WILLIAM F. MAAG LIBRARY
YOUNGSTOWN STATE UNIVERSITY

COMPREHENSIVE CHEMICAL KINETICS

COMPREHENSIVE

Section 1. THE PRACTICE AND THEORY OF KINETICS
(3 volumes)

Section 2. HOMOGENEOUS DECOMPOSITION AND
ISOMERISATION REACTIONS (2 volumes)

Section 3. INORGANIC REACTIONS (2 volumes)

Section 4. ORGANIC REACTIONS (6 volumes)

Section 5. POLYMERISATION REACTIONS (3 volumes)

Section 6. OXIDATION AND COMBUSTION REACTIONS
(2 volumes)

Section 7. SELECTED ELEMENTARY REACTIONS (1 volume)

Section 8. HETEROGENEOUS REACTIONS (4 volumes)

Section 9. KINETICS AND CHEMICAL TECHNOLOGY (1 volume)

Section 10. MODERN METHODS, THEORY, AND DATA

CHEMICAL KINETICS

EDITED BY

R. G. COMPTON

M. A., D. Phil. (Oxon.)
*University Lecturer in Physical Chemistry
and Fellow, St. John's College, Oxford*

VOLUME 31

MECHANISM AND
KINETICS OF ADDITION
POLYMERIZATIONS

ELSEVIER
AMSTERDAM-OXFORD-NEW YORK-TOKYO
1992

Scientific Editor

Ing. Jan Šebenda, DrSc.

Reviewer

Ing. Jaroslav Bartoň, DrSc., Corresponding Member of the Slovak
Academy of Sciences

Published in co-edition with Academia, Publishing House of the Czechoslovak Academy of
Sciences, Prague

Distribituon of this book is being handled by the following publishers

for the U.S.A. and Canada
Elsevier Science Publishing Company, Inc.
655 Avenue of the Americas
New York, NY 10010

for the East European Countries, China, North Korea, Cuba, Vietnam and Mongolia
Academia, Publishing House of the Czechoslovak Academy of Sciences,
Prague, Czechoslovakia

for all remaining areas
Elsevier Science Publishers
25 Sara Burgerhartstraat
P. O. Box 211, 1000 AE Amsterdam, The Netherlands

ISBN 0-444-98795-9 (Vol. 31)
ISBN 0-444-41631-5 (Series)

Library of Congress Cataloging-in-Publication Data

Kučera, Miloslav, 1927-
  Mechanism and kinetics of addition polymerizations / (Miloslav Kučera).
    p. cm. --(Comprehensive chemical kinetics; v. 31)
  Translated from the Czech.
  Includes bibliographical references.
  ISBN 0-444-98795-9
  1. Addition polymerization. 2. Chemical reaction, Rate of. I. Title. II. Series.
  QD501.B242 vol. 31
  (QD281.P6)
  541.3'94 s-dc20
  (541.3'93)

1st edition © Miloslav Kučera, 1984
2nd revised edition © Miloslav Kučera, 1992
Translation © Danica Doskočilová, 1992
All rights reserved. No part of this publication may be reproduced, stored in a retrieval system,
or transmitted in any form or by any means, electronic mechanical, photocopying, recording or
otherwise, without the prior written permission of the copyright owners.

Printed in Czechoslovakia

COMPREHENSIVE CHEMICAL KINETICS

ADVISORY BOARD

Professor C. H. BAMFORD

Professor S. W. BENSON

Professor LORD DAINTON

Professor G. GEE

Professor G. S. HAMMOND

Professor W. JOST

Professor K. J. LAIDLER

Professor SIR HARRY MELVILLE

Professor S. OKAMURA

Professor N. N. SEMENOV

Professor Z. G. SZABO

Professor O. WICHTERLE

Volumes in the Series

        Section 1.   **THE PRACTICE AND THEORY OF KINETICS**
                        (3 volumes)

Volume 1   The Practice of Kinetics
Volume 2   The Theory of Kinetics
Volume 3   The Formation and Decay of Excited Species

        Section 2.   **HOMOGENEOUS DECOMPOSITION AND ISOMERISATION REACTIONS** (2 volumes)

Volume 4   Decomposition of Inorganic and Organometallic Compounds
Volume 5   Decomposition and Isomerisation of Organic Compounds

        Section 3.   **INORGANIC REACTIONS** (2 volumes)

Volume 6   Reactions of Non-metallic Inorganic Compounds
Volume 7   Reactions of Metallic Salts and Complexes, and Organometallic Compounds

        Section 4.   **ORGANIC REACTIONS** (6 volumes)

Volume 8   Proton Transfer
Volume 9   Addition and Elimination Reactions of Aliphatic Compounds
Volume 10   Ester Formation and Hydrolysis and Related Reactions
Volume 12   Electrophilic Substitution at a Saturated Carbon Atom
Volume 13   Reactions of Aromatic Compounds

        Section 5.   **POLYMERISATION REACTIONS** (3 volumes)

Volume 14   Degradation of Polymers
Volume 14A   Free-radical Polymerisation
Volume 15   Non-radical Polymerisation

        Section 6.   **OXIDATION AND COMBUSTION REACTIONS**
                        (2 volumes)

Volume 16   Liquid-phase Oxidation
Volume 17   Gas-phase Combustion

        Section 7.   **SELECTED ELEMENTARY REACTIONS** (1 volume)

Volume 18   Selected Elementary Reactions

        Section 8.   **HETEROGENEOUS REACTIONS** (4 volumes)

Volume 19   Simple Processes at Gas–Solid Interface
Volume 20   Complex Catalytic Processes
Volume 21   Reactions of Solids with Gases
Volume 22   Reactions in the Solid State

        Section 9.   **KINETICS AND CHEMICAL TECHNOLOGY** (1 volume)

Volume 23   Kinetics and Chemical Technology

        Section 10.   **MODERN METHODS, THEORY, AND DATA**

Volume 24	Modern Methods in Kinetics
Volume 25	Diffusion-limited Reactions
Volume 26	Electrode Kinetics: Principles and Methodology
Volume 27	Electrode Kinetics: Reactions
Volume 28	Reactions at the Liquid–Solid Interface
Volume 29	New Techniques for the Study of Electrodes and their Reactions
Volume 30	Electron Tunneling in Chemistry. Chemical Reactions over Large Distances
Volume 31	Mechanism and Kinetics of Addition Polymerizations

Contributor to Volume 31

M. KUČERA   Research Institute of Macromolecular Chemistry, Tkalcovská 2, 656 49 Brno, Czechoslovakia

To my grandmother and parents

A good theory accelerates progress more than hundreds of inaccurate experiments; an unambiguous measurement is more reliable than an admirable theory

Preface to the second (English) edition

From the submission of the manuscript of the first, Czech, edition of this book to the editor, roughly eight years have passed. A long enough time to substantially broaden our knowledge of polymerizations. New findings sometimes confirm previous experience, sometimes they shift our knowledge from the realm of postulates and hypotheses to the rank of theory; rules and laws are derived from the theories. In previously less explored fields, new findings permit the formulation of hypotheses, in other disciplines they oppose facts previously accepted as safely established. All new findings must be included in the text of a book aspiring to be modern.

The first edition of the book was drafted with respect to the limited extent required by the publishers. This requirement is also the case for the second edition. Therefore the insertion of new results dictated the omission of some sections, e. g. of a part of the text specific for CSFR, of the chapter on the sources and toxicity of monomers, and of whole part 9 on the methods of polymer investigation. The general concept of the book has not changed; its aim is to inform on all important findings, to facilitate the search for the common features of any of the various basic polymerization processes, without distracting the reader's attention by a too detailed description of some special cases isolated in the context.

I believe that the general concept of this book remains modern, making its study useful even for a reader with a wider choice of books of similar orientation.

Miloslav Kučera

Introduction to the first (Czech) edition

The best known products of macromolecular chemistry are plastics, synthetic rubber and fibres. The world average per capita consumption of plastics exceeded 8 kg (44 kg in the USA and in Japan). The production of synthetic fibres and rubber exceeds the production of the natural materials. A large proportion of these substances is produced by polymerization.

Polymer synthesis is a very particular process, sensitive even to slight variations of both chemical and physical parameters. It would probably be impossible to produce a good quality polymer by purely empirical methods, or else the development of the technology would be very expensive. On the other hand, if every production phase should have to be supported by scientific understanding, such production could never be started. Therefore only a reasonable proportion of scientifically supported facts and suitable empirical procedures leads to technologically feasible results. Unfortunately we have no recipe for determining the sufficient amount of scientific data necessary for the realization of an important technological development.

In this book, the basic principles of polymerization theory are treated. It cannot be expected that everything in it could withstand the assault of new findings. However, well selected and well performed experiments are pillars that can outlive even an explosive development of the field. Their interpretation by hypotheses or theories may change sooner or later. Correct considerations will become a special case (i. e. part) of more general theories. Incorrect explanations will disappear. Therefore the reader should regard the text of this book with a suitable dose of healthy scepsis.

I have tried to arrange the chapters of this book in the following order: exposition of the problem, present interpretation, documentation, plus possibly an example and an indication of importance for further development. It is possible that the selected manner is not the best in all cases. It is probable that the selected documentation does not include the first published reports.

For the former I wish to apologize to the readers, for the latter to the authors of those first measurements. In the explanation it is often instructive to include the main stages in the development of present ideas. If our understanding of some feature is not yet unambiguous, only the main explanations are included. The proofs are selected from the pool of existing documents. The most important proof is again the experiment. The closing of the sequence experiment, interpretation, generalization usually leads to a further experiment, and one turn of the helix is closed. Quite often, however, the proof is missing, and is substituted by various procedures. It happens that we can devise an experiment that would be helpful for the continuity of the development. However, the state of science or of experimental techniques does not permit its immediate realization. In such a case we remain stuck in the realm of theories or hypotheses, and we are waiting impatiently for the progress of science, often in related disciplines, enabling us to carry on, either directly or by a by-pass.

Experiment and theory must coexist, similarly as the scientific and empirical approaches. The ability to choose a suitable proportion of these components in research work probably belongs to the most exacting requirements of scientific and technical work. Whoever possesses this art, or is able to learn it, may be successful.

Scientists are usually sensitive to due recognition of their achievements. New, healthy ideas are rather rare, and their occurrence depends on talents and the amount of work. Naturally authors feel discrimination when their contribution is not connected with their name, or even is ascribed to someone else. It is therefore customary to cite the authors of original scientific communications.

I have tried to meet this custom without substantially interrupting the clear flow of the text. Those interested in direct information can easily find the original communications in the cited literature (and in the papers cited therein).

In this book I am trying to present a close review of our overall knowledge of polymerizations at the turn of the decade (1979–1981). No book can fulfil the needs of a specialist in his own field. A specialist should be familiar with everything in the books, and he has to follow all periodical literature of his discipline. He should also be sufficiently informed on the progress in related fields. Thus the depth of specialization, and the quality and quantity of contacts with related disciplines again has to result from a compromise. As any extreme, even here excess to either side is detrimental. With this in mind I have tried to draft this book. A specialist in a certain discipline, e. g. in radical polymerizations, certainly will not regard the text concerning his field as exhausting. I hope, however, that he will find some interest in reading the chapter on ionic or coordination polymerizations. I also believe that the book could be useful to students at all levels of schools of chemical orientation.

I assume that the book will be read and studied by people with some knowledge of elementary physical and organic chemistry and mathematics. I have tried to lower this threshold level by detailed derivations and frequent footnotes. The text of the book is printed in two types. The descriptive and documentary parts are thus roughly separated. Naturally the separation could not be sharp, and in no case should it be regarded as stressing importance.

In writing this book I took advantage of the experience gathered during my own research work and through contacts with both undergraduate and graduate students. Without the necessary support both in the family and at my workplace the book could not have been written. I am greatly obliged to all who have helped me directly or indirectly, and I wish to express my sincere thanks to all of them: the first place to the management of the Research Institute of Macromolecular Chemistry for the understanding and approval of the preparation of this publication; particularly to Director of the Institute of Macromolecular Chemistry of the Czechoslovak Academy of Sciences, the Corresponding member of the Czechoslovak Academy of Sciences, Professor K. Friml, and to the Deputy Director of this Institute, Ing. J. Trekoval, DrSc. Both of them showed constant interest in the nascent book, and their enthusiasm was a strong moral stimulus for me. I also wish to express my sincere thanks to my long-time coworker PhDr. K. Majerová for exceptionally careful and devoted copying of my manuscript.

I am indebted to Ing. J. Bartoň, DrSc., and to the scientific editor Ing. J. Šebenda, DrSc. for highly qualified reviewing of the manuscript, for their notes and comments. They have contributed to the overall quality of the book.

Miloslav Kučera

Acknowledgments

In this book the experience of many authors is summarized. Individual observations are sorted and evaluated, and from these, generalized findings and comprehensive concepts are drawn. By an overwhelming majority all these forms of cognition are published; a verbal tradition essentially does not exist. I have used well established journals and books as sources of information. Some data included in Tables and Figures are subject to copyright. I am greatly indebted to the authors and to the publishers of journals and books who have granted their permission for the publication of these data in my book. This concerns particularly the Tables: $5\text{-}2^a$, $6\text{-}2^b$, $1\text{-}4^f$, $5\text{-}4^i$, $1\text{-}5^f$, $2\text{-}5^l$, $2\text{-}6^a$, $1\text{-}8^a$, $2\text{-}8^a$ and the Figures: $5\text{-}4^e$, $6\text{-}4^e$, $3\text{-}5^b$, $10\text{-}5^b$, $11\text{-}5^b$, $12\text{-}5^b$, $13\text{-}5^g$, $14\text{-}5^k$, $28\text{-}5^a$, $29\text{-}5^a$, $31\text{-}5^c$, $1\text{-}6^a$, $2\text{-}6^a$, $3\text{-}6^d$, $4\text{-}6^d$, $8\text{-}6^b$, $11\text{-}6^b$, $2\text{-}7^j$, $3\text{-}7^b$, $4\text{-}7^c$, $5\text{-}7^c$, $6\text{-}8^a$, $7\text{-}8^a$, $8\text{-}8^h$, $9\text{-}8^c$, $10\text{-}8^c$, $11\text{-}8^c$, $12\text{-}8^c$.

[a] – Makromol. Chem., Hüthig & Wepf Verlag Basel
[b] – J. Polym. Sci., Periodicals Division, Interscience Publishers, J. Wiley & Sons, Inc.
[c] – Polymer, IPC Business Press Ltd. ©
[d] – Macromolecules, Copyright (1977) American Chemical Society, New York, Washington
[e] – Rendiconti della Classe di Scienze fisiche, matematiche e naturali, Serie VIII, Vol. 26, Accademia nazionale dei Lincei, Roma
[f] – P. E. M. Allen and C. R. Patrick: Kinetics and Mechanisms of Polymerisation Reactions, Ellis Horwood Limited, Chichester 1974
[g] – Academic Press, Inc., New York
[h] – Royal Society of Chemistry, Chem. Commun. London
[i] – Royal Society's Proceedings, London
[j] – La Chimica e l'industria, Milano
[k] – Marcel Dekker, Inc., New York
[l] – W. J. Bailey

An exact reference to each source, together with the authors' names, is cited at the respective Table or Figure. I sincerely thank the publishers and authors of the cited communications.

Miloslav Kučera

Contents

Preface xi
Introduction xii
Acknowledgments xv
List of symbols 1

Chapter 1

Polymerization 11

   1. Basic processes and the initiator 12
      1.1 Separation of basic processes 12
      1.2 The initiator 12
   2. Classification according to type of active centre 14
   3. Classification according to reactivity of active centres 14
   4. Classification according to number of monomers 14
   5. Classification according to number of phases at the beginning and during reaction 16
   6. Liquid phase polymerization 18
   7. Classification according to the arrangement of chains formed . . . 21
   8. Kinetic factors 22
   9. Special kinds of polymerizations 24
      9.1 Ring opening polymerization 24
      9.2 Cyclopolymerization 24
      9.3 Polymerization by activated monomer mechanism 25
      9.4 Isomerization polymerization 25
      9.5 Topochemical polymerization 25
      9.6 Popcorn polymerization 26
      9.7 Group transfer polymerization 26
References 26

Chapter 2

Monomers 37

   1. Classification of monomers 37
      1.1 Chemical classification 37
         1.1.1 Vinyl and vinylidene monomers 28
         1.1.2 Dienes 29

1.1.3 Alkynes	31
1.1.4 Cyclic monomers	31
1.1.5 Other monomers	38
1.2 Relationship between monomer structure and its polymerization mode	40
1.3 Donor-acceptor properties of monomers	42
1.3.1 Monomers as electron donors	43
1.3.2 Monomers as electron acceptors	43
2. Properties of polymerizing monomers	44
2.1 Physical properties of monomers	44
2.2 Chemical properties of monomers	45
3. Reactivity of monomers	45
4. Monomer and other components of the system	61
4.1 Physical relationship	62
4.2 Chemical interactions	62
4.2.1 Monomer complexes	63
5. Two, three or more types of monomer	65
5.1 Non-complexing monomers	65
5.2 Monomers forming donor-acceptor complexes	65
6. Models of monomers	68
References	69

Chapter 3

Initiation … 75

1. Formation of primary radicals	75
1.1. Thermal initiation	75
1.2 Initiators	78
1.2.1 Peroxides and their decomposition	79
1.2.2 Azo-initiators and their decomposition	85
1.3 Photo-initiation	88
1.3.1 Sensitization	89
1.4 Initiators produced in situ	91
1.4.1 Effect of oxygen	93
1.5 Other methods leading to radical formation	93
1.6 Non-typical initiators	95
1.6.1 Multifunctional initiators	95
1.6.2 Polymeric initiators	96
1.7 Behaviour of primary radicals	98
1.7.1 Cage effect	98
1.7.2 Efficiency of initiation	99
1.8 Reaction of radical with unsaturated molecule	101
1.8.1 Initiation of radical polymerization	101
2. Formation of initiating anion	102
2.1 Properties of metal-carbon bond	103
2.1.1 Initiation by alkylmetals	105
2.2 Properties of metal-oxygen and metal-nitrogen bonds	112
2.2.1 Initiation by alkoxides	112
2.3 Electron transfer from donor to monomer	114
2.3.1 Initiation by alkali metals	115
2.3.2 Initiation by radical anions	116

2.4 Other methods leading to initiating anions	118
2.4.1 Multifunctional carbanions	118
2.4.2. Macroanions	119
2.5 Reaction of anion with monomer	119
2.5.1 Initiation of anionic polymerization	121
2.5.2 Activators of anionic polymerization	123
3. Formation of initiating cation	124
3.1 Initiators of cationic polymerizations	124
3.1.1 Brönsted acids	125
3.1.2 Lewis acids	126
3.2 Special kinds of initiators	129
3.2.1 Initiation by HI/I_2	129
3.2.2 Multifunctional initiators	130
3.2.3 Macrocations	131
3.3 Reaction of cation with monomer	132
4. Formation of complexes initiating coordination polymerizations	133
4.1 Ziegler–Natta catalysts	133
4.1.1 Supported catalysts	136
4.1.2 Cocatalysts of ZN catalysts	137
4.1.3 Internal and external donors	139
4.1.4 Dual function catalysts	139
4.1.5 Reaction of Ziegler–Natta centres with monomers	140
4.2 Transition metal catalyst active in absence of alkylmetals	141
5. Initiating donor-acceptor complexes	143
5.1 Formation of radical-ions and their reaction with monomers	143
5.2 Polymerizations initiated by thermally activated donor-acceptor complexes	145
5.2.1 Polycombination of zwitterions	147
5.3 Monomers as ion pairs (monomer salts)	148
6. Other types of initiation	149
6.1 Electroinitiation	149
6.2 Mechanochemical initiation	150
6.3 Initiation in plasma	151
7. Initiation of non-polymerizing compounds	151
8. Initiation kinetics	153
References	154

Chapter 4

Active centres of polymerizations 163

1. Active centres of radical polymerizations	164
1.1 Reactivity of radicals	165
1.2 Reactivity of radical centres during polymerization	172
1.3 "Hot" radicals	174
1.4 Intervention in the structure of centres	176
1.5 Some radicals of particular importance	179
2. Ionic active centres	180
2.1 Ions, ion pairs and their associates	180
2.2 Centres of anionic polymerization	183
2.2.1 Carbanions	184
2.2.2 Other anions	188

	2.3 Centres of cationic polymerizations	191
	2.3.1 Carbocations	191
	2.3.2 Carboxonium centres	193
	2.3.3 Oxonium centres	193
	2.3.4 Covalent centres	195
	2.3.5 Centres of cationic polymerization with activated monomer	195
	2.3.6 Cations derived from elements other than C and O	196
	2.4 Zwitterions	197
3. Radical-ions		199
4. Coordination centres		203
	4.1 Coordination centres of Ziegler–Natta polymerizations	203
	4.1.1 Solid Ziegler–Natta catalysts	203
	4.1.2 Soluble Ziegler–Natta catalysts	208
	4.1.3 Properties of coordination centres of Ziegler–Natta polymerizations	211
	4.2 Other coordination centres	213
5. Transformation of centres		214
	5.1 Transformation anion to cation	215
	5.2 Transformation cation to anion	216
	5.3 Direct transformation anion to cation and cation to anion	216
	5.4 Transformation anion to radical	217
	5.5 Transformation with the participation of coordination centres	217
	5.5.1 Anion to coordination centre	217
	5.5.2 Coordination centre to anion	218
	5.5.3 Coordination centre to cation	218
	5.5.4 Coordination centre to radical	218
	5.5.5 Anion to centre of metathesis	219
6. Reaction of coordination centres during polymerization		219
7. Living and dormant centres		221
References		222

Chapter 5

Propagation 231

	Theoretical conditions for propagation	231
	1.1 Thermodynamic conditions	231
	1.1.1 Ceiling (floor) temperature	232
	1.1.2 Heat of reaction and entropy changes during polymerization	235
	1.1.3 Influence of pressure on propagation	235
	1.2 Chemical influence	240
	1.2.1 The presence of impurities	240
	1.2.1 Propagation and polycondensation	241
	1.3 Kinetic influence	242
	1.3.1 Stationary polymerization	242
	1.3.2 Non-stationary polymerization	243
	1.3.3 Temperature of polymerization	244
2. Environment surrounding active centre (external influence)		244
	2.1 Medium	245
	2.2 Consequences of polymer formation	247
	2.2.1 Viscosity increase	247
	2.2.2 Solid phase formation	250

2.2.3 Active centres and polymer	251
2.2.4 The matrix (template) effect	252
3. General features of propagation	256
3.1 Conversion	256
3.2 Transfer of heat during polymerization	259
3.3 Volume contraction of the polymerizing system	260
4. Controlled propagation	261
4.1 Stereospecific growth	262
4.1.1 Chain configuration and statistics of stereochemical propagation	262
4.1.2 General factors controlling the mode of addition	264
4.1.3 Enantioelective (stereoelective) and enantioselective (stereoselective) polymerization	268
4.1.4 Propagation on the transition metal-carbon bond	270
4.2 Polymerization in micelles (emulsion polymerization)	280
4.2.1 Initiation in the aqueous phase	281
4.2.2 Polymerization in emulsion initiated by an insoluble initiator	288
4.2.3 Inverse emulsion polymerization	288
4.2.4 Ionic emulsion polymerization	289
5. Copolymerization	289
5.1 Copolymerization equation	290
5.2 Copolymerization parameters	293
5.2.1 The experimental determination of copolymerization parameters	295
5.2.2 The theory of copolymerization parameters	300
5.2.3 The distribution of monomeric units in the copolymer	306
5.3 Ionic copolymerization	310
5.4 Coordination copolymerization	311
5.5 Multicomponent copolymerization	312
5.5.1 The penultimate effect	312
5.5.2 Copolymerization with monomers and their complexes	313
5.5.3 Ternary copolymerization and copolymerization of more than three monomers	314
5.6 Thermodynamics of copolymerization	320
5.6.1 Copolymerization at temperatures above T_c of one of the monomers	326
5.7 Copolymerization kinetics	328
5.8 Copolymerization types	332
5.8.1 Polymerizations producing periodic copolymers	332
5.8.2 Block copolymers	335
5.8.3 Graft copolymers	336
6. Mechanism of propagation	338
6.1 Addition	338
6.2. Insertion	339
6.2.1 Coordination insertion	339
6.3 Ring-opening polymerizations	341
6.3.1 The formation of macrocycles during ring-opening polymerization	346
6.3.2 Metathesis	348
6.3.3 Polymerization on activated ligands	349
6.4 Cyclopolymerization	350
6.5 Polymerization with activated monomer	352
6.6 Isomerization polymerization	352
6.7 Group-transfer polymerization	352
6.8 Branching, cross-linking	356

7. Depropagation	360
8. Propagation kinetics	362
8.1 Propagation in living systems	363
8.1.1 Determination of propagation rate on free ions and ion-pairs in living polymerizations	363
8.2 Rate constant of propagation	369
References	369

Chapter 6

Termination . . . 383

1. Decay of radical activity	383
1.1 Combination, disproportionation of radicals	383
1.1.1 The effect of pressure on termination	390
1.1.2 The effect of polymer concentration and chain lenghts	393
1.2 Termination of and by primary radicals	394
1.3 Systems with retarded termination	395
1.3.1 The gel effect	396
1.3.2 Trapped radicals	399
1.4 Inhibition, inhibitors	401
1.4.1 Autoinhibition	404
1.5 Non-stationary states of radical polymerizations	405
1.5.1 The pre-effect of radical polymerization accompanied by mutual radical termination	406
1.5.2 The decay of radical polymerization due to mutual termination of radicals	408
1.5.3 Rotating sectors	408
1.5.4 Spatially intermittent polymerization	414
2. Termination of ionic polymerizations	417
2.1 Termination involving counter ion	417
2.1.1 Shift of equilibria between ions, ion pairs, and esters	417
2.1.2 Combination of ions	418
2.1.3 Termination by unstable counter-ion	419
2.2 Transformation of an active centre to an inactive species or to an ion of low activity	420
2.2.1 Aggregation (association) of centres	420
2.2.2 Polymerization-inactive complexes of ions	422
2.2.3 Autoinhibition (autotermination)	424
2.2.4 Termination due to improper addition of monomers	426
2.2.5 Termination by impurities	426
2.3 Combination of macroions	427
3. Termination of coordination polymerizations	428
3.1 Termination by donors	428
3.2 Termination by breaking metal-polymer bond in active centre	428
3.3 Termination by deactivation of the organometallic component	429
3.4 "Spontaneous" termination	429
4. Use of termination	430
4.1 Macromers (macromonomers)	430
4.2 Desactivation of centres	431
4.2.1 Determination of the number of centres	431
5. Kinetics of termination	435
References	438

Chapter 7

Transfer . . . 443

1. General characteristics of transfer reactions . . . 443
   1.1 Retardation . . . 444
      1.1.1 Degradative transfer . . . 450
   1.2 Branching (cross-linking) by transfer . . . 451
      1.2.1 Branching by the "wrong" addition of monomer or initiator . . . 452
      1.2.2 Multifunctional transfer agents . . . 453
      1.2.3 Transfer to polymer . . . 453
   1.3 Reversible transfer . . . 454
2. Transfer in radical polymerizations . . . 454
   2.1 Mechanism of some transfers . . . 455
3. Transfers in anionic polymerizations . . . 457
   3.1 Transfers of H^-, H^+ . . . 457
4. Transfers in cationic polymerizations . . . 459
   4.1 Transfer during polymerization of hydrocarbon monomers and of monomers containing heteroatoms in side chain . . . 460
   4.2 Transfers during polymerization of heterocyclic monomers . . . 461
5. Transfers in coordination polymerization . . . 462
   5.1 Transfers by organometallics . . . 462
   5.2 Transfers by molecular H_2 . . . 463
   5.3 Other transfers of coordination polymerizations . . . 464
6. Molecular mass of polymerization products . . . 464
7. Copolymerizing transfer agents . . . 469
8. The use of transfers . . . 473
   8.1 Control of molecular mass . . . 473
   8.2 Grafting by transfer process . . . 473
   8.3 Formation of macromers by transfer . . . 474
      8.3.1 Inifer, iniferter techniques . . . 475
References . . . 476

Chapter 8

Kinetics . . . 481

1. Polymerization rate . . . 481
   1.1 Rate of initiation . . . 482
   1.2 Radical polymerization . . . 484
      1.2.1 Overall reaction rate of ideal polymerization . . . 485
      1.2.2 Reactions of radicals of various lengths . . . 490
      1.2.3 Non-ideal radical polymerization . . . 492
      1.2.4 Example: kinetic analysis of vinyl chloride polymerization . . . 497
   1.3 Ionic polymerization . . . 502
      1.3.1 Living polymerization with slow initiation . . . 502
      1.3.2 Livig equilibrium copolymerization . . . 505
      1.3.3 Non-stationary polymerization . . . 511
      1.3.4 Example: propagation of heterocycles. Periodic changes of growing centre reactivity . . . 514
   1.4 Coordination polymerization . . . 515
      1.4.1 Kinetic model of Ziegler–Natta polymerizations . . . 515
      1.4.2 Example: polymerization of ethylene . . . 523

2. The degree of polymerization of the product 526
   2.1 Molecular mass distribution of the product 527
      2.1.1 Number and weight (mass) average degrees of polymerization 529
      2.1.2 Rate of living polymerizations and molecular mass distribution
           curves of products 531
      2.1.3 Molecular mass distribution in products of radical polymerizations . 535
      2.1.4 Example: changes of molecular mass distribution during
           polymerization of heterocycles 539
References 544

Chapter 9

Conclusion 547

1. Activation energy of polymerization and thermodynamics 547
2. Replacement of active centers by a high frequency magnetic field 549

Index 552

List of Symbols

A	acceptor
A, A_1, A_2	dilatometric constant, conversion factor
AC	active centre
Ac	cation, cationic part of an acid
AIBN	2,2′-azo-bis-isobutyronitrile
A_s^{-*}	solvated anion radical
A*	excited acceptor
a_p	surface area of a particle of volume V
acac	2,4-pentanedione-3yl (acetylacetonate)
[al]	concentration of a monomeric organoaluminium compound at the surface of a catalyst
B	anion, base
B	proportionality constant
B_2, B_3	virial coeficients
Bu	butyl
C	proportionality constant
C	Coulomb charge
CT	charge-transfer (donor—acceptor) complex
C_{tr}	k_{tr}/k_p, apparent transfer constant
$C_{tr,1}$, $C_{tr,2}$	apparent transfer constant of monomer 1 or monomer 2, respectively
$°C_{tr,1}$, $°C_{tr,2}$	apperant transfer constant of standard monomer 1, or standard monomer 2, respectively
c	concentration
cp	cyclopentadienyl
D	permittivity of medium
D	electron donor

D^*	excited donor
DA	donor—acceptor complex
DBP	dibenzoylperoxide
DMF	dimethylformamide
$^1(DA)^*$, $^3(DA)^*$	excited DA complex in singlet or triplet state, respectively
D_s^{+*}	solvated cation radical
(D_s^{+*}, A_s^{-*})	excited DA complex with solvated ion radicals
d, d'	length of the illuminated and dark zones, respectively, in SIP reactor
$\bar{d}$	d'/d
d_m	density of monomer
d_p	density of polymer
E, E_A	activation energy
E	contribution of primary radical destruction in a cage
$^{1/2}E$	half-wave potential
$[E]$	concentration of emulsifier
E_i, E_{iso}, E_p	activation energy of initiation, isomerization, propagation
EPR, ESR	electron paramagnetic (spin) resonance
E_s	Taft steric substituent constant
Et	ethyl
e	unit charge (of electron)
e	electron
e	Alfrey—Price e value
e_n	bond electronegativity
^Ae_n	electronegativity of element A
^Be_n	electronegativity of element B
F_1, F_2	mole fractions of monomers 1 and 2 in copolymer
$F_{(n)}, X_n$	fractions of molecules with degree of polymerization n
$F_1(T, \xi_0)$	function describing the dependence of k_t on the temperature T and friction constant ξ
$F_2(\alpha, N)$	function describing the dependence of k_t on the expansion coefficient α and the degree of polymerization of macroradicals N_A and N_B
f	efficiency of initiator
f_1, f_2	mole fractions of monomers 1 and 2 in monomer batch
$f_{1(az)}$	mole fraction of monomer 1 at azeotropic point
ΔG_p	change of Gibbs energy (free enthalpy). Indices characterize the conditions of the process
$\Delta G^{\neq}$	Gibbs thermodynamic activation potential

$\Delta G_{dis,el}$	contribution of electrostatic interactions to molar change in the Gibbs energy in ion pair dissociation
GPC	gel permeation chromatography
g	monomer mass
g	shear stress
$\Delta H, \Delta H^{\neq}$	enthalpy change, activation enthalpy
h	Planck constant
I	total number of moles of initiator
I	rate of radical generation by photoinitiation
I'	rate of radical generation by thermal initiation
$\bar{i}$	mean number radicals in a particle
$J_k(x)$	Bessel function of the first kind
K	constant of the Mark—Houwink equation
K_c	critical concentration of polymer
K_d	equilibrium dissociation constant
k	Boltzmann constant

Rate constants:

k	general
k_-, k_{+-}	propagation on free anions or ion pairs
$k_{(+-)c}$	propagation on contact ion pairs
$k_{(+-)s}$	propagation on separated ion pairs
$k'_{(+-)}$	depropagation on ion pairs
k_*	propagation on radical centres
k_a, k_d	absorption or desorption of primary radicals by particles or from particles
k_d	decomposition of initiator
^1k_d	decomposition of primary radicals
^ik_d	induced decomposition of initiator
k_i	initiation
$k_{i,tr}$	initiation after transfer
$k_{p,ef}$	propagation, effective
k_p	propagation, mean
$k_{p\star}$	propagation on hot radicals
k_q	inhibition
k_t	termination
$\bar{k}_t$	first-order termination

$\bar{k}_t$	second-order termination
$^P k_t$	termination in a polymer-rich particle
$^M k_t$	termination in monomer-rich phase
$k_{t(N_A, N_B)}$	termination among radicals with degree of polymerization N_A, N_B
$k_{t(\text{rel})}$	termination, relative
$k_{t(p)}$	termination, at pressure p
$k_{t(0)}$	termination at atmospheric pressure
$k_{t,\text{pr}}$	termination by primary radicals
k_{tr}	of transfer
$k_{tr,M}$, $k_{tr,P}$	of transfer with monomer, or with polymer
$k_{t,Q}$	termination with inhibitor
$k_{tr,Y}$	of transfer with Y
$k_{t,z}$	termination of entangled radicals
k_u	escape of radicals from particle (cm h^{-1})
k_z	combination of active centre with spin trapping agent
L	electron affinity
L	$-\mathrm{d}\ln D_r/\mathrm{d}T$; factor expressing change in relative permittivity (dielectric constant) with temperature
L	localization energy
LC	living centre
LCAO-MO	linear combination of atomic orbitals
M	relative molecular mass of monomer
M	monomer, substrate
M	number molecular mass
$\bar{M}$, $\bar{M}_n = \bar{P}M$	number average molecular mass
$\bar{M}_w$	weight average molecular mass
M_E	electrophilic monomer
M_N	nucleophilic monomer
$[M]_\infty$	equilibrium monomer concentration
M°	chain carrying an active centre (radical, ionic or coordination) at one end
°M°	chain carrying active centres at both ends
Me	methyl
$M_{j,X}$	molecule of unreactive polymer after transfer with agent XT
$[M_M]$	monomer concentration in monomer-rich phase
$[M_P]$	monomer concentration in polymer-rich particle
M_q	mass of monomer in a monomer—polymer particle
M_t	metal atom
$[M_N^*]$	concentration of radicals with degree of polymerization N

$[M^*]_Z$	concentration of active centres decayed by degradative transfer
m_1	amount of monomer M_1 in terpolymer
m_P	number of macromolecules
m_P	mass of polymer
$m_{P,i}$	mass of macromolecule with degree of polymerization i
N	Avogadro's number
N	non-radical product
N	number of polymer particles swollen by monomer
$[N]$	concentration of macromolecules
NMR	nuclear magnetic resonance
N_A, N_B	degree of polymerization of macroradicals A and B
N_i	number of moles of macromolecules with degree of polymerization i
n	number of monomer units in the chain of an intermediate or in a macromolecule
n	number of molecules which have dissociated from active centres ($= 1/2$ of end groups)
n	$d[M_1]/(d[M_2]$
n_c	critical chain length
P	polymerization number
$\bar{P}$	number average degree of polymerization
$\bar{P}_c$	mean degree of polymerization at critical polymer concentration
${}^1\hat{P}_n$, ${}^2\hat{P}_n$	degree of polymerization of sequence of monomer M_1 or M_2, respectively
Ph	phenyl
P_i	chain of degree of polymerization i
P_j	inactive chain of degree of polymerization j
$\bar{P}_w$	weight average degree of polymerization
P_{12}	reaction probability of particles 1 and 2
p	surface area occupied by 1 g of emulsifier
p	pressure
pK_b	negative logarithm of dissociation constant of weak bases
ppm	parts of impurity per million parts of basic substance
p_s	packing factor
Q	Price's Q value
Q	inhibitor
Q	$({}^Mk_t/{}^Pk_t)^{1/2}$
Q	$[R_p^*]/[R_m^*] = k_a/k_d$
$\sim Q^\star$	hot radical after reaction with inhibitor

q, ε	Schwan–Price characteristics of radicals and monomers in copolymerization
$^Mq, {}^Pq$	monomer or polymer mass, respectively, in monomer-polymer particle
$\boldsymbol{R}$	gas constant, $\boldsymbol{R} = 8.314$ joule/grad $(= 1.987$ cal/grad$)$
R	radius of hypothetical sphere in which a radical chain end can move
R_1, R_2	substituents
$R°$	growing active centre (radical, ionic or coordination) $R° = M°$; the symbol $R°$ occurs in relations that have originally been derived for the radical case, but which are of a more general character
R^*	radical
$^1R^*$	primary radical
R^{-*}	anion radical
R^{+*}	cation radical
R_M^*, R_P^*	number of radicals in monomer- or polymer-rich phase
$[R^*]_{s,i}$	hypothetical „stationary" concentration of radicals during the induction period
r_1, r_2	copolymerization parameters
$\bar{r}, \bar{r}_0$	mean distance between chain ends
r	$^M\varrho/{}^P\varrho$
r_g, r_l	distance between ions in the gas or liquid phase
$r_{R\cdots M}$	distance $R\cdots M$
$\langle r^2 \rangle$	mean square end-to-end distance in undisturbed polymer random coil
S	entropy
S	solvent
S	$v/[M]^2$
$\bar{S}, \bar{S}_0$	mean radius of gyration
$\Delta S, \Delta S^{\neq}$	entropy change, entropy of activation
$\Delta S_{dis,el}$	contribution of electrostatic interactions to molar entropy change in ion pair dissociation
s	kind of moment of macromolecular size distribution
T	temperature (K)
T_c	ceiling temperature
T_f	floor temperature
T_i	isokinetic temperature
T_g	glass transition temperature

^1t	start of darkness period
^2t	end of darkness period
t_n	retardation of polymerization caused by a pre-effect
Δt_m	duration of mixing of initiator with monomer
t_{inh}	length of inhibition period
t_{ind}	length of inhibition and induction periods
Δt_{ind}	length of induction period
U	energy
U	intercept of the stationary phase of the conversion curve with the ordinate
UV	ultraviolet radiation
u	flow rate in SIP
V	volume of a particle
V	molar volume
$V, \Delta V$	volume, volume change
ΔV_{visc}	activation energy of viscous flow
V_P	molar volume of monomer unit in polymer
V_M	molar volume of free monomer
$^M V$	total volume of monomer-rich phase
$^P V$	total volume of precipitated polymeric particles
ΔV_P	volume change in polymerization
v	contribution of autoacceleration in VC polymerization
v	overall polymerization rate
v_{init}, v'_{init}	rate of initiation
$^M v_{init}, {}^P v_{init}$	rates of initiation in monomer- or polymer-rich phases, respectively
v_p	rate of polymerization
$\bar{v}_p$	mean polymerization rate
$v_{p,o}$	rate of polymerization in the absence of transfer agent
$v_{p,s}$	standard polymerization rate
$(v_{pol})_s$	polymerization rate per unit catalyst surface area and time (g polymer m^{-2} s^{-1})
v_s	overall polymerization rate in the stationary state
v_t	termination rate
$v_{t,P}$	rate of termination by combination, generating a macromolecule with degree of polymerization P
W	substituent
w	$v_p/v_{p,o}$
w_i	mass fraction of ith polymer fraction
X_n	fraction of molecules with degree of polymerization n

XT	transfer agent
x	relative fraction of living centres
x_{disp}	mole fraction of chains terminated by disproportionation
x_j	mole fraction of component j
y	$[M_1]/[M_2]$
y	coordinate along the long axis of SIP reactor
Z	substituent
Z	spin trapping agent
Z	contribution of thermal polymerization in the dark, $\bar{v}_p(dark)/v_s$
Z_j	fraction of jth type of centre
$Z_{t,n}$	rate of all kinds of termination of R_n radicals
z	$(^2t - {}^1t)/{}^1t$
α	expansion coefficient $(\partial V/\partial T)_p (1/V)$, (K^{-1})
α	exponent of the Mark–Houwink equation
α	$[M_k^*]/[M_{k-1}^*]$
β	exponent. For the model of closely packed spheres, $\beta = 1/2$; for entanglements $\beta = 1$
γ_1, γ_2	fraction of radicals generated by transfer, consumed by reaction with growing radicals, or inactivated by mutual collisions
δ	reaction constant in the Taft equation
δ_{C_B}	chemical shift in NMR spectra
$\delta_{C\beta}$	^{13}C chemical shift
ε	Schwan ε value
$\varepsilon_a, \varepsilon_r$	resonance energy of the attacking or resulting radical
ε_s	resonance energy of the substrate
η, η_0	viscosity of solution and solvent
η_{rel}	relative viscosity η/η_0
θ	efficiency of catalyst, m_p/n_K
θ	theta temperature
ϑ_1	partial molar volume of solvent
$\varkappa$	conversion, $([M]_0 - [M])/[M]_0$
$\varkappa_i$	conversion of initiator $([I]_0 - [I])/[I]_0$
$\varkappa_i, \varkappa$	conversion of initiator at monomer conversion $\varkappa$
λ	wavelength
μ, μ_0	chemical potential of solution and solvent
μ	growth rate of monomer–polymer particles

μ'_r	rth moment of a statistical distribution
ν	mean kinetic chain length
ν	momentary length of polymer produced from unentangled radicals
ζ_0, ζ_k	friction coefficient of a polymeric segment and of a segment of spherical shape
π	Ludolf number
π	osmotic pressure
ϱ	excursion radius of a radical chain end
ϱ	rate of radical generation in the aqueous phase
ϱ	constant in relations describing the pressure dependence of k_t
ϱ	reaction constant in the Taft equation
ϱ_A	entrance rate radicals into all particles in an emulsion (mol h^{-1} cm^{-3})
$^M\varrho, {}^P\varrho$	density of monomer or polymer, respectively
σ	Taft polar substituent constant
σ	statistical moment referred to the centre of a distribution
τ	mean lifetime of radicals
ϕ	volume fraction of polymer at critical polymer concentration
$^M\phi, {}^P\phi$	volume fraction of monomer and polymer, respectively, in a polymer-rich particle
$\varphi_1(\varphi_2)$	volume fraction of component 1(2)
φ_m	volume fraction of monomer in a particle
$\varphi(t)$	efficiency of radical generation in the light and in the dark
χ	$(S_{12} + S_{21}) - (S_{11} + S_{22})$
$\chi^{\pm}$	$(S_{12}^{\pm} + S_{21}^{\pm}) - (S_{11}^{\pm} + S_{22}^{\pm})$
χ	Flory–Huggins interaction parameter
ψ	randomness coefficient, $(P_{21}/F_1) = (P_{12}/F_2)$
ψ_1	thermodynamic solution parameter
ψ^*, ψ	wavefunction
●	transition metal atom
○	designation of an unspecified active centre
—	designation of an anionic active centre
+	designation of a cationic active centre
*	designational of a radical active centre
★	hot radical
□	vacancy in the coordination sphere of a transition metal; coordination centre

*	designation of an excited molecule (atom);
*	designation of a complex conjugate function
$-^{\bullet}$	anion radical
$+^{\bullet}$	cation radical
× × × ×	crystalline polymer
$\sim$ or $\wwww$	polymer chains of undefined length

Chapter 1

Polymerization

*Even liquid water is a
kind of polymer*
A finding documenting our dependence on polymers

Macromolecules are formed by polycondensation and polymerization. Both reactions have been and are precisely defined and differentiated. For formal reasons, the term polymerization is superior to polycondensation (both lead to the formation of polymers) and polycondensation is then designated by the compromise term condensation polymerization[†]. In my opinion, due to the differences in the mechanism and kinetics of polymerizations and polycondensations, a separate treatment of each of these polyreactions is justified.

The trivial scheme of polymerization is simple

$$IM_n + M \underset{k_d}{\overset{k_p}{\rightleftarrows}} IM_{n+1} \qquad (1)$$

However, it comprises an enormous number of possible ways by which the monomer M can be added to the initiator I or to the chain IM_n growing on the reactive end group, usually called the active centre. In addition, the real situation is complicated by the occurrence of competing and consecutive reactions. To this day, a general theory of polymerizations suitable for a unified explanation of observed phenomena and predicting the behaviour of real polymerizing systems has not been formulated. Truly speaking, present findings are so varied that so far nobody has attempted to unify them.

The progress of our knowledge has led to the description of many kinds of polymerization. Each kind is composed of a large number of special cases but it always exhibits some common features. Let us present some such sets as examples without claiming completeness.

[†] Until recently it has been customary to classify synthetic polyreactions as polymerizations, polycondensations and polyadditions [1]. According to the IUPAC nomenclature [2] the term polymerization is superior to the concepts of addition polymerization and condensation polymerization. Both these subordinate terms are very broadly defined and so far the classification of reactions bearing the features of polymerization and polycondensation is rather arbitrary. Only addition polymerizations are treated in this book. Processes of the polycondensation type are not described.

References p. 26

1. Basic processes and the initiator

Each polymerization has a chain character and is therefore a manifestation of four basic processes: (a) initiation, (b) propagation (growth), (c) termination (ending) and (d) transfer. These four processes determine the course of polymerization and the quality of the product. When they proceed simultaneously, a very complicated situation occurs. The better our knowledge of each part of the overall process, the better our chance of rapid and rational control.

1.1 Separation of basic processes

A complete separation of the elementary processes is possible only in very special and rarely occurring cases (e. g. in living polymerizations with rapid initiation). Even a "mere" change in the velocity of partial steps is of great importance. Such change can be induced in many ways, the most important of which are physical effects (temperature, pressure, viscosity, etc.), choice of initiator and purity of the medium.

The activation energy of propagation is usually lower than that of transfer. Thus by the lowering the polymerization temperature, transfer is relatively suppressed and a product with a higher degree of polymerization is obtained[†]. In viscous medium, the mobility of growing chains is lower than that of the small monomer molecules. The mutual reactions of growing chain ends are limited whereas the transport of monomer to them is unrestricted. A soluble or solid initiator can be selected for ionic polymerizations with a stable or less stable counter-ion. If we do not allow the transfer agent to contaminate the batch, then transfer with this agent will not occur, etc.

1.2 The initiator

Until recently, "catalysts" were discussed in connection with the start of polymerization but today we know that the compounds initiating chain reactions do not satisfy the definition of a catalyst. A catalyst and a substrate form a transition complex which is decomposed to the product and the regenerated catalyst. The path to the product over the catalyst—substrate complex crosses a lower energy barrier (has a lower activation energy) than

---

† The degree of polymerization is the number of monomeric units composing the macromolecule. Chains consisting of large number of monomeric units are designated as high polymers. Their opposite is a low polymer, with an oligomer composed of a few units at the most.

the substrate transformation without a catalyst. Initiating compounds form aggregates with the monomer (or other substances), leading to polymers along a path with a lower peak. However, they are incorporated in the chains and are not generally regenerated. Therefore they are called initiators, and the compounds supporting their effect are called co- initiators.

Only exceptionally in ionic polymerizations are the initiators partly regenerated. Only a fragment of the original molecule or a particle resulting from the reaction of the initiator with the monomer, macromolecule or solvent, etc., may be incorporated in the chain.

Some hesitation still persists over the coordination "catalysts". We know that they are not "real" catalysts but nevertheless the term is used [e. g. Ziegler–Natta (ZN) catalysts]. The nomenclature in the field of initiation is far from being unified. In descriptions of the start of coordination polymerizations, the term catalyst is so familiar that to change it would result in considerable difficulty. Therefore, since many examples of the acceptance of incorrect terms can be found (even the theory of equilibria is called thermodynamics), the usual custom will be observed in this volume.

Active centers are formed by reaction(s) of the initiator and/or of its complex with the first molecule of monomer.

2. Classification according to type of active centre

An important criterion for classification is the type of active centre and depending on its type we classify polymerizations as radical, ionic (which are further classified as anionic or cationic) and coordination.

According to this classification, the polymerization type can usually be easily determined. The structure of the initiator, the manner of its reaction with the monomer, the effects of the medium and last, but not least, sensitive spectroscopic or resonance methods usually, but not always, provide sufficiently convincing information. We know systems containing radical ions. Several years ago it was sometimes assumed that stereospecific polymerizations (now classified as coordination polymerizations) proceed by a radical or cationic mechanism.

It is the radical processes that have the longest tradition. They are the best described and already seem to be approaching completion. The remaining polymerization types are at present at the stage of collecting and sorting of information.

References p. 26

3. Classification according to the reactivity of active centres

The rate of polymeric chain generation used to help in the classification of polymerization processes, and to this day chain polymerizations and step polyadditions are distinguished. This classification was introduced long ago and is mainly based on the observation that the molar mass of the product is sometimes roughly independent of time (conversion) whereas in other cases it increases considerably. Strictly speaking, every polymerization is a chain polymerization and, at the same time, each is a polyaddition.

During the polymerization of isobutyl vinyl ether, macromolecules grow about 10^6 times more rapidly than the chains of poly(tetramethylene oxide) (PTHF) from tetrahydrofuran (THF) (at the same temperature and with the same initiator, $Ph_3C^+SbCl_6^-$). Therefore it is possible, even by relatively rough methods, to record the change in lenght of PTHF macromolecules during the reaction, whereas with poly(isobutyl-vinyl-ether) this is not possible, even by sensitive and rapid methods. Nevertheless, both chain types grow by stepwise monomer addition to carboxonium or oxonium centres, respectively.

The borderline between chain and step polyadditions has never been strictly defined. In fact it is mobile, and if the molar mass change with time is accepted as a criterion, this borderline is expected to change with the development of experimental methods. Polymerizations proceeding on the multiple bonds of monomers or by opening of their rings need not be classified in this way [†].

4. Classification according to number of monomers

The number of monomers and the manner of their alternation in the polymer permits the classification of polymerizations as homopolymerization and copolymerization.

[†] As a process characterizing name, the term step polyaddition is preserved for cases where a monomer with multiple bonds is "copolymerizing" with saturated monomers. The formation of polyurethanes is an example of such a process

$$n\ O{=}C{=}N{-}R{-}N{=}C{=}O\ +\ n\ HO{-}R'{-}OH\ \rightarrow$$

$$\rightarrow\ O{=}C{=}N{-}R{-}NH{-}\underset{O}{\overset{\|}{C}}{-}\left[{-}O{-}R'{-}O{-}\underset{O}{\overset{\|}{C}}{-}NH{-}R{-}NH{-}\underset{O}{\overset{\|}{C}}{-}\right]_{n-1}{-}O{-}R'{-}OH.$$

Reactions of this type resemble polycondensations rather than polymerizations. The above example belongs to the group of polyreactions called migration polymerizations (the H atom migrates from O to N).

Copolymerization is a variant of polymerization where macromolecules composed of two or more kinds of monomer are formed. According to the frequency of entry of various monomers into the chains, copolymers (and even the reactions by which they are formed) are sometimes more closely specified by special attributes. Thus copolymers may be unspecified, statistical, random, periodic, alternating, block or graft.

The building of copolymers even with an unspecified distribution of monomeric units in the macromolecules certainly obeys some rules. Either these rules are as yet unknow or else it remains unclear which of the known rules have been operating; this kind of copolymer is designated poly(M_1-co--M_2 ...) [3] where M_1, M_2, M_3 ... are the various monomers.

In statistical copolymers the sequence of monomeric units obeys some known statistical law, e. g. Markov statistics of zero order (Bernoulli), or of the first, second or higher orders. Such copolymers are designated as poly(M_1-*stat*-M_2-*stat*-M_3 ...) [3].

Random copolymers are a special case of statistical copolymers. The probability of finding a given monomeric unit at any place in the chain is independent of the nature of the neighbouring units (Bernoulli distribution). For such a copolymer the probability of finding the sequence $M_1M_2M_3$, $P[\sim M_1M_2M_3\sim]$ is given by the relation

$$P[\sim M_1M_2M_3\sim] = P[M_1]P[M_2]P[M_3]... = \prod P[i]$$
$$i = M_1M_2M_3...$$

$P[M_1]$, $P[M_2]$, $P[M_3]$ ... are the unconditional probabilities of the occurrence of various monomeric units. Only in these cases may the copolymer be called random and designated as poly(M_1-*ran*-M_2-*ran*-M_3 ...) [3].

In periodical copolymers the monomers are arranged in certain ordered sequences:

$\sim M_1M_2M_3M_1M_2M_3\sim$, $(M_1M_2M_3)_n$;
$\sim M_1M_2M_2M_1M_2M_2M_1M_2M_2\sim$, $(M_1M_2M_2)_n$;
$\sim M_1M_1M_2M_2M_1M_1M_2M_2\sim$, $(M_1M_1M_2M_2)_n$;
$\sim M_1M_2M_1M_3M_1M_2M_1M_3\sim$, $(M_1M_2M_1M_3)_n$;

etc. These copolymers are symbolized respectively by the expressions poly(M_1-*per*-M_2-*per*-M_3); poly(M_1-*per*-M_2-*per*-M_2); poly(M_1-*per*-M_1-*per*-M_2--*per*-M_2); poly(M_1-*per*-M_2-*per*-M_1-*per*-M_3); etc. [3].

A special case of periodical copolymers are the alternating copolymers. They are composed of two monomers, strictly alternating in the chain, $M_1M_2M_1M_2M_1M_2$ ... They are designated poly(M_1-*alt*-M_2) [3].

In the formation of a block copolymer, a second monomer is attached to the growing homopolymer chain, and further growth of the macromolecule proceeds only by homopolymerization of this second monomer. This process

References p. 26

can be repeated, so that the polymeric chain may be composed of two or more sections of varying structure

$$M_1M_1M_1M_1M_1M_2M_2M_2M_2M_2M_2M_2M_3M_3M_3M_3M_3M_3M_3M_3M_1M_1M_1M_1M_1 \sim$$

Various blocks may even be statistical or periodical copolymers. Block copolymers are also formed by the combination of suitable end groups of homopolymers of various monomers. This may occur either by direct combination or by means of a bonding group. According to the IUPAC nomenclature [3] these copolymers are designated as polyM_1-*block*-polyM_2-*block*-polyM_3-*block*-polyM_1.

When the end of one macromolecule reacts with atoms inside the chain of another homopolymer, we speak of grafting.

$$\sim M_1M_1M_1M_1M_1M_1M_1M_1M_1M_1M_1 \sim$$
$$|$$
$$M_2$$
$$|$$
$$M_2$$
$$|$$
$$M_2$$
$$|$$
$$M_2$$
$$\wr$$

A graft copolymer may, of course, also be formed by polymerization of the monomer on an active centre placed on some atom group inside the chain built of other monomer(s). These are designated as polyM_1-*graft*-polyM_2 [3].

Copolymerizations are of considerable theoretical importance. They enable us compare the reactivity of various monomers, the reactivity of active centres derived from various monomers, as well as reactivity changes caused by substituents and physical effects. The practical importance of copolymers is great. They form a considerable part of the industrially produced plastics. One of the advantages of copolymers is the chemical bond between their constituent units, i. e. separation of components cannot occur.

Homopolymers are sometimes modified by a mechanical admixture of another homopolymer. As only about 5 % of pairs of all known polymers are mutually miscible, compatibility may be a problem in mixtures (blends). Copolymerization is technically applied to overcome, for example, the brittleness of polystyrene, polypropylene and PVC. It is also applied for improving the curing properties and modifying the viscoelastic properties of rubbers. By copolymerization, the relation between the hydrophobic and hydrophilic properties of macromolecules can also be modified. Their resistance to solvents may be enhanced.

5. Classification according to the number of phases at the beginning and during reaction

The polymerizing system may consist of one or more phases, According to this we classify polymerizations as (a) homogeneous, (b) homogeneous changing to heterogeneous; precipitating (slurry polymerization) and (c) heterogeneous.

When the monomer is a solvent both of the corresponding polymer and of the initiator, polymerization proceeds in a single phase; it is homogeneous. The dissolving role of the monomer may sometimes be taken over by the solvent. Of course, homogeneous polymerizations are not limited to the liquid phase. Solid pure trioxane can be transformed into a highly crystalline polymer by γ radiation.

Studies of the mechanism and kinetics of homogeneous polymerizations in the liquid phase are simpler than in the other cases. Therefore they are preferred when possible, or at least the partial problems of inhomogeneous polymerizations are modelled in the homogeneous phase.

Polymers that are insoluble in their own monomers or in the monomer— —solvent mixture separate as a solid phase from the homogeneous solution in the course of the reaction. Such polymerizations are designated as precipitating. Their course is complicated by sorption of the initiator, growth centres, monomer and solvent on the solid surface. A detailed investigation of the consequences of the whole complex of factors is very difficult.

Heterogeneous polymerizations proceed in two or more phases. Heterogeneity may be caused by the presence of a solid or of a gaseous phase or else the liquid monomer may be dispersed in another liquid with which it does not dissolve. Very important are the systems (a) with a solid initiator and (b) of two practically immiscible liquids. The former is useful for producing stereospecific polymers which are usually formed by a coordination mechanism. The latter makes possible an elegant and efficient removal of the heat of polymerization and it is applied technically with radical polymerizations in suspension or emulsion.

Polymerization from the gaseous phase (disregarding dimer to tetramer formation) is an example of a heterogeneous reaction where the active centres are present in the condensed phase and the monomer in the gaseous phase. Polymerization does not, of course, proceed in the gaseous state but on the surface of the component carrying the active centres, i. e. also in the condensed phase. These polymerizations are of industrial importance.

The systems suitable for the polymerization of the most common monomers are summarized in Table 1.

The homogeneity of polymerizing systems is closely connected with the physical state of the monomer. In the liquid state, several polymerization variants are possible, each of which deserves to be mentioned in some detail.

References p. 26

TABLE 1

Standard methods of polymerization of common monomers

Monomer	Abbreviation[a]	Homogeneous	Precipitation	Heterogeneous	
				Insiluble initiator	Initiator in solution
Ethylene	E		+	+	
Propene	P			+	
1-Butene	1-Bu	+		+	
2-Methyl propene (isobutene)	IB	+	+		
1,3-Butadiene	B	+		+	+
Styrene	S	+		+	+
Vinyl chloride	VC		+		+
Vinylidene chloride	VDC		+		+
Vinyl acetate	VAC	+			+
Acrylonitrile	AN		+		+
Vinylidene cyanide	VDN		+		+
Methyl acrylate	MA	+			+
Methyl methacrylate	MMA	+			+
Tetrafluoroethylene	TFE		+		+
Formaldehyde	Fd		+		
Trioxane	TOX		+		
Oxirane	EOX	+		+	
6-Hexanelactam	A	+		+	
Tetrahydrofuran	THF	+			
Octamethylcyclotetrasiloxane	D_4	+			

[a] The abbreviations have not been internationally recognized.

6. Liquid-phase polymerization

All polymers are formed in a condensed phase. In the liquid state polymerizations start (a) in bulk, (b) in solution, (c) in suspension and (d) in emulsion.

With a few exceptions, polymerizations are strongly exothermic (see Chap. 5, Sect. 1.1). Therefore removal of the heat of polymerization is the main problem affecting the construction of polymerization reactors and the technology of the process. The kinetics of the process also determines the technology, with the stringent requirement of the purity of medium. In chemistry perhaps only the preparation of semiconductors poses higher requirements on the purity of raw materials and on the prevention of contamination than does macromolecular syntheses. The removal of the heat of polymerization and the requirements concerning the purity of the medium and product thus determine the method of polymerization in the liquid phase.

A very pure polymer is produced by homogeneous polymerization in bulk. In this case, the heat of reaction can be removed only by the reactor walls. With a wall area large relative to the reactor volume, kilogram batches can be slowly polymerized with success. In this way, plates of optically pure polymer are produced from methyl methacrylate. In precipitating polymerizations, the generated heat can be used for evaporating part of the monomer or solvent. The vapours of these compounds are condensed outside the reactor and the liquid is recycled. The cooling area is inereased in this way, maintaining isothermal conditions even with large batches.

The solvent may, of course, also be used for kinetic or technological reasons. Monomer concentration is reduced and polymerization is usually slower. The transport of solutions is often easier than the transport of the solid polymer and some polymers are used as solutions, e. g. paints and coatings. In such cases, a suitably adjusted polymerization mixture may by the final product.

However, the introduction of the solvent into the polymerization medium poses new problems. The solvents must be pure, without inhibiting and transfer agents. Every solvent takes part in the polymerization process; its effect is almost never limited to the mere physical dilution of the monomer. It solvates the active centres; it participates in processes connected with energy and impulse transfer; often it serves as a transfer agent (so that the degrees of polymerization of solution-polymerized products are usually lower compared with bulk-polymerized polymers); it may form complexes with some component of the system; it modifies initiation efficiency by the cage effect; etc.

In most cases, the solvent must be removed from the polymer. After polymerization is complete, either the product is precipitated from the solution or the solvent is distilled off. With precipitation polymerizations the separation is automatic. Residual solvents, of course, must always be removed, mostly by a suitable drying process.

An elegant way of removing the heat of reaction occurs in suspension or emulsion polymerizations. Suspension polymerization is kinetically simpler. It really proceeds in bulk, as every monomer–polymer drop of the suspension is an individual "reactor". These particles are small (100–150 μm), they have a large surface area, and the heat is effectively transferred by water to the cooling jacket. The polymer is contaminated by the tenside used for suspension stabilization. Therefore it must be washed, and even so it is sometimes less suitable for high-performance electrotechnical applications than a polymer prepared in bulk. For the suspension process, the initiator must be soluble in the monomer.

References p. 26

Emulsion polymerization is more complicated. It leads to higher polymers which are formed more rapidly than under otherwise comparable conditions in suspension. The difference lies in the localization of initiation in the aqueous phase (the initiator must be insoluble in the monomer), in the solubilization of the monomer in micelles, and usually in the presence of not more than one growing radical in a particle (see Chap. 5 Sect. 4.2). Water is usually removed by evaporation from the polymer emulsion. The polymer is contaminated by non-volatile substances. In large-scale production, always in big reactors, emulsion polymerizations are usually continuous, suspension polymerizations discontinuous. Today, the 80 m^3 reactors for the polymerization in suspension are used, but larger reactors up to 200 m^3 are being developed. To the best of the authors knowledge, all existing industrial suspension and emulsion polymerizations use water as the dispersing phase. In principle, non aqueous media could be used.

TABLE 2

Most frequently used methods of polymerization of common monomers in the liquid phase

Monomer		Bulk	Solution	Suspension	Emulsion
Abbreviation[a]	Name				
E	Ethylene		+		
P	Propene	+	+		
1-Bu	1-Butene		+		
IB	Isobutene	+	+		
B	1, 3-Butadiene		+	+	+
S	Styrene	+	+	+	+
VC	Vinyl chloride	+	+	+	+
VDC	Vinylidene chloride	+	+	+	+
VAC	Vinyl acetate	+	+	+	+
AN	Acrylonitrile	+	+	+	+
VDN	Vinylidene cyanide	+	+	+	+
MA	Methyl acrylate	+	+	+	+
MMA	Methyl methacrylate	+	+	+	+
TFE	Tetrafluoroethylene	+	+	+	+
Fd	Formaldehyde	+	+		
TOX	Trioxane	+	+		
EXO	Oxirane		+		
A	6-Hexanelactam	+	+		
THF	Tetrahydrofuran	+	+		
D$_4$	Octamethylcyclotetrasiloxane	+	+		

[a] The abbreviations have not been internationally recognized.

Ionic and coordination polymerizations are inhibited by the presence of a certain amount of water. In this case, the amount of water means tens to hundreds parts per million (ppm). Therefore the emulsion and suspension processes in water are limited to monomer polymerizing by the radical mechanism. The most frequently used methods of liquid-phase polymerization of the more conventional monomers are summarized in Table 2.

7. Classification according to the arrangement of chains formed

The manner of monomer addition determines the arrangement of the growing chains. According to this, we distinguish the polymerization mechanisms (a) uncontrolled propagation and (b) stereospecific.

An irregular[†] or atactic[††] chain is generated by non-stereospecific polymerization with undirected monomer addition to the growing centre. The placement of monomers in the chain varies statistically, head-to-tail and head-to-head for vinyl monomers and 1, 2 and 1, 4 (*cis* or *trans*) for the dienes. In fact, each generated chain is a separate chemical individuum and, strictly speaking, ist segments are not mutually similar. The chain cannot, in practice, be folded into a lamella and therefore such polymers do not crystallize. A completely irregular or atactic chain is, of course, a fiction. Even when polymerization proceeds in the absence of explicitly expressed directing factors, fluctuations occur which always cause a certain regularity in the arrangement of some parts of the macromolecules. Most polymers formed in real systems are able to crystallize partly. The opposite case is also known. Polymerizations which have been considered as strictly stereospecific produce small amounts of atactic polymers.

Stereospecific polymerizations were discovered at the beginning of the 1950s by Ziegler [4]. By the interaction of $TiCl_4$ with organoaluminium compounds, Ziegler obtained a product initiating rapid polymerization of ethylene to high-molecular weight, high-density polyethylene. The reaction proceeds at low temperature and low pressure. No wonder it attracted great

---

[†] According to the IUPAC definition, an irregular polymer is one whose macromolecules are composed of several types of constituonal units in a varying sequential order [2].
[††] An atactic polymer is a regular polymer with macromolecules composed of a certain number of statistically distributed configurational units. The constitutional unit is a type of atom or group of atoms composing the macromolecule (e. g. $-[CH_2-CHPh]-$ or $-[CHPh-CH_2]-$ in polystyrene). The configurational unit is a constitutional unit with one or several stereoisomeric centres. These definitions would require a more detailed explanation. In this volume they will only rarely be used, the stereochemistry of polymers is a special branch of macromolecular chemistry. More information can be found in the original literature [2].

References p. 26

attention. Many workers have taken up its investigation, and important rules were soon revealed. Transition metals of Groups IV–VI and VIII together with organometallic compounds of Groups I–IV yielded catalysts for the polymerization of ethylene, 1-alkenes and dienes. The most suitable transition metal compounds were oxyhalides and halides of Ti, V and Cr in a higher oxidation state.

Natta et al. [5] have shown that compounds of transition metals in the lower oxidation state together with organometallic compounds yield catalysts suitable for the preparation of crystalline polyalkenes with a high melting point. The similarity of the enumerated catalyst types is evident. Today they form a specific group in the class of coordination catalysts: the Ziegler–Natta (ZN) catalysts. The original ideas, considering the solid surface of ZN catalysts as the determining factor for the generation of structurally ordered polymers, were soon abandoned. Today both solid and soluble ZN catalysts are known; stereospecific polymerization can proceed even in homogeneous medium.

The importance of ZN catalysts in industrial production is enormous. They enable us to produce polymers with stereoregular or tactic structures–isotactic (di-isotactic) or syndiotactic (Chap. 5, Sect. 4). They are used in the synthesis both of plastics and of elastomers.

8. Kinetic factors

Some features of polymerizations are manifested in the kinetic factors and may be stationary (in special cases either living or immortal) or non-stationary.

The character of polymerization is determined by a number of factors. In a first approximation, let us regard concentration and reactivity of active centres as the most important. When both these quantities remain constant, the polymerization is stationary. In such process the reaction rate, and consequently also the amout of heat of polymerization released per unit time, change only in their dependence on monomer concentration (provided the reaction is not of zero order with respect to M). A special case of stationary reactions are the living polymerizations discovered in 1956 by Szwarc [6]. Under certain conditions, a living polymerization is a mere propagation without termination and transfer; it exhibits stationarity of the second order. Most radical reactions are (pseudo)stationary over the greater part of their duration. In special cases, even radical polymerizations may be living. The living mechanism is more frequent in anionic and (less often) in cationic polymerizations.

Very recently, a paper on immortal polymerization [7] has been published. This is a special (and so far rare) kind of polymerization where active centres do not decay but rather participate in transfer, yielding macromolecules of uniform length. The described exemplary polymerization of methyloxirane in the presence of (tetraphenylporphinato) aluminium chloride proceeds even in the presence of CH_3COOH or HCl.

Non-stationary polymerization are complicated from the kinetic point of view. The changing concentrations of active centres, of monomer and possibly even of further components produce conditions unsuitable for an analysis of the process. Even technical and technological difficulties occur. Nevertheless, these have to be solved as most known coordination and cationic, and a considerable number of anionic, polymerizations are non-stationary. Information on the polymerization mechanisms of the more conventional monomers are summarized in Table 3.

TABLE 3

Usual mechanism of polymerization of common monomers

Monomer	Polymerization[a]			
	Radical	Coordination	Cationic	Anionic
E	2	3		
P		3		
1-Bu		3		
IB			3	
B	2	3	3	1, 3
S	2(1)	1, 3	1, 3	1, 3
VC	2			3
VDC	2			
VAC	2			
AN	2			1, 3
VDN	2			3
MA	2(1)			1, 3
MMA	2(1)			1, 3
TFE	2			
Fd			3	3
TOX			3	
EOX		3	3	1, 3
A			3	3
THF			1, 3	
D_4			3	1, 3

[a] 1 = living; 2 = stationary; 3 = non-stationary

9. Special kinds of polymerization

Some kinds of polymerization with a characteristic structure of the monomer or polymer or with a characteristic initiation or propagation mechanism have special names. These reactions will be treated in greater detail in Chap. 5.

9.1 Ring-opening polymerizations (ROP)

These generally differ from the more conventional processes proceeding on multiple bonds. As indicated by their name, they occur by the breaking of some (usually polar) bond between two atoms of a cyclic monomer, and the product of this opening, original cyclic but now linear, forms a part of the growing chain

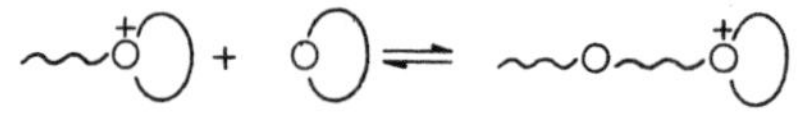

By ring-opening polymerization, polymers can formed from cycloalkenes, but mostly from simple and more complicated heterocycles containing heteroatoms such as O, N, S, P and Si. Ring-opening polymerizations are mostly initiated by the ionic mechanism.

Metathesis is a special kind of ROP and also of disproportionation. The monomers are cycloalkenes, i. e. compounds which are both cyclic and contain a double bond. Both linear and cyclic, but always unsaturated, macromolecules are generated. The similarity of disproportionation and metathesis is indicated by the scheme

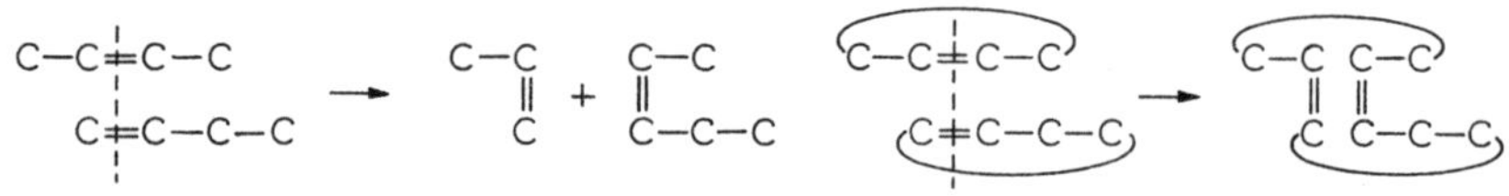

Disproportionation   Metathesis

Metatheses mostly proceed by a coordination mechanism on catalysts of the Ziegler–Natta type.

9.2 Cyclopolymerizations

Cyclopolymerizations are a type of reverse ROP. Non-cyclic monomers, mostly conjugated or non-conjugated dienes, yield rings composing the resulting macromolecule. For example

Cyclopolymerization may proceed by a radical, ionic or coordination mechanism, of course, according to the conditions but usually only by one of these.

9.3 Polymerization by activated monomer mechanism

The active centre usually appears simultaneously as the end group of the growing chain. In some cases, the Gibbs energy of the active centre–monomer system is more negative for the chain–activated monomer pair

$$\sim\!\!\sim\!\!O + M \rightleftharpoons \sim + M^O$$

The equilibrium in these cases is shifted to the right; the monomer yields a formation carrying charge (energy), and only the monomer activated in this way is added to the end of the (usually) neutral chain. This type of addition is typical for the ionic polymerization of lactams.

9.4 Isomerization polymerizations

These are manifested by the formation of a polymer with a structure which can only be explained by monomer isomerization in the course of addition. An active centre rich in energy is sometimes transformed to a more stable form by a transfer of atoms or atom groups, and further monomer is added only to this more stable form

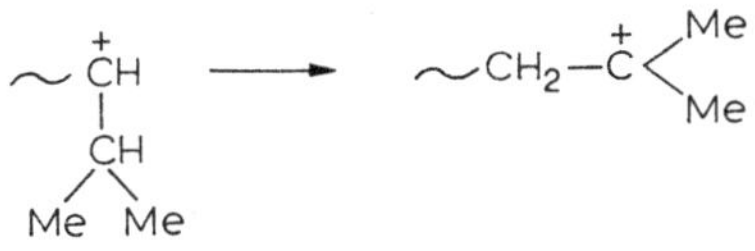

$$\sim \overset{+}{C}H\!-\!\underset{\underset{Me\;Me}{\diagdown\!\!\diagup}}{CH} \longrightarrow \sim CH_2\!-\!\overset{+}{C}\!\!<\!\!\overset{Me}{\underset{Me}{}}$$

9.5 Topochemical polymerization

A great majority of polymerizations are simultaneously affected by many physical and chemical factors, and their course is the result of a superposition of these effects. Only in rare cases does one of these factors dominate and the polymerization is formally simplified. In topochemical polymerizations, the growth of macromolecules is governed by forces in the crystal lattice of the monomer. Solid-state polymerization of trioxane (trioxacyclohexane) is a typical example of topochemical polymerization.

References p. 26

9.6 Popcorn polymerization

This is a kind of heterogeneous polymerization yielding a cross-linked, insoluble, weakly swelling, sponge-like polymer with many voids, of white opalescence. The inhomogeneities are an inherent part of the process; they are not due to gases or vapours. Popcorn (proliferous, cauliflower) polymers are often formed spontaneously together with the glassy part of cross-linked polymers and copolymers; as such they may be the cause of technological difficulties. Pronounced manifestations of popcorn polymerization may be observed in radical copolymerization of butadiene with methyl acrylate [8] (cross-linking occurs by transfer with polymer), in copolymerization of styrene with divinylbenzene, and in many special cases.

9.7 Group transfer polymerization

This has recently been described as a new method of polymerizing α, β-unsaturated esters, amides and nitriles with organosilicon derivatives of ketene as initiators and Lewis acids or bases as catalysts. The products formed have the characteristics of living polymers [9].

References

1 See, for example, Terminology of Plastics and Rubber, Czechoslovak State Standard ČSN 640 001.
2 IUPAC, Macromolecular Division, Nomenclature Commission, Basic Definitions of Terms Relating to Polymers, Pure Appl. Chem., 40 (1974) 477.
3 IUPAC, Macromolecular Division, Commission on Macromolecular Nomenclature. Source-Based Nomenclature for Copolymers, Pure Appl. Chem., 57 (1985) 1427.
4 K. Ziegler, BE 534, 792 and 534, 889 (1955); BE 540, 459 (1956).
5 G. Natta, P. Pino and M. Farina: Int. Symp. Macromol. Chem., Milan, 1954; Ric. Sci. Suppl. A, 25 (1955) 120.
6 M. Szwarc, Carbanions, Living Polymers and Electron Transfer Processes, Interscience, Wiley, New York, 1968.
7 S. Asano, T. Aida, and S. Inoue, J. Chem. Soc., Chem. Commun., 17 (1985) 1148.
8 J. W. Breitenbach and O. F. Olaj, in R. Houwink and A. T. Staverman (Eds.) Chemie und Technologie der Kunststoffe, Vol. I, Akademische Verlagsgesellschaft, Leipzig, 4th edn., 1962, p. 320.
9 W. B. Farnham and D. Y Sogah, Polym. Prepr. Am. Chem. Soc. Div. Polym. Chem. 27(I) (1986) 1670.

Chapter 2

Monomers

> *It is a crime to make poor products
> from valuable raw materials*
>
> Common experience

By suitable activation, relatively stable compounds, e.g methane or benzene, can be made to polymerize. However, the term "monomer" is usually used to designate unsaturated compounds or molecules with reactive groups, but always those that predominantly react by addition to an active centre and give rise to the same kind of centre in the reaction[†]. The subject of interest in this volume are polymerizing, not polycondensing, monomers. Even so, there are very many of them. An overwhelming majority of these monomers are organic compounds composed of C, H, O, N, S, P and also Si.

1. Classification of monomers

The great number of monomers and the diversity of their behaviour make necessary their rational classification. The various systems will reflect the profession of the classifier. Thus, an organic chemist is likely to classify the monomers in a different way from a physicist. Macromolecular chemistry operates on the basis of information resulting from a whole complex of scientific disciplines. A discussion of the set of monomers from several points of view is indispensable.

1.1 Chemical classification

Each monomer is placed in the corresponding category according to the type and arrangement of atoms in its molecule. The generalized experience concerning the behaviour of the other members of the group is also valid for the newly placed monomer.

[†] In copolymerizations, several kinds of active centre can alternate in a certain order.

References pp. 69–74

1.1.1 VINYL AND VINYLIDENE MONOMERS

This group has been best investigated theoretically, and is the most used in large-scale production. The basic members are ethylene and its derivatives. According to the type of substituent, the group contains two subgroups, alkenes (olefins) and polar vinyl monomers.

The number and position of substituents are important. One bulky group, e.g. naphthyl, or two substituents on a single olefinic carbon do not usually hinder polymerization. On the other hand, an ethylene derivative with a substituent on each of the two carbons does not undergo radical polymerization at conventional pressures and temperatures. Usually, however, it can be copolymerized with a suitable monomer or it can be polymerized by an ionic mechanism. With a larger number of substituents, the tendency to polymerization is further limited. The small fluorine atom represents an exception, as do the cases discussed in Sects. 1.3 and 5.2.

(a) *Alkenes*

The most important representatives are the lowest 1-alkenes, ethylene and propene. Ethylene is not particularly easily polymerized by radical or ionic mechanisms. Its importance as a monomer was greatly enhanced by the discovery of coordination polymerizations. Propene is oligomerized by radical and ionic initiators. This explains the importance of Natta's modification [1] of Ziegler [2, 3] catalysts, enabling "inferior" raw materials to yield high-quality polymers.

Similarly, 2-methylpropene (isobutene) is an important monomer. It only polymerizes by a cationic mechanism, and its copolymers with dienes are known as butyl rubber. Higher 1-alkenes (1-butene, 1-hexene, 1-octene) are important copolymerization components [4, 5]; they produce tailored branching of some polyethylene types prepared by a coordination mechanism. Longer-chain alkenes (C_{10}, C_{12}, C_{16}) are also sometimes used as comonomers [6] but more often they are employed as oligomers with outstanding viscosity characteristics in modern lubricating oils. They are in great demand as raw materials for some organic chemistry products (higher alcohols, detergents ...). Other alkenes have not gained technical importance so far. Nevertheless they remain a subject of theoretical interest, e.g. the cationically polymerizing 3-methyl-1-butene or 4-methyl-1-pentene [7].

Of great importance are the ethylene derivatives with aromatic substituents. Styrene (vinylbenzene) is one of the monomers produced industrially in large volume. Polystyrene and styrene copolymers still belong to the important representatives of modern plastics and rubbers. Styrene can be polymerized by any of the known procedures. It has suitable physical properties, and therefore it is one of the most frequently studied monomers. It also often serves to model situations observed in work with other, less accessible or less manageable, monomers.

Ethylenes carrying a substituent even larger than phenyl are also easily polymerized. Disubstituted derivatives such as 1,1-diphenylethylene and 1,2-diphenylethylene, on the other hand, can only be dimerized. The behaviour of alkenes of this group can hardly be described by a single scheme. Exceptions to elaborately derived rules often occur. The behaviour of specific representatives is, of course, known to some extent. So α-methylstyrene is polymerized cationically just as easily as styrene whereas the rate, under otherwise similar conditions, is down to one half for *trans-β*-methylstyrene and to one third for the *cis* derivative. The type of initiator employed is very important. Of two polymerizing monomers, the slower one may become quicker by a change of initiator of the same polymerization type [8]. These monomers are evidently very sensitive to side effects, and a "small" change may result in large kinetic or stereochemical consequences.

(b) *Vinyl and vinylidene monomers with polar substituents*

These monomers usually polymerize by classical methods, i.e. radical or ionic, more readily than ethylene. On the other hand, they are too good as electron donors; in coordination polymerizations they act as catalytic poisons. The group of polar vinyl monomers is very large. Mostly these compounds are of "only" theoretical interest. Many of them are, however, technically and socially important, and the exploitation of others is anticipated.

Reactive substituents can undergo chemical transformations [9]. Particularly in ionic polymerizations, the number of possible reactions of the monomeric molecules is broadened considerably. This limits the possibility of employing some polymerization types. So, for example, vinyl chloride (VC) cannot be polymerized on cationic or carbanionic centres because these would be destroyed by reaction with Cl— bonds in the polymer. Technically, the most important representatives of simple vinyl monomers are VC, methyl methacrylate (MMA) and acetonitrile (AN). Other monomers of this class, though important and interesting, need not be mentioned at this stage. They will be discussed in the pertinent parts of the text.

1.1.2 DIENES

Dienes are very important monomers. The molecule of any diene may carry various substituents. In spite of the great number of possible variants, fewer are employed in production and science than in the case of vinyl monomers. The mutual position of double bonds, cumulative, conjugated or isolated, is of great importance.

References pp. 69–74

(a) *Conjugated dienes*

These are the most important. The two double bonds mutually activate each other; conjugation is essentially not destroyed by addition to the growing chain end. Therefore the conjugated dienes are difunctional monomers. They are polymerized by a relatively "simple" mechanism. Of all the polymers generated in living tissues, we have so far been able to imitate most closely natural rubber, poly-*cis*-1,4-isoprene. Butadiene, isoprene and chloroprene are the dienes most often employed in macromolecular chemistry.

The contribution of conjugation in electron transfer during addition to an active centre is qualitatively equal for butadiene and, for example, acrylate:

$$\underset{|}{\overset{|}{C}}=\underset{|}{\overset{|}{C}}-\underset{|}{\overset{|}{C}}=\underset{|}{\overset{|}{C}}, \quad \underset{|}{\overset{|}{C}}=\underset{|}{\overset{|}{C}}-\overset{|}{C}=O.$$

Due to oxygen electronegativity and substituent effects, the two compounds need not react in the same way. Nevertheless, some tendency to similar behaviour persists.

(b) *Non-conjugated dienes with isolated double bonds*

These react as if each double bond is independent. They react as tetrafunctional monomers yielding branched or cross-linked macromolecules[†].

Divinylbenzene (DVB) also behaves as a non-conjugated diene; it can remain difunctional only upon the interruption of conjugation in the aromatic ring. This, of course, does not occur. Non-conjugated dienes are used in crosslinking copolymerizations for the preparation of insoluble three-dimensional macromolecules (especially DVB for ion-exchanger backbones) and for scientific purposes.

(c) *Compounds with cumulative double bonds*

These compounds are not used as monomers. Propadiene is a strong catalytic poison in coordination polymerizations, similar to the configurationally related ketene. On the other hand, CO_2, with a similar bond configuration, can be copolymerized under certain conditions (see Chap. 5, Sect. 5.8).

[†] Cross-linked macromolecules are sometimes designated as three-dimensional. Strictly speaking each macromolecule is three-dimensional, even when regarded in hypothetical, exactly linear conformation, not spatially ordered in a crystal or in a random coil. With non-branched macromolecules, the transversal dimensions are sometimes neglected with respect to length. All three dimensions of cross-linked (network) macromolecules are mutually comparable.

1.1.3 ALKYNES

These do not, so far, constitute industrially important monomers. Nevertheless, they do have some technical importance which is documented by the number of published studies of their polymerization. Acetylenes yield chains with a conjugated system of double bonds with semiconducting properties. It is probably just this possibility of conjugation in the generated chain that prevents formation of three-dimensional structures. Acetylene as such is a potentially tetrafunctional monomer.

Even the polymerization of di-ynes need not lead to branching. Behaviour according to the scheme

$$n\ R_1-C\equiv C-C\equiv C-R_2 \longrightarrow$$

was in fact observed [10]. The degree of polymerization is not usually high. The properties of the acetylene monomer can be affected by substitution. The polymer chains with conjugated multiple bonds are, of course, coloured.

1.1.4 CYCLIC MONOMERS

A double bond can be formally regarded as a limiting case of a two-membered ring. Molecules with three or more atoms in a ring can also be polymerized. The mechanism and course of the reaction will, indeed, be different, and strongly dependent on the type and number of ring atoms, their arrangement and substitution. The driving force of the polymerization of cyclic monomers consists of steric and acidobasic contributions. Less strained rings are more stable and their polymerization is driven by entropy.

(a) *Cycloalkanes*

These are not frequently used monomers. There exists a clear connection between the strain of various members in the series of cyclic hydrocarbon molecules and their heats of combustion (see Table 1). The high heats of combustion of the first members are the consequence of the C—C bond angle deviation from 109°28'. In cyclohexane, the most stable cycloalkane which can exist in the chair conformation, the C—C bond angle value deviates very little from that observed in unstrained compounds. Cyclopentane exhibits the smallest deviation of the C—C bond angle from the theoretical value. Its higher heat of combustion is due to steric interactions of pairs of neighbouring hydrogen atoms. A similar situation is observed with cycloheptane [12a].

TABLE 1

Strain in cycloalkanes

Number of ring members	$\Delta H_c/n$ [a]	Angular strain[b,c]
3	697.4	24°44'
4	686.3	9°44'
5	664.3	0°44'
6	658.9	−5°16'
7	662.6	−9°51'
8	663.9	
9	664.7	
10	663.9	
11	663.1	
12	660.1	
15	659.3	
∞[d]	658.9	0°

[a] Heat of combustion, ΔH_c (kJ mol^{-1}), for a methylene group.
[b] Defined as $(1/2)(109°28' -$ respective bond angle) assuming a strictly planar conformation.
[c] Atoms in the cycle are arranged in space, therefore the angular strain is not a measure of ring strain. (Scientific editor's note.)
[d] Linear n-alkane.

Cyclopropane and cyclobutane can be opened by aggressive agents. The same agents also attack the hydrocarbon chain [13], so that three- and four-membered hydrocarbon rings are not suitable monomers. Cyclic compounds with a higher number of ring atoms are even less suitable because of the lower ring strain.

Bicyclic systems with a deformed carbon-carbon bond are more easily polymerized with the opening of one or both cycles. So far, however, such monomers are rather rare.

(b) *Cyclic alkenes, non-conjugated and conjugated cyclic dienes*

These are used more often than cycloalkanes; nevertheless they are far from being "conventional monomers". They polymerize either as 1, 2-disubstituted alkene derivatives [14] (without ring opening) or else the cyclic monomer is split, yielding a macrocycle or a linear chain (metathesis).

Probably the most often polymerized member of the first group is indene, and often-studied representative of the second group is norbornene [15].

(c) *Aromatic compounds*

Aromatic compounds polymerizing by substitution on the aromatic ring are found among conventional compounds; e.g. benzene [16]. So far they have not gained importance as monomers. Unsaturated compounds with aromatic substituents belong to a different class.

(d) *Heterocycles*

Typical bond lengths and bond angles in heterocyclic molecules are shown in Table 2. Introduction of an electronegative atom into a ring built of carbon or silicon atoms greatly increases the susceptibility to attack by nucleophilic or electrophilic agents. Such compounds often polymerize by an ionic mechanism. Many are easily accessible, and are therefore widely employed both industrially and in theoretical studies.

TABLE 2

Bond lengths and angles in heterocycles[a]

Bond type	Length (nm)	Angle
C–C–C	0.154	109°28'
C–N–C	0.147	109°
C–O–C	0.144	111°
C–S–C	0.182	100°
Si–O–Si		134° [b]

[a]Ref. 17.
[b]Ref. 18.

There exists an enormous number of heterocyclic compounds; even a rough classification and characterization aspiring to completeness would take too much space. Therefore only the most common ones will be mentioned here. Discussion of their properties will indicate a great deal, even about those that have been omitted.

(i) *Oxiranes*. The molecular structure of oxirane is [12b. 19, 20]

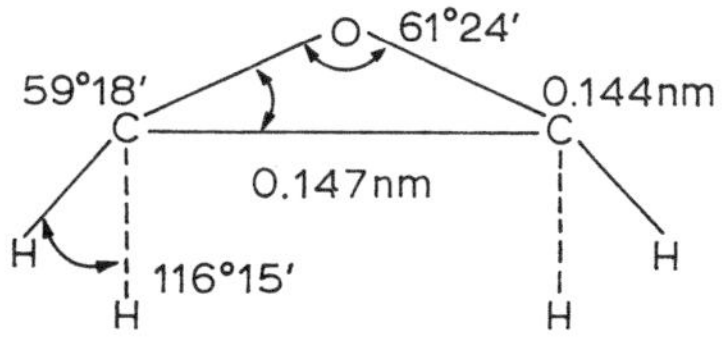

The strain amounts to 113,8 kJ mol^{-1} [21]. The polar C—O bond is generally broken by ionic processes. However, some exceptions exist. In hydrogenation, the ring is opened by a radical mechanism [22]; in oxirane with very polar substituents (tetracyano-oxirane) even the C—C bond may be broken [23].

Acids, HX, react with oxiranes in several steps [24].

$$H^+ + CH_2\text{—}CHR \xrightleftharpoons[]{K} CH_2\text{—}CHR \xrightarrow{k_1, S_N1} HOCH_2CHR^+$$

with k_2, S_N2 branch with $+HX$ giving

$$HOCH_2CHRX \text{ (or } HOCHRCH_2X) + H^+$$

The oxirane ring is probably broken by both the S_N1 and S_N2 processes, depending on the reaction conditions [25–28]. In acid-initiated reactions of mutually competing heterocycles of the ether type (e.g in copolymerization), the reactivity of epoxides is lower than would be expected from their high internal strain [29], and it depends more on their basicity [30].

Bases also break the oxirane ring in several steps [31].

$$BI^- + CH_2\text{—}CHR \xrightarrow[S_N2]{k_4} B\text{—}CH_2\text{—}CHOI^- \text{ (or } BCHCH_2OI^-)$$
$$\qquad\qquad\qquad\qquad\qquad\qquad\qquad\; |\qquad\qquad\qquad\; |$$
$$\qquad\qquad\qquad\qquad\qquad\qquad\qquad R\qquad\qquad\qquad R$$

$$BCH_2\text{—}CHOI^- + HB \rightleftharpoons BCH_2CHOH + BI^-$$
$$\qquad\; |\qquad\qquad\qquad\qquad\qquad\qquad\; |$$
$$\qquad R\qquad\qquad\qquad\qquad\qquad\qquad R$$

The rate-determining step is the S_N2 reaction; oxiranes with a more electronegative substituent are more reactive.

From the point of view of macromolecular chemistry, coordination of oxiranes to metal compounds is an important reaction

$$CH_2\text{—}CHR + \overset{\delta+}{M_t}\text{—}\overset{\delta-}{Y} \rightleftharpoons CH_2\text{—}CHR \longrightarrow M_tOCH_2CHRY$$

where M_t = Li, Mg, Al, Sn, Fe; Y = CO_3^{2-}, —OH, —OR, —Br, —H, ... The electron density on the oxirane oxygen is lowered by coordination; this facilitates attack of ether by a nucleophilic ligand.

(ii) *Oxetanes, oxolanes and higher cyclic ethers* [32]. Four-membered rings are opened easily, five-membered less readily, while the six-membered rings do not polymerize at all. Larger cycles can again be polymerized, though with some difficulty. In four-membered oxetanes, the driving force results from mutual hydrogen atom repulsion rather than from ring strain. The effect of substituents (such as —CH_2Cl, —CH_3, —OR, —CN) is reflected in the quantitative parameters of the reaction; qualitatively it remains unchanged.

Unsubstituted tetrahydrofuran does polymerize, probably due to hydrogen repulsion.

Substituted THFs do not polymerize under normal conditions. Similarly polymerization has not been observed with six-membered cyclic ethers which assume an unstrained conformation, thus lacking the driving force for polymerization. With larger cycles, polymerization can occur, in spite of lack of strain, due to the low probability of closure in rings that have once been opened [33] (see entropic term, Chap. 5 Sect. 1.1).

(iii) Other heterocycles. In cyclic formals of the general formula

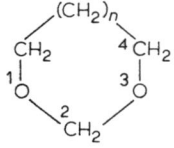

the tendency to polymerization is again a function of ring strain and of hydrogen or substituent interactions. The six-membered formals all polymerize; the twelve-membered rings are, relatively, the least ready to polymerize [34] and the eight- and nine-membered are the most easily polymerized (up to high conversions) [35]. The tendency to polymerization is generally more suppressed by substituents in positition 2 than in position 4 [36]. The cyclic oligomers of formaldehyde (trioxane, tetraoxane, etc.) are readily polymerized and cyclic oligoacetaldehydes only with difficulty. Their ceiling polymerization temperature is low (see Chap. 5 Sect. 1.1).

The opening of lactone rings of various sizes does not directly depend on ring strain [37]; indirect dependences can, however, be observed [38].

(a) With five- and six-membered lactones, the ability to polymerize depends on ring structure.

(b) Four-, seven- and eight-membered rings polymerize in all cases.

(c) Alkyl or aryl substituents always suppress reactivity.

Among the simple lactones, only the five-membered γ-butyrolactone does not polymerize under conventional conditions. It can, however, be polymerized under very high pressure [39].

Lactams with seven-membered or larger rings can be polymerized or hydrolytically condensed yielding long chains. Six-membered and smaller rings can only be polymerized by the ionic mechanism [40]. Lactams as monomers are strongly resonance-stabilized and exceedingly temperature-resistant. Non-ionic initiators are generally only effective at temperatures of about 450 K or higher [41].

Among the multitude of lactams, only a few are important; their importance is, however, quite extraordinary, both from the theoretical and from the technical point of view. ε-Caprolactam is the most frequently used.

Alkyleneimines can undergo ring opening, simultaneously polymerizing or aminoalkylating the ring-opening agent. With the exception of N-substituted alkyleneimines, they can also be alkylated or acylated by suitable agents. Most reactions in which the alkyleneimine ring is opened belong to nucleophilic substitutions where the C—N bond is broken and a new C—N bond is generated. The mechanism of this reaction is complicated; it depends considerably on the structure of the alkyleneimine, of the attacking nucleophile and on reaction conditions [42].

Ethyleneimine (aziridine) and carbon-substituted alkyleneimines are relatively weakly basic compounds [43, 44]. This is ascribed to π-electron delocalization in the ring [44]. By protonation of the imine cycle this "aromaticity" disappears, and iminium rings are much more readily broken than imines [45]. Also the N-alkylethyleneimines are only weakly basic, and they are also more resistant to ring-opening nucleophilic reactions [46].

Alkyleneimines are so far of only theoretical importance; as monomers they find very little industrial use.

Of the cyclic sulphides, mostly only thiiranes, thietanes, sulpholanes and some cyclic di- and trisulphides, as well as compounds with sulphur or oxygen in the ring, serve as monomers. Thiiranes are much more reactive than their oxygen analogues [47].

Polymerization of 2-oxa-6-thio (3.3) spiroheptane at 270 K yields a soluble crystalline polymer with oxetane rings

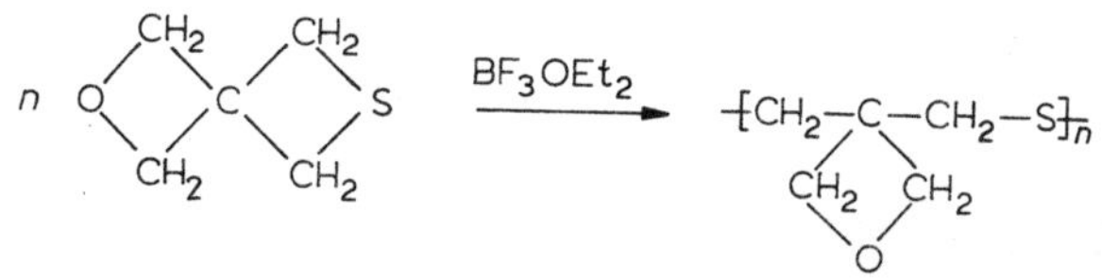

Thus under the given conditions, the thietane ring is more reactive than oxetane [48]. On the other hand, the polymerization of 1,3-oxathiolane

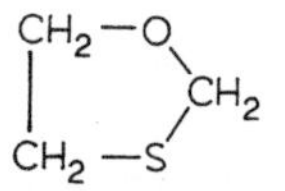

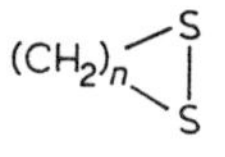

is slower than that of 1,3-dioxolane, and 1,3-dithiolane does not polymerize at all [49]. This is supposed to be caused by the lower basicity of sulphur derivatives. Trithiane and higher cyclic sulphides can be polymerized in similar way to their oxygenated analogues. The mechanism and kinetics of these reactions are, however, different. Cyclic disulphides can also be polymerized [50] by the action of water. Rings with 4, 5, 6, 8, 10 and 12 atoms are the most reactive, while rings with 7, 9 and 11 atoms are opened less readily. This variation in reactivity in larger rings is caused by the compression of the volume needed for a methylene group in compounds with an even number of ring atoms. The ease of opening in a five-membered derivative is unexpected and it has been explained by hindered rotation about the —S—S— bond leading to ring strain [51].

Cyclic siloxanes are very important monomers, both from the theoretical and the practical points of view. Organocyclosiloxanes were discovered and characterized by Kipping [52]. The most frequently studied and practically applied organosilicon monomers are cyclic compounds of the type $\{Si(R_1R_2)\text{—}O\}_n$[†]. The homologues with $n = 3$—7 are the starting compounds for macromolecular synthesis and are the best known. Much larger cycles composed of 25 and more siloxane units have, however, been isolated [53].

Most cyclic organosiloxanes have only a very small ring strain; the highest, 12—20 kJ mol^{-1}, occurs in the trimer [54]. Their polymerization is thermodynamically controlled, by the entropy factor. Substituent type and ring size determine the polymerization mechanism and kinetics. Practically all polymerizations are ionic [55]. The most frequent substituents, R, are —CH$_3$, —C$_6$H$_5$, —CH=CH$_2$, and for special applications —CH$_2$CH$_2$CN, —CH$_2$CH$_2$—CF$_3$ and others.

The character of the organosilicon cyclic siloxanes differ considerably, of course, from all the monomers described so far. This is due to the properties of silicon. The siloxane bond has a much larger valence angle than its carbon analogue (see Table 2), the segments of the siloxane chain being extremely mobile. Silicon is much more electropositive than carbon (according to Pauling, the electronegativity of carbon is 2.5 while that of silicon is 1.8), and it has a coordination number 6. Siloxane monomers are attacked by nucleophilic agents on silicon, and by electrophilic molecules on oxygen. The monomeric rings are thus opened by different mechanisms.

Silazanes, particularly hexamethylcyclotrisilazane, octamethylcyclotetrasilazane and other alkyl- and aryl-substituted cyclosilazanes, can also be polymerized [55]. Organosilazanes are more easily cyclized than siloxanes;

[†] In the literature on silicons, D is used to indicate the difunctional siloxane unit

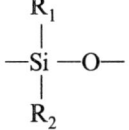

they are also more easily hydrolyzed. The tendency of the —Si—N— bond to hydrolysis decreases with the volume of the substituents on N and Si. The compounds with this bond are more stable to bases than to acids [56]. Silazanes do not belong to the group of common monomers. However, the properties of their polymers are being carefully studied.

Anhydrides of N-carboxy-α-amino acids represent a separate class of monomers. They were discovered at the beginning of this century [57] (Leuchs anhydrides) and are of the general formula

$$\begin{array}{c} \text{RCH}-\text{C} \overset{\displaystyle\diagup\text{O}}{\diagdown\text{O}} \\ | \qquad\qquad | \\ \text{HN}-\text{C} \diagdown_{\text{O}} \end{array}$$

They are polymerized by ring opening with separation of CO_2, yielding poly(amino acids) which are a useful protein model. More than 100 of these anhydrides, their polymers and copolymers are known [58]. All these compounds are very sensitive to moisture and to the effects of chemical agents.

1.1.5 OTHER MONOMERS

They are very numerous [59] and are mainly compounds which have been specifically synthesized for the purpose and used as models, either as monomers of the required properties, with the studied effect stressed by a substituent or arrangement of atoms, or in the polymeric form as models of chains with a particular selected property.

Of the monomers not discussed in the preceding paragraphs, formaldehyde is the most important, both theoretically and practically. It is one of the few monomers undergoing both polymerization and polycondensation, although each under different conditions. Formaldehyde is a thermally stable compound, it is practically undecomposed below 573 K [60]. Its molecules readily undergo mutual reaction by various processes (the Canizzaro and Tishchenko reactions, aldol condensation, etc.). With other compounds it is condensed with equal ease. Its ability to polymerize spontaneously is proverbial. Formaldehyde is difficult to purify because of its high reactivity. Depolymerization of spontaneously formed or synthetic low-molecular-weight polyformaldehyde produces the cyclic trimer (trioxane) in high yield. Trioxane is much less reactive, it is more easily purified and can be both polymerized and copolymerized. Today high-molecular-weight polyformaldehyde is prepared both from monomeric formaldehyde and from trioxane.

Higher aldehydes are not nearly so important as monomers. Acetaldehyde and its cyclic trimer, 2,4,6-trimethyl-1,3,5-trioxane, can yield a high polymer.

However, its ceiling temperature (see Chap. 5 Sect. 1.1) is low, and this is one of the reasons why it has not found technical applications. Acetaldehyde is a stable isomer of vinyl alcohol. Under the conditions investigated so far, the equilibrium of these two isomers is shifted towards acetaldehyde to such an extent that poly(vinyl alcohol) can only be prepared indirectly. By substitution of hydrogen with chlorine, i.e. by transformation of acetaldehyde into chloral, the electron density on the C=O bond is reduced. Chloral polymerization is used practically though not at a large scale. Aromatic aldehydes cannot be polymerized, probably because of the conjugation of the carbonyl double bond with the aromatic ring [61]. The lowest unsaturated aldehydes, e.g. acrolein, can be used as monomers with two double bonds, reacting by different mechanisms.

This monomer type enables us to prepare linear chains by a specific mechanism. The second functional group remains unreacted in the polymer (copolymer), and it can be used to modify the properties of the polymer chain in a second reaction step.

The polymerization of ketones was studied for theoretical reasons. Acetone has an unusually low ceiling temperature and yields high-molecular-weight polymers only at low temperatures (ca. 70 K); its polymer depolymerizes at temperatures of about 100 K.

Carbon monoxide and carbon dioxide cannot be polymerized. However, they can be copolymerized with some monomers. The same is true of sulphur dioxide. Carbon suboxide can be polymerized under certain conditions [62].

Even compounds which themselves do not polymerize, or do so only with great difficulty, can be used as monomers by complexing with suitable agents.

The polymerizations of pyridine and quinoline may serve as an example (without exhausting the existing possibilities) to illustrate this. Polymerization of these compounds would give rise to nitrogen-containing chains with conjugated double bonds. Neither of these compounds can be directly transformed to a high polymer. By complexation of pyridine or quinoline with $ZnCl_2$, and also with many other coordinationally unsaturated salts and acids, at temperatures exceeding 550 K, the conditions necessary for the generation of conjugated polymers are established. The N—C bond in the ring is probably weakened by the immobilization of the electron pair on nitrogen, due to coordination. Heterolytic splitting of this bond is thus made possible [63].

Complex formation as a means of changing the electronic structure of compounds to be polymerized is now being increasingly applied. Suitable complexes enable us to employ compounds which otherwise do not readily polymerize. In this way the polymerization mechanism and kinetics of a large number of conventional monomers can also be widely modified.

An overwhelming majority of monomers are of lower density than the corresponding polymer. For some applications, monomers of the opposite characteristics would be more suitable, i.e. compounds which expand on polymerization (see Chap. 5 Sect. 3.3). Such compounds tend to be relatively complicated, e.g. spiro *ortho*-carbonates, spiro *ortho*-esters, bicyclic *ortho*-esters [64] and many other [65, 66].

Phosphorus-containing polymers, which can exhibit biological activity, have aroused considerable interest. Other such polymers are excellent fire retardants.

Many authors are seeking suitable monomers based on, for example, 1,3-dioxo-2-phospholanes

and even more complicated compounds [67, 68].

Another type of phosphorus-containing monomers are the phosphazenes

which can be polymerized by ring opening at high temperatures (ca. 520 K) yielding linear chains with conjugated phosphorus—nitrogen bonds [69, 70].

With the exception of the silicones, macromolecules built of covalently bound atoms other than carbon, nitrogen and oxygen are rare.

1.2 The relationship between monomer structure and its polymerization mode

By drastic means, using extremely strong acids such as FSO_3H—SbF_5, even methane can be made to polymerize [61], and even N_2 and H_2O to copolymerize in plasma [71, 72]. Only low-molecular-weight and poorly defined products are, of course, formed. Long chains with desirable properties can only be synthesized when the probability of disturbing side reactions is low. Therefore exceedingly energetic initiation methods which lead to strongly reactive and poorly controlled centres must be avoided.

In polymerization, monomers capable of addition are counterparts of the active centres. An important factor is the electron distribution in the monomer molecule, especially between the atoms on a double bond (or on a polar bond in heterocycles). The manner of electron sharing is given by the type of atoms and by their arrangement in the molecule. Also the electron shifts caused by the induction effects of strong electrostatic fields in the vicinity of ions are determined by monomer structure.

Almost all compounds polymerizing by the radical mechanism belong to the classical monomers with a double or triple bond. Radicals of relatively low reactivity formed from the initiators do not usually attack the bonds of electron-rich atoms (with an excess of electrons). They react readily with electron-deficient atoms. Thus the anionically polymerizing monomers usually also polymerize by a radical mechanism. Typical cationic monomers do not undergo radical polymerization. The quite neutral ethylene forms a transition between the two groups. It polymerizes reluctantly by the radical and ionic mechanisms; cationically it only yields oligomers.

The combination of a multiple bond with an electronegative substituent usually produces an anionically polymerizing monomer; a substituent of donor character increases the probability of efficient monomer addition to a cation. Compounds of the type

$$CH_2=CH-\underset{A}{C}=\bar{\underline{O}} \leftrightarrow {}^+CH_2-CH=\underset{A}{C}-\underline{\bar{O}}|^-$$

are therefore expected to undergo anionic polymerization. Monomers with the possibility of conjugation

$$CH_2=CH-\underline{\bar{O}}R \leftrightarrow {}^-|CH_2-CH=\overset{+}{\underline{O}}-R$$

should be polymerized by the cationic mechanism [73]. Exceptions to these rules occur, especially in copolymerizations.

Heterocycles form a specific class of monomers. They do not usually undergo radical polymerization, and the kind of ionic polymerization mechanism is determined by the kind of heteroatom, substituent and ring size. Oxiranes and, aziridines are polymerized by both ionic mechanisms. With the exception of lactone, four-membered and larger heterocycles with oxygen and with substituted nitrogen can only be polymerized cationically; heterocycles with unsubstituted nitrogen can also be polymerized anionically.

The centres of coordination polymerizations should be able to form only a weak and reversible coordination bond with the monomer. Compounds irreversibly solvating the coordination centres (under the given conditions) act as catalytic poisons. Thus monomers with unscreened polar substituents and most heterocycles cannot be polymerized on Ziegler–Natta coordination centres (see Chap. 3 Sect. 4.1).

References pp. 69–74

1.3. The donor–acceptor properties of monomers

An important property of the reacting compounds is their ability to donate or accept electrons, i.e. to behave as reducing or oxidizing agents. For a long time, monomers were not considered from this point of view. Only in recent years has the role of electron transfer in macromolecular chemistry been

TABLE 3

Half-wave potentials of some monomers
Values referred to standard calomel elctrode at 298 K.

Monomer	$^{1/2}E$ (V)	
	In dimethoxy-ethane[a]	In dioxan-water mixture[b]
AN	−2.09	−2.01
MAN[c]	−3.13	
MA	−2.95	−1.95
MMA	−3.07	−2.01
B	−3.27	−2.60
S	−3.19	−2.35
IZ[d]	−3.43	−2.70
α-MeS[e]	−3.28	−2.39
VAC	f	
VC	f	

[a] Refs. 75 and 76.
[b] Ref. 77.
[c] Methacrylonitrile.
[d] Isoprene.
[e] α-Methylstyrene.
[f] No reduction wave.

appreciated. Quantitative measures of the monomer redox properties are their electron affinity and ionization potential. Both quantities define the energy requirements of the molecule (atom) on the loss or addition of an electron (electrons) in the gas phase. They are also measured in the gas phase or calculated for it. For polymerizations which occur exclusively in the condensed phase, the values of true ionization potentials and electron affinities are of only indirect importance. The energy balance is usually completely changed by solvation. For a quantitative placement of a monomer on a scale of redox properties, the value of, for example, the half-wave potential would be suitable. Also the relative reduction potentials obtained in potentiometric titrations can be very useful. Both methods yield values generally correlating with the electron affinity [74]. The half-wave potentials of the most common monomers are summarized in Table 3. Known values of half-wave and reduction potentials are scarce, especially for the media in which the monomers are really polymerized. Often we must get along with a mere distinction of monomers as electron donors or acceptors.

1.2.1 MONOMERS AS ELECTRON DONORS

In a broader sense, each cationically polymerizing monomer has a nucleophilic character. To the best of the author's knowledge, monomers have not so far been ordered according to their electron-donating ability.

The following monomers have been classified as donors: styrene [78, 79], vinyl ethers and vinyl sulphides [78], vinyl acetate [80]; vinylcarbazole [81], aziridine [82], 2-oxazoline [83], 2-benzyliminotetrahydrofuran [84], five- and six-membered cyclic iminoethers, cyclic amines, cyclic phospholanes and Schiff bases [85]. Trialkoxyethylenes [86]

$$\text{RO}\diagdown_{\diagup}\!\!=\!\!\diagup^{\text{OR}}_{\diagdown\text{OR}}$$

were described as electron-rich monomers.

The ability to donate electrons is a very important property of monomers, especially in copolymerizations. It may be expected that monomer characterization from this point of view by a suitable quantitative method under various conditions would considerably extend our knowledge of the addition mechanism in polymerizations.

1.2.2 MONOMERS AS ELECTRON ACCEPTORS

Every anionically polymerizing monomer has some electrophilic character.

The ability to accept electrons from donors is particularly pronounced in acrylic acid derivatives [85]: its alkyl esters [78, 87, 88], acrylonitrile [88], acrylamide, hydroxylacrylates [85], and further in styrenes substituted with an electronegative atom or group: m-nitrostyrene, 2,6-dichlorostyrene [86], β-nitrostyrene [89]; bicyclobutane-1-carbonitrile [89]; lactones: β-propiolactone [85], sulfolactone; vinyl ketones [87]; unsaturated dicarboxylic acids and their derivatives: diethyl fumarate, fumaronitrile [90], ROOC—N=N—COOR [86], cyclic anhydrides of diacids [91], particularly maleic anhydride [78, 92]; ethylenes substituted with electronegative groups [93, 95]

$$\text{NC}\diagdown_{\text{NC}\diagup}\!\!\text{C}\!=\!\text{CHCN} \qquad \text{EtOOC}\diagdown_{\text{NC}\diagup}\!\!\text{C}\!=\!\text{C}\diagup^{\text{H}}_{\diagdown\text{CN}} \qquad \text{MeOOC}\diagdown_{\text{CN}\diagup}\!\!\text{C}\!=\!\text{C}\diagup^{\text{H}}_{\diagdown\text{COOMe}}$$

References pp. 69–74

These monomers can be copolymerized by a radical mechanism. The unfavourable effect of the 2-substituent is compensated by two 1-substituents strongly stabilizing the electron-deficient transition radical.

Under certain circumstances, the electron-accepting ability of monomers can be greatly enhanced by their complexation with Lewis acids.

For example, the reactivity of acrylic and methacrylic acids is considerably changed in the presence of $ZnCl_2$, $SnCl_4$, triethylaluminium, etc. [95, 96] (see Sect. 4.2).

2. Properties of polymerizing monomers

The monomer assumes several functions in the polymerizing medium. They are proportionately manifested in all the physical and chemical parameters of the polymerizing mixture. The monomers for macromolecular synthesis must be pure. The requirements concerning the permissible amount of impurities are ever more stringent; this trend is uncompromising and unequivocal.

Not long ago, ethylene and propene polymerization on ZN catalysts required monomers with a maximum of 10 ppm of H_2O. This standard was found unsatisfactory. Modern facilities operate with water concentrations of units of ppm or less. Ionic processes are not less sensitive to H_2O.

2.1 Physical properties of monomers

The physical properties of the monomers participate in determining the mechanism and kinetics of polymerizations, and thus also the type of laboratory and industrial process employed. For example the vapour pressure determines the range of pressures to be considered in the selected the temperature interval; the melting point defines the minimum polymerization temperature of pure liquid monomer; solubility determines the selection of solvents and concentrations; surface tension affects the formation of heterogeneous and microheterogeneous systems; viscosity affects diffusion-controlled processes, mixing and the removal of the heat of reaction. The heat of evaporation of monomers is often used for removing the heat of polymerization in industrial processes. The permittivity of monomers directly affects ionic equilibria; these equilibria are also a function of the solvating efficiency of monomers and they may be complicated by selective solvation. In strong electromagnetic fields (in the vicinity of ions), conformations of molecules, and thus also their dielectric properties, may be altered.

To the author's knowledge, data on all physical properties of monomers have not been collected in a single treatise. The existing sources are limited to the basic data and the remaining information has to be sought in original articles. Some values of physical properties are summarized in Table 4. If more detailed data are needed in the following text, they will be included at the appropriate place.

2.2 Chemical properties of monomers

Only a compound whose addition to the active centre proceeds virtually without side reactions can be a monomer for macromolecular synthesis; it is very important that active centre inactivation does not occur. Other criteria are not so evident at first sight.

Many difficulties in polymer processing are caused by chain branching[†]. The macromolecules of technical polymers consist of thousands of monomeric units. The ratio of the rates of regular addition and of the branching reaction should be $>10^3$ for strictly linear polymers.

Allyl-type monomers do not yield high polymers. The substituent on the carbon in the β position with respect to the double bond is easily eliminated (especially hydrogen, halogenides, etc.). The generated radical is resonance-stabilized. It reacts much more readily with growing radicals than with the monomer. The low probability of long chain formation is a consequence of these terminating and transfer reactions.

There exists a large number of monomers. Each kind is characterized by a specific behaviour, different for each polymerization procedure. The derivation of some kind of general rule for the chemical behaviour of monomers would require a lot of space, and is not within the scope of this volume. The chemical reactions of selected monomers will be described in the following text, especially in the paragraph on the propagation mechanism (see Chap. 5, Sect. 6).

3. Reactivity of monomers

This is a relative value derived from measurements of the rate of monomer addition to a specific active centre.

[†] Low density polyethylene is branched, and it is more easily processed than the linear high density polyethylene. This is caused by the special kind of branching: short branches regularly spaced along the chain backbone act as an internal plasticizer.

References pp. 69–74

TABLE 4

Basic physical properties of common monomers[a]

Trivial name (systematic)	Formula	Molar mass	Density [g cm^{-3}] at (K)	B. p. (K)	M. p. (K)	Vapour pressure Pa 10^{-3} at (K)
Acetylene (ethyne)	CH≡CH	26.04	0.618 1 (191)	189.4	191.2	6 047.8[b] (308.6)
Acrolein (acrylaldehyde)	CH$_2$=CHC(=O)H	56.06	0.840 6[c]	425.7	186.1	28.5[d]
Acrylamide (propenamide)	CH$_2$=CHCONH$_2$	71.08	1.122 (323)	498	458	0.018 7 (313.2)
Acrylic acid (propenic acid)	CH$_2$=CHCOOH	72.06	1.047 2[e]	514.5	284.5	0.413[d]
Acrylonitrile (propenenitrile, vinyl cyanide)	CH$_2$=CHCN	53.06	0.807 5[e]	350.5	189.6	11.06[d]
1,3-Butadiene	CH$_2$=CH—CH=CH$_2$	54.09	0.621 1[d]	268.8	164.3	4 321.6[b]
1-Butene	CH$_2$=CH—CH$_2$—CH$_3$	56.10	0.595 1[d]	266.9	87.8	8.4 (310.9)
Chloroprene (2-chloro--1,3-butadiene)	CH$_2$=CH—CCl—CH$_2$	88.54	0.958 3[c]	332.6		
Ethylene (ethene)	CH$_2$=CH$_2$	28.05	0.566 (171.2)	169.3	103.8	49 069.6[b] (282.4)
Vinylidene chloride (1,1-dichlorethene)	CH$_2$=CCl$_2$	96.95	1.212 9[d]	304.9	151.1	0.029 (273.2)

Name	Formula	MW				
1,1-Diphenylethene	$CH_2=CPh_2$	180.24	1.020 6 (295.2)	550.2	282.2	2.1 (420.2)
Fumaric acid (*trans*-butenedioic acid)	HOOCCH=CHCOOH	116.07	1.635[d]	563.2[f]	559	0.23 (338)
Maleic adic (*cis*-butenedioic acid)	HOOCCH=CHCOOH	116.07	1.609[d]		403.7	
Maleic anhydride	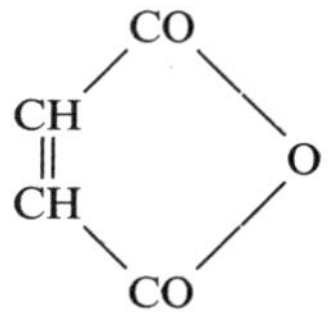 (structure)	98.06	1.300 1 (337.8)	470.2	325.8	0.45[b] (333.1)
Isoprene (2-methyl-1,3-butadiene)	$CH_2=CH-C(CH_3)=CH_2$	68.11	0.680 6[d]	307.2	127.2	53.3 (288.5)
Methacrylic acid (2-methylpropenic acid)	$CH_2=C(CH_3)COOH$	86.09	1.015 3[d]	434.2	288.7	1.3[h] (334)
Methyl methacrylate	$CH_2=C(CH_3)COOCH_3$	100.11	0.936[d]	374	225.0	4.27[d]
Propylene (propene)	$CH_2=CHCH_3$	42.07	0.513 9 (253.2)	225.4	88.0	30.18 (235)
Isobutene (2-methyl propene)	$CH_2=C(CH_3)_2$	56.10	0.626 6 (266.6)	266.6	132.2	1.33 (191)
Styrene (ethenyl benzene, vinyl benzene)	$CH_2=CHPh$	104.14	0.907 5[c]	418.0	142.6	0.67[d]
1-Methylstyrene (isopropenyl benzene)	$CH_2=C(CH_3)$ Ph	118.17	0.916 5	436.6	250	2.53[d]
Vinyl acetate	$CH_2CHOCOCH_3$	86.09	0.933 8[e]	345.7	173.0	11.82[d]
Vinyl chloride (chloroethene)	$CH_2=CHCl$	62.50	0.991 76 (257.2)	259.8	119.4	13.3 (117.4)

TABLE 4 (*continued*)

Trivial name (systematic)	Formula	Molar mass	Density [g cm^{-3}] at (K)	B. p. (K)	M. p. (K)	Vapour pressure Pa 10^{-3} at (K)
Vinyl fluoride (fluoroethene)	CH$_2$=CHF	46.04		222.2	112.2	133.3 (207)
Butyl vinyl ether	CH$_2$=CHOC$_4$H$_9$	100.16	0.780 3^c	367.5	160.5	5.60^d
Ethylene oxide[g] (oxirane)	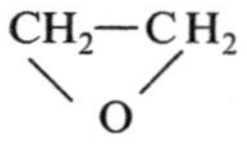 CH$_2$—CH$_2$ \\ O	44.05	0.892 2 (279.2)	283.9	162.2	66.3^b (273.2)
Ethylenimine[g] (aziridine)	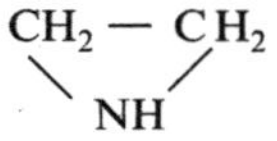 CH$_2$—CH$_2$ \\ NH	43.07	0.832^d		329.2	26.7^h (296.5)
Ethylene sulphide[g] (thiirane)	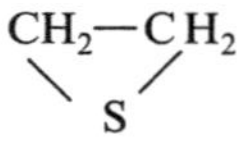 CH$_2$—CH$_2$ \\ S	60.11	1.004 6^d	328.2		26.7^h (293)
Oxacyclobutane (oxetane)	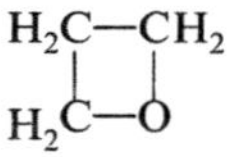 H$_2$C—CH$_2$ \\ H$_2$C—O	58.08				
Tetrahydrofuran[g] (oxolane)	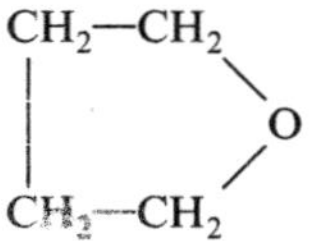 CH$_2$—CH$_2$ \\ CH$_2$—CH$_2$ \\ O	72.11	0.889 2^d	338.2	164.7	26.7^h (302.8)

Compound	Structure			
δ-Valerolactone[g]	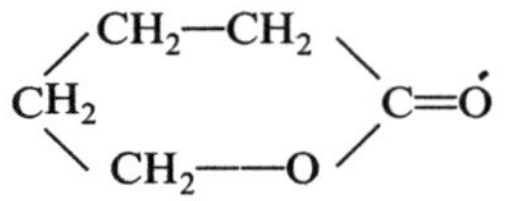 CH₂-CH₂-C(=O)-O-CH₂-CH₂ (ring)	100.12		492.2	261.7	1.3[h] (353.4)
6-Hexanelactam[g]	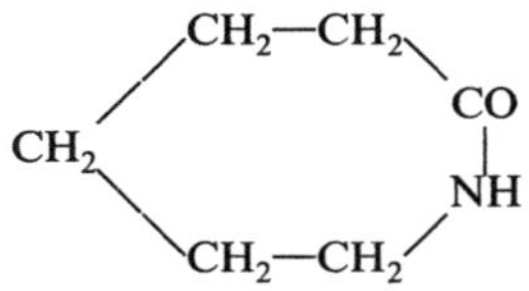 CH₂-CH₂-CO-NH-CH₂-CH₂-CH₂ (ring)	113.16		413.2[i]	335.2	1.3[h] (307.2)
1,3,5-Trioxane[g]	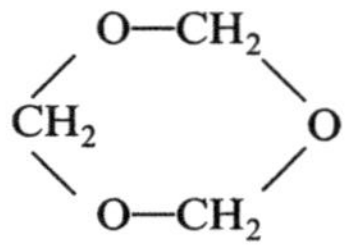 (O-CH₂)₃ ring	90.08	1.17 (338.2)	387.7[j]	336.2	1.3[h] (288)
Octamethylcyclotetrasiloxane	[−Si(Me)₂−O−]₄	296.62	0.955 8 (293)	447.5	290.4	1.3[h] (338.2)

[a] Taken from ref. 97.
[b] Critical conditions [98].
[c] 273 K/273 K.
[d] 293 K.
[e] 293 K/293 K.
[f] Sublimes.
[g] Taken from ref. 99.
[h] Taken from ref. 100.
[i] At 2×10^3 Pa.
[j] At 101.2×10^3 Pa

Many attempts have been made to define monomer reactivity[†] by some numerical value, and to find relations suitable for relating this value to monomer structure. These attempts have always resulted in only short series, based almost always on experience. They are very important and must be regarded as an indispensable quantitative basis for a future general theory of monomer reactivity.

Most data were obtained from copolymerization studies. The copolymerization parameter r (see Chap. 5, Sect. 5.2) is the rate constant ratio for the addition of two different monomers to the same active centre. The inverse values of r_1 determined for the copolymerization of a series of monomers with the monomer M_1, define the relative reactivities of these monomers with the active centre from the first monomer, M_1°. Thus it is possible to order monomers according to their reactivities in radical, anionic, cationic and coordination polymerizations from the tabulated values of copolymerization parameters [101–103].

Such a series naturally calls for an explanation of why a specific monomer assumes the experimentally found position in the sequence. Many authors are attempting to find a relation between reactivity and some well-defined physico-chemical properties of monomers. The most valuable are the correlations indicating the type of electron distribution around the atoms on the double bond.

Hatada et. al [104] made use of the finding that the ^{13}C NMR spectrum of vinyl groups is controlled by π-electron density [105, 106]. Therefore they

TABLE 5

^{13}C chemical shifts of vinyl groups of some monomers [90]
Chemical shift is referred to CS_2 as external standard. A shift towards higher frequencies is defined as positive.

Monomer	β-Carbon (ppm)	α-Carbon (ppm)
Benzyl vinyl ether	105.6	41.6
p-Methylbenzyl vinyl ether	105.9	41.3
Methyl methacrylate	68.6	56.6
Methyl α-ethyl acrylate	70.5	50.7
1,1-Diphenylethylene	78.6	43.0
Vinyl chloride	76.0	67.5
Vinylidene chloride	77.7	64.0

[†] Reactivity is a qualitative, general and relative term, similar to acidity or basicity. It can be quantitatively expressed by the value of rate constant in the corresponding reaction of the evaluated component with a partner in a given medium.

compared the ^{13}C NMR shifts of various vinyl monomers with their reactivities (see Table 5). For radical polymerizations they found a linear dependence between the polarities of monomer double bonds, and also of their radical end groups (as characterized by e values, see Chap. 5, Sect. 5.2) and $\delta_{C\beta}$. The linear relation between the proportionality constants and $\delta_{C\beta}$ is considered to prove proportionality between the substituent-induced polarity of the monomer double bond and the polarity of the resulting radical. The square of the difference $(\delta_{C\beta M2} - \delta_{C\beta M1})^2$ can serve as a measure of the alternation effect in copolymerization. In radical polymerization of para-substituted styrenes [107], the logarithm of the propagation rate constant k_p is inversely proportional to $\delta_{C\beta}$ [104] (k_p decreases in the series —Br > —Cl > —H > —CH$_3$ > —OCH$_3$). Therefore the lower the electron density on the β carbon, the easier will be the electron transfer from the growing radical to the monomer, i.e. the more reactive will the monomer be. The same relation is valid for anionic polymerization. This also proves that the carbanion attack on the monomer is the rate-determining step in copolymerizations with styrene. The reverse is true in cationic polymerizations. The higher the electron density on the β carbon of substituted styrenes, the higher the monomer reactivity. Thus even here the attack of the carbenium ion on the β carbon of the monomer is the rate-determining step.

Natta et al. [108] have studied the copolymerization of styrene with substituted styrenes on ZN catalysts. They came to the conclusion that monomer coordination to the electron-poorest part of the catalytic complex is one of the basic characteristics of this stereospecific copolymerization. The relation between $\log r_1^{-1}$ and $\delta_{C\beta}$ is typically cationic [104]. This is a direct proof of Natta's ideas. This also means that the vinyl group of the monomer is coordinated to the catalytic complex by the β carbon and by its nearest neighbours.

Another possibility of monomer classification is offered by the correlation of copolymerization parameters with Hammett or Taft constants.

Taft [109] postulated that the rate of aliphatic ester hydrolysis is a function of substituent type. Using methyl as a standard substituent, he compared, for example, the effects of larger alkyls, and interpreted the results by means of two constants characterizing the steric and polar behaviour of the alkyls. In copolymerization theory, the Taft equation was used many times to describe the relative reactivities (See Chap. 5, Sect. 5.2) in homologous polymer series

$$\log r_1^{-1} = \varrho\sigma + \delta E_s$$

where σ and E_s are Taft polar and steric substituent constants; ϱ and δ are the corresponding reaction constants. The relative reactivities of the alkyl methacrylate [110] and alkylvinylketone [111] series are affected only by the polar factors and the reactivities of methyl-α-acrylates [112] by steric factors.

Van Der Meer et al. [113] have found that vinyl ester reactivity increases with the electron-withdrawing ability of the ester group. All the measured ester radicals prefer addition of their own monomer to that of ethylene. The observed relative reactivity is mostly affected by polar factors; resonance stabilization plays only a minor role. Vinyl ester reactivity grows in the series vinyl iso-butyrate < vinyl butyrate < vinyl propionate < vinyl acetate < vinyl formate ~ vinyl pivalate.

Szwarc and co-workers [114] measured the apparent[†] rate constant of the addition of some substituted styrenes to polystyrylsodium

$$(\sim CH_2\text{—}CH|^-, Na^+)$$
$$\quad\quad\quad\quad |$$
$$\quad\quad\quad\quad Ph$$

in THF solutions. Addition, mostly to free ions was observed, and the authors proved the validity of the Hammett equation ($\log k_{\sim S^-} = \sigma \varrho$). In the reverse case, for the addition of styrene to the carbanion ends of poly(substituted styrenes), the observed rate constants could not be correlated with the electronic effect of the substituents. This behaviour is connected with the different tendency to dissociation of the ion pairs on the ends of various substituted styrenes, and this must be included in the considerations. Tobol-

TABLE 6

Propagation rate constants in anionic polymerizations of nitrostyrenes[a]

Substituent	k_p (dm^3 mol^{-1} s^{-1})	k_p/k_0
p-Methoxy	0.38	0.42
m-Methyl	0.57	0.63
p-Methyl	0.63	0.71
H [b]	0.90	1.0
m-Methoxy	1.3	1.4
p-Chloro	0.73	0.82
p-Bromo	1.4	1.6
m-Bromo	1.8	2.0
m-Nitro	2.2	2.5

[a] Polymerization temperature 301 K in mixture of ethanol/THF (2:1 by volume); initiator sodium ethanolate.
[b] Unsubstituted β-nitrostyrene as reference.

[†] The attribute overall, effective or apparent rate constant (or activation energy etc.) is used for non-elementary processes. Such a constant may be any kind of a (mostly unknown) function of the elementary rate and equilibrium constants.

TABLE 7

Reactivity of monomers with various types of conjugation [118]

Monomer	k_p [a]	r_1^{-1}	$\bar{\varepsilon}$ [b]	$\bar{q}$ [w]
Monomers with π–σ interaction				
2-Methylpropene	N	0.00	−8.8	19
1-Hexene	N	0.00	−8.3	13
Ethylene	242[d]	0.01	−0.1	9
Allyl chloride	N	0.03	+0.8	25
Monomers with π–n interaction				
Ethyl vinyl ether	N	0.01	−10.0	3
Methyl vinyl sulphide		0.20	−9.0	81
Vinyl chloride	9 500[c]	0.06	−0.8	35
Monomers with π–n–π interaction				
Vinyl pyrrolidone	4 760[c]	0.12	−10.1	52
Vinyl acetate	9 500[c]	0.02	−2.3	23
N-Vinyl carbazole		0.17	+6.3	32
Monomers with π–π interaction				
Butadien	100[c]	1.3	−5.0	116
Chloroprene	229[f]	19.0	−3.1	135
p-Methoxystyrene	71[g]	0.86	−7.7	106
Styrene	106[g]	1.00	−5.0	100
p-Chlorostyrene	150[g]	1.35	−2.9	93
p-Nitrostyrene		5.26	+2.7	113
2-Vinyl pyridine	96[h]	1.60	+1.9	94
Monomers with π-π interaction (with heteroatoms)				
Methyl methacrylate	512[h]	1.90	+1.5	87
Methyl acrylate	1 580[h]	2.20	+3.8	81
Butyl acrylate	3 800[i]	2.50	+7.1	87
Acrylonitrile	14 500[h]	3.50	+7.1	87
Methyl vinyl ketone		3.50	+4.0	90

[a] Cited in original work without experimental detail.
[b] $\bar{\varepsilon}_2 = \bar{\varepsilon}_1 \pm B(-RT \ln r_1^{-1} r_2^{-1})^{1/2}$
[c] $\bar{q}_2 = RT \ln r_1 + \bar{q}_1 - A\varepsilon_1(\varepsilon_1 - \varepsilon_2)$
A, B = const.; $\bar{\varepsilon}$ = polar factor in % of elementary charge [119],
$\bar{q}$ = factor of resonance stabilization referred to styrene (%) [119],
r_1^{-1} = relative monomer reactivity towards styrene macroradical at 333 K.
[d] 356 K. [e] 333 K. [f] 313 K. [g] 303 K. [h] 298 K. [i] 300 K. N = low activity towards own radical.

References pp. 69–74

sky et. al. [115, 116] measured the reactivities of styrene derivatives towards polystyryllithium, also in THF solution. Monomer reactivity decreased in the series styrene > p-methylstyrene > p-methoxystyrene.

β-Nitrostyrenes are easily polymerized anionically, basically due to the strong electron-withdrawing effect of the nitro group on the double bond, both by induction and resonance effects. The double bond is electron-deficient and susceptible to nucleophilic attack. Carter et al. [117] synthesized and polymerized a number of substituted nitrostyrenes and measured their reactivity. Their results are summarized in Table 6.

The interactions of double bond electrons with substituents strongly depend on the type of mutually interacting electrons. This criterion may be used for the classification of monomer reactivity [118], and three groups can be defined: interactions π—σ; π—n, π—n—π, and π—π. The advantage of this classification is the possibility of generalizing the properties of chemically

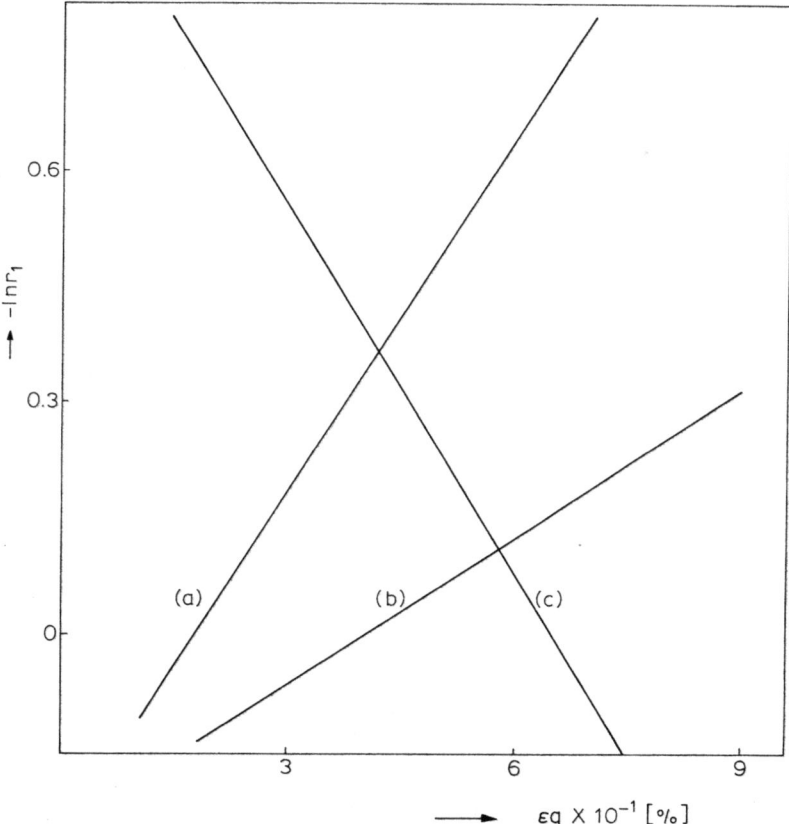

Fig. 1. Reactivity of vinyl monomers (a) with electronegative macroradicals at constant ε, (b) with electropositive macroradicals at constant q and (c) with electronegative macroradicals at constant q.

differing monomers yielding a transition complex and active centre with very similar arrangements of valence electrons. Data on the characteristic representatives of all three groups are summarized in Table 7. For each monomer, the corresponding quantities have been derived from polymerizations and copolymerizations.

It appears that within each group, the reactivity of monomers towards the styrene radical increases with both q and ε (see Chap. 5, Sect. 5.2). Higher q values correspond to greater resonance stabilization of the newly formed radical; growth in ε is connected with the interaction of the easily polarizable macroradical with the β carbon of the monomer whose electronegativity is increased.

An contradictory change of both quantities is manifested by competing effects.

In the monomer groups, the value of q increases in the order $\pi-\sigma < \pi-n < \pi-\pi$. Monomer reactivity towards the macroradical increases

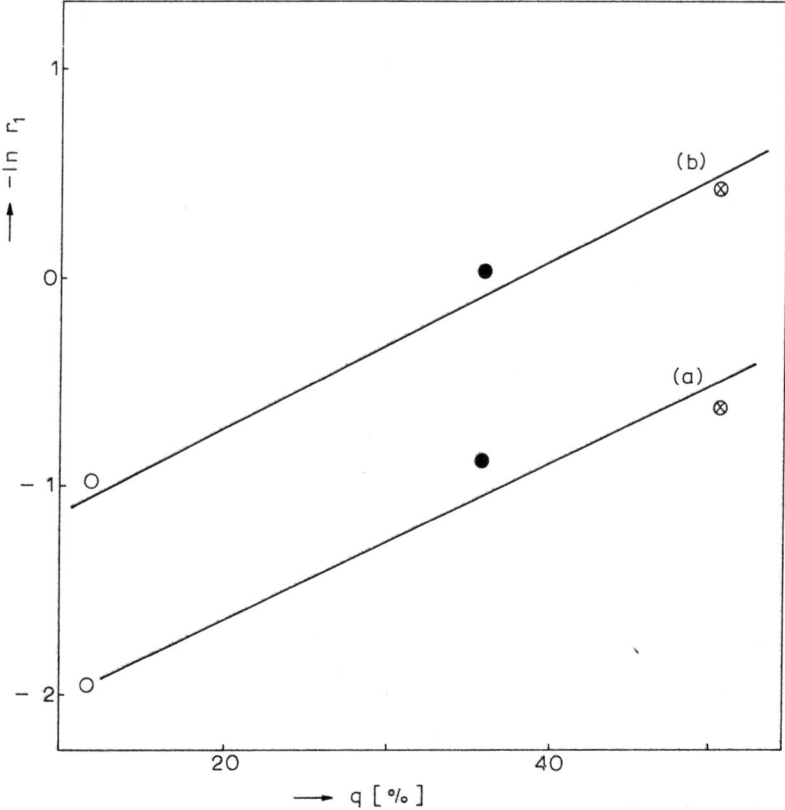

Fig. 2. Reactivity of monomers with $\pi-n$ conjugation (a) with styrene macroradical and (b) with methacrylate macroradical. ○, Ethyl vinyl sulphide; ●, N-vinylcarbazole; ⊗, methyl vinyl sulphide.

References pp. 69–74

and the reactivity of the macroradical decreases in the same order. The sign of its partial charge does not change the character of the monomer group activity [118, 120, 121]. Monomer reactivity in copolymerization is quantitatively characterized in Figs. 1–3. In homopolymerization, the monomers with conjugated bonds $(\pi\text{—}n, \pi\text{—}\pi)$ are the most active. Within each monomer group, reactivity grows with ε. This is again due to the polarizability of the macroradical end group in the transition complex during chain growth [122].

The classification of monomers according to the character of vinyl group-substituent conjugation makes possible a deeper analysis of experimentally found kinetic data and a more detailed analysis of the addition mechanism.

Szwarc [123] observed that the reactivity of the methyl radical with various monomers is comparable with the reactivity of polyradicals with the same monomers. Similar results concerning the phenyl radical were obtained independently by Bevington and Ito [124] and by Pryor and Fiske [125]. On

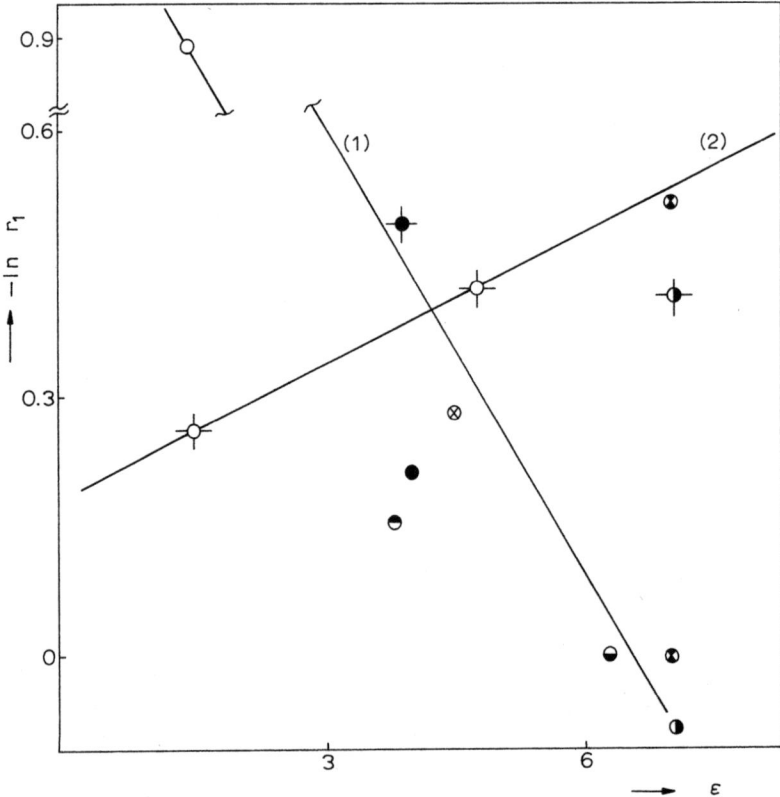

Fig. 3. Reactivity of monomers with $\pi\text{–}\pi$ conjugation (1) with styrene macroradical, (2) with acrylonitrile macroradical (points with vanes). ○, MMA; ●, methyl vinyl ketone; ⊗, acrolein; ◑, butyl acrylate, ⊗ AN; ⊖ methyl acrylate; ⊖ acrylamine; ✧ methacrylonitrile.

the other hand, the reaction of the benzoyloxy radical with vinyl monomers is mostly controlled by the polar character of the monomers used [126]. Sato and Otsu [127] compared the preceding results with the reactivity of the vinyl monomer—alkoxy radical [specifically $(CH_3)_3CO^*$] pair. The relative reactivities of vinyl derivatives towards various radicals are summarized in Table 8.

TABLE 8

Relative rate of monomer addition to selected radicals

Radicals	Monomer reactivity
Me*, Ph*	AN > MMA > S ≫ VAC
PhC(=O)—O*	S > VAC > MMA ≫ AN
$(CH_3)_3C-O^*$	S > MMA > AN > VAC

Me* and Ph* are donors of electrons; during addition the resonance effect prevails over polar one. Benzoyl-oxyl is an electron-acceptor radical; the addition step is controlled merely by the polar effect [128]. During the addition of monomers to trimethylmethoxyl, both polar and resonance properties of their electron configuration are involved.

Substitution of the α proton by methyl in acrylic monomers is not accompanied by reduced reactivity of the methylated derivative. Bolshakov et al. have shown that the low rate of, for example, methacrylate or methacrylamide polymerizations at temperatures of 100–120 K in melting ethanol is caused by the low reactivity of the resulting active centres [129].

The group of *n*-alkyl cyanoacrylates

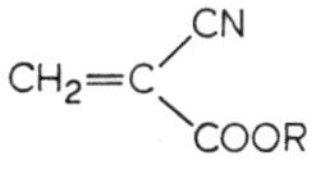

$$CH_2=C\begin{smallmatrix}CN\\COOR\end{smallmatrix}$$

are the most reactive monomers. Unless stabilized by acid (sulphonic, SO_2, etc.) they immediately "spontaneously" polymerize on almost any surface [130]. They are used as adhesives, even in medical applications. Polymerization is initiated by traces of weak bases, e.g. water (i.e. HO^-) but even by CH_3COO^-, CN^-, I^-, Br^- and by covalent bases such as R_3N and pyridine. It proceeds in the presence of O_2, CO_2, H_2O or MeOH, which would safely stop any anionic polymerization of monomers yielding less stable anions. The polymerizing ability of monomers is thus also a function of the stability of their own active centre.

References pp. 69–74

In contrast to the monomers that polymerize easily are the 1, 2-substituted ethylenes, especially those with bulky substituents. They exhibit a low to zero rate of addition to their own radical, and they do not directly produce homopolymers. Alfrey and Young [131] have shown that the greatest barrier to autoaddition is the steric hindrance preventing the existence of a suitable transition complex. They do, however, copolymerize with sterically less shielded monomers.

Acenaphthylene

is an exceptional monomer from this point of view as it does polymerize readily [132]. This is explained by the large strain in the five-membered ring, part of which is relieved by linear polymer formation [133].

The reactivity ratios of vinyl groups in multifunctional monomers are largely unknown. This is mainly due to the formation of insoluble products which are difficult to analyze. In very special cases this difficulty can be circumvented [134].

Divinylbenzene finds industrial use on a small scale. The reaction of the *para* derivative with the dienyllithium carbanion is more rapid than that of the *meta* isomer because of a more favourable arrangement of the vinyl substituents. The *meta* derivative should be more reactive than styrene. Quantum calculations indicate a higher electron density in the styrene than in the divinylbenzene ring [135].

In agreement with these findings, Worsfold [136] has shown that the reaction of the first double bond of *p*-divinylbenzene with styryllithium is about ten times more rapid than that of styrene.

Monomer reactivity is a broad concept, and it can not always be limited only to reactions of the double or triple bonds of a vinyl or acetylene group. Weakly polar monomers, such as styrene or butadiene, react almost exclusively by their double bonds. The anionic polymerization of polar monomers, such as α, β-unsaturated esters and nitriles, is accompanied by many side reactions. A fairly large amount of oligomers and side products are formed, and these may affect the active centres, thus indirectly modifying propagation.

TABLE 9

Relative reaction rates of some oxirane derivatives [138]

Oxirane	Initiator	
	HClO$_4$	EtONa
Chloromethyloxirane (epichlorohydrin)	0.065	4.8
2,3-Epoxy-1-propanol	0.41	2.0
Oxirane	1	1
Methyloxirane	54	0.50
Trans-2,3-epoxybutane	119	0.03
Cis-2,3-epoxybutane	238	0.06
2,2-Dimethyloxirane	5 100	0.06
2-Phenyloxirane	> 10^4	0.14

TABLE 10

Basicity of cyclic ethers [139, 140]

Monomer		pK_b[a]
Oxetane		3.13
2-Methyltetrahydrofuran		4.56
Tetrahydrofuran		5.00
δ-Valerolactone		5.10[b]
ε-Caprolactone		5.31[b]
Tetrahydropyran		5.42
3,3-Bis (chloromethyl) oxetane		5.65
1,4-Dioxan		5.71
γ-Butyrolactone		6.12[b]
Methyloxirane		6.94
1,3-Dioxolane		7,55
2-Vinyl-4, 5-dimethyl-1, 3-dioxolane	trans	8.76[c]
	cis	8.75[c]
2-Isopropenyl-4,5-dimethyl-1,3-dioxolane	trans	8.52[c]
	cis	8.00[c]
2-Propenyl-4,5-dimethyl-1,3-dioxolane	trans	8.20[c]
	cis	6.41[c]
2-Phenyloxirane		8.30
Chloromethyloxirane (epichlorohydrin)		8.84
Trioxane		10.00
β-Propiolactone		10.06[b]
β-Methyl-β-propiolactone		5.59[b]

[a] 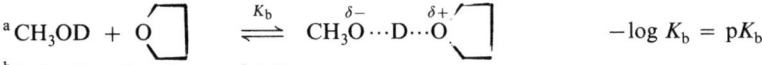 $-\log K_b = pK_b$.

[b] Ref. 141; taken from ref. 142.
[c] Ref. 143.

In the polymerization of methacrylonitrile initiated by butyllithium, the organometal is bound to nitrile groups [137]

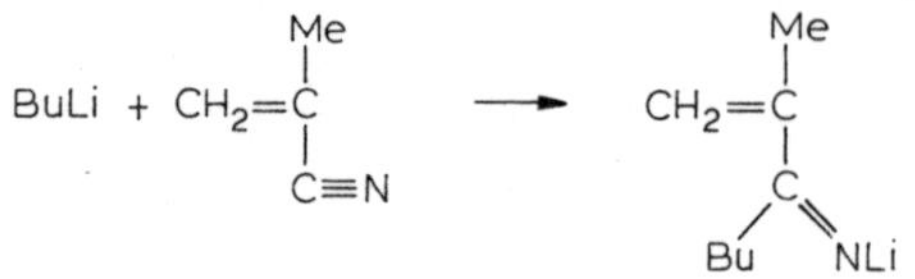

$$BuLi + CH_2=C(Me)(C\equiv N) \longrightarrow CH_2=C(Me)-C(Bu)=NLi$$

At 198 K in toluene, 10–12 % of initiator reacts in this way.

The section on monomer reactivity would not be complete without mentioning heterocycles.

The relative rates of oxirane ethanolysis with acid and basic catalysts are summarized in Table 9. It can be seen that the relative reactivity of the monomer in cationic polymerization is controlled by the basicity of the cyclic ether.

According to Table 10, basicity depends on constitution, ring size and substituents. Basicity changes with ring size in the order $4 > 5 > 6 > 3$. In five-membered rings, the basicity decreases from ether through lactone to formal. Basicity is increased by methyl substitution and decreased by chloromethyl substitution for hydrogen. The effect of configuration is complicated and cannot be concisely formulated.

The formation of an amide bond in a lactam molecule occurs by overlap of p_x atomic orbitals of nitrogen and carbon. In addition, another bond between these two atoms can be formed by sharing of the two electrons of nitrogen penetrating into the $2p_z^2$ atomic orbital of carbon.

According to Chaitin and Volf [144], this delocalization of the free electron pair on nitrogen is the reason why the amide bond is so stable to radiolysis [145] and high temperature [146] in the absence of polymeriza-

TABLE 11

Value of positive change on N and C atoms of the amide bond in lactams [147]

Lactam	Number of ring members	Atomic charge[a]	
		N	C
Azetidinone	4	0.503 5	0.89
α-Pyrrolidone	5	0.346	0.052 6
α-Piperidone	6	0.12	0.309 2
6-Hexanelactam	7	1.43	0.946
ξ-Enantholactam	8	1.907	0.697 4

[a] Referred to elementary charge.

tion initiators. The position of the delocalized pair with respect to the C and N atoms depends on the ring size, on the substituents and on their position. The polarity of the amide bond changes with the increasing number of methylene groups (see Table 11). The values of the partial charges on nitrogen and carbon as a function of the lactam ring size pass through a minimum. In agreement with the course of this function, the strength of the amide bond decreases [148] in the order α-pyrrolidone > α-piperidone > ε-caprolactam.

The change in the positive charge on N is followed by a change in polymerizability, which according to Yoda and Miyake [149] decreases in the order ζ-enantolactam > ε-caprolactam > α-pyrrolidone > α-piperidone. The interaction of the $2p_z$ electrons of nitrogen with the π electrons of the carbonyl group is also a function of ring size. It affects the character of electron distribution in the amide bond, and it is also manifested by a change in ionization potential [150].

The reactivity of lactams towards bases and acids is greatly affected by substituents. For example, in the polymerization of C-monomethyl-ε--caprolactams in the presence of water, the rate decreases with the methyl approaching the $\diagup$C—O group [151]. Also, the 7-alkyl-ε-caprolactams (7-ethyl-, 7-propyl-) can be polymerized, though less easily with increasing substituent size [152]. With increasing size of substituent in position 5, the polymerizability of lactams decreases (the equilibrium concentration of 5-ethyl-ε-caprolactam is 65 %) or is completely lost (5-propyl-, 5-phenyl-, 5-n-heptyl-ε-caprolactam). The influence of a substituent with +I effect on nitrogen is very large. N-Methyl-ε-caprolactam does not polymerize. The main reason of this is the disappearance of the cis-character of the amide bond, reducing the heat of polymerization by 5.9 kJ mol^{-1}, i.e. by the enthalpy difference of the cis and trans configurations of the amide bond [153]. In consequence of this, only lactams with a greater ring strain can be polymerized, e.g. with four- [154] and eight-membered [155] or larger rings [156]. The presence of a substituent with −I effect increases the reactivity of lactams.

4. Monomer and other components of the system

Polymerization is only rarely limited to the interaction of a monomer with the initiator, transfer agent or impurity. Each component of the system, including the so-called inert materials, participates in polymer formation. The mutual interaction of various components with the monomer may vary from a simple physical process to a complicated chemical reaction. In all cases intermolecular forces are involved, as manifested by the formation of solvates, associates and complexes. Sometimes it is difficult to determine the poorly defined boundary between physical and chemical processes.

References pp. 69–74

4.1 Physical relationships

The viscosity of polymerizing systems is a very important factor because the rate of bimolecular termination may become diffusion-controlled (see Chap. 6, Sect. 1.1). In cases where the monomer is a solvent for its own polymer, highly viscous solutions are usually formed, modifying the polymerization mechanism and kinetics. Suspension and emulsion polymerization is, of course, possible only when the monomer is insoluble in the dispersing medium. For emulsion polymerization, however, the monomer cannot be quite insoluble in this medium. This would hinder, or even prevent its transfer from the monomer droplets into the micelles. The forces between the monomer molecules, or the monomer and solvent molecules, determine the quality of the cage enclosing the primary radical immediately after its formation; thus they considerably affect initiation efficiency. Even the monomer makes its contribution to the overall polarity of the medium. In this way it cooperates in the establishment of ionic equilibria, and thus also in all elementary processes. One of the most interesting properties of monomers is their ability to solvate selectively certain groups of atoms, due to which the monomers concentrate in the vicinity of some kinds of particle. Unfortunately the solvation processes have so far not been studied quantitatively to any extent. Nevertheless it is beyond any doubt that solvation is one of the most important factors in polymerizing systems. It affects both energetic and thermodynamic factors as well as kinetics.

In special cases, by no means rare, more or less stable chemical individual complexes, are formed instead of the poorly defined solvates.

4.2 Chemical interactions

In addition to the formation of active centres and participation in elementary processes, the discussion of which forms the main topic of this volume, monomers very often react with some component(s) of the polymerizing medium under complex formation. This reaction is very important. Complex formation lowers the effective monomer concentration, and changes in the polymerization rate usually occur. When the complex is much more active than the monomer, it may react preferentially with the active centre. This, of course, changes the addition mechanism and kinetics. When the monomer and complex also compete, the macrokinetics need not necessarily change. Usually, however, the mechanism of the whole process is greatly complicated, and a kind of copolymerization occurs.

4.2.1 MONOMER COMPLEXES

Bamford et al. [157] were the first to observe the acceleration of vinyl monomer polymerization caused by inorganic salt addition. They polymerized acrylonitrile dissolved in dimethyl-formamide with 2,2'-azobisisobutyronitrile. The reaction rate was increased by the addition of LiCl. The observed effect was ascribed to the increase of the rate constant and interpreted by complex formation between lithium chloride and the nitrile group of the radical.

This work aroused considerable attention, and many other authors began to study similar effects. They observed several types of interaction, with considerable kinetic and stereochemical consequences in radical polymerizations [158].

(a) The complex of the monomer and/or radical with the metal compound is formed by the interaction of the free electron pair of the heteroatom in the monomer with the unoccupied orbital of the metal [157].

(b) The monomer and/or radical is bound to a suitable compound by a hydrogen or ionic bond [159].

(c) A complex is formed by interaction of the unoccupied metal orbital with the monomer π orbitals (e.g. ethylene—Ag^+) [160].

Radical polymerizations or copolymerizations of polar vinyl monomers are accelerated by Lewis acids [161]. In particular, monomers with conjugated nitrile or carbonyl groups form complexes easily. This increases the ability of the monomer to accept electrons [169]. The extent of electron delocalization is, of course, not the only factor determining monomer reactivity. A polymerization may even be slowed down by complex formation. Acrylonitrile, methacrylonitrile, methyl acrylate and methyl methacrylate polymerizations at 323 K in bulk or in toluene solution with $ZnCl_2(1:1)$ are accelerated by complexation. The complexes of the same monomers with $SnCl_4$, Et_2AlCl and $EtAlCl_2(1:1)$ are less active than the uncomplexed monomers, in spite of greater electron delocalization [163]. The polymerizability of complexes is probably also controlled by their polar interaction with macroradicals. Complex formation leads to an increase in the partial positive charge on the C^β carbon of the monomer vinyl group and on the C^α carbon of the macroradial (162, 163]; the addition rate is lowered. $ZnCl_2$ is a weaker acid than $SnCl_4$ and the alkylaluminium chlorides. In addition, the polar effects of its complexes are weaker.

Uncomplexed acrylates do not copolymerize with ethylene; only polyacrylates (e.g. polyethyl acrylate) are formed. BF_3 is a strong Lewis acid used as an initiator in cationic polymerizations. With acrylic monomers it forms a complex which can be relatively easily copolymerized with ethylene and propene. This process is similar to radical reactions; it requires an initiator [96].

Alkylaluminium chlorides form complexes with methyl acrylate which can react with styrene yielding alternating copolymers [164]. The donor–acceptor complex of acrylonitrile (donor) and potassium persulphate (acceptor) makes possible the homopolymerization of acrylonitrile in polar media (in water, dimethylformamide, dimethylsuphoxide, dioxan) at low temperatures [165]. The initiating radicals are formed according to the scheme

$$CH_2=\underset{CN}{CH} + K_2S_2O_8 \rightleftharpoons [(CH_2=CH-CN^*)^+ KSO_4^-] + KSO_4^*$$
$$\updownarrow$$
$$KHSO_4 + CH_2=\overset{*}{C}-CN$$

Acrylamide, 2-methyl-5-vinylpyridine and N-vinylpyrrolidone can be polymerized under similar conditions, and also after decomposition of a monomer–peroxide complex. On the other hand, styrene, methyl methacrylate, isoprene, methyl acrylate, vinyl acetate and ascorbic acid do not polymerize under these conditions. Complex formation between persulphate and these monomer donors is more favourable energetically [165]. The complex is more stable, it is not decomposed into initiating radicals and polymerization does not occur.

The complexes of monomers with transfer agents play an important role in the reaction with the polymer radical. According to type, they may either accelerate or retard the transfer reaction. Monomers form complexes with halogenated compounds, e.g. styrene with CCl_4 [166] and styrene and vinyl acetate with CBr_4 [167]. Phosphorus trichloride and phosphorus organohalogenides of the PCl_2Ph, $PClPh_2$ type are electrophilic compounds. They form complexes with methyl methacrylate and they may slow down its polymerization [168, 169]. Similar complexes are not formed with styrene [168]. Many authors assume the existence of donor–acceptor interactions between the polar methyl methacrylate or vinyl pyridine and the transfer agents [170].

Monomer complexes play an important role even in non-radical processes. In coordination polymerizations, the interactions of monomers with catalysts are evidently of greatest importance; without them this type of addition would not be possible. The formation of unstable complexes between the electrophilic initiator and nucleophilic monomer is also necessary in cationic polymerizations. The idea that under certain conditions the formation of stable complexes between initiator and monomer may prevent polymerization [171] is now frequently accepted [172–174].

In anionic polymerization, monomer complexes are less important. ε-Caprolactam forms a strongly bound complex with LiCl, with a pronounced effect on the polymerization kinetics [175]. The formation of similar complexes is probably typical for all lactams.

Complexes are frequently encountered in macromolecular chemistry. They always have the character of more or less labile compounds which can only rarely be isolated. Direct proof of their existence is difficult. By indirect proof only some of their properties are revealed, for example the ability to compete with other components during addition, a shift or change in the character of a reaction path, contribution to the overall thermodynamical parameters of the process, etc. In the author's opinion, the question of the existence of monomer complexes has not yet been fully appreciated. Work in this field should lead to important discoveries for elucidating polymerization mechanisms and kinetics, with corresponding consequences for industrial production.

So far, the question of complex formation between two or more monomers has been deliberately avoided. These complexes are of quite extraordinary importance for the synthesis of alternating copolymers and they will be briefly treated in Sect. 5.2.

5. Two, three or more types of monomers

Two or more monomers in the charge are only encountered in copolymerizations. At this point it will be useful to discuss the manner of mutual monomer interactions. Their kinetic consequences and modification of polymerization mechanisms will be treated in Chap. 5, Sect. 5.5.

5.1 Non-complexing monomers

In this case, each monomer behaves independently. Either no polymeric product is formed or a copolymer or mixture of homopolymers may result. Selective solvation of some monomer at the active centres may occur; aggregates of the monomer molecules may be formed. Specific information on the physical properties and the resulting chemical interactions in monomer mixtures, especially in the possible presence of solvent, is not available. Authors do not generally pay much attention to this problem (it is not a simple problem and a rigorous analysis of the existing relations is often beyond the possibilities current at the time) and all these effects are usually summed up in the rate constant, sometimes even in the „elementary" one.

5.2 Monomers forming donor–acceptor complexes

When a monomeric donor of suitable structure is encountered by a monomeric acceptor, a donor–acceptor (DA) complex is formed. It is characterized by a reorganized electron structure accompanied by charge shift. Therefore such formation are also designated as charge-transfer (CT) complexes. Until recently, the tendency to the formation of alternating copolymers has been ascribed to a special kind of monomer reactivity and of the resulting

References pp. 69–74

radicals. The attempts to better investigate the basis of this kind of reactivity have met with ever-increasing obstacles. Only departure from established ideas has brought a solution. Bartlett and Nozaki [176] interpreted the deviations from the simple kinetic scheme of Mayo and Lewis and Alfrey and Goldfinger (see Chap. 5, Sect. 5) by the existence of complexes that are added to the centre as a whole. Their hypothesis was confirmed by further studies [177].

This explanation was supported by the following observations

(a) the presence of complexes during polymerization could be proved (see, for example, refs. 178–182);

(b) the existence of a significant tendency to alternation at considerably differing concentrations of the two monomers in the charge;

(c) a high polymerization rate in the vicinity of an equimolar monomer concentration in the charge;

(d) spontaneous initiation of copolymerization in some cases.

Yoshimura et al. [183] have used a similar scheme for the quantitative interpretation of the initial rates in some, essentially alternating, copolymerizations.

Perhaps the most common monomer pair forming donor–acceptor complexes is styrene with maleic anhydride. Tsuchida et al. [184] have measured the copolymerization rate in various solvents. They reached the conclusion that even 1:1 styrene–maleic anhydride complexes are capable of propagation. They presented a spectroscopic proof of the existence of DA complexes and measured the equilibrium constant of their generation. Similar conclusions on the role of complexes in the copolymerization of α-methylstyrene with maleic anhydride were presented by Seymour and Garner [185]. The problem has not yet been settled. Dodgson and Ebdon [186] have studied the dependence of copolymer composition on the ratio of monomer concentrations in the feed. They do not consider copolymer formation by styrene– maleic anhydride complex propagation as safely proved. Ledwith [187] also regards sceptically the role of DA complexes in copolymerizations. He argues that, with the exception of extremely strong donors and acceptors, with an ionic or polarized ground state, the intermolecular forces of organic molecules are weak, with a minimum charge-transfer contribution. On the other hand, Otsu et al. [188] consider the concept of donor–acceptor complexes as proved, and they provide further evidence of its validity. By means of the spin-trapping technique they proved the existence of the styrene radical $C_6H_5CH=CH^*$ and of the radical of the complex in the system styrene–maleic anhydride in ethyl benzene. In the presence of an initiator, the same system was a source of the radicals

$$(CH_3)_3CO-CH_2-\overset{*}{\underset{\underset{Ph}{|}}{CH}} \qquad (CH_3)_3CO-CH_2-\underset{\underset{Ph}{|}}{CH}-\underset{\underset{CO}{|}}{CH}\underset{\underset{O}{\diagdown\diagup}}{}\underset{\underset{CO}{|}}{CH}-CH_2-\overset{*}{\underset{\underset{Ph}{|}}{CH}}$$

$$AB$$

[$(CH_3)_3CO-$ is an initiator residue]. With copolymerization of free monomers, they should have observed an increasing A/B ratio according to the method used; with complex propagation, A/B should remain constant. The authors observed both cases. They concluded that maleic anhydride with a monomeric donor, like styrene, yields a DA complex by a reversible reaction, with an equilibrium constant of 10^{-1} to 10^{-2} dm^3 mol^{-1}. The initiating radical is formed from the complex, and the copolymerization is in fact a terpolymerization involving the two free monomers and their complex. These authors have applied the same technique in a study of the type of radicals formed in copolymerization of maleic anhydride with vinyl sulphides. Even in this case they provided evidence of the existence of a complex.

An electron acceptor such as maleic anhydride forms complexes with many donors, amongst which is vinyl acetate [80, 189, 190]. Its existence at 363 K was proved by UV spectroscopy, ^{1}H NMR and by the formation of an alternating copolymer [80]. The complex is not formed above 363 K. From the two monomers, a statistical copolymer is formed, its composition depending on the ratio of initial monomer concentrations.

Further monomers with which maleic anhydride produces donor–acceptor complexes are conjugated dienes [191–193], vinyl ethers [194a], furan, thiophene, indole [195, 196], β-isopropenylnaphthalene [197], 4-vinylpyridine [194b], 2-vinyl-1,3-dioxolane [198], cycloalkenes [199] and other complicated vinylic or acrylic monomers [200]. Maleic anhydride complexes have even been observed in some terpolymerizations [201].

Donor–acceptor complexes of p-methoxystyrene with trisubstituted ethylenes

$$ACH=C\underset{\diagdown B}{\overset{\diagup Cl}{}}$$

(where A = B = CH_3COO-; A = $NC-$; B = CH_3COO- and A = CH_3CO-, B = $NC-$) were observed by Hall et al [93, 94]. An indication of the existence of these complexes is the appearance of light colour after mixing of the components. Styrene and ring-halogenated styrenes do not yield complexes [202]. A review on the participation of donor–acceptor complexes in polymerizations was published by Gaylord [203].

In donor–acceptor complexes of strongly nucleophilic and electrophilic monomers, charge transfer within the product is pronounced, and the complex will be of zwitterion (amphiion) character. This situation was described by Saegusa et al. [85, 204].

In this case, monomeric donors were, for example, cyclic phosphorus compounds: 2-phenyl-1, 3, 2-dioxaphospholane and 2-phenoxy-1, 3, 2-dioxaphospholane

Monomeric acceptors were α-ketoacids: pyruvic acid (MeCCOOH with $\overset{\text{O}}{\overset{\|}{}}$) or phenylglyoxalic acid (PhCCOOH with $\overset{\text{O}}{\overset{\|}{}}$).

6. Models of monomers

Most information on each elementary polymerization step was derived from the course of polymerization reactions. In some cases it is useful to model a reaction so as to make the initial compounds, intermediates and products best amenable to analysis. In such cases the application of a non-polymerizing monomer may be indicated.

Busson and van Beylen [205] studied the role of the cation and of the carbanionic part of the active centre during anionic polymerization in non polar media. They were interested in the problem of complex formation between the cation and the monomer double bond [206] and they therefore measured the reaction of various 1,1-diphenylethylenes with Li^+, K^+ and Cs^+ salts of living polystyrene in benzene and cyclohexane at 297 K. Diphenylethylene derivatives were selected for two reasons.

(a) After addition to the living polystyrene end, they do not further polymerize for steric reasons [207]. The interpretation of the results is much simpler compared with studies of substituted styrene homopolymerizations [208]. Special experimental conditions and techniques for the isolation of the first addition step need not be applied.

(b) The reaction can be easily studied by spectrophotometric methods because the absorption maxima of the living polystyrene, at 335 nm, and of the resulting adduct (living 2-polystyryl-1,1-diphenylethylene, ca. 440 nm) are sufficiently separated and permit an accurate measurement of the changes in optical density.

The authors reached the conclusion that, in the absence of strongly solvating solvents (e.g. ethers), a characteristic feature of the anionic polymerization is monomer complexation with the positive ion or the polarization of the monomer double bond by this ion, especially by Li^+.

The above example illustrates the application of non-polymerizing monomers. Not many compounds exist that are suitable for modelling monomer behaviour during addition to the active centre. 1,1- and 1,2-diphenylethylenes are the most important representatives.

References

1. G. Natta, SPE J., 5 (1959) 373.
2. K. Ziegler, BE 534, 792 and 534, 888 (1955), BE 540, 459 (1956).
3. K. Ziegler, E. Holzkamp, H. Breil and H. Martin, Angew. Chem., 67 (1955) 426.
4. G. E. Weismantel, Chem. Eng. News, 8 (1981) 47.
5. N. Platzer, Ind. Eng. Chem. Prod., Res. Dev., 22(1) (1983) 158.
6. J. V. Seppala, Macromolecules, 18 (1985) 2409.
7. J. P. Kennedy and J. E. Johnston, Adv. Polym. Chem., 19 (1975) 57.
8. See, for example, T. Sao and T. Otsu, Makromol. Chem., 178 (1977) 194.
9. See, for example, S. Bywater, in C. H. Bamford and C. F. H. Tipper (Eds.), Comprehensive Chemical Kinetics, Vol. 15, Elsevier, Amsterdam, 1976, p. 40.
10. D. Day and H. Ringsdorf, J. Polym. Sci., Polym. Lett. Ed., 16 (1978) 205.
11. E. L. Eliel, Stereochemistry of Carbon Compounds, McGraw-Hill, New York, 1962, p. 188
12. K. C. Frisch and S. L. Reegen (Eds.), Ring Opening Polymerization, Dekker, New York, 1969, (a) p. 2; (b) p. 15.
13. H. K. Hall, Jr., H. Tsuchiya, P. Ykman, J. Otton, V. Papanu, S. C. Snider and A. Deutschman, Jr., Polym. Prepr. Am. Chem. Soc. Div. Polym. Chem., 18(1) (1978) 104.
14. J. P. Kennedy, Cationic Polymerization of Olefins, Wiley, New York, 1975, p. 37.
15. J. Kress, J. A. Osborn, R. M. E. Gregne, K. J. Ivin and J. J. Rooney, J. Am. Chem. Soc., 109 [1987] 899.
16. G. G. Engstrom and P. Kovacic, J. Polym. Sci. Polym. Chem. Ed., 15 (1977) 2453.
17. A. Macoll, Prog. Sterochem., I (1954) 361.
18. K. Yamasaki, A. Kotera, M. Yokoi and Y. Ueda, J. Chem. Phys., 18 (1950) 1414.
19. T. E. Turner and J. A. Howe, J. Chem. Phys., 24 (1956) 924.
20. M. Igarashi, Bull. Chem. Soc. Jpn., 26 (1953) 330.
21. J. K. Stille and J. A. Empen, J. Polym. Sci. Part A-1(5) (1967) 273.
22. W. Höckel and F. S. Bollig, Chem. Ber., 86 (1953) 1137.
23. W. J. Linn, J. Am. Chem. Soc., 87 (1965) 3659.
24. Y. Yshii and S. Sakai, in K. C. Frisch and S. L. Reegen (Eds.), Ring Opening Polymerization, Dekker, New York, 1969, p. 16.
25. F. A. Long, J. G. Pritchard and F. E. Stafford, J. Am. Chem. Soc., 79 (1957) 2362; 80 (1958) 4162.
26. C. C. Tung and A. J. Speziale, J. Org. Chem., 28 (1963) 2009.
27. J. Koshikalis and E. Whalley, Trans. Faraday Soc., 59 (1959) 809.
28. R. E. Parker and N. S. Isaacs, Chem. Rev., 59 (1959) 737.
29. M. Tameres, S. Searles and J. M. Goodenow, J. Am. Chem. Soc., 86 (1964) 3934.
30. K. Matyjaszewski, J. Macromol. Sci. Rev. Macromol. Chem. Phys. C, 26(1) (1986) 1–32.

31 D. R. Boyd and E. R. Marle, J. Chem. Soc., 105 (1914) 2117.
32 P. Dreyfuss and M. P. Dreyfuss, in K. C. Frisch and S. L. Reegen (Eds.), Ring Opering Polymerization, Dekker, New York, 1969, p. 112.
33 H. K. Hall, Jr., Polym. Prepr. Am. Chem. Soc. Div. Polym. Chem., 6(2) 1965 535.
34 J. W. Hill and W. H. Carothers, J. Am. Chem. Soc., 57 (1935) 925.
35 A. A. Strepikheev and A. V. Volokhina, Dokl. Akad. Nauk SSSR, 99 (1954) 407.
36 M. Okada, Y. Yamashita, Y. Ishii, Makromol. Chem., 80 (1964) 196.
37 J. I. Goldfarb and L. I. Belenkii, Russ. Chem. Rev., 29 (1960) 214.
38 H. K. Hall, Jr. and A. K. Schneider, J. Am. Chem. Soc., 80 (1958) 6409.
39 F. Korte and W. Glet, J. Polym. Sci. Part B, 4 (1966) 685.
40 H. K. Reimschuessel, in K. C. Frisch and S. L. Reegen (Eds.), Ring Opening Polymerization, Dekker, New York, 1969, p. 303.
41 M. Rothe, Polym. Prepr. Am. Chem. Soc. Div. Polym. Chem., 18(1) (1977) 45.
42 M. Hauser, in K. C. Frisch and S. L. Reegen (Eds.), Ring Opening Polymerization, Dekker, New York, 1969, p. 222.
43 H. C. Brown and M. Gerstein, J. Am. Chem. Soc., 72 (1950) 2926.
44 C. E. O'Rourke, L. B. Clapp and J. O. Edwards, J. Am. Chem. Soc., 78 (1956) 2159.
45 G. D. Jones, J. Org. Chem., 9 (1944) 125.
46 J. E. Earley, C. E. O'Rourke, L. B. Clapp, J. O. Edwards and B. C. Lawes, J. Am. Chem. Soc., 80 (1958) 3458.
47 P. Sigwalt, in K. C. Frisch and S. L. Reegen (Eds.), Ring Opening Polymerization, Dekker, New York, 1969,
48 E. J. Goethals, Int. Symp. Macromol. Chem., Prague, 1965. Preprint 334, Abstract 117.
49 O. C. Dermer, Wadc Tech. Rep. 55–447, ASTIA Doc. AD 110, 496 (1956). J. Lal, in P. H. Plesch, (Ed.), The Chemistry of Cationic Polymerization, Pergamon Press, New York, 1963, Chap. 13.
50 F. O. Davis and E. M. Fettes, J. Am. Chem. Soc., 70 (1948) 2611.
51 J. G. Affleck and G. Dougherty, J. Org. Chem., 15 (1950) 864.
52 F. S. Kipping, J. Chem. Soc., 101 (1912) 2138.
53 J. F. Brown Jr. and G. M. J. Slusarczuk, J. Am. Chem. Soc., 87 (1965) 931.
54 W. A. Piccoli, G. G. Haberland and R. L. Merker, J. Am. Chem. Soc., 82 (1960) 1883.
55 C. Eaborn, Organosilicon Compounds, Academic Press, New York, 1960, pp. 339–350.
56 E. E. Bostick, in K. C. Frisch and S. L. Reegen (Eds.), Ring Opening Polymerization, Dekker, New York, 1969, p. 352.
57 H. Leuchs, Chem. Ber., 39 (1906) 857.
58 Y. Shalitin, in K. C. Frisch and S. L. Reegen (Eds.), Ring Opening Polymerization, Dekker, New York, 1969, p. 421.
59 See, for example R. W. Lenz, Organic Chemistry of High Polymers, Wiley, New York 1967, pp. 473, 564. T. Endo and T. Ogasawara, Netsu Kokasei Jushi, 5(2) (1984) 98 Chem. Abstr., 102 (1985) 46258 t.
60 J. F. Walker, Formaldehyde, American Chemical Society Monograph Series, New York, 2nd edn., 1953.
61 O. Wichterle, Organic chemistry (Organická chemie), Přírodovědecké vydavatelství, Prague, 1952, p. 249.
62 A. W. Snow, H. Haubenstock and N. L. Yang, Macromolecules, 11(1) (1978) 77.
63 V. A. Kargin and V. A. Kabanov in K. C. Frisch and S. L. Reegen (Eds.), Ring Opening Polymerization, Dekker, New York, 1969, p. 360.
64 M. S. Cohen, C. Bluestein and M. Dunkel, Proc. 8th Int. Conf. Radiat. Curing, 1984, pp. II/1–II/12.
65 T. Endo, Kobunshi Kako, 33 (5) (1984) 222.

66 T. Endo and T. Ogasawara, Netsu Kokasei Jushi, 5(2) (1984) 98 Chem. Abstr. 102 (1985) 46258 t.
67 T. Saegusa, S. Kobayashi and Y. Kimura, Macromolecules, 10 (1977) 64.
68 T. Saegusa and S. Kobayashi, Polym. Prepr. Am. Chem. Soc. Div. Polym. Chem., 20(1) (1979) 138.
69 H. R. Alcock, J. L. Schmutz and K. M. Kosydar, Macromolecules, 11 (1978) 179.
70 D. R. Tur, V. V. Korshak, S. V. Vinogradova, S. S. A. Pavlova, G. I. Timofeeva, Ts. A. Goguadze and N. O. Alikhanova, Dokl. Akad. Nauk SSSR, 291(2) (1986) 364.
71 H. Yasuda, H. C. Marsch and J. Tsai, J. Appl. Polym. Sci., 19 (1975) 2157.
72 H. Yasuda and T. Hirotsu, J. Polym. Sci. Polym. Chem. Ed., 15 (1977) 2749.
73 O. Wichterle, Technology of Plastics (Technologie Plastických Hmot), Part I, SPN, Prague, 1952, p. 80.
74 M. Szwarc, Progr. Phys. Org. Chem., 6 (1968) 323.
75 N. Yamazaki, S. Nakahawa and I. Tanaka, T. Macromol. Sci. Chem., A2 (1971) 1121.
76 N. Yamazaki, Fortschr. Hochpolym. Forsch., 6 (1969) 377.
77 T. Fuero, K. Asada and T. Furukawa, J. Polym. Sci., 40 (1959) 511.
78 T. Otsu, T. Sato and M. Abe, Polym. Prepr. Am Chem. Soc. Div. Polym. Chem., 17(2) (1976) 615.
79 M. H. Litt and J. Radovic-Wellinghoff, Prepr. Am. Chem. Soc. Div. Polym. Chem., 17(2) 1976) 596.
80 R. B. Seymour, D. P. Carner, G. A. Stahl and L. J. Sanders, Polym. Prepr. Am. Chem. Soc. Div. Polym. Chem., 17 (2) (1976) 660.
81 S. R. Turner, M. Stolka and J. C Wilson, Polym. Prepr. Am. Chem. Soc. Div. Polym. Chem., 17 (2) (1976) 602.
82 A. Takahashi, Polym. Prepr. Am. Chem. Soc. Div. Polym. Chem., 18 (1) (1977) 751.
83 T. Saegusa, H. Ikeda and H. Fujii, Macromolecules, 5 (1972) 354.
84 T. Saegusa, Y. Kimura, K. Sano and S. Kobayashi, Macromolecules 7, (1974) 546.
85 Cited in T. Saegusa, T. Yokohama, Y. Kimura and S. Kobayashi, Macromolecules, 10 (1977) 791.
86 H. K. Hall, Jr. and V. Papanu, Polym. Prepr. Am. Chem. Soc. Div. Polym. Chem., 18(1) (1977) 648.
87 T. Saegusa, S. Kobayashi and Y. Kimura, Macromolecules, 10 (1977) 68.
88 W. R. Cabaness and C. H. Chiang, Polym. Prepr., Am. Chem. Soc. Div. Polym. Chem., 17 (2) (1976) 632.
89 M. E. Carter, J. L. Nash, Jr., J. W. Drueke, Jr., J. W. Schwietert and G. B. Buttler, J. Polym. Sci. Polym. Chem. Ed., 16 (1978) 937.
90 M. Yoshimura, Y. Shirota and H. Mikawa, Polym. Prepr. Am. Chem. Soc. Div. Polym. Chem., 17 (2) (1976) 590.
91 A. Takanashi, Polym. Prepr. Am. Chem. Soc. Div. Polym. Chem., 18 (1) (1977) 751.
92 S. N. Bhandáni and J. Prasad, Makromol. Chem., 178 (1977) 1651.
93 H. K. Hall, Jr and R. C Daly, Macromolecules, 8 (1975) 22.
94 H. K. Hall, Jr. and P. Ykman, Polym. Prepr. Am. Chem. Soc. Div. Polym. Chem., 17 (2) (1976) 654.
95 W. Kuran, S. Pasynkiewicz, R. Nadir and Z. Florjanczyk, Makromol. Chem., 178 (1977) 1881.
96 A. L. Logothetis and J. M. McKenna, J. Polym. Sci. Polym. Chem. E., 15 (1977) 1431.
97 Y. P. Castille, in T. Brandsup and E. H. Immergut (Eds.), Polymer Handbook, Wiley-Interscience, New York, 1967, pp. VIII-1–27.
98 J. Timmermans, Physico-chemical Constants of Pure Organic Constants, Elsevier, Amsterdam, 1950.

99 M. Večeřa, J. Gasparič, J. Churáček and J. Borecký, Chemical Tables of Organic Compounds (Chemické tabulky organických sloučenin) SNTL, Prague, 1975.
100 J. Dykyj and M. Repáš, Saturated Vapor Pressure of Organic Chemicals (Tlak nasycené páry organických sloučenin), Veda, Vydav. SAV, Bratislava, 1979.
101 K. Veselý, Polyreactions (Polyreakce), SNTL, Prague, 1955, pp. 87–94.
102 H. Mark, B. Immergut, E. H. Immergut, L. J. Young, and K. I. Beynon, in G. E. Ham (Ed.), Copolymerization, Interscience, New York, 1964. pp. 695–843.
103 B. Vollmert, Elements of Macromolecular Chemistry (Grundriss der makromolekularen Chemie, Springer-Verlag, Berlin, 1962.
104 K. Hatada, K. Nagata, T. Hasegawa and H. Yuki, Makromol. Chem., 178 (1977) 2413.
105 M. Karplus and A. Pople, J. Chem. Phys., 38 (1963) 2803.
106 G. E. Maciel, J. Phys. Chem., 69 (1965) 1947.
107 M. Imoto, M. Kinoshita and N. Nishigaki, Makromol. Chem., 86 (1965) 217.
108 G. Natta, F. Danusso and D Sianesi, Makromol. Chem., 30 (1959) 238.
109 R. W. Taft, in Steric Effects in Organic Chemistry, M. S. Newman, (Ed.), Wiley, New York, 1956, p. 556.
110 T. Otsu, T. Ito, M. Imoto, J. Polym. Sci. Part B, 3 (1965) 113.
111 T. Otsu and H. Tanaka, J. Polym. Sci. Polym. Chem. Ed., 13 (1975) 2605.
112. G. G. Cameron and G. P. Kerr, Eur. Polym. J., 3 (1967) 1.
113 R. Van der Meer, E. H. M. Van Gorp and A. L. German, J. Polym. Sci. Polym. Chem. Ed., 15 (1977) 1489.
114 M. Shima, D. N. Bhattacharyya, J. Smid and M. Szwarc, J. Am. Chem. Soc., 85 (1963) 1306.
115 A. V. Tobolsky and A. J. Boudreau, J. Polym. Sci., 51 (1961) 553.
116 B. D. Philips, T. L. Hanlon and A. V. Tobolsky, J. Polym. Sci. A-1, (2) (1964) 4 241.
117 M. E. Carter, J. L. Nash, Jr., J. W. Drueke, Jr., J. W. Schwietert and G. B. Butler, J. Polym. Sci. Polym. Chem. Ed., 16 (1978) 937.
118 A. V. Chernobai, Vysokomol. Soedin. Ser. B, 19 (1977) 329.
119 T. C. Schwan and C. C. Price, J. Polym. Sci., 40 (1959) 457.
120 R. Houwink and A. Stavermann, Chemistry and Technology of Polymers (Chimiya i Technologiya Polimerov), Chimiya, Moscow, 1963, p. 271.
121 C. Walling, Svobodnye Radikaly v Rastvore (Free Radicals in Solution, Wiley, New York 1957.) Izd. Inostr. Lit., Moscow, 1960, p. 98.
122 A. V. Chernobai, Vysokomol. Soedin. Ser. A, 16 (1974) 2217.
123 M. Szwarc, J. Polym. Sci., 16 (1955) 367.
124 J. C. Bevington and T. Ito, Trans. Faraday Soc., 64 (1968) 1329.
125 W. A. Pryor and T. R. Fiske, Trans. Faraday Soc., 65 (1969) 1865.
126 J. C. Bevington, Radical Polymerization, Academic Press, London, 1961.
127 T. Sato and T. Otsu, Makromol. Chem., 178 (1977) 1941.
128 J. C. Bevington and C. S. Brooks, J. Polym. Sci., 22 (1956) 257.
129 A. I. Bolshakov, A. I. Michailov, I. M. Barkalov, Vysokomol. Soedin., 20 (1978) 1820.
130 E. F. Donnelly, D. S. Johnson, D. C. Pepper and D. J. Dunn, J. Polym. Sci. Polym. Lett. Ed., 15 (1977) 399.
131 J. Alfrey and L. J. Young, in G. E. Ham (Ed.), Copolymerization, Interscience, New York, 1964, pp. 67–87.
132 K. Dziewanski and Z. Lylso, Chem. Ber., 47 (1914) 1679.
133 J. Ballesteros, G. J. and L. Teasdale, J. Macromol. Sci. Chem., A11(1) (1977) 29.
134 I. M. Bolbit and S. J. Frenkel, Vysokomol. Soedin., 20 (1978) 294.
135 R. N. Young and L. J. Fetters, Polym. Prepr. Am. Chem. Soc. Div. Polym. Chem., 18 (1) (1977) 693.

136 D. J. Worsfold, Macromolecules, 3 (1970) 514.
137 H. Vankerckhoven and M. Van Beylen, Eur. Polym. J., 4 (1978) 189.
138 R. E. Parker and N. S. Isaacs, Chem. Rev., 59 (1959) 737. G. Gee, W. C. E. Higginson, P. Levesley and K. J. Taylor, J. Chem. Soc., 86 (1964) 762.
139 Y. Yamashita, T. Tsuda, M. Okada and S. Iwatsuki, J. Polym. Sci. Part A-1, (4) (1966) 2121.
140 H. E. Wirth and P. L. Slick, J. Phys. Chem., 66 (1962) 2277.
141 T. Tsuda, T. Shimizu and Y. Yamashita, Kogyo Kagaku Zasshi, 68 (1963) 2473.
142 R. D. Lundberg and E. F. Cox, in K. C. Frisch and S. L. Reegen (Eds.), Ring Opering Polymerization, Dekker, New York, 1969, p. 279.
143 Z. J. Jedliński and T. Łukaszczyk, Macromolecules, 8 (1975) 700.
144 B. Sh. Chaitin and L. A. Volf, Vysokomol. Soedin., 20 (1978) 1580.
145 W. Dameran, G. Labmann, H. G. Thom, J. Phys. Chem., 223 (1963) 59.
146 P. H. Hermans, D. Heikens and P. F. Van Velden, J. Polym. Sci., 30 (1958) 81.
147 D. Marrel, S. Catle and D. Tedder, Valentnost (Valence), Mir, Moscow, 1968, p. 495.
148 N. Ogata, Bull. Chem. Soc. Jpn., 31 (1961) 245.
149 N. Yoda and A. Miyake, J. Polym. Sci., 43 (1960) 117.
150 B. K. Potanov, A. F. Fimochkina, D. N. Shigorin and G. A. Ozerova, Dokl. Akad. Nauk SSSR, 180 (1968) 398.
151 L. E. Volinsky and H. R. Mighton, J. Polym. Sci., 49 (1961) 217.
152 O. B. Salamatina, A. K. Bonetskaya, S. M. Skuratov, B. P. Fabrickii, P. F. Shelavina and J. L. Goldfarb, Vysokomol. Soedin., 7 (1965) 485.
153 M. S. Gutowski and C. H. Holm, J. Chem. Phys., 25 (1956) 1228.
154 T. Kagyia, H. Kishimoto, S. Narisawa and K. Fukui, J. Polym. Sci. Part A-1 (3) (1965) 145.
155 R. S. Muromova, A. A. Strepikheev, Z. A. Rogovin, Vysokomol. Soedin., 5 (1963) 1096.
156 B. Masař and J. Šebenda, Collect. Czech. Chem. Commun., 39 (1974) 110; 40 (1975) 93.
157 C. H. Bamford, A. D. Jenkins and R. Johnston, Proc. R. Soc. London Ser. A, 364; J. Polym. Sci., 28 (1958) 355.
158 V. A. Kabanov, in Kinetics and Mechanism of Polyreactions, IUPAC Symposium on Macromolecular Chemistry, Akadémiai Kiadó, Budapest, 1971, p. 437.
159 N. A. Vengerova, V. R. Georgieva, V. P. Zubov, V. A. Kabanov and V. A. Kargin, Vysokomol. Soedin., 125 (1970) 46.
160 T. Otsu, Y. Kinishita and A. Nakamachi, Makromol. Chem., 115 (1968) 275.
161 W. Kuran, S. Pasynkiewicz, R. Nadir, Makromol. Chem., 178 (1977) 411.
162 J Furukawa, E. Kobayashi, S. Nagata, T. Moritani, J. Polym. Sci. Polym. Chem. Ed., 12 (1974) 1799.
163 W. Kuran, S. Pasynkiewicz, Z. Florjanczyk and E. Lusztyk, Makromol. Chem., 177 (1976) 2627.
164 M. Hirooka, J. Polym. Sci. Part B, 10 (1972) 171.
165 S. I. Trubicina, I. Ismailov and M. A. Askarov, Vysokomol. Soedin. Ser. A, 19 (1977) 495.
166 J. W. Breitenbach, O. F. Olaj and N. Liaris, Monatsh. Chem., 103 (1972) 990.
167 G. Gleixner, J. W. Breitenbach, O. F. Olaj, Makromol. Chem., 178 (1977) 2249.
168 H. Uemura, T. Taninaka and Y. Minoura, J. Polym. Sci. Polym. Lett. Ed., 15 (1977) 493.
169 J. Pellon, J. Polym. Sci., 43 (1960) 537.
170 A. B. Riabov, L. A. Smirnova and T. G. Cvieshnikova, Proceedings in Chemistry and Chemical Technology (Trudy po Khimii i Khimicheskoi Tekhnologii), Gorkii, 2nd edn., 1974, p. 20.
171 M. Kučera, International Symposium on Cationic Polymerization, Rouen, C21–1, 1973; J. Macromol. Sci. Chem., A7(8) (1973) 1611.
172 M. Chmelíř, International Symposium on Cationic Polymerization, Rouen, C7-1, 1973.
173 P. H. Plesch, Makromol. Chem., 175. (1974) 1 065.

174 D. C. Pepper, J. Polym. Sci. Polym. Symp., 50 (1975) 51.
175 A. Ciferri and S. Russo, Polym. Prepr. Am. Chem. Soc. Div. Polym. Chem., 18 (1) (1977) 87.
176 P. D. Bartlett and K. Nozaki, J. Am. Chem. Soc., 68 (1946) 1495.
177 J. A. Seiner and M. Litt, Macromolecules, 4 (1971) 308, 312, 314, 316.
178 E. N. Gur'yanova, I. P. Gol'dshtein and L. P. Romm, The Donor-Acceptor Bond, Wiley, Chichester, 1975.
179 V. Perdec, Proc. 28th, IUPAC Macromol. Symp., 1982, p. 891.
180 A. M. Mosin, R. M. Aizatullova and V. G. Genchel, Deposited Doc. 1982, SPSTL 943 Khp-D82; Chem. Abstr. (1984) 145323a.
181 Y. Shirota, Encycl. Polym. Sci. Eng., 3 (1985) 327.
182 M. J. Costigan and I. E. Craven, J. Macromol. Sci. Chem., A23(5) (1986) 647.
183 M. Yoshimura, T. Nogami, M. Yokayama, H. Mikawa and Y. Shirota, Macromolecules, 9 [1976] 211.
184 E. Tsuchida, T. Tomono and H. Sano, Makromol. Chem., 151 (1975) 245 and preceding papers.
185 R. B. Seymour and D. P. Garner, Polymer, 17 (1976) 21.
186 K. Dodgson and J. R. Ebdon, Eur. Polym. J., 13 (1977) 791.
187 A. Ledwith, Polym. Prepr. Am. Chem. soc. Div. Polym. Chem., 17 (2) (1976) 614.
188 T. Otsu, T. Sato and M. Abe, Polym. Prepr. Am. Chem. Soc. Div. Polym. chem., 17 (2) (1976) 615.
189 C. Caze and C. Loucheux, J. Macromol. Sci. Chem., A9 (1) (1975) 29.
190 R. B. Seymour and D. P. Garner, J. Paint Technol., 48 (1976) 41.
191 Y. Yamashita, S. Iwatsuki and T. Tokubo, J. Polym. Sci. Part C, 23 (1968) 753.
192 N. G. Gaylord, O. Solomon, M. Stolka and B. K. Patnaik, J. Macromol. Sci. Chem., A8 (1974) 981.
193 S. V. Shulindin, K. G. Sanatulin, R. Z. Shakirov and B. E. Ivanov, Vysokomol. Soedin. Ser. B 16 (1974) 599.
194 M. L. Hallensleben, Makromol. Chem., (a) 144 (1971) 267; (b) 142 (1971) 303.
195 E. J. Goethals, G. Humbrecht and C. Verschuere, Eur. Polym. J., 10 (1974) 847.
196 B. Kamo, I. Morita, S. Horie and S. Furusawa, Polym. J. (Jpn.), 6 (1974) 121.
197 M. L. Hallensleben and I. Lumme, Makromol. Chem., 144 (1971) 261.
198 J. L. Acosta and J. L. Mateo, Rev. Plast. Mod., 24 (1972) 562.
199 T. Yamaguchi, K. Nagai and T. Ono, Kobunshi Ronbunshu, 31 (1974) 361.
200 See, for example, Z. M. Rzaev, L. V. Bryksina and S. I. Sadyk-Zade, J. Polym. Sci. Part D 42 (1973) 519.
201 T. Tobuko, S. Iwatsuki and Y. Yamashita, Macromolecules, 3 (1970) 518.
202 W. Ring, Angew. Makromol. Chem., 21 (1972) 149.
203 N. G. Gaylord, J. Polym. Sci. Polym. Symp. 56 (1976) 497.
204 T. Saegusa, S. Kobayashi, Y. Kimura and T. Yokohama, Polym. Prepr. Am. Chem. Soc. Div. Polym. Chem., 18 (1) (1977) 125.
205 R. Busson and M. Van Beylen, Macromolecules, 10 (1977) 1320.
206 S. S. Medvedev, Russ. Chem. Rev., 37 (1968) 834.
207 A. G. Evans and D. B. George, J. Chem. Soc., (1961) 4653.
208 H. Hirohara, M. Nakayama and N. Ise, J. Chem. Soc. Faraday Trans. 1, 68 (1972) 58.

Chapter 3

Initiation

> *Whatever you search for, you shall find*
> Sophocles

By "initiation" we understand the reaction or set of reactions generating the active centre. The number of monomer molecules consumed in initiation is equal to the number of active centres generated during the whole polymerization process[†]. Each monomer and each type of polymerization condition require a certain kind and method of initiation. Several kinds are available, and each can be applied in many variants.

1. Formation of primary radicals

Radicals are formed by a collision of two monomer molecules of sufficient potential or kinetic energy, by decomposition of an excited molecule after the absorption of a light quantum, by initiator decomposition, and in many other ways. These are called the primary radicals because generally they exhibit a different reactivity and behaviour from the radical active centres generated from them.

1.1 Thermal initiation

Very carefully purified styrene polymerizes slowly [1]. Initiation is evidently due to the monomer itself. In 1936, Flory had already advanced the hypothesis on the formation of a diradical from two monomer molecules [2]

$$2\ CH_2{=}CHPh \rightarrow Ph{-}\overset{H}{\underset{*}{C}}{-}CH_2{-}CH_2{-}\overset{H}{\underset{*}{C}}{-}Ph \qquad (1)$$

Doubt has been cast on the initiating ability of this type of diradical. In would tend rather to cyclization; the probability of propagation is negligible [3]. However, it appears again in more recent theories.

[†] In certain cases, depending on the initiation mechanism, this number could be doubled.

References pp. 154–162

Barr et. al. [4] have studied the thermal initiation of styrene at 330–410 K in the presence of a stable radical, 1,1-diphenyl-2-picrylhydrazyl (DPPH≡Ph_2N-N^* [$C_6H_2(NO_2)_3$]). From the decay of DPPH absorbance at 525 nm, the rate of radical formation from styrene can be calculated because each newly formed radical rapidly reacts with DPPH. The adduct absorbs light of different wavelength and cannot initiate polymerization. In the presence of DPPH, styrene is not thermally polymerized and we speak of inhibition. The rate of radical formation can be calculated from the DPPH concentration and from the length of the inhibition period. The spectral and kinetic methods did not yield the same result. A rate higher by a factor of ≈ 7 was determined spectroscopically. The discrepancy is ascribed by the authors to the much higher reactivity of the trinitrobenzene derivatives with styrene diradicals than with the monoradicals. The results support the diradical mechanism of thermal initiation. The effectiveness of the whole process is low and its activation energy is 121 kJ mol^{-1}. On the basis of own results and of previous suggestions [5] the authors formulated the scheme

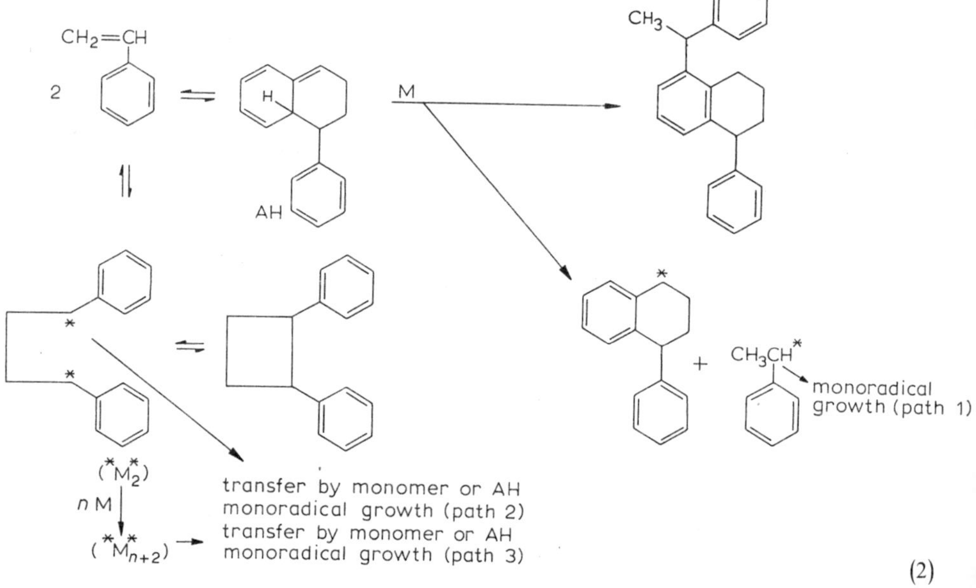

(2)

The idea of the formation of Diels–Alder intermediates from two styrene molecules was first postulated by Mayo [5] and has recently been revived by several authors. Sato et. al. [6] had described path 1 [see reaction (2)] before Barr [4]; they did not consider polymerization on diradicals. Kauffmann [7] supported the postulation of the formation of the 4π–2π Diels–Alder adduct (5-phenylbicyclo-[4,4,0]-deca-1,7,9-triene) spectrophotometrically. Kinetic analysis led him and his collaborators [8] to the conclusion that the adduct is formed in two stereoisomeric forms, with axial (a) and equatorial (b) position of the phenyl ring

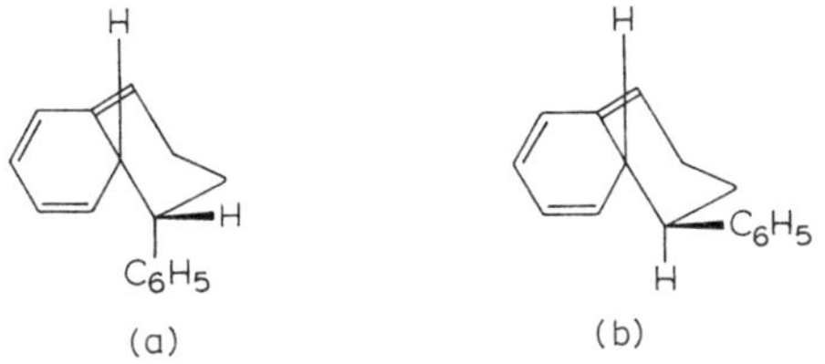

(3)

For mainly energy reasons, the isomer (a) can transfer its hydrogen to styrene and to the propagating radical $\sim C^*$; it is chemically more reactive than the stereoisomer (b). Form (a) leads to the formation of the monoradical and to transfer [9]† [see eqn. (4)]; form (b) yields the trimer exclusively. As long as the radicals formed from the monomer and adduct (a) do not sufficiently separate (see the cage effect, Sect. 1.7), they can recombine forming the trimer. Thus trimer formation competes with the initiation reaction. The whole process is illustrated in scheme (4) where a, b is the Diels–Alder adduct, PhT represents 1-phenyltetraline, T the styrene trimers, $[R_1^* \cdots R_2^*]$ the radical pair in the solvent cage, R_1^*, R_2^* polymerizable monoradicals and P* the propagating radical [7].

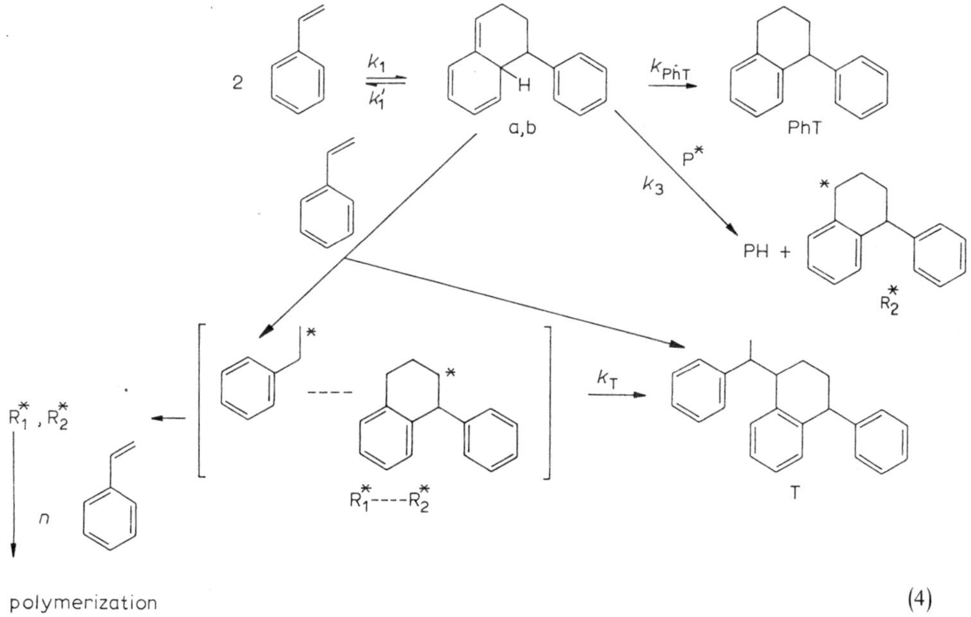

(4)

---

† The stationary concentration of the adduct (a) is 5.2×10^{-6} mol dm^{-3} at 353 K and 8.5×10^{-6} mol dm^{-3} at 333 K [7].

References pp. 154–162

To date, the role of the diradical in the thermal initiation of styrene has not been conclusively proved, and even less is known about its quantitative contribution to the initiation rate. Hara et. al. [10] have studied the products of the initial stages of thermal styrene polymerization by the method of field desorption mass spectrometry. They confirmed the formation of the trimer generated by the reaction of radicals with the surrounding monomer molecules forming a cage. Based on kinetic data Ito [11] assumes that the thermal initiation of styrene is a third-order reaction with respect to monomer; this observation might be an important contribution to our understanding of this problem. The generation of monoradicals does not at present raise any controversy. Moreover it is assumed that other vinyl monomers with an aromatic substituent (2-vinylfuran, 2-vinylthiophene [12]), may yield initiating radicals by a reaction analogous to that of styrene.

Methyl methacrylate may also give rise to initiating radicals [13]. A collision of two monomer molecules leads to the formation of linear dimers, trimers, two cyclobutane derivatives and a diradical [14]

$$\begin{array}{c} \text{MeO}-\overset{\overset{\displaystyle O}{\|}}{\underset{\underset{\displaystyle \|}{\text{MeO}-\text{C}}}{\text{C}}}-\overset{\displaystyle *}{\underset{\displaystyle }{\text{C}}}-\text{CH}_2-\text{CH}_2-\overset{\displaystyle *}{\underset{\displaystyle }{\text{C}}}\begin{array}{c}\overset{\displaystyle O}{\|}\\ \text{C}-\text{OMe}\\ \\ \text{C}-\text{OMe}\\ \underset{\displaystyle O}{\|}\end{array} \end{array}$$

The diradical should be in the singlet state with a tendency to recombination. Because of the high rotation barrier [15], its *cis* and *trans* forms should be discriminated, leading to the formation of *cis*- and *trans*-cyclobutane derivatives [13]. According to Bagdasarjan, about 0,1% of the singlet diradicals are inactivated by an intersystem crossing yielding the relatively stable triplet with a much lower rotation barrier [15]. The diradical in the triplet state is an initiating particle.

The rate of the thermally initiated methyl methacrylate polymerization amounts to only about 1% of the rate measured with styrene. It can be increased by the presence of heavy metal atoms [16] which can change the multiplicity of the diradical and thus also its reactivity.

Attempts at the thermal polymerization of vinyl chloride [15], vinyl acetate [17] and methyl acrylate [18] were unsuccessful.

1.2 Initiators

Almost all industrial radical polymerizations are initiated by labile compounds decomposing to reactive free radicals. These unstable compounds are called initiators. As practically used substances, they have been closely studied, especially since each monomer and each set of polymerization conditions require a different, strictly defined pattern of behaviour from the initiator. The half-time of initiator decomposition must correspond to the polymerization temperature; the order of the rate-determining step of the decomposition should be as low as possible (zero in the ideal case) in order to keep the initiation rate changes at a minimum during the course of polymerization; the initiator should normally have no transfer activity; its residue should form the required end group of the macromolecule; it should become inactive or be easily removable (extractable) at the end of the polymerization process, or it should be able to form strong complexes. Even properties not directly connected with polymerization are not without importance, for example solubility in polar and/or non-polar solvents, storage life, handling safety, the colour of the initiator itself and of its decomposition products, etc.

Such a collection of properties cannot be found even in a group of chemically related compounds, not to speak of a single chemical entity. A great number of compounds able to form radicals and suitable for polymerization initiation have already been investigated. The most important of these will now be discussed.

1.2.1 PEROXIDES AND THEIR DECOMPOSITION

According to Pauling, the bond between the two oxygen atoms in a peroxide has an energy of about 146 kJ mol^{-1}. Thus it easily undergoes thermal splitting with the formation of sufficiently reactive free radicals. Peroxide decomposition is rarely limited to the formation of only two fragments. Each radical can rapidly decompose further to radical and non-radical products. The primary and secondary radicals and non-radical products can affect the reactions of the original peroxide. Induced peroxide decomposition differs, of course, from its monomolecular decomposition; the polymerization kinetics therefore becomes more complicated due to the occurrence of radicals generated by secondary decomposition. One of the most frequently used initiators is dibenzoylperoxide. We shall use it as an example for the illustration of the above statements.

Berger et al. [19] have studied polymerization of styrene initiated by dibenzoyl peroxide (DBP) with the carbonyl groups labelled by the isotope ^{14}C. DBP can react in several ways [20, 21]

(a) primary decomposition

$$\text{Ph-C(=O)-O-O-C(=O)-Ph} \xrightarrow{2fk_d} 2 \text{ Ph-C(=O)-O}^* = 2R_1^*$$

(5a)

References pp. 154–162

(b) secondary decomposition

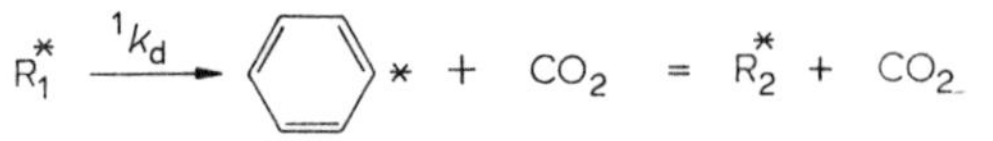

$$R_1^* \xrightarrow{^1k_d} \bigcirc {}^* + CO_2 = R_2^* + CO_2 \qquad (5b)$$

(c) decomposition iduced by macroradical P*

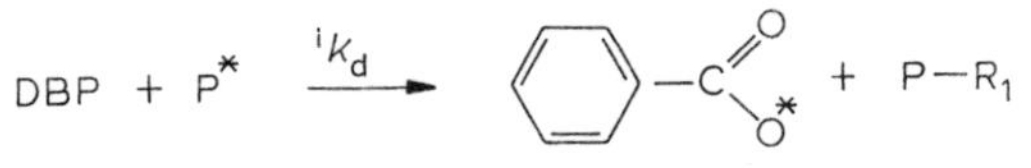

$$DBP + P^* \xrightarrow{^ik_d} \bigcirc-C\overset{O}{\underset{O^*}{\diagdown}} + P-R_1 \qquad (5c)$$

Both R_1 and R_2 are sufficienty reactive. Their mutual populations in the initiation reaction will depend on the mean lifetime of the benzoyloxy radical. This is about 10^{-9} s so that there is sufficient time for the secondary dissociation [22]. The macromolecules generated are terminated by both fragments at a ratio depending on the polymerization temperature [19, 23].

The rate constants of eqns. (5) can be expressed by the relations [19, 24]

$$\ln(2fk_d) = 26.39 - 107.6 \text{ (kJ mol}^{-1}) (RT)^{-1} \; ^\dagger$$

$$\ln \frac{^1k_d}{k_{i,1}} = -13.30 + 38 \text{ (kJ mol}^{-1}) (RT)^{-1} \; ^{\dagger\dagger}$$

$$\ln(^ik_d) = 21.86 - 52.84 \text{ (kJ mol}^{-1}) (RT)^{-1} \qquad (6)$$

$2fk_d$ and ^ik_d grow, whilst $^1k_d/k_{i,1}$ falls with increasing temperature. With increasing temperature, the fraction of chains initiated by the radicals

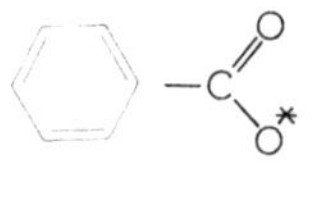

$$\bigcirc-C\overset{O}{\underset{O^*}{\diagdown}}$$

was found to grow, reaching 2/3 at 313 K.

Decomposition of DBP in the polymerization of vinyl acetate can also be expressed by eqns. (5) [25]. The medium has a considerable effect on the values of various constants; in the presence of a not very reactive monomer such as vinyl acetate the half-time of peroxide breakdown by induced decom-

---

† f is used to designate the constant characterizing the efficiency of initiation. The selection of this symbol is unfortunate, because the same symbol is widely used to designate mathematical functions. However, it has been established in the world literature, and the introduction of another symbol might lead to confusion.

†† $k_{i,1}$ is the rate constant in the reaction of R_1^* with monomer, i. e. of initiation.

TABLE 1

Peroxides used to initiate polymerization of ethylene to low-density polyethylene [28]

Peroxide	Formula	Active oxygen contents (%)	Temperature for a half-life of 1 min (K)
Isobutyryl peroxide	$\left[\begin{array}{c}\text{Me} \\ | \\ \text{HC}-\overset{\overset{\text{O}}{\|\|}}{\text{C}}-\text{O}- \\ | \\ \text{Me}\end{array}\right]_2$	9.2	363
Isopropyl peroxydicarbonate	$\left[\begin{array}{c}\text{Me} \\ | \\ \text{HC}-\text{O}-\overset{\overset{\text{O}}{\|\|}}{\text{C}}-\text{O}- \\ | \\ \text{Me}\end{array}\right]_2$	7.8	368
Di-n-butyl peroxydicarbonate	$\left[n\text{-Bu}-\text{O}-\overset{\overset{\text{O}}{\|\|}}{\text{C}}-\text{O}-\right]_2$	6.8	363
Di-sec.butyl peroxydicarbonate	$\left[sec.\text{-Bu}-\text{O}-\overset{\overset{\text{O}}{\|\|}}{\text{C}}-\text{O}-\right]_2$	6.8	363
Dicyclohexyl peroxydicarbonate	$\left[\text{C}_6\text{H}_{11}-\text{O}-\overset{\overset{\text{O}}{\|\|}}{\text{C}}-\text{O}-\right]_2$	5.6	363
Di(2-ethylhexyl) peroxydicarbonate	$\left[\text{C}_4\text{H}_9-\overset{\overset{\text{Et}}{|}}{\text{CH}}-\text{CH}_2-\text{O}-\overset{\overset{\text{O}}{\|\|}}{\text{C}}-\text{O}\right]_2$	4.6	363
Tert.butyl perneodecanoate	$\text{C}_9\text{H}_{19}-\overset{\overset{\text{O}}{\|\|}}{\text{C}}-\text{O}-\text{O}-\text{C}(\text{CH}_3)_3$	6.5	373
Tert.butyl perpivalate	$(\text{CH}_3)_3-\text{C}-\overset{\overset{\text{O}}{\|\|}}{\text{C}}-\text{O}-\text{O}-\text{C}(\text{CH}_3)_3$	9.2	383
Bis(3,5,5-trimethylhexanoyl) peroxide	$\left[iso\text{-C}_8\text{H}_{17}-\overset{\overset{\text{O}}{\|\|}}{\text{C}}-\text{O}\right]_2$	5.1	388
Dioctanoyl peroxide	$\left[\text{C}_7\text{H}_{15}-\overset{\overset{\text{O}}{\|\|}}{\text{C}}-\text{O}-\right]_2$	5.6	393

References pp. 154–162

TABLE 1 (continued)

Peroxide	Formula	Active oxygen contents (%)	Temperature for a half-life of 1 min (K)
Didecanoyl peroxide	$[C_9H_{19}-\underset{\underset{O}{\|}}{C}-O-]_2$	4.7	393
Dipropyonyl peroxide	$[C_2H_5-\underset{\underset{O}{\|}}{C}-O-]_2$	11.0	388
Dilauroyl peroxide	$[C_{11}H_{23}-\underset{\underset{O}{\|}}{C}-O-]_2$	4.0	393
Tert.-butyl perisobutyrate	$H-\underset{\underset{CH_3}{\|}}{\overset{\overset{H_3C}{\|}}{C}}-\underset{\underset{O}{\|}}{C}-O-O-C(CH_3)_3$	10.0	403
Tert.-butyl per-2-ethylhexanoate	$C_4H_9-\underset{\underset{C_2H_5}{\|}}{CH}-\underset{\underset{O}{\|}}{C}-O-O-C(CH_3)_3$	7.4	403
Tert.-butyl peracetate	$CH_3-\underset{\underset{O}{\|}}{C}-O-O-C(CH_3)_3$	12.1	433
Tert.-butyl per-3,5,5-trimethylhexanoate	$iso\text{-}C_8H_{17}-\underset{\underset{O}{\|}}{C}-O-O-C(CH_3)_3$	6.9	433
Tert.-butyl perbenzoate	$Ph-\underset{\underset{O}{\|}}{C}-O-O-\underset{\underset{Me}{\|}}{\overset{\overset{Me}{\|}}{C}}-Me$	8.2	433
Di-tert.-butyl peroxide	$Me-\underset{\underset{Me}{\|}}{\overset{\overset{Me}{\|}}{C}}-O-O-\underset{\underset{Me}{\|}}{\overset{\overset{Me}{\|}}{C}}-Me$	10.9	463
Tert.-butyl hydroperoxide	$Me-\underset{\underset{Me}{\|}}{\overset{\overset{Me}{\|}}{C}}-O-O-H$	17.8	533[a]

[a] Measured with pure peroxide. The half-life of the commercially available product is much lower.

position is drastically reduced [26]. Secondary and induced splitting have also been observed with dialkyl peroxides [27] and peroxyesters. Generally, therefore, the decomposition mechanism of the peroxide initiator and the effect of the medium cannot be neglected in studies of polymerization kinetics.

The induction and steric effects of substituents in the vicinity of the peroxide bond correspondingly affect the rate of its splitting and of the consecutive reactions.

This can be documented by Table 1 which summarizes data on the temperatures where just one half of some initiators are decomposed per minute.

The medium also has a strong effect on the peroxides. Acids induce heterolysis of the O—O bond and migration of the aryl or alkyl group to the electron-deficient oxygen [30].

Tert.-butyl hydroperoxide and di-*tert.*-butyl peroxide are decomposed into free radicals, initiating the polymerization of methyl methacrylate and styrene at room or lower temperature in the presence of $HClO_4$, H_2SO_4, HCl, CF_3SO_3H, CF_3COOH or BF_3OEt_2 [31]. Not every peroxide is decomposed into initiating radicals by acids. Cumyl hydroperoxide does not initiate the polymerization of methyl methacrylate in the presence of acid. This efect is ascribed to the cationic reactions of peroxide with the benzene ring [32]. The fraction of cumyl hydroperoxide decomposed into radicals is very small. The mechanism of the whole process is so far unclear; the polymerization might be inhibited by phenol generated during the decomposition [31].

The decomposition of some peroxides is accelerated by two to three orders of magnitude by some transition metals [33–36]. This effect is discussed in the following section on redox initiation.

Redox initiation

In many reactions, free radicals are formed along with the intermediates, although their existence is not evident from the material balance of initial compounds and products. The overall reaction of ferrous ions with hydroperoxide is expressed as [37]

$$H_2O_2 + 2 Fe^{2+} \xrightarrow{k} 2 Fe^{3+} + 2 HO^- \qquad (7)$$

More detailed investigations have led to the conclusion that the above reaction proceeds in two steps

(a) electron transfer to the peroxide

$$Fe^{2+} + H_2O_2 \rightarrow Fe^{3+} + HO^- + HO* \qquad (7a)$$

(b) reduction of the radical with a further Fe^{2+} ion

$$Fe^{2+} + HO* \rightarrow Fe^{3+} + HO^- \qquad (7b)$$

The radical *OH lives only for a short time. Its addition to a monomer that is possibly present leads to the formation of a radical active centre that is generally more reduction-resistant than HO*, and polymerization of the monomer occurs. The presence of Fe^{2+} accelerates the decomposition of hydroperoxide several-fold. Electron transfers between the oxidizing and reducing agents make possible the formation of a relatively high concentration of active centres, thus accelerating the polymerization process.

Redox initiation was discovered by Kern [38] in 1937. Today a great number of pairs are known, producing radicals at very much higher rates than the peroxides themselves. They enable vinyl and diene monomers to be polymerized at low temperatures without the use of spontaneously rapidly decomposing (and therefore dangerous) peroxides. On the other hand, the use of the peroxide-reducing agent pair introduces some complications. A non-stationary polymerization process may result.

To this day, redox systems are selected empirically. In spite of considerable efforts devoted to the generalization of past experience with redox initiation, neither the ionization potentials, electron affinities nor other molecular constants can be used at the moment as a reliable criterion for characterizing radical formation. The mechanism of interaction, the intermediates and the medium are all of great importance. The same is true of the monomer, which may even replace the reducing component in special cases [39]. General schemes are therefore of little importance and it will be more useful to discuss specific cases in greater detail.

Probably the most studied systems are those using a peroxide—reducing agent combination as represented by reaction (7). Its rate constant [40] $k = 4.45 \times 10^8 \exp(-9,400/RT)$ dm^3 mol^{-1} s^{-1}. The pair hydroperoxide–Fe^{2+} has been used for polymerizing many monomers, mainly methyl acrylate, methyl methacrylate and acrylonitrile. In addition to the organic peroxides, the persulphate ion is also frequently used as an oxidizing component. The reducing components used are perhaps even more numerous than the peroxides: hydrazine, hydroxylamine or hydrogen sulphide, $S_2O_4^{2-}$, salts of Ag^+, Cu^+ [41–43] and many others. The effectiveness of two-component organic redox systems may be enhanced by a so-called promoter, a small amount of transition metal ions.

(a) *Transition metal ion–reducing agent systems*

Nayak et al. [44] polymerized methyl methacrylate using Mn^{3+} with fructose in acid medium. They proposed the following mechanism of radical formation

$$Mn^{3+} + FR \underset{}{\overset{K}{\rightleftharpoons}} \text{complex}$$

$$\text{complex} \xrightarrow{k_T} R^* + Mn^{2+} + H^+ + F$$

$$R^* + Mn^{3+} \xrightarrow{k_0} NR$$

$$R^* + M \xrightarrow{k_i} RM_1^* \tag{8}$$

where FR is fructose, F its residue after complex decomposition, and NR are non-radical products. The mechanism includes complex formation between Mn^{3+} and fructose. The initiating radicals are generated by complex decomposition (they can also be inactivated by reaction with the metal ion). For the constant, the values $K_{293} = 80 \text{ dm}^3 \text{ mol}^{-1}$ and $k_T = 2.2 \times 10^{-4} \text{ s}^{-1}$ were found by the authors. A significant effect of donor–acceptor complexes in a considerable number of redox initiations of this type was proved by Bartoň [35, 36].

Many systems of this type exist, examples of which are: Cr^{6+} or $Fe^{3+}-$ -thiourea (tested in polymerizations of acrylonitrile, methyl acrylate and acrylamide [44]; Ce^{3+} with alcohols, aldehydes, amines, phosphates and carboxylic acids [45]; V^{5+} with glycols, Mn^{3+} with dicarboxylic acid and their derivatives [46]; Fe^{3+} with benzoin [47]; etc.

(b) *Other redox systems*

Diazonium salts are decomposed by metal ions

$$[R-N\equiv N]^+ + Fe^{2+} \rightarrow [RN\equiv N]^* + Fe^{3+}$$

$$[R-N\equiv N]^* \rightarrow R^* + N_2 \qquad (9)$$

Both radicals, $[R-N\equiv N]^*$ and R^*, can initiate polymerization [47]. In the system Cu^2–hydrazine hydrate, the reaction

$$N_2H_4 + Cu^{2+} \rightarrow N_2H_3^* + Cu^+ + H^+ \qquad (10)$$

occurs. The radical $N_2H_3^*$ initiates polymerization of methyl acrylate [49]. Although oxygen is normally a radical inhibitor, in this case its presence considerably accelerates polymerization; it probably re-oxidizes Cu^+. The system $NaClO_3$ with Na_2SO_3 initiates the polymerization of acrylonitrile and acrylamide in acid medium [50]. The system BrO_3^-–thiourea produces initiating radicals for methyl methacrylate polymerization, again in acid medium [51].

1.2.2 AZO INITIATORS AND THEIR DECOMPOSITION

The most frequently used initiator of this type is 2,2'-azobisisobutyronitrile (AIBN). It was first prepared by Thiele and Hauser [51] who, at the same time, correctly identified its decomposition products. Analysis of the products was refined by Bickel and Waters [53]. According to them, the resulting compounds are the tetramethyldinitrile of succinic acid (84%), isobutyronitrile (3%) and 2,3,5-tricyanohexane (9%).

Talât-Erben and Bywater [54] proved the existence of unstable intermediates. They proposed the mechanism

$$R-N=N-R \rightarrow 2\ R^* + N_2$$

$$R^* + R^* \begin{matrix} \nearrow \\ \\ \searrow \end{matrix} \begin{matrix} (CH_3)_2C-C(CH_3)_2 \\ |\quad\ | \\ CN\ CN \\ \\ (CH_3)_2CH(CN) + (CH_3)(CN)C=CH_2 \end{matrix} \qquad(11)$$

for the decomposition of AIBN [R = $(CH_3)_2$ C(CN)–]. Dimethylketene-cyanoisopropylimine $(CH_3)_2$ C=C=NC$(CH_3)_2$ CN is also a decomposition product of AIBN. The isobutyronitrile radical is therefore probably resonance-stabilized

$$(CH_3)_2 \overset{*}{C}(CN) \longleftrightarrow (CH_3)_2C=C=N^*$$

and the product mentioned above is formed by its "anomalous" combination.

Complications caused by secondary decomposition of AIBN need not be considered. Its initiating effectiveness, f, is practically independent of temperature, but it increases with pressure. The combination rate of the radicals $(CH_3)_2$ C*(CN) in the toluene "cage" also increases with pressure; the process therefore is evidently diffusion-controlled [55]. AIBN decomposition can be accelerated by the addition of Lewis acids. In this respect some formal similarity therefore exists between the behaviour of peroxy and azo initiators.

Lakhinov et al. [56] observed an 8.5-fold increase in the rate constant of AIBN decomposition in methyl methacrylate caused by the presence of $ZnCl_2$[†]. The initiation rate increase more than fourfold (see Figs. 1 and 2), and the degree of polymerization falls more slowly with increasing concentration of initiator. The complexes of BF_3 with AIBN decompose to radicals 14 times more rapidly than the azo initiator itself [59], thus making possible its use at around room temperature.

Other azo initiators, e. g. 1,1′-diacetoxy-1,1′-diphenylazoethane [60], 2-cyano-2-propylazoformamide [61] and many others [62], are little used. The reasons for this are that they are less easily available, their decomposition may be complicated and affected by many factors, and their initiating effectiveness may not be very high. Photoactive groups have been attached to the ends of

---

[†] Other authors did not observe catalysis of radical decomposition of AIBN in the presence of $ZnCl_2$. See, for example, ref. [57]. This discrepancy was explained by Bartoň; it is due to the use of various solvents, forming complexes of differing stability with $ZnCl_2$ [34].

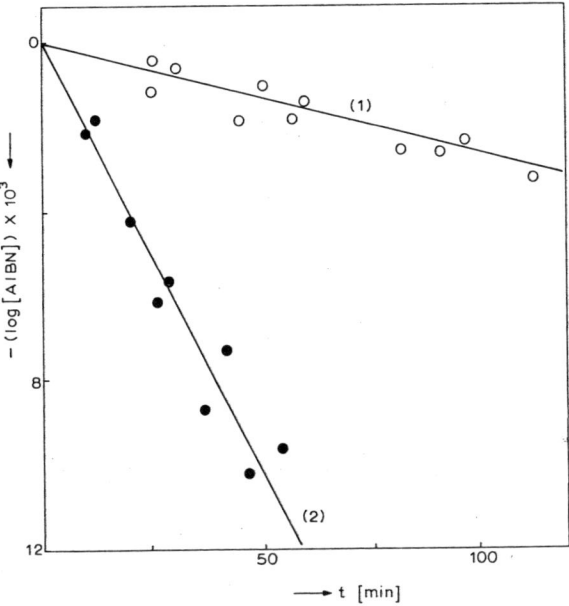

Fig. 1 Decomposition of AIBN at 323 K. Concentration $[\text{AIBN}] = 1.5 \times 10^{-3}$ mol dm^{-3}: (1) in MMA; (2) in MMA–ZnCl$_2$ at $[\text{ZnCl}_2]/[\text{MMA}] = 0.1$.

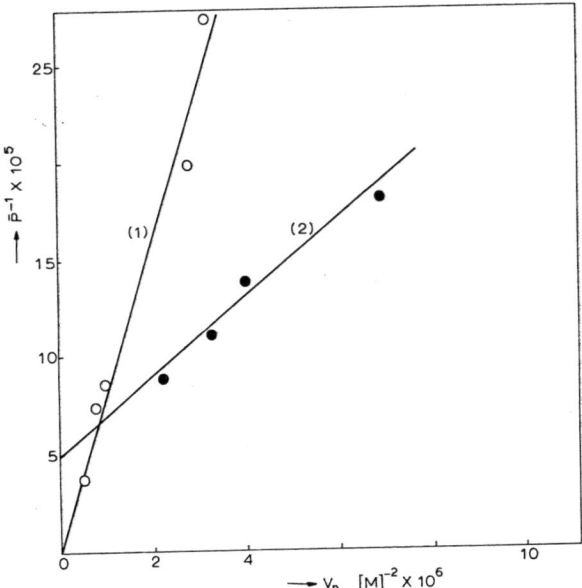

Fig. 2 Dependence of $(1/\bar{P})$ on $v_p/[\text{M}]^2$. Polymerization of (1) MMA and (2) MMA–ZnCl$_2$ at $[\text{ZnCl}_2]/[\text{MMA}] = 0.1$; 323 K, AIBN.

References pp. 154–162

macromolecules by means of some azo initiators, but so far only on a laboratory scale. An example of such an application is $N-$[4,4'-azobis (4-cyanopentanoyl)]-bisdibenz [b, f] azepine [63].

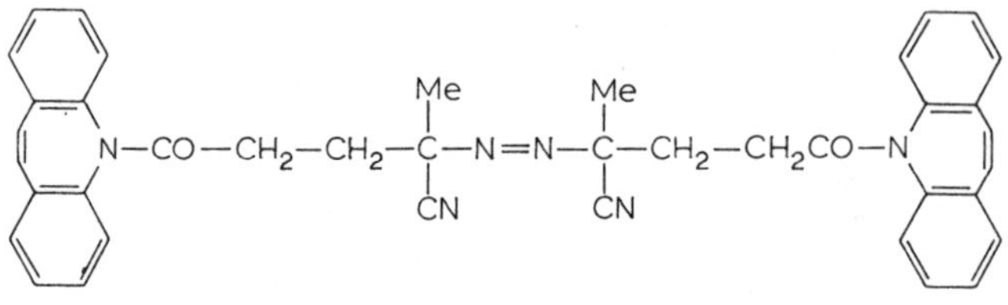

$$\text{Dibenzazepine-N−CO−CH}_2\text{−CH}_2\text{−C(Me)(CN)−N=N−C(Me)(CN)−CH}_2\text{−CH}_2\text{CO−N-Dibenzazepine}$$

1.3 Photoinitiation

This is a typical laboratory method for starting polymerization. From the research point of view it has the invaluable advantage of giving an immediate start to radical generation, which can be varied at will, and stopped by simply switching the light on or off. However, it has not been found suitable for technological applications so far. Polymerizing systems are rarely optically pure and isotropic. The intensity of light falls with the distance from the source and it also depends on the conversion. The production of radicals, and therefore also the initiation rate, thus depends on the coordinates of the place at which they are generated. In the present large-volume reactors it is practically impossible to evade this drawback. Nevertheless, the method is technologically important for some special applications, for example in the curing of films [64].

Only light of wavelengths (λ) that are absorbed by the monomer can be used for direct initiation. Almost all monomers absorbing radiation of wavelengths $\lambda \leq 390$ nm are degraded with the formation of radicals initiating polymerization. Polymerization of styrene and vinyl acetate is induced by photons $\lambda = 250$ nm [65, 66]; the quantum yield is, however, small. Actually, only a few vinyl monomers can give a good yield when polymerized with direct photoinitiation [67][†]. Much more frequently, photosensitizers which enhance the utilization of the light input, are used.

[†] Except for MMA, these are rather uncommon monomers, such as allyl-, cinnamyl-, 2-hydroxyethyl-, and methyl methacrylate; barium-, 2-ethylhexyl-, and 2-hydroxypropylacrylate; diallyl phthalate; diallyl isophalate; diallyl terephthalate; N, N'-methylenebisacrylamide; pentaerythritoltetramethyl acrylate; tetraethyleneglycol dimethacrylate; N-vinylcarbazole; vinyl-2-furyl acrylate.

1.3.1 SENSITIZATION

This is usually accomplished by the addition of molecules inducing initiation by various mechanisms.

(a) *Initiation by energy transfer*

The sensitizer is excited by light to the triplet state and, in the next step, the excitation energy is transferred to the monomer. From the monomer in the triplet state, the initiating radicals are formed. The polymerization of methyl methacrylate and acrylamide in the presence of uranyl ions is a classic example [68, 69].
Venkatarao and Santappa assume the initiation mechanism[†]

$$UO_2^{2+} + h\nu \xrightarrow{k_1} {}^3UO_2^{2+} \tag{12a}$$

$${}^3UO_2^{2+} \xrightarrow{k_2} UO_2^{2+} + \Delta \tag{12b}$$

$${}^3UO_2^{2+} \xrightarrow{k_3} UO_2^{2+} + h\nu' \tag{12c}$$

$${}^3UO_2^{2+} + M \xrightarrow{k_4} UO_2^{2+} + {}^3M \tag{12d}$$

$${}^3M + M \xrightarrow{k_i} 2R_i^* \tag{12e}$$

$${}^3M + M \xrightarrow{k_s} M + \Delta' \tag{12f}$$

$$R_i^* + M \xrightarrow{k_5} RM^* \tag{12g}$$

The first reaction describes the excitation of uranyl ions. The excited sensitizer can lose the energy Δ by a non-radiative process (12b), by emission (12c) or by energy transfer in monomer excitation to the triplet state (12d). Radicals are formed by reaction (12e). The detailed mechanism of step (12e) is so far unknown. Electron transfer probably occurs, with radical cation and radical anion formation; these can recombine by their oppositely charged ends. The products retain their radical character. Step (12g) corresponds to propagation and step (12f) to inactivation of the excited monomer by collision with another molecule. The photosensitized initiation and polymerization of methacrylamide [69] probably proceeds according to scheme (12). Ascorbic acid and β-carotene act as sensitizers of isoprene photoinitiation in aqueous media [70], and diacetyl (2, 3-butenedione) as sensitizer of vinylidene chloride photopolymerization in a homogeneous medium (*N*-methylpyrrolidone was used as solvent) [71].

[†] The triplet is usually formed by intersystem crossing from the excited singlet [see for example, scheme (13)], not by direct irradiation. Thus, eqn. (12a) is more complicated.

References pp. 154–162

(b) *Photosensitization by electron transfer*

In the excited state, some sensitizers have a tendency to electron transfer with radical formation. The whole process is somewhat reminiscent of redox systems.

As an example, the polymerization of vinyl monomers by anthraquinone sulphonates (AS) in the presence of reducing agents [72] is presented.

$$AS \underset{-h\nu}{\overset{+h\nu}{\rightleftarrows}} {}^SAS \qquad (13a)$$

$$^SAS \longrightarrow {}^TAS \qquad (13b)$$

$$^TAS + Cl^- \longrightarrow AS^{-*} + Cl^* \qquad (13c)$$

$$AS^{-*} + H^+ \longrightarrow ASH^* \qquad (13d)$$

$$Cl^* + M \longrightarrow M_1^* \qquad (13e)$$

Antraquinone sulphonate absorbs a quantum of light yielding an excited singlet (13a), which is transformed to a triplet by an intersystem crossing (see ref. [73] for explanation of this term) (13b). The latter is reduced by a chloride ion. The generated chlorine atom initiates polymerization (13e). In this process, antraquinone sulphonate is reduced to an anthraquinol derivative.

The system benzophenone—triethylamine produces ketyl and amine radicals by electron transfer between the photoactivated benzophenone and the amine in the ground state. The generated radicals initiate the polymerization of methyl methacrylate [74].

(c) *Initiation by fragmentation*

Some radical initiators decomposing by heat can also be decomposed by UV radiation. Photopolymerization of acrylonitrile in the presence of AIBN or hydrogen peroxide, or of other initiators [75–78] has been reported. The quantum yield of AIBN decomposition is 0.4 at 298 K and 0.6 at 318 K. Photopolymerization of methyl methacrylate, styrene, and vinyl acetate can be initiated by tetramethylsilane, methylchlorosilanes, and halides of Group IV metals [79]. We assume that the radicals are formed by homolytic splitting of the covalent bond

$$Si(CH_3)_3Cl \xrightarrow{h\nu} {}^*Si(CH_3)_3 + Cl^*$$

$$^*Si(CH_3)_3 + M \longrightarrow (CH_3)_3SiM^* \qquad (14)$$

(d) *Initiation by decomposition products of donor—acceptor complexes*

Decomposition of initiators can be strongly affected by DA complexes of the initiator with a further component of the system, e. g. the solvent [80]. Photopolymerization of acrylonitrile is greatly accelerated by the addition of isobutyl vinyl ether [81]. Spectroscopic measurements have shown that acrylonitrile and isobutyl vinyl ether form a donor—acceptor complex which is stable in the dark. When illuminated, it initiates polymerization.

The details of the initiation mechanism are not yet known. Polymerization proceeds exclusively by the radical mechanism, and the increased rate is ascribed to the enhanced rate of initiation by photoexcitation of the complex (DA designates the complex) [67]

$$DA + h\nu \rightarrow (^{\delta+}D\cdots^{\delta-}A)^*$$

$$(^{\delta+}D\cdots^{\delta-}A)^* \rightarrow DA + \Delta$$

$$\rightarrow D^{+*} + A^{-*} \quad (15)$$

$$\rightarrow D^+ - A^-$$

$$\rightarrow {}^*D - A^*$$

The accelaration of styrene photopolymerization by oxygen is also explained by excitation of the DA complex of these two substances [82]. A copolymer is produced which decomposes upon illumination [83]. Polymerization of methyl methacrylate is initiated by the photoexcited complex of the monomer with triethylaluminium [84]. Methyl methacrylate, acrylonitrile and acrylates in general readily produce unstable DA complexes which decompose to products quite different from the initial components. Methyl methacrylate, for example, polymerizes in the presence of quinoline and bromine. With the monomer, these pairs yield a DA complex which is unstable upon illumination [85a]

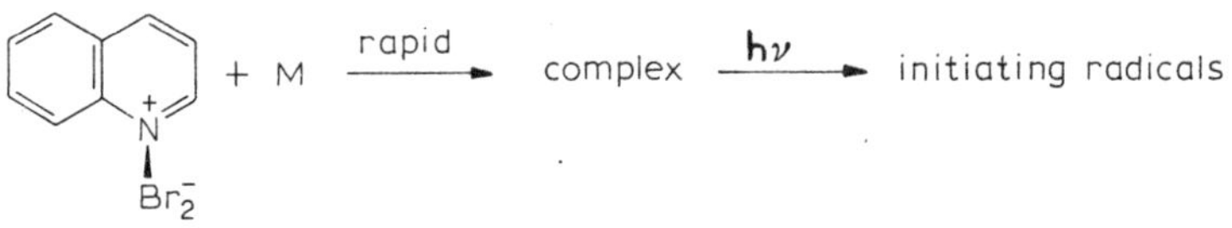

$$\text{(quinoline-N-Br}_2^-\text{)} + M \xrightarrow{\text{rapid}} \text{complex} \xrightarrow{h\nu} \text{initiating radicals} \quad (16)$$

The effectiveness of initiation also depends on external conditions, e. g. on type of solvent [85b].

Numerous complexes have been described which decompose in light with the formation of initiating radicals. It appears that the participation of the monomer plays an important role in the generation of complexes and exciplexes (see Sect. 5.1).

References pp. 154–162

The importance of donor—acceptor complexes for the initiation of radical (and similarly of ionic) polymerizations is ever increasing. Therefore this topic will be treated separately in Sect. 6 where the reactions of DA complexes producing active polymerization centres will be more generally described.

1.4 Initiators produced in situ

Technical reasons sometimes require an enhancement of the initiation rate, for example when we wish to carry out the polymerization at low temperature, to shorten or nullify the inhibition period [86] (and thus increase the productivity of the polymerization equipment), or to paralyze the consequences of degradative transfer [87]. When a redox system cannot be used, only a single classical way is left, the selection of a rapidly decomposing initiator. The main drawback of this solution is the high risk in manipulating compounds of this type. Smith [88] advanced the idea of synthesizing the initiator directly in the polymerization medium. Barter and Kellar [86] prepared the initiating diacyl peroxide of an acid anhydride with H_2O_2 or organic peracid.

For the polymerization of vinyl chloride in aqueous medium, they used isobutyric acid anhydride and H_2O_2 as diisobutyryl peroxide precursors

$$(C_3H_7\overset{\overset{O}{\|}}{C})_2O + H_2O_2 + HCO_3^- \rightarrow C_3H_7\overset{\overset{O}{\|}}{C}OOH + C_3H_7\overset{\overset{O}{\|}}{C}-O^- + H_2CO_3$$

$$C_3H_7\overset{\overset{O}{\|}}{C}OOH + (C_3H_7\overset{\overset{O}{\|}}{C})_2O + HCO_3^- \rightarrow$$

$$\rightarrow C_3H_7\overset{\overset{O}{\|}}{C}OO\overset{\overset{O}{\|}}{C}C_3H_7 + C_3H_7\overset{\overset{O}{\|}}{C}O^- + H_2CO_3 \qquad (17)$$

This system enabled them to polymerize vinyl chloride at 303 K at a rate comparable with that obtained with dialkylperoxydicarbonates at 323 K.

Even dialkyl dicarbonates can be produced in the polymerizing medium. This technique was applied by Ravey and Waterman [87] for the initiation of vinyl chloride—propene copolymerization.

1.4.1 EFFECT OF OXYGEN

Addition of oxygen to an alkyl radical produces the relatively unreactive peroxyradical

$$R^* + [^*\underline{O}-\underline{O}^*] \longleftrightarrow [|\underline{O}-\overset{*}{\underline{O}}|] \rightarrow ROO^* \tag{18}$$

Thus in radical polymerizations, elementary oxygen is always an inhibitor. Its inhibiting effects in the polymerizations of styrene, vinyl acetate, vinyl chloride, the acrylates and methacrylates, were described some years ago [89, 90].

Oxygen bound in peroxides or complexes can initiate or accelerate polymerizations. Some monomers, e. g. vinyl acetate, are oxidized to peroxidic compounds. Initiating radicals are liberated by thermal decomposition of peroxides.

The number of possible reactions between oxygen and the components of the polymerizing system is considerable. Both in laboratory work and in industrial processes, oxygen must always be removed with the greatest care prior to polymerization.

1.5 Other methods leading to radical formation

In Sect. 1.3, photoinitiation by means of donor—acceptor complexes was described. In some cases, these complexes may play an important role even without the contribution of light energy. In the presence of aliphatic amines and CCl_4, methyl methacrylate is polymerized at temperatures ≤ 300 K. In polar solvents (N,N-dimethylformamide, dimethylsulphoxide, chloroform), interaction of aliphatic amines as donors with methyl methacrylate as acceptor produces complexes [91] which yield initiating radicals with CCl_4 [92]

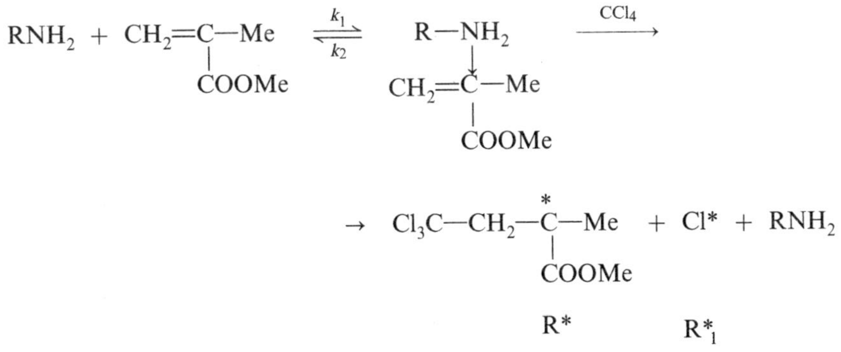

$$RNH_2 + CH_2=\underset{COOMe}{\overset{|}{C}}-Me \underset{k_2}{\overset{k_1}{\rightleftarrows}} \underset{\underset{COOMe}{\overset{|}{CH_2=C-Me}}}{\overset{|}{R-NH_2}} \xrightarrow{CCl_4}$$

$$\rightarrow Cl_3C-CH_2-\underset{COOMe}{\overset{*}{\overset{|}{C}}}-Me + Cl^* + RNH_2$$

$$R^* R_1^*$$

$$R^*, R_1^* + M \rightarrow R_{(1)}M^*$$

References pp. 154–162

In aliphatic amines, the free electron pair on N is placed in an sp^3 orbital, whereas in pyridine it is in an sp^2 orbital [91]. The sp^2 electrons are more strongly attracted by the nucleus. Thus pyridine and aromatic amines are less suitable for donor—acceptor interactions. Methyl methacrylate does not polymerize with these compounds. The electron density on the styrene double bond is higher than in methyl methacrylate; under the described conditions, styrene does not polymerize even in the presence of aliphatic amines [92].

An interesting method of initiating radical polymerization (although so far only theoretical) is offered by plasma. For polymerizations, plasma is usually generated by electrical glow discharge at radio frequencies [94]. Collisions between electrons with high kinetic energy and monomer molecules lead to the formation of ions, excited molecules, and radicals. In inorganic media, the concentration of radicals exceeds that of ions by 3–6 orders of magnitude [95]. Polymerization tests indicate that radicals also predominate in organic media. [82]. Even molecules lacking the properties of conventional monomers are polymerized in plasma. Therefore we discriminate between polymerization in plasma, eqns. (20), and plasma-initiated polymerization, eqns. (21) [96–98].

$$M_i \rightarrow M_i^*$$
$$M_k \rightarrow M_k^*$$
$$M_i^* + M_k^* \rightarrow M_{i+k} \tag{20}$$

$$M_i^* + M_k \rightarrow M_{i+k}^*$$
$$M^* + M \rightarrow MM^*$$
$$M_n^* + M \rightarrow M_{n+1}^* \tag{21}$$

γ-radiation is also used in initiation studies. Derivatives of acrylic and methacrylic acids are polymerized by γ rays (^{60}Co radiation) of intensity 5×10^4 J kg^{-1} h^{-1} at 100 K [99]. Under certain conditions, polymerization-initiating radicals are also produced by electroinitiation [100] (see Sect. 6.1). Reactive radicals are also formed by the breaking of covalent bonds by mechanochemical reactions or on transition from the liquid to the solid phase and vice versa (especially by freezing and melting of aqueous suspensions) [101]. Mechanochemical initiation can be of some importance in the industrial production of block and graft copolymers. The remainder of the these methods of radical generation are not much used in technical applications.

1.6 Non-typical initiators

The requirements of macromolecular syntheses can sometimes only be satisfied by special initiators. In addition to the conventional demands[†], for example concerning solubility in a certain medium, vapour pressure, physical state, etc., other initiator properties may be of importance.

1.6.1 MULTIFUNCTIONAL INITIATORS

The initiating radical forms the end group of the chain. A diradical, formed by the decomposition of difunctional initiators, is built into the interior of the macromolecule. This may be of importance for some polymer properties, e. g. stability to thermo-oxidation, chain flexibility, etc.

Glazomitskii et. al. [102] have described a difunctional initiator generated by ozonization of the end groups of oligomers of acrylic acid esters. It has peroxidic functional groups. A similar initiator was described by Galubei et. al. [103]; the peroxy end groups are connected by a polyoxirane chain

$$\text{ROO}\overset{\text{O}}{\underset{\|}{\text{C}}}\text{R}'\overset{\text{O}}{\underset{\|}{\text{C}}}\text{O}(\text{CH}_2\text{CH}_2\text{O})_n\overset{\text{O}}{\underset{\|}{\text{C}}}\text{R}'\overset{\text{O}}{\underset{\|}{\text{C}}}\text{OOR}$$

The main advantage of this type of initiator is the production of significantly longer polymer chains without reduction of the polymerization rate (compared with monofunctional initiators) [104]. When both functional groups are part of the same peroxide, they are both decomposed to the initiating radicals at practically the same rate. When the difunctional initiator is composed of peroxides decomposing at various rates, this property may be utilized for the synthesis of block copolymers.

Bylina et al. [105] have studied the bulk polymerization of styrene initiated by diacyl diperoxides

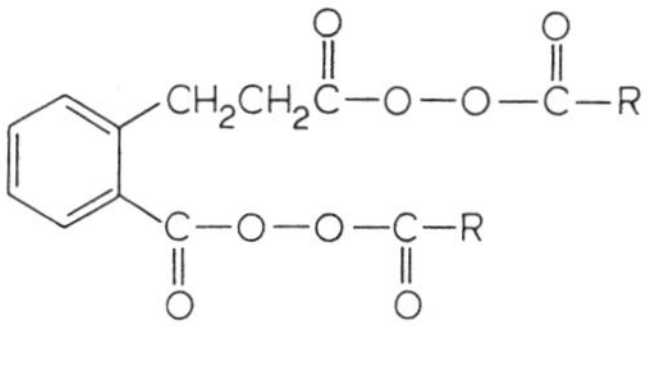

R = Me or C_6H_5

[†] The principal requirement upon the initiator is controllable rate and reproducible mode of decomposition, as well as small variations in the reactivity of the generated radicals.

References pp. 154–162

Each peroxide group behaves roughly as a monofunctional diacyl initiator. By a suitable selection of the substituent R, polystyrene with long chains and low active oxygen content in peroxide groups, or polystyrene of roughly half the molecular weight and high peroxide content can be obtained. These peroxide groups can be used for the initiation of another monomer polymerization.

1.6.2 POLYMERIC INITIATORS

As the title indicates, the initiating group is bound to a polymer chain. Some initiators from the preceding paragraph evidently also have the character of polymeric initiators. The azo initiators also have their polymeric analogues.

Piirma and Gunesin [106] polymerized methyl methacrylate in bulk with di-*tert.*-butyl-4,4'-azobis (4-cyanoperoxyvalerate) at 298 K in the presence of amine as a reducing agent

$$R''OOC-RN=NR-COOR'' + R'NH_2 \rightarrow$$
$$\rightarrow {}^*OC-RN=NR-CO^* + NR$$

$$^*OC-RN=NR-CO^* + n\,CH_2=CMe \rightarrow$$
(with C=O groups and COOMe)

$$\rightarrow PMMA\sim OC-RN=NR-CO\sim PMMA \quad (1)$$

where NR indicates non-radical products. The generated macromolecules, 1, were used to initiate isoprene or butyl acrylate polymerization

$$1 \xrightarrow{333\,k} 2\,PMMA\sim OCR^* + N_2$$

$$PMMA\sim OCR^* \xrightarrow{monomer} PMMA\sim OCR\sim\!\!\sim\!\!\sim\!\!\sim^* \quad (22)$$

Another macroinitiator of this type is azonitrile formed by the condensation of 4,4'-azobis-4-cyanovaleryl chloride with polyoxirane

$$m \text{ ClCCH}_2\text{CH}_2\overset{\text{Me}}{\underset{\text{CN}}{\overset{|}{\underset{|}{\text{C}}}}}-\text{N}=\text{N}-\overset{\text{Me}}{\underset{\text{CN}}{\overset{|}{\underset{|}{\text{C}}}}}\text{CH}_2\text{CH}_2\text{CCl} +$$

(with C=O groups as indicated)

$$+ m \text{ HO}-(\text{CH}_2-\text{CH}_2-\text{O})_n\text{CH}_2\text{CH}_2\text{OH} \xrightarrow[-(2m-1)\text{HCl}]{}$$

$$\rightarrow \text{Cl}\left[\text{CCH}_2\text{CH}_2\overset{\text{Me}}{\underset{\text{CN}}{\overset{|}{\underset{|}{\text{C}}}}}-\text{N}=\text{N}-\overset{\text{Me}}{\underset{\text{CN}}{\overset{|}{\underset{|}{\text{C}}}}}\text{CH}_2\text{CH}_2\text{C}-\text{O}-(\text{CH}_2\text{CH}_2-\text{O})_n\text{CH}_2\text{CH}_2\text{O}\right]_m\text{H}$$

(23)

Thermal decomposition of —N=N— groups in the presence of monomers (for example vinyl chloride) leads to the formation of block copolymers [107]. When the decomposing —N=N— groups is bound to a polymeric chain as a substituent, graft copolymers are produced. An example of such an initiator may be the copolymer of styrene with (4-vinylphenylazo)-2-methylmalononitrile [108]

(structure showing styrene copolymer with pendant —N=N—C(CN)(CN)—Me group decomposing to give $N_2 + {}^*\text{C(CN)}_2\text{Me}$ and a polymer radical)

(24)

Multifunctional polymeric initiators are also formed by the reaction of ozone with polyalkenes [109, 110]

$$\text{RH} + \text{O}_3 \rightarrow \text{R}^* + {}^*\text{OH} + \text{O}_2$$

$$\text{R}^* + \text{O}_2 \rightarrow \text{ROO}^*$$

$$\text{ROO}^* + \text{RH} \rightarrow \text{ROOH} + \text{R}^* \qquad (25)$$

Scheme (25) is the most general symbolic representation of thermo-oxidative degradation of hydrocarbon polymers. The generated hydroperoxide or dialkyl peroxide groups can be used to initiate radical polymerization of many monomers. Homopolymers, graft or block copolymers are produced depending on the reaction conditions. The fixation of the initiating radical to a certain position in the chain (which may be a component of the solid phase or in solution) is manifested by specific phenomena during polymerization (see Chap. 5, Sect. 2.2).

References pp. 154–162

From the preceding lines it is evident that multifunctional and polymeric initiators have been the object only of laboratory work so far; they have not yet progressed to larger scale production. Nevertheless, requirements of specialized polymers may dictate their wider application in the future.

1.7 Behaviour of primary radicals

The dissociation of initiator molecules can be represented by two reaction types

$$R-R \rightarrow R^* + R^* \tag{26}$$

$$R-X-R \rightarrow R^* + R^* + X \tag{27}$$

where X is a stable molecule, for example N_2, in the decomposition of azo compounds. Radicals generated by the splitting of a covalent bond, are always formed in pairs.

1.7.1 CAGE EFFECT

Radical generated according to reactions (26) and (27) cannot be mutually independent immediately after their formation. According to Noyes [111], the surrounding molecules form a kind of cage hindering their separation.

$$\text{initiator} \rightleftharpoons \{R^* \cdots R^*\} \rightleftharpoons \text{\guillemotright} R^* \cdots\cdots R^* \text{\guillemotleft} \rightarrow 2R^* \tag{28}$$

In the first step, the generated radicals are still in contact. Such highly reactive particles as radicals cannot be in close vicinity without reacting. The initiator molecule need not always be regenerated in this process; products may be formed which do not produce further radicals. In the next stage, the radicals have escaped from the range of immediate contact, but they remain close enough for renewed contact and reaction to occur with non-negligible probability.

The occurrence of molecule X during radical formation according to reaction (27) is a strongly complicating factor. The primary cage is affected by intervention of the fragment X which can separate the radical pair and thus reduce the probability of their mutual reaction.

The last step is the further separation of the two radicals to a distance approaching the average radical spacing in the system. The probability of mutual reaction is small for a pair that has passed the third stage. Radicals that have escaped immediate contact (the primary cage) can react with the monomer or with their twin or with the radical product of the reaction of the twin with the monomer. While the primary cage $\{R^* \cdots R^*\}$ can be physically defined, the state represented by the symbol $\{R*\cdots\cdots R*\}$ cannot be defined. The difference between the latter state and 2 R* is a matter of convention based on the probability of radical reactions [112a].

The process of mutual radical separation in the pair should be diffusion-controlled. Therefore the rate of radical liberation from the cage depends on the viscosity of the medium. Actually, some correlation between the two quantities has been observed but it is not simple.

The cage effect is collective designation of a number of processes manifested in various ways. We understand its meaning as a whole; the individual processes, however, have not yet been sufficiently elucidated.

1.7.2 EFFICIENCY OF INITIATION

Not all the radicals generated by the dissociation of the initiator can initiate polymerization. The effectiveness of initiation, f, is defined by the ratio of the initiation rates and initiator decomposition

$$f = \frac{v_{init}}{-d[I]/dt}, \qquad (29)$$

Thus f is the standardized fraction of initiating radicals referred to all radicals generated in the system.

The rate constants of radical reactions with vinyl monomers exhibit values around 10^2 mol^{-1} dm^3 s^{-1} and often even much higher at 298–350 K. The data of flash photolysis indicated values of up to 10^8 mol^{-1} dm^3 s^{-1} for a reaction of low-molecular-weight radicals with monomer [113]. The concentration of monomer usually does not exceed the range 1–10 mol dm^{-3}. The interval preceding the reaction of the radical with the monomer is therefore $\leq 10^{-3}$ s, but rarely less than 10^{-6} s. The mean residence time of the radicals in the cage is 10^{-9} to 10^{-10} s. The mean life time of combining radicals is of the same order. Some data on the effectiveness of initiation, f, are summarized in Table 2.

The fraction of radical reactions in the cage is not necessarily the same for photochemical and thermal dissociation of the initiator. Photochemically generated radicals are usually excited and so they have less tendency to produce stable molecules. They may be flung away from each other by excess energy at the moment of generation, and the probability of mutual collision is reduced [124].

The cage effect has a considerable influence on the course of radical polymerizations. It is held responsible for many kinetic anomalies.

References pp. 154–162

TABLE 2

Initiation efficiency, f, and the fraction of mutually reacting primary radicals $(1 - F)$ [112b]

Initiator	Solvent	Decomposition mode	Temperature (K)	$(1 - F)^a$	f^a	Method	Ref.
Azocumene	Toluene	b	313	0.26		d	114
Azocumene	Toluene	c	298	0.35		f	114
2,2-Azobisisobutyronitrile	Styrene	b	333		0.67	e	115
	Styrene/DMF	b	333–353		0.73	f	116
	Acrylonitrile/DMF	b	333		0.77	f	117
	Methacrylonitrile/DMF	b	333		0.72	f	117
	MMA/DMF	b	333		0.67	f	117
					0.70	f	116
	MA/DMF	b	333		0.67	f	117
	DMF	b	333		0.67	f	117
			313–343		0.70	f	119
Azobis-α-phenylethane	Ethylbenzene	b	378	0.29		d	115
Azodibenzoyl	MMA	b	363		0.75	e	120
Acetylperoxide	Isooctane	b	353	0.53		d	114
Hexafluoroazomethane	Various	c	373–473	0.1–0.7		d	121
Azomethane	Various	c	373–473	0.5–0.8		d	122
Di-tert.-butyl peroxyoxalate	Various	b, c	318	0.03–0.54		d	123

[a] $(1 - F)$ is the fraction of mutually reacting radicals and f is the fraction of radicals reacting with the inhibitor or with another molecule. The values of $(1 - F)$ and $(1 - f)$ shuld be approximately equal.

[b] Thermal.

[c] Photo.

[d] Analysis of the radical–inhibitor reaction product.

[e] Determination of the isotopically labelled initiator fragment bound in the polymer.

[f] Comparison of inhibitor decay rate (usually $FeCl_3$, but also $CuCl_2$ or $CuBr_2$) with rate of N_2 generation.

1.8 Reaction of radicals with unsaturated molecules

Primary radicals are unstable particles. Their ability to add a monomer can be classified by various degrees. So, for example, the reaction of the benzoyloxy radical with styrene exhibits a significantly higher activation energy than the addition of the butyronitrile radical to the same monomer [19]. Therefore the initiation rate differs from case to case.

1.8.1 INITIATION OF RADICAL POLYMERIZATION

The active centres of polymerization are produced by the addition of the primary radical to the monomer, i. e. to a π electron system. Only rarely is this simple process, and almost all branches of theoretical chemistry and chemical physics have contributed to its elucidation. The addition is a bimolecular reaction interpreted kinetically as a second-order reaction [125]. Unfortunately, most studies have been concerned with reaction in the gaseous phase. In the condensed phase, the probability that the excess energy of the reaction product will be removed by collision with a third molecule is very much higher; thus the results obtained in the gaseous phase need not be valid generally.

The rate of hydrogen atom addition to alkenes grows in the order ethylene < propene $\leqq$ 1-butene < 2-methylpropene.

The rate of H* addition to 1-pentene is roughly equal to the addition to 1-butene, of the H* addition to *cis* and *trans* isomers of 2-pentene as to *cis*- and *trans*-butenes. Cycloalkenes add a H* atom in a similar way to simple alkenes of comparable structure. H* attacks either the terminal or the internal C atom of 1,3-butadiene; the first way predominates, probably due to allylic or hyperconjugative stabilization of the generated radical.

The addition rate of a methyl radical to ethylene obeys the equation [126]

$$\log k = \frac{8.52 - 32\,200\,[\text{kJ mol}^{-1}]}{(2.3RT)}$$

An attack of methyl radical on propene produces predominantly butyl* (90%), but also the isobutyl radical [127]. In additions to higher alkenes, neither of the two C atoms of the double bond is preferred by the methyl radical, which lacks electrophilic character. The relative reactivity of methyl with respect to ethylene, propene, 1-butene, and 2-methylpropene is roughly equal [128].

Concerning reactivity and the way of attacking alkenes, the ethyl radical is roughly comparable with the methyl radical. Its addition to some vinyl monomers is characterized by the Arrhenius parameters summarized in Table 3.

References pp. 154–162

TABLE 3

Rate constants of ethyl radical addition to some monomers

Monomer	log k	Temperature (K)	E (kJ mol^{-1})	log A	Ref.
Acrylonitrile	5.80	373	14.2	7.79	129
Methacrylonitrile	5.57	356	19.3	8.39	130
Vinyl acetate	3.70	360	28.9	7.89	129
Vinyl n-butyl ether	3.87	369	25.5	7.39	129

The hydroxyl radical is an electrophilic particle. The rate of its addition to propene is six times higher than to ethylene [131]. Also, hydroperoxyl is electrophilic. The possibilities of its reactions are already very diverse [132]. This makes more difficult the evaluation of addition.

All the data discussed above are only an approximation or analogy of the real initiation of radical polymerization. The detailed investigation of the kinetics and mechanism of actual initiation processes is often our future task. Only rarely it is possible to obtain quantitative data on initiation, for example the value of the initiation rate constant, without simplifying assumptions. [133]. Some further information on this problem will be presented in Sect. 8.1 and in Chap. 5, Sect. 4.2 and in Chap. 8, Sect. 1.1.

2. Formation of initiating anions

The dissociation of bases, acids and salts to ions in melts and suitable solvents has been known for a long time. In order that anions formed in this way can initiate polymerization, stringent conditions must be met. The bond between the anion and the monomer must not be solvolyzed by the medium; solvation of the anion must not prevent penetration of the monomer; the acidity of the monomer must be higher than that of the eventual solvent; the nucleophilic properties of the anion must not be reduced by resonance below a certain lower limit, the level of which depends on monomer acidity; the medium must not contain an acceptor which could neutralize the anion. For this reason, strongly polar solvents are almost completely excluded as polymerization media†. In non-polar media, the formation of anions by dissocia-

---

† Only monomers with pronounced electron acceptor properties can polymerize in water (e. g. derivatives of nitrostyrenes, cyanocrylates, vinylidene cyanide). Generally, only oligomers are formed.

tion is very limited. In addition to the enumerated requirements, some practical factors also have to be considered (e. g. ease of purification, manipulation, price, etc.). Of the large number of anion-yielding elements and compounds, only a small fraction are used. These include some metals, organometallic compounds, strong mineral and organic bases and certain donor–acceptor complexes.

2.1 Properties of metal–carbon bonds

The properties of metal–carbon bonds are mainly determined by the electronegativities of the constituting atoms. Mulliken [134] defined the electronegativity of an element as the sum of its first ionization potential and its electron affinity divided by a constant. There exist several methods of electronegativity determination, each of which gives somewhat different results. Probably the most frequently used values are those derived from thermodynamic data by Pauling [135]; some of these are summarized in Table 4. By means of electronegativity Pauling calculated a further useful quantity, the ionic character of a bond. This defines the polarity of a bond in

TABLE 4

Mean electronegativity valeus, e_n, according to Pauling calculated from thermodynamic data [136]

	e_n		e_n		e_n
F	3.98	Ge	2.01	Ca	1.00
O	3.44	Sn	1.96	Li	0.98
Cl	3.16	Si	1.90	Sr	0.95
C	2.55	Al	1.61	Na	0.93
H	2.20	Be	1.57	Ba	0.89
As	2.18	Ti	1.54	K	0.82
Sb	2.05	Zr	1.33	Rb	0.82
B	2.04	Mg	1.31	Cs	0.79

percent between the limits of a totally non-polar and a perfect ionic bond. It is calculated by means of the empirical equation [137]

$$\% \text{ionic character} = 16(^A e_n - {}^B e_n) + 3.5(^A e_n - {}^B e_n)^2 \tag{30}$$

Bonds with over 50% ionic character are designated as ionic; the rest are covalent. The polarities of some C–metal bonds are given in Table 5.

The calculation of the bond ionic character in the above way is, of course, inaccurate. But in spite of all these limitations, the simple approach described

References pp. 154–162

TABLE 5

Ionic character of the carbon bond with some metals [138]

Bond	Ionic character (%)	Bond	Ionic character (%)
C–Cs	57	C–Ti	18
C–Rb	52	C–Sn	15
C–K	52	C–Si	12
C–Na	47	C–As	6
C–Li	43	(C–Cl)	6
C–Mg	34	(C–H)	4
C–Al	22		

enables us to estimate the properties of molecules containing a metal atom. Organocaesium compounds will evidently be ionic, non-volatile and almost insoluble in hydrocarbons. Organolithium compounds are much more soluble, and also non-volatile. The solubilities of alkyl-and aryllithium compounds are connected with their association. "Micelles" containing Li atoms in a core surrounded by hydrocarbon groups are formed [139]. Non-aqueous solutions of the alkyl derivatives of all alkali metals are conductive. Alkylmagnesium compounds are also practically non-volatile. Grignard reagents, RMgX (X = halogen), yield conducting solutions in ethers and amines; even in these compounds, the ionic character of the metal bond to the rest of the molecule is preserved. On the other hand, alkylsilicon and alkyltin compounds are volatile liquids of unlimited solubility in benzene; they do not dissociate in hydrocarbons.

An electropositive metal in organic compounds of alkali metals is replaced by a more electropositive one in series of reversible reactions. More electronegative, i. e. more acid, hydrocarbon groups or whole molecules replace those which are less acid [140]. Caesium replaces lithium in ethyllithium. Benzene, which is a stronger acid than ethane, replaces ethyl in ethyllithium. Toluene and H_2 are more acid than benzene, and they can therefore replace phenyl in phenylsodium [141, 142].

The tendency of the aluminium atom to complete its electron octet is manifested as the Lewis acidity of organoaluminium compounds. This is demonstrated by the formation of 1:1 complexes of trialkylaluminium with anions or bases such as ethers and amines. Organoaluminium compounds are generally more acid than the organoderivatives of boron, indium and gallium. For steric reasons, complex formation is more difficult in compounds of boron compared with aluminium (the covalent radius of boron is 0,08 nm and aluminium is 0.126 nm).

The Lewis acidity of organoaluminium compounds is the reason of their association. The association of alkylaluminium molecules and aluminium hydrides proceeds by way of electron-deficient bonds [143a]. The associates of dimethylberyllium, dimethyl- and diethylmagnesium, methyl- and ethyllithium are of the same type; each alkyl group is simultaneously bound to two or three metal atoms [143b].

A great majority of organometallic compounds, especially of those which can be used as initiators, are easily solvolyzed by polar molecules. Inactive or weakly active products are formed. If the properties of the metal–carbon bond are to be useful for initiating anionic polymerizations, the organometallic compound must be able to yield a carbanion, either free or bound to the counter ion.

2.1.1 INITIATION BY ALKYLMETALS

In non-polar (hydrocarbon) solvents, only the organometallic compounds of lithium are sufficiently soluble. In solution they are present as aggregates ("micelles"). In benzene the hexamers $(EtLi)_6$ and $(n\text{-BuLi})_6$ (also in cyclohexane), the tetramers of iso-PrLi, sec.-BuLi, tert.-BuLi [144, 145] and the dimers of benzyllithium have been observed [146]†. The reaction of hydrocarbon monomers with alkyllithium compounds yielding active centres is slow [148]. Therefore the monomer is consumed simultaneously by the initiation and propagation reactions, and initiation is only complete at high conversion. The growing chains form mixed aggregates with the initiator [149], thus modifying the reactivity of alkyllithium.

Bywater and Worsfold [145] have measured the initiation rate in the polymerization of styrene with n-BuLi and sec.-BuLi in benzene. The active centres

$$BuCH_2\text{—}\underset{\underset{Ph}{|}}{CH^-}, Li^+$$

---

† So far, the problem of the number of alkylmetal molecules in aggregates has not been reliably settled. This is reflected in the most recent studies. It generally holds that the number of molecules in the aggregates decreases with increasing solvating ability and temperature of the solvent, with decreasing concentration of the dissolved organolithium compound, with increasing possibilities of steric hindrance in its substituents, and with the degree of electron delocalization in the substituents [147].

References pp. 154–162

are formed more rapidly from sec.-BuLi than from n-BuLi by about two orders of magnitude. The order of the initiation is 0.25 with respect to sec.-BuLi and 0.166 with respect to n-BuLi. The authors proposed the simple scheme

$$(BuLi)_n \xrightleftharpoons{K} n\,BuLi \qquad K = \frac{[BuLi]^n}{[(BuLi)_n]}$$

$$BuLi + M \xrightarrow{k_i} BuMLi \qquad (31)$$

For the initiation rate, this yields

$$v_{init} = k_i[K(BuLi)_n]^{1/n}[M]$$

The experimental verification of this relation indicates a rapid establishment of the deaggregation equilibrium, which is shifted far to the left (the concentration of free BuLi must be very small). The degradation intermediates are also inactive. Table 6 shows that similar dependences have also been observed with the polymerization of isoprene and the dimerization of 1,1-diphenylethylene in fluorene [150].

TABLE 6

Parameters of the reaction of alkyllithium with styrene and isoprene in benzene at 303 K [151][a].

Initiator	Monomer	$k_i K^{1/n}$
n-BuLi	Styrene	2.3×10^{-5} [b]
Sec.-BuLi	Styrene	9.8×10^{-3}
Sec.-BuLi	Isoprene	2.8×10^{-3}
Tert.-BuLi	Isoprene	3.5×10^{-3}

[a] $(BuLi)_n \xrightleftharpoons{K'} n\,BuLi$; When practically all organometal is aggregated, the concentration of $(BuLi)_n$ is n times less than the analytical total BuLi concentration. It then holds that $v_{init} = k_i(K'/n)^{1/n}[BuLi]^{1/n}[M]$.
[b] Apparent activation energy $E_c = 75.3$ kJ mol^{-1}.

In aliphatic solvents, centres are formed more slowly than in benzene (see Fig. 3). Evidently the deaggregation mechanism is different from that in the aromatics. The dependence of the initiation rate on the type of initiating alkyl is preserved. The apparent orders with respect to initiator are higher, around 1. In this case, initiation may be the result of direct contact between monomer and aggregate, yielding a mixed associate [152]

$$(sec.-BuLi)_4 + M \rightarrow [sec.-BuLi)_3\,(sec.-Bu-MLi)_1] \qquad (32)$$

which reacts more rapidly than the initiator with the monomer. At the beginning, the formation of active centres is of autocatalytic character. This soon makes it impossible to analyze exactly the initiation of non-polar monomer polymerizations in aliphatic hydrocarbons.

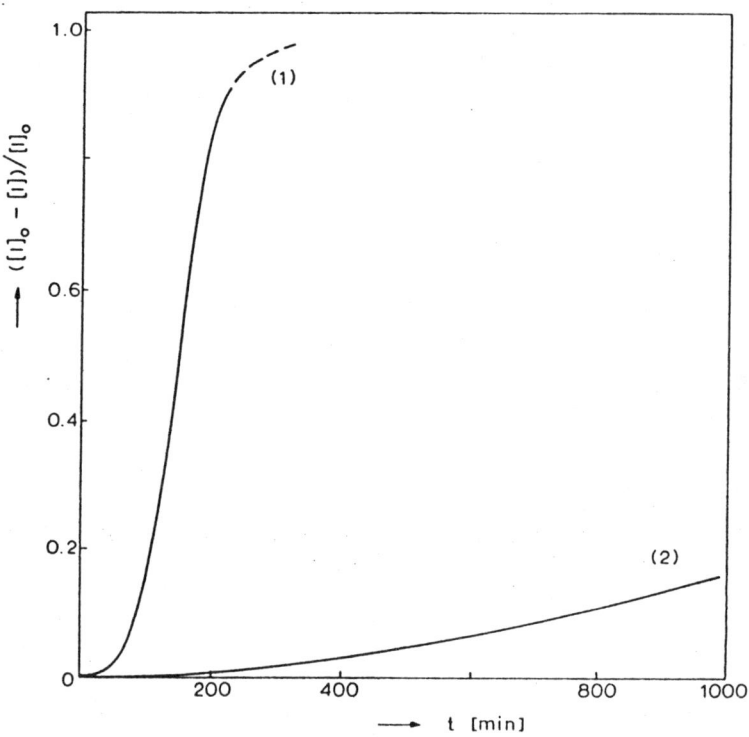

Fig. 3 Kinetics of active centre formation (see ref. 151, p. 13). Reaction of butyllithium with isoprene $(0.04 \text{ mol dm}^{-3})$ in hexane at 303 K. (1) sec.-BuLi $(7.9 \times 10^{-5} \text{ mol dm}^{-3})$; (2) n-BuLi $(1.0 \times 10^{-4} \text{ mol dm}^{-3})$.

The addition of lithium alkoxides to an alkyllithium initiator causes the induction period of active centre formation to disappear [152].

Any knowledge of the mode of action of lithium alkoxides is important as a certain amount of lithium alkoxides is always present. They are formed by reaction of the initiator with traces of oxygen. Another possible impurity is LiOH (hydrolysis product), which leads to a reduction in the actual RLi concentration [153]

$$\text{RLi} + \text{LiOH} \rightarrow \text{RH} + \text{Li}_2\text{O} \tag{33}$$

Very little information exists on the initiation of non-polar monomers in polar media. A rare exception is the description of the dimerization of 1,

References pp. 154–162

1-diphenylethylene in THF solutions initiated by alkyllithium compounds. Greater, although still only marginal, attention has been paid to essentially non-polar systems containing small amounts of ethers or amines. During the interaction of these donors with the ~C—metal bond (particularly for Li), a significant polarization of this bond occurs with decomposition of the organometal associates to product composed of a smaller number of entities.

Kmínek et al. [154–156] established that the electron transfer ability of donors decreases in the order N, N, N', N'-tetramethylethylenediamine > THF > 1, 2-dimethoxyethane > triethylamine and the electron accepting ability of organometals in the order BuLi > oligostyryllithium > oligoisoprenyllithium.

In THF, the alkyllithium compounds are aggregated [157] and the situation is reminiscent of the conditions in hydrocarbon solutions. At high concentrations, the association number (i. e. the number of molecules in the aggregate) decreases. This anomaly is explained by the existence of aggregate—solvent complexes, for example $(MeLi)_4$ 8THF. Benzyllithium and its polymeric analogue polystyryllithium are not associated. Phenyllithium is mostly present as a dimer or monomer. Both forms are in equilibrium and are solvated. Only the monomeric form of the initiator is active. In practice, benzyllithium reacts only in the form of an ion pair. The fraction of the free benzyl anion must be very small [151c].

These effects considerably complicate the kinetics of polymerization. It is even harder to isolate initiation from the collection of outward manifestations of all partial processes. Such an analysis has not been attempted so far.

All this does not prevent attempts to utilize all established pieces of knowledge for practical purposes. For example, styrene–isoprene block copolymers have been successfully synthesized in this rather complicated medium [158].

We possess more information on systems containing polar monomers. Organolithium and organomagnesium compounds initiate the polymerization of a number of monomers with an electron-withdrawing substituent. These polymerizations are rarely of the living type. The initiator usually reacts not only with the double bond of the monomer, but also with the polar substituent (both on the monomer and the polymer) yielding inactive products.

Vinyl chloride [159], vinylidene chloride and chloroprene are polymerized in this way at ambient temperature [160]. Polymerization ceases before all the monomer has been exhaused. The chemistry of these processes is little known. Vinyl chloride [161] can be metallised by alkyllithium compounds and radicals are formed by the interaction of n-BuLi with alkyl and aryl halides [162]. Thus a contribution of the radical mechanism in polymerizations of chlorine-containing monomers cannot be excluded. The generation of high polymers indicates such a process, whereas the small influence of atmospheric oxygen and inhibition by tetrahydrofuran [159] bears evidence to the contrary.

Other vinyl monomers, such as acrylonitrile, methacrylonitrile, *tert.*-butyl vinyl ketone and methyl isopropenyl ketone, polymerize at 203 K, i. e. most probably by non-radical mechanisms. Even here, conversion of monomer to polymer is not complete, and utilization of the initiator is low. Only the polymerization of acrylate momomers proceeds to full monomer consumption at low temperatures. Additional monomer, even when introduced after some delay, is also polymerized. This indicates that a part of the active centres remains living for some time. However, the number of high-molecular-weight chains is lower than the number of added initiator molecules. At the same time, initiation is very rapid [163].

The best studied process is probably the initiation of methyl methacrylate [see, for example, refs. 164–166 and the review ref. 151(d)]. Trekoval [167] summarized the known facts into the following scheme. The polymerization effectiveness of alkyl- and arylmetallic compounds with the same hydrocarbon ligands decreases in the order K > Na > Li > Be > Mg > Zn > > Cd > B. 1,2 addition of conjugated systems to these initiators decreases in roughly the same order. Thus in the above series, the tendency to 1,4 addition increases from left to right. Magnesium compounds tend to add monomer in the 1,4 manner, whereas sodium compounds add almost exclusively by the 1,2 route. Lithium derivatives usually add monomer by both mechanism simultaneously, but 1,2 addition predominates. Alkyl derivatives of Group IA and IIA metals initiate anionic polymerization and alkyl compounds of Group IIB and IIIA metals cationic polymerization. In the methyl methacrylate molecule, the π electrons of the C=C and C=O bonds are conjugated. In the interaction of butyllithium with this monomer, addition can occur in three ways: 3,4, 1,4 and 1,2.

(a) 3,4 addition

$$\text{BuLi} + \text{CH}_2=\underset{\underset{\underset{\text{OMe}}{|}}{\overset{\text{CO}}{|}}}{\overset{\overset{\text{Me}}{|}}{\text{C}}} \rightarrow \text{Bu—CH}_2-\underset{\underset{\underset{\text{OMe}}{|}}{\overset{\text{CO}}{|}}}{\overset{\overset{\text{Me}}{|}}{\text{C}^-}} \text{Li}^+ \quad (34)$$

(b) 1,4 addition

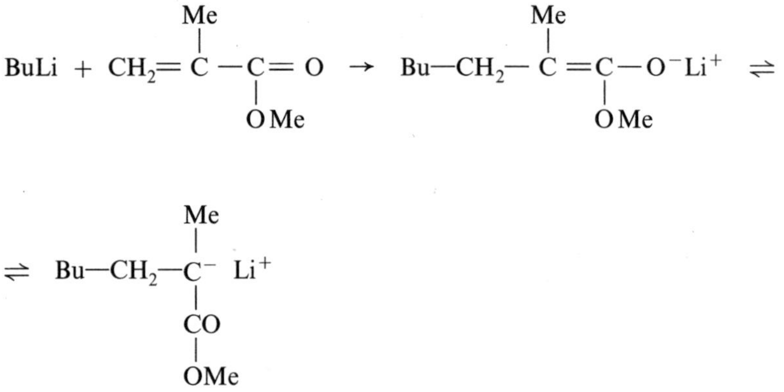

$$BuLi + CH_2=\underset{\underset{OMe}{|}}{\overset{\overset{Me}{|}}{C}}-C=O \rightarrow Bu-CH_2-\underset{\underset{OMe}{|}}{\overset{\overset{Me}{|}}{C}}=C-O^-Li^+ \rightleftharpoons$$

$$\rightleftharpoons Bu-CH_2-\underset{\underset{OMe}{\underset{|}{CO}}}{\overset{\overset{Me}{|}}{C^-}} Li^+ \qquad (35)$$

(c) 1,2 addition

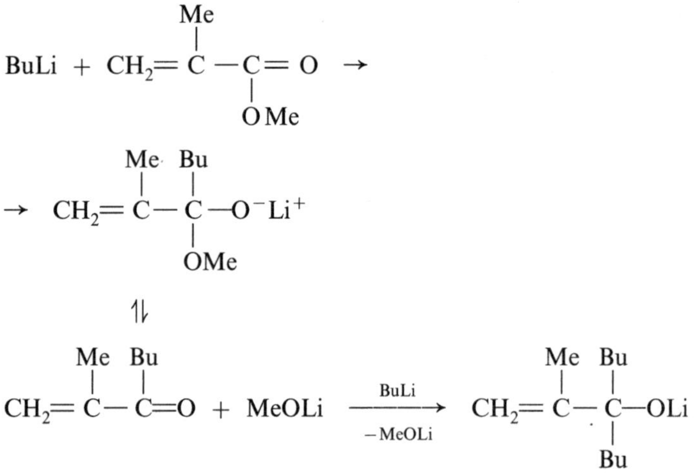

$$BuLi + CH_2=\underset{\underset{OMe}{|}}{\overset{\overset{Me}{|}}{C}}-C=O \rightarrow$$

$$\rightarrow CH_2=\overset{\overset{Me}{|}}{C}-\underset{\underset{OMe}{|}}{\overset{\overset{Bu}{|}}{C}}-O^-Li^+$$

$$\updownarrow$$

$$CH_2=\overset{\overset{Me}{|}}{C}-\overset{\overset{Bu}{|}}{C}=O + MeOLi \xrightarrow[-MeOLi]{BuLi} CH_2=\overset{\overset{Me}{|}}{C}-\underset{\underset{Bu}{|}}{\overset{\overset{Bu}{|}}{C}}-OLi \qquad (36)$$

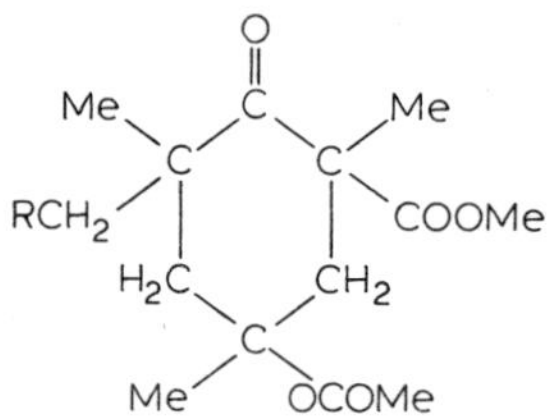

The 1,4 and 3,4 additions are the proper initiation reactions leading to the same active centre, though perhaps at different rates. Addition to the carbonyl group yields products whose participation in polymerization is still subject to controversy. Some authors assume that reaction (36) yields inactive compounds [165]. A considerablle part of the initiator is also consumed in the formation of very short chains [168, 169]. Trimers are probably formed by an attack of the carbanion on a carbonyl group [168], i. e. by a reaction analogous to reaction (36). In this process, the terminal carbanion attacks the third monomeric unit in the chain, as counted from the anion. The formation of the low-molecular-weight fraction of the product has also been explained by the existence of non-propagating "internal" complexes [168] [see chap. 6 eqn. (84)].

None of the described reactions has been proved so far. It is, however, beyond doubt, that at the start of the polymerization an oligomer of molar mass 500–800 is formed at an extreme rate; this oligomer can be isolated from the final product.

The structure of the hydrocarbon group also affects reactions leading to the start of polymerization. Initiation by butyllithium leads to the rapid formation of a relatively large amount of lithium methoxide, whereas by the reaction of diphenylhexyllithium with methyl methacrylate only a small amount of methoxide is formed slowly; in both cases, the reaction proceeds according to scheme (36) or by cyclization or termination of the active centres [170] by 1,2 addition [i. e. again in analogy to reaction (36)]

$$\sim \underset{\underset{OCOMe}{|}}{\overset{\overset{Me}{|}}{C^-}} Li^+ + O=\underset{\underset{OMe}{|}}{\overset{\overset{Me}{|}}{C}} - C=CH_2 \rightarrow \sim \underset{\underset{OCOMe}{|}}{\overset{\overset{Me}{|}}{C}} - \overset{\overset{O}{\|}}{C} - \overset{\overset{Me}{|}}{C}=CH_2 + MeOLi \qquad (37)$$

The described behaviour of both initiators can be interpreted by the preferential reaction of diphenylhexyllithium according to scheme (34) and of butyllithium according to scheme (36). The subsequent phases of macromolecule formation are affected by the liberated methoxide.

The initiation of methyl methacrylate polymerization by alkyl- and aryllithium compounds is a very complex process which has not been completely elucidated so far. Even less information is available on the effects of organometals not containing lithium. Recently, the interest of some authors has centred on the application of Grignard reagents for the initiation of anionic polymerizations, especially of MMA [171, 172].

For further elucidation of the initiation mechanism in polymerizations of polar monomers with organometals, a method for determining the instantaneous concentrations of active centres has to be found, as well as an initiating system acting without side reactions.

Alkyl- and arylmetals in the classical form are little used for initiating anionic polymerizations of heterocycles.

For the initiation of caprolactam polymerization, Grignard reagents [173] have been used as a source of lactamate salt. Tsuchiya and Tsuruta have described an interesting case of methyloxirane polymerization initiated with diethylzinc. The oxirane ring is split at the CH$_2$—O or O—CH bonds, depending on the ratio [H$_2$O]/[Et$_2$Zn].

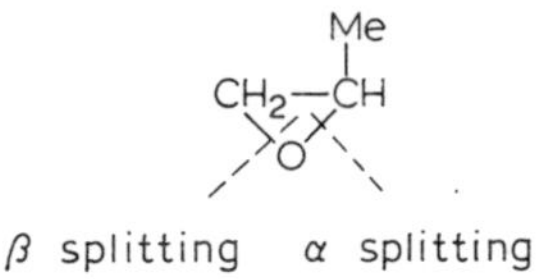

$$\underset{\beta \text{ splitting} \quad \alpha \text{ splitting}}{\overset{\displaystyle CH_2\!-\!\overset{\displaystyle Me}{\underset{\displaystyle }{CH}}}{\diagdown\!O\!\diagup}}$$

For $0 < ([H_2O]/[Et_2Zn]) < 1$, β splitting is characteristic, designated as cationic by the authors. For $1 < ([H_2O]/[Et_2Zn]) < 2$, α splitting (anionic) occurs [174].

Quirk and Seung used Bu$_3$NaMg for initiating the copolymerization of oxirane with styrene. They obtained block copolymers while with poly(styryl)lithium or 1,1-diphenylhexyllithium, oxirane did not polymerize at all [175].

Organometals, however are gaining ever increasing attention in the form of delocalized anions (see Sect. 2.4).

2.2 Properties of metal–oxygen and metal–nitrogen bonds

Both types of bond are relatively stable. Only the bonds of the most electropositive metals of Groups IA and IIA of the periodic table with the electronegative oxygen or nitrogen atoms exhibit sufficient ionic character for the dissociation of the respective salts to ions and ion pairs at low temperatures and in relatively non-polar media. In this way, initiating anions are formed which are mostly more stable (and therefore also less reactive) than carbanions. Alkali metal hydroxides and amides have often been used in the past.

The classical examples are the polymerization of styrene in NH$_3$, initiated by KNH$_2$ [176], and the polymerization of octamethylcyclotetrasiloxane initiated by alkali metal hydroxides [177]. At the present time, simple strong mineral bases are no longer used for the initiation of anionic polymerizations.

Probably the most important initiators of the group of oxygen-containing anions are the alkoxides.

2.2.1 INITIATION BY ALKOXIDES

During the initiation of methyl methacrylate polymerization by alkyllithium, lithium alkoxide is formed [see eqn. (37)]. This compound directly affects the subsequent course of the reaction. It has aroused the interest of scientists who started to used various lithium alkoxides directly as initiators

[178], those with a branched hydrocarbon group proving to be the most suitable. With their use, polymerization can proceed at ambient temperature to complete monomer consumption at a relatively low rate. Side reactions between the alkoxide and the carbonyl group of the monomer do not occur and the system is relatively insensitive to moisture and oxygen.

In solvents of various solvating strength, the *tert.*-alkoxides of alkali metals are associated to various degrees. For initiation, the aggregated alkoxide has to be activated by the monomer or by a saturated ester of suitable structure [179].

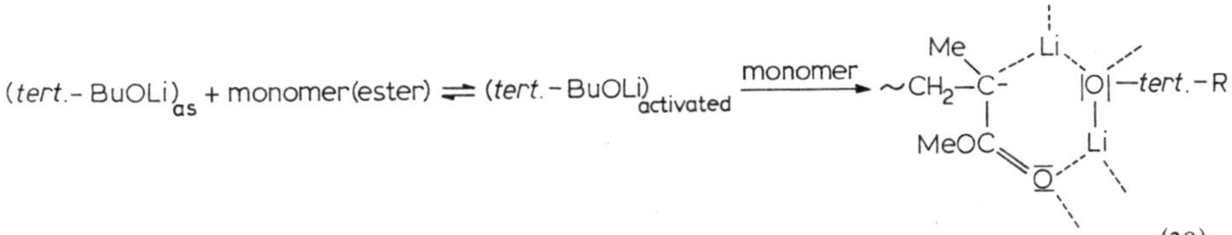

$$(tert.\text{-BuOLi})_{as} + monomer(ester) \rightleftharpoons (tert.\text{-BuOLi})_{activated} \xrightarrow{monomer} \sim CH_2-\underset{MeOC}{\overset{Me}{\underset{|}{C}}}-\underset{\bar{O}}{\overset{Li}{|O|}}-tert.\text{-R}$$

(38)

Only about 10% of *tert.*-ROLi molecules yield polymer chains. The remaining alkoxide, coordinated to the growth centres, forms a protective envelope. It suppresses the basicity of the centres and thus limits their ability to react with the ester group of the monomer. When the reaction conditions are selected so that almost no *tert.*-ROLi is left for the formation of the protective envelope, propagation occurs simultaneously with termination[†]. Polymerization ceases prior to the consumption of all monomer and a strongly branched polymer results.

The polymerizations of many other monomers are initiated by alkoxide ions.

The anionic polymerization of β-nitrostyrenes may serve as an example. It is initiated by alkoxide ions in solvents able to loose a proton (alcohols in this case)

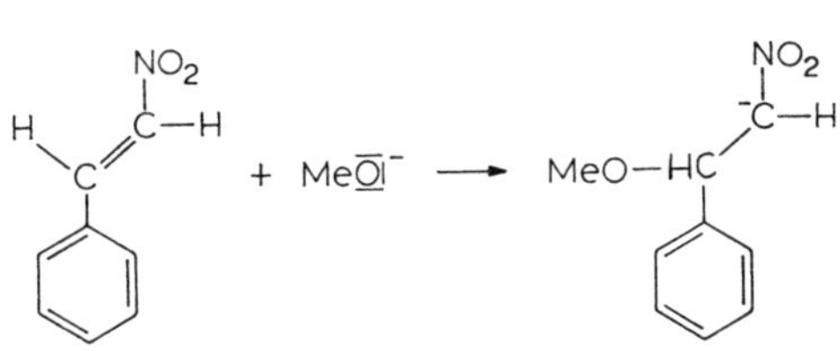

$$\underset{\text{Ph}}{\overset{H}{\underset{}{C}}}=\underset{H}{\overset{NO_2}{\underset{|}{C}}}-H + Me\bar{O}|^- \longrightarrow MeO-HC-\underset{Ph}{\overset{NO_2}{\underset{|}{\bar{C}}}}-H$$

[†] This situation occurs in the presence of very low concentrations of *tert.* ROLi or of special activators (e. g. methyl or butyl isobutyrate). It is also induced in the presence of strongly solvating solvents (e. g. THF).

p-β-Dinitrostyrene, on the other hand, yields an ion with a free electron pair on the α carbon

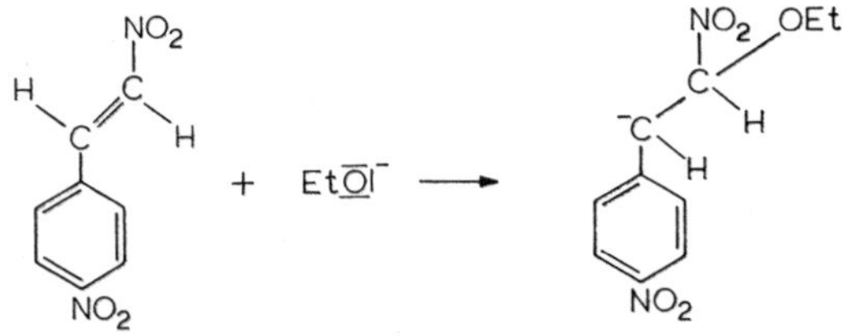

High-molecular-weight polymers of low polydispersity are formed [180]. BuONa initiates the isomerization polymerization of acrylamide, producing poly (*β*-alanine) (see also Chap. 5, Sect. 6.6). This initiation can be described as addition of the $CH_2=CHCO\bar{N}H$ anion to the monomer double bond [181]. The anion is formed by the reversible reaction

$$CH_2=C\underset{CONH_2}{|} + Bu\bar{O}| \rightleftharpoons CH_2=CH\underset{CO\bar{N}H}{|} + BuOH$$

All initiator molecules yield active centres at the very beginning of the polymerization.

2.3 Electron transfer from donor to monomer

In aromatic hydrocarbons, some substituted alkenes, dienes, substituted acetylenes and ketones, one half of the π orbitals are empty and an electron can easily be placed in these antibonding orbitals. The capture of an electron by the acceptor molecule is an exothermic process because the energy of the antibonding orbitals lies below the level of the ionization potential of the acceptor radical anion. Many radical anions formed from unsaturated molecules are themselves stable; they do not decompose and may exist indefinitely under suitable experimental conditions [182a]. On the other hand, they react easily with other molecules.

Electron transfer from donor to monomer is a frequently used process initiating anionic polymerizations. An electropositive metal, mostly a member of Group IA in the periodic table, is usually the donor. The acceptor is either the monomer direct or an aromatic molecule which is used to mediate the electron transfer from the metal to the monomer. For initiation by electron transfer, a more polar medium is typical (e. g. THF, dimethoxyethane, etc.). The solvation energy represents a major part of the overall balance of the driving force of the process.

2.3.1 INITIATION BY ALKALI METALS

This is based on the existence of the equilibrium [182b]

$$\text{metal} + \text{acceptor} \rightleftharpoons \text{metal}^+ + \text{acceptor}^{-*} \tag{39}$$

The equilibrium between alkali metals and many acceptors is often shifted to the far right and, because of this, experimental studies of the equilibrium constant are very difficult. Such studies are only possible under certain conditions, including knowledge of the qualities of the metal, of the acceptor and of the solvent (see, for example, ref. 183).

The polymerization of dienes by sodium metal has been known since the beginning of this century [184]. Ziegler et al. [185] postulated dianion formation from butadiene

$$CH_2=CH-CH=CH_2 + 2\,Na \rightarrow Na^{+\,-}CH_2-CH=CH-CH_2^-Na^+$$

Vinyl monomers with an aromatic group in the molecule are also efficient acceptors. Styrene and its derivatives undergo several transformations under the effect of alkali metals

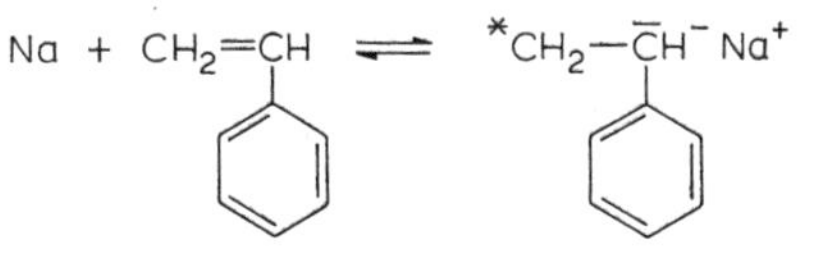

$$Na + CH_2=CH(Ph) \rightleftharpoons {}^*CH_2-\bar{C}H^-(Ph)\,Na^+ \tag{40}$$

$$2\ {}^*CH_2-CH^-(Ph)\,Na^+ \rightleftharpoons Na^+{}^-CH(Ph)-CH_2-CH_2-\bar{C}H^-(Ph)\,Na^+ \tag{41}$$

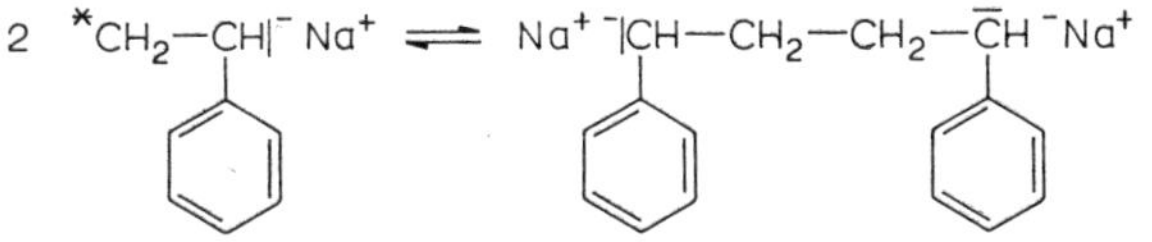

In strongly solvating solvents (e. g. THF, dimethoxyethane, etc.), styrene yields a radical ion (or, more precisely, a radical ion pair). The radical ends combine yielding diions (diion pairs). Reactions (40) and (41) are of general validity even for other alkali metals and monomers. Dimers can be produced according to reaction (41) only with weakly reactive monomers (e. g. α-methylstyrene). Due to the high propagation rate on ion pairs (see Chap. 5, Sect. 8.1), much more often a dianion with a larger number of monomer units is formed.

The relatively slow reaction (40) is the reason why a polymer with various chain lengths is produced. New centres continue to be generated during the course of the polymerization with less and less monomer available. This difficulty is circumvented by experimental separation of initiation from propagation. A solution of oligomeric diions is added to the system prepared for polymerization. This technique requires special measures to prevent

References pp. 154–162

contamination, of course. It can, however, maintain living polymerization in homogeneous medium with a very narrow molar mass distribution in the product.

2.3.2 INITIATION BY RADICAL ANIONS

In some cases it may be useful not to separate initiation from propagation and to work in homogeneous medium. It is then necessary to create conditions for the rapid transfer of an electron from the metal to a suitable molecule, and to use the generated radical ion for initiation. The simplest adical ion is the solvated electron.

Alkali metals dissolve in liquid ammonia (THF, etc.)

$$Li + nNH_3 \rightleftharpoons Li^+[(NH_3)_n]^{-*} \qquad (42)$$

The resulting complexes can initiate polymerization [186] of vinyl and heterocyclic [187] monomers.

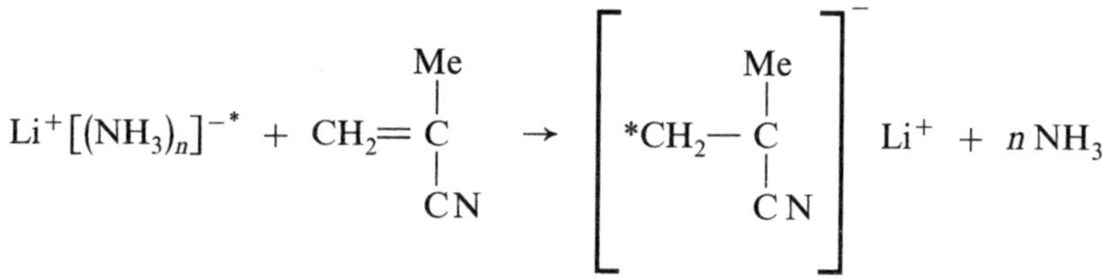

$$Li^+[(NH_3)_n]^{-*} + CH_2=\underset{CN}{\overset{Me}{C}} \rightarrow \left[*CH_2-\underset{CN}{\overset{Me}{C}} \right]^- Li^+ + nNH_3 \qquad (43)$$

$$2\left[*CH_2-\underset{CN}{\overset{Me}{C}} \right]^- Li^+ \rightarrow Li^+|\underset{CN}{\overset{Me}{C}}{^-}-CH_2-CH_2-\underset{CN}{\overset{Me}{C}}|Li^+ \qquad (44)$$

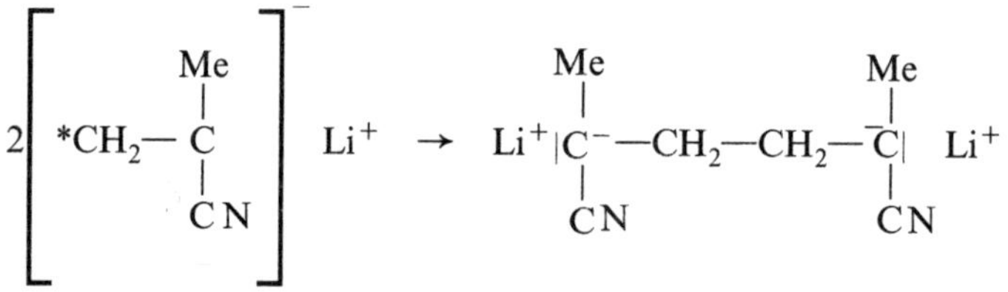

Electron transfer from alkali metals to aromatics is very easy in suitable solvents [182]. The radical anions produced in this way do not dimerize. The formation of a covalent C—C bond would be accompanied by the loss of resonance stability in the aromatic system. Paul et. al. [188] have shown that the unpaired electron is placed in the lowest unoccupied orbital of the molecule, and that the stability of the radical anion increases in the order diphenyl < naphthalene < phenanthrene < anthracene.

The formation of radical anions of this type can be schematically described as follows (with Na and naphthalene as examples of donor and acceptor) [189].

$$\text{(naphthalene)} + Na \rightleftharpoons [\text{(naphthalene)}]^{-*} Na^+ \qquad (45)$$

The course of the reaction strongly depends on solvent. In dimethyl ether, the equilibrium (45) is shifted strongly to the right; in the less basic diethyl ether, interaction does not occur. The most used solvents are again THF, dimethoxyethane, etc. The electron is transferred to the monomer by the green solution of naphthalenesodium

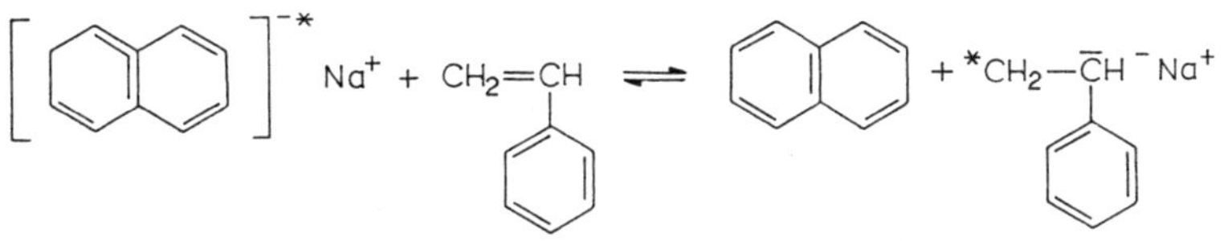

$$[\text{naphthalene}]^{-*} \text{Na}^+ + \text{CH}_2=\text{CH}(\text{Ph}) \rightleftharpoons \text{naphthalene} + {}^*\text{CH}_2-\bar{\text{CH}}(\text{Ph})\ \text{Na}^+ \quad (46)$$

The monomeric radical anion is dimerized [reaction (41)], and polymerization takes place on the ions or ion pairs. The equilibrium (46) is very rapidly established. Radical defects decay by combination; thus naphthalenesodium is used quantitatively for practically instantaneous initiation. This method can therefore also be used to prepare polymer chains of uniform length. The regenerated naphthalene does not usually cause trouble. In some cases, the aromatic residue can be incorporated into the chains [190] (the metal cation is omitted for simplicity).

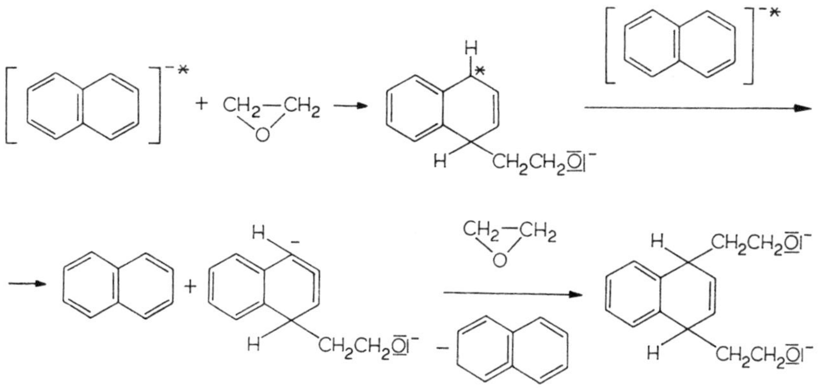

(47)

A scheme of this variant of electron transfer from the naphthalene radical anion was first proposed by Morton et al. to describe the initiation of octamethylcyclotetrasiloxane polymerization [191]. Initiation according to scheme (47) is rapid with oxirane and slow with octamethylcyclotetrasiloxane.

Even molecules with heteroatoms can be used as electron acceptors. The initiating anion radicals are formed from aromatic ketones [192] (e. g. benzophenone) or nitriles [193] (e. g. benzonitrile, naphthonitrile). It appears that both ketones and nitriles can react in several ways, and the evidence concerning the correctness of the proposed mechanisms does not at present seem to be conclusive [194].

References pp. 154–162

2.4 Other methods leading to initiating anions

Ions are formed by the action of particle beams, decomposition of molecules excited by photons, donor–acceptor interactions of suitable compounds and by the effects of an electric current on solutions of supporting electrolytes[†]. Some special polymerization processes can be initiated by anions and cations generated in this way. The practical importance of all the enumerated methods is so far not large. Nevertheless, these processes continue to be the subject of intense study. A short description of the most interesting of these methods is contained in Sects. 5 and 6.1. Polymerization-initiating anions can even be formed from cations (see Chap. 4, Sect. 5).

2.4.1 MULTIFUNCTIONAL CARBANIONS

In order to fulfil certain special requirements concerning the polymer architecture, initiators with several anions in their molecule are needed.

(a) *Delocalized carbanions*

Reactions of the Lochmann base (a mixture of butyllithium with potassium *tert*.-butoxide in pentane [195]), with conjugated polyenes yield fully delocalized carbanions in high yield at a high rate [196]

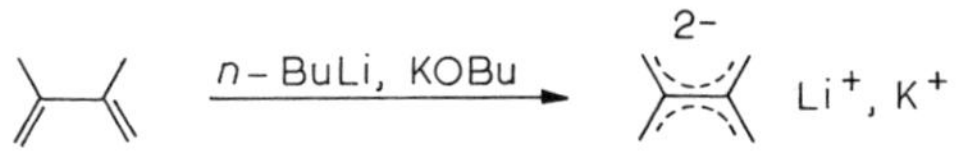

A large number of compounds of this type have been prepared [197], for example (counter-ions omitted)

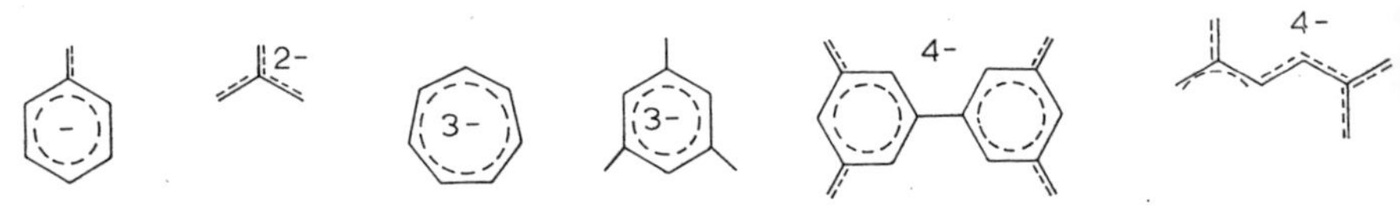

They can be used to initiate the polymerization of monomers on several centres simultaneously, i. e. for the preparation of star polymers. Delocalized carbanions of a different type are formed by the deprotonation of acetylene copolymers [198]

[†] Naturally the appearance of ions in electrolyte solutions is not caused by the electric current. However, the current affects initiation and/or propagation. The mechanism of this process is presently being elucidated.

Mono- and dianions can, of course, be used for the initiation of linear chain propagation. Gordon and Loftus observed interesting differences in styrene polymerization when initiated with BuLi and with delocalized carbanions. In the latter case, the anions have various counter-ions (Li^+ and K^+); at each of these, propagation proceeds at a different rate. When the initiator contains a greater number of delocalized carbanions, insolubility of the product and non-uniformity of charge on the anions cause some problems [199].

(b) *Carbanions with separated charges*

The initiator anions need not, of course, be part of a conjugated system. Charge separation is equivalent to the existence of several, mutually largely independent, active centres in a single molecule. One such initiator, the frequently applied dianion of α-methylstyrene tetramer, has already been treated in Sect. 2.3. Even molecules with more than two isolated anions can be prepared. It appears, however, that in their application, difficulties (mainly concerning solubility) dominante over their advantages. With the exception of some sporadic applications (as described in the patent literature) for the synthesis of star polymers, they have not aroused much practical interest.

2.4.2 MACROANIONS

In the preceding sections and paragraphs, anion quality was mainly discussed. However, even the rest of the molecule to which the anion is bound is not without importance. Longer or shorter molecules, carrying anionic growth centres at one or both ends, are of special value. Such macroanions, in effect representing a kind of initiator with separated charges, can be easily prepared by living polymerization. From the kinetic point of view, they are not particularty important as initiators. They are, however, indispensable for the preparation of block and graft copolymers [200, 201]. Also the variant of polymeric initiators with growth or active centres inside the chains is of considerable consequence for the preparation of graft copolymers. They serve as sites for the propagation of grafts from suitable monomers (grafting from).

2.5. Reaction of anion with monomer

Multiple bonds between the atoms in the molecules of conventional monomers may possess a relative excess or deficiency of electrons. In principle, only a few of these bond types exist: nitrile, aldehyde, carbonyl, carboxyl, ester, vinyl and acetylene. In macromolecular chemistry, the reactions of anions with oxiranes, the amide and ester (in rings) and the siloxane bond are also of importance.

References pp. 154–162

Anion addition to saturated aldehydes, esters and ketones usually occurs at the 1,2 position [202]

$$\sim\!\overset{|}{\underset{|}{C}}{}^{-} + R_2C=O \rightarrow \sim\!\overset{|}{\underset{|}{C}}-C(R_2)-\bar{O}|^{-} \qquad (48)$$

Under certain condittions, deviations from this scheme may occur.

Ketones with hydrogen in the α position can react as Brönsted acids [203]

$$BuLi + Me_2C=O \rightarrow LiOC(Me)=CH_2 + BuH \qquad (49)$$

The reactivity of organometals increases in the order Li < Na < K; alkali metal organometals are more efficient than magnesium and aluminium compounds and the reaction of aryllithium organometals with ketones is more rapid than that of alkyllithium compounds [202]. Wittig and Bub [204, 205] postulated the formation of transition complexes with simultaneous nucleophilic attack on the carbonyl and electrophilic attack on the oxygen by the metal.

$$Ph_2C=O + Ph_3Al \rightleftharpoons Ph_2\overset{+}{C}-O\cdots\overset{-}{AlPh_3} \rightarrow Ph_3COAlPh_2 \qquad (50)$$

It may be expected that, under suitable conditions, carbanions will substitute the alkoxide and be added to the remaining carbonyl.

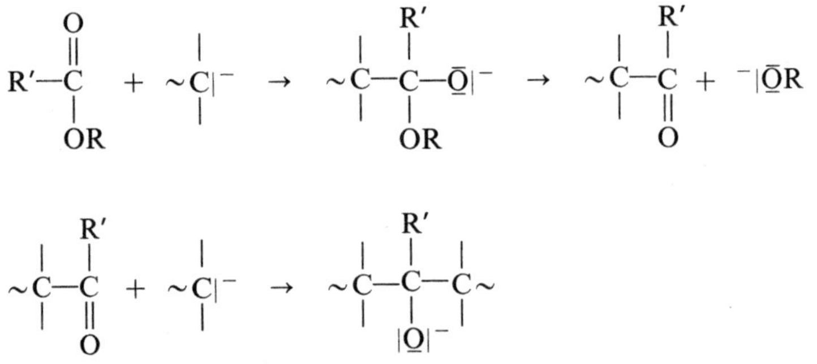

$$R'-\underset{OR}{\overset{O}{\underset{\|}{C}}} + \sim\!\overset{|}{\underset{|}{C}}|^{-} \rightarrow \sim\!\overset{R'}{\underset{OR}{\overset{|}{\underset{|}{C}}-\overset{|}{\underset{|}{C}}-\bar{O}|^{-}}} \rightarrow \sim\!\overset{R'}{\underset{O}{\overset{|}{\underset{\|}{C}}-\overset{|}{\underset{|}{C}}}} + {}^{-}|\underline{O}R$$

$$\sim\!\overset{R'}{\underset{O}{\overset{|}{\underset{\|}{C}}-\overset{|}{\underset{|}{C}}}} + \sim\!\overset{|}{\underset{|}{C}}|^{-} \rightarrow \sim\!\overset{R'}{\underset{|\underline{O}|^{-}}{\overset{|}{\underset{|}{C}}-\overset{|}{\underset{|}{C}}-\overset{|}{\underset{|}{C}}\sim}} \qquad (51)$$

Various steps of this reaction may be intramolecular within the complex. The oxygen atoms are probably coordinated to the counter-ions of the anions. The character of the nitrile bond is similar to that of the carbonyl. The product of anion addition to the nitrile carbon

$$\sim\!\overset{|}{\underset{|}{C}}|^{-} + \overset{R}{\underset{|}{C}}\equiv N| \rightarrow \sim\!\overset{R}{\underset{|}{\overset{|}{C}}-\overset{|}{C}=\bar{N}^{-}} \qquad (52)$$

is not further split in the sense of reaction (51). Addition of a further anion does not occur [206].

Reactions (36), (48) and (52) are only examples illustrating mainly undesirable complications of anionic polymerization.

2.5.1 INITIATION OF ANIONIC POLYMERIZATION

The active centre of anionic polymerization is formed by the addition of an anion to a

$$>\!C\!=\!C\!<\quad (-C\equiv C-)$$

bond of the monomer. The small tendency of the carbanion to side reactions enables us to prepare active centres outside the actual polymerization medium. This is of advantage when initiation is slow and the active centres are stable. Monomers which not polymerize on the C=C bond present active centres by different reactions. Formaldehyde produces a centre by a scheme analogous to reaction (48). Nucleophilic attack on an oxirane ring leads to its opening

$$\sim\!\underline{\overline{O}}|^- + CH_2\!\!-\!\!CH_2 \longrightarrow \sim\!\underline{\overline{O}}\!-\!CH_2\!-\!CH_2\!-\!\underline{\overline{O}}|^- \qquad (53)$$

Larger cyclic ethers are more basic and therefore more stable toward anions. Even these can be opened by strongly nucleophilic agents [207]

$$Ph_3CMgBr + \underset{O}{\triangleleft} \longrightarrow Ph_3C(CH_2)_4OMgBr$$

Tetrahydrofuran is a frequently used solvent which appears to be not quite inert. The opening of its ring by anions is not infinitely slow. In its presence, polymerization temperatures should be kept below 273 K.

In polymerizations of cyclic siloxanes, because of the higher coordination number of silicon the active centre can be formed by the decomposition of the transition complex with pentavalent silicon [208]

$$H\underline{\overline{O}}|^- + \underset{|}{\overset{|}{Si}}\!\!\!\underset{O}{\overset{O}{\diagdown}} \rightleftharpoons HO\!-\!\overset{|}{\overset{-}{Si}}\!\!\!\underset{O}{\overset{O}{\diagdown}} \rightleftharpoons HO\!-\!\overset{|}{\underset{|}{Si}}\!\sim\!\overset{|}{\underset{|}{Si}}\!-\!\underline{\overline{O}}|^- \qquad (54)$$

The opening of lactam rings by strong bases and their anionic polymerization proceed by means of a complicated disproportionation [173]. The reaction is usually initiated by an organometallic lactam derivative with the alkali

References pp. 154–162

metal on nitrogen. Initiation is characterized by an induction period during which the activator (acyllactam) is formed

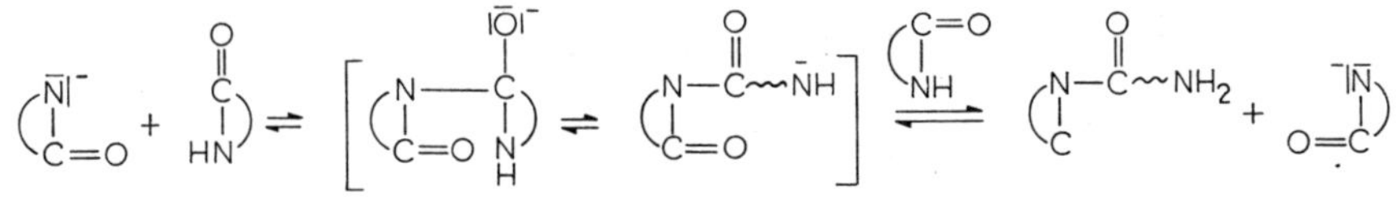

$$(55)$$

Reaction (55) exhibits a higher activation energy than the addition of the caprolactam anion to acyllactam. An addition of acyllactam as activator strongly reduces the induction period and makes possible polymerization at lower temperatures.

Initiation of ε-caprolactam polymerization in the presence of acyllactam can be described by the scheme

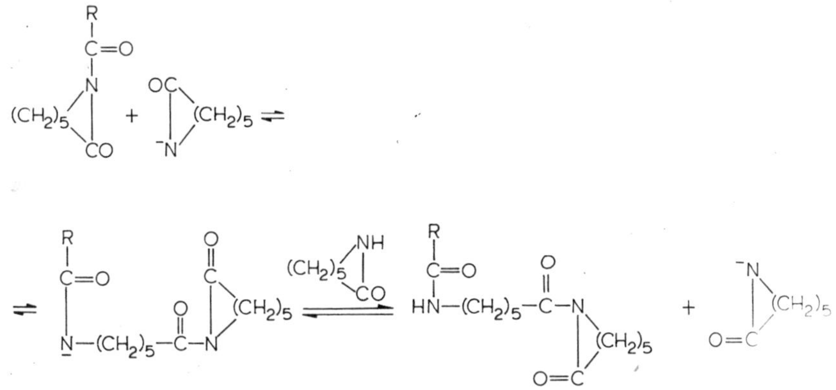

$$(56)$$

Initiation according to reaction (55) is generally accepted, although other mechanisms have also been proposed. Khaitin and Volf [209] consider anion attack on the amide carbon. Further additions would occur with cation participation (Ac^+ is the cationic and B^- the anionic part of initiator AcB).

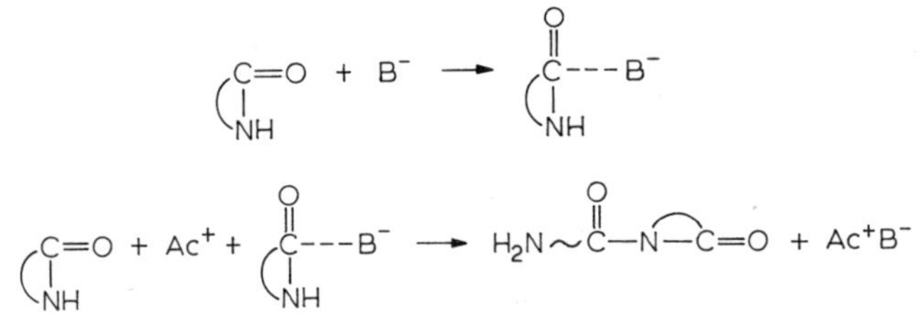

$$(57)$$

2.5.2 ACTIVATORS OF ANIONIC POLYMERIZATION

Many authors have observed an acceleration of caprolactam polymerization in the presence of N-acetyl-ε-caprolactam, isocyanates, thiolactones, organic sulphates, aromatic amides, benzyldioxime carbonate, arylenedicarbamoyllactams, phosphorus oxychlorides and phosphorus pentachloride. Activator effectiveness increases with growing substituent electronegativity in N-substituted lactams.

Highly active molecules are therefore of the type [210] phenylphosphonyl--N-N'-biscaprolactam.

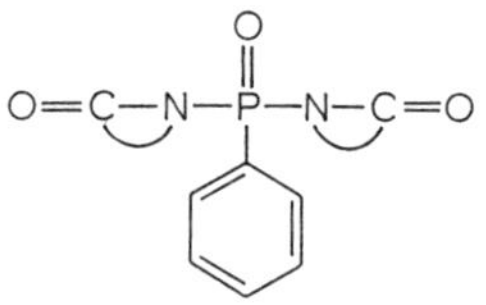

Polymerization of 2-pyrrolidone is also activated by various N-acyllactams, especially by N-acetyl-2-pyrrolidone, lower lactones, CO_2 and N-benzoyllactams [211].

The activation described is due to an increase in the concentration of active centres. The polymerization rate can, of course, also be enhanced by increased reactivity of the centres. To solvent-separated ion pairs and to free ions, monomers in general are added more rapidly than to contact ion pairs. When we succeed in separating the active centre ions by the addition of a strongly solvating compound, the initiation and propagation rates will be enhanced, and changes affecting the mode of addition may appear [212, 213].

For cation solvation, macrocyclic polyethers, crown ethers [214] of type

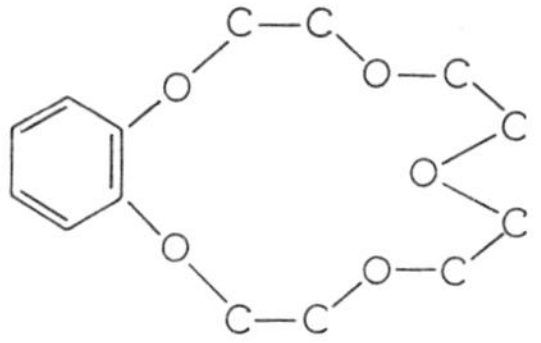

are particularly suitable (monobenzo-15-crown-5; the first number corresponds to the total number of atoms and the second to the number of ring oxygen atoms). Crown ethers significantly increase the solubility of ionic compounds in non-polar media and yield crystalline complexes with many salts [215]. Cryptands exhibit a similar effect [216, 217] e. g. 4, 7, 13, 16, 21, 24-hexaoxa-1, 10-diazabicyclo [8, 8, 8] hexacosan (Kryptofix [222].

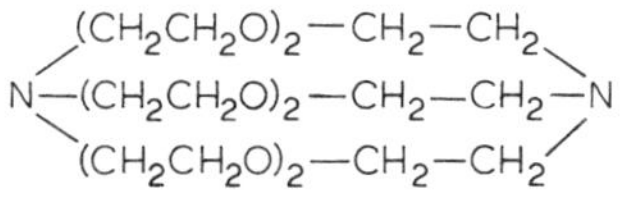

References pp. 154–162

By the addition of crown ethers or cryptands, the rate of the anionic polymerization of methylthiiran, β-lactones, and other monomers may be enhanced by up to two orders of magnitude [218–220].

The initiators of anionic polymerization can also be activated by some types of chelates. Magnon et al. [221] have described ethylene oligomerization with butyllithium activated by tetramethylethylenediamine

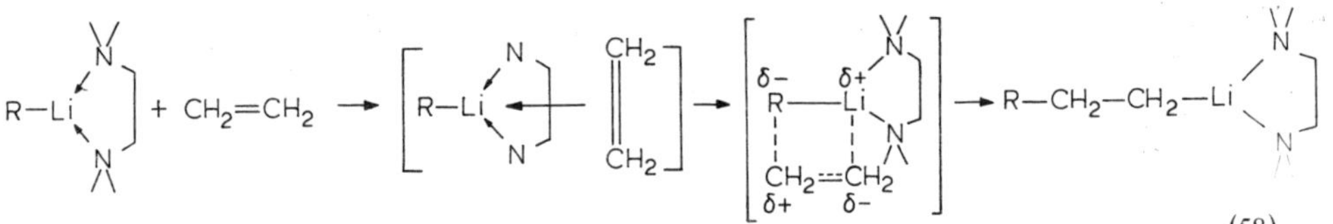

$$R-Li\begin{pmatrix}N\\N\end{pmatrix} + CH_2{=}CH_2 \rightarrow \left[R-Li\begin{pmatrix}N\\N\end{pmatrix}\cdots\begin{matrix}CH_2\\\|\\CH_2\end{matrix}\right] \rightarrow \left[\begin{matrix}\stackrel{\delta-}{R}-\!\!\!-\stackrel{\delta+}{Li}\\ ||\\ \underset{\delta+}{CH_2}{\cdots}\underset{\delta-}{CH_2}\end{matrix}\right] \rightarrow R-CH_2-CH_2-Li\begin{pmatrix}N\\N\end{pmatrix} \tag{58}$$

The above examples adequately support the statement that valuable results may be obtained by active centre modifications. Complexes on active centres enable us to control the kinetics and mechanism of subsequent polymerization stages.

3. Formation of initiating cation

Cationic polymerizations are presently overshadowed by other methods of polymer synthesis. The main reason is the extraordinary complexity of the chemical processes in cationic polymerizations, and the consequent difficulty in controlling technological problems. Some monomers cannot, however, be polymerized in any other way. This is a sufficient reason for studying the generation and reactions of carbocations (and also carboxonium and oxonium ions).

3.1 Initiators of cationic polymerizations

Cationic polymerization are initiated by acids, i. e. by electrophilic agents with a pronounced ability to bind electrons offerred by the donor. The initiator nomenclature has not been established at the present time. Most authors of original studies use the word initiator to designate the acid which is responsible for cation formation (without which the cation would not be formed).

The compound producing the initiating cation by splitting is called the "initiator" by Kennedy [222]. In contrast to other authors [223], he designates the molecule responsible for this reaction as the "co-initiator"[†]

$$AlCl_3 + RCl \rightleftharpoons R^+ + AlCl_4^- \tag{59}$$

In the subsequent text, the older definitions, designating $AlCl_3$ as initiator and RCl as co-initiator, will be used.

The initiators of cationic polymerizations must produce sufficiently reactive cations which are able to yield active centres with the monomer. It is therefore useful to discuss each of the two main initiator types separately.

3.1.1 BRÖNSTED ACIDS

By dissociation, strong mineral acids liberate a proton-initiating polymerization

$$HClO_4 \rightleftharpoons H^+ + ClO_4^- \tag{60}$$

Not many initiators belong to this class even though the halogenoacetic, fluorosulphonic, and other acids are included. A detailed analysis of their polymerization mechanism is obscured by complex formation with monomer and with solvent, by the occurrence of aggregates, and by anion reactions in acids with an unstable anion. In spite of its apparent simplicity, initiation by Brönsted acids has not yet been investigated in detail. The "pseudo-cationic" polymerization of styrene is an instructive example.

Gandini and Plesch [224] have published their observations on the polymerization of styrene in methylene chloride with $HClO_4$. At low temperatures (<200 K), this polymerization proceeds in three stages. The first short stage may be observed immediately after breaking the ampoule containing $HClO_4$ in the styrene solution. It is actually only a flash of rapid polymerization; the presence of ions is indicated by conductivity measurements. Soon, however, a long "stationary" state is established where the reaction is relatively slow, conductivity drops to zero, and the rate is not affected by the presence of water. Ions could not be detected even by sensitive methods. At higher conversions, the polymerization again becomes more rapid; when the styrene concentration drops below four times the $HClO_4$ concentration, the ionic character of the reaction reappears together with all the accompanying

---

[†] The Kennedy's nomenclature has been derived for straightforward cases of initiating H^+ or R^+ generation. When the structure of the initiating particle is unknown, the same logical approach cannot be used to discern between initiator and co-initiator. It is even more important, according to this nomenclature, cationic polymerizations would be initiated by Lewis bases, and only co-initiated by Lewis acid. This clearly contradicts all experience. Many systems are known that can be cationically polymerized in the presence of a Lewis acid, without any co-initiator.

effects, and the remaining monomer is rapidly transformed to polymer. This is the third phase. The authors have called the middle "stationary" phase pseudo-cationic, assuming an active centre with an ester structure. This work arouse the interest of other authors (see for example, refs. 225–229), many of whom do not consider the ester structure of the active centre to be safely proved. In a broader context, formations with propagation proceeding on a covalent bond have actually been found (see Chap. 4, Sect 2.3).

3.1.2 LEWIS ACIDS

Lewis acids are more frequently used as initiators than are Brönsted acids. The mechanism of their function is largely unknown. They are generally more active and sometimes indispensable. The application of the halogenides of almost all metals of Groups IIIA–VIII has been described. Organometals, mainly of aluminium, are also efficient cationic initiators under certain conditions.

Several hypotheses have been proposed for explaining the mechanism of the formation of initiating cations by means of Friedel–Crafts "catalysts".

(a) Direct initiation [230, 231]

$$AlBr_3 + CH_2=C(Me)_2 \rightleftharpoons {}^-AlBr_3-CH_2-(CH_3)_2C^+ \tag{61}$$

(b) Co-initiation by solvent [223] [see reaction (59)].
(c) Autodissociation [232]

$$Al_2Cl_6 \rightleftharpoons 2\,AlCl_3 \rightleftharpoons AlCl_2^+ + AlCl_4^- \tag{62}$$

A very interesting variant of this kind of initiation, leading to living cationic polymerizations of vinyl ethers, is the dissociation [233]

$$2\,I_2 \rightleftharpoons I^+ + I_3^- \tag{63}$$

(d) Co-initiation by a protogenic molecule [234]

$$BF_3 + H_2O \rightleftharpoons BF_3OH^- + H^+ \tag{64}$$

(e) Decomposition of donor–acceptor complexes [235] (see Sect. 5).

All these hypotheses have been known for a very long time. There is no doubt that each of them is justified in a particular system and under certain conditions. Simultaneous participation of several mechanisms of ionic particle generation is not excluded.

Co-initiation

In 1946, Evans et al. [236] observed that 2-methylpropene polymerization with BF_3 only starts in the presence of a small amount of water. By analogy to enzyme–co-enzyme terminology, they called water the cocatalyst. The term co-initiator was later established. Reaction (64) and its modifications play an important role in cationic polymerizations. The presence of a small amount of a compound witch a separable proton sometimes greatly accelerates polymerization. The problem of trace impurities has been the subject of bitter controversy for a long time. To this day, the existence of a co-initiating impurity at the start of a cationic polymerization cannot be safely excluded. The situation is further obscured by the formation of initiator and co-initiator complexes with monomers and solvents.

Plesch [237] noticed that the concentration of co-initiator can change in the course of polymerization. This affects the whole reaction. Three cases may occur.

(a) The contact of Lewis acid with monomer does not result in polymerization or else polymerization does occur but it is slow and stops at low conversion. The reaction can be resumed by the addition of water. This behaviour is ascribed to traces of a co-initiator, usually H_2O, which is consumed in the course of the reaction. Thus in these systems, the presence of a special co-initiator is necessary; it cannot be substituted by any of the components of the main system. This behaviour is oberved, for example with 2-methylpropene in the presence of $TiCl_4$ in $MeCl_2$ at 213–253 K or with the system isobutyl vinyl ether–BF_3 in THF.

(b) Enhanced efficiency of drying the polymerization components leads to a considerable reduction in polymerization rate. Reproducibility remains poor but the reaction continues to proceed until the monomer is completely consumed. The author assumes that this situation is also caused by residual water but that, in this case, water is not consumed in the medium. This behaviour has been observed for styrene with $SnCl_4$ in nitrobenzene and CCl_4 or in a mixture of these solvents.

(c) Careful drying leads to rapid, reproducible polymerization. The rate can be affected by the addition of water. It is very improbable that this kind of behaviour could be due to impurities. This behaviour is typical for styrene in methylene chloride as solvent; polymerization with $SnCl_4$ at ~300 K is accelerated by the addition of water; with $TiCl_4$ at 183 K, water does not affect the rate. Addition of water to the system 2-methylpropene–$AlBr_3$–heptane at 211–295 K reduces the polymerization rate. Initiation may occur by reactions (59), (61), (64), by complex decomposition and by other processes.

Situations (a) and (b) used to be considered easily understandable but case (c), on the other hand, was difficult to explain. It now appears that even the behaviour in cases (a) and (b) is not simple. Biddulph et al. [238] and especially Cheradame and Sigwalt [239] have observed only partial polymerization of 2-methylpropene with $TiCl_4$ at low temperatures. After prepolymerization, the reaction could not be revived by further addition of a $TiCl_4$ solution. However, polymerization proceeded to completion when either the whole charge or at least TiCl$_4$ was passed through the gas phase by distillation. If, after prepolymerization, the TiCl$_4$–2-methylpropene mixture was held at a higher temperature for several hours, partial polymerization occurred after cooling. Conversion could be slightly increased by the addition of indene or α-methylstyrene to prepolymerized 2-methylpropene. The monomers polymerized completely after cooling (∼100–150 K).

Co-initiation by water (and by protogenic compounds) exhibits many unclarified features. Water may be indispensable for the start of polymerization [240], it may act as an accelerator [241], a retardant or an inhibitor [242]; it may accelerate or retard polymerization depending on concentration [241, 243, 244] or it may be totally without effect [245]. The co-initiating effect of water depends on the initiator, monomer, solvent, temperature, and concentration of water. In spite of the fact that water is today considered to be the dominant co-initiator, its indispensability for the start of polymerization is often overestimated. In can be shown that it is always possible to bring a selected monomer–initiator pair to polymerization (possibly slow) in a non-aqueous medium, for example by suitable choice of solvent, temperature or other conditions [246]. Ion formation by reaction (64) has not actually been proved; the initiating cation is formed in a more complicated way. This is indicated by the inhibition periods observed during the polymerization of styrene with SnCl$_4$ [244], by the inhibiting effect of water on 2-methylpropene polymerization in the presence of AlBr$_3$ [242] and by other observations.

All the enumerated examples indicate the insufficiency of reactions (59)–(64) to explain completely the initiation of cationic polymerizations. An inseparable aspect of initiation is the activation of the primary products produced by ionization or dissociation. Several kinds of ion pair of various reactivities are known to exist. The solvate envelope of free ions must affect the frequency of active ion–monomer collisions, i.e. the initiation rate. In the author's opinion, the key to our understanding of some co-initiation effects in cationic polymerization is a suitable interpretation of the Winstein dissociation scheme [247]

$$AB \overset{S}{\rightleftharpoons} A^+B^- \overset{S}{\rightleftharpoons} A^+ \| B^- \overset{S}{\rightleftharpoons} A^+ + B^- \qquad (65)$$

In eqn. (65), A^+B^- is a contact (externally solvated) ion pair, $A^+\|B^-$ is a solvent-separated ion pair (with solvent molecules between the ions), and A^+, B^- are "free" (solvated) ions; S stands for solvent. The initiating efficiency of a Lewis acid AB generally increases with a shift of the equilibria (65) from left to right. In the system monomer–initiator–(solvent) the concentration of active formations suitable for initiation and propagation is a function of the equilibrium position. In non-polar, poorly solvating media, only weakly active or completely inactive forms of the initiator are present and polymerization is either slow or does not occur at all. The addition of a substance affecting the equilibrium must be reflected in the initiation and polymerization rates. By shifting the equilibrium to the right, compound S either activates or accelerates initiation. Even a small amount of a polar, strongly solvating compound can significantly increase the concentration of polar formations. The equilibria can be shifted towards active initiator forms by a suitable solvent or by the monomer itself. In this case, the system polymerizes even without a "co-initiator". When the polar properties of compound S profoundly differ from the properties of the monomer (for example water and styrene), then a sufficient amount of S reduces the initiation and polymerization rates, sometimes down to zero. The solvate (hydrate) envelope prevents efficient contact between cation and monomer[†]. Although direct proof is not always available, considerations concerning shifts of solvation equilibria and the existence of relatively stable monomer complexes may plausibly explain many experimentally observed effects [241–244].

The rate of monomer consumption on free cations is higher by one to two orders of magnitude than on ion pairs [248, 249]. Physical shifts of dissociation equilibria naturally cannot explain all unclarified initiation problems in cationic polymerizations. It would nevertheless be rash to neglect these evidently existing processes intentionally and to limit the explanations to only chemical interactions [228].

The equilibria considered so far only as a result of van der Waals' intermolecular attractive forces are, of course, strongly affected by complex formation, for example of the donor–acceptor type (see Sect. 5).

3.2 Special kinds of initiators

Polymer synthesis can proceed smoothly or with difficulty with the medium, the reaction conditions, and the initiator playing important roles. This is particularly true when polymers of specific, predetermined structure are to be prepared.

3.2.1 INITIATION BY HI/I_2

In a series of papers (see for example, ref. 250), Higashimura and Sawamoto have described the living cationic polymerization of vinyl ethers at low

---

[†] Most solvents and impurities are of nucleophilic character. They preferentially solvate the cation. Conditions are generated for the occurrence of a collection of poorly defined reactions at the active centres, thus impeding analysis of cationic polymerizations.

temperatures initiated by the system HI and I_2 in which the route of the propagation reaction is determined by the polarity of the medium.

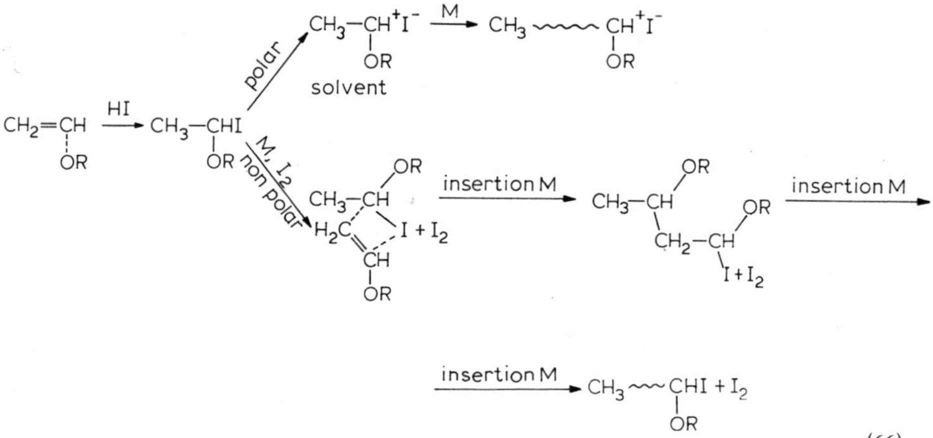

$$CH_2=CH(OR) \xrightarrow{HI} CH_3-CHI(OR) \begin{cases} \xrightarrow{polar\ solvent} CH_3-CH^+I^-(OR) \xrightarrow{M} CH_3\sim\sim CH^+I^-(OR) \\ \xrightarrow{M,\ I_2\ non\ polar} \underset{\text{CH(OR)}}{\overset{\text{CH}_3-\text{CH(OR)}}{\text{H}_2\text{C}\cdots\text{I}+I_2}} \xrightarrow{insertion\ M} CH_3-CH(OR)-CH_2-CH(OR)\cdots I+I_2 \xrightarrow{insertion\ M} \end{cases}$$

$$\xrightarrow{insertion\ M} CH_3\sim\sim CHI(OR) + I_2$$

(66)

3.2.2 MULTIFUNCTIONAL INITIATORS

Molecules with two or more cationic active centres are very useful for certain macromolecular syntheses. The first dicationic initiator was used in our laboratory [251–253]. It is formed by hydrolysis and condensation of dimethyldichlorosilane in the presence of a strong acid, e. g. $HClO_4$. The condensation is an equilibrium process. When its volatile products are removed by evacuation, a siloxonium dication is finally formed.

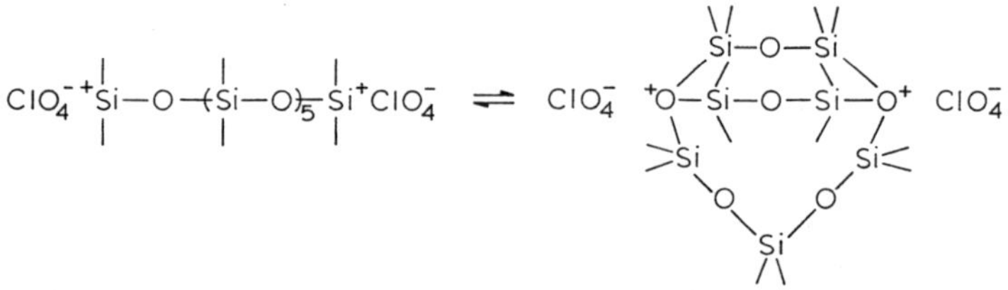

$$ClO_4^- {}^+Si-O-(Si-O)_5 Si^+ ClO_4^- \rightleftharpoons ClO_4^- {}^+O-Si-O-Si-O^+ ClO_4^- \text{ (cyclic siloxane)}$$

(67)

A somewhat complicated method for the preparation of another dication (octamethylene-bis-dioxolenyliumperchlorate)

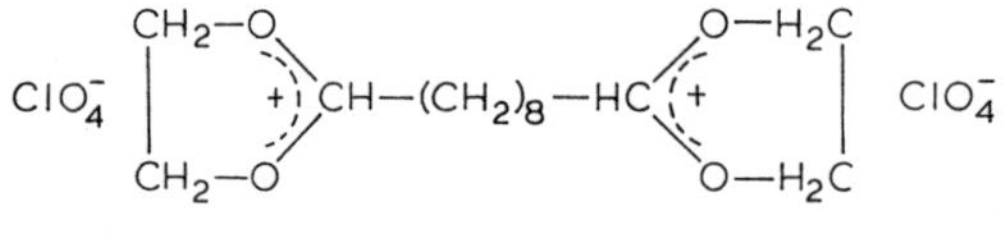

$$ClO_4^- \quad \underset{CH_2-O}{\overset{CH_2-O}{\diagup}}\!\!\!\!{}^+\!\!CH-(CH_2)_8-HC^+\!\!\!\!\underset{O-H_2C}{\overset{O-H_2C}{\diagdown}} \quad ClO_4^-$$

was described by Yamashita et al. [254]. Smith and Hubin polymerized tetrahydrofuran with the aid of fluorosulphonic acids and their anhydrides

[255]. Chains with active centres at both ends, i. e. dications, were formed from $(CF_3SO_2)_2O$ [255]. A very similar dication

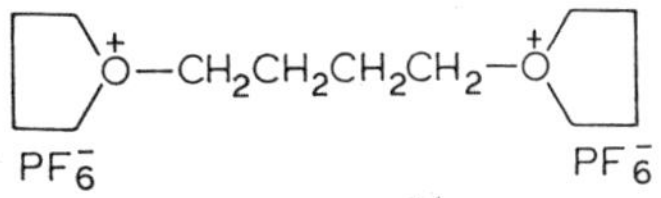

$$\overset{+}{\underset{PF_6^-}{\boxed{}\!\!\!O}}-CH_2CH_2CH_2CH_2-\overset{+}{\underset{PF_6^-}{O\!\!\!\boxed{}}}$$

is formed from THF and PF_5 [256].

Burgess et al. [257] made use of the reaction of alkylhalogenides with the silver salts of superacids.

$$R{-}Br + AgClO_4 \rightarrow R^+ClO_4^- + AgBr \tag{68}$$

Cations on silicon are formed from chlorosilanes by a similar reaction [258].

$$Ph_3SiCl + AgPF_6 \rightarrow Ph_3Si^+PF_6^- + AgCl$$

Molecules with two halogens in the α and ω positions yield a dication.

A reaction analogous to reaction (68) was first described by Kennedy and Melby [259] who obtained the initiating carbocation by a reaction of alkyl halogenide with Et_3Al or Et_2AlCl. They have not prepared dication (multication) initiators directly nevertheless the route to them is evident.

Higashimura and Sawamoto [250] prepared a difunctional initiator by a reaction analogous to reaction (66)

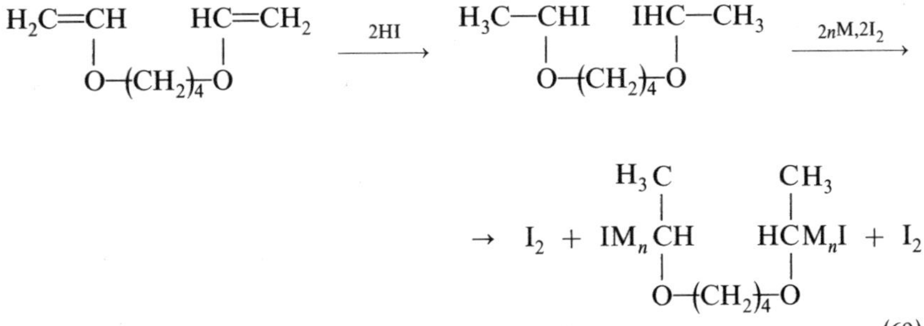

$$\begin{array}{c} H_2C{=}CH \quad\quad HC{=}CH_2 \\ | \quad\quad\quad\quad | \\ O{-}(CH_2)_4{-}O \end{array} \xrightarrow{2HI} \begin{array}{c} H_3C{-}CHI \quad IHC{-}CH_3 \\ | \quad\quad\quad\quad | \\ O{-}(CH_2)_4{-}O \end{array} \xrightarrow{2nM, 2I_2}$$

$$\rightarrow \quad I_2 + IM_n\underset{\underset{O{-}(CH_2)_4{-}O}{|}}{\overset{\overset{H_3C}{|}}{C}H} \quad \underset{\phantom{O{-}(CH_2)_4{-}O}}{\overset{\overset{CH_3}{|}}{H}C}M_nI + I_2 \tag{69}$$

3.2.3 MACROCATIONS

Similar to their anionic counterpart (see Sect. 2.4), even with cationic polymerizations the structure and size of the molecule to which the active centre is bound plays an important role. The required macromolecules with one or two active ends are formed by living polymerizations. Modern macromolecular syntheses use them as agents, especially for the preparation of

References pp. 154–162

block and graft copolymers [200, 201]. The cationic path is more convenient for the systhesis of hydrophilic chains (from heterocyclic monomers); the anionic reaction, on the other hand, is more suitable for the synthesis of hydrophobic blocks. By a convenient combination of both these reactions, very desirable polymers with both hydrophobic and hydrophilic portions can be obtained.

Macrocations are also formed by the attack of strong acids on polyalkene macromolecules [260].

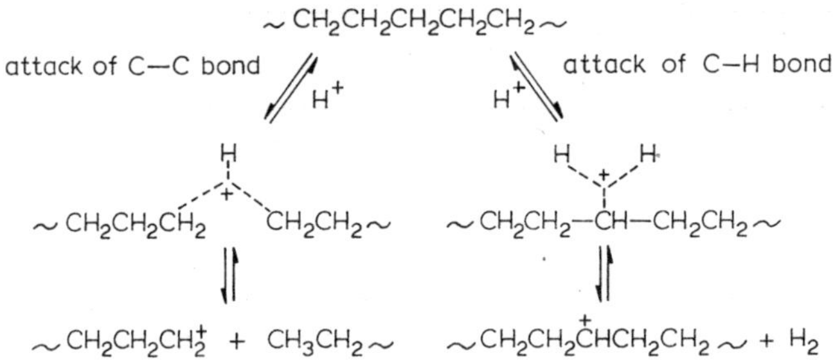

$$\sim CH_2CH_2CH_2CH_2CH_2 \sim$$

attack of C—C bond $\quad H^+ \quad\quad H^+ \quad$ attack of C—H bond

$\sim CH_2CH_2\overset{H}{\overset{|}{\overset{+}{C}}H_2} \quad \overset{+}{C}H_2CH_2 \sim \qquad \sim CH_2CH_2-\overset{H\underset{+}{}H}{\overset{|}{C}H}-CH_2CH_2 \sim$

$\sim CH_2CH_2\overset{+}{C}H_2 + CH_3CH_2 \sim \qquad \sim CH_2CH_2\overset{+}{C}HCH_2CH_2 \sim + H_2$

The active centres formed in this way can again be used for the preparation of block and/or graft copolymers [261–263].

3.3 Reaction of cation with monomer

Interaction of a cation with a monomer yields an active centre; this is initiation in its most proper sense. The proton and the most simple carbocations are attached to the β atom of an alkene. This is documented both by experimental findings and by the corresponding calculations of the localization energies of the respective structures (Hückel) and of the electron densities on the alkene carbon atoms [264].

Also with heterocyclic monomers, simple cations are attached to the atom with the highest electron density in the valence sphere, i. e. to the heteroatoms. The reactions of more complicated carbocations are evidently of greater complexity. The triphenylmethyl cation separates a hydride ion from tetrahydrofuran, producing triphenylmethane and an oxonium ion

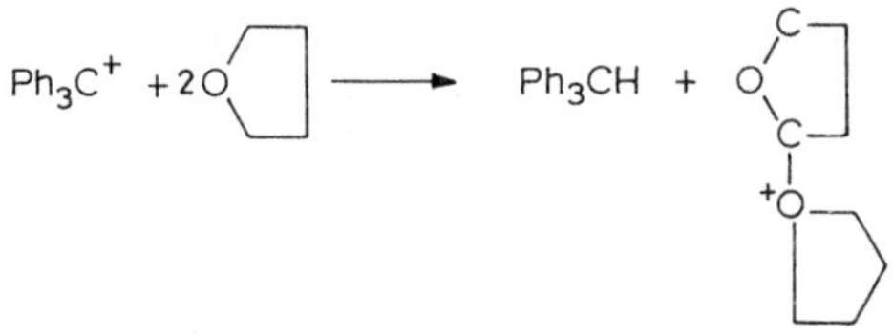

$$Ph_3C^+ + 2 \underset{}{\bigcirc\!\!\!\!|} \longrightarrow Ph_3CH + \underset{\overset{+}{O}\!\!\!\!\bigcirc}{\overset{\bigcirc\!\!\!\!|}{}} \tag{70}$$

With lactams, a tautomeric equilibrium is assumed to exist between the compounds formed by the addition of a simple primary cation to oxygen or nitrogen. With a proton as the cation, an acyllactam ammonium ion is finally produced [265, 266].

$$\left[\begin{array}{c} C=\overset{+}{O}H \\ | \\ NH \end{array} \rightleftharpoons \begin{array}{c} C-OH \\ \| \\ \underset{+}{NH} \end{array} \rightleftharpoons \begin{array}{c} \overset{+}{C}-OH \\ | \\ NH \end{array} \right] \rightleftharpoons \begin{array}{c} C=O \\ | \\ \underset{+}{NH_2} \end{array}$$

$$\begin{array}{c} C=O \\ | \\ \underset{+}{NH_2} \end{array} + \begin{array}{c} O=C \\ | \\ HN \end{array} \rightleftharpoons \begin{array}{c} C=O \\ | \\ N-C\sim\overset{+}{NH_3} \\ \| \\ O \end{array} \xrightleftharpoons[-M]{+M} \begin{array}{c} C=O \\ | \\ N-C\sim NH_2 \\ \| \\ O \end{array} + \begin{array}{c} C=O \\ | \\ \underset{+}{NH_2} \end{array} \quad (71)$$

As for every reaction of ions, even initiation is affected by the medium. It is essentially a rapid process. Of course, in various systems the solvation and desolvation, aggregation and deaggregation, and coordination processes occur to various degrees. Their outward manifestation is a greater or smaller hindrance of the whole process.

Very rapid initiations are known, manifested by an instantaneous start to the polymerization after which the number of active centres is not further increased. Polymerizations with slow initiation are also quite frequent, starting only after some inhibition and/or induction period. In the course of these polymerizations, the concentration of active centres is not usually constant. A stationary state is not excluded, of course but it occurs much less frequently than with radical polymerizations.

4. Formation of complexes initiating coordination polymerizations

The unoccupied d orbitals of transition metals are suitable for monomer coordination. A certain structure of the complexes of these metals can result in an extremely useful link between space-oriented monomer coordination and polymerization.

4.1 Ziegler–Natta catalysts

In chemical technology, no other catalyst type can be found that approaches the Ziegler–Natta (ZN) catalysts in their versatility and universal application. Since their discovery, enormous energy has been devoted to their

References pp. 154–162

exploitation and to the elucidation of their operating mechanism. The amount of the accumulated information is enormous (see, for example, the monograph by Boor [267a]). However, a unique consistent theory of their action has not been presented to date.

Any attempt to describe concisely the structure of ZN catalysts must necessarily suffer from oversimplification. ZN catalysts can be used in soluble, insoluble or coloidal form, either free or attached to a support. In principle, they consist of mixtures of salts of Group IV–VIII metals with alkylmetallic compounds of Group I–III metals (cocatalysts).

Actually, the number of practically important compounds is much smaller than would correspond to the above definition. Organoaluminium and organomagnesium compounds are the most frequently used alkylmetals. Alkylmetallic compounds of Zn, Be, and Li are studied to a lesser extent. Of the transition metals, the salts of Ti, V, Cr, Co, Ni, and W are of importance. For the polymerization of alkenes (especially propene and 1-butene) to isotactic polymers, $TiCl_3$ activated by milling (or its mixed crystals with $AlCl_3$) was the most important until recently, together with Et_2AlCl. These systems also contain a third component of donor character (e. g. diethyleneglycol) and they are usually called ZN catalysts of the first generation. Syndiotactic polypropylene, less important than isotactic so far, contaminated by other configurations, is only obtained with a soluble vanadium catalyst (of the VCl_4MeOPh + Et_2AlCl type) at temperatures < 223 K [268].

The polymerization of conjugated dienes to products with a controlled structure usually occurs in the presence of alkylaluminium compounds. The choice not only the transition metal but also of its ligands is of importance. Some systems produce a certain kind of stereochemical structure irrespective of external conditions. So, for example, vanadium compounds yield predominantly the *trans*-1,4 structure whereas cobalt salts yield the *cis*-1,4 structure. Other catalysts are very sensitive, and a small external effect completely changes their stereochemical activity [267b] [e. g. $Cr(acetylacetonate)_3$–R_3Al]. Examples of several catalytic systems are summarized in Table 7.

Typical of the second-generation catalysts is a modified solid phase: $TiCl_3$ or $TiCl_3 \cdot \frac{1}{3}AlCl_3$, adjusted during production or later by a donor, usually diethyl ether. These catalysts are more active with more efficient control of monomer addition to yield an isotactic product so that the unwanted atactic fraction only amounts to about 2%.

The active centre of all these catalysts is the Ti–C bond surrounded by suitable ligand fields, mainly from the direction of the transition metal. Even though milling of the solid phase greatly increases the surface and thus also the number of transition metal atoms potencially able to form active centres with suitable cocatalysts, the amount of transition metal inside the crystals remains relatively large. These atoms have no chance of becoming active

TABLE 7

Polymerization of some alkenes and dienes with ZN catalysts [267b]

Alkenes and dienes	Catalysts	Product characteristics/m. p. (K)
Ethylene	$Et_3Al + TiCl_4$	XXX/410
	CMO^a, $BuEtMg$, $TiCl_4$, $EtAlCl_2$	> 1 t PE/g Ti[b]
	Al, Mg, Si, Ti[c]	$> $ t PE/g Ti, h[d]
Propene	$Et_2AlCl + VCl_4$ (> 230 K)	syndio; XXX/403
	$\begin{Bmatrix} EtBuMg, Et_3Al, BuCl, EtOH \\ TiCl_4EB^e \rightarrow K \end{Bmatrix}$	
	K, Et_3Al, Et_2AlCl, MT^f	95,8 % iso[g]
Propene and higher 1-alkenes	$Et_2AlCl + TiCl_3$	iso, XXX/443
RS-3-Methylpentene and similar racemic 1-alkenes	$Bu_3Al + TiCl_4$	Optically inactive
Vinyloxirane	$Et_2AlCl + VCl_3$	
Cyclobutene	$Et_3Al + VCl_3$	Elastomer
	$Et_2AlCl + V(acac)_3$	XXX
2-Butene	$Et_2AlCl + V(acac)_3$ (≤ 240 K)	XXX
Butadiene	$Et_2AlCl +$ soluble Co salt	cis-1,4; elastomer
	$Et_3Al + VCl_3$	trans-1,4; XXX/418
	$Et_3Al + Ti(OnBu)_4$	1,2-iso; XXX/400
	$Et_3Al + Cr(acac)_3$	1,2-syndio; XXX/423
Isoprene	$Et_3Al + TiCl_4$ Al/Ti ≈ 1	cis.-1,4; elastomer
1,5-Hexadiene	$Et_2AlCl + V(acac)_3$	Elastomer
	$Bu_3Al + TiCl_3$	XXX/408

[a] Chloromethyloxirane.
[b] Ref. 269.
[c] Contents of catalyst components (metals and organic residues) cannot be expressed by simple stoichiometric ratios.
[d] Ref. 270.
[e] Ethylbenzoate.
[f] Methyltoluate.
[g] Ref. 271.

References pp. 154–162

centres but they are a ballast in the product. As variable valency metals, they catalyze the decomposition of peroxides (formed by the interaction of accidentally present radicals with oxygen) and thus contribute to rapid polymer degradation. The radicals formed from these peroxides attack hydrocarbon chains, yielding further radicals which produce new peroxides with oxygen. Residual transition metals, when present at a concentration of more than a few ppm, must therefore be washed out of the polymers. Technologically this is always a difficult and expensive operation. This is the reason that a tendency has developed to attach a thin layer of the transition metal compound on a suitable carrier, and thus to utilize its greater readiness to form active centres. In this way, the unwanted metal residue in the polymer can be reduced by about an order of magnitude; an acceptable concentration may be reached in this way so that the difficult washing procedure can be eliminated. Many porous materials with a large surface area have been used as carriers: SiO_2, Al_2O_3. Mg(OH) Cl, various polymers, and others. In fact, highly efficient catalysts have been obtained but unfortunately their iso-specific regulating activity was insufficient. This type of catalyst is sometimes also designated as a second generation catalyst [272].

The third generation of ZN catalysts is recent. It comprises extremely active complexes formed by the interaction of $TiCl_4$ with $MgCl_2$ or with organomagnesium compounds modified by Lewis bases. They operate together with organoaluminium cocatalysts and exhibit not only high activity and productivity, but also a high efficiency in regulating stereospecificity (see below).

4.1.1 SUPPORTED CATALYSTS

The reasons that led to the development of supported catalysts were explained above. Of the great number of alleged suitable carriers, including natural zeolites, kaolin, perlite, metal oxides, graphite, many polymers and copolymers [e. g. poly (styrene-co-divinylbenzene)], hydroxylated polybutadiene, polyethylene-graft-poly(4-vinylpyridine), and many others (see, for example, ref. 273), evidently only synthetic SiO_2 (silica), Al_2O_3 (alumina), and $MgCl_2$ are really important. While in classical ZN catalysts only a few parts per thousand of transition metal atoms are utilized for active centre formation [274], in the supported catalysts this fraction grows to several tenths [275] and even to over 75% of the total amount of transition metal.

In an overwhelming majority of cases, the transition metal atom is bound to the carrier surface by an oxygen bridge

$$\text{Mg(Cl) OH} + \text{TiCl}_4 \rightarrow \text{Mg(Cl)}-\text{O}-\text{TiCl}_3 + \text{HCl} \tag{72}$$

$$-\underset{|}{\overset{|}{\text{Si}}}-\text{OH} + \text{CrO}_2\text{Cl}_2 \rightarrow -\underset{|}{\overset{|}{\text{Si}}}-\text{O}-\text{CrO}_2\text{Cl} + \text{HCl} \tag{73}$$

The Solvay catalyst, formed according to reaction (72), exhibits high activity and is used for the preparation of the ethylene–propene copolymer and for butadiene homopolymerization. The catalysts of the Union Carbide Company (UCC) [reaction (73)] are formed by the anchoring of organosilicon Cr derivatives on a silica surface with large internal surface area and pores of certain dimensions. After contact of both components, Cr is reduced with hydrogen (or better with CO) to a lower oxidation state. The UCC supported catalysts are used for initiating ethylene polymerization. A modern Solvay catalyst for propene polymerization is a highly porous $TiCl_3$ formed from $TiCl_4$ by reduction with Et_2AlCl at low temperature and washed with a mixture of diisoamyl ether and n-hexane. The solid product formed is once more treated with $TiCl_4$ in n-hexane to yield the catalytic complex.

The Solvay catalysts are activated by aluminium organometals, particularly Et_2AlCl, $(isoBu)_3 Al$ or $(isoBu)_2 AlH$ [267a].

No metal is so inert as not to exert any physical or chemical effect on an anchored transition metal. Consequently it also affects the properties of active centres. Anhydrous $MgCl_2$ assumes a quite exceptional position among the supports. In fact its properties have made possible the construction of third-generation catalysts. The Ti and Mg ions are of approximately the same dimensions; Ti easily becomes part of a cocrystalline lattice in $MgCl_2$. Originally the superactive catalysts were obtained by jointly milling $MgCl_2$ and $TiCl_4$. Because of the low reproducibility of the results obtained with the systems prepared by milling, $MgCl_2$ is now more often obtained from organomagnesium compounds directly during catalyst synthesis by means of compounds readily liberating chlorine. The effect of Mg is not limited to accommodation of the transition metal in its crystal lattice. The properties (activity) of the centre are significantly affected by Mg atoms neighbouring with Ti; the centres are diluted and stabilized [275]. Their activity is also much greater than that of classical ZN centres (roughly by an order of magnitude) [276]; together with the better utilization of the transition metal, this explains the high productivity of third-generation catalysts.

$MgCl_2$ exhibitits the highest activating effect but other salts have some effect on the activity of ZN catalysts too, e. g. in the system $(C_{17}H_{35}COO)_3$-Cr/Et_2AlCl/metal chloride, activity decreases in the order [277] $MgCl_2 > MnCl_2 > CoCl_2 > 0 > AlCl_3 \approx NbCl_5$.

4.1.2. COCATALYSTS OF ZN CATALYSTS

Probably all accessible organometals, metaloid compounds, and many other substances have been tested as cocatalysts. A cocatalyst must fulfil a number of requirements:

(i) it must be able to alkylate the transition metal atoms (when the starting compound of this metal does not already contain a suitable metal–carbon or metal–H bond);

(ii) it must maintain the transition metal oxidation state preferentially at the most suitable level. For example, propene and higher alkenes, or dienes only polymerize on Ti^{3+}–C, ethylene also on Ti^{4+}–C and Ti^{2+}–C [278]. Therefore in copolymerizations of ethylene with higher alkenes, the copolymer is only formed on Ti^{3+}–C centres; in Phillips catalysts, Cr^{3+} is more active than Cr^{2+};

(iii) the cocatalyst should not react with donors to modify the properties of active centres, or at least it should yield modifying or inert products;

(iv) it should protect the active centres from deactivation by impurities; and

(v) the transition metal complex with the cocatalyst must exhibit some degree of basicity. More acid centres tend to transfer to monomer and are more active (this may be of advantage), but they have a lower tendency to produce a tactic polymer [280].

The requirements (iii) and (iv) are evidently in conflict and also the requirements (ii) and (iv) may be antibatic. ZN catalysts are therefore very sensitive to the component ratio, to impurities, and to temperature, not to speak of monomer type.

(a) *Aluminoxanes*

Homogeneous catalysts have significant advantages compared with heterogenous catalysts concerning both theoretical studies and technological applications. Reports on the application of cyclic and linear poly (aluminoxanes)

$$[\overset{\frown}{Al-O}]_n \quad\quad R+\underset{R}{Al}-O+_n AlR_2$$
$$|$$
$$R$$

represent an important contribution in the field of ZN catalysts. They were first used in connection with the soluble titanium component Cp_2TiRCl [281] and more recently with other transition metals (Hf), but especially with Zr. These systems are extremely active, able to produce $> 5 \times 10^6$ g of polyethylene per 1 g of Zr per h [282]. Methylaluminoxane cocatalysts are more active than ethylaluminoxanes. They induce polymerization not only of ethylene but also of propene as well as the copolymerization of higher 1-alkenes. Shortly after discovery they would only yield atactic polypropylene [283, 284] but later, with the centres modified by a Zr ligand using racemic ethylene-bis (tetrahydroindenyl) zirconium dichloride, even isotactic, optically active polypropylene and polybutene could be obtained.

4.1.3 INTERNAL AND EXTERNAL DONORS

For ethylene polymerization, the iso- (syndio-) specific catalytic activity is not needed. For higher 1-alkene polymerizations it is, of course, indispensable; crystalline polymers are in greater demand. For some time, third-generation catalysts could not be widely used because of their insufficient ability to induce tactic chain propagation. This drawback was overcome by the revamped discovery of the donor effect of Lewis bases [285].

According to our present somewhat simplified ideas, the isotactic centre has a single vacancy at which monomer insertion proceeds at a very high rate. An atactic centre has two vacancies, and it is slower by about two orders. A small amount of suitable donors [of which ethyl benzoate, its derivatives or dicarboxylic acid esters (dibutylphtalate etc.) appear to be important] results in preferential coordination of the donor to the free vacancy of the more acid (atactic) centre, which is thus transformed to a more rapid isotactic one [280]. Therefore the catalyst with a suitable amount of donor may be enhanced, and the iso-specific effect of the catalyst is significantly improved. Another suggestion is that steric control is a function of the asymmetry around the metal–polymer bond in the active centre. According to this hypothesis, the Lewis base mainly hinders coordination of $TiCl_4$ to $MgCl_2$ (which would yield mainly non-stereospecific centres), being itself coordinated, together with $TiCl_4$, to $MgCl_2$ [286]. Lewis bases which react in this sense, usually during the preparation of the catalyst solid phase, are called internal donors.

External donors are introduced together with the organometallic cocatalyst. In the absence of external donor, the organoaluminium component (e. g. Et_3Al) rapidly reacts with the internal donor. The external donor reduces the rate of internal donor consumption and/or substitutes its loss [286]. Chemically, external donors are very similar to internal donors, and they may even be identical.

4.1.4 DUAL FUNCTION CATALYSTS (DIMERIZATION–COPOLYMERIZATION)

Branched polyethylene, prepared by radical polymerization at high pressures, has good processing properties and even some physical properties which are useful in practical applications. Its production, however, is more expensive than the low-pressure processes, yielding a highly crystalline polymer of high density. Branched polyethylene can also be prepared by a low-pressure process by copolymerization with higher 1-alkenes. In order to avoid additional purification, storage, dosage, and possibly even regeneration of the comonomer, a catalyst consisting of two mutually compatible components was developed. The first of these induces dimerization of ethylene to 1-butene at suitable rate and the other component (the polymerization catalyst proper) induces copolymerization of 1-butene with ethylene. Ethylene is dimerized on $(RO)_4Ti$ with Et_3Al; the efficiency of the reaction decreases in the order $(iso\text{-}PrO)_4 Ti > (n\text{-}BuO)_4 Ti > (EtO)_4 Ti$.

The other component is $TiCl_4$ anchored on $MgCl_2$; once more with Et_3Al for copolymerization. This system produces polyethylene with 20–30 branches per 1000 C atoms. A change in the component ratio is accompanied by a systemmatic, predictable change in the properties of the resulting polyethylene [287].

4.1.5 REACTION OF ZIEGLER–NATTA CENTRES WITH MONOMERS

Present views concerning the operation mechanism of ZN catalysts are not conclusive. Cossee [288, 289] assumes that, in the first step, donor–acceptor interaction occurs between the transition metal and the monomer. A σ bond is formed by the overlap of the monomer π orbital with the $d_{x^2y^2}$ orbital of the transition metal. A second π bond is formed by reverse (retrodative) donation of electrons from the d_{yz} orbital of the transition metal into the antibonding π^* orbital of the monomer. In the following phase, a four-centre transition complex is formed with subsequent monomer insertion into the metal–carbon bond. This, in principle, monometallic concept is criticized by the advocates of the necessary presence of a further metal in the active centre. According to them, the centre is bimetallic. Monometallic centres undoubtedly exist; on the other hand, technically important ZN catalysts are multicomponent systems in which each component has its specific and non-negligible function in active centre formation. The non-transition metal in these centres is their inherent component, and most probably the centre is bimetallic. Even present ideas concerning the structural difference in centres producing isotactic and atactic polymers are not united.

According to Cossee, the isotactic centre has one and the atactic two vacancies. According to Rodriguez, the organometal is coordinated into one of the two vacancies of the isotactic centre. The addition of donors (diethyl ether, triethylamine, pyridine, etc.) greatly affects the polymerization rate (catalyst activity) as well as the ratio of the stereoregular and atactic product components. Donors affect the structure of active centres and modify it.

According to present ideas, the initiation mechanism of diene polymerization on ZN centres does not substantially differ from that of alkene initiation. Experiments support the idea according to which the chain is propagated on the transition metal atom; it may be bound to the metal either by a σ, or by a π allylic bond

$$\bullet\!\!\leftarrow\!\!-\!\!-\!\!-\!\!\begin{array}{c}CH_2\\ \|\\ CH\\ \|\\ CH\\ |\\ R\end{array} \quad\rightleftharpoons\quad \bullet\!-\!CH_2\!-\!CH\!=\!CH\!-\!R$$

$$\pi \qquad\qquad\qquad\qquad \sigma \tag{74}$$

Transition metals from the left-hand side of the periodic table (Ti, V) prefer a σ bond, whereas the π allylic structure is preferred by metals from the right-hand side (Ni, Co). The existence of an equilibrium between the two bond types has also been assumed, with growth on the σ bond; the position of the equilibrium would be a function of the type of ligands on the transition metal atom. The presence of an organometal, such as Et_2AlCl, is not necessary for the formation of a highly stereoregular product. If present, the organometal may affect the kinetics and stereospecific action of the catalyst, either by alkylation of the transition metal (this is probable with Ti, V, and Cr), or by its reduction (Ni and Co); this makes possible the oxidative addition of diene, yielding an active centre with a π allylic bond [267d].

All the findings collected to date do not enable us to formulate a detailed initiation mechanism for the polymerization of alkenes and dienes on ZN centres. The initiation and growth mechanisms are usually combined in a single scheme which will be discussed in detail in Chap. 5, Sect. 4.

4.2 Transition metal catalysts active in the absence of alkylmetals

Many transition metals and their compounds with organic ligands initiate the polymerization of alkenes and/or dienes. Some of them do not need any special treatment to this end while others require the presence of some organic or mineral compound or a special physical modification. In contrast to ZN catalysts, they are active without an organometal of Groups I–III. They are commonly known as metal alkyl free (MAF) catalysts. Many of their features are, of course, in common with ZN catalysts. MAF catalysts initiate stereoselectively controlled polymerization. Even less is known of their operating mechanism than that of ZN catalysts. It is assumed that propagation also occurs on the transition metal-carbon bond.

The Philips catalyst [290] is probably the most important. It is still used for producing over 60% of the world's polyethylene. The most frequently used kind is prepared by impregnating a support (silica, silicoalumina, etc. with a surface area $> 200 \, m^2 \, g^{-1}$) with Cr compounds, drying and activating with a stream of dry air (with addition of CO) at 700–1100 K. Chromates are assumed to be formed, but only a small part forms active centres

$$\begin{array}{c} \text{OH} \quad \text{OH} \\ | \quad\quad | \\ -Si-O-Si- \end{array} + CrO_3 \longrightarrow \begin{array}{c} ^-O \quad\quad O^- \\ \diagdown \quad \diagup \\ Cr^{2+} \\ \diagup \quad \diagdown \\ O \quad\quad O \\ | \quad\quad | \\ -Si-O-Si- \end{array} + H_2O \tag{75}$$

The catalyst contains Cr in several oxidation states. The mechanism of the Cr—C bond formation has not been safely explained so far; these bonds in particular, together with their immediate vicinity, are assumed to be the active centres.

There exists a large number of MAF catalysts and only their general features can be briefly mentioned here. When the cocatalyst (alkylmetal) is omitted in a typical ZN system, the remaining transition metal salt is rarely active. Some transition metals, especially Ti, V, Cr, Co, Ni, Zr, Nb, Mo, W, Pd, Rh, and Ru, are however, active, often after special treatment. Generally alkenes are more readily polymerized by transition metals from the left-hand side of the periodic table and dienes by metals from the right-hand side.

Exceptions do exist. In active catalysts the ligand is a halogen or a group bound to the metal through a heteroatom, or a hydrocarbon group connected to the metal by a σ or π allylic bond. Sometimes the transition metal carries two ligands. The oxidation state of the metal in the active centre is no longer a matter of discussion, and the optimum oxidation level of transition metals in these systems is more or less known. Myers and Lunsford have found that, in the supported catalyst Cr^{3+}/SiO_2 during ethylene polymerization, chromium in the active centre is in the oxidation state 3. Reduction of Cr^{3+} to Cr^{2+} is accompanied by loss of activity [291]. Both soluble and insoluble forms of MAF are known. An additional component often enhances activity and

TABLE 8

Composition of some transition metal salts initiating polymerization of alkenes or dienes [267e]

Salt	Monomer
$TiCl_2$	Ethylene, propene
$(\pi\text{-Allyl})_3ZrBr$	Butadiene
$(\pi\text{-Allyl})_2CoI$	Butadiene
$(Ph_3SiO)_2CrO_2$	Ethylene
$Pd(CN)_2$	Propene
NiO on support	Ethylene
$(\pi\text{-Allyl})Cr$	Ethylene, butadiene
$(\pi\text{-Crotyl})NiCl$	Butadiene
$(\pi\text{-Methallyl})NiCl$	Butadiene
$(\pi\text{-2,6,10-Dodecatriene})NiCl$	Butadiene
cp_2Cr	Ethylene
Ph_2Cr	Ethylene
$PhCH_2)_4Cr$	Ethylene, propene
cp_4Ni	Ethylene

changes the stereoregulating activity of the catalyst. Their properties can often be affected by physical means, for example by milling. They can also be modified by anchoring on a mineral or organic carrier. Of the monomers polymerizing on this type of catalysts mainly ethylene, propene, butadiene, but also acetylene, cyclobutene and other cycloalkenes, butenes and isoprene should be named. The composition of some MAF catalysts is illustrated in Table 8.

5. Initiating donor–acceptor complexes

Mutual donor–acceptor interaction is affected by the changes of electron density on the atoms and bonds of interacting particles. The formation of a donor–acceptor bond between molecules is caused by electron transfer from the highest occupied orbital of the donor to the lowest unoccupied orbital of the acceptor molecule. The properties and structure of the complex depend (inter alia) on the type of the participating orbitals. Usually the complex is of zwitterion character, easily decomposing under the influence of external forces to sets of atoms with properties differing from those of the original donor or acceptor. These zwitterions or their decomposition products can initiate polymerizations.

5.1 Formation of radical ions and their reaction with monomers

With a donor molecule of low ionization energy and an acceptor of high electron affinity, electron transfer may occur even in the ground state of the complex. When the differences in electron affinity are not pronounced, a structure without bond predominates in the ground state. Electron transfer can be considerably boosted by excitation, by a photon for example. A strong polarization of the donor–acceptor complex occurs [292], leading to its dissociation to radical ions in a solvent of suitable polarity. When a convenient monomer (e. g. N-vinylcarbazole [293], styrene [294], α-methylstyrene [295], etc.) acts as donor, it can be polymerized by the generated radical ions.

Absorption of a quantum by a DA complex in the ground state is not the only way of generating an excited donor–acceptor complex (called exciplex). According to Weller [296] and Ledwith [297] in can also be formed by interaction of an excited component of the complex with a non-excited

References pp. 154–162

component, even in systems where complex formation in the ground state could not be proved.

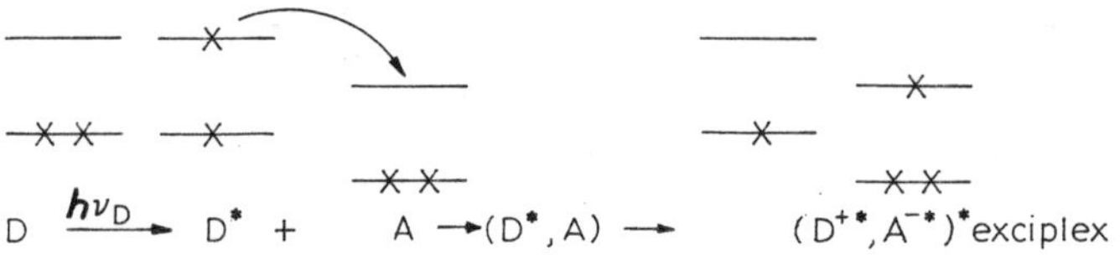

$$D \xrightarrow{h\nu_D} D^* + A \rightarrow (D^*, A) \rightarrow (D^{+*}, A^{-*})^* \text{exciplex} \tag{76}$$

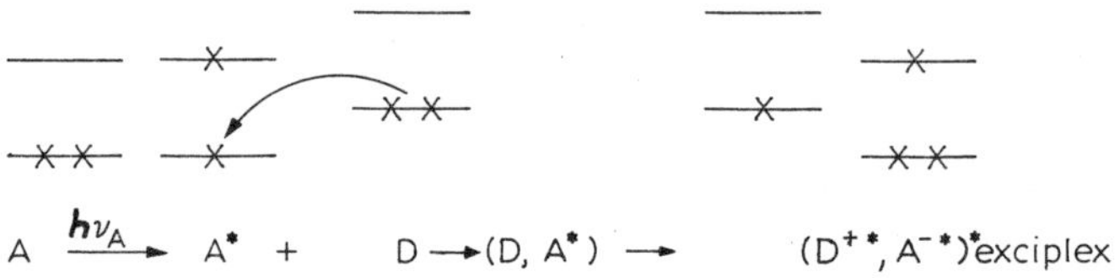

$$A \xrightarrow{h\nu_A} A^* + D \rightarrow (D, A^*) \rightarrow (D^{+*}, A^{-*})^* \text{exciplex} \tag{77}$$

The energy of the highest occupied orbital is characterized by its ionization energy, I_D, and the energy of the lowest unoccupied molecular orbital by its electron affinity E_A. The energy necessary for ionization from the donor and acceptor ground states to D^{+*} and A^{-*} is $I_D - E_A$. When starting from the excited states A^* and D^*, the ionization energy is reduced by $-h\nu$. At the equilibrium distance of the separated radical ions D^{+*} and A^{-*}, the total exciplex energy is reduced by the electrostatic contribution $-e^2/r$, resulting in the relation [298]

$$\Delta E = I_D - E_A - h\nu - \frac{e^2}{r} \tag{78}$$

A molecule in the excited state is a better electron donor and a better electron acceptor than an unexcited molecule [292, 298]. The occurrence of exciplexes generated by donor–acceptor interaction of a photoexcited component is frequent. The exciplex thus formed is not necessarily identical to the exciplex formed by excitation of DA complex from its ground state [293]. The relations between DA complexes, excited DA complexes, and exciplexes (from the same components) are illustrated in the scheme [297]

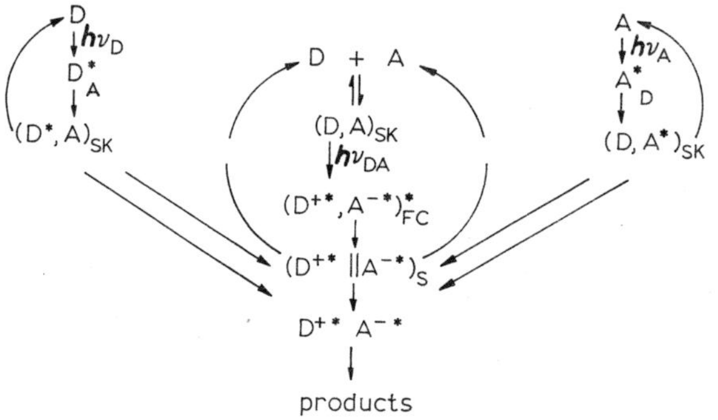

$$
\begin{pmatrix}
D \\
\downarrow h\nu_D \\
D^* \\
\downarrow A \\
(D^*, A)_{SK}
\end{pmatrix}
\quad
\begin{pmatrix}
D + A \\
\updownarrow \\
(D, A)_{SK} \\
\downarrow h\nu_{DA} \\
(D^{+*}, A^{-*})^*_{FC} \\
\downarrow \\
(D^{+*} \| A^{-*})_S \\
\downarrow \\
D^{+*} \quad A^{-*} \\
\downarrow \\
\text{products}
\end{pmatrix}
\quad
\begin{pmatrix}
A \\
\downarrow h\nu_A \\
A^* \\
\downarrow D \\
(D, A^*)_{SK}
\end{pmatrix}
\tag{79}
$$

Here $(D^*, A)_{SK}$, $(D, A^*)_{SK}$ and $(D, A)_{SK}$ are used to designate a collision donor–acceptor complex, $(D^{+*}, A^{-*})_{FC}$ the photoexcited donor–acceptor complex (of the Franck–Condon type), and $(D^{+*} \parallel A^{-*})_S$ a radical ion pair where both components are separated by solvent. The frequency v, of the exciting radiation is generally different for the donors, D, complexes, DA, and acceptors, A. Each step in scheme (79) is probably reversible, and the position and rate equilibruim establishment (of which the solvation and desolvation processes are the slowest) determine the rate and effectiveness of product formation. A review of donor–acceptor interactions and of the properties of DA bonds was published, for example, in refs. [299 and 300] and in the original communication [301].

A specific, greatly simplified form of the general scheme (79) is the initiation of 2-methylpropene polymerization [302]. It occurs in the presence of VCl_4 (Ti halogenides, etc.) and photons. The donor–acceptor complex of 2-methylpropene with VCl_4 forms an intermediate

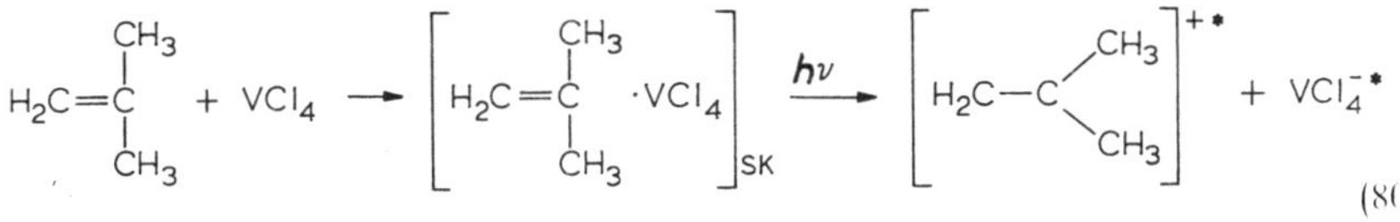

$$H_2C=\underset{CH_3}{\overset{CH_3}{C}} + VCl_4 \longrightarrow \left[H_2C=\underset{CH_3}{\overset{CH_3}{C}} \cdot VCl_4 \right]_{SK} \xrightarrow{h\nu} \left[H_2C-C\underset{CH_3}{\overset{CH_3}{\diagup}} \right]^{+\bullet} + VCl_4^{-\bullet}$$

(80)

Polymerization is initiated by the radical cations

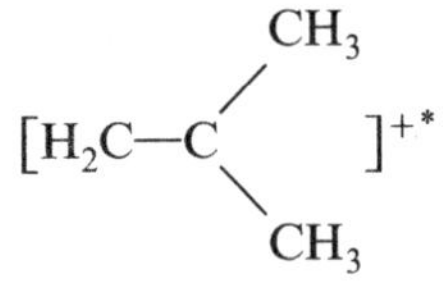

$$\left[H_2C-C\underset{CH_3}{\overset{CH_3}{\diagup}} \right]^{+*}$$

or by dications formed by their combination

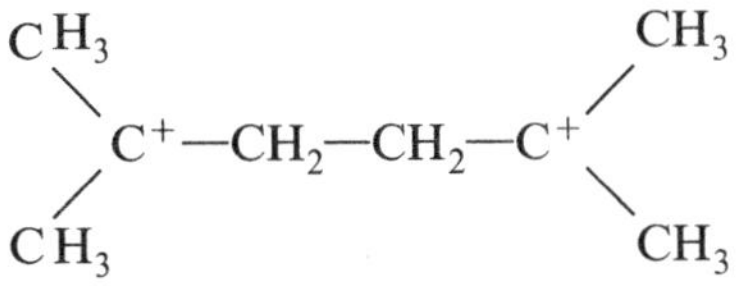

$$\underset{CH_3}{\overset{CH_3}{\diagdown}}C^+-CH_2-CH_2-C^+\underset{CH_3}{\overset{CH_3}{\diagup}}$$

5.2 Polymerization initiated by thermally activated donor–acceptor complexes

Complexes of monomers with acceptors can initiate polymerizations of the respective monomers even without photoexcitation. The properties of DA complexes and the conditions of the thermal activation are determined by the type of donor (monomer), acceptor, and solvent [300, 303]. Ion radicals from the monomer, dipolar intermediates (zwitterions), as well as the relatively

References pp. 154–162

stable products of the monomer–acceptor reaction may act as initiating particles

$$D + A \rightleftharpoons [DA]_{CT} \underset{\text{deactivation}}{\overset{\text{activation}}{\rightleftharpoons}} [(D^{+*}A^{-*}) \rightleftharpoons D^{+*} + A^{-*}] \rightarrow$$

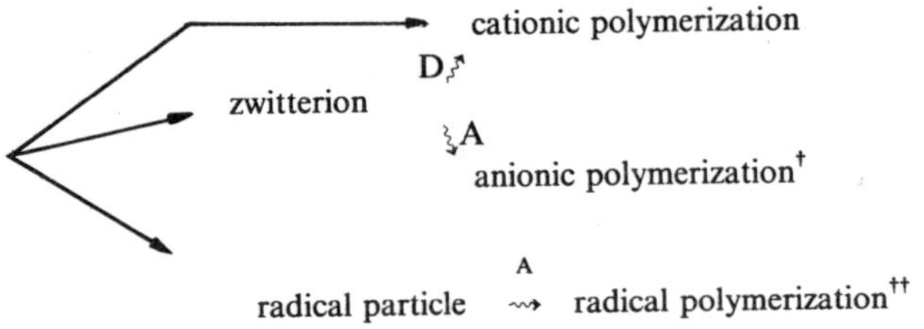

$$\begin{array}{l} \rightarrow \text{cationic polymerization} \\ \quad D^{+*} \\ \rightarrow \text{zwitterion} \\ \quad A^{-*} \\ \quad \text{anionic polymerization}^{\dagger} \\ \\ \rightarrow \text{radical particle} \overset{A}{\rightsquigarrow} \text{radical polymerization}^{\dagger\dagger} \end{array} \qquad (81)$$

Many specific examples of reaction type (81) have been described, and their number has been increasing. In many of these, vinyl carbazole is the donor. Other monomers can also react in a similar way [304].

Hall and co-wokers [305] formulated a general scheme of initiating the polymerization of vinyl monomers [scheme (82a)] and cyclic ethers [scheme (82b)] using alkenes with electrophilic substituents (A is an electrophilic, D a nucleophilic substituent, and X a dissociable group):

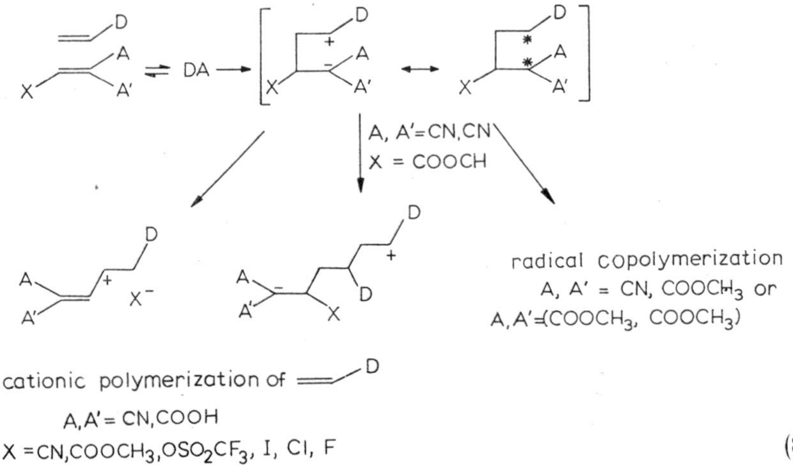

(82a)

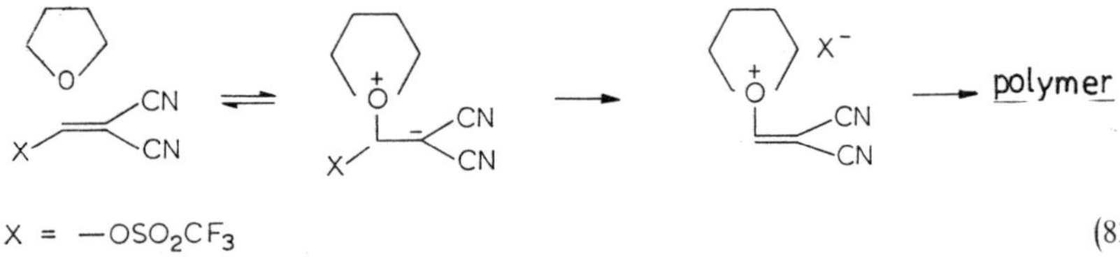

(82b)

The tendency of vinyl ethers to polymerization decreases with the lowering of the alkyls inductive effect, i. e. *tert.*-Bu > iso-Pr > Et ≈ *n*-Bu, iso-Bu. Polymerization only takes place in strongly polar solvents such as acetonitrile or nitromethane.

5.2.1 POLYCOMBINATION OF ZWITTERIONS

A special case of donor–acceptor interactions occurs when two monomers come into contact such that one exhibits a signicant donor and the other an acceptor character. The donor–acceptor complex then has the structure of a zwitterion resulting from the combination of radical ions after electron transfer from donor to acceptor

$$M_D + M_A \rightleftharpoons [M_D^{+*} M_A^{-*}] \rightleftharpoons {}^+M_D{-}M_A^- \qquad (83)$$

When the equilibrium (83) is shifted sufficiently to the right, zwitterion concentration in the medium is so high that the end groups of opposite charge react before the monomer has had a chance to become attached to a suitable ion. A strictly alternating copolymer is formed (see also Chap. 5, Sect. 5.8)

$${}^+M_D{-}M_A^- + {}^+M_D{-}M_A^- \rightarrow {}^+M_D{-}M_A{-}M_D{-}M_A^- \xrightarrow{n(+M_D{-}M_A^-)}$$

$$\rightarrow {}^+M_D{-}(M_A{-}M_D)_{n+1}{-}M_A^- \qquad (84)$$

In systems where the equilibrium is shifted less to the right of eqn. (83), monomer addition to a suitable end can also occur. As the addition rates of monomers M_A and M_D are generally different, the product need not have a strictly alternating structure.

These copolymerizations start spontaneously without the addition of initiator. They were discovered by Saegusa et al. [306]. Originally only a few monomer pairs were known which yielded zwitterions by way of DA complexes. Representatives of M_D are

cyclic iminoethers

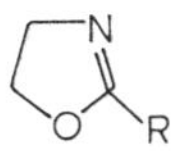

◀
―――

† When an anionically polymerizing monomer is the acceptor.
†† When a radically polymerizing monomer is the acceptor.

References pp. 154–162

cyclic exoiminoethers

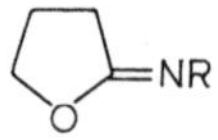

and azetidine

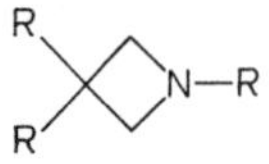

and examples of M_A are

β-propiolactone

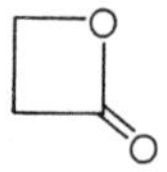

cyclic anhydrides

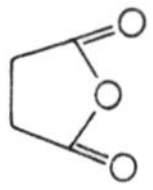

and sulpholactone

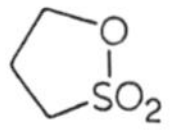

Their number is steadily increasing.

5.3 Monomers as ion pairs (monomer salts)

An interesting possibility of development in ionic polymerizations was described by Salamone et al. [307, 308]. An ion pair is formed from a pair of suitable monomers, probably by way of a donor–acceptor complex as intermediate. Both the cationic and the anionic components of the pair can polymerize individually

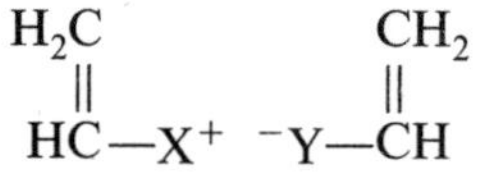

$$\begin{array}{cc} H_2C & CH_2 \\ \| & \| \\ HC-X^+ & ^-Y-CH \end{array}$$

Each vinyl particle acts as a counter-ion of the vinyl particle of opposite charge. These salts may be regarded as ionic active centres with a polymerizable counter-ion.

The authors have described the preparation of three salts of this type:

4-vinylpyridine with vinylsulphonic acid, 2-acrylamido-2-methyl-propanesulphonic acid and *p*-styrenesulphonic acid

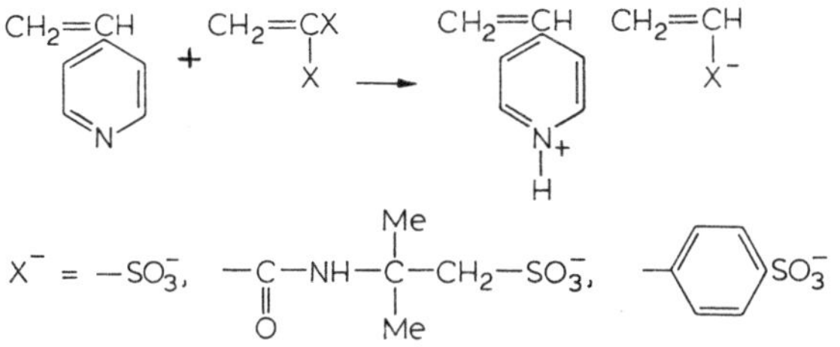

$$\underset{N}{\underset{|}{C_5H_4N}}-CH=CH_2 + CH_2=CX\!-\!X \longrightarrow \text{[pyridinium·X}^-\text{ adduct]}$$

$$X^- = -SO_3^-,\ -\underset{O}{\overset{\|}{C}}-NH-\underset{Me}{\overset{Me}{\underset{|}{C}}}-CH_2-SO_3^-,\ \text{–}\!\!\bigcirc\!\!\text{–}SO_3^-$$

Detailed data on this type of initiation are not yet available. The published idea is a stimulating challenge deserving additional investigation.

6. Other types of initiation

A description of all special and often very specific methods of initiation would be beynd the scope of this part of the volume. Nevertheless, some initiation methods of more general importance must be mentioned.

6.1 Electroinitiation

An electric field between two electrodes creates conditions suitable for initiation and/or modifications of other elementary polymerization steps.

Electroinitiated polymerization is induced by electron transitions between the electrodes and the molecules of the electrolyzed solution. The following classification is based on the features of the transfer reaction [309].

a) Direct electroinitiation; electron transfer occurs between the electrodes and the monomer, yielding active centres.

b) Indirect electroinitiation; electrons are transferred between the electrode and some component of the solution (other than the monomer), yielding a particle suitable for initiating polymerization.

Generally, it is difficult to prove direct electroinitiation experimentally due to the instability and low concentration of the radical ions formed from the monomer. When the medium between the electrodes contains only the monomer which undergoes polymerization at high field intensities, direct electroinitiation is clearly involved [310].

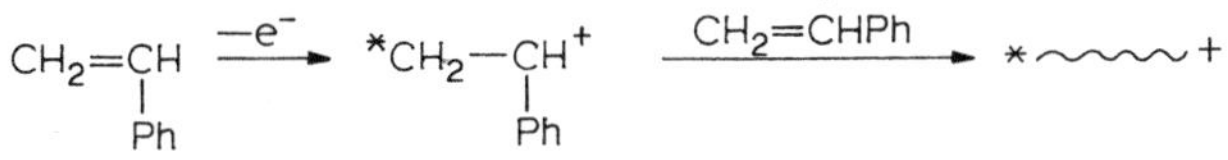

$$CH_2=\underset{Ph}{\underset{|}{CH}} \xrightarrow{-e^-} {}^*CH_2-\underset{Ph}{\underset{|}{CH}}{}^+ \xrightarrow{CH_2=CHPh} {}^*\!\!\sim\!\!\sim\!\!\sim\!\!+ \tag{85}$$

References pp. 154–162

When other compounds are present in addition to the monomer, the electron affinities and ionization energies of the components are important. When the reduction potential of the monomer is higher (more positive) than that of the cations from the supporting electrolyte and solvent, polymerization is started by direct monomer reduction yielding a radical anion. When the oxidation potential of the monomer is higher (more negative) than that of the anions from the supporting electrolyte and solvent, polymerization is started by direct monomer oxidation yielding a radical cation.

Indirect electroinitiation occurs by electrolysis of the supporting electrolyte in the presence of monomer when conditions for direct initiation are not fulfilled. Cerrai et al. [311] electrolyzed $R_4N^+I^-$ and $R_4N^+I_3^-$ in 1,2-dichloroethane in the presence of isobutyl vinyl ether. They found iodine formed by the oxidation of anions

$$2\,I^- \rightarrow I_2 + 2\,e^-$$

$$2\,I_3^- \rightarrow 3\,I_2 + 2\,e^-$$

The tetraalkylammonium cation is not reduced. The solvent is decomposed on the cathode, yielding H_2, ethylene, vinyl chloride, and carbanions. Similar processes also occur in the cathode space with the electrolyte $R_4N^+ClO_4^-$. $HClO_4$ is formed around the anode and the monomer is polymerized in the anode space; so far it is not known whether this is only by the effect of $HClO_4$ or whether cations from the supporting electrolyte are also involved.

Systematic studies of electrochemical initiation are just beginning [312–314]. It has not yet been applied industrially. Even at present, however, it can be very useful from the theoretical point of view.

6.2 Mechanochemical initiation

This type of initiation has been little studied, and has not been technically applied. Covalent bonds where mechanical stress is concentrated can split with radical formation by kneading [315], milling [316], vibration milling [317], vigorous stirring of polymer solutions, treatment of polymers and their solutions by ultrasound [318] or by freezing and melting of polymer emulsions [319, 320]. The generated radicals react with accessible atoms: with polymer C—H bonds, with oxygen, and with monomers (if present). Initiation of mechanochemically generated radicals leads to modification of the stressed polymer, most frequently to block or graft copolymers.

Initiating particles may also be formed by processes other than direct strain of covalent bonds. It appears that the mechanical energy transferred to the reactor walls by the milled particles in a vibrating mill leads to excitation of the metal atoms at the wall surface. Sorption of monomer (styrene) on the excited surface enables electrons to be transferred from metal to monomer, yielding an anion radical attached to the wall by its anionic end. The electron from the radical end can be transferred to a suitable acceptor which can initiate chain propagation [319].

6.3 Initiation in plasma

Plasma itself, as generated for example by glow or radiofrequency discharge, is a mixture of ions and radicals. It is not surprising that monomers polymerize under its effect. Plasma, however, is a non-specific source of poorly defined initiating particles, potentially able to induce electron defects even in stable molecules. Interaction of plasma ions and radicals with monomers carrying double bonds can start a process called plasma-initiated polymerization [see scheme (21)]. This method of initiation is being intensively studied, and is even beginning to penetrate into industrial applications [321]. In combination with plasma polymerization [see scheme (20)], it is employed for the application of various kinds of layers on the surface of glass for optical purposes or resist layers in electrical engineering, to catalyst supports, etc.

The energy of the plasma is sufficient to produce poorly defined branched or even cross-linked products. In the presence of polymerizing monomers, even N_2, H_2O, CO, CO_2, etc. can be incorporated in polymer chains [322].

7. Initiation of non-polymerizing compounds

Initiation can be most effectively studied in systems with the smallest number of elementary steps in the polymerization reaction. This requirement is fulfilled in living systems (without termination and transfer) with monomers which do not generate polymer chains. By eliminating propagation (or suppressing it to dimerization) kinetically relatively simple systems can be obtained which are suitable for the application of convenient analytical methods.

Schulz and Höcker [323] have studied the behaviour of compounds of the general formula

$$\text{Ph}-\underset{CH_2}{\overset{\|}{C}}-R-\underset{CH_2}{\overset{\|}{C}}-\text{Ph}$$

during contact with alkali metals. The composition of the product strongly depends on the type of the bridge R connecting the two styrene units. When

R permits conjugation (for example 1,4-phenylene), electron addition to the double bond leads to an increase in electron density even at the other double bond. The negative charge is distributed over the whole molecule which may be regarded as a more complicated allylic ion. Under these conditions electron transfer to the other double bond is strongly hindered; with Li it does not occur at all while with Na and K it takes place mainly only after dimerization of the monoradical anion.

When conjugation between the two double bonds is interrupted by a substituent R, electron transfer to the other double bond is only insignificantly hindered. Important data on donor–acceptor interactions between monomer and metal, and on the effects of R, solvents and temperature on electron transfer [scheme (86)] can be derived from the rate of generation and composition of the products.

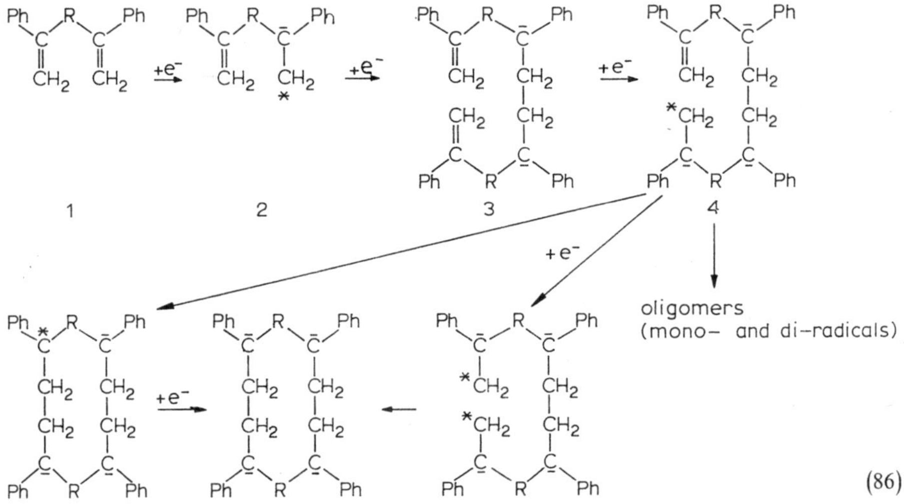

$$(86)$$

Lithium is a weak donor. It leads to the generation of only low radical concentrations where intramolecular combinations (cyclization) predominate. With increasing donor properties of the alkali metal, the generation of anion radicals is accelerated and their concentration increases. The probability of intermolecular reactions increases and more oligomers and polymers are formed.

In cationic polymerizations, the occurrence of living systems is limited. Therefore the search for a medium where only cationic initiation would take place is even more difficult. It is usually possible to exclude, or at least limit, propagation; transfer and termination can rarely be excluded. Even so, some simplification can be achieved, and it has been exploited.

Kennedy and Castner [324] have studied the reaction of tert.-BuCl with Me_2cpAl (cp = cyclopentadiene). They assumed the formation of the ion pair tert.-$Bu^+Me_2cpAlCl^-$, yielding either neopentane or tert.-Bu-cyclopentadiene. Alkylation of *tert.*-BuCl by Me_2cpAl thus models cationic initiation and termination without propagation. The fraction of *tert.*-Bu-cyclopentadiene increases with decreasing temperature.

The possibilities of studying the initiation of coordination polymerizations with the aid of non-polymerizing monomers are very limited. Monomer coordination to the transition metal atom actually determines the whole polymerization path. A modification of monomer structure (to prevent polymerization) changes its coordinating ability to such an extent that a completely different reaction would be studied instead of initiation.

8. Initiation kinetics

It is rarely possible to measure the rate of active centre formation directly. Sufficiently sensitive methods, which readily react to changes in centre concentration have seldom been developed to date. We have to rely mainly on indirect data, often based on assumptions. The simplest kinetic scheme of initiation consists of two reactions

$$I \underset{}{\overset{k_d}{\rightleftharpoons}} R° \tag{87}$$

$$R° + M \xrightarrow{k_i} M° \tag{88}$$

where I is the initiator, k_d the rate constant of initiator decomposition, $R°$ the primary initiating particle, M the monomer, k_i the rate constant of initiation, and $M°$ the active centre.

The solution for specific cases is greatly simplified when one of the reactions (87) or (88) is much slower than the other and thus controls the initiation rate. [In radical polymerizations, this is usually reaction (87).] We know, of course, that reaction (87) can be reversible, that $R°$ can decay by secondary decomposition to $R_1°$ (the reactivity of which generally differs from that of $R°$), and both reactions can only be a part of a much more complicated set of interactions, especially in ionic and coordination polymerizations. An exact kinetic analysis must be based on a proved scheme with identified intermediate transition states and products, and a knowledge of the rate constants and of the rates of various initiation stages. Such a complete and complex analysis does not yet exist.

In all branches of science it is customary to simplify processes which cannot be treated exactly to a manageable level at the price of (acceptably) lowered accuracy, while the requirement may become more stringent with the advancement of knowledge. In macromolecular chemistry, the rate of initiation has become a very useful piece of information. Disregarding the detailed history of centre formation, it can be determined experimentally under cer-

References pp. 154–162

tain circumstances, for example with the aid of inhibitor addition, by comparison of the rates of propagation and/or termination with the rate of monomer consumption in the initial polymerization stages, etc.

Even information on the relation of the initiation rate to those of propagation and termination is very useful. When initiation is very rapid, active centres will be formed practically immediately from all the initiator. In such a case we only have to determine their fate, i. e. if each of the centres outlives the whole polymerization process (this leads to a great simplification of the kinetics). If part or all of the centres decay, the situation is more complicated; the polymerization process is non-stationary. When initiation is slow, the further phase (propagation) may be stationary or non-stationary, depending on whether the termination rate is equal to or less than the initiation rate (at termination rate higher then initiation rates, long chains cannot be generated). With slow initiation, polymerization always starts slowly, and the rate of monomer consumption increases with time. Conversion curves exhibit induction periods. The length of the generated chains and its distribution is likewise a function of initiation rate and of the relative rates of all basic polymerization processes.

A more detailed discussion of initiation rates presumes a knowledge of termination and transfer and it has therefore been included in Chap. 8.

References

1. C. Walling, E. R. Briggs and F. R. Mayo, J. Am. Chem. Soc., 68 (1946) 1145.
2. P. J. Flory, J. Am. Chem. Soc., 59 (1937) 241.
2. B. H. Zimm and J. K. Bragg, J. Polym. Sci., 9 (1952) 476.
4. N. J. Barr, W. I. Bengough, G. Beveridge and G. P. Park, Eur. Polym. J., 14 (1978) 245.
5. F. R. Mayo, J. Am. Chem. Soc., 90 (1968) 1289.
6. T. Sato, M. Abe and T. Otsu, Makromol. Chem., 178 (1977) 1061.
7. H. F. Kauffmann, Makromol. Chem., 178 (1977) 3007.
8. O. F. Olaj, H. F. Kauffmann and J. W. Breitenbach, Makromol. Chem., 178 (1977) 2707.
9. H. F. Kauffmann, O. F. Olaj and J. W. Breitenbach, Makromol. Chem., 177 (1976) 939, 3065.
10. J. Hara, N. Teramae, J. Saito and S. Tanaka, Kobunshi Ronbonshu, 41(8) (1984) 453; Chem. Abstr., 101 (1984) 152416g.
11. K. Ito, Polym. J. (Tokyo), 18(11) (1986) 877.
12. C. Aso, T. Kunitake and H. Miyazaki, J. Polym. Sci. Part A-1, (7) (1969) 1497.
13. C. Walling and E. R. Briggs, J. Am. Chem. Soc., 68 (1946) 1141.
14. J. Lingnau, G. Meyerhoff and M. Stickler, in Lüderwald and R. Weis (Eds.), Makro Mainz, Preprints of Short Communications, Vol. 1, 1979, p. 86.
15. L. M. Stephenson and J. I. Brauman, J. Am. Chem. Soc., 93 (1971) 1988.
16. J. Lingnau and G. Meyerhoff, Makromol. Chem., 185(3) (1984) 587.
17. K. S. Bagdasarjan, Theory of Free-Radical-Polymerization, Jerusalem, 1968.

18 J. W. Breitenbach and W. Thury, Experientia, 3 (1947) 281. C. Cuthbertson, G. Gee and E. K. Rideal, Nature (London), 140 (1973) 889. J. W. Breitenbach and R. Raff, Chem. Ber., 69 (1936) 1107.
19 K. C. Berger, P. C. Deb and G. Mayerhoff, Macromolecules, 10 (1977) 1075.
20 W. A. Pryor and J. H. Coco, Macromolecules, 3 (1970) 500.
21 M. M. Koton, T. M. Kiseleva and M. I. Bessenov, Dokl. Akad. Nauk SSSR, 96 (1954) 85.
22 P. C. Deb and I. D. Gaba, Makromol. Chem., 179 (1978) 1549.
23 J. C. Bevington, Radical Polymerization, Academic Press, London, 1961.
24 K. C. Berger, Makromol. Chem., 176 (1975) 3575.
25 P. C. Deb and S. Ray, Eur. Polym. J., 13 (1977) 1015.
26 C. G. Swain, W. H. Stockmayer and J. T. Clarke, J. Am. Chem. Soc., 72 (1950) 5426.
27 G. A. Razuvaev, J. A. Oldekon and E. I. Fedotova, Usp. Khim., 21 (1952) 378.
28 G. Luft, H. Bitsch and H. Seidl, J. Macromol. Sci. Chem., A11 (1977) 1089.
29 See, for example, R. A. Wolf and W. Rozich, Tetrahedron Lett., 14 (1975) 1273 and papers cited therein.
30 R. Hiatt, in D. Swern, (Ed.), Organic Peroxides, Vol. 2, Wiley, New York, London, 1971.
31 E. Rizzardo and D. H. Solomon, J. Macromol. Sci., Chem., A11(9) (1977) 1697.
32 N. C. Deno, W. E. Billups, K. E. Kramer and R. R. Lastomirsky, J. Org. Chem., 35 (1970) 3080.
33 M. S. Kharash and A. Fono, J. Org. Chem., 24 (1959) 606.
34 J. K. Kochi and P. E. Mocadlo, J. Org. Chem., 30 (1965) 1134.
35 J. Bartoň, D. Sc. Thesis, SAV, Bratislava, 1980.
36 J. Bartoň and E. Borsig, Complexes in Free-Radical Polymerizaation, Elsevier, Amsterdam, 1988, pp. 91–100.
37 N. Uri, Chem. Rev., 50 (1952) 375.
38 W. Kern, Angew. Chem., 61 (1949) 471.
39 S. I. Trubitsina, I. Ismailov and M. A. Askarov, Vysokomol. Soedin. Ser. A, 19 (1977) 495.
40 W. G. Barb, J. H. Baxendale, P. George and K. R. Hargrave, Trans. Faraday Soc., 47 (1951) 462.
41 F. Rodriguez and R. D. Givey, J. Polym. Sci., 55 (1961) 713.
42 S. G. Palit and T. Guha, J. Polym. Sci., 34 (1959) 243.
43 E. S. Roskin, Zhur. Prikl. Khim., 30 (1957) 1030.
44 P. L. Nayak, R. K. Samal and N. Baral, J. Macromol. Sci. Chem., A11(5) (1977) 1971. P. L. Nayak and R. K. Samal, J. Polym. Sci. Polym. Chem. Ed., 15 (1977) 2603 and papers cited therein.
45 M. M. Husain and A. Gupta, J. Macromol. Sci. Chem., A11(12) (1977) 2177.
46 P. L. Nayak, R. K. Samal and M. C. Nayak, Eur. Polym. J., 14 (1978) 287.
47 B. Boutevin and Y. Pietrasanta, Makromol. Chem., 186(4) (1985) 817.
48 W. Cooper, Chem. Ind., (1953) 407.
49 C. C. Menon and S. L. Rapor, J. Polym. Sci., 54 (1961) 45.
50 T. J. Sven, Y. Jen and J. Lockwood, J. Polym. Sci., 31 (1958) 481.
51 D. D. Dash, T. R. Mohanty and P. L. Nayak, J. Macromol. Sci. Chem., A11(11) (1977) 2029.
52 J. Thiele and K. Heuser, Annalen, 290 (1896) 1.
53 A. F. Bickel and W. A. Waters, Rec. Trav. Chim., 69 (1950) 1940.
54 M. Talât–Erben and S. Bywater, J. Am. Chem. Soc., 77 (1955) 3710, 3712.
55 M. J. Botnikov, V. M. Zhulin, L. G. Bubnová and. G. A. Stashina, Vysokomol. Soedin., 19 (1977) 229.
56 M. B. Lakhinov, V. P. Zubov and V. A. Kabanov, J. Polym. Sci. Polym. Chem. Ed., 15 (1977) 1777.

57 B. L. Funt and G. Pawelchak, J. Polym. Sci. Polym. Lett. Ed., 13 (1975) 451.
58 See J. Bartoň and E. Borsig, Complexes in Free-Radical Polymerization, Elsevier, Amsterdam, 1988, p. 129.
59 A. L. Logothetis and J. M. McKenna, J. Polym. Sci. Polym. Chem. Ed., 15 (1977) 1431.
60 A. Mozhe, I. Vizovishek and S. Lapanie, Makromol. Chem., 178 (1977) 3051.
61 J. C. Bevington and A. Wahid, Polymer, 4 (1963) 129.
62 See, for example, C. H. Bamford, W. G. Barb, A. D. Jenkins and P. F. Onyon, The Kinetics of Vinyl Polymerization by Radical Mechanisms, Butterworths, London, 1958.
63 C. H. Bamford, A. Ledwith and Y. Yagci, Polymer, 19 (1978) 354.
64 L. R. Gatechair and D. Wostratzky, Tech. Pap. Soc. Manuf. Eng. Ser. FC, (1982) FC 82–279.
65 O. H. Wheeler and C. B. Covarrubias, Can. J. Chem., 40 (1962) 1224.
66 H. W. Melville, Proc. Soc. (London) Ser. A, 237 (1956) 149.
67 S. S. Labana, J. Macromol. Sci. Rev. Macromol. Chem., C11 (1974) 299.
68 V. A. Krongauz, Teor. Eks. Khim. An SSSR, 1 (1965) 47.
69 K. Venkatarao and M. Santappa, J. Polym. Sci. Part A–1, (8) (1970) 3429.
70 K. Fujimoto, J. Tominaga and H. Okuno, Br. Pat. 1,290,020 (1972).
71 K. Matsuo, G. W. Nelb, R. G. Nelb and W. H. Stockmayer, Macromolecules, 10 (1977) 654.
72 Q. Anwaruddin and M. Santappa, J. Polym. Sci. Part. A–1, (7) (1969) 1315.
73 W. Moore, Physical Chemistry, Prentice-Hall, Englewood Cliffs, NJ, 4th edn. 1972.
74 M. R. Sandner, C. L. Osborn and D. J. Trecker, J. Polym. Sci Part. A–1, (10) (1972) 3173.
75 H. Miyama, H. Harumiya and A. Takeda, J. Polym. Sci. Part. A–1, (10) (1972) 943.
76 K. Tsuda and K. Kosegaki, Makromol. Chem. 161 (1972) 267.
77 E. A. Lissi and A. Zanocco, J. Polym. Sci. Polym. Chem. Ed., 21 (1983) 2197.
78 H. J. Timpe, V. Schikowsky and R. Vergara, Acta Polym., 35(3) (1984) 208.
79 Y. Minoura and T. Toshima, J. Polym. Part. A–1, (8) (1970) 273.
80 S. E. Morsi, A. B. Zaki and M. A. El-Khyami, Eur. Polym. J., 13 (1977) 851.
81 S. Tazuke and S. Okumara, J. Polym. Sci. Part A–1, (7) (1969) 715.
82 T. Kodaira and K. Hayashi, J. Polym. Sci. Part B,9, (1971) 907.
83 T. Kodaira, K. Hayashi and T. Ohnishi, Polym. J., 4 (1973) 1.
84 N. Gaylord, S. S. Dixit and B. K. Patnaik, J. Polym. Sci. Part B, 9 (1971) 927.
85 P. Ghosh and P. S. Mitra, J. Polym. Sci. Polym. Chem. Ed., (a) 13 (1975) 921; (b) 15 (1977) 1743.
86 J. A. Barter and D. E. Kellar, J. Polym. Sci. Polym. Chem. Ed., 15 (1977) 2545.
87 M. Ravey and J. A. Waterman, J. Polym. Sci. Polym. Chem. Ed., 15 (1977) 2521.
88 E. S. Smith, Goodyear Tire Rubber Co., U. S. Pat. 3,022,281, (1962).
89 F. A. Bovey and J. M. Kolthoff, J. Am. Chem. Soc., 69 (1947) 2143.
90 C. E. Barnes, J. Am. Chem. Soc., 67 (1945) 217.
91 D. A. Seanor, in A. D. Jenkins (Ed.), Polymer Science, Vol. 2, North-Holland, 1972, p. 1247.
92 S. Hussain, S. D. Baruah and N. N. Dass, J. Polym. Sci. Polym. Lett. Ed., 16 (1978) 167.
93 S. M. Mukherjee and S. P. Singh, Reaction Mechanism in Organic Chemistry, Macmillan Company of India, 1975, p. 55.
94 Y. Osada, A. T. Bell and M. Shen, Polym. Prepr. Am. Chem. Soc. Div. Polym. Chem., 19(1) (1978) 639.
95 L. C. Brown and A. T. Bell, Ind. Eng. Chem. Fundam., 13 (1974) 210.
96 H. Yasuda and T. Hirotsu, J. Polym. Sci. Polym. Chem. Ed., 15 (1977) 2749.
97 G. K. Vinogradov, Khim. Vys. Energ., 20(3) (1986) 195.
98 Y. Osada, M. Takase and Y. Iriyama, Polym. J. (Tokyo), 15(1) (1983) 81.

99. A. I. Bolshakov, A. I. Mikhailov and I. M. Barkalov, Vysokomol. Soedin. Ser. A, 20 (1978) 1820.
100. S. N. Bhandani and Y. K. Prasad, Makromol. Chem., 178 (1977) 1841.
101. A. Casale and R. S. Porter, Adv. Polym. Sci., 17, (1975) 64.
102. K. L. Glazomitskii, Y. I. Poliakov, R. F. Smirnov, T. I. Yurchenkova, A. S. Chegolya and E. S. Roskin, Vysokomol. Soedin. Ser. A, 19 (1977) 2483.
103. V. I. Galubei, L. V. Dubnik, T. A. Tolnygina, A. B. Petrova and V. I. Sokolova, Vysokomol. Soedin. Ser. A, 19 (1977) 1321.
104. I. S. Tsvetkov and R. F. Markovskaya, Vysokomol. Soedin. Ser. A, 16 (1974) 1936.
105. G. S. Bylina, M. S. Matveentseva and Y. A. Oldekon, Vysokomol. Soedin. Ser. B, 19 (1977) 555.
106. I. Piirma and B. Gunesin, Polymer Prepr. Am. Chem. Soc. Div. Polym. Chem., 18(1) (1977) 687.
107. J. L. Laverty and Z. G. Gardlun, J. Polym. Sci. Polym. Chem. Ed., 15 (1977) 2001.
108. R. Kerber, O. Nuyken and R. Steinhausen, Makromol. Chem., 178 (1977) 1833.
109. P. Citovický, D. Mikulášová, V. Chrástová and G. Beňo, Eur. Polym. J., 13 (1977) 655.
110. J. Yamauchi, K. Ikemoto and A. Yamaoka, Makromol. Chem., 178 (1977) 2483.
111. R. M. Noyes Prog. React. Kinet. 1 (1961) 129.
112. P. E. M. Allen and C. R. Patrick, Kinetics and Mechanisms of Polymerization Reactions, Horwood, Chichester, 1974, (a) pp. 109–111; (b) pp. 112–113.
113. O. Ito, K. Nogami and M. Matsuda, J. Phys. Chem., 85 (1981) 1365.
114. S. F. Nelsen and P. D. Bartlett, J. Am. Chem. Soc., 88 (1967) 143.
115. J. C. Bevington, Trans. Faraday Soc., 51 (1955) 1392. J. C. Bevington and D. J. Stamper, Trans. Faraday Soc., 66 (1970) 688.
116. J. Betts, F. S. Dainton and K. J. Ivin, Trans. Faraday Soc., 58 (1982) 1203.
117. C. H. Bamford, A. D. Jenkins and R. Johnston, Trans. Faraday Soc., 58 (1962) 1212.
118. W. I. Bengough and W. H. Fairservice, Trans. Faraday Soc., 61 (1965) 1206.
119. W. I. Bengough and W. H. Fairservice, Trans. Faraday Soc., 67 (1971) 414.
120. J. W. Taylor and J. C. Martin, J. Am. Chem. Soc. 89 (1967) 6904.
121. O. Dobis, J. M. Pearson and M. Szwarc, J. Am. Chem. Soc., 90 (1968) 278.
122. K. Chakravorty, J. M. Pearson and M. Szwarc, J. Am. Chem. Soc., 90 (1968) 283.
123. H. Kiefer and T. G. Traylor, J. Am. Chem. Soc., 89 (1967) 6663.
124. R. M. Noyes and L. F. Meadows, J. Am. Chem Soc., 82 (1960) 1872.
125. P. I. Abell, in C. H. Bamford and C. F. H. Tipper (Eds.), Comprehensive Chemical Kinetics, Elsevier, Amsterdam, 1976, pp. 111–165.
126. J. A. Kerr and M. J. Parsonage, Evaluated Kinetic Data on Gas Phase Additions, Butterworths, London, 1972.
127. M. Miyoshi and R. K. Brinton, J. Chem. Phys., 36 (1962) 3019.
128. R. J. Cvetanovic and R. S. Irwin, J. Chem. Phys., 46 (1967) 1694.
129. D. G. L. James and D. McCallum, Can. J. Chem., 43 (1965) 633.
130. D. G. L. James and T. Ogawa, Can. J. Chem., 43 (1965) 640.
131. E. D. Norris, Jr., O. H. Stedman and H. Niki, J. Am. Chem. Soc., 93 (1971) 3570.
132. L. I. Avramenko, L. N. Evlashkima and R. V. Kolesnikova, Izv. Akad. Nauk. SSSR Ser. Khim., (1967) 259.
133. V. I. Galubei and Yu. K. Epimakhov, Ukr. Khim. Zh. (Russ. Ed.), 49(8) (1983) 871.
134. R. S. Mulliken, J. Chem. Phys., 2(1934) 272; 3 (1935) 573.
135. L. Pauling, J. Am. Chem. Soc., 54 (1930) 3570.
136. A. L. Allred, J. Inorg. Nucl. Chem., 17 (1961) 215.
137. N. B. Hannay and C. P. Smyth, J. Am. Chem. Soc., 68 (1946) 171.
138. L. Pauling, The Nature of the Chemical Bond, Cornell University Press, Ithaca, NY, 1940.

139 T. L. Brown, Adv. Organomet. Chem., 3 (1965) 365.
140 E. G. Rochov, D. T. Hurd and R. N. Lewis, The Chemistry of Organometallic Compounds, Wiley, New York, 1957, p. 70.
141 J. B. Conant and G. W. Wheland, J. Am. Chem. Soc., 54 (1932) 1212.
142 H. Gilman, A. L. Jacoby and H. Ludeman, J. Am. Chem. Soc., 60 (1938) 2336.
143 T. Mole and E. A. Jeffery, Organoaluminium Compounds, Elsevier, Amsterdam, 1972, (a) p. 2; (b) pp. 95–96.
144 H. L. Lewis and T. L. Brown, J. Am. Chem. Soc., 92 (1970) 4664.
145 S. Bywater and D. J. Worsfold, J. Organomet. Chem., 10 (1967) 1.
146 T. L. Brown, Acc. Chem. Res., 1 (1968) 23.
147 R. N. Young, R. P. Quirk and L. J. Fetters, Advances in Polymer Science, Vol. 56, Springer-Verlag, Berlin, Heidelberg, New York, Tokyo, 1984, p. 7.
148 E. N. Kropacheva, B. A. Dolgoplosk and E. M. Kuznetsova, Dokl. Akad. Nauk SSSR, 130 (1960) 1253.
149 M. Morton, R. A. Pett and L. J. Fetters, Macromolecules, 3 (1970) 333.
150 R. A. H. Cassling, A. G. Evans and N. H. Rees, J. Chem. Soc. B, (1966) 519.
151 S. Bywater, in C. H. Banford, C. F. H. Tipper (Eds.), Comprehensive Chemical Kinetics, Vol. 15, 1976, (a) p. 11; (b) p. 15; (c) p. 28; (d) pp. 40–46.
152 J. E. L. Roovers and S. Bywater, Macromolecules, 1 (1968) 328.
153 H. L. Hsieh, J. Polym. Sci. Part A–1, (8) (1970) 533.
154 I. Kmínek, Thesis, Institute of Macromolecular Chemistry, Czechoslovak Academy of Science, Prague, 1980.
155 I. Kmínek, M. Kašpar and J. Trekoval, Collect. Czech. Chem. Commun., 46 (1981) 1124.
156 I. Kmínek, M. Kašpar and J. Trekoval, Collect. Czech. Chem. Commun., 46 (1981) 2317.
157 R. Waack, M.–A. Doran: J. Am. Chem. Soc. 91 (1969) 2456.
158 J. Podešva, P. Špaček, A. Sikora and A. F. Podoľskii, J. Polym. Sci. Polym. Chem. Ed. 22 (1984) 3343.
159 J. Furukawa, T. Tsuruta, A. Fujita and A. Kawasaki, J. Chem. Soc. Jpn. Ind. Chem. Soc. 63 (1960) 645.
160 B. L. Erusalimskii, I. G. Krasnoselskaya and I. V. Kulevskaya, Russ. Chem. Rev., 37 (1968) 874.
161 V. Jíšová, M. Kolínský and D. Lím, J. Polym. Sci. Part A–1 (8) (1970) 1525.
162 G. A. Russell and D. W. Lamson, J. Am. Chem.Soc., 91 (1969) 3967. J. F. Garst, R. H. Cox, J. T. Barbas, R. D. Roberts, J. I. Morris and R. C. Morrison, J. Am. Chem. Soc., 92 (1970) 5761.
163 B. J. Cottam, D. M. Wiles and S. Bywater, Can. J. Chem., 41 (1963) 1905.
164 F. H. Owens, W. L. Myers and F. E. Zimmerman, J. Org. Chem, 26 (1961) 2288.
165 N. Kawabata and T. Tsuruta, Makromol. Chem., 86 (1965) 231.
166 A. A. Korotkov, S. P. Mitsengendler and V. N. Krasulina, J. Polym. Sci., 53 (1961) 217 and previous papers.
167 J. Trekoval, D. Sc. Thesis, Institute of Macromolecular Chemistry, Czechoslovak Academy of Science, Prague, 1978.
168 D. L. Glusker, I. Lysloff and E. Stiles, J. Polym. Sci., 49 (1961) 315.
169 W. E. Goode, F. H. Owens and W. L. Myers, J. Polym. Sci., 47 (1960) 75.
170 H. Schreiber, Macromol. Chem., 36 (1959) 86.
171 P. E. M. Allen, C. Mair and E. H. Williams, Eur. Polym. J., 20(2) (1984) 119.
172 Z. K. Cao, K. Ute, T. Kitayama, Y. Okamoto and K. Hatada, Kobunshi Ronbunshu, 43(7) (1986) 435.
173 H. K. Reimschuessel, in K. C. Frisch and S. L. Reegen, (Eds.), Ring Opening Polymerization, Dekker, New York, 1969, p. 314.

174 S. Tsuchiya and T. Tsuruta, Makromol. Chem., 110 (1967) 123.
175 R. P. Quirk and N. S. Seung, Polym. Prepr. Am. Chem. Soc. Div. Polym. Chem. 25(1) (1984) 206.
176 W. C. E. Higginson and N. S. Wooding, J. Chem. Soc., (1952) 760.
177 D. T. Hurd, R. C. Osthoff and M. L. Corrin, J. Am. Chem. Soc., 76 (1954) 249.
178 J. Trekoval and D. Lím, J. Polym. Sci. Part C, 4 (1964) 333.
179 J. Trekoval, D. Sc. Thesis, Institute of Macromolecular Chemistry, Czechoslovak Academy of Sciencees, Prague, 1978, pp. 139–157.
180 M. E. Carter, J. L. Nash, Jr., J. W. Drueke, Jr., J. W. Schwietert and G. Butler, J. Polym. Sci, Polym. Chem. Ed., 16 (1978) 937.
181 G. Camino, S. L. Lim and L. Trossarelli, Eur. Polym. J., 13 (1977) 479.
182 M. Szwarc, Carbanions, Living Polymers and Electron Transfer Processes, Wiley-Interscience, New York, 1968, (a) p. 298; (b) p. 328.
183 A. I. Shatenshtein, E. S. Petrov and E. A. Yakoleva, J. Polym. Sci. Part C, 3 (1976) 1729.
184 G. Heublein, Zum Ablauf ionischer Polymerisations-reaktionen, Akademie Verlag, Berlin, 1975, p. 193.
185 K. Ziegler, L. Jakob, H. Wollthan and A. Wenz, Liebigs Ann. Chem., 511 (1934) 34.
186 C. G. Overberger, E. Pearce and N. J. Mayes, J. Polym. Sci., 34 (1959) 109.
187 Z. Jedlinski, M. Kowalczuk and P. Kurcok, Makromol. Chem., Macromol. Symp., 3 (1986) 277.
188 D. E. Paul, D. Lipkin and S. I. Weissman, J. Am. Chem. Soc., 78 (1956) 116.
189 M. Morton, in G. E. Ham (Ed.), Vinyl Polymerization, Vol. I, Dekker, New York, 1969, pp. 211–229.
190 I. M. Panayotov, J. V. Berlinova and Kh. B. Tsvetanov, Eur. Polym. J., 7 (1971) 127.
191 M. Morton, A. Rembaum and E. E. Bostick, J. Polym. Sci., 32 (1958) 530.
192 A. Zilkha, P. Neta and M. Frankel, J. Chem. Soc., (1960) 3357.
193 G. Greber and G. Egle, Makromol. Chem., 92 (1966) 180.
194 I. M. Panayotov and Kh. B. Tsvetanov, Monatsh. Chem., 101 (1970) 1672.
195 L. Lochmann, J. Pospíšil and D. Lím, Tetrahedron Lett., 2 (1966) 257.
196 J. J. Bahl, R. B. Bates and B. Gordon III, J. Org. Chem., 44 (1979) 2290.
197 B. Gordon III, J. E. Loftus and J. Clark, Polym. Prepr. Am. Chem. Soc. Div. Polym. Chem. 27(1) (1986) 96.
198 L. M. Tolbert, J. A. Schomaker, Polym. Prepr. Am. Chem. Soc. Div. Polym. Chem., 27(2) (1986) 197.
199 B. Gordon III and J. E. Loftus, Polym. Prepr. Am. Chem. Soc. Div. Polym. Chem., 27(1) (1986) 353.
200 M. Kučera, Chem. Listy, 77 (1983) 1083.
201 M. Kučera, Chem. Listy, 78 (1984) 626.
202 T. Eichler, in S. Patai (Ed.), The Chemistry of the Carbonyl Group, Interscience, London, 1966, p. 621.
203 Y. Yasuda, N. Kawabata and T. Tsuruta, J. Org. Chem. 32 (1967) 1720.
204 G. Wittig and O. Bub, Annalen, 566 (1949) 113.
205 G. Wittig, Angew. Chem., 70 (1958) 65.
206 O. Wichterle, Organic chemistry, Vol. II, Přírodovědecké nakladatelství, Prague, 1952, p. 289.
207 W. J. Bailey and F. Marktscheffel, J. Org. Chem., 24 (1959) 874.
208 W. T. Grubb and R. C. Osthof, J. Am. Chem. Soc., 77 (1955) 1405.
209 B. Sh. Khaitin and L. A. Volf, Vysokomol. Soedin. Ser. A, 20 (1978) 1580.
210 S. W. Shalaby and H. K. Reimschuessel, J. Polym. Sci. Polym. Chem. Ed., 15 (1977) 1349.
211 J. Roda, M. Kusková and J. Králíček, Makromol. Chem., 178 (1977) 3203.

212 J. Jarrin, F. Dawans, E. Marechal and S. Boileau, J. Polym. Sci. Polym. Chem. Ed., 22 (1984) 2345.
213 D. Bremner, Chem. Br., 22 (1986) 194.
214 C. J. Pedersen, J. Am. Chem. Soc., 89 (1967) 7017; 92 (1970) 391.
215 J. Smid, in M. Szwarc (Ed.), Ion Pairs in Organic Reactions, Wiley, New York, 1972, p. 127.
216 J. M. Lehn, J. P. Sauvage and B. Dietrich, J. Am. Chem. Soc., 92 (1970) 2916.
217 S. Hubert, C. Momtaz, P. Hemery and S. Boileau, Polym. Prepr. Am. Chem. Soc. Div. Polym. Chem., 27(1) (1986) 134.
218 S. Boileau, B. Kaempf, J. M. Lehn and F. Schue, J. Polym. Sci. Polym. Lett. Ed., 12 (1974) 203
219 A. Deffieux and S. Boileau, Macromolecules, 9 (1976) 369.
220 S. Slomkovski and S. Penczek, Macromolecules, 9 (1976) 367.
221 H. Magnon, F. Rodriguez, M. Abadie and F. Schue, J. Macromol. Sci. Polym. Chem. Ed., 15 (1977) 897.
222 J. P. Kennedy, Cationic Polymerization of Olefins: A Critical Inventory, Wiley, New York. 1975.
223 See, for example, A. Ledwith and D. C. Sherrington, in C. H. Bamford and C. F. H. Tipper (Eds.), Comprehensive Chemical Kinetics, Vol. 15, Elsevier, Amsterdam, 1976, pp. 67–127.
224 A. Gandini and P. H. Plesch, Proc. Chem. Soc. London, (1966) 625.
225 D. C. Pepper and P. J. Reilly, J. Polym. Sci., 58 (1962) 639.
226 D. C. Pepper, 23rd Congress IUPAC, Boston, 1971, Preprints, p. 98.
227 D. C. Pepper, J. Polym. Sci. Polym. Symp., 56 (1976) 39.
228 A. Gandini and H. Cheradame, Advances in Polymer Science, Springer-Verlag, Berlin, Heidelberg, New York, 1980, pp. 76–79.
229 M. Szwarc, Macromolecules, 17(10) (1984) 1993.
230 W. H. Hunter and R. U. Yohé, J. Am. Chem. Soc., 55 (1933) 1248.
231 A. R. Gantmakher and S. S. Medvedev, Dokl. Akad. Nauk SSSR, 106 (1965) 1031.
232 D. W. Grattan and P. H. Plesch, J. Chem. Soc. Dalton, Trans., (1977) 1734.
233 D. D. Eley, F. L. Isack and C. H. Rochester, J. Chem. Soc. A, (1968) 1651.
234 A. G. Evans and G. W. Meadows, Trans. Faraday Soc., 46 (1950) 327.
235 H. Scott, G. A. Miller and M. M. Labes, Tetrahedron Lett., (1963) 1073.
236 A. G. Evans, G. W. Meadows and M. Polanyi, Nature (London) 158 (1946) 94.
237 P. H. Plesch, in J. C. Robb and F. W. Peaker (Eds.), Progress in High Polymers, Vol. II, Illife, London, 1968, pp. 137–228.
238 R. H. Biddulph, P. H. Plesch and P. P. Rutherford, J. Chem. Soc., (1965) 275.
239 H. Cheradame and P. Sigwalt, Bull. Soc. Chim. Fr., 3 (1970) 843 and other communications of Sigwalt's group.
240 D. Clark, in P. H. Plesch (Ed.), Proceedings of a Conference held at the University College of North Staffordshire, Heffer, Cambridge, 1953, p. 99.
241 R. D. Colclough and F. S. Dainton, Trans. Faraday Soc., 54 (1958) 886, 894, 898, 901.
242 J. M. Solich, M. Chmelíř and M. Marek, Collect. Czech. Chem. Commun., 34 (1969) 2611.
243 M. Kučera and E. Spousta, Makromol. Chem., 82 (1965) 60.
244 M. Kučera, J. Švábík and K. Majerová, Collect. Czech. Chem. Commun., 37 (1972) 2004, 2708.
245 A. Gandini and P. H. Plesch, Eur. Polym. J., 4 (1968) 55.
246 M. Kučera, J. Macromol. Sci. Chem., A7 (1973) 1611.
247 S. Winstein and G. C. Robinson, J. Am. Chem. Soc., 80 (1958) 169.
248 J. M. Sangster and D. J. Worsfold, Macromolecules 5 (1972) 229.
249 B. M. Mandal, Polym. Bull., 2 (1980) 625.
250 T. Higashimura and M. Sawamoto, Makromol. Chem. Suppl., 12 (1985) 153.

251　M. Kučera, J. Láníková J. and E. Spousta, CS 110, 064 (1962).
252　M. Kučera and E. Spousta, J. Chem. Soc., (1965) 1478.
253　M. Kučera and H. Kelblerová, Collect. Czech. Chem. Commun. 44 (1979) 542.
254　Y. Yamashita, M. Hirota, K. Nobutoki, Y. Nakamura, A. Hirao, S. Kozawa, K. Chiba, H. Matsui, G. Hattori and M. Okada, J. Polym. Sci Part B, 8 (1970) 481.
255　S. Smith and A. J. Hubin, Polym. Prepr. Am. Chem. Soc. Div. Polym. Chem. 13(1) (1972) 66; J. Macromol. Sci. Chem. A7(7) (1973) 1399.
256　R. Hoene and K. H. W. Reichert, Makromol. Chem., 177 (1976) 3545.
257　F. J. Burgess, A. V. Cunliffe, D. H. Richards and D. Thompson, Polymer, 19 (1978) 334.
258　A. Guiot, Polym. Prepr. Am. Chem. Soc. Div. Polym. Chem., 26(1) (1985) 46.
259　J. P. Kennedy and E. G. Melby, J. Polym. Sci. Polym. Chem. Ed., 13 (1975) 29.
260　Y. K. Mo, J. Polym. Sci. Polym. Lett., 13 (1975) 651.
261　M. Kučera, Z. Salajka, K. Majerová and M. Kunz. CS 220, 722 (1981).
262　M. Kučera, Z. Salajka and. K. Majerová, CS 238 488 (1987)
263　M. Kučera, D. Kimmer and K. Majerová, CS 240 627 (1987)
264　T. Yonezawa, T. Higashimura, K. Katagiri, K. Hayashi, S. Okamura and K. Fukui, J. Polym. Sci., 27 (1957) 311.
265　J. Šebenda, in C. H. Bamford and C. H. F. Tipper (Eds.), Comprehensive Chemical Kinetics, Elsevier, Amsterdam, 1976, pp. 379–471.
266　M. Rothe, G. Reinisch, W. Jaeger and I. Schopov, Makromol. Chem., 54 (1962) 183.
267　J. Boor, Jr., Ziegler–Natta Catalysts and Polymerizations, Academic Press, New York, 1979, (a) pp. 2–5; (b) p. 140; (c) pp. 162–163; (d) p. 380; (e) p. 286.
268　G. Natta, I. Pasquon and A. Zambelli, J. Am. Chem. Soc., 84 (1962) 1488.
269　G. K. Lund (Dow Chemical), V. S. Pat. 4,604,374 (1986).
270　K. Takatani (Asahi Chemical Ind. Co. Ltd.), Jpn. Pat. 61,133,206 (1984).
271　K. Sasaki, T. Kujiva and A. Ito (Mitsui Toatsu Chemical Inc.), Jpn. Pat. 61, 207, 403 (1985).
272　K. Y. Choi and W. H. Ray, J. Macromol. Sci. Rev. Macromol. Chem. Phys., C25(1) (1985) 1–97.
273　J. Lieto, D. Milstein, R. L. Albright. J. V. Minkiewicz and B. C. Gates, Chemtech 13(1) (1983) 46.
274　A. D. Caunt, P. J. T. Tait and S. Davies, Transition Met. Catal. Polym. Alkenes Dienes, Part A), MMI Press Symp. Sci., 1983, pp. 149–169.
275　F. J. Karol, Polym. Prepr. Am. Chem. Soc. Div. Polym. Chem. 24(1) (1983) 107.
276　N. Kashiwa and J. Yoshitake, Makromol. Chem. Rapid Commun., 4 (1983) 41.
272　K. Soga, S. I. Chen, Y. Doi and T. Shiono, Macromolecules, 19(12) (1986) 2893.
278　K. Soga, M. Ohtake, R. Ohnishi and Y. Doi, Makromol. Chem., 186 (1985) 1129.
279　B. Rebenstorf and R. Larsson, J. Catal., 84(1) (1983) 240.
280　P. Pino, B. Rotzinger and E. v. Achenbach, Makromol. Chem. Suppl., 13 (1985) 105.
281　J. Cihlář, Thesis, Research Institute of Macromolecular Chemistry, Brno, 1978.
282　W. Kaminsky, M. Miri, H. Sinn and R. Woldt, Makromol. Chem. Rapid Commun., 4(6) (1983) 417.
283　E. Gianetti, G. M. Nicoletti and R. Mazzocchi, J. Polym. Sci. Polym. Chem. Ed., 23 (1985) 2117.
284　W. Kaminsky, K. Kuelper and S. Niedoba, Makromol. Chem. Macromol. Symp., 3 (1986) 377.
285　R. Vilím, J. Ambrož and O. Hamřík, CS 99, 922 (1960).
286　V. Busico, P. Corradini, L. De Martino, A. Proto, V. Savino and E. Albizzati, Makromol. Chem., 186(6) (1985) 1279.
287　Y. V. Kissin and D. L. Beach, J. Polym. Sci. Polym. Chem. Ed., 24 (1986) 1069.
288　P. Cosee, J. Catal., 3 (1964) 80.

289 P. Cosee, Rec. Trav. Chim. Pays Bas 85 (1966) 1152.
290 J. P. Hogan and R. L. Bank, (Philips Petroleum Co.), BE 530, 617 (1955); U. S. Pat., 2, 825, 721 (1958).
291 D. L. Myers and J. H. Lunsford, J. Catal., 92 (1985) 260.
292 S. Tazuke, Pure Appl. Chem., 34 (1973) 329.
293 A. Ledwith, in M. Gordon and W. R. Ware (Eds.), The Exciplex, Academic Press, New York, 1975, p. 209.
294 E. B. Lyudvig, A. P. Gantmakher and S. S. Medvedev, Dokl. Akad. Nauk SSSR, 119 (1950) 90.
295 M. Irie and K. Hayashi, Progr. Polym. Sci. Jpn., 8 (1975) 106.
296 A. Weller, Pure Appl. Chem., 16 (1968) 115.
297 A. Ledwith, J. Polym. Sci. Polym. Symp., 56 (1976) 483.
298 H. Leonhardt and A. Weller, Ber. Bunsenges. Phys. Chem., 67 (1963) 791.
299 E. N. Gur'yanova, I. P. Gol'dshtein and I. P. Romm, The Donor–Acceptor Bond, Willey, Chichester, 1975.
300 Y. Shirota, in J. I. Kroschwitz, Encyclopedia of Polymer Science Engineering, Vol. 3, Wiley, New York, 1985, pp. 327–363.
301 A. M. Mosin, R. M. Aizatullova and V. G. Genchel, Deposited Document, 1982, SPSTL, 943-Khp–D82, Chem. Abstr. 100 (1984) 145323.
302 M. Marek, L. Toman and J. Pilař, J. Polym. Sci. Polym. Chem. Ed., 13 (1975) 1565.
303 Y. Shirota and H. Mikawa, J. Macromol. Sci. Rev. Macromol. Chem., C16(2) (1977–1978) 129.
304 M. Matsuda and Y. Ishiroshi, J. Polym. Sci. Part A–1, (8) (1970) 387.
305 M. Abdelkader, A. B. Padias and H. K. Hall, Jr., Polym. Prepr. Am. Chem. Soc. Div. Polym. Chem., 27(1) (1986) 90.
306 See, for example, T. Saegusa, S. Kobayashi, Y. Kimura and H. Ikeda, J. Macromol. Sci. Chem., A9(5) (1975) 641.
307 J. C. Salomone, A. C. Watterson, T. D. Hsu, C. C. Tsai and M. U. Mammud, J. Polym. Sci., Polym. Lett. Ed. 15 (1977) 487.
308 J. C. Salomone, A. C. Watterson, L. Quach and M. K. Raheja, Polym. Prepr. Am. Chem. Soc. Div. Polym. Chem., 26(1) (1985) 196.
309 P. Giusti, J. Polym. Sci. Polym. Symp., 50 (1975) 133.
310 H. Lambla, R. Koenig and A. Banderet, Eur. Polym. J., 8 (1972) 1.
311 P. Cerrai, P. Giusti, G. Guerra and M. Tricoli, Eur. Polym. J., 10 (1974) 1195.
312 L. Toppare, S. Eren, L. Tuerker and U. Akbulut, Polymer, 25(11) (1984) 1655.
313 U. Akbulut, L. Toppare and Yurttas, J. Polym. Sci. Polym. Lett. Ed., 24(4) (1986) 185.
314 B. L. Funy, in J. I. Kroschwitz, (Ed.), Encyclopedia of Polymer Science Engineering, Vol. 5, Willey, New York, 1986, pp. 587–601.
315 R. J. Ceresa, Trans. J. Plastic. Inst., 28 (1960) 178, 202.
316 A. A. Berlin, G. S. Petron and V. F. Prosvirkina, Khim. Nauk Prom., 2(4) (1957) 522.
317 C. V. Oprea and M. Popa, Acta Polym., 37(3) (1986) p. 177 and previous papers.
318 A. Casale and R. S. Porter, Adv. Polym. Sci., 17 (1975) 1.
319 P. Kruus, Ultrasonics, 21(5) (1983) 201.
320 C. V. Oprea and F. Weiner, Angew. Makromol. Chem., 126 (1984) 89.
321 G. K. Vinogradov, Khim. Vys. Energ., 20(3) (1986) 195.
322 H. Yasuda and T. Hirotsu, J. Polym. Sci. Polym. Chem. Ed., 15 (1977) 2749.
323 G. G. H. Schulz and H. Höcker, Makromol. Chem., 178 (1977) 2589.
324 J. P. Kennedy and K. F. Castner, Polym. Prepr. Am. Chem. Soc. Div. Polym. Chem. 18(1) (1977) 655.

Chapter 4

Active centres of polymerizations

*Imagination is better
than knowledge*

Albert Einstein

For a macromolecule to be generated, a certain group of atoms suitable for electron transfer and bond rearrangements between itself and the monomer must exist. Such a group usually forms the end of the growing molecule and is generally very reactive. It is one of the main factors determining the manner and rate of macromolecule formation. When the addition step is not strictly selective, we speak of active centres (active centres of polymerizations, active polymerization centres.). Only when propagation is the only reaction occurring do we designate them as living centres. Thus living centres are a special case of active centres.

The properties of the centre are determined by the type and arrangement of the atoms of which it is composed. The structure of the centre determines the extent of the electron defect delocalization and the polarity of the whole formation, and thus also its energetic state and ability to form associates and complexes. Steric effects are also a result of the structure of the centre. A complete, quantitative and exact description is, however, almost always beyond the possibilities of the present state of science. Only in rare cases can they be constructed exactly and prepared according to the requirements of macromolecular syntheses. In spite of that, active centres are subject to strict requirements. The reaction of the centre with the monomer must be straightforward with a minimum of side reactions. Monomer addition must be neither too slow nor explosive. The possibility of regulating the rate and method of addition is of great advantage. As our knowledge of the consequences of an intentional modification of the structure of the centre is still very limited, the overall polymerization rate is usually regulated by an adjustment of the concentrations of the reacting components and/or the reaction temperature.

References pp. 222–229

1. Active centres of radical polymerization

Polymers generated by radical polymerization and having a backbone built of atoms other than carbon are so far unknown. Therefore the active centre in radical polymerizations almost always contains a $-\overset{|}{\underset{|}{C}}*$ atom with an unpaired electron. This structure should be planar.

However, evidence has been presented which indicates that the atoms bound to the $-\overset{|}{\underset{|}{C}}*$ radical are not always coplanar, even in the simplest cases. H_3C* is not strictly planar, and the carbons of

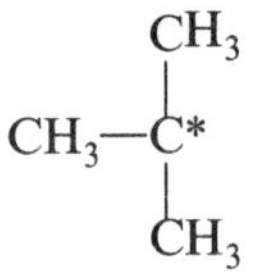

$$CH_3-\overset{CH_3}{\underset{CH_3}{C*}}$$

lie at the apices of a tetrahedron [1–3]. This may be caused by a highly excited vibrational state of the radicals [4]. In the absence of substrate, $CH_3CH_2^*$, decays by disproportiation, and to a small extent by combination yielding ethane, ethylene and butane [5]. $PhCOCH_2^*$ generated by the reaction of diacetyl peroxide with acetophenone, does not dimerize. It was shown that it has the structure of a cyclopropanonespirocyclohexadienyl [6]

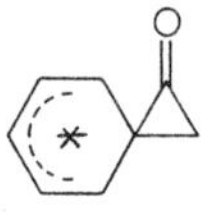

β-Thioalkyl radicals and β-haloalkyl radicals exhibit a conformation with a distortion at the β carbon, shifting S (or the halogen) towards the orbital with the unpaired electron [7]. The spin density on the α atom of $(RS)_3C*$ is about 0.7 [8]; the unpaired electron is in an orbital with 8–9 % s character [4]. The mean lifetime of vinyl radicals considerably increases when the centres of these radicals are sterically hindered by bulky groups such as Me_3C-, Me_3Si- [9, 10]. The stabilization energy of

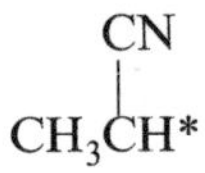

$$CH_3\overset{CN}{\underset{|}{CH*}}$$

is (21 ± 10) kJ mol^{-1} [11].

These few examples illustrate the difficulties met in attempts to describe the character of an active centre quantitatively. The two-dimensional graphically described structure of the centre is almost always simple, $\sim CH_{(2-n)} F_n-CZW^*$ (where Z and W designate mono- or polyatomic substitutents.) Attempts to calculate the energetic state of the centre are greatly complicated by conjugation, hyperconjugation, induction and inductomeric effects, by steric hindrance and many other factors. For this reason also, a quantitative description of the abilitiy of the centre to react with various substrates is impossible. A calculation of the energetic states of the centre-substrate system along the reaction path, informing us of the probability of the occurrence of side reactions, is a future goal. Our present methods of determining the reactivity of the centre are based on the concepts of quantum chemistry but the procedure itself is always empirical.

1.1 Reactivity of radicals

The rate of the reaction of radical R* with substrate S is given by the product of the rate constant and the concentrations (better activities) of the reacting components

$$v = k[R^*][S] \qquad (1)$$

The interactions of various radicals with the same S, under otherwise identical conditions, are not equally rapid. Thus the value of the rate constant is a measure of the reactivity of the radical with the given substrate. According to Arrhenius [12] and the Eyring [13] transition state theory, it holds (in standard notation) that

$$k_{R^*S} = A \exp\left(-\frac{E}{RT}\right) = \frac{RT}{Nh} \exp\left(\frac{\Delta S^{\ddagger}}{R} - \frac{\Delta H^{\ddagger}}{RT}\right) \qquad (2)$$

The calculation of the rate constant from molecular parameters is based on an estimate of the properties of the hypothetical transition complex. We assume that the structure of the complex forms a transition between the structure of the reacting molecule (or molecules) and the product, but it may be nearer either to the former or to the latter.

Effects modifying the stability of the transition complex usually also affect the activation energy. Many of these factors, especially the steric effects, also affect $\Delta S^{\ddagger}$. In some cases, a certain factor, j, can modify the activation energy and $\Delta S^{\ddagger}$ so that the induced effects counteract each other. At a certain

References pp. 222–229

temperature (the isokinetic temperature T_I) the effects are mutually compensated

$$T_I(\Delta S_0^{\ddagger} - \Delta S_j^{\ddagger}) = E_e - E_j$$

so that the reaction rate is unaffected by the corresponding factor, $k_0/k_j = 1$. For a series of reactions obeying the isokinetic relation, the plot of E_j vs. $\Delta S_j^{\ddagger}$ is a straight line of slope T_I. At $T < T_I$, the factor j mostly affects the activation energy; above T_I changes in $\Delta S^{\ddagger}$ are more important.

It is evident that a calculation of the rate constant neglecting $\Delta S^{\ddagger}$ will only be correct for a limited number of cases. Series of reactions are known in which $\Delta S^{\ddagger}$ really remains approximately constant. Other series obey the isokinetic relation; in others the change in $\Delta S^{\ddagger}$ is independent of E [14]. Nevertheless, for the time being, our only choice is to consider only the energy, and compare the result with experiment. A procedure for calculating E at variable $\Delta S^{\ddagger}$ in a generally applicable form is not yet available.

The determination of the activation energy in the case where the structure of the transition complex resembles the product (Evans et al. [15] and Ogg and Polanyi [16]) is based on the assumption that the activation energy difference between reaction (3) and a standard reaction (4) is proportional to the difference in the heats of reaction

$$\text{RCHX}^* + \text{CH}_2\!\!=\!\!\text{CHY} \xrightarrow{k} \text{RCHXCH}_2\text{CHY}^* \qquad (3)$$

$$\sim\!\text{CH}_2^* + \text{CH}_2\!\!=\!\!\text{CH}_2 \xrightarrow{k'} \sim\!\text{CH}_2\text{CH}_2\text{CH}_2^* \qquad (4)$$

Thus the difference in E is determined by the resonance energies of the attacking and resulting radicals (ε_a and ε_r), of the substrate (ε_s) and by a proportionality constant α

$$-\Delta H - (-\Delta H') = \varepsilon_a + \varepsilon_s - \varepsilon_r \qquad (5)$$

$$\Delta E = E - E' = \alpha(\Delta H - \Delta H') = \alpha(\varepsilon_a + \varepsilon_s - \varepsilon_r) \qquad (6)$$

The reaction coordinate can be roughly approximated by a discontinuous curve composed of two branches†. The ascending branch represents the dissociation of a bond generated in the course of the reaction, and the descending branch represents repulsion between the attacking radical and substrate. A case of the interaction of two substrates attacked by a radical is illustrated in Fig. 1. The activation energy difference $\Delta E = \alpha(\varepsilon_r - \varepsilon_s)$. When n and m are the slopes of the non-bonding and bonding branches, α is given

---

† In a more general approximation, potential energy surfaces would have to be considered. The submitted curves represent intersections of these surfaces with a selected plane.

by the ratio $\alpha = n/(m - n)$. This procedure can rarely be used. Its more recent variant is the correlation of relative substrate reactivity with the energy necessary for the transition of its molecule from the ground state to the hypothetical excited state. The theory is based on the concept of radical

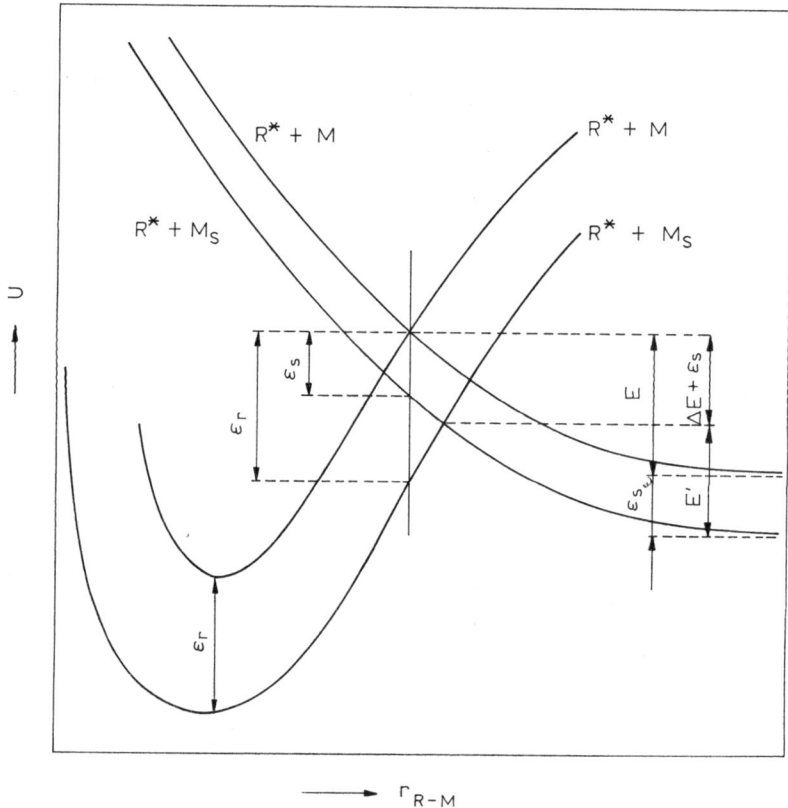

Fig. 1. Dependence of potential energy on the distance of the reacting components. M_s is a standard monomer and M is the monomer used for testing the reactivity of the radical R*. The meanings of the other symbols and of the curves are described in the text.

attack on a π-electron system [17–20]. The excited state is represented by a polarization where one of the π electrons is placed in an orbital of the attacked atom. The remaining electrons are distributed among the other C atoms; σ electrons do not participate in the process. According to Brown [19], the energy necessary for ground-state polarization is called the localization energy L. These energies can be computed by means of the molecular orbital theory LCAO-MO [21–23]. The direct proportionality between the computed localization energy and the logarithm of the relative rate constant

References pp. 222–229

(which is a function of the activation energy) for radical attack on a series of substrates was expressed geometrically by Szwarc and Binks [20, 24]

$$\ln \frac{k}{k'} = \text{const.} - \Delta L \frac{n}{m+n} (RT)^{-1}$$

$$\Delta E = \frac{\Delta L n}{m+n} = \Delta L \, \alpha \tag{7}$$

In this approach, the activation energy is again determined from the points of intersection of the curve (more generally surface) of the radical-excited substrate bond potential energy with the curves of non-bonding interactions between the attacking radical and substrate or standard (Fig. 2).

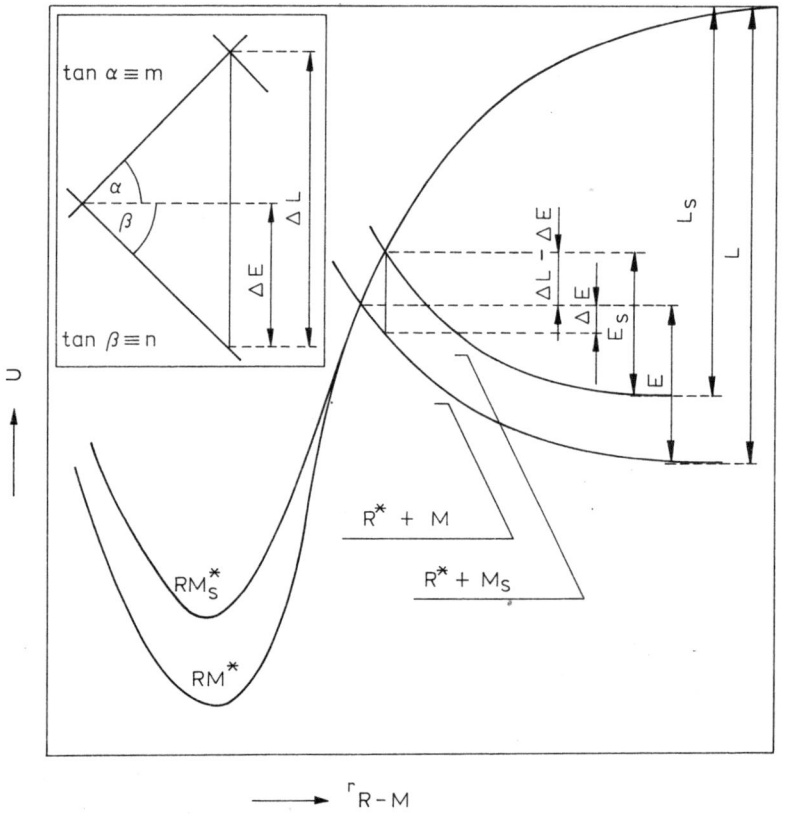

Fig. 2. Determination of α and ΔE from the plot of potential energy vs. the distance between the reacting components. L and L_s are the localization energies of the monomer and of the standard monomer, E and E_s the activation energies in the reaction of the substrates with the tested radical, and m and n the slopes of the bonding and non-bonding curves.

The methods of geometrical analysis described are only a very rough solution because the intersection of the potential energy surfaces is not identical with the energy profile of the reaction path. Considerations of resonance splitting of potential energy surfaces in the transition complex range lead to further refinement [25–27].

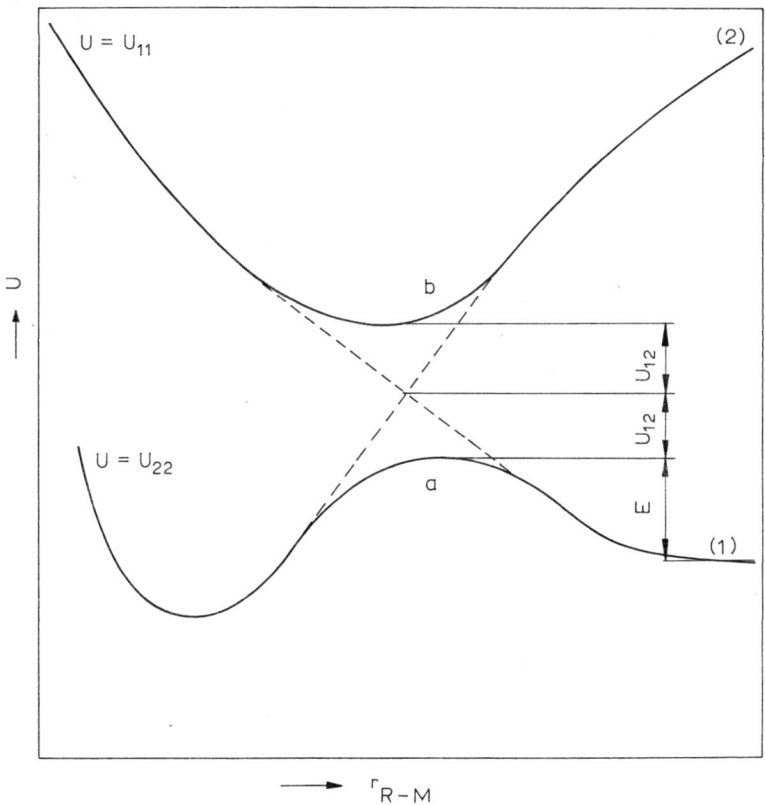

Fig. 3. Splitting of the potential surfaces for the system R* + M in two states.

Let $U(r_{R^*M}, \chi)$ be an operator describing the interaction of radical R* with substrate M as a function of the distance r_{RM} and of a set of electron coordinates χ in the system. For $r_{RM} = \infty$, the system has two corresponding states 1 and 2 (Fig. 3). Curve (1) represents the initial situation. The reacting components are in the ground state, and the perturbation energy is given (to a first approximation) by the relation

$$U_{11}(r_{RM}) = \int \psi_1^* U \psi_1 \, d\chi \qquad (8)$$

In state 2, M is excited. Therefore for curve (2) we write

$$U_{22}(r_{RM}) = \int \psi_2^* U \psi_2 \, d\chi \tag{9}$$

The ordinate of the point of intersection of the two curves represents the peak of the reaction path only in the case when states 1 and 2 are exactly described for the final values of r_{RM}. This condition is, of course, not fulfilled, and the eigenfunctions ψ_1 and ψ_2 are therefore not directly applicable. The energy can, however, be calculated by means of their linear combination

$$\psi = a\psi_1 + b\psi_2 \tag{10}$$

This procedure leads to two permitted energy values of the system at all values r_{RM}. They correspond to the states a and b and the corresponding curves do not intersect. For small r_{RM} values (for the product) a converges to curve (2) and b to curve (1). At the intersection of these converging curves (1) and (2) ($U_{11} = U_{22}$), the energies of the two real states are $U_a = U_{11} - U_{12}$ and $U_b = U_{11} + U_{12}$, where

$$U_{12} = \int \psi_1 U \psi_2 \, d\chi$$

When the reaction between R and M proceeds adiabatically from the initial state, then curve a represents the correct profile of the reaction path. ΔE is smaller than U_{12} in the preceding considerations.

In radical reactions, U_{12} is not usually negligible. However, for some interaction series it exhibits little change or it may be proportional to the rate constant. In such cases it is possible to obtain relatively easily rate constant values of good accuracy [28] according to eqns. (6) and (7).

Fukui and co-workers [29] called attention to the fact that the radical-substrate interaction is affected by the π configuration. The second-order perturbation energy, ΔU_{rs}, between the rth π orbital of the radical and the sth π orbital of the substrate is related to the activation energy of the reaction between the r and s positions

$$E = C - \Delta U_{rs} \tag{11}$$

where C is a term containing the contributions of the σ orbital, and which, for series of comparable reactions, is considered as constant. ΔU_{rs} was computed by the LCAO-MO method using the concept of frontier orbitals [30, 31][†]. By means of eqn. (11), head-to-tail addition has been correctly

---

[†] The highest occupied orbital in the transition complex, and the lowest unoccupied orbital in the ground state.

TABLE 1

Values of the pre-exponential factor, A, activation energy, E, activation entropy, $\Delta S^{\neq}$, and standard entropy $\Delta S°$, for the addition of alkyl radicals to ethylene

Radical	$\log\left(\dfrac{A}{dm^3\,mol^{-1}\,s^{-1}}\right)^b$	E^b (kJ mol^{-1})	$-\Delta S^{\neq}$ ^{d,e} (JK^{-1} mol^{-1})	$-\Delta S°$ ^{d,e} (JK^{-1} mol^{-1})
CH_3^*	8.5	32	125	133
CH_3^{*a}	8.2	31	131	142
$C_2H_5^*$	8.3	31	129	139
$CH_3CH_2CH_2^*$	7.3^c	25^c	(148)	138
$(CH_3)_2CH^*$	7.8	29	(138)	101
$CH_3(CH_2)_2CH_2^*$	7.4^c	28^c	(146)	142
$(CH_3)_3C^*$	7.5	30	(145)	113

^{a} Reaction of radical with propene.
^{b} Values recommended in ref. 33.
^{c} Tentative values from ref. 33.
^{d} Values from ref. 14, p. 376.
^{e} At 298 K and 0.098 MPa.

predicted for radical polymerizations of styrene, butadiene, acrylonitrile, and vinyl chloride as well as their alternating tendency in copolymerization [32].

We have some indirect evidence on the structure of the transition complex between the reacting radical and the unsaturated compound from kinetic data. A comparison of $\Delta S^{\neq}$ with the measured value $\Delta S°$ (see Table 1) leads to the conclusion that the greater part of the entropy change accompanies the formation of the transition complex. This indicates that the bond between radical and substrate is more or less completed in this state, and that the transition state resembles the product.

The contribution of theoretical procedures to rate constant calculations is beyond any doubt. This concerns first of all the determination of relative radical reactivities with a series of substrates of gradually changing structure, and predictions of the reaction path direction. The agreement of theory with experiment is usually better with non-polar radicals than with the polar radicals. In spite of this, experimental findings are still the main source of our knowledge of radical reactivities. Theoretical concepts are helpful in the classification and interpretation of these findings. Quantum chemistry will be able to triumph at the time when it is capable of producing a recipe for the preparation of an active centre of optimum properties, and for the conditions of its reaction with the substrate along a selected reaction path.

References pp. 222–229

Experimental methods for determining the reactivity of a radical active centre are based on kinetic studies; reactivity can be estimated from rate constant values, e.g. in propagation and transfer (see Chap. 8, Sect. 1.2).

1.2 Reactivity of radical centres during polymerization

Most kinetic studies are based on the implicit or explicit assumption that the reactivity of a centre is independent of the length of the chain to which it is attached. It appears that this assumption is still acceptable for common situations. It is evidently not valid for the extreme situations, i.e. for chains that are just beginning to grow, for chains with conjugated multiple bonds, and for polymerizations affected by physical factors. The reactivity change of the centre is caused either by spontaneous changes of its structure, or by physically induced changes of its surroundings or of the whole system.

An active centre is formed by the addition of a primary radical (for example from an initiator) to the monomer. A polar primary radical may considerably affect the electronic configuration of the centre, and this effect will decay with the number of added monomers. The first additions may proceed by a different mechanism and may even be more rapid or slower than additions at later stages, when the inductive effect has been suppressed.

TABLE 2

Dependence of activation energy difference on the ratio of pre-exponential factors of transfer (tr) and propagation (p) for the addition of MMA to a radical with the $F_3C\sim\overset{|}{\underset{|}{C}}*$ end group

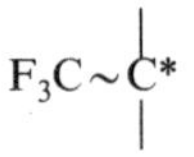

n^a	$(E_{tr,n} - E_{p,n})$ (kJ mol^{-1})	$A_{tr,n}/A_{p,n}$
1	12.4 ± 1.3	10.3 ± 4.5
2	17.3 ± 2.0	176 ± 55
3	18.0 ± 1.4	340 ± 163
4	18.9 ± 1.4	473 ± 147
5	19.7 ± 2.0	620 ± 280

[a] n is the degree of polymerization of the radical

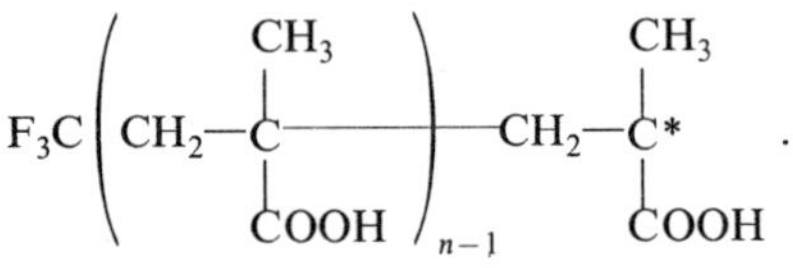

$$F_3C\left(CH_2-\underset{\underset{COOH}{|}}{\overset{\overset{CH_3}{|}}{C}}\right)_{n-1} CH_2-\underset{\underset{COOH}{|}}{\overset{\overset{CH_3}{|}}{C}}*$$

Barson and Ensor [34] have studied the temperature dependence of the apparent transfer constant in telomerization of MMA with CBr_4. They found dramatic differences in the Arrhenius parameters between the first, second, etc. monomer additions to the CBr_3^* radical (Table 2.) The authors explain these results by electron transfer from radical to monomer, yielding the combining ions

$$Br_3C-CH_2-\underset{\underset{OCH_3}{|}}{\underset{\underset{C=O}{|}}{\overset{\overset{CH_3}{|}}{C^+}}} \quad \underset{\underset{OCH_3}{|}}{\underset{\underset{C=O}{|}}{\overset{\overset{H}{|}}{\underset{|}{C}}-\overset{\overset{CH}{|}}{\underset{|}{C^*}}}} \quad ; \quad Br_3C-CH_2-\underset{\underset{OCH_3}{|}}{\underset{\underset{C=O}{|}}{\overset{\overset{CH_3}{|}}{C^+}}} \quad \underset{\underset{OCH_3}{|}}{\underset{\underset{C-\bar{O}|^-}{\|}}{\overset{\overset{H}{|}}{\underset{|}{*C}}-\overset{\overset{CH_3}{|}}{\underset{|}{C}}}}$$

In this way, the activation energy of addition can be reduced with respect to the reaction which is unaffected by the polar end. The rate of reaction of the active centre with the telogen† CBr_4 also changes with chain length, of course.

Another type of reactivity change of the centres was observed in the polymerization of acetylene and its derivatives. The generated chain with conjugated double bonds makes possible a far-reaching delocalization of the unpaired electron. The reactivity of the active centre decreases with chain length [35, 36], so that the number of monomer additions is limited.

Some authors suggest that the accommodation of a conjugated chain in space does not favour far-reaching electron delocalization [37]. It appears, however, that the low degree of polymerization and relatively narrow molar mass distribution in the product of radical polymerization of acetylenic monomers are indeed caused by inter- and intramolecular delocalization of the unpaired electron in the generated conjugated macroradical [38]. The final experimental proof of this statement depends on the application of sufficiently sensitive separation methods. The reaction of alkyne monomers is not straightforward. Oligomers (including cyclic trimers) are formed during polymerization, hindering sufficiently exact characterization of the polymerization products.

A change in physical conditions in the vicinity of the active centre in the course of polymerization may sometimes be the cause of processes which can formally be accounted for by a change in centre reactivity.

Rate constants change in the course of MMA and VAC polymerizations [39, 40] (Tables 3 and 4). The considerable decrease in k_t at high conversion is caused by a switch of the rate-determining step from the kinetic domain to a diffusion-controlled process [41] (see Chap. 6, Sect. 3.1)

---

† The effective transfer agent (the cause of the formation of a low molecular weight polymer), the telomer.

TABLE 3

Change of rate constants k_p and k_t with conversion during MMA bulk polymerization[a]

Conversion, $\varkappa$	k_p (dm^3 mol^{-1} s^{-1})	$2 \times 10^{-4} k_t$ (dm^3 mol^{-1} s^{-1})
0.0	384	44 209
0.1	259	2 730
0.2	334	726
0.3	434	142
0.4	615	89
0.5	515	40
0.6	185	5
0.7	54	0.56
0.8	6	0.076

[a] Values are taken from ref. 39. Polymerization temperature 295.5 K, photochemical initiation. Rate constants were determined from the stationary polymerization rate and from the non-stationary pre-effect. The necessary conditions for the calculation were not quite fulfilled; the rate constant values are not exact but they adequately illustrate a characteristic situation.

TABLE 4

Change of rate constants k_p and k_t with conversion during VAC bulk polymerization[a]

Conversion, $\varkappa$	k_p (dm^3 mol^{-1} s^{-1})	$10^{-5} k_t$ (dm^3 mol^{-1} s^{-1})	E_p (kJ mol^{-1})	E_t (kJ mol^{-1})
0.04	900	240	18	4.2
0.23	1 290	126	18	6.2
0.46	1 980	90	15	14
0.57	555	6.7	36	49
0.65	30	1.2	48	54

[a] Values are taken from ref. 40. Polymerization temperature 298 K. Rate constants were determined from the stationary polymerization rate and from the non-stationary pre-effect.

1.3 "Hot" radicals

The energetic state of an active centre is determined by an equilibrium between the chemical and physical factors within the centre itself and in its immediate vicinity. Usually the equilibrium is rapidly established so that, for the subsequent chemical reaction, the centres are ready in the same energy state, i.e. they are equally reactive. Situations cannot, however, be excluded where the rate of energy equilibration is comparable with the rate of the successive reaction. In such a system, radicals of various reactivity react with the substrate. This effect was noticed by Tüdös [42] who developed the theory of "hot" radicals.

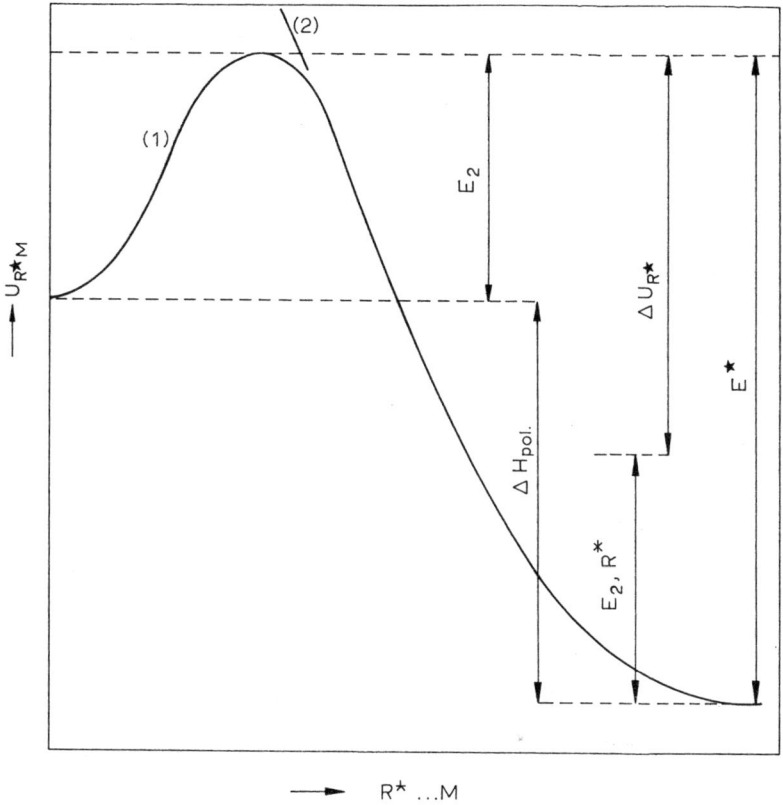

Fig. 4. Profile of the reaction path. (1) Radical with monomer; (2) hot radical with monomer.

At the moment of formation, the reaction product contains the heat from the activation energy of the exothermic elementary reaction, and it is in a highly excited state. The energy diagram of the reaction

$$R^* + M \rightarrow R^\star$$

is given by the reaction cordinate (Fig. 4.) The vibrationally excited "hot" radical $R^\star$ collides with the reaction system components and is deactivated in a series of collisions. According to the classical theory of chain reactions, the "hot" radical is first deactivated, and the reaction can only proceed after the substrate has accummulated sufficient activation energy, E_2, by molecular collisions. In the theory of "hot" radicals, it is assumed that fresh active

References pp. 222–229

centres with excess energy $\left(E^\star \geqq E_2 + \Delta H = 75 \text{ to } 120 \text{ kJ mol}^{-1}\right)$ are capable of direct reaction

$$R^\star + M \rightarrow R_1^\star$$

The activation energy is supplied by the excess energy of "hot" radicals, therefore the process does not require any external activation.

In real systems, some effective rate constant $\tilde{k}_p$ is measured; its value should lie in the interval $k_p^* \leqq \tilde{k}_p \leqq k_p^\star$. By diluting the substrate, the value of $\tilde{k}_p$ should asymptotically approach some value which is independent of the solvent. According to this theory, a chemically inert solvent is a negative "catalyst" because it lowers both the rate and the rate constant. The situation described has been observed in some cases, for example in the polymerization of styrene [43]. Some effects, connected with inhibition reactions, are supposed to confirm the theory.

1.4 Intervention in the structure of centres

Bolshakov et al. [44] observed a large difference in the reactivities of active centres in MMA and MA polymerizations at low temperatures (100–130 K). Substitution of the α proton with a $-CH_3$ group results in steric hindrance of the centre and lower reactivity of the macroradical. At higher temperatures, steric hindrance is less severe, due to increased methyl mobility, and the reactivities of the radicals

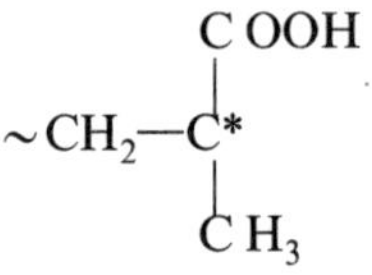

$$\sim CH_2-\underset{\underset{CH_3}{|}}{\overset{\overset{COOH}{|}}{C^*}}$$

and

$$\sim CH_2-\underset{\underset{H}{|}}{\overset{\overset{COOH}{|}}{C^*}}$$

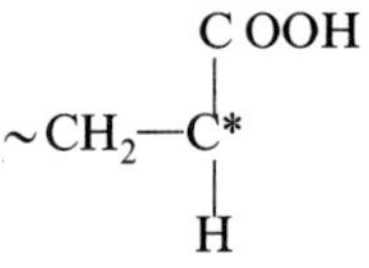

are equalized.

Important information on active centres can be obtained from measurements of polymerization and copolymerization rates of related monomers under various conditions. When the kinetic properties of a centre can be changed by a substituent, its properties can also be changed by a molecule of a non-polymerizing substance (or molecules of non-polymerizing substances) with which the centre forms an associate or complex. Aromatic compounds and metals are known to operate in this way.

Henrici-Olivé and Olivé were the first to put forward the hypothesis that complexes are sometimes formed between the active centre and the monomer and or/solvent [45]. As only the complex with monomer is capable of propagation, part of the centres is inhibited and the polymerization rate is reduced. This theory was found to be valid with styrene [46], but not with MMA [47]. Burnett called attention to the important circumstance that radicals solvated in various ways may react differently, or at least at different rates [47]. His conclusions were based on kinetic studies of MMA polymerization in various halogenated aromatics. In the copolymerization of butyl vinyl ether with methacrylates, complex formation between the active centre and condensed aromatics prior to monomer addition was observed by Shaikhudinov et al. [48]. The growing polymer forms a stable donor-acceptor complex with naphthalene, described by the formula.

$$\sim CH_2-\underset{\underset{COOCH_3}{|}}{\overset{\overset{CH_3}{|}}{C^*}} + \bigcirc\!\!\bigcirc \longrightarrow \left[\sim CH_2-\underset{\underset{COOCH_3}{|}}{\overset{\overset{CH_3}{|}}{C}}{}^{-}\bigcirc\!\!\bigcirc^{+*} \right]$$

Studies of radical-solvent interactions have not as yet been the centre of interest of macromolecular chemists (with some exceptions [49–51]). It is, however, beyond any doubt that the products of such interactions are responsible for most observed "solvent effects". The observations of Bamford and his co-workers [52] called attention to the possibility of modifying the properties of a radical active centre by its interaction with mineral salts [53].

In addition to the cases discussed earlier (see Chap. 2, Sect. 4.2), several others deserve to be mentioned.

The pH dependence of the polymerization rate of acrylic acid [54] in the presence of various neutralizing agents does not exhibit a course. Between pH 2 and 6, the rate drops abruptly; afterwards it grows, depending on the degree of neutralization and the kind of neutralizing agent. The rate increase is connected with the presence of a cation which restricts the repulsion between the anions of the active centre and the monomer

$$\sim CH_2-\underset{\underset{O\diagup\diagdown O^-}{\overset{|}{C}}}{\overset{\overset{H_3C}{|}}{C^*}}\cdots\cdots\cdots Ac^+ \cdots\cdots\underset{\underset{{}_{-}O\diagup\diagdown O}{\overset{|}{C}}}{\overset{\overset{CH_3}{|}}{C}}=CH_2$$

After neutralization with ethylenediamine at pH 6.6, the active centre is not dissociated, and ethylenediamine adds one proton. The increase in polymerization rate is caused by complex formation between these two molecules. The transition complex can be represented schematically as

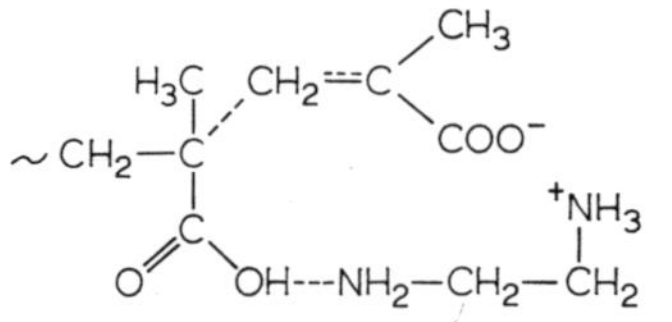

$$\sim CH_2-\underset{\underset{O}{\overset{\displaystyle C}{\|}}\overset{\displaystyle }{}}{\overset{H_3C}{\underset{}{C}}}\cdots \overset{CH_2=C\overset{\displaystyle CH_3}{\diagup}}{}\underset{}{\overset{}{COO^-}}$$
$$ OH\text{---}NH_2\text{--}CH_2\text{--}CH_2\text{--}{}^+NH_3$$

The importance of the ion pair formed at the active centre or monomer can be also demonstrated in the copolymerization of 1,2-dimethyl-5-vinylpyridinium sulphate [55]

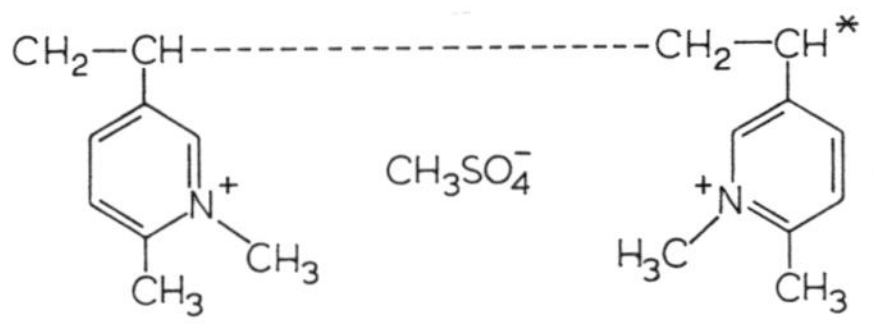

In methanol it is built into the copolymer much more rapidly than the non-ionized 2-methyl-5-vinylpyridine

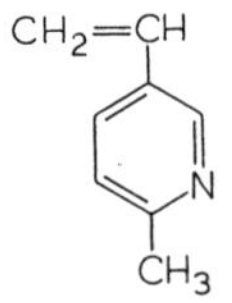

On the other hand, in a more polar medium, 2-methyl-5-vinylpyridine is faster.

The effect of salts on radical polymerizations is the subject of growing attention. It appears that the addition of salts to the active centre may affect not only the rate, but also the method of monomer addition, i.e. the regularity or tacticity of the product. The indicated possibilities have not been technically exploited so far.

1.5 Some radicals of particular importance

The allylic radical plays an extraordinary role in polymerizations. Qualitatively its reactivity can be estimated from the series of increasing particle stability [56]

$$H^* < CH_3^* < Ph^* < RCH_2^* < R_2CH^* < CCl_3^* < R_3C^* < CBr_3^* <$$

$$< RCH^*(COR) < RCH^*(CN) < RCH^*(COOR) < CH_2=CH-CH_2^* < PhCH_2^* < Ph_2CH^* <$$

$$< Cl^* < Br^* < Ph_3C^* < I^*$$

The resonance energy of the allylic radical is 48.8 kJ mol^{-1} [57, 58]. It is easily formed and its reactivity is low. It reacts reluctantly with monomers. This is why propene, 2-methylpropene, 1-butene (and higher members of the homologous series) do not polymerize by the radical mechanism

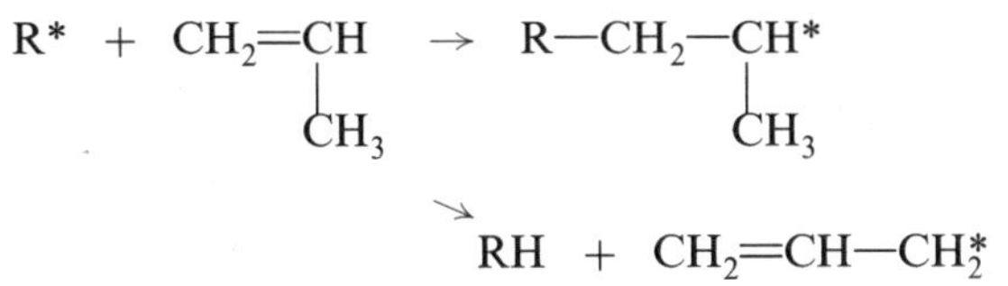

$$R^* + CH_2=CH(CH_3) \rightarrow R-CH_2-CH^*(CH_3)$$

$$\searrow RH + CH_2=CH-CH_2^*$$

Compounds which readily yield the allyl radical are potent transfer agents. The reaction of the generated radical with monomer is slow and the probability of its decay by some other reaction is large; polymerization dies out†.

The chain rarely grows on two or more active centres. Multifunctional active centres are sometimes intentionally formed [59] but more often they result from transfer to polymer. Their practical importance is not great so far. Not much is known about how reactivity is mutually affected in multifunctional radicals. We assume that centres mutually isolated by several methylene groups (several monomer units) behave as individual centres.

Rarely, the unpaired electron of the active centre can belong to an atom other than carbon.

Copolymer formation from vinyl monomers and PhPCl$_2$ has been described [60–62]

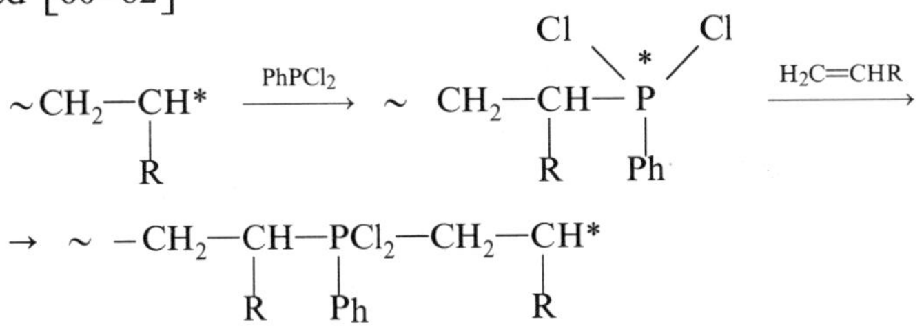

$$\sim CH_2-CH^*(R) \xrightarrow{PhPCl_2} \sim CH_2-CH(R)-P^*(Cl)(Cl)(Ph) \xrightarrow{H_2C=CHR}$$

$$\rightarrow \sim -CH_2-CH(R)-PCl_2(Ph)-CH_2-CH^*(R)$$

Under some conditions, ethylene and vinyl compounds (vinyl chloride, styrene, acrylic acid, chloroprene, etc.) can be copolymerized with SO_2 [63–65]. During synthesis of this type of polysulphone, carbon radical centres alternate with

$$\sim CH_2-\underset{X}{CH}-SO_2^*.$$

The latter tend to depropagation, especially at higher temperatures (>273 K). The properties of centres with an unpaired electron on a heteroatom are little known.

Radical active centres are unstable formations with mean lifetimes mostly of the order of 0.1–10 s. Because of their reactivity, the concentration of the centres cannot be increased much above 10^{-7} mol dm^{-3}. Only the most sensitive of modern ESR instruments are able to detect and measure their signal quantitatively [66]. Therefore in those cases in which it is necessary, some authors allow these radicals to react with suitable inhibitors to yield stable radicals that can be followed by ESR [67].

2. Ionic active centres

Ions are formed by the dissociation of salts and heteropolar splitting of covalent bonds. The rules of ion formation and behaviour have been studied in detail, and for aqueous solutions they are fairly well known. Descriptions of ions, of their immediate vicinity, and of their reactions in less polar systems (e.g. in MeOH) are less clear. The available information on ion behaviour in non polar or weakly polar media (of relative permittivity 2–10) is even more limited. In non-polar systems, ions are much more reactive than even the most reactive radicals. Their electric charge is the cause of mutual ion associations, of ion solvation by the molecules of various compounds, and of many other effects.

2.1 Ions, ion pairs, and their associates

In solution, the conditions for the existence of equilibria between free ions and ions more tightly bound together by electrostatic forces are generally fulfilled. Particles whose motion in the medium is independent of other ions or ion groups are designated as free ions. An ion pair represents a couple of ions of opposite charge forming an electrically neutral particle. A group of a greater number of ions, forming a single entity for a shorter or longer time, is called an ionic associate. An associate can either carry a charge or be electrically neutral.

Two ions, A and B, carrying Z_A and Z_B charge with centres at a distance r exhibit an electrostatic interaction energy [68]

$$U_{el,r} = \frac{Z_A Z_B e^2}{4\pi r D_0 D_r} \quad (J) \tag{12}$$

where e is the proton unit charge $(1.602\ 1 \times 10^{19} C)$, D_0 is the permittivity of vacuum $(8.854\ 187\ 82 \times 10^{-12}\ C^2\ J^{-1}\ m^{-1})$, and D_r is the relative permittivity of the medium.

For small values of r (less than ca. 1 nm), the ion interaction also includes the contribution of short-distance repulsion forces and attractive (dispersion) forces. The repulsion forces are little known; for simple spherical molecules their contribution can be expressed by the product br^{-12}. The attractive forces are better defined; by analogy to the repulsion forces they can be expressed as cr^{-6}, where b and c are constants characterizing the particles. To a first approximation the overall energy can be described by

$$U_{el,r} = \frac{Z_A Z_B e^2}{4\pi r D_0 D_r} - cr^{-6} + br^{-12} \tag{13}$$

For a pair of ions of opposite charge, the value of $U_{el,r}$ exhibits a minimum for a very short inter-ion distance r. For NaCl in vacuum [69], the interaction energy is about 500 kJ mol^{-1}[†].

Heterolytic dissociation of a covalent bond in the gas phase (the term ionization was recommended for this process [68a]) requires energy of the same order. The calculation is, however, subject to considerable error, and experimental verification is not easy.

In solution, the energy necessary for the separation of an ion pair to free ions is lower by the factor D_r^{-1} compared with vacuum. The stumbling block of these considerations is the determination of D_r in the immediate vicinity of the ions. The solvent molecules are oriented and their properties, (e.g. their conformation) are changed by the strong electrostatic field of the ions. These changes can affect D_r: the values of r_g and r_l, corresponding to the minimum of interaction energy $(dU/dr = 0)$ in the gaseous and liquid phases, respectively, are not equal, and their difference is often neglected. This reduces the accuracy in calculating $U_{el,r}$.

Orientation of solvent molecules in the neighbourhood of the ion, and their coordination to the ion, is known as ion solvation. The manner of solvation is very important. If the surface of the ion pair as a whole is surrounded by

---

[†] The energy necessary to separate two NaCl ions from a distance corresponding to an ion pair to infinity is about 500 kJ mol^{-1}.

References pp. 222–229

solvent molecules, we speak of a contact pair. When one or several solvent molecules penetrate between the two ions of the pair, we have a solvent-separated ion pair. Both these forms are thermodynamically stable particles, and their existence has been proved experimentally [70]. Systematic studies of reversible reactions leading to the dissociation of contact ion pairs in ether solvents [71] had led to the discovery of a larger number of contact ion pair types. On the basis of present knowledge it may be assumed that, in every electrolyte solution, several kinds of ion pairs, differing by type of solvation, can exist under suitable conditions [68b].

Attempts to derive theoretical relations describing various dissociation degrees are based on the contribution of electrostatic forces to the Gibbs energy, enthalpy, and change in dissociation entropy. These contributions can be calculated [68c]

$$-\Delta G_{dis,\,el} = \frac{NZ_A Z_B e^2}{4\pi D_r D_0 r_{AB}} \tag{14}$$

Dissociation is accompanied by a change in entropy

$$\Delta S_{dis,\,el} = -\frac{\partial \Delta G_{dis,\,el}}{\partial T}$$

$$= \frac{NZ_A Z_B e^2}{4\pi D_r^2 D_0 r_{AB}} \left(-\frac{\partial D_r}{\partial T}\right) \tag{15}$$

The change in relative permittivity, D_r, is usually described by the relation (D'_r and L characterize the liquid medium, in the simplest case the solvent)

$$D_r = D'_r \exp(-LT)$$

thus

$$\frac{\partial \ln D_r}{\partial T} = -L$$

Then

$$\Delta S_{dis,\,el} = \frac{NZ_A Z_B e^2}{4\pi D'_r D_0 r_{AB}} L \exp(LT) = -L \Delta G_{dis,\,el} \tag{16}$$

From eqns. (14) and (16)

$$\Delta H_{dis,\,el} = \Delta G_{dis,\,el}\,(1 - TL) \tag{17}$$

The electrostatic contribution of the Gibbs energy (free enthalpy) is usually negative; the entropy change is always negative. However, the electrostatic factors alone cannot completely describe the behaviour of electrolytes in solution [68d, 72–75]. This is caused by the neglect of permittivity changes, and perhaps also of other properties of the immediate neighbourhood of the ion [76]. Nevertheless, log K_{dis} calculated for salts in various solvents varies with D_r^{-1} in agreement with theory [77].

Winstein et al. [78] found that any form of an ion pair can participate in a reaction, specifically in monomer addition. Ionization[†] and dissociation of the electrolyte is described by the equilibria [see Chap. 3. eqn. (65)]

$$AcB \underset{}{\overset{S}{\rightleftharpoons}} Ac^+B^- \underset{}{\overset{S}{\rightleftharpoons}} Ac^+ \| B^- \underset{}{\overset{S}{\rightleftharpoons}} Ac^+ + B^- \tag{18}$$

The rate of monomer addition usually (but not always) increases with a shift of equilibrium (18) to the right, from the non-ionized forms towards free ions. The difference in the values of the rate constants of addition often amounts to several orders of magnitude (three and even more).

2.2 Centres of anionic polymerizations

These centres are formed by the addition of monomer to a suitable anion. They are almost always simpler than their cationic reverse part. The counter ion is usually a metal cation[††] able to interact with the electrons of the growing end of the macromolecule, and to bind in its ligand sphere monomer or solvent molecules or parts of the polymer chain. This changes the properties of the whole centre. Therefore, by selection of the metal, the stability of the centre, the tendency of the centres to aggregation, the position of the equilibrium between the contact and solvent-separated ion pairs and free ions, and the stereoselectivity of the centre [the ability to produce polymers with an ordered structure (tacticity, see Chap. 5, Sect. 4.1)] are predetermined. The chemical reactions of the metal cations are, however, very limited. Most solvents and potential impurities are of nucleophilic character. They readily solvate the cation, leaving the anion relatively free. The determination

---

[†] In this case, ionization means the transition of a covalent bond between the atoms of a certain compound in solution to an ionic bond.
[††] There are rate exceptions, for example R_4N^+, R_4P^+ may act as counter-ions instead of the metal.

References pp. 222–229

of the energy state of the anionic centre is similar to the case of the radical centre, but contributions from ionization, dissociation and solvation have to be considered.

2.2.1 CARBANIONS

The most important representatives of anionic polymerization centres are formed from vinyl and diene monomers. The trivial schematic representation of a carbanion

$$\sim\overset{|}{\underset{|}{C}}\,|\,^-$$

never reflects reality. The free electron pair always interacts with the n, π, and even σ electrons of nearby atom groups. The configuration of electrons in the centre is always, of course, also affected by the type of counter-ion and by its distance from the centre of negative charge distribution. Carbanions often absorb visible light, and in such cases they are coloured.

The best studied active centre is the carbanion derived from styrene. With Na^+ as a counter-ion, at temperatures <273 K in THF it represents a real living centre. Monomer addition is the only chemical reaction proceeding on it[†]. The number of possible reactions of the centre increases with increasing Stokes radius[††] of the counter-ion and with increasing temperature. With Li^+, the centre is strictly living only at temperatures <233 K.

The kinetic characteristics of the centres has been well studied in living systems. The rate of styrene addition to a free carbanion is about a thousand times more rapid than to a contact ion pair. The transition from the well-solvating THF to poorly solvating hydrocarbon solvents leads to complications; the centres associate.

An alkali metal and α-methylstyrene easily yield a thermodynamically stable dianionic tetramer, by analogy to reaction (41) in Chap. 3. The electrons of the terminal carbon are sp^2 hybridized, and Li^+ in the contact ion pair interacts with the aromatic ring; thus Li^+ is not placed on the α C atom but in the space above the Ph–C bond [79].

[†] Strictly speaking, this statement is not quite true (see Chap. 6, Sect. 2.2 and Chap. 7 Sect. 3.
[††] Including the solvation sphere.

It appears that two tetramer structures are possible

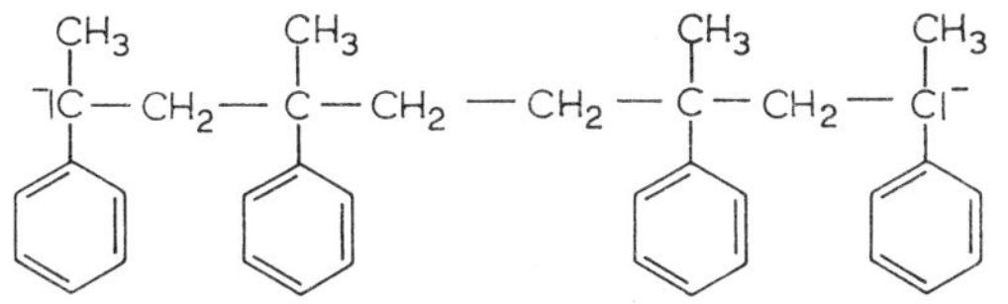

[Structure a: tetramer with four phenyl-substituted units connected by CH₂ groups, terminated with Cl⁻]

a

[Structure b: alternative tetramer arrangement with phenyl-substituted units, terminated with Cl⁻]

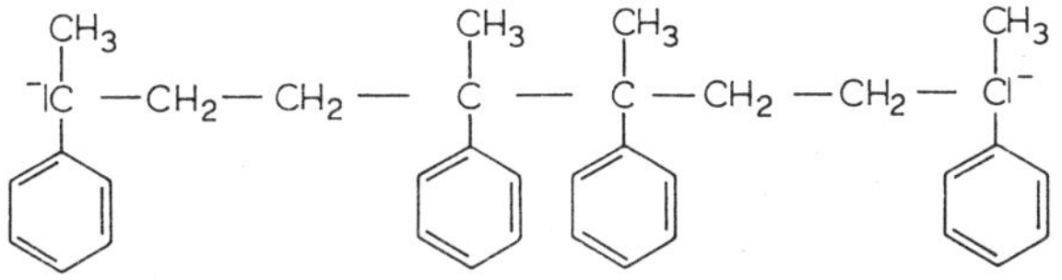

b

Formation of the dianions *a* or *b* depends on the conditions and also, among other things, on the ratio of the reacting components and on type of alkali metal [80–83]. According to Malhotra [83], in the presence of monomer, the tetramer *a* grows more rapidly than the dianion *b*. This results in the generation of chains of unequal lengths, and the final effect resembles chain growth on ion pairs of various activity, for example on contact and solvent-separated pairs (see Chap. 5, Sect. 8.1).

The structure and chemistry of delocalized anions were studied by Yasuda and Nakamura [84]. In pentadienyl, K^+ is surrounded by a horseshoe (U)-shaped structure (*cis, cis*; Z, Z structure)

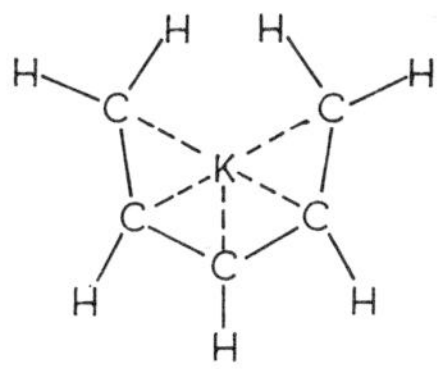

[Diagram of U-shaped pentadienyl structure with K⁺ at center]

In the Li^+ derivative, the hydrocarbon anion is of shape (E, E)

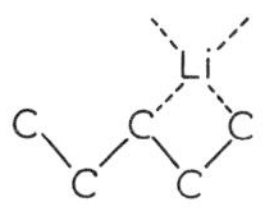

[Diagram of W-shaped pentadienyl structure with Li⁺]

These planar U and W forms are more stable than other conformations of these two anions. The geometry of the anions strongly depends on the medium. These findings enable us to better understand initiation and propagation processes. In diene, 1; and substituted alkene (acrylate, 2; 2-vi-

References pp. 222–229

nylpyridine, 3) polymerizations, the active centres are planar (or almost planar) systems with delocalised electrons [85].

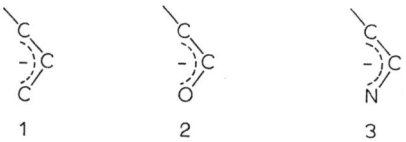

The rotation barrier about the β–γ bond is not negligible, so that the centres exist in either the *cis* or *trans* forms (Z, E); mutual transitions between these two forms only occur under certain conditions and at a certain rate. In diene polymerizations in hydrocarbon medium with Li^+ as counter-ion, monomer addition transforms a *cis* centre into a *cis* segment of the chain, and a *trans* centre into a *trans* polymer (the former case is typical for isoprene polymerization; in THF the equilibrium is shifted towards the *cis* form over the whole of the available temperature range). A *cis* $\rightleftharpoons$ *trans* transition is, of course, not excluded (for example with butadienyllithium), and when it is more rapid than addition, chain stereospecificity is reduced.

In polar solvents, the *cis* form is generally more stable, but in the activated complex the *trans* conformation is preferentially formed and can be preserved [86]. Therefore data on the isomerization of ion pairs in active centres are of great importance, especially for controlled propagation to stereospecific (stereoregular) polymers.

The location of the cation with respect to the growing end also plays an important role. It is most likely placed in the range of the π electron cloud, above or below the plane of the anion. Cation oscillations from one to the other position may therefore occur. When the cation jumps are more rapid than monomer addition, then the structure of the centre is insufficiently defined, so that it can form a *meso* or racemic diad with the penultimate unit; this configuration cannot be changed by further monomer addition. In this case, the structure of the generated diad will also be affected by the spatial disposition of the pair centre – approaching monomer [86]. Under certain conditions (yet to be found) the rate of the cation oscillations is expected to be reduced, and the directing effect of the centre should be enhanced. When the direct interaction between centre and cation is interrupted, for example by selective solvation of the metal ion, then the stereospecific effect of the centre as well as its reactivity and selectivity with respect to various monomers will be modified. Polymerization of butadiene and its copolymerization with styrene, initiated with alkylsodium, leads to different products in the presence of various ethers. A low polymer is formed with aliphatic ethers; with crown ethers a high polymer and a copolymer of different composition are produced.

Crown ethers, e.g. bicyclohexyl-18-crown-6

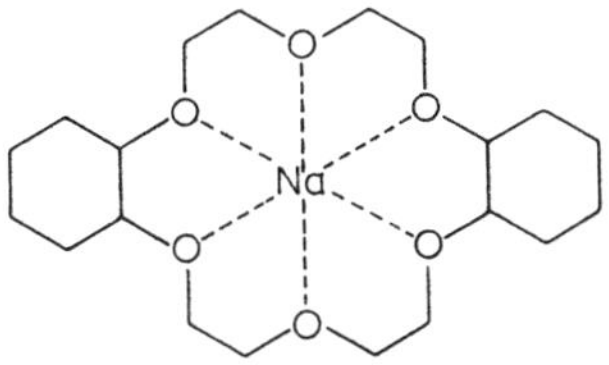

enhance anion reactivity by complexation of the counter-ion [87].

When the electrons of the monomer vinyl group are withdrawn by a substituent less stable than methyl or phenyl, then the centres have the opportunity not only to add monomer, but also to react with it in a different way. Instead of the vinyl double bond, the centre can attack the nitrile group of methacrylonitrile

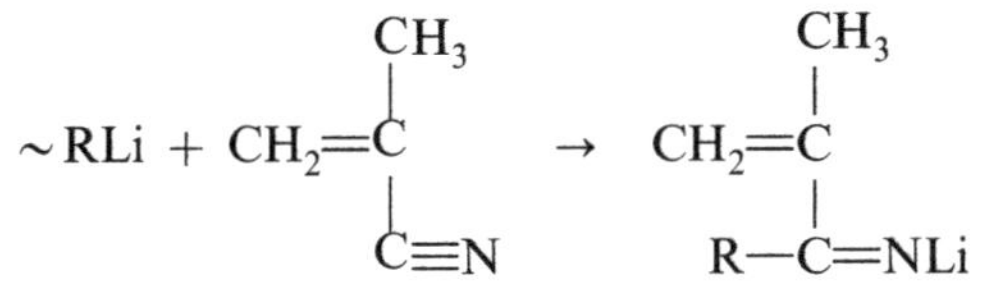

$$\sim RLi + CH_2=\underset{\underset{C\equiv N}{|}}{\overset{\overset{CH_3}{|}}{C}} \rightarrow CH_2=\underset{\underset{R-C=NLi}{|}}{\overset{\overset{CH_3}{|}}{C}}$$

or it can add to six-membered heterocyclic rings [88] formed from $-C\equiv N$ during polymerization [89]. Similar reactions are, of course, also possible with acrylonitrile [90]. The active centres are transformed by these reactions to less reactive groups and polymerization dies out. The ester groups of acrylate and methacrylate monomers and polymers are similarly a source of side reactions, leading to activity decay; for example, the strongly basic C anion is transformed to the less basic O anion [91]

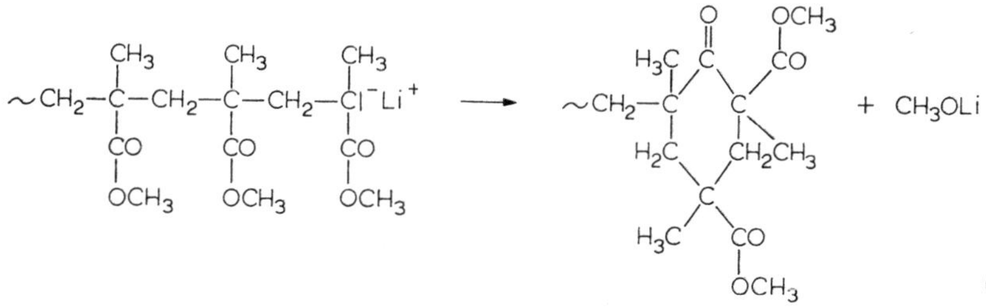

(20)

[see also eqns. (34)–(36) in Chap. 3].

The polymerization of methyl methacrylate is an elegant example of the success of careful studies resulting in a useful tailored modification of the active centre. The typically non-stationary methacrylate polymerization in hydrocarbon medium can be transformed to a stationary process by the

References pp. 222–229

addition of tertiary alkoxides of alkali metals. The active centre is protected by the alkoxides [see eqn. (38) in Chap. 3]. The generated formation reacts almost exclusively with the vinyl groups of the monomer [92, 93].

On special centres (complexes of Grignard reagents) selectively chosing first the R then the S antipode, two different polymer chains can be obtained from a racemic mixture of α-methylbenzyl-methacrylate [94]. The immediate vicinity of the centre is undoubtedly of great importance for monomer addition. This is documented by the effect of solvents. When the centre is sorbed on a polyelectrolyte matrix, the distance of the matrix ions will affect the length and quality of the counter-ion bond of the centre, and thus also the rate of polymerization [95].

These are examples of findings which, when generalized, should lead to a solution of the problem of structure-reactivity relations in anionic centres.

2.2.2 OTHER ANIONS

Of great importance are the silanolate active centres

$$\sim \overset{|}{\underset{|}{Si}}-O^- Ac^+$$

on which polysiloxane chains from cyclic monomers are propagated on an industrial scale. These centres are also interesting from a theoretical point of view. They represent one of the cases where Ac^+ may be an alkali metal ion, but also a tetraalkylamonium or phosphonium cation. Because Si has a coordination number 6, these centres can easily associate [96, 97].

The electrophilic silicon can even bind nucleophilic compounds (ethers, amines, etc.) yielding complexes which initiate polymerizations of vinyl compounds [98]

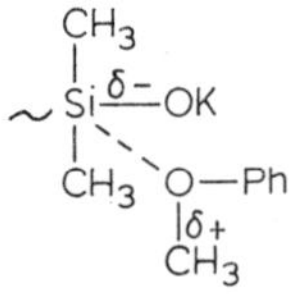

$$\sim \overset{CH_3}{\underset{CH_3}{\overset{|}{Si}}}\overset{\delta-}{-}OK$$
$$\qquad \overset{}{\underset{}{}} O-Ph$$
$$\qquad \overset{}{\underset{\delta+}{|}}$$
$$\qquad CH_3$$

By interaction of alkyllithium with 2-methylthiirane, active centres are produced in the form RS^-Li^+. With monomer these centres yield polypropylenesulphide. In the presence of lithium tert.-alcoholates, complexes of the shape

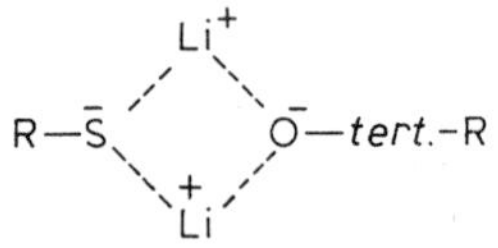

$$R-\bar{S}\overset{Li^+}{\underset{\underset{Li}{+}}{\diamondsuit}}\bar{O}-tert.-R$$

are produced, followed by

$$\begin{array}{c} \text{Li}-\text{O}-tert.\text{-R} \\ -\text{CH}-\text{S}-\text{Li} \\ \text{CH}_3\text{Li}-\text{O}-tert.\text{-R} \end{array}$$

on which polypropylene disulphide is formed by the overall reaction [99]

$$2n\ \underset{S}{\underset{\diagdown\ \diagup}{\text{CH}_2-\text{CH}-\text{CH}_3}} \rightarrow \underset{\text{CH}_3}{\underset{|}{+\text{CH}_2-\text{CH}-\text{S}-\text{S}+_n}} + n\ \text{CH}_2=\text{CH}-\text{CH}_3$$

Thus the modification of centres by tertiary alcoholates is not limited to MMA polymerization.

The use of the silyl anion derivative $(\text{CH}_3\text{O})_n (\text{CH}_3)_{3-n}\text{Si}^-\text{Na}^+$ [100] in the role of an active centre is as yet only possibility. The usefulness of theoretical studies of some centres of this kind, for example of those formed during phosphazene polymerization (see Chap. 2, Sect. 1.1), should increase with time.

(a) *Centres of ring-opening polymerization*

Anionic polymerizations of heterocycles differ considerably from polymerizations of unsaturated monomers.

In heterocycles, the bonds of heteroatoms to neighbouring atoms are the least resistant to ion attack. By anion force, non-symmetrically substituted monomers can yield centres of various structure following α or β cleavage [101]

$$\underset{\beta\alpha}{\underset{\diagdown\ \diagup}{\underset{\text{O}}{\overset{\text{Me}}{\underset{|}{\text{CH}_2-\text{CH}}}}}} + \text{B}^- \diagup \begin{array}{c} \overset{\text{Me}}{\underset{|}{\text{BCH}}}\diagup\text{CH}_2\diagdown_{\text{O}^-} \\ \\ \text{BCH}_2\diagdown\underset{|}{\overset{\text{CH}}{\underset{\text{Me}}{}}}\diagdown_{\text{O}^-} \end{array} \quad (21)$$

During lactone polymerization, the O-acyl or O-alkyl bonds are broken yielding carboxylate or alcoholate anions. Each monomer addition need not result in a single opening type; centres formed by initiation can be transformed to the other anion structure during propagation [102]. Penczek and Slomkowski have found both types in some heterocycles [103] so that, for example, oxirane grows on the centre $\sim\text{CH}_2\text{O}^-$; phenyloxirane on $\sim\text{CH}_2\text{CH(Ph)}\text{-O}^-$ and $\sim\text{CH(Ph)CH}_2\text{-O}^-$; γ-propiolactone on

$$\sim\text{CH}_2\text{C}\underset{\text{O}}{\overset{\text{O}}{\diagup}} -$$

but γ-caprolactone on $\sim\text{CH}_2\text{O}^-$, and methylthiirane on $\sim\text{CH}_2\text{CH(Me)S}^-$.

References pp. 222–229

In these centres, the negative charge is concentrated on the heteroatom. The centre-counter-ion interaction is much stronger than with carbanion centres. This is manifested by a tendency to aggregate and solvate formation. The reactivity of this kind of ion pair may exceed the reactivity of free ions, due to the unequal contribution of the desolvating energies of ion pairs and free ions [103].

In polymerizations of heterocycles, the reactivity of anionic centres is considerably affected by cation solvation with the polymer [104]. Schematically

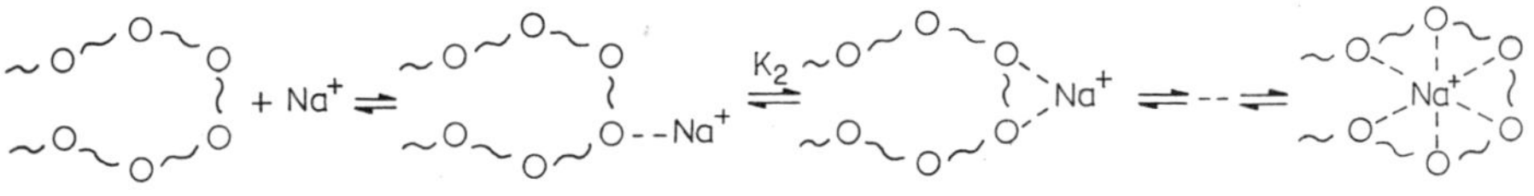

$$(22)$$

In this way, the polymer may to some extent assume the role of crowns and cryptands (see Chap. 3, Sect. 2.5), and thus affect the polymerization process.

(b) *Centres of anionic polymerization with activated monomer*

The interpretation of the mechanism of anionic lactam polymerization based on the conventional scheme (ionic active centre with approaching monomer) could not exhaustively explain all the observed effects. Agreement could only be obtained when the acido-basic properties of lactams and polyamides had been respected. The equilibrium

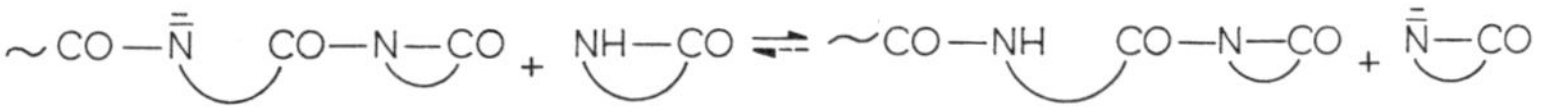

$$(23)$$

is shifted strongly to the right, and it is established more rapidly than the addition itself. A small part of the monomer exists in the ionic, i.e. highly active form; it can react in several ways [105] of which only one increases the degree of polymerization of some chains by one unit. The activated monomer is almost always written as a free anion. It is of course evident that participation of ion pairs or even of aggregates of a larger number of particles cannot be excluded, even at the relatively high permittivity of lactams and their polymers ($D = 10$–33 [105, 106]). The position of equilibrium (23) is a function of temperature, lactam acidity (decreasing in rings from 3 to 7 members but changing little in larger rings [107]) and permittivity of the medium.

Lactam polymerization with anionically activated monomer has its counterpart in the cationic processes of lactam polymerization. This type of mechanism has also been observed recently in some polymerizations of oxygen-containing heterocycles (see Chap. 4, Sect. 2.3)

2.3 Centres of cationic polymerization

These centres take the form of free cations or ion pairs, rarely of associates of a larger number of ions. The counter-ion is always of a weakly basic character. A cationic centre is characterized by an electron gap, formally situated in the valence sphere of one of the constituting atoms, and it is strongly electrophilic. The acidity of the centres is usually higher than that of concentrated H_2SO_4; they belong to the class of superacids. Even when the electron defect is delocalized to a larger number of atoms (as in radicals), it is useful to name the cations according to the element from which the electrons have been removed.

2.3.1 CARBOCATIONS

This common name covers two cation groups [108]
(a) trivalent ("classical") carbenium ions with an sp^2 hybridized C atom. In the absence of conformational effects and steric interference they are almost planar.
(b) penta- or tetracoordinated ("non-classical") carbonium ions with three two-electron covalent bonds and a fourth two-electron, three-centre bond

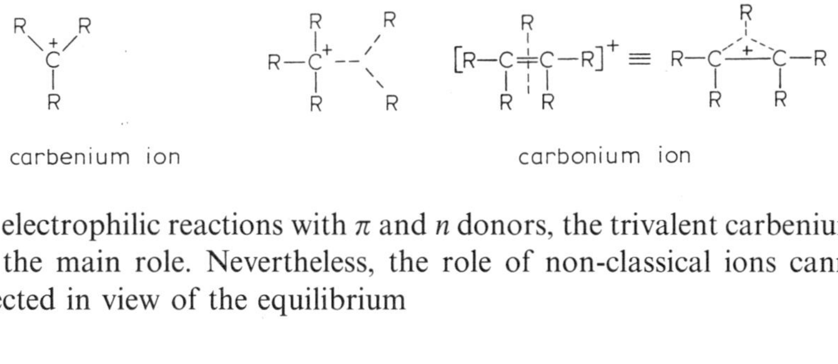

carbenium ion   carbonium ion

In electrophilic reactions with π and n donors, the trivalent carbenium ions play the main role. Nevertheless, the role of non-classical ions cannot be neglected in view of the equilibrium

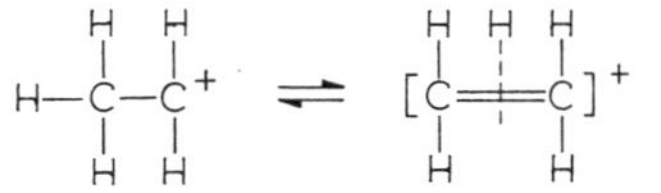

$$\text{H---C---C}^+ \rightleftharpoons [\text{C}\doteq\text{C}]^+ \tag{24}$$

According to Olah et al. [109], the ethylium carbenium ion is more stable than the carbonium ion by 40 kJ mol^{-1}. Thus the cationic polymerization centre of hydrocarbon monomers (in free ion form) has the trivial shape

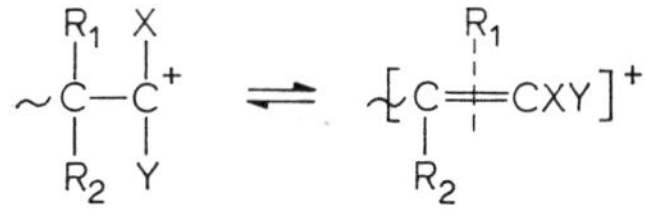

$$\sim\text{C}-\text{C}^+ \rightleftharpoons [\text{C}\doteq\text{CXY}]^+ \tag{25}$$

The equilibrium (25) has been known for some time, but its consequences have been neglected, even in relatively modern monographs [110] (see Chap. 7, Sect. 4.1).

So far attempts to calculate exactly the cation-substrate energy state (cation reactivity) have been no more successful than in the case of radical reactivities, perhaps even less so. For some reactions, however, satisfactory agreement between theory and experiment has been obtained by semiempirical methods [111]. The applied methods are not simple, and they have not been applied to polymerizations.

Cationic active centres rarely react with the monomer directly without side reactions. Their energy state permits several modes of interaction, as with other molecules. One of the complicating factors is ion solvation, which changes the concentration relationship in the immediate vicinity of the centre. A theory quantitatively describing the reasons and consequences of solvation is at present being developed [112–121]. Reactions of a counter-ion with a cation may be irreversible and hard to control. Even a hypothetical isolation of the centre from all the mentioned effects would not guarantee its preservation as a single chemical entity. Under certain circumstances, a less stable carbocation can be rearranged to a more stable form.

So, for example, in the polymerization of 3-methyl-1-butene, monomer addition to the active centre yields a cation on the secondary carbon I, which is then isomerized to II. The energy change in the transition from I to II is about 40 kJ mol^{-1}.

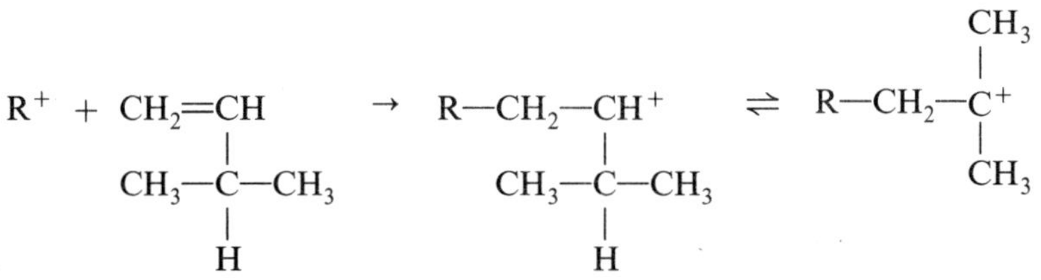

$$R^+ + \underset{\underset{H}{\overset{|}{\underset{|}{CH_3-C-CH_3}}}}{CH_2=CH} \rightarrow \underset{\underset{H}{\overset{|}{\underset{|}{CH_3-C-CH_3}}}}{R-CH_2-CH^+} \rightleftharpoons \underset{\underset{CH_3}{|}}{\overset{\overset{CH_3}{|}}{R-CH_2-C^+}}$$

$$\text{I} \qquad\qquad \text{II}$$

The conditions of this isomerization have been roughly established [121].

All these circumstances cause cationic polymerization to be a very sensitive process, despite the fact that it may also be living, i.e. without being subject to transfer and termination [123, 124]. Monomer addition to carbocations usually yields an atactic polymer.

2.3.2 CARBOXONIUM CENTRES

Reactions of carbocations with acetal bonds, aldehydes, ketones, etc. yield carboxonium ions. Various structures have been assigned to these ions. They most probably actually exist in various variants, mutually connected by equilibria. The proportion and structure of the predominant form depend on the structure of the original particles and on their neighbourhood.

Trioxane polymerization proceeds on the active centres [125, 126]

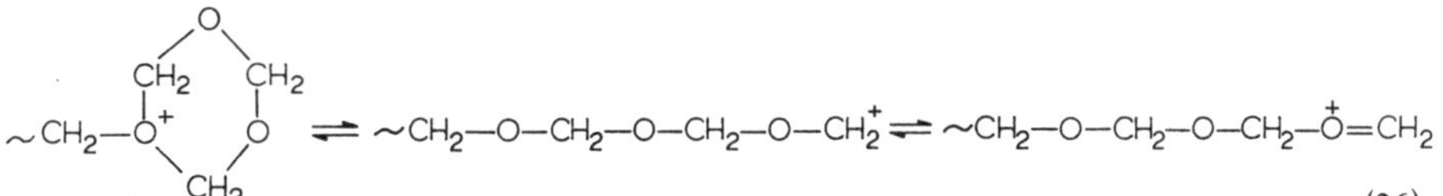

$$\sim CH_2-\overset{+}{O}\underset{CH_2}{\overset{CH_2\diagdown O\diagup CH_2}{\diagdown \diagup}} \rightleftharpoons \sim CH_2-O-CH_2-O-CH_2-O-\overset{+}{CH_2} \rightleftharpoons \sim CH_2-O-CH_2-O-CH_2-\overset{+}{O}=CH_2 \quad (26)$$

Immediately after the start of dioxolane polymerization, initiated by the triethyloxonium salt $Et_3O^+B^-$, the carboxonium centres present are of the form [127]

$$\sim CH_2-O^+=CH_2$$
$$B^-$$

and

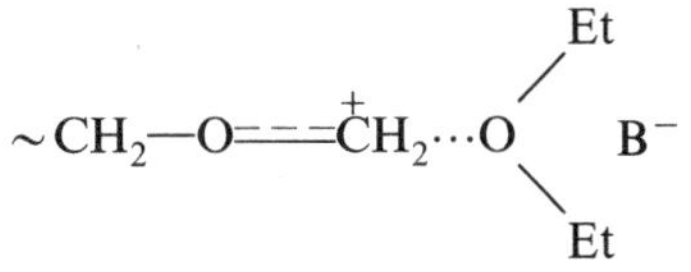

$$\sim CH_2-\overset{+}{O}===CH_2\cdots O\diagup^{Et}_{\diagdown Et} \quad B^-$$

At sufficiently high conversions, the more basic stabilizing diethyl ether is replaced by the less basic but more abundant polymeric chain, and the generated centres are of the form III [128]

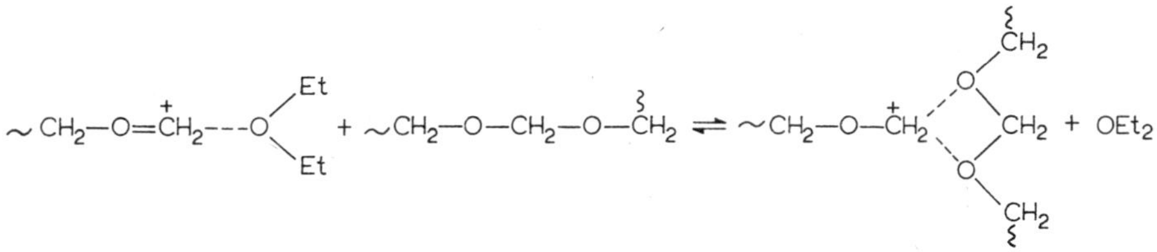

$$\sim CH_2-\overset{+}{O}=CH_2---O\diagup^{Et}_{\diagdown Et} + \sim CH_2-O-CH_2-O-\overset{?}{CH_2} \rightleftharpoons \sim CH_2-O-\overset{+}{CH_2}\diagup^{O-CH_2}_{\diagdown O-CH_2} CH_2 + OEt_2$$

$$III$$

Eizner and Yeruzalimski [129] observed that the bond between O and the terminal methylene group in $CH_3\overset{+}{O}===CH_2$ is of double bond character.

The carboxonium centres are resonance-stabilized [130], and therefore much less reactive than carbenium centres.

References pp. 222–229

In β-lactone

$$(O=\overset{\frown}{C}OCH_2CH_2)$$

polymerization, both acylium

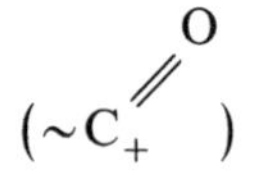

$$(\sim \underset{+}{C}\!\!\nearrow^{O})$$

and carboxonium

$$(\sim CH_2O\!\!\overset{+}{\cdots}\!\!C\!\!\overset{O}{\underset{\diamondsuit}{\diagup}})$$

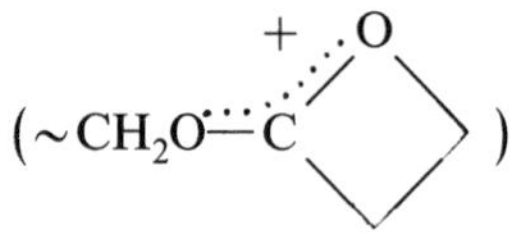

centres are formed during initiation; their ratio depends on the kind of initiator. With increasing chain length, the acylium centres disappear, and further growth occurs on carboxonium centres [131]. In ε-caprolactone polymerization, acylium centres also participate in propagation and side reactions [132].

Analogous to carboxonium centres are the siloxonium centres (see Chap. 3, Sect. (3.2) [133, 134]

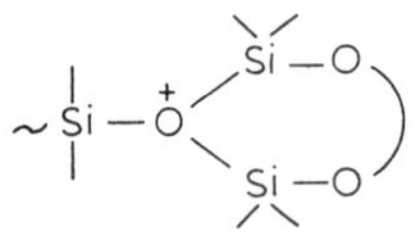

$$\sim \underset{|}{\overset{|}{Si}}-\overset{+}{O}\!\!\underset{Si-O}{\overset{Si-O}{\diagup\!\!\!\diagdown}}$$

These are excellent initiators and centres of the cationic propagation of polysiloxane chains.

2.3.3 OXONIUM CENTRES

Oxonium centres are formed by the reaction of a cation with the oxygen of an ether-type monomer. They are even more stable than carboxonium centres. For example the reactivity of an oxonium ion is not sufficient for the separation of a hydride ion from the monomer [128]. The behaviour of the centre strongly depends on the stability of the counter- ion (see Chap. 6, Sect. 2.1).

The reduced probability of side reactions enables some heterocycle polymerizations to proceed by a living mechanism.

The structure of oxonium centres is described by the general formula

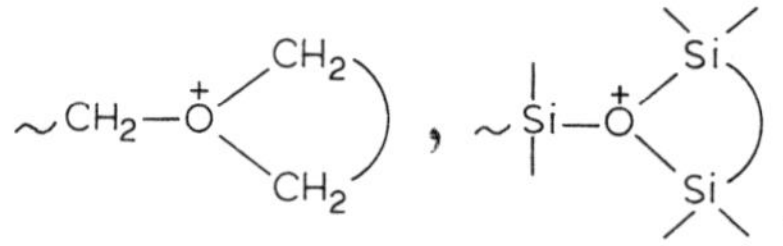

$$\sim CH_2-\overset{+}{O}\!\!\underset{CH_2}{\overset{CH_2}{\diagup\!\!\!\diagdown}} \quad , \quad \sim \underset{|}{\overset{|}{Si}}-\overset{+}{O}\!\!\underset{Si}{\overset{Si}{\diagup\!\!\!\diagdown}}$$

Which illustrates only one of the possible (limiting) forms; actually only this form, in the shape given above or in connection with a counter-ion, represents an oxonium centre. In reality they exist in several forms (as esters, ion pairs, free ions) connected by equilibria.

Ion pairs and free ions are specifically solvated by monomer. This is the reason for the small difference in the monomer addition rates to ion pairs or free ions [135, 136].

2.3.4 COVALENT CENTRES

The existence of centres with non-ionic character has already been suspected in studies of polymerizations which are supposed to proceed on carbocations; the theory of pseudo-cationic polymerization was proposed [137] (see Chap. 3, Sect. 3.1). The transformation of an ion pair to a covalent compound will evidently be easier for acid centres with heteroatoms, i.e. in heterocycle or vinyl ether polymerizations. Propagation on covalent bonds has actually been observed, first in the studies of oxazoline polymerization [138] and later even with THF [139, 140] and with other monomers (see, for example, refs. 131, 141 and 142).

Ionic and covalent centres usually coexist [143], and their equilibrium proportions are a function of temperature, counter-ion type, quality of solvent and conversion.

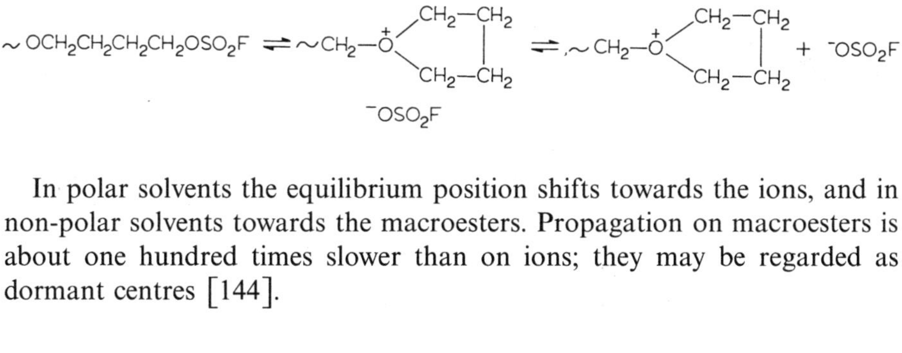

$$\sim OCH_2CH_2CH_2CH_2OSO_2F \rightleftharpoons \sim CH_2 - \overset{+}{O} \begin{pmatrix} CH_2-CH_2 \\ | \\ CH_2-CH_2 \end{pmatrix} \rightleftharpoons \sim CH_2 - \overset{+}{O} \begin{pmatrix} CH_2-CH_2 \\ | \\ CH_2-CH_2 \end{pmatrix} + {}^-OSO_2F$$
$${}^-OSO_2F$$

In polar solvents the equilibrium position shifts towards the ions, and in non-polar solvents towards the macroesters. Propagation on macroesters is about one hundred times slower than on ions; they may be regarded as dormant centres [144].

2.3.5 CENTRES OF CATIONIC POLYMERIZATION WITH ACTIVATED MONOMER

When the basicities of a monomer and of some atom group in a polymer differ considerably, a situation may occur where the propagating chain end remains neutral for most of its propagation time, and a part of the monomer

References pp. 222–229

is activated by the cation (usually proton). This case occurs when reaction (27a) is more rapid than (27b).

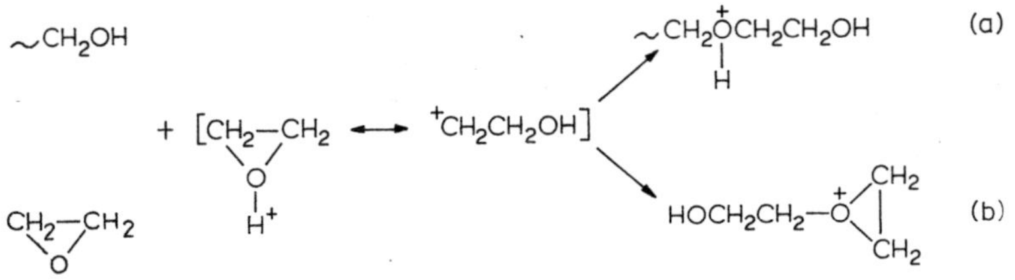

$$\sim\!CH_2OH + [CH_2\!-\!CH_2 \rightleftharpoons {}^+\!CH_2CH_2OH] \longrightarrow \begin{cases} \sim\!CH_2\overset{+}{\underset{H}{O}}CH_2CH_2OH & (a) \\[6pt] HOCH_2CH_2\!-\!\overset{+}{O}\!\!\diagup\!\!\overset{CH_2}{\underset{CH_2}{|}} & (b) \end{cases} \quad (27)$$

$$CH_2\!-\!CH_2\diagdown\!O\diagup\,\,H^+$$

In order to satisfy the conditions for polymerization with activated monomer, $\sim\!CH_2OH$ must be more basic (nucleophilic) than oxirane. As

$$\sim\!CH_2\overset{+}{\underset{H}{O}}CH_2OH$$

is, on the contrary, much more acidic than the monomer, it rapidly eliminates a proton which activates a further oxirane, and the process is repeated; for the overall reaction we can write

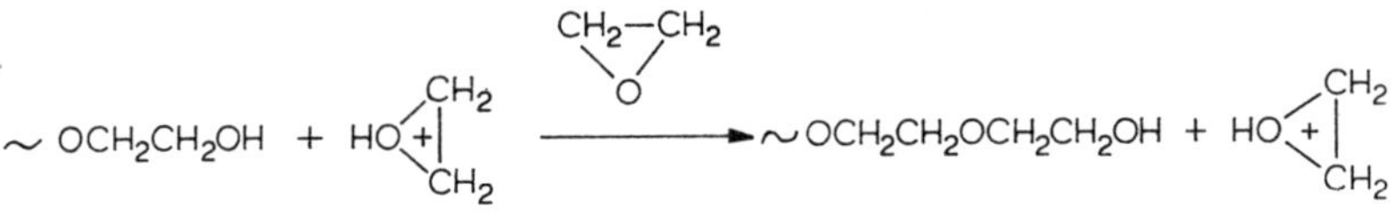

$$\sim\!OCH_2CH_2OH + HO\!\overset{+}{\diagup\!\!\diagdown}\!\!\overset{CH_2}{\underset{CH_2}{|}} + \overset{CH_2\!-\!CH_2}{\underset{O}{\diagdown\!\!\diagup}} \longrightarrow \sim\!OCH_2CH_2OCH_2CH_2OH + HO\!\overset{+}{\diagup\!\!\diagdown}\!\!\overset{CH_2}{\underset{CH_2}{|}} \quad (28)$$

So far, not many systems are known to meet the conditions of the growth mechanism with activated monomer. Polymerizations of oxirane, chloromethyloxirane [145, 146] and of lactams [147] belong to this class. The existence of an electroneutral propagating macromolecule and of an activated monomer results in reduced probability of back-biting reactions and in easier preparation of macromers [145] by means of exchange reactions.

2.3.6 CATIONS DERIVED FROM ELEMENTS OTHER THAN C, Si AND O

To date these have not been very important as active centres of polymerizations, mainly because of the limited number of conventional monomer types suitable for building such a centre. Nitrogen cations are an exception. Cyclic imines polymerize on centres [148, 149]

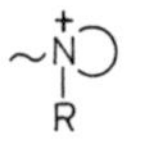

$$\sim\!\overset{+}{\underset{R}{N}}\!\!\diagup\!\!\diagdown$$

lactams either on growing chain centres (for example $-NH_2$) with an activated monomer [147, 150, 151]

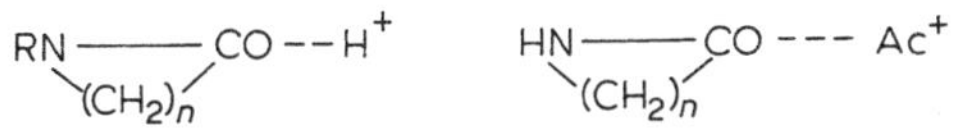

$$RN\underset{(CH_2)_n}{\diagdown}CO--H^+ \qquad HN\underset{(CH_2)_n}{\diagdown}CO---Ac^+$$

or by means of conjugated (stable) cations in a classical manner with a charge at the growing end [147]

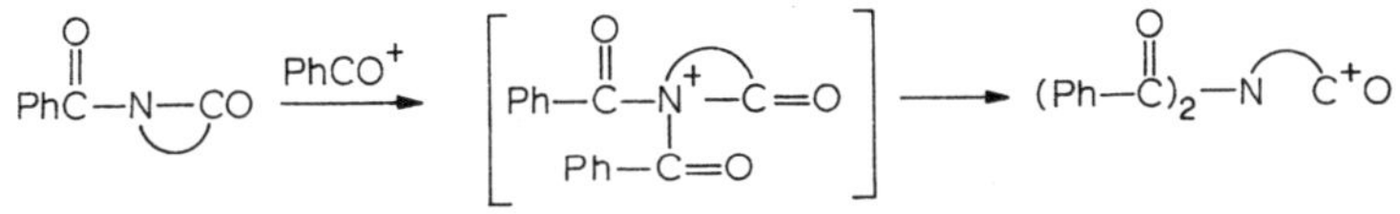

$$PhC(=O)-N-CO \xrightarrow{PhCO^+} \left[\begin{array}{c} Ph-C(=O)-N^+-C=O \\ | \\ Ph-C=O \end{array}\right] \longrightarrow (Ph-C(=O))_2-N\overset{+}{\frown}C^+O$$

Centres of this kind might even gain technical importance in the future because N-substituted lactam derivatives can only be polymerized by a cationic process.

Many centres with a formal positive charge on nitrogen have already been described, especially with polymerizations of more or less exotic monomers. 4,5-Dihydro-1,3-oxazine is transformed to polymer by the overall reaction

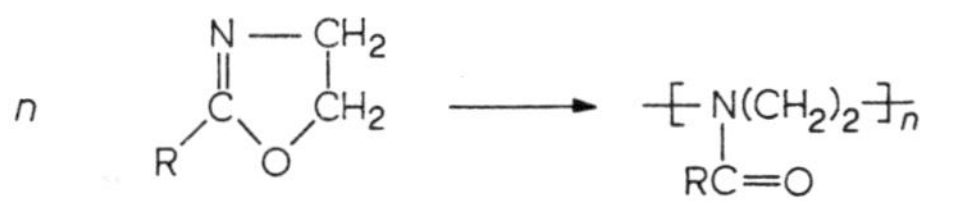

$$n \underset{R}{\overset{N=CH_2}{\underset{O}{\diagup\diagdown}}}\overset{CH_2}{\underset{}{|}} \longrightarrow -\!\!\!\left[N(CH_2)_2\right]_n\!\!\!-\\ \qquad\qquad\qquad\qquad\qquad RC=O$$

both on ions (ion pairs) or on a covalent active centre, according to the counter-ion [152] (X = I, Y = Cl)

$$\sim\overset{\bar{N}}{\underset{R}{\diagup\diagdown_O}}X^- \qquad \sim NCH_2CH_2Y \\ \qquad\qquad\qquad\quad\; |\\ \qquad\qquad\qquad RC=O$$

A fair amount of data on the structure of centres suitable for the polymerization of phosphazenes, phosphonites [153] and sulphur-containing monomers [154] can be found in the literature.

2.4 Zwitterions

There exist compounds the molecules of which exhibit the character of donor-acceptor complexes with separated centres of positive and negative charge. Initiator and monomer sometimes do not produce a classical ion pair, but a molecule carrying both types of opposite charge. In such a case, the generated zwitterions (amphiions) represent a special class of active centres.

References pp. 222–229

The mechanism of monomer addition to these centres exhibits some special features (see Chap. 5, Sect. 5.8).

Some examples will be given by way of illustration. Zwitterion character is typical of compounds with an S-heterocycle connected to a phenol ring by means of the sulphur atom [155].

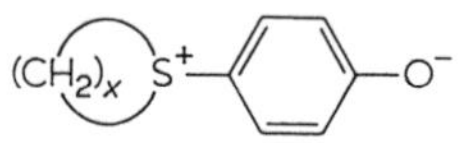

$$(CH_2)_x\overset{+}{S}-\underset{}{\bigcirc}-O^-$$

These internal salts are stabilized by hydration. Removal of the water of crystallization causes polymerization. The nucleophilic phenolic anion attacks the α carbon of the sulphonium ion, generating a linear chain

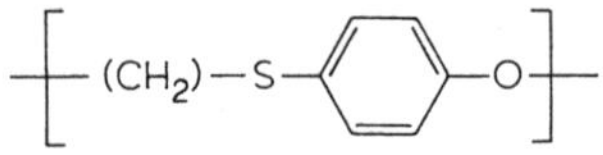

$$\left[(CH_2)-S-\underset{}{\bigcirc}-O \right]$$

The S—Ph bond is not broken [156].

According to Meerwein et al. [157], the result of the interaction between 2-chloromethyloxirane and BF_3OEt_2 is an unstable intermediate of zwitterionic character

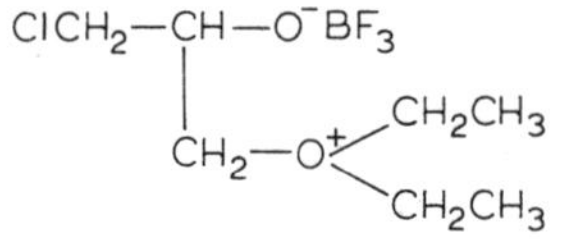

$$\begin{array}{c} ClCH_2-CH-O^-BF_3 \\ | \\ CH_2-O\overset{+}{\underset{}{<}}\begin{array}{c}CH_2CH_3\\CH_2CH_3\end{array} \end{array}$$

This intermediate reacts with another molecule of BF_3OEt_2 forming a neutral molecule and $EtO_3{}^+BF_4{}^-$. Komratov et al. [158] discovered macrozwitterions, apparently analogous to the type shown above, on the polymerization of THF through BF_3 in the presence of methyloxirane. Depending on the chain length, temperature and polarity of the medium [157], both ends will either be free or will form an ion pair; the latter may be of the contact or solvent-separated type. At 293 K, most active centres exist in the form of cyclic zwitterions [158].

By reaction with covalent organic bases (Et_3N, Me_2PhN, pyridine) N-alkyl-2-cyanoacrylates can be transformed to zwitterions of the type

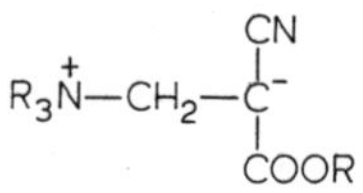

$$R_3\overset{+}{N}-CH_2-\underset{\underset{COOR}{|}}{\overset{\overset{CN}{|}}{C^-}}$$

and further molecules of N-alkyl-2-cyanoacrylates polymerize at their anionic end. As a weak base is sufficient for monomer ionization, the reaction is not suppressed by traces of O_2, CO_2, H_2O, and CH_3OH (which would effectively terminate a conventional anionic polymerization process) [159–161].

Saegusa et al. observed that a monomer able to transfer electrons can be oxidized by an electrophilic monomer yielding a donor-acceptor complex with significantly separated charge. The generated zwitterions combine spontaneously, producing a strictly alternating copolymer. No initiator is needed for this reaction, which was designated by the authors as oxidation-reduction copolymerization (not to be confused with redox initiation) [162].

The donors used are, for example, 1,3,2-dioxaphospholane [163], 2-phenylimino-1,3-dioxolane [164]. Vinyl monomers with electronegative substituents (acrylic acid, acrylamide, methacrylonitrile, etc.), 3-hydroxy-1-propanesulphonic acid anhydride, β-propiolactone [163], etc. may be used as acceptors.

The choice of monomers for this type of polymerization is steadily increasing (see, for example, refs. 165 and 166 and Chap. 5, Sect. 5.8).

3. Radical ions

Radical ions are, in the main, not very important as active centres of polymerizations. In media suitable for the existence both of radicals and of ions, the latter are usually more reactive. Moreover, the radicals decay by combination; their contribution to chain propagation is usually negligible. Radical ions are more important as precursors of active centres, as intermediates generated from initiators and monomers; through their radical ends they can combine (disproportionate) yielding active centres, frequently diions. Studies of radical ion behaviour contribute to our knowledge of the processes connected with electron transfer from molecule to molecule. These oxidation-reduction processes are very important in macromolecular chemistry.

Our knowledge of the energy states of a biatomic molecule capturing an electron in the gas phase is sufficiently precise [167]. The electron affinities

References pp. 222–229

and ionization potentials of many compounds have been tabulated. In a more complicated molecule, the energy of the more reactive parts may be dissipated into the „cooler" parts. This favours electron capture. The situation is different in solutions where the heat of reaction is considerably affected by the contribution of solvation energy. The solvent molecules assist in dissipating the liberated energy.

A solvated electron may be regarded as the simplest radical anion. The chemistry of solvated electron reactions is qualitatively similar to radical anion reactions but the physical properties of electron solutions are very complicated [167]. They are not suitable as a model of radical anions.

For some types of anionic polymerization, the equilibrium between the alkali metal and the aromatic hydrocarbon is important

$$M_t + \text{acceptor} \rightleftharpoons M_t^* + \text{acceptor}^{-*}$$

Conjugated hydrocarbons can also be reduced by Ca or Mg amalgams [168].

The heat of this reaction and the change in the Gibbs energy (for gas ions, solid metal D and crystalline acceptor A) are given by the relations [167c]

$$\Delta H = \Delta H_{\text{subl}(D)} + Ip_{(D)} + \Delta H_{\text{subl}(A)} - L_{(A)}$$

$$\Delta G = \Delta G_{\text{subl}(D)} + Ip_{(D)} + \Delta G_{\text{subl}(A)} - L_{(A)}$$

where ΔH_{subl} is the heat of sublimation, $Ip_{(D)}$ is the ionization potential of the alkali metal atom in the gas phase, and $L_{(A)}$ is the electron affinity of the hydrocarbon. In solutions, the solvation of the particles D^+ and A^{-*} has to be considered. Therefore the overall heat of reaction will be reduced because the heats of solvation are negative. When only pairs are generated instead of free ions, enthalpic members corresponding to ion association must be introduced into the above relations.

The equilibrium

$$Na + (Ph—Ph)_1 \rightleftharpoons Na^+, (Ph—Ph)_1^{-*}$$

has been carefully measured experimentally. The obtained experience is instructive (the overall reaction is exothermic and strongly dependent on the type of solvent). It cannot, however, be directly applied to systems with monomers (for example styrene and its derivatives). When monomers are used as acceptors, the interpretation of the results is obscured by rapid polymerization.

Electron transfer from a radical anion to an acceptor is formally similar to the case referred to above. In this case, however, an equilibrium between two radical ions R_1^{-*} and R_2^{-*} and between their parent molecules M_1 and M_2 has to be considered.

$$R_1^{-*} + M_2 \underset{}{\overset{K_{RM,i}}{\rightleftharpoons}} M_1 + R_2^{-*} \tag{29}$$

When the radical ions participate in the reaction in the form of ion pairs

$$R_1^{-*}, Ac^+ + M_2 \underset{}{\overset{K_{RM,p}}{\rightleftharpoons}} M_1 + R_2^{-*}, Ac^+ \tag{30}$$

the corresponding equilibrium constant will generally have a different value because

$$K_{RM,p} = \frac{K_{RM,i} K_{dis\, R_1^{-*}Ac^+}}{K_{dis\, R_2^{-*}Ac^+}} \tag{31}$$

$K_{disR^{-*}Ac^+}$ are the dissociation constants of the ion pairs $R_1^{-*}Ac^+$ and $R_2^{-*}Ac^+$. Under suitable conditions, the equilibria (29)–(31) can be followed by spectrophotometric methods [167c, 169]. There exist some very important specific reactions of the type shown in eqns. (29) and (30) which are poorly characterized. This concerns, for example, the electron transfer from naphthalene^{-*} metal$^+$ (Szwarc initiator) to styrene or other monomers [see Chap. 3, eqn. (46)]. The rapid consecutive reactions of the styrene radical ion make a direct measurement of the equilibrium impossible. Indirect data are not reliable.

Radical anions from a vinyl (vinylidene) monomer undergo dimerization or they can add a molecule of the parent monomer to yield a dimeric radical ion. Using styrene as an example, these reactions can be illustrated by the equations

$$2\, CH_2\text{—}CHPh^{-*} \rightleftharpoons {}^-CHPhCH_2CH_2CHPh^-$$

$$CH_2\text{–}CHPh^{-*} + CH_2.CHPh \rightleftharpoons {}^-CHPhCH_2CH_2CHPh^*$$

In principle, dimerization is reversible and the equilibrium is rapidly established. The equilibrium concentration of radical ions derived from vinyl monomers is so small that it cannot be detected by ESR (for monomer concentrations 10^{-1} mol dm^{-3} it is $<10^{-7}$ mol dm^{-3}). The rate constant of dissociation of the dimeric dianion of α-methylstyrene (αMeS)

$$^-\alpha MeS - \alpha MeS^- \rightarrow 2\, \alpha MeS^{-*}$$

in THF at 198 K is 6×10^{-8} s$^-$. i.e. small, as could be expected. On the other hand, relatively slow dimerization was also observed, for example in studies of dibenzylidine (tolane PhC≡CPh) radical anions.

These data indicate the necessary existence of a certain concentration of radical centres next to ionic centres. Considerable efforts have been exerted to determine quantitatively the proportion of radical propagation next to anionic growth [167e, 170, 171]. The participation of radical ends in polymerization cannot be excluded but reliable proof of such a process has not been presented so far.

References pp. 222–229

When an alkene molecule loses an electron, a cation radical is formed. The very reactive cation radical $(CH_3)_2\overset{+}{C}-CH_2^*$ is generated from 2-methylpropene in light in the presence of $TiCl_4$. It can be detected by ESR in the frozen parent compound at 123 K [172]. We assume that at higher temperatures these formations are dimerized to dications. The existence of a donor-acceptor complex is a necessary condition for the mechanism generating cation radicals (see Chap. 3, Sect. 5). α-Methylstyrene is cationically polymerized when illuminated in the presence of tetracyanobenzene in methylene chloride. From the two compounds, of which α-methylstyrene is the donor (D) and tetracyanobenzene the acceptor (A), the donor-acceptor complex is generated in the singlet and triplet states; it dissociates to solvated ion radicals [173]

$$D + A \rightleftharpoons (DA) \xrightarrow{h\nu} {}^1(DA)^* \rightarrow {}^3(DA)^*$$
$$\searrow \qquad \searrow$$
$$(D_S^{+*}, A_S^{-*}) \ (D_S^{+*}, A_S^{-*}) \rightleftharpoons D_S^{+*} + A_S^{-*}$$

The properties of the ion radical pair depend on the amount of energy in the exciplex. Ions decay more rapidly from the singlet than from the triplet state. This difference is explained by the concept of spin conversion [174].

Cation radicals are readily formed from vinylcarbazole. Their further behaviour is a sensitive function of external conditions.

In the presence of O_2, trans-1,2-9-dicarbazoylcyclobutane

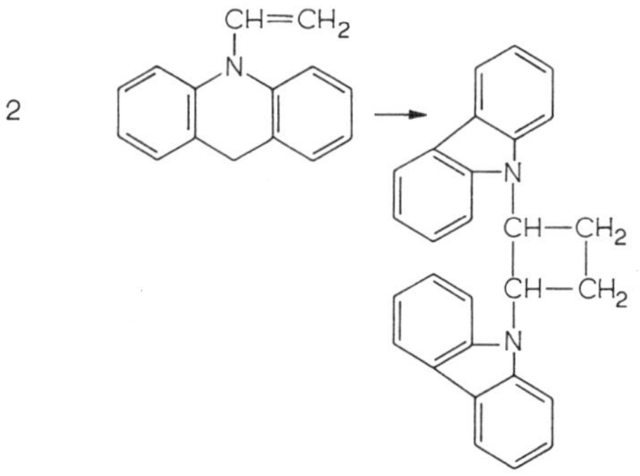

is the product of cation radical interaction with monomer [175]. In the absence of O_2, polyvinylcarbazole is predominantly formed. Both reactions are affected by the solvent in a complicated way. Polymerization and dimerization in benzonitrile is stopped both by radical and by cationic inhibitors. In the presence of nitrobenzene, only cationic vinylcarbazole polymerization occurs [176], even in the presence of O_2.

Only in recent years have the properties of cation radicals been intensively studied. Considering the important role played by these particles in organic reactions in general [177, 178], and particularly in macromolecular syntheses [179, 180], it may be expected that the interest in them will increase. This is also true of anion radicals.

4. Coordination centres

Strictly speaking, some kind of coordination is a prerequisite for any ionic polymerization. Some active centres can bind the monomer prior to its controlled attachment to the end of a propagating macromolecule. Chains of a regular or tactic polymer are thus formed. Such processes are designated as coordination polymerizations proper. At the present time, the centres of alkene coordination polymerizations and the precursors of such centres are of greatest importance.

4.1 Coordination centres of Ziegler–Natta polymerizations

Ziegler et al. [181] discovered that the mixture of $TiCl_4$ with $R_{3-n}AlCl_n$ can polymerize ethylene at low temperatures and pressures. The product is linear, macromolecular polyethylene.

Natta [182] has shown that a pure compound of a transition metal in a lower oxidation state (e.g. $TiCl_3$, VCl_3) in mixture with some organometals produces an efficient catalytic system for the preparation of crystalline polyalkenes with high melting points. The Ziegler and Natta pairs are evidently related, and they are currently considered as a single group called Ziegler–Natta (ZN) catalysts.

4.1.1 SOLID ZIEGLER-NATTA CATALYSTS

Modifications of $TiCl_3$ are still an important component of ZN catalysts. The purple crystals of α-$TiCl_3$ are composed of alternating layers of titanium and chloride ions; the β modification is brown, of fibrillar structure (Figs. 5 and 6) and is prepared by the reduction of $TiCl_4$ with alkylmetals. γ-$TiCl_3$ is also purple and has a layer structure like the α form (with a different arrangement). The δ form is prepared by milling α-$TiCl_3$ with $AlCl_3$; it exhibits a mixed $(\alpha + \gamma)$ structure, and is a very active catalyst [185].

Modern modifications of classical ZN systems, represented by the Solvay catalyst, are based on β-$TiCl_3$. $TiCl_4$ is mixed with Et_2AlCl at 274 K, the reduction temperature is slowly increased up to 338 K and maintained at this value for 1 h. From the mixed $TiCl_3 \cdot \frac{1}{3} AlCl_3$ crystals, $AlCl_3$ is washed out with diisoamyl ether; in this process, $\beta - \delta$ recrystallization takes place, and δ-$TiCl_3$

References pp. 222–229

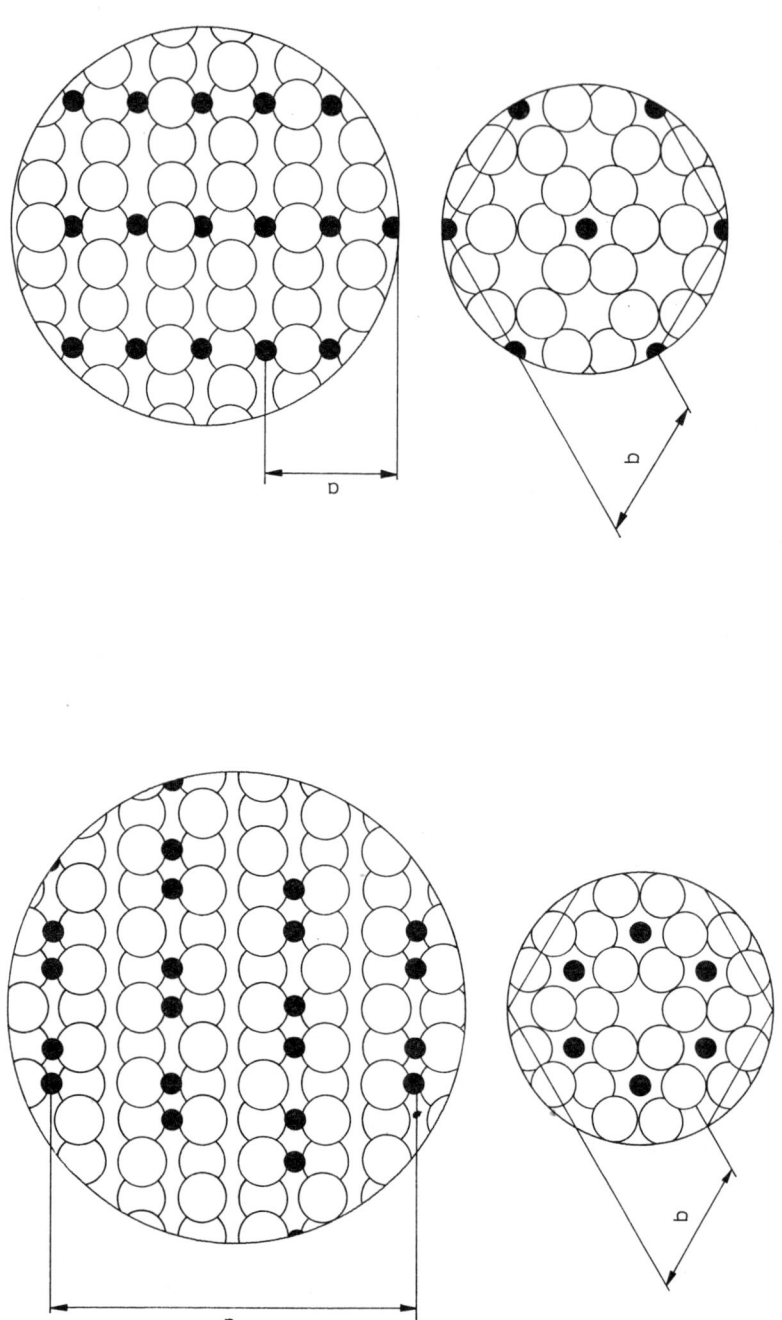

Fig. 5. Placement of Ti (●) and Cl (○) atoms in the α modification of TiCl$_3$ [183]. The drawing is an adaptation of an illustration in ref. 184.

Fig. 6. Placement of Ti (●) and Cl (○) atoms in the β modification of TiCl$_3$ [183]. The drawing is an adaptation of an illustration in ref. 184.

is reacted with $TiCl_4$ in *n*-hexane [186]. The resulting solid phase is 3–6 times more active than conventional δ-$TiCl_3$ because of the microporous structure and small crystal dimensions. Thus it contains more Ti atoms with one or two vacancies suitable for active centre formation [187].

According to Nielsen [187], the following Ti types are placed at the corners of $TiCl_3$ crystallites: Ti atoms with one vacancy and one pendant Cl atom (the pendant Cl is bound only to this one Ti atom, and to no other in the crystal lattice); Ti with two vacancies and two pendant Cl atoms; and Ti with one vacancy and three pendant Cl atoms. From these atom groups, the active centres are generated (Cl atoms bound to other Ti atoms are only indicated by valence lines).

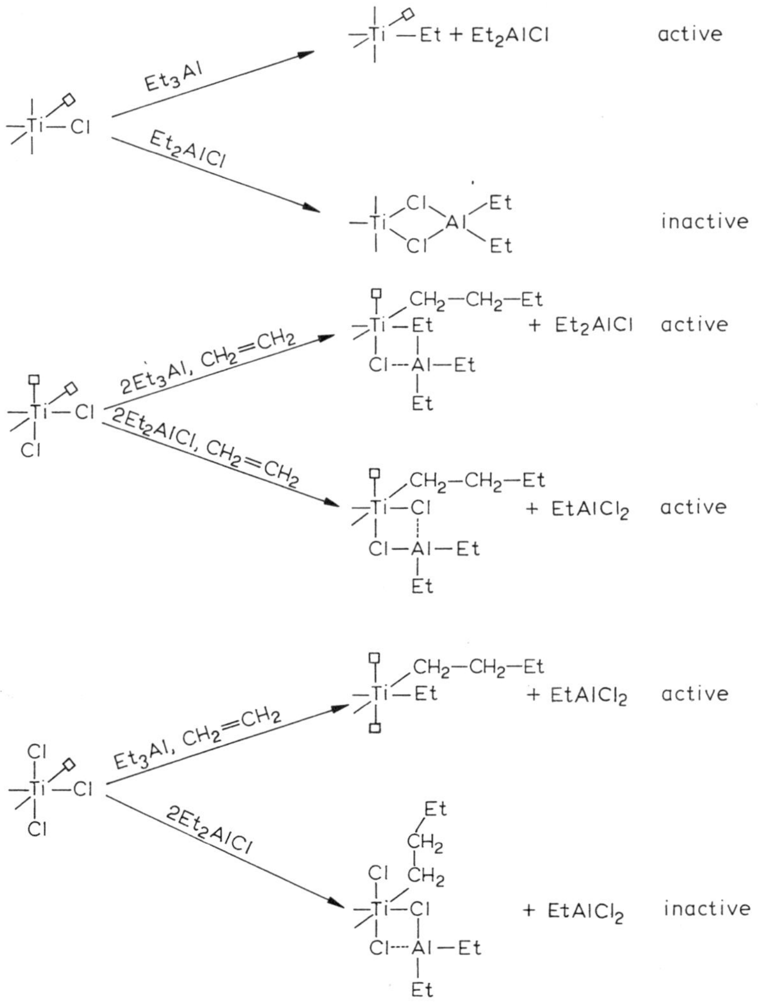

With Solvay-type catalysts, Et_2AlCl is used as cocatalyst.

It has not yet been unambiguously decided whether the active centre is formed by the transition metal atom with its ligands (monometallic centre), I, or by the bimetallic complexes II or III; most authors now favour the idea that the centre is bimetallic.

```
         R                    R
          \                    \
           Al                   Al
          /  \                 /  \
      R  Cl   R           Cl  Cl   R
       \ |   /             \ |    /
  —Ti    Ti                 Ti
   /\    /\                 /\

    I          II                III
```

The two ideas are not incompatible. The Ti—C bond is activated by monomer coordination to the unoccupied d orbital of alkylated Ti, perhaps also with a contribution of the bimetallic complex with the organometal. Catalytic activity as well as the regularity, tacticity or stereoregularity (see Chap. 5, Sect. 4) of the polymer produced depend on the structure and concentration of the organometallic compound used. Its change affects the degree of alkylation; other reactions, e.g. reduction of $TiCl_3$, may also occur.

Data on the reduction of halides of other transition metals by Et_3Al and Et_2AlCl are very numerous [a frequently used organometal is $(iso-Bu)_3Al$, and there are many others], nevertheless a closed concept concerning the fine structure and function of the generated active centres cannot be formulated at present [188].

(a) *Supported Ziegler-Natta catalysts*

Gas-phase polymerization technologies exclusively using supported catalysts are being employed more extensively in large-scale polyalkene production. Supported catalysts are also penetrating into the production of speciality polymers. They must therefore be regarded as the most perspective variant.

A logical way of increasing the fraction of transition metal atoms used for active centre formation seems to be the application of the metal compound to a polymer surface [189]. The carrier could be identical with, or similar to, the product. This approach is being intensively studied, and various polymers are being tested as carriers, for example poly(styrene-*co*-divinylbenzene) [190], chloromethylated poly(styrene-*co*-divinylbenzene) [191], polyethylene-*graft*-poly(4-vinylpyridine) [192], and many others.

So far, more active catalysts have been prepared from mineral supports which should be inert towards the product (polyalkene), e.g. pure SiO_2 or Al_2O_3. Even this way has not been easy. Hardly any commercial silica or alumina product is really inert towards the polymer (metal atoms present as impurities and which are bound to O and Si or Al exhibit very strong Lewis acid character at elevated temperatures). Their effect on the properties of the supported centres is not simple nor has it yet been sufficiently characterized.

Of great industrial importance for ethylene polymerization are the derivatives of transition metal oxides. The oxides are usually anchored on silica or alumina surface, and thus they belong to the class of supported catalysts. So far we possess only rough information on the principles and mechanism of their operation.

For example, chromium trioxide reacts with an —OH group pending from the support surface yielding a centre precursor [193]. This is activated by partial reduction of Cr^{6+} with agents without detachable H (CO, monomer etc.) [194].

$$\begin{array}{c} \sim\text{—Si—OH} \\ | \\ \text{O} \\ | \\ \sim\text{—Si—OH} \end{array} + Cr^{6+} \xrightarrow{-H_2O} \begin{array}{c} -Si-O \\ | \quad\quad\;\; \diagdown \\ \text{O} \quad\quad Cr^{6+} \\ | \quad\quad\;\; \diagup \\ -Si-O \end{array} \xrightarrow{\text{reduction}}$$

$$\xrightarrow{\text{reduction}} \begin{array}{c} -Si-O \\ | \quad\quad\;\; \diagdown \\ \text{O} \quad\quad Cr^{2+,3+} \\ | \quad\quad\;\; \diagup \\ -Si-O \end{array} \xrightarrow[\text{with monomer}]{\text{alkylation}} \begin{array}{c} -Si-O \\ | \quad\quad\;\; \diagdown \quad\quad | \\ \text{O} \quad\quad Cr^{x+}\!\!-\!\!C- \\ | \quad\quad\;\; \diagup \quad\quad | \\ -Si-O \end{array}$$

In this scheme

$$\begin{array}{c} | \\ (-SiO)_2\, Cr_2O_5 \\ | \end{array}$$

may be the precursor. Reduction with CO yields coordinationally unsaturated compounds with Cr^{3+} and Cr^{2+} ions which can be alkylated with monomer. An active catalyst can be generated directly by an exchange reaction of

$$\begin{array}{c} | \\ -Si-OH \\ | \end{array}$$

with a complex containing Cr in the necessary oxidation state [195].

The number of positive charges on a Cr atom in an active centre is a subject of some controversy; most authors suppose that centres with Cr^{2+} are more active than those with Cr^{3+} [196–198]; arguments in favour of the opposite view can also be found [199]. It appears that even Cr^+ [196], Cr^{4+} and Cr^{5+} [200] may occur in active centres. In order to interact with the monomer they must, however, always be coordinationally unsaturated.

References pp. 222–229

Ethylene is not sufficient for the reduction of molybdenum and tungsten catalysts containing metals in high oxidation states. Hydrogen, Na or NaH must be used. We assume that these centres and their reactions are similar to centres with chromium [201].

No support can be regarded as inert with respect to the active centres. By its universally positive effect on the activity of centres, $MgCl_2$ is superior to any other support. In spite of the great technical importance of Mg in active centres, generally not much is known of their structure in third-generation catalysts (or perhaps because of its positive effects; all the important producers have published hundreds of patents, but the crucial factors may still be kept secret). It is suspected that the separation (dilution) of transition metal atoms by a barrier of Mg atoms enables the majority of transition metals to become part of the active centres; on these centres, the polymer grows more rapidly than on centres without Mg. Mutual contact of the centres is hindered, bimolecular termination of centres (transition metal reduction to a less active oxidation state) is limited, and the centres live longer.

Quite recently, combinations of classical supports (SiO_2) with $MgCl_2$ have tended to appear, often precipitated from solutions of organomagnesium compounds on $SiO_2(Al_2O_3)$ surfaces by means of chlorine (halogen) donors; or the quality of the centres is modified by an organomagnesium compound together with an organoaluminium or even organosilicon component together with Lewis bases, without Cl donors. These systems appear to be the most promising variants of ZN catalysts for alkene polymerizations from the gas phase.

4.1.2 SOLUBLE ZIEGLER–NATTA CATALYSTS

Soluble ZN catalysts were found shortly after the Ziegler discovery. The specific features of homogeneous polymerizing systems can be technologically exploited. The main importance of soluble ZN catalysts lies in the possibility of employing them as models of insoluble centres. Solutions are generally more accessible to physical and physicochemical studies of the interesting particle structure. Ethylene and conjugated dienes polymerize well on homogeneous centres, but higher alkenes only very reluctantly. In homogeneous ZN catalysts, the main role again belongs to the transition metal atom (Ti, V, Cr, Co, etc.) to which two or more stable subsituents are bound; the latter may be hydrocarbons such as cyclopentadienyl (cp), or may contain heteroatoms, for example 2,4-pentanedione-3-yl (acac). Even homogeneous derivatives of catalytic systems require the presence an organometallic compound as cocatalyst. The role of the organometal is again to alkylate the transition metal or even to activate the metal-carbon bond which is formed.

Cyclopentadienyl compounds of Ti as polymerization catalysts were discovered by Natta et al. [202] and their properties were studied in greater detail by Breslow and Newburg [203]. The authors described the interaction of both catalytic components in the absence of monomer by the equations
component complexation

$$2 \text{ cp}_2\text{TiCl}_2 + (\text{Et}_2\text{AlCl})_2 \underset{}{\overset{\text{rapid}}{\rightleftharpoons}} 2 \text{ cp}_2\text{TiCl}_2.\text{AlEt}_2\text{Cl}$$

I

alkylation of Ti component

$$\text{cp}_2\text{TiCl}_2.\text{AlEt}_2\text{Cl} \underset{}{\overset{\text{slow}}{\rightleftharpoons}} \text{cp}_2\text{TiEtCl}.\text{AlEtCl}_2$$

II

reduction of Ti component

$$2 \text{ cp}_2\text{TiEtCl}.\text{AlEtCl}_2 \underset{}{\overset{\text{very slow}}{\rightleftharpoons}} 2 \text{ cp}_2\text{Ti}^{3+}\text{Cl}.\text{AlEtCl}_2 + \\ + \text{ CH}_3\text{CH}_3 + \text{CH}_2.\text{CH}_2$$

III

The polymerization rate reflected the time change of the concentration of complex II. This complex, or a very similar compound, was assumed to be the active centre. Complex II is also formed immediately and quantitatively on contact of the alkylated Ti component with EtAlCl_2 [204].

$$\text{cp}_2\text{TiEtCl} + \tfrac{1}{2}(\text{EtAlCl}_2)_2 \rightleftharpoons \text{cp}_2\text{TiEtCl}.\text{AlEtCl}_2$$

Henrici–Olivé and Olivé [205] assume complex II to be octahedral with the metal atoms connected by two Al—Cl—Ti links, with one vacancy (suitable for ethylene coordination), and a reactive ethyl group in the *cis* position

This notion was refined by Armstrong et al. [206]. The octahedral complex is only formed during the reaction with the monomer whose π electrons overlap the Ti d orbital yielding a σ bond. Since Ti^{4+} has no d electron, reverse donation cannot occur. The submitted hypotheses clearly have the character of a simple model describing the structure of the active centre, but it is evidently too simple to explain some observed facts, e.g. the growth of the reaction rate with increasing $\text{EtAlCl}_2/\text{cp}_2\text{TiEtCl}$ ratio, bimolecular ter-

References pp. 222–229

mination, etc. Even the ionic character of the centres, resulting in complex II dissociation

$$cp_2TiRCl.AlRCl_2 \rightleftharpoons [cp_2TiR]^+ + [RAlCl_3]^-$$

has not been widely accepted. The active centre should have the form of a cation $[cp_2TiR]^+$ [207] or of a zwitterion [208]

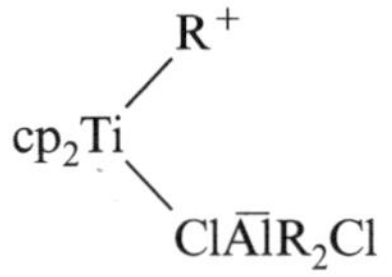

$$cp_2Ti\begin{matrix} \nearrow R^+ \\ \searrow Cl\overline{Al}R_2Cl \end{matrix}$$

When only a shift of the centres of charge is considered, i.e. ionization instead of dissociation, then this hypothesis becomes very similar to the coordination mechanism.

The active centres are greatly affected by the cocatalyst which, in the case discussed above, is the organo-aluminium component. As an acceptor part of the centre, it reduces the electron density on the donor (titanium). This facilitates monomer addition by insertion. Thus the activity of the centre is a function of the acidity of the aluminium component. The partly hydrolyzed organo-aluminium compound of general formula

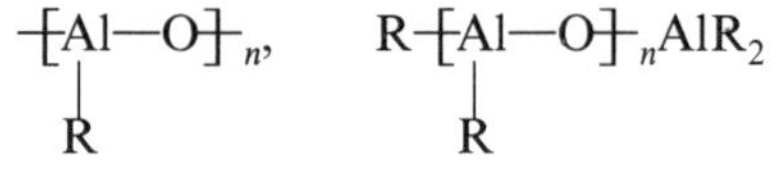

$$\begin{matrix} +Al-O+_n, \\ | \\ R \end{matrix} \qquad \begin{matrix} R+Al-O+_nAlR_2 \\ | \\ R \end{matrix}$$

where (R = Me, Et and n = 10–20) containing highly polarized Al—O bonds

$$\left\{ \overline{Al} \overset{\overset{+}{O}}{\diagup} \diagdown Al \right\} \longleftrightarrow \left\{ Al \diagup \overset{\overset{+}{O}}{\diagdown} \overline{Al} \right\}$$

yields centres of reactivity enhanced by almost two orders compared with the non-hydrolyzed cocatalyst [209]. Aluminoxanes also greatly increase the activity of centres formed from vanadium compounds [210] and especially of cp_2ZrCl_2 [211, 212]. These catalysts can produce 25–30 × 10^6 g of polyethylene per 1 g of Zr per hour. Aluminoxanes with chiral racemic ethylene-bis (tetrahydroindenyl) zirconium dichloride can polymerize > 300 kg of propene on 1 g of Zr per hour to isotactic polypropylene.

According to present ideas, heterogeneous and homogeneous ZN centres have different structures. The semiconducting properties of the $TiCl_3$ crystal are also reflected in the character of the surface Ti atom forming the centre. The difference between the two types of centre is clearly proved by the fact that aluminoxanes do not accelerate polymerization of the lowest alkenes on solid $TiCl_3$ [213].

Another type of homogeneous centre is the polybutadiene or polystyrene-*block*-polybutadiene (10 000 : 162 M_r; M_r = relative molecular mass) chains with an $\sim \bar{C}Li^+$ anionic end, reacted with $TiCl_4$

$$\sim \bar{C}Li^+ + \tfrac{2}{3}TiCl_4 \rightleftharpoons \tfrac{1}{3}(\sim CTiCl_2, TiCl_3, 3LiCl) + \tfrac{2}{3} \sim C^*$$

The macroradicals $\sim C^*$ undergo termination, mostly by disproportionation. The polymeric chain renders the complex soluble in non-polar solvents (hexane). According to the authors, the active centres have the form [214]

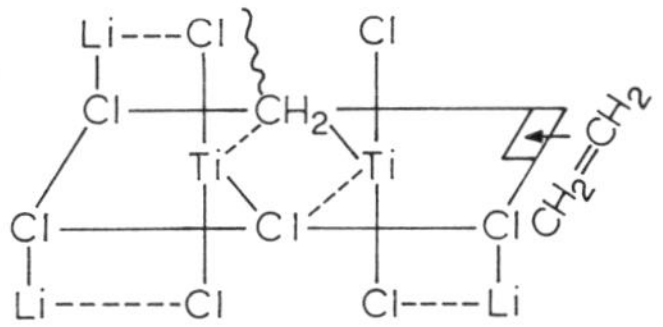

and ethylene polymerization on them has the character of a living system. Block copolymers are produced [215].

4.1.3 PROPERTIES OF COORDINATION CENTRES OF ZIEGLER–NATTA POLYMERIZATIONS

Much empirical data on the behaviour of ZN centres have been accummulated. An external effect does not usually change just a single elementary reaction (as we would wish in order to determine the structure of the centre), but affects a host of chemical and physical processes, resulting in an overall measured effect.

It is rather difficult to isolate individual changes and to estimate their size. Experimental insufficiency, underestimation of the importance of some variables or trivial negligence may lead to situations where, in addition to the measured effect, other factors are also operating. Varied and often even controversial information then appears in the original literature. (This is, of course, true even of phenomena other than the characterisation of active centres.) In spite of all the difficulties, however, many properties of various centres are mutually similar and can be generalized.

General features of ZN centres

One of these properties is the replication of the catalyzing particle shape by the generated polymer. The product flakes produced are 15–20 times larger than the catalyst particles [216]. Various opinions exist concerning monomer permeability through the polymer layer towards the polymerization centres.

Hitherto it has been widely accepted that monomer diffusion through the polymer limits the possibility of its coordination to the active site; this opinion is still maintained by many authors (see, for example, refs. 217 and 218), even though Chien [219] and Doi et al. [220] have presented results showing that the reduction in polymerization rate is not caused by hindered monomer diffusion to the centres. The size of the particles is evidently also of importance [221]. The highest oxidation state of the transition metal is never the most active. Its coordination unsaturation is very important. Magnesium-containing centres generally yield a product with a narrow distribution curve [222]. Classical ZN catalysts ($TiCl_3$, Et_2AlCl), on the other hand, yield a broad distribution. Chromium-containing catalysts produce polymers (polyethylene) with a relatively high double bond content. Increasing the Lewis acidity of centres leads to higher activity [223], transfer to monomer is easier, and the iso-specific effect of the centres is reduced [224].

As yet we have no explanation of the fact that the most active centres are formed specifically by Zr with aluminoxanes, and why Ti with Mg is the best pair of metal components in third-generation catalysts.

Valuable data on the properties of active centres are obtained from kinetic measurements. They reveal the simultaneous existence of several centre types. The stable centres are active during the whole course of polymerization; in addition, some fraction of decaying centres is also present. Isotactic centres exhibit stereoregulating ability and are, moreover, extremely active. Centres may oscillate between active and inactive (dormant) forms and some centres selectively polymerize enantiomers from a racemate. External effects, caused by specific properties of centres, will be discussed in subsequent chapters. In addition, the centres on which dienes are polymerized will be treated in Chap. 5, Sect. 4. The structure of these centres is a function of the coordinated diene, and it is therefore better presented together with propagation.

In my opinion, an active centre of alkene polymerization in the liquid phase is not a single chemical entity to be visualized by a single (and simple) chemical formula. Probably a set of compounds, of complexes with variable composition, a dynamic system where the effects of individual components are mutually complementary or overlapping is really in play. The same macroscopic effect (centres of equal activity and iso-specific regulating ability) can be obtained with various starting organometals and donors. In such a system, subsystems may exist each of which is externally manifested as an individual active centre (rapid or slow, isotactic, with a tendency to transfer or termination, or living, etc.) [225].

With supported centres of low mobility, exchange reactions are hindered, and therefore these centres may be more active and/or longer living.

4.2 Other coordination centres

Many systems are known where centres of coordination polymerizations are generated. Since it would require a volume the size of the present one to discuss all of these, only a few examples can be presented here in order to illustrate the possibilities in this field.

Monomers such as methyloxirane produce a chain the end of which cannot control the addition of further monomer. When the generation of a stereoregular, crystalline polymer is required, the mode of methyloxirane addition to the chain must be regulated by a catalyst. Hydrolysis of mixed Meerwein alkoxides [226] yields the compound

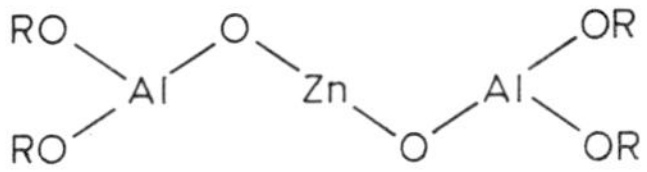

$$\text{RO}_2\text{Al}-\text{O}-\text{Zn}-\text{O}-\text{Al}(\text{OR})_2$$

The empty coordination sites on Al and Zn are reversibly filled by electron pairs from alkoxide groups with intra- or intermolecular associate formation. The degree of association depends on the solvent, the type of metal M_t^{2+}, and the OR group in $[(RO)_4Al_2O_2M_t^{2+}]_n$; in benzene, n varies from 1 to 8. We assume that the monomers are inserted, with simultaneous ring opening, into the

$$-\overset{|}{\text{Al}}-\text{OR}$$

bond [227]. Not only oxiranes, but also heterocycles with a larger number of ring members are polymerized on the

$$-\overset{|}{\text{Al}}-\text{OR}$$

group [228].

Copolymerization of CO_2 with oxiranes proceeds with high yield in the presence of catalysts produced by reaction of Et_2Zn with polyphenols [229]. The prerequisite of high activity is the presence of an internal coordination bond and the existence of an empty coordination site on the Zn atom in the

$$-\text{Zn}-\text{O} \rightarrow \text{Zn}-$$

sequence. The highest activity is exhibited by compounds such as [230]

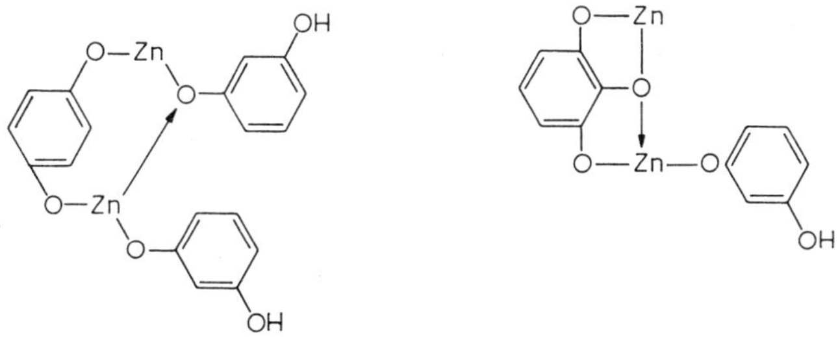

Active centres of coordination polymerizations can be formed by unusual combinations of elements. We have observed slow polymerizations of VC, MMA, S and AN yielding products of high melting point on complexes such as [231]

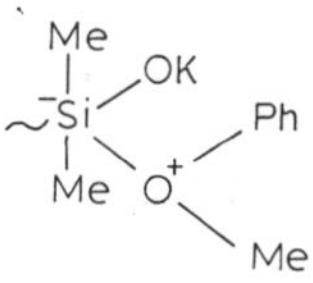

$$\begin{array}{c} \text{Me} \\ | \\ \sim\text{Si} \\ | \\ \text{Me} \end{array} \begin{array}{c} \text{OK} \\ \diagdown \\ \diagup \\ \text{O}^+ \end{array} \begin{array}{c} \\ \text{Ph} \\ \diagdown \\ \text{Me} \end{array}$$

Oligomerization of THF can be initiated by unusually arranged complexes of $TiCl_4$ with chloromethyloxirane.

We found [232] that polymers could not be produced by means of the relatively stable complexes

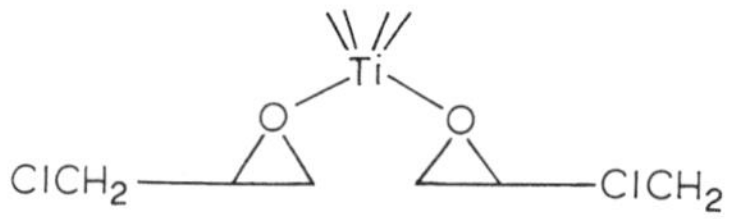

$$ClCH_2-\underset{O}{\triangle}\underset{}{\overset{Ti}{\diagup\diagdown}}\underset{O}{\triangle}-ClCH_2$$

Only the anomalous complexation

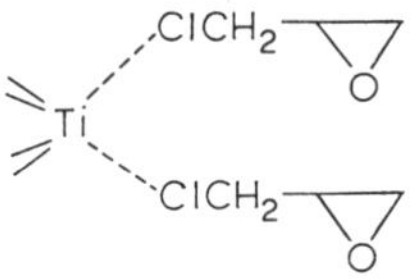

$$\overset{Ti}{\diagup\diagdown}\begin{array}{c}\cdot\cdot\cdot ClCH_2-\triangledown \\ \cdot\cdot\cdot ClCH_2-\triangledown\end{array}$$

makes possible THF coordination to activated ligands, with the opening of its ring (see Chap. 5, Sect. 6.3).

This chapter as well as the whole section on coordination active centres is typically characterized by a large number of observations, results, and hypotheses. Very complicated processes are evidently concerned which lack a general theory so far.

5. Transformation of centres

For full control of macromolecular syntheses, it is necessary to master the art of the design and synthesis of active centres, as well as of their modification during polymerization, and not only from the quantitative point of view, concerning their number and reactivity. The qualitative aspect is also of importance, the change of the basic character of the centres, for example from the anionic to the radical type. Modification of centres is at present the subject of many theoretical studies, and it is also of extraordinary practical importance.

Sokolskii et al. [233] noticed that ZN coordination centres spontaneously change to cationic centres during polymerization. In our laboratory we have attempted to obtain quantitative data on a similar reaction by means of models [232]. Tsuchyia and Tsuruta [234] pointed out the possibility of choice between anionic and cationic polymerization of methyloxirane with diethylzinc —H_2O (see Chap. 3, Sect. 2.1).

5.1 Transformation anion to cation

A complicated way of transforming living polymeric anions to cations was used by Yamashita [235].

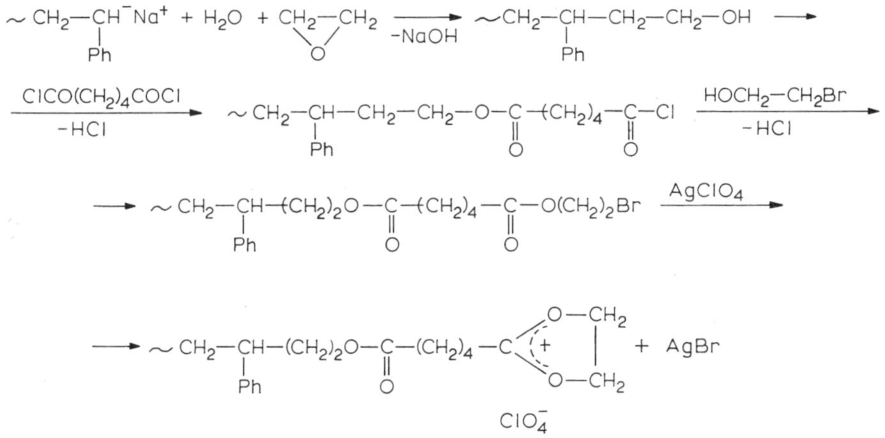

$$\sim CH_2-\underset{Ph}{CH^-}Na^+ + H_2O + \underset{O}{CH_2-CH_2} \xrightarrow[-NaOH]{} \sim CH_2-\underset{Ph}{CH}-CH_2-CH_2-OH \longrightarrow$$

$$\xrightarrow[-HCl]{ClCO(CH_2)_4COCl} \sim CH_2-\underset{Ph}{CH}-CH_2-CH_2-O-\underset{O}{\overset{\|}{C}}-(CH_2)_4-\underset{O}{\overset{\|}{C}}-Cl \xrightarrow[-HCl]{HOCH_2-CH_2Br} \longrightarrow$$

$$\longrightarrow \sim CH_2-\underset{Ph}{CH}-(CH_2)_2O-\underset{O}{\overset{\|}{C}}-(CH_2)_4-\underset{O}{\overset{\|}{C}}-O(CH_2)_2Br \xrightarrow{AgClO_4}$$

$$\longrightarrow \sim CH_2-\underset{Ph}{CH}-(CH_2)_2O-\underset{O}{\overset{\|}{C}}-(CH_2)_4-C\overset{O-CH_2}{\underset{O-CH_2}{\big\langle^+\big|}} + AgBr$$
$$ClO_4^-$$

A simpler process was found by British authors [236–238].

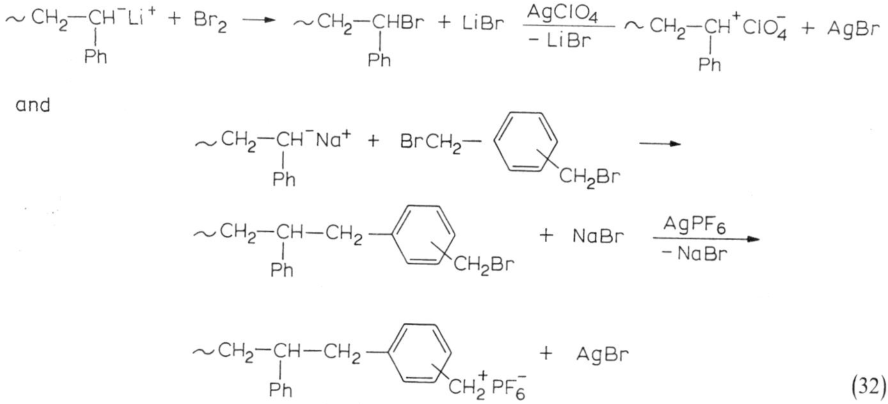

$$\sim CH_2-\underset{Ph}{CH^-}Li^+ + Br_2 \longrightarrow \sim CH_2-\underset{Ph}{CHBr} + LiBr \xrightarrow[-LiBr]{AgClO_4} \sim CH_2-\underset{Ph}{CH^+}ClO_4^- + AgBr$$

and

$$\sim CH_2-\underset{Ph}{CH^-}Na^+ + BrCH_2-\underset{CH_2Br}{\bigcirc} \longrightarrow$$

$$\sim CH_2-\underset{Ph}{CH}-CH_2-\underset{CH_2Br}{\bigcirc} + NaBr \xrightarrow[-NaBr]{AgPF_6}$$

$$\sim CH_2-\underset{Ph}{CH}-CH_2-\underset{\overset{+}{CH_2}PF_6^-}{\bigcirc} + AgBr \qquad (32)$$

5.2 Transformation cation to anion

A transformation of a macrocation to an anion with low efficiency was performed by Abadie et al. [239].

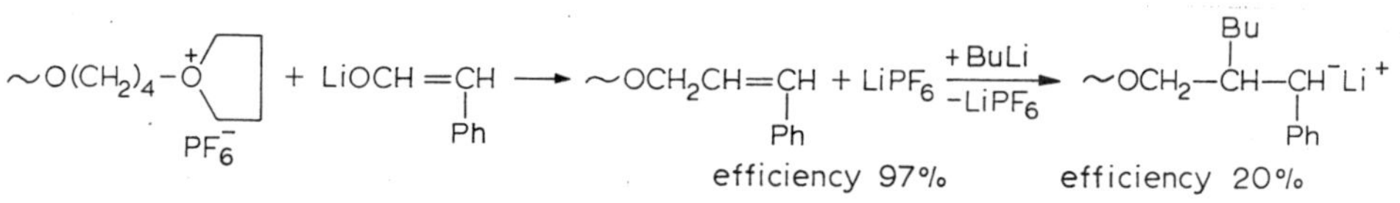

$$\sim\!O(CH_2)_4\!-\!\overset{+}{O}\!\!\diagup\!\!\diagdown \;\; PF_6^- \;\;+\; LiOCH\!=\!\underset{Ph}{CH} \longrightarrow \sim\!OCH_2CH\!=\!\underset{Ph}{CH} + LiPF_6 \xrightarrow[-LiPF_6]{+BuLi} \sim\!OCH_2\!-\!\underset{Ph}{\overset{Bu}{CH}}\!-\!CH^-Li^+$$

$$\text{efficiency 97\%} \qquad \text{efficiency 20\%}$$

Of higher efficiency is the change of an oxonium centre to a secondary amine, which can be transformed to a carbanion by alkyllithium [240]

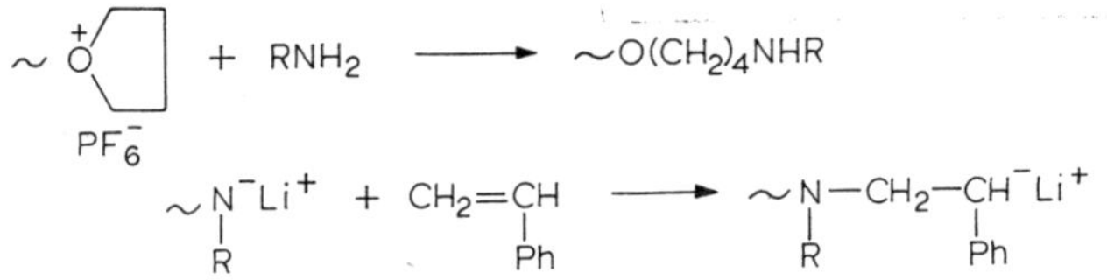

$$\sim\!\overset{+}{O}\!\!\diagup\!\!\diagdown \;\; PF_6^- \;\; + \; RNH_2 \longrightarrow \sim\!O(CH_2)_4 NHR$$

$$\sim\!\underset{R}{N^-}Li^+ \;+\; CH_2\!=\!\underset{Ph}{CH} \longrightarrow \sim\!\underset{R}{N}\!-\!CH_2\!-\!\underset{Ph}{CH^-}Li^+$$

5.3 Direct transformation anion to cation and cation to anion

It appears that a combination of polymeric ions with difunctional initiators should be of some advantage over the transformation processes described above [241]. We have found that the combination of a macroanion with a dicationic initiator and of a macrocation with a dianionic initiator proceeds with high efficiency [242]. Macroions can therefore be transformed by difunctional initiators

$$\sim\!C^-M_t^+ \;+\; B^-{}^+Si\!\sim\!\!\sim\!Si^+B^- \longrightarrow \sim\!C\!-\!Si\!\sim\!\!\sim\!Si^+B^- + M_tB$$

$$\sim\!\!\sim\!{}^+O\!\!\diagup\!\!\diagdown \;\; B^- \;\;+\; M_t^+{}^-C\!\sim\!C^-M_t^+ \longrightarrow \sim\!\!\sim\!\sim\! C\!-\!C\!\sim\!C^-M_t^+ + M_tB$$

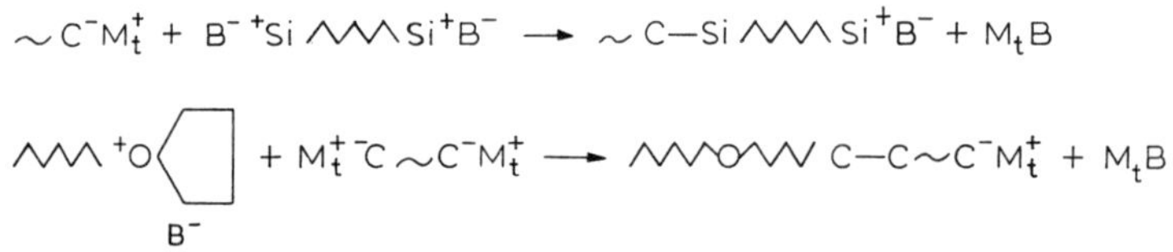

or generally

$$\sim\!\!\sim^- \;+\; ^+\!\sim\!\!\sim\!\!\sim^+ \longrightarrow \sim\!\!\sim\!\!\sim\!\!\sim^+$$

$$\sim\!\!\sim\!\!\sim^+ \;+\; ^-\!\sim\!\!\sim^- \longrightarrow \sim\!\!\sim\!\!\sim^-$$

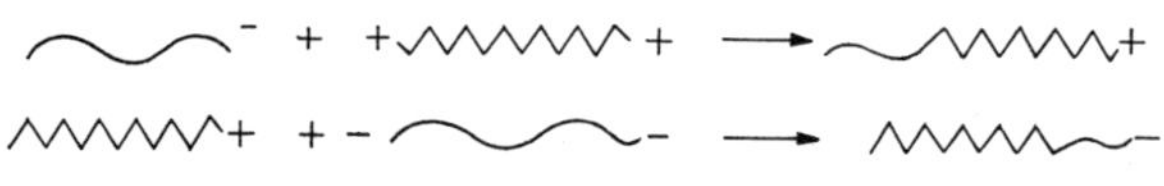

In this way, anionic centres are generated from cationic centres and vice versa [243]. In order to prevent the formation of a multiblock copolymer and to limit the combination in practice only to transformation, it is useful to maintain a high relative concentration of the transforming di-ion. This can be realized by the introduction of the macro-ion solution into the initiator.

5.4 Transformation anion to radical

An anionic active centre can be transformed to a radical centre by peroxyalkyl halides [244] or by molecular oxygen [245]. It appears, however, that the most elegant way so far was found by the British workers [246, 247]. They make use of the decomposition of unstable alkylsilver compounds generated from alkyllead salts and silver halides. The transformation thus proceeds in two steps. In the first step, a suitable Pb-containing end group is formed on the macroanion

$$\sim M^- Na^+ + Et_3PbCl \rightarrow \sim MPbEt_3 + NaCl$$

and this is decomposed in the second step

$$\sim PbEt_3 + AgX \rightarrow \sim MAg + Et_3PbX$$

$$\sim CAg \rightarrow \sim C^* + Ag$$

In order to prevent rapid generation of radicals in large concentrations (a situation which leads to their considerable decay by combination and disproportionation), the authors recommend the addition of solid $AgClO_4$ to the solution of the $\sim MPbEt_3$ polymer. Radicals are generated slowly from the slowly dissolving salt. Mercury, iron and perhaps even other metals can be used instead of silver.

5.5 Transformation with the participation of coordination centres

5.5.1 ANION TO COORDINATION CENTRE

The transformation of a carbanionic to a ZN coordination centre was described by Richards and co-wokers [248].

$$3 \sim^- Li^+ + AlCl_3 \rightarrow (\sim)_3 Al + 3 LiCl$$

$$\beta TiCl_3 + (\sim)_3 Al \rightarrow \sim TiCl_2 + (\sim)_2 AlCl$$

$$\sim TiCl_2 \xrightarrow{nE} \sim (CH_2CH_2)_n TiCl_2$$

References pp. 222–229

A carbanion can be transformed to a ZN centre also by nucleophilic substitution of a transition metal ligand [249]

$$\sim CH_2-\underset{Ph}{CH^-}Li^+ + (BuO)_4Ti \rightarrow \sim CH_2-\underset{Ph}{CH}-Ti(OBu)_3 +$$

$$+ BuOLi \xrightarrow[-BuOLi]{nCH\equiv CH} poly(styrene)\text{-}block\text{-}poly(acetylene)$$

By a similar reaction, described in Sect. 4.1, coordination centres are also generated from carbanions and $TiCl_4$ [214, 215].

5.5.2 COORDINATION CENTRE TO ANION

This kind of transformation is mentioned in the patent literature [250]. After mixing with RLi or ROLi, 1-alkenes polymerizing on ZN catalysts yield macroanions of the poly(1-alkene). In the presence of a suitable monomer, a block copolymer is formed

$$n\text{ (1-butene)} \xrightarrow[Et_2Zn]{TiCl_3,\ Et_3Al} \xrightarrow[S]{LiOCH_3} poly(1\text{-butene})\text{-}block\text{-}poly(styrene)$$

5.5.3 COORDINATION CENTRE TO CATION

Iodine readily reacts with polymers bound to coordination centres, yielding macromolecules containing iodide end groups. The $\sim$C—I bond can react with silver salts of superacids yielding carbocations essentially in the sense of the last steps of eqn. (32). Doi et al. [251, 252] have used this reaction for the synthesis of the copolymer poly(propene)-*block*-poly(tetramethylene oxide) with blocks of syndiotactic polypropylene.

5.5.4 COORDINATION CENTRE TO RADICAL

Vanadium-containing coordination centres producing syndiotactic polypropylene at 195 K can be transformed to radical centres simply by raising the temperature to 298 K [252]. In this way, Japanese authors have prepared the copolymer poly(propene)-*block*-poly(methylmethacrylate). The radical end is probably formed by homolytic splitting of the $\sim$C—V bond, and it can be stabilized by the V ion. The authors state that, in this way, two-component blocks of polypropylene with various polymers propagating by the radical mechanism can be prepared.

5.5.5 ANION TO CENTRE OF METATHESIS

Substitution of the donor Cl in WCl_6 (tungsten is the most frequently used transition metal in centres of metathesis) by a carbanion produces a grouping which is capable of further cycloalkene polymerization by metathesis [253]

$$-\overset{|}{\underset{|}{W}}-Cl + Li^+ ^-\underset{\underset{Ph}{|}}{CH}-CH_2\sim \longrightarrow \sim CH_2-\underset{\underset{Ph}{|}}{CH}-\overset{|}{\underset{|}{W}}- + LiCl$$

$$\sim CH_2-\underset{\underset{Ph}{|}}{CH}-\overset{|}{\underset{|}{W}}-Cl \longrightarrow \sim CH_2-\underset{\underset{Ph}{|}}{C}=\overset{|}{\underset{|}{W}}- + HCl$$

Reaction block copolymers, e.g. poly(styrene)-*block*-poly(pentenamer), have also produced by this method.

The possibilities of coordination centre transformations are being further intensively studied [254, 255].

6. Reaction of active centres during polymerization

The transformation of monomer to polymer (the consumption of monomer) is the consequence of propagation. Knowing the number of active centres, the rate of monomer consumption, and the structure of macromolecules, growth can be defined with sufficient accuracy. However, chains rarely grow on a single type of centre. Actually this only occurs in simple radical homopolymerizations; in ionic reactions, the contribution of weakly reactive centres can also sometimes be neglected. The spin-conserving method (see Chap. 6, Sect. 4.2), with the reagent preferentially trapping the more reactive radicals, may be of some help.

Pichot et al. applied the spin-trapping technique to study the copolymerization of acrylonitrile with vinyl chloride or vinyl acetate [256]. Macroradicals terminated by the latter two units are trapped preferentially. The authors also noted suppression of cyclization by the spin-trapping agent; cyclization is otherwise very common in acrylonitrile copolymerizations.

In ionic polymerizations, the fraction of free ions and of both solvent-separated and contact ions pairs is determined spectrophotometrically by the measurement of electrical conductivity and by NMR chemical shifts.

Direct observation of more than one type of ion pair was described by Hogen–Esch and Smid [257] for the fluorenyl salt. Differences in spectra enabled them to differentiate between contact and solvent-separated ion pairs. Buncel and Menon measured the equilibrium contact $\rightleftharpoons$ separated ion pairs of diphenylmethyl-and triphenylmethylcarbanions with Li^+ and K^+ in

References pp. 222–229

TABLE 5

Characteristic wavelengths of light absorbed by contact and solvent separated ion pairs of diphenylmethyl- and triphenylmethylcarbanions in ethers
Effect of solvent and counter-ion. Temperature 298 K.

$\sim C^- M_t^+$	Solvent	λ_{max} (nm)		Fraction $\sim C^- \parallel M_t^+$
		$\sim C^-, M_t^+$	$\sim C^- \parallel M_t^+$	
$Ph_2CH^- Li^+$	Et_2O	407		0.0
	THF	418	448	0.4
	MeOEtOMe		448	0.8
$Ph_2CH^- K^+$	Et_2O	432		0.0
	THF	440		0.0
	MeOEtOMe	441		0.0
$Ph_3C^- Li^+$	Et_2O	446/390		0.1
	THF		500/435	1.0
	MeOEtOMe		496/432	1.0
$Ph_3C^- K^+$	Et_2O	476/414		0.0
	THF		486/420	0.6
	MeOEtOMe		494/430	0.9

ethers, and spectrophotometrically in the visible and near UV ranges [258]. The results are summarized in Table 5.

Pticyna et al. measured the electrical conductivity of polyoxirane terminated by $\sim OEtO^- K^+$ active centres in tetrahydrofuran. At concentrations of $> 10^{-3}$ mol dm^{-3} these centres associate. The dissociation constant of low-molecular-weight alcoholates increases with growing chain length. Poly(oxyethylene) (in a similarly way to crowns and cryptands) solvates the cation and thus promotes charge separation [259]. Littlejohn et al. quantified $EtOCH_2^+ SbCl_6^-$ dissociation in $MeCl_2$ by specific conductivity and permittivity determinations [260].

Very low conductivities of organic solutions can be measured quite easily. However, the character of the results is not simple, and their connection with the partial polymerization processes is not straightforward.

Very rapid reactions are studied by means of the stopped-flow technique. In some cases this technique makes direct observation of active centres possible, as well as the determination of their immediate concentrations even in non-stationary polymerizations [261–264].

An important reaction of active centres is transfer which sometimes leads to the insertion of the transfer agent into the polymer. Determination of the amount of such fragments in macromolecules, possibly even of the structure and mode of the fragment binding are very valuable for describing the transfer reaction. To facilitate such analysis (often to make it possible at all), it is sometimes useful to label the transfer agent either by a radiating atom or by a chemical group of conspicuous properties [265], for example optical. The initiator or the terminating agent can also be labelled in a similar way, so that a polymer is obtained in which all end groups can easily be counted [266, 267].

With the progress of experimental techniques and analytical methods, polymer research is increasingly concerned with the knowledge and control of side reactions. They often determine the course of the polymerization process and the quality of the product. It does not pay to underestimate these reactions, and therefore they are the subject of due attention. However, no systematic review of this field has so far been published, and specific cases have to be sought in original papers [268].

7. Living and dormant centres

Active centres rarely represent a single chemical entity. Macromolecules almost always grow on formations which mutually differ to a greater or lesser extent, by structure or by energy content, and therefore also by reactivity. Addition to a hot radical, to a free ion or separated ion pair is more rapid than to a radical or contact ion pair, to an ester or aggregate. Sometimes the difference in the reactivity of two forms of one active centre is large. The rate constant on the more active form is by three, four or even more orders higher than on its less active counterpart. In such case, we speak of living and dormant centres[†].

Mutual transition of a dormant form into a living form and vice versa must be spontaneous, without external intervention. The concentrations of living and dormant forms are usually connected by an equilibrium. During macromolecule propagation, the centre can oscillate between the active and inactive states; generally, the two forms have different mean life times.

Dormant centres should not be confused with the inactive form of chain ends to which active centres are sometimes transformed by termination and which can be reversed to living forms only by slow re-initiation. Centres whose reactivity has been lost by termination are dead. The existence of living, dormant and dead centres is manifested in the mechanism and kinetics of the whole polymerization. The consequences of their occurrence will be discussed in the appropriate parts of the subsequent text (see Chap. 5, Sect. 8.1).

[†] These terms were introduced by Szwarc [167]. They have now become customary in technical literature, although originally they were criticized as being unduly taken over from the world of living matter.

References pp. 222–229

References

1. J. F. Kochi, in H. Williams (Ed.), Advances in Free-Radical Chemistry, Vol. 5, Elek, London, 1975, p. 189.
2. T. Koenig, T. Balle and W. Snell, J. Am. Chem. Soc., 97 (1975) 662.
3. R. V. Lloyd and D. E. Wood, J. Am. Chem. Soc., 97 (1975) 5986.
4. A. K. Butler and M. J. Perkins (Eds.), Organic Reaction Mechanism, Wiley, London, 1975.
5. G. E. Ham, in G. E. Ham (Ed.), Vinyl Polymerization, Vol. 1, Part 1, Dekker, New York, 1967, p. 2.
6. H. C. McBay, J. Org. Chem., 40 (1975) 1883.
7. Y. K. Irino and R. W. Fessender, J. Phys. Chem. 79, (1975) 834.
8. H. B. Stegmann, K. Scheffler and D. Seebach, Chem. Ber., 108 (1975) 64.
9. D. Griller, J. W. Cooper and K. V. Ingold, J. Am. Chem. Soc., 97 (1975) 4269.
10. J. W. Cooper, D. Griller and K. V. Ingold, J. Am. Chem. Soc., 97 (1975) 233.
11. K. D. King and R. D. Goddard, J. Am. Chem. Soc., 97 (1975) 4504.
12. S. Arrhenius, Z. Phys. Chem., 4 (1889) 226.
13. See for example, K. J. Laidler, Chemical Kinetics, McGraw-Hill, New York, London, 1965.
14. P. E. M. Allen and C. R. Patrick, Kinetics and Mechanisms of Polymerization Reactions, Horwood, Chichester, 1974, p. 263. W. Linert and R. F. Jameson, Chem. Soc. Rev. 18 (1989) 477.
15. M. G. Evans, J. Gergely and E. C. Seaman, J. Polym. Sci., 3 (1948) 866.
16. R. A. Ogg and M. Polanyi, Trans. Faraday Soc., 31 (1935) 1375.
17. G. W. Wheland, J. Am. Chem. Soc., 64 (1942) 900.
18. C. A. Coulson, J. Chem. Soc., (1955) 1435.
19. R. D. Brown, Q. Rev., 6 (1952) 63.
20. M. Szwarc and J. H. Binks, IUPAC Kekule Symposium 1958, Theoretical Organic Chemistry, Butterworths, London, 1959, pp. 262–290.
21. K. F. O'Driscoll and T. Yonezawa, Rev. Macromol. Chem., 1 (1966) 1.
22. A. Streitwieser, Jr., Molecular Orbital Theory for Organic Chemistry, Wiley, New York, 1961.
23. J. D. Roberts, Molecular Orbital Calculations, Benjamin, New York, 1962.
24. J. H. Binks and M. Szwarc, J. Chem. Phys.,, 30 (1959) 1494.
25. W. Kauzmann, Quantum Chemistry, Academic Press, New York, 1957, p. 536.
26. K. J. Laidler, Chemical Kinetics of the Excited State, Oxford University Press, Oxford, 1955, pp. 28–36.
27. V. N. Kondratev, Kinetic of Chemical Reactions in the Gas Phase (Kinetika Khimicheskikh Gazovykh Reaktsii), Akad. Nauk SSSR, Moscow, 1958, p. 185.
28. See P.E.M. Allen and C.P. Patrick, Kinetics and Mechanisms of Polymerization Reactions, Horwood, Chichester, 1974, p. 275.
29. K. Hayashi, T. Yonezawa, C. Nagata, S. Okamura and K. Fukui, J. Polym. Sci., 14 (1954) 312.
30. K. Fukui, T. Yonezawa, C. Nagata and M. Shingu, J. Chem. Phys., 20 (1952) 722.
31. L. Salem, Molecular Orbital Theory of Conjugated Systems, Benjamin, New York, 1966, p. 327.
32. K. Hayashi, T. Yonezawa, C. Nagata, S. Okamura and K. Fukui, J. Polym. Sci., 20 (1956) 536.
33. T. A. Kerr and M. J. Parsonage, Evaluation of Kinetic Data on Gas-phase Addition Reactions: Reactions of Atoms and Radicals with Alkanes, Alkynes and Aromatic Compounds, Butterworths, London, 1972.
34. C. A. Barson and R. Ensor, Eur. Polym. J., 13(2) (1977) 113.

35 J. Manassen and R. Rein, J. Polym. Sci. Part A-1 (8) (1970) 1403.
36 J. Kříž, J. Polym. Sci. Part A-1 (10) (1971) 615.
37 C. H. Bamford, W. G. Barb, A. D. Jenkins and P. F. Onyon, Kinetics of Vinyl Polymerization by Radical Mechanisms, Butterworths, London, 1958, p. 220.
38 S. Amdur, A. T. Y. Cheng, C. J. Wong, P. Ehrlich and R. D. Allendoerfer, J. Polym. Sci. Polym. Chem. Ed., 16 (1978) 407.
39 P. Hayden and H. Melville, J. Polym. Sci., 43 (1960) 201.
40 W. I. Bengough and H. W. Melville, Proc. R. Soc. (London) Ser. A, 230 (1955) 429.
41 See P.E.M. Allen and C. R. Patrick, Kinetics and Mechanisms of Polymerization Reactions, Horwood, Chichester, 1974, p. 102.
42 F. Tüdös, Hung. Acta Chim., 43 (1965) 397; 44 (1965) 403.
43 G. M. Burnett, G. G. Cameron, S. N. Joiner, J. Chem. Soc. Faraday Trans. 1, (1973) 69, 322.
44 A. I. Bolshakov, A. I. Mikhailov and I. M. Barkalov, Vysokomol. Soedin., 20 (1978) 1820.
45 G. H. Henrici-Olivé and S. Olivé, Makromol. Chem., 68 (1963) 219. Z. Phys. Chem. (Frankfurt), 47 (1965) 286; 48 (1966) 35, 51.
46 G. H. Olivé-Henrici and S. Olivé, Makromol. Chem., 96 (1966) 221.
47 G. M. Burnett, Kinetics and Mechanisms of Polyreactions, IUPAC International Symposium on Macromolecular Chemistry, Akademiai Kiadó, Budapest, 1969, p. 403.
48 E. M. Shaikhudinov, B. A. Zhubanov, S. R. Rafikov and S. Kh. Khusainova, Vysokomol. Soedin. Ser. A 19 (1977) 1861.
49 V. P. Zubov and V. A. Kabanov, Vysokomol. Soedin. Ser. A, 13 (1971) 1305.
 V. A. Kabanov and D. A. Topchiev, Vysokomol. Soedin. Ser. A, 13 (1971) 1324.
50 C. H. Bamford and S. Brumby Makromol. Chem., 105 (1967) 122.
51 E. Tsuchida and T. Tomono, Makromol. Chem., 141 (1971) 265.
52 C. H. Bamford, A. D. Jenkins and P. Johnston, Proc. R. Soc. (London) Ser. A, 241 (1957) 364.
53 V. A. Kabanov, Kinetics and Mechanisms of Polyreactions, IUPAC International Symposium on Macromolecular Chemistry, Akademiai Kiadó, Budapest, 1969, p. 435.
54 D. A. Topchiev, V. G. Popov and V. A. Kabanov Kinetics and Mechanisms of Polyreactions, IUPAC International Symposium on Macromolecular Chemistry, Akademiai Kiadó, Budapest, 1969, p. 452.
55 V. R. Georgieva, V. P. Zubov, V. A. Kabanov and V. A. Kargin, Dokl. Akad. Nauk SSSR, 190 (1970) 1128.
56 G. E. Ham (Ed.), Vinyl Polymerization, Vol. I, Part I. Dekker, New York, 1967, p. 50.
57 G. Levin and W. A. Goddard, Org. React. Mech., (1974) 74.
58 G. Levin and W. A. Goddard, J. Am. Chem. Soc., 97 (1975) 1649.
59 K. L. Glazomickii, J. N. Polyakov, R. F. Smirnov, T. N. Jurchenkova, A. S. Cegolya and E. S. Roskin, Vysokomol. Soedin. Ser. A, 19 (1977) 2483.
60 E. L. Gefter, Phosphorus containing Monomers and Polymers (Fosfororganicheskie monomery i polimery), Izd. Akad. Nauk SSSR, Moscow, 1960, p. 207.
61 W. B. McCormack, U.S. Pat. 2,671,077 (1954); Chem. Abstr., 48 (1955) 6737h.
62 A. Boryniec and B. Laszkiewicz, J. Polym. Sci. Part A-1 (1963) 1963.
63 F. S. Dainton and K. J. Ivin, Q. Rev., 12 (1958) 61.
64 K. Soga, L. Hattori, J. Kinoshita and S. Ikeda, J. Polym. Sci. Polym. Chem. Ed., 15 (1977) 745.
65 R. E. Cais and J. H. O'Donnell, J. Polym. Sci. Polym. Lett. Ed., 15 (1977) 659.
66 W. Lau, D. G. Westmoreland and R. W. Novak, Macromolecules, 20(2) (1987) 457.
67 Y. Maeda, P. Schmid, D. Griller and K. V. Ingold, J. Chem. Soc. Chem. Commun., 13 (1978) 525.

68 P. E. M. Allen and C. R. Patrick, Kinetics and Mechanisms of Polymerization Ractions, Horwood, Chichester, 1974, p. 39; (a) p. 41; (b) p. 42; (c) p. 43; (d) p. 46.
69 E. A. Guggenheim and R. Fowler, Statistical Thermodynamics, Cambridge University Press, London, 1949.
70 See, for example, T. E. Hogen-Esch and J. Smid, J. Am. Chem. Soc., 88 (1966) 307.
71 N. Hirota, J. Am. Chem. Soc., 90 (1968) 3603.
72 J. T. Denison and J. B. Ramsay, J. Am. Chem. Soc., 77 (1955) 2615.
73 J. B. Ramsay and E. L. Colichman, J. Am. Chem. Soc., 69 (1974) 3041.
74 F. H. Healey and A. E. Martell, J. Am. Chem. Soc., 73 (1951) 3296.
75 C. Carjaval, K. S. Tolle, J. Smid and M. Szwarc, J. Am. Chem. Soc., 87 (1965) 5584.
76 See for example, R. Phillippe and A. M. Piette, Bull, Soc. Chim. Belg., 64 (1955) 600; F. H. Healey, K. H. Stern and A. E. Martell, J. Chem. Phys., 19 (1951) 1114.
77 See for example, H. K. Bodengek and J. B. Ramsay, J. Phys. Chem., 69 (1965) 543.
78 S. Winstein and G. C. Robinson, J. Am. Chem. Soc., 80 (1958) 169 and many previous papers on this topic.
79 R. Wicke and K. F. Elgert, Makromol. Chem., 178 (1977) 3063.
80 M. Szwarc, Makromol. Chem., 35 (1960) 132.
81 A. Vrancken, J. Smid and M. Szwarc, Trans. Faraday Soc., 58 (1962) 2036.
82 A. Vrancken, J. Smid and M. Szwarc, J. Am. Chem. Soc., 83 (1961) 2772.
83 S.L. Malhotra, J. Macromol. Sci. Chem., A15(4) (1981) 533.
84 H. Yasuda and A. Nakamura, Polym. Prepr. Am. Chem. Soc. Div. Polym. Chem., 27(1) (1986) 136.
85 S. Bywater, Polym. Prepr. Am. Chem. Soc. Div. Polym. Chem., 27(1) (1986) 149.
86 A. Garton and S. Bywater, Macromolecules, 8 (1975) 694.
87 T. C. Cheng and A. F. Halasa, J. Polym. Sci. Polym. Chem. Ed. 14 (1976) 583.
88 H. Vankerckhoven and M. Van Beylen, Eur. Polym. J., 4 (1978) 189.
89 See, for example, B. L. Jerusalimski and A. V. Novoselova, Faserforsch. Textiltech. 26(4) (1975) 293.
90 W. Berger and H. − J. Adler, Makromol. Chem. Macromol. Symp., 3 (1986) 301.
91 M. Völker, A. Neumann and V. Baumann, Makromol. Chem., 63 (1963) 182.
92 J. Trekoval, D. Sc. Thesis, Institute of Macromolecular Chemistry, Czechoslovak Academy of Science, Prague, 1978, (a) p. 41; (b) p. 144.
93 J. Trekoval, Collect. Czech. Chem. Commun., 42 (1977) 1529.
94 Y. Okamoto, K. Ohta and H. Yuki, Macromolecules, 11 (1978) 724.
95 G. Weill and A. Blumstein, Polym. Prepr. Am. Chem. Soc. Div. Polym. Chem., 17(2) (1976) 479.
96 M. Kučera and M. Jelínek, Vysokomol. Soedin., 2 (1960) 1860.
97 M. Kučera J. Polym. Sci., 58 (1962) 1263.
98 M. Kučera, Vysokomol. Soedin., 6 (1963) 938.
99 A. D. Aliev, I. I. Solomatina, A. J. Koshevnik, Zh. Zhumabaev and B. A. Krencel, Vysokomol. Soedin. Ser. A, 19 (1977) 197.
100 H. Watanabe, K. Higuchi, M. Kobayhashi, M. Hara, Y. Koike, T. Kitahara and Y. Nagai, J. Chem. Soc. Chem. Commun., (1977) 534.
101 S. Tsuchiya and T. Tsuruta, Makromol. Chem., 110 (1967) 123.
102 A. Hofman, R. Szymański, S. Słomkowski and S. Penczek, Makromol. Chem., 185 (1984) 655.
103 S. Penczek S. Słomkowski, Polym. Prepr. Am. Chem. Soc. Div. Polym. Chem., 27(1) (1986) 171.
104 D. K. Dimov, I. M. Panayotov, V. N. Lazarov and Kh. B. Tsvetanov, J. Polym. Sci. Polym. Chem. Ed., 20 (1982) 1389.

105 J. Šebenda, Makromol. Chem. Macromol. Symp., 6 (1986) 1.
106 M. A. Bagirov, A. A. Aliev, Yu. N. Gazaryan and V. P. Malin, Plast. Massy, 12 (1982) 32.
107 B. Valter, M. T. Terekhová, E. S. Petrov, J. Stehlíček and J. Šebenda, Collect. Czech. Chem. Commun., 50 (1985) 834.
108 G. A. Olah, J. Am. Chem. Soc., 93 (1972) 808.
109 G. A. Olah, J. D. DeMember, R. H. Schlosser and Y. Halpern, J. Am. Chem. Soc., 94 (1972) 156.
110 J. P. Kennedy, Cationic Polymerization of Olefins, Wiley, New York, 1975.
111. I. Koyano, in C. H. Bamford and C. F. H. Tipper (Eds.), Comprehensive Chemical Kinetics, Vol. 18, Elsevier, Amsterdam, 1976, Chap. 6.
112 H. L. Friedman, in J. O. M. Bockris and B. A. Conway (Eds.) Modern Aspects of Electrochemistry, Vol. 6, Plenum Press, New York, 1971.
113 R. Zwanzig, J. Chem. Phys., 52 (1970) 3625.
114 R. W. Guerney, Ionic Processes in Solution, Mc Graw-Hill, New York, 1962.
115 C. V. Krishnan and H. L. Friedman, J. Phys. Chem., 75 (1971) 388, 3608.
116 F. H. Stillinger and A. Rahman, J. Chem. Phys., 57 (1972) 1281.
117 B. Ya. Simkin and I. I. Sheikhet, J. Mol. Liq., 27(2) (1983) 79.
118 L. G. Gorb, I. A. Abronin, N. H. Kharchevnikova and G. M. Zhidomirov, Zh. Fiz. Khim., 58(1) (1984) 9.
119 R. L. Perry and J. P. O'Connell, Mol. Phys., 52(1) (1984) 137.
120 D. Chandler, J. Phys. Chem., 88(16) (1984) 3400.
121 R. R. Dogonadze, E. Kalman, A. A. Kornyshev and J. Ulstrup (Eds.), The Chemical Physics of Solvation, Part A: Theory of Solvation, Elsevier, Amsterdam, 1985.
122 J. P. Kennedy and J. E. Johnston, Adv. Polym. Sci., 19 (1975) 57.
123 P. K. Bossaer, E. J. Goethals, P. J. Hackett and D. C. Pepper, Eur. Polym. J., 13 (1977) 489.
124 T. Enoki, M. Sawamoto and T. Higashimura, J. Polym. Sci. Polym. Chem. Ed., 24 (1986) 2261.
125 M. Kučera and E. Spousta, J. Polym. Sci. Part A-1 (1964) 3431
126 M. Kučera, D.Sc.Thesis, Research Institute of Macromolecular Chemistry, Brno, 1967.
127 Y. Yokoyama, M. Okada and H. Sumimoto, Makromol. Chem., 178 (1977) 529.
128 S. Penczek, Makromol. Chem., 175 (1974) 1217.
129 Y. E. Eizner and B. L. Erusalimski, Vysokomol. Soedin. Ser. A, 12 (1970) 1614.
130 B. C. Ramsey and R. W. Taft, J. Am. Chem. Soc., 88 (1966) 3058.
131 S. Penczek, Polym. Prepr. Am. Chem. Soc. Div. Polym. Chem., 25(1) (1984) 222.
132 A. Hofman, R. Szymanski, S. Słomkovski and S. Penczek, Makromol. Chem., 185(4) (1984) 655.
133 M. Kučera, J. Láníková and E. Spousta, Czech. Pat. 110, 064 (1962).
134 M. Kučera and H. Kelblerová, Collect. Czech. Chem. Commun., 44 (1979) 542.
135 S. Kobayashi, H. Danda and T. Saegusa, Macromolecules, 7 (1974) 415.
136 A. M. Buyle, K. Matyjaszewski and S. Penczek, Macromolecules, 10 (1977) 269.
137 P. H. Plesch, IUPAC Symposium on Macromolecules, Prague, 1965, Butterworths, London, 1966, p. 117.
138 T. Saegusa, Makromol. Chem., 175 (1974) 1193.
139 K. Matyjaszewski and S. Penczek, J. Polym. Sci. Polym. Chem. Ed., 12 (1974) 1905.
140 K. Matyjaszewski, T. Diem and S. Penczek, Makromol. Chem., 180 (1979) 1917.
141 K. Matyjaszewski, R. Szymanski, P. Kubisa and S. Penczek, Acta Polym., 35(1) (1984) 14.
142 T. Saegusa and S. Kobayashi, Polym. Prepr. Am. Chem. Soc. Div. Polym. Chem., 26(1) (1985) 32.
143 M. U. Mahmud, G. Wegner, W. Kern, J. B. Bland and Y. Osada, J. Macromol. Sci. Chem., A11 (1977) 2233.

144 S. Penczek, Makromol. Chem. Suppl., 3 (1979) 17.
145 S. Penczek, P. Kubisa and R. Szymanski, Makromol. Chem. Macromol. Symp., 3 (1986) 203.
146 M. Wojtania, P. Kubisa and S. Penczek, Makromol. Chem. Macromol. Symp., 6 (1986) 201.
147 R. Puffr and J. Šebenda, Makromol. Chem. Macromol. Symp., 3 (1986) 249.
148 P. A. Gembitskii, A. I. Khmarin, N. A. Kleshcheva and O. S. Zhuk, Vysokomol. Soedin. Ser. A, 20 (1978) 1505.
149 E. J. Goethals, E. H. Schacht, P. Bruggeman and P. Bossaer, Polym. Prepr. Am. Chem. Sec. Div. Polym. Chem., 18(1) (1977) 1.
150 M. Rothe, Polym. Prepr. Am. Chem. Soc. Div. Polym. Chem., 18(1) (1977) 45.
151 G. Bertalan, I. Rusznák and P. Anna, Paper presented at Microsymposium: Ring-opening Polymerization, M8, Karlovy Vary 1980.
152 S. Kobayashi and T. Saegusa, Makromol. Chem. Macromol. Symp., 3 (1986) 179.
153 D. R. Tur, V. V. Korshak, S. V. Vinogradova, S. A. Pavlova, G. I. Timofeeva, Ts. A. Goguadze and N. O. Alikhanova,, Dokl. Akad. nauk SSSR, 291(2) (1986)) 364.
154 S. Florquin and E. J. Goethals, Paper presented at Microsymposium: Ring-opening Polymerization, M17, Karlovy Vary, 1980.
155 D. N. Bhattacharya, C. L. Lee, J. Smid and M. Szwarc, Polymer, 5 (1964) 54.
156 P. Sigwalt, J. Polym. Sci. Polym. Symp., 50 (1975) 95.
157 H. Meerwein, E. Battenberg, H. Gold, E. Pfeil and G. Willfang, J. Prakt. Chem., 154 (1940) 83.
158 G. N. Komratov, R. A. Barzykina and G. V. Korovina, Vysokomol. Soedin. Ser. A 20 (1978) 608.
159 F. Leonard, J. A. Collins and H. A. Porter, J. Appl. Polym. Sci.,, 10 (1966) 1617.
160 E. F. Donnelly, D. S. Johnston and D. C. Pepper, J. Polym. Sci. Polym. Lett. Ed., 15 (1977) 399.
161 D. S. Johnston and D. C. Pepper, Makromol. Chem., 182 (1981) 393.
162 T. Saegusa, T. Yokohama, Y. Kimura and S. Kobayashi, Macromolecules, 10 (1977) 791.
163 T. Saegusa, S. Kobayashi and Y. Kimura, Macromolecules, 10 (1977) 64, 73.
164 T. Saegusa, S. Kobayashi and K Hayashi, Macromolecules, 11 (1978) 360; Polym. Prepr. Am. Chem. Soc. Div. Polym. Chem. 19 (1978) 1.
165 G. Odian, P. A. Gunatillake and D. Tomalia, Macromolecules, 18 (1985) 605.
166 I. J. McEwen, Prog. Polym. Sci., 10(4) (1984) 317.
167 M. Szwarc, in Carbanions, Living Polymers and Electron Transfer Processes, Interscience, Wiley, New York, 1968, (a) p. 301; (b) p. 348; (c) p. 328; (d) p. 320; (e) p. 373; (f) p. 23.
168 A. Mathias and E. Warburst, Trans. Faraday Soc., 58 (1962) 948.
169 D. E. Paul, D. Lipkin and S. I. Weissman, J. Am. Chem. Soc., 78 (1956) 116.
170 K. F. O'Driscoll, R. J. Boudreau, and A. V.Tobolsky, J. Polym. Sci., 31 (1958) 115; D. B. George and A. V. Tobolsky, J. Polym. Sci. Part B, 2 (1964) 1.
171 C. G. Overberger and N. Yamamoto, J. Polym. Sci. Part A-1, (4) (1966) 3101.
172 M. Marek, J. Polym. Sci. Polym. Symp. 56 (1976) 149, 496.
173 K. Hayashi, J. Polym. Sci. Polym. Symp. 56 (1976) 490.
174 M. Irie, M. Masuhara, K. Hayashi and N. Nagata, J. Phys. Chem., 78 (1974) 341.
175 A. Ledwith, J. Polym. Sci. Polym. Symp., 56 (1976) 483.
176 S. Tagawa, A. Sugawara and Y. Tabata, Polym. Prepr. Am. Chem. Soc. Div. Polym. Chem., 17(2) (1976) 636.
177 A. J. Bard, A. Ledwith and H. J. Shine, Adv. Phys. Org. Chem., 13 (1976) 156.
178 L. Eberson, Adv. Phys. Org. Chem., 12 (1976) 1.
179 E. Oberrauch, T. Salvatori and S. Cesca, J. Polym. Sci. Polym. Lett. Ed., 16 (1978) 345.
180 B. L. Funt, W. Severs and A. Glasel, J. Polym. Sci. Polym. Chem. Ed., 14 (1976) 2763.

181 K. Ziegler, E. Holzkamp, H. Breil and H. Martin, Angew. Chem., 67 (1955) 426.
182 G. Natta, SPE J. 5 (1959) 373.
183 G. Natta, P. Corradini and G. Allegra, Rendiconti della Classe di Scienze Fisiche, Matematiche e Naturali, Ser. VIII, Vol. 26, Academia Nazionale dei Lincei, Rome, 1959, p. 159.
184 J. Boor, Jr., Ziegler–Natta Catalysts and Polymerizations, Academic Press, New York, 1979, pp. 100, 101.
185 J. Boor, Jr., Macromol. Rev., 2 (1967) 124.
186 J. P. Harmans and P. Henrioulle, (Solway et Cie), U.S. Pat. 4, 210, 738 (1980).
187 R. P. Nielsen, in R. P. Quirk, H. L. Hsieh, G. B. Klingensmith and P. J. T. Tait (Eds.), Transition Metal Catalyzed Polymerization, Alkenes and Dienes, Part A, MMI Press, Symposium Series, Vol. 4, Horwood, London, New York, 1983, pp. 47–78.
188 W. Cooper, in C. H. Bamford and C. F. H. Tipper (Eds.), Comprehensive Chemical Kinetics, Vol. 15, Elsevier, Amsterdam, 1976, pp. 133–145.
189 See, for example, J. Lieto, D. Milstein, R. L. Albright, J. V. Minkiewicz and B. C. Gates, Chemtech., 13(1), (1983) 46.
190 S. A. Mrktychyan, B. A. Uvarov, V. I. Tsvetkova, Yu. M. Tovmasyan, S. O. Chistyakov, G. F. Rachinskij and F. S. D'yachkovskii, Vysokomol, Soedin, Ser. A, 28(10) (1986) 2108.
191 S. A. Benell, W. M. Coleman III and W. R. Howell Jr. (Dow Chemical Co.), U.S. Pat. 4, 623, 707 (1986).
192 L. A. Rishina, E. I. Vizen and L. N. Sosnovskaya, Polimery 31(10) (1986) 380.
193 J. Hogan, J. Polym. Sci. Part A-1 (8) (1970) 2637.
194 J. I. Ermakov and V. A. Zakharov, Coordination Polymerization, Academic Press, New York, 1975, p. 107.
195 L. K. Przheval'skaya, V. A. Shvets and V. B. Kazanskii, Kinet. Katal. 15 (1974) 180, 430.
196 M. P. McDaniel, in R. P Quick, H. L. Hsieh, G. B. Klingensmith and P.J.T. Tait (Eds.) Transition Metal Catalyzed Polymerization, Alkenes and Dienes, Part B, MMI Press Symposium Series, Vol. 4, Horwood, London, New York, 1983, pp. 713–755.
197 S. N. Gan, S. Chen, R. Ohnishi and K. Soga, Makromol. Chem. Rapid Commun., 5 (1984) 535.
198 K. Soga, S. Chen, T. Shiono and Y. Doi, Polymer, 26(12) (1985) 1888.
199 B. Rebenstorf, Z. Anorg. Allgem. Chem., 513 (1984) 103.
200 Z. Tvaruzkova, and B. Wichterlova, J. Chem. Soc. Faraday Trans. 1, 79(7) (1983) 1591.
201 H. N. Friedlander, J. Polym. Sci., 38 (1958) 91.
202 G. Natta, P. Pino, G. Mazzanti, V. Giannini, E. Mantica and M. Peraldo, J. Polym. Sci., 26 (1957) 120.
203 D. S. Breslow and N. R. Newburg, J. Am. Chem. Soc., 79 (1957) 5072; 81 (1959) 81. W. P. Long and D. S. Breslow, J. Am. Chem. Soc., 82 (1960) 1953.
204 F. L. Borisova, E. A. Fushman, E. I. Vizen and N. M. Chirkov, Eur. Polym. J., 9 (1973) 953.
205 G. Henrici-Olivé and S. Olivé, J. Polym. Sci. Part B, 8 (1970) 271 and previous papers.
206 D. R. Armstrong, P. G. Perrins and J. J. Stewart, J. Chem. Soc. Dalton Trans., (1972) 1972.
207 A. E. Shilov, A. K. Shilova and B. N. Bobkov, Vysokomol. Soedin., 4 (1962) 1688.
208 F. S. D'Yachkovskii, Vysokomol. Soedin., 7 (1965) 114.
209 J. Cihlář. Thesis, Research Institute of Macromolecular Chemistry, Brno, 1978.
210 I. N. Meshkova, T. M. Ushakova, J. L. Lelyukhina, N. N. Korneev and F. S. D'Yachkovskii, Vysokomol. Soedin. Ser. B, 19 (1977) 849.
211 W. Kaminski, Polym. Prepr. Am. Chem. Soc. Div. Polym. Chem., 26(2) (1985) 373.
212 W. Kaminski, K. Kuelper and S. Niedoba, Makromol. Chem. Macromol. Symp., 3 (1986) 377.
213 J. Cihlář, J. Mejzlík, O. Hamřík, P. Hudec and J. Majer, Makromol. Chem. 181 (1980) 2549.
214 A. Siove and M. Fontanille, J. Polym. Sci. Polym. Chem. Ed., 22 (1984) 3877.

215 M. Fontanille and A. Siove, in R. P. Quirk, H. L. Hsieh, G. B Klingersmith and P. J. T. Tait (Eds.), Transition Metal Catalyzed Polymerization, Alkenes and Dienes, Part A, MMI Press Symposium Series, Vol. 4, Horwoord, London, New York, 1983, pp. 313–321.
216 Karol, F. J., Jacobson, F. I.: Stud. Surf. Sci. Catal. 25 (1986) 323.
217 H. Meyer and K. H. Reichert, Makromol. Chem., 57 (1977) 211.
218 E. J. Nagel, V. A. Kirillov and W. H. Ray, Ind. Eng. Chem. Prod. Res. Dev., 19 (1980) 372.
219 J. C. W. Chien, J. Polym. Sci. Polym. Chem. Ed., 17 (1979) 2555.
220 Y. Doi, M. Murata, K. Yano and T. Keii, Ing. Eng. Chem. Prod. Res. Dev., 21 (1982) 580.
221 S. Floyd, K. Y. Choi, T. W. Taylor and W. H. Ray, J. Appl. Polym. Sci., 31(7) (1986) 2231.
222 K. Soga, S. Chen, T. Shiono and Y. Doi, Makromol. Chem., 187 (1986) 351.
223 N. M. Karayannis and S. S. Lee, Makromol. Chem., 184 (1983) 2275.
224 P. Pino, B. Rotzinger and E. v. Achenbach, Makromol. Chem. Suppl., 13 (1985) 105.
225 M. Kučera, Chem. Listy 83 (1989) 730, 829; Chem. Listy, 83 (1989) 936.
226 H. Meerwein and T. Bersin, J. Prakt. Chem., 147 (1937) 203.
227 P. Teyssié, J. P. Bioul, A. Hamilton, J. Heuschen, L. Hocks, R. Jerome and T. Ouhadi, Polym. Prepr. Am. Chem. Soc. Div. Polym. Chem., 18(1) (1977) 65.
228 T. Saegusa, H. Imai and S. Matsumoto, J. Polym. Sci. Part A-1, (6) (1968) 459.
229 M. Kobayashi, Y. L. Tang, T. Tsuruta and S. Inoue, Makromol. Chem., 169 (1973) 69.
230 W. Kuran, S. Pasynkiewicz and J. Skupinska, Makromol. Chem., 178 (1977) 47.
231 M. Kučera, Vysokomol. Soedin., 5 (1963) 938.
232 M. Kučera, A. Zahradníčková and K. Majerova, Polymer, 17 (1976) 519, 528, 535.
233 D. V. Sokolskii, N. D. Lavorokhin and M. V. Favorskaya, Dokl. Akad. Nauk SSSR, 187 (1969) 1325.
234 S. Tsuchyia and T. Tsuruta, Makromol. Chem., 110 (1967) 123.
235 Y. Yamashita, Polym. Prepr. Am. Chem. Soc. Div. Polym. Chem., 13(1) (1972) 539.
236 F. J. Burgess, A. V. Cunliffe, D. H. Richards and D. C. Sherrington, J. Polym. Sci. Polym. Lett. Ed., 14 (1976) 471.
237 F. J. Burgess, A. V. Cunliffe, J. R. MacCallum and D. H. Richards, Polymer, 18 (1977) 719, 726.
238 F. J. Burgess, A. V. Cunliffe, J. W. Dawkins and D. H. Richards, Polymer, 18 (1977) 733.
239 M. J. M. Abadie, F. Schue, T. Souel, D. B. Hartley and D. H. Richards, Polymer, 23 (1982) 445.
240 P. Cohen, M. J. M. Abadie, F. Schue and D. H. Richards, Polymer, 23 (1982) 1105.
241 M. Kučera, K. Majerová and F. Božek, CS 190, 983 (1977).
242 M. Kučera, F. Božek and K. Majerová, Polymer, 20 (1979) 1013.
243 M. Kučera, Chem. Listy, 77 (1983) 1083.
244 G. Reiss and F. Palacin, IUPAC, Helsinki, Preprint 1, 1972, p. 123.
245 J. Brossas, J. M. Catala, G. Clouet and Z. Gallot, C. R. Acad. Sci. Ser. C, 278 (1974) 80.
246 M. J. M. Abadie, F. J. Burgess, A. V. Cunliffe and D. H. Richards, J. Polym. Sci. Polym. Letters E., 14 (1976) 477.
247 M. J. M. Abadie, F. Schué, T. Souel and D. H. Richards, Polymer, 22 (1981) 1076.
248 P. Cohen, M. J. M. Abadie, F. Schué and D. H. Richards, Polymer, 22 (1981) 1316.
249 M. E. Galvin and G. E. Wnek, Polym. Bull., 13(2) (1985) 109.
250 C. L. Willis (Shell Oil Co.), U.S. Pat. 4, 480,075 (1984).
251 Y. Doi, Y. Watanabe, S. Ueki and K. Soga, Makromol. Chem. Rapid Commun., 4 (1983) 533.
252 Y. Doi, T. Kayma and K. Soga, Makromol. Chem., 186(1) (1985) 11.
253 A. J. Amass, S. Bas, D. Gregory and M. C. Mathew, Makromol. Chem., 186 (1985) 325.
254 F. S. D'Yachkovskii, Makromol. Chem. Macromol. Symp., 3 (1986) 331.
255 I. V. Ikonitskii, N. A. Buzina, L. S. Bresler, M. D. Shul'diner and V. A. Kormer, Vysokomol. Soedin. Ser. A, 28 (1986) 1078.

256 C. Pichot, R. Spitz and A. Guyot, J. Macromol. Sci. Chem., A11 (1977) 251.
257 T. E. Hogen-Esch and J. Smid, J. Am. Chem. Soc., 87 (1965) 699; 88 (1966) 307.
258 E. Buncel and B. Menon, J. Chem. Soc. Chem. Commun., (1978) 758.
259 N. V. Pticyna, V. K. Kazakievich and K. S. Kazanskii, Vysokomol. Soedin Ser. A, 19 (1977) 2787.
260 M. A. Littlejohn, C. C. Ma, H. Garreau, D. R. Squire and V. T. Stannett, J. Macromol. Sci. Chem., A11 (1977) 1603.
261 M. De Sorgo, D. C. Pepper and M. Szwarc, J. Chem. Soc. Chem. Commun., (1973) 419.
262 J. P. Lorimer and D. C. Pepper, Proc. R. Soc. (London) Ser. A, 351 (1976) 551.
263 M. Sawamoto and T. Higashimura, Macromolecules, 11 (1978) 328.
264 C. V. Schulz, L. L. Böhm, M. Chmelíř, G. Löhr and B. J. Schmitt, IUPAC Symposium on Macromolecular Chemistry, Akadémiai Kiadó, Budapest 1969, pp. 225–226.
265 Y. Ikada, H. Iwata and S. Nagaoka, Macromolecules, 10 (1977) 1364.
266 A. K. Alimoglu, C. H. Bamford, A. Ledwith and S. U. Mullik, Macromolecules, 10 (1977) 1081.
267 S. Szakács and O. Ocs, Makromol. Chem., 179 (1978) 1649.
268 J. Šebenda in F. Ciardelli and P. Giusti (Eds.), Structural Order in Polymers, Pergamon Press, Oxford, 1981, p. 95.

Chapter 5

Propagation

> *...first of all we should define the contents of the subject, classify it exactly, present various representatives and describe their properties*
>
> V. V. Korshak

The chains of industrially produced polymers contain hundreds to thousands of monomeric units. The properties of polymers strongly depend on their structure, on the manner of attachment of the monomeric units in the chains, and on the distribution of chain lengths and shapes. Branched macromolecules in solution or melt exhibit completely different rheological properties compared with their linear counterparts. When the length and number of branches exceed some critical limit, difficulties in their processing usually occur.

Thus the addition of the monomer to the active centre must be repeated about one thousand times without interruption by another reaction. In other words, propagation must be about one thousand times more rapid than termination or any of the many possible transfer processes.

1. Theoretical conditions for propagation

Chain growth can only occur under suitable thermodynamic conditions. For the chains to have required properties, some chemical conditions must also be fulfilled. The possibility of the technical control of propagation depends on the rates of the respective reactions, i. e. on the kinetics of the process.

1.1 Thermodynamic conditions

As any other process, propagation can spontaneously proceed only when accompanied a decrease in the Gibbs energy (free enthalpy) of the system

$$\Delta G = \Delta H - T \Delta S \tag{1}$$

where the symbols have their conventional meaning.

Propagation is a reversible reaction

$$\mathsf{\sim\!\!\sim\!\!M^\circ + M \xrightleftharpoons{K} \sim\!\!\sim\!\!MM^\circ} \quad (2)$$

There must therefore exist some conditions where the rate of propagation is just equal to the rate of depropagation, i. e. where $\Delta G = 0$. The temperature at which the rates of reaction (2) are equal in both directions is of special importance for polymerizations.

1.1.1 CEILING (FLOOR) TEMPERATURE

Equations (1) and (2) indicate that ΔH and ΔS must both be either positive or negative

$$T_c = \frac{\Delta H}{\Delta S} > 0 \quad (3)$$

The equilibrum constant K of eqn. (2) is given by the ratio of the propagation and depropagation rate constants, k_p k_d

$$K = \frac{k_p}{k_d} = \frac{A_p \exp\left[-E_p/(RT)\right]}{A_d \exp\left[-E_d/(RT)\right]} = \frac{A_p}{A_d} \exp\left[-\frac{1}{RT}(E_p - E_d)\right] \quad (4)$$

The temperature derivative of the logarithmic form of eqn. (4) yields the temperature dependence of the equilibrium constant (Van't Hoff isobar)[†]

$$\frac{d \ln K}{dT} = \frac{E_p - E_d}{RT^2} = \frac{\Delta H}{RT^2} \quad (5)$$

As the activation energies E_p and E_d are always positive, (5) yields for $\Delta H > 0 \Rightarrow E_p > E_d$ and for $\Delta H < 0 \Rightarrow E_p < E_d$. The acceleration of exothermic propagation with increasing temperature is slower than the growth of the rate of depropagation; the opposite is true for endothermic propagation[††]. The consequences of this are shown schematically in Figs. 1 and 2 where the rates of propagation $k_p[M^\circ][M]$ and of depropagation $k_d[M^\circ]$ are compared. Polymer chains will only be formed at higher propagation rates, i. e. only in those cases where the sum curve of both rates lies above the temperature axis. The intersection point of the sum curve with the abscissa represents the

---

[†] To a first approximation, the pre-exponential factors A_p and A_d are considered to be independent of temperature.
[††] Most known propagations are exothermic processes.

ceiling temperature of polymerization (T_c) for cases with $\Delta H < 0$, and the floor temperature of polymerization (T_f) for cases with $\Delta H > 0$. In the thermodynamically permitted range, polymer formation and quality are determined by kinetic and chemical factors.

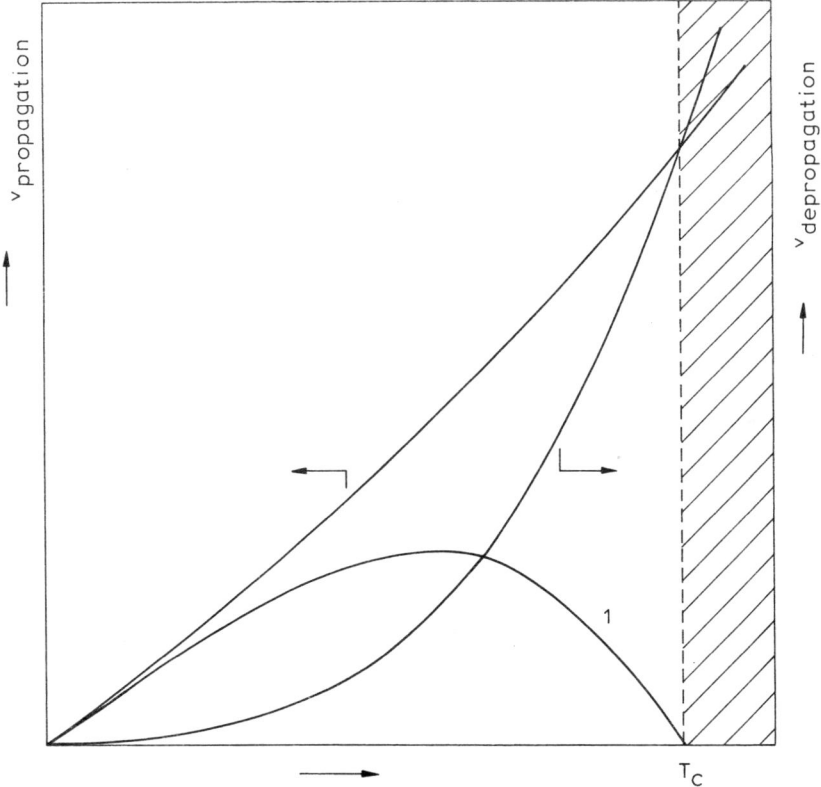

Fig. 1 T_c. The difference in the rates of propagation and depropagation (curve 1) determines the temperature range where long chains can be generated. In the hatched section, polymerization does not occur.

The concept of the limiting polymerization temperature has been elaborated by Dainton and Ivin[1]. Originally they only considered T_c, and they took the name from previous studies [2]: "ceilling temperature T_c". In the literature it is still usually designated by this symbol. Eisenberg and Tobolsky [3] are the authors of the broader concept of polymerization equilibria. Due to fluctuations in monomer addition, both k_p and k_d vary, and T_c is statistical in character. It is rather a temperature interval of some width. Sometimes it is defined as the temperature above which the formation of a high polymer $(\bar{P} > 100)$ is excluded[†]

[†] For example, T_c is highest for oligomers.

For a given monomer, T_c and T_f depend on the initial and final states. This follows from eqn. (3). The value of ΔS is not constant for all cases, of course[†]. It depends on the size of the various entropy contribution. Evidently it is always necessary to characterize the initial and final states in order to prevent misunderstanding.

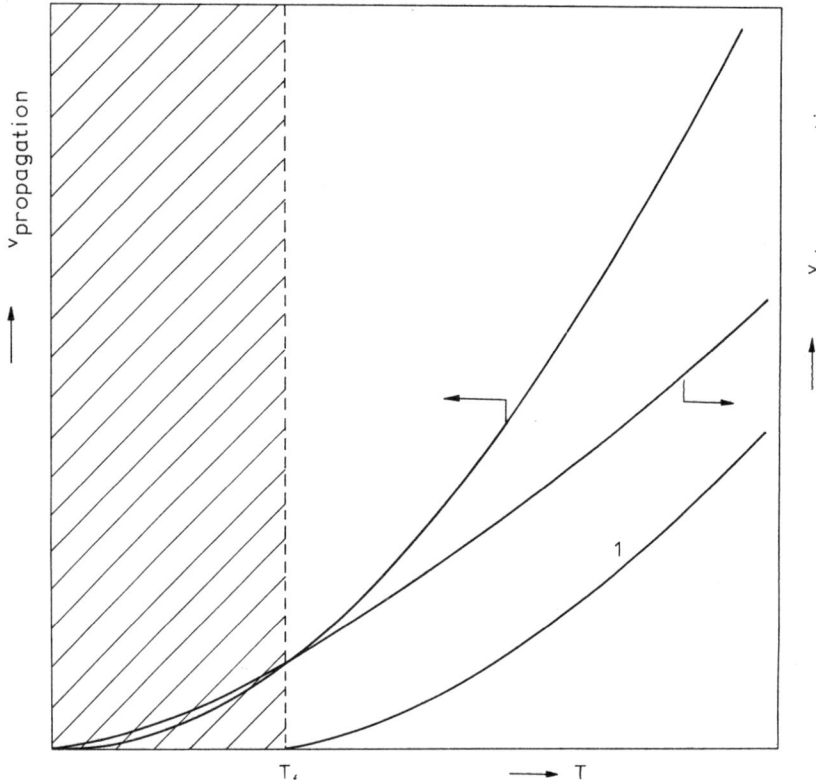

Fig. 2. T_f. See legend to Fig. 1.

From eqn. (2) it follows that

$$K \approx [M]_\infty^{-1} \qquad (6)$$

Therefore for the standard state we can write

$$\Delta H^\circ - T \Delta S^\circ = RT \ln [M]_\infty \qquad (7)$$

$$T_c = \frac{\Delta H}{\Delta S^\circ + R \ln [M]_\infty} \qquad (8)$$

[†] Even ΔH is not constant for the propagation of a given monomer (see Sect. 1.1).

Equation (8) describes the dependence of T_c on dilution; $\Delta S°$ here corresponds to the standard entropy change in polymerization of 1 mol of monomer.

Actually there exist many exceptions to eqn. (6) caused by non-uniform monomer attachment to the growing chain ends. Equations (7) and (8) are even less accurate as they operate with concentrations instead of activities, which should be used rigorously. Equations (3), (6), and (8) can therefore be used only for rough orientation, and the proper values T_c and T_f should be determined experimentally.

The approach of Dainton and Ivin [1] is general, simple, and formally quite correct. Practically it really yields only a limited amount and quality of information on the polymerizing systems. The mechanistic approach of Eisenberg and Tobolsky [3] is more specialized; it only applies to living systems. However, it yields information not only on monomer–polymer equilibria but also on the equilibrium distribution of molecular mass. The work of Tobolsky was extended by Wheeler et al. who further refined equilibria calculations in homopolymerizations [4, 5]; a general solution of equilibrium copolymerizations in living media was developed by Szwarc [6]. These latter developments are not based on formal thermodynamics.

1.1.2 HEAT OF REACTION AND ENTROPY CHANGES DURING POLYMERIZATION

For a hydrocarbon monomer whose double bond is not affected by substituents (ethylene), the heat of polymerization should be about -84 ± 8 kJ mol^{-1}, corresponding to the difference between the mean energy of a C=C bond (610 kJ mol^{-1}) and that of two C—C bonds in the polymer ($2 \times 347 = 694$ kJ mol^{-1}).

Several procedures exists for calculating the heats of polymerization for various monomers but they are not very important. It is still more useful to consider the experimentally measured values as the best basis both for theoretical studies and for practical calculations. For ethylene derivatives, the value of $-\Delta H_{lc}$ (from liquid monomer to amorphous solid polymer) ranges from 33.5 kJ mol^{-1} for α-methylstyrene, through ca. 96 kJ mol^{-1} for vinyl chloride, to 174 kJ mol^{-1} for tetrafluoroethylene.

Unequal ΔH values are usually caused by [7]

(a) differences in resonance stabilization energies of the monomer and polymer effected by conjugation or hyperconjugation or by a change in bond type (hybridization);

b) steric strain in the polymer and opposite strain in the monomer; and

c) various degrees of van der Waals force effects between monomer molecules, polymer molecules and other system components (e. g. solvation).

These will be discused below.

(a) Let us consider ethylene as standard. Propene exhibits a ΔH_{lc} value which is lower by ca. 7 kJ mol^{-1} (i. e. -81.6 kJ mol^{-1}) [8]; this is ascribed to hyperconjugation[†]. The heat of styrene polymerization and of its ring-substituted derivatives is considerably lower, by about 20 kJ mol^{-1}. This is caused by conjugation of the aromatic phenyl ring with the double bond. Weaker monomer stabilization was observed with conjugated dienes (by about 12 kJ mol^{-1}), vinyl-pyridines, and other monomers of this type. ΔH_{lc} is also lowered by conjugation with a carboxyl or carbonyl group in the respective monomers (acrylates and their derivatives). A similar, but unconjugated substituent (as in vinyl acetate and in higher esters) does not affect ΔH_{lc} so that for these monomers the heat of polymerization is again equal to about -88 kJ mol^{-1} as for ethylene.

Interesting cases where propagation is promoted by resonance stabilization of the polymer have been discussed by Orr [9]. Acetylene derivatives yield polymers with backbone atoms bound by a system of conjugated double bonds. The resonance energy contributing to polymer formation was estimated to amount to ca. 40 kJ per half mole of $(-CH=CH-CH=CH-)_n$ units; with this contribution the heat of acetylene polymerization is high at about -193 kJ mol^{-1}. Resonance stabilization sometimes leads to active centre isomerization, to the formation of a different polymer structure from that resulting from the given monomer by simple propagation (see Sect. 6.6).

Ethylene derivatives with electronegative substituents exhibit high heats of polymerization: vinyl chloride, nitroethylene $(-\Delta H_{lc} = 91$ kJ mol$^{-1})$, vinylidene fluoride $(-\Delta H_{lc} = 130$ kJ mol$^{-1})$, and tetrafluoroethylene. According to Mortimer, the repulsion interactions of non-bonding electrons from electronegative groups are reduced by polymerization [10]. For vinyl chloride, Flitcroft and Skinner measured a heat of hydrogenation higher by ca. 8 kJ mol^{-1} compared with ethylene. According to these authors, this is caused by the polarity of the C—Cl bond which has the character of a "stunted" double bond (4 %) which can participate in hypeconjugation in vinyl chloride [11]. The third factor is the formation of a weak hydrogen bond —C—H···Cl— [12]. The polymer chains are more firmly bound by these bonds than are the monomer molecules. The heat of polymerization is increased by the heat of macromolecule association [7].

(b) Steric hindrance in polymers or strain in monomers are the result of interatomic bond length or bond angle deformations and they considerably affect the heats of polymerization. Even interactions of non-

---

[†] It appears that even the higher ΔH of isoprene compared with butadiene can be explained by hyperconjugation, mainly due to the stabilization of the remaining double bond in polyisoprene.

bonded atoms are not negligible; they usually result in repulsive forces. Their contribution decreases the $-\Delta H$ values of exothermic and increases the ΔH values of endothermic polymerizations. Ferstanding and Goodrich calculated the non-bonding energy contributions of all combinations of the interacting groups of propene approaching the growing end of polypropylene [13]. The interaction energy exhibits a minimum at a distance of 0.26 nm with the monomer twisted by 40° about the axis of approach (it amounts to about 29–38 kJ mol^{-1}). It increases sharply on further approach. By similar calculations for methyl methacrylate, Bawn et al. [14] found an interaction energy of 21 kJ mol^{-1} between the ultimate and penultimate units. Moreover, end group free rotation is hindered by an energy barrier of over 80 kJ mol^{-1}. Only a part of these repulsion forces (from 20 to ca. 80 kJ) contributes to a reduction of the heat of polymerization. During addition to a growing chain end, the monomer must assume the energetically most favourable position. This means that the ultimate and penultimate units are usually mutually twisted by a certain angle. Probably even the penultimate and several further units twisted about the C—C bond to the preceding unit so as to generate a minimum energy conformation at the terminal chain segment. Thus the penultimate unit concept is not limited to copolymerizations for which it was originally proposed [15].

All structural effects decreasing the heat of polymerization are cumulated in α-methylstyrene: resonance of the double bond with the aromatic ring, —CH$_3$ hyperconjugation stabilizing the monomer, and polymer destabilizing 1,1-disubstitution. The same is true of methacrylic acid and of all its derivatives (ΔH_{lc} between 54 and 59 kJ mol^{-1}) and partly even of vinylidene chloride ($\Delta H_{lc} = 75.4$ kJ mol^{-1}).

Older ideas have been modified by recent progress. 1,2-Disubstituted ethylenes are now described as reluctant, rather than unable, to form homopolymers. They can be polymerized by new initiation procedures under more suitable external conditions of pressure, etc. (e. g. fumaric acid esters [16] and the *tert.* - Bu ester of crotonic acid, $(CH_3)HC=CHCOO$ *tert.* - Bu [17]). The cyclic compounds, polymerizing on the double bond without ring opening, also belong to this class of monomers. They are polymerized even more easily than the preceding group. Both substituents are firmly connected, facilitating an attack of the active centre on the double bond. The stabilizing effects of conjugation or induction effects are mutually compensated by the symmetry of the substituents. Every ring has some strain. Binding of cyclic monomers in a polymer chain usually reduces the ring strain [18] thus increasing the heat of polymerization.

Saturated cyclic monomers which are opened during polymerization exhibit heats of polymerization very close to ring strain values [7].

(c) In the sum of the contributions constituting the heat of polymerization, the effect of hydrogen bonds and of other van der Waals bonding forces is usually not dominant. These forces strongly depend on medium effects, for example the solvent. Strong hydrogen bonds affect monomer association more than that of polymer chains; with weak hydrogen bonds, the reverse is true. A detailed description of this kind of energy contribution for specific systems is a highly specialized topic.

The second thermodynamic quantity controlling propagation is the entropy. Changes of phase transitions, of concentration, and addition itself

TABLE 1

Values of thermodynamic polymerization parameters for some monomers [19]

Monomer	State[a]	$\Delta H°$ (kJ mol^{-1})	$\Delta S°$ (J K^{-1} mol^{-1})	Method[b]
Ethylene	gc	−103.3	−154.8	C
Propene	gc	−94.0	−181	C
2-Methylpropene	gc	−78.2	−192	C
Styrene	lc	−72.8	−103.8	R
α-Methylstyrene	ss	−33.5	−120.3	R
VC	gc	−130.1		E
VDC	lc̄	−75.3		E
TFE	gc̄	−171.1	−188.3	E
VAC	lc	−89.1		H
Acrylamide	ss	−82.8		H
MA	lc	−78.2		H
MMA	lc	−56.1	−116.3	R
MAN	ss	−64.0	−142.3	R
AN	lc	−72.5		H
Fd	gc̄	−55.2	−174.9	R
Cyclopropane	lc	−113.0	−69.0	C
Cyclobutane	lc	−105.0	−55.2	C
Cyclopentane	lc	−21.8	−42.7	C
3,3′-Bischloromethyloxetane	lc	−84.5		H
β-Propiolactone	lc	−80.5		E
Pivalolactone	lc	−92.0		E
THF	lc	−12.5	−41.0	R
1,3-Dioxolane	ss	−21.3	−77.8	R
1,3,5-TOX	lc̄	−28		C
6-Hexanelactam	lc	−12.5	+ 5.4	E
ζ-Enantholactam	lc	−23.8	+ 16.7	E
2-Pyrrolidone	lc	−5.4	−30.2	C

[a] g – gas, pressure 9 . 81 × 10^4 Pa; l – one-component liquid; s – solution of concentration 1 mol dm^{-3}; c – amorphous polymer; c̄ – crystalline polymer.
[b] C – from calculated or estimated heats of formation or entropies of the monomer; E – from experimentally determined heats of formation or entropies of the monomer; H – experimentally determined heat of polymerization; R – from equilibria.

contribute to its overall change. Further contributions are connected with solvation and desolvation processes, transitions from amorphous to crystalline product, etc. All these contributions are part of the overall entropy of polymerization which is measurable.

The results are not always reliable; they depend on the method of measurement (monomer ⇌ polymer equilibrium, calorimetric data) and mainly on the definition of the initial and final states.

So far nobody has attempted to isolate the entropy of propagation from the measured data.

The entropy change at polymerization has the tendency to increase with growing complexity of monomer structure [19]. The values of the heats of polymerization and of the entropy changes of some common monomers are summarized in Table 1.

1.1.3 INFLUENCE OF PRESSURE ON PROPAGATION

From the relations for general thermodynamic functions, the change of Gibbs energy with pressure is given by

$$\left(\frac{\partial G}{\partial p}\right)_T = V \tag{9}$$

In most cases, the growth of polymeric chains is accompanied by volume contraction. Therefore external pressure tends to shift the monomer ⇌ polymer equilibrium in favour of the polymer or, in other words, it increases the ceiling temperature of polymerization (lowers T_f). This analysis can be refined by means of the known thermodynamic relations. The change in enthalpy with pressure is described by the thermodynamic equation of state

$$\left(\frac{\partial H}{\partial p}\right)_T = -T\left(\frac{\partial V}{\partial T}\right)_p + V \tag{10}$$

and the change in entropy with pressure by the relation

$$\left(\frac{\partial S}{\partial p}\right)_T = -\left(\frac{\partial V}{\partial T}\right)_p \tag{11}$$

Equations (10) and (11) can be written in the form of eqns. (12) and (13) using the expansion coefficient α [$\alpha = (\partial V/\partial T)_p (1/V)$] for monomer ($\alpha_M$) and polymer ($\alpha_p$). For polymerizations in liquid medium, the changes in enthalpy and entropy with pressure are a function of the molar volume of the free monomer, V_M, of the molar volume of the monomeric unit in the polymer, V_p, of the expansion coefficients α_M, α_P, and of temperature

$$\left(\frac{\partial \Delta H}{\partial p}\right)_T = \Delta V_P - T(\alpha_P V_P - \alpha_M V_M) \tag{12}$$

where ΔV_p is the volume change in polymerization, and

$$\left(\frac{\partial \Delta S}{\partial p}\right)_T = \alpha_M V_M - \alpha_P V_P \tag{13}$$

As $\Delta V_P < 0$ (but for minor exceptions which will be mentioned in Sect. 3.3), entropy changes will decrease with growing pressure. The effect of pressure on the entropy of propagation is relatively larger than on the enthalpy. Thus pressure should favour propagation, in agreement with our observations. For transitions of liquid unsaturated monomers to solid polymer, the usual values in propagation are $\Delta V_P = -20 \text{ cm}^3 \text{ mol}^{-1}$, $V_M = 100 \text{ cm}^3 \text{ mol}^{-1}$, and $\alpha_M = 10^{-3} \text{ K}^{-1}$. For polymers above the glass transition temperature (T_g), $\alpha_P = 5 \times 10^{-4} \text{ K}^{-1}$ and below T_g, $\alpha_P \approx 2 \times 10^{-4} \text{ K}^{-1}$; for crystalline polymers $\alpha \approx 2 \times 10^{-4} \text{ K}^{-1}$. Using the approximate assumption that α_P and V_P are independent of pressure, Allen and Patrick [19] estimated the changes of thermodynamic parameters: at a pressure of 98.1 (196.2, 490.5, and 981) MPa, ΔS_{lc} will be more positive by 7 (15, 42, and 85) $J \text{ K}^{-1} \text{ mol}^{-1}$ and roughly independent of temperature up to the critical temperature of the monomer. At about 300 K, the changes in ΔH with pressure will be >0 but less than 5 kJ mol^{-1} in the range 0.1–981 MPa.

1.2 Chemical influences

Structural features predisposing compounds to react as monomers have been briefly treated in Chap. 2, Sec. 1. Chemical factors affecting polymerizations are, in fact, the subject of all parts of this volume. Therefore only certain aspects will be mentioned in this chapter.

1.2.1 THE PRESENCE OF IMPURITIES

The condition for well-defined monomer addition to the active centre during propagation is the removal from the polymerizing system of all molecules that either react in a differential way or that change the manner of addition. Monomers, solvents, and initiators always contain molecules which will react readily with the active centres: water, oxygen, compounds related to the monomers, their oxidation or hydrolytic products, and many others. In view of the competing processes

$$\sim Ac° \underset{X, k_x}{\overset{M, k_p}{\rightleftarrows}} \begin{matrix} \sim MAc° \\ \sim AcX° \end{matrix} \qquad (14)$$

the requirement of uninterrupted thousand-fold monomer addition† will only be fulfilled when the monomer concentration (strictly activity) is $10^3 \, k_x/k_p$ times greater than the concentration of the impurity X. The value of k_x is often greater than that of k_p, sometimes even by several orders. The requirements of polymeric syntheses concerning the purity of the medium are thus well justified.

Modern coordination polymerizations of alkenes proceed in systems containining less than 10 ppm of water. Only a single water molecule may be present among 5×10^4 molecules of propene. There exist sufficient reasons for tightening of this norm by at least an order of magnitude.

1.2.2 PROPAGATION AND POLYCONDENSATION

In some cases the required polymer can be obtained by both basic polyreaction types.

Polydimethylsiloxane used to be prepared by polycondensation of dimethyldichlorosilane

$$\text{HO}|\text{H Cl}|-\underset{\underset{\text{Me}}{|}}{\overset{\overset{\text{Me}}{|}}{\text{Si}}}-|\text{Cl} + \text{H}|\text{O}|\text{H Cl}|-\left[\underset{\underset{\text{Me}}{|}}{\overset{\overset{\text{Me}}{|}}{\text{Si}}}-|\text{Cl H}|\text{O}|\text{H Cl}|-\underset{\underset{\text{Me}}{|}}{\overset{\overset{\text{Me}}{|}}{\text{Si}}}\right]-|\text{Cl} + \text{H}|\text{OH} \longrightarrow$$

$$\longrightarrow \text{HO}-\underset{\underset{\text{Me}}{|}}{\overset{\overset{\text{Me}}{|}}{\text{Si}}}-\text{O}-\left[\underset{\underset{\text{Me}}{|}}{\overset{\overset{\text{Me}}{|}}{\text{Si}}}-\text{O}-\underset{\underset{\text{Me}}{|}}{\overset{\overset{\text{Me}}{|}}{\text{Si}}}\right]_n-\text{OH} + 2(n+2) \text{ HCl}$$

This is a smooth reaction which proceeds easily. It requires, however, very pure dichlorosilane. Methyltrichlorsilane impurity leads to inadmissible chain branching. Technical separation of the trichloro from the dichloro derivative is dificult; the difference in the boiling points of the two compounds

---

† $\sim AcX°$ need not be active or it may cause some other unwanted effect.

is only 2 K and even rectification on columns with 100 theoretical plates is not satisfactory. It is therefore preferable to condense dimethyldichlorosilane with water so as to yield six-, eight-, and ten-membered cyclic molecules (hexamethylcyclotrisiloxane, etc.). These cycles are much easier to separate from the linear condensation products which contain compounds with trifunctional siloxane units. Pure cyclic polysiloxanes are ionically polymerized to yield the macromolecular component of silicone rubber. The residue is for the preparation of lubricants or laquers. In syntheses of shorter or branched and even crosslinked molecules, the requirements concerning the purity of the medium are not as stringent.

It is generally much easier to purify cyclic monomers than their linear low-molecular - weight precursors.

1.3 Kinetic influences

Whenever propagation does not contradict the laws of thermodynamics, the respective polymer may be produced. From both the theoretical and the technical points of view, the rate at which it is produced is, of course, extremely important. Very slow polymerizations, proceeding for days or weeks, would hardly be of great industrial interest. The same is true of the other extreme, of very rapid polymerizations. In their outward manifestations, the latter may approach explosions. Therefore information on the rate of propagation, on the controlling step, on rate changes in the course of the reaction, and on the methods of rate regulation is of greatest importance in macromolecular chemistry.

1.3.1 STATIONARY POLYMERIZATION

When the concentracion of active centres does not change during propagation, we speak of stationary polymerization. All previous and the majority of the presently valid analyses of radical polymerization kinetics assume stationary (or, better, pseudo-stationary [20, 21]) states. The conditions under which the stationary state assumption is a valid and useful approximation, can be expressed as follows [22]:

(a) The higher the value of the rate constant of termination, the more rapid is the approach of the polymerizing system to the stationary state.

(b) For the reacting system to exist predominanty in the stationary state, the time required to establish a concentration of active centres equal to one half of their stationary concentration ($[R_s^\circ]/[R^\circ] = 2$) must be short compared with the half-time of monomer decay. When long chains are formed, this condition is fulfilled for propagation half-times that are about ten times longer than the half-times of initiation.

(c) The stationary state is more easily established and proved when the respective rate constants and initiation rates are time-independent. A stationary state may exist even if this is not fulfilled. However, in the period of increase in the active centre concentration, the change in these quantities must amount to only a fraction of their original value.

The existence of stationary states in polymerizations has been proved by a comparison of assumptions with experimental findings. A direct proof is mostly impossible because of the too low concentration of reactive intermediates.

Living polymerization with fast initiation

This is a special kind of stationary polymerization. The requirement $d[LC]/dt = 0$ is realized by very rapid transformation of the initiator to living growth centres (LC) and the absence of termination and transfer. Each centre continues to grow until the establishment of the polymerization—depolymerization equilibrium [eqn. (2)]. When all growing centres are equally reactive and when a sufficient suply of monomer to all of them is provided, chains of uniform length are generated. The whole polymerization consists only of propagation. Kinetically this is the simplest case, even when complications connected with reactivity differences among growth centres and monomer diffusion variations occur. Many anionic polymerizations proceed by the living mechanism (as long as they are carried out in kinetically pure media, i. e. without impurities); examples of living cationic (mostly with heterocycle monomers), radical, and coordination polymerizations are also known.

1.3.2 NON-STATIONARY POLYMERIZATION

When the active centre concentrations change during propagation, the whole polymerization is non-stationary. Kinetically the process becomes more complicated and often even experimental control of the process becomes more difficult. On the other hand, a non-stationary condition can be utilized in studies of the elementary polymerization steps. To this end, the non-stationary phases of radical polymerizations are suitable, where outside these phases the process is essentially stationary [23–25]. Hayes and Pepper [26] called attention to the existence and solution of a simple non-stationary case caused by slower decay of rapidly generated cationic centres. In more complicated cases, exact analysis of the causes of a non-stationary condition is often beyond present possibilities. Information from the process kinetics is often not conclusive. It should be mentioned that, even when the condition $d[Ac]/dt = 0$ is strictly valid, polymerizations may be non-stationary, particularly in those cases when during propagation the more active form of the centres is slowly transformed to the less active form or vice versa.

References pp. 372–382

1.3.3 TEMPERATURE OF POLYMERIZATION

The rate of propagation shows little temperature dependence. The activation energy usually amount to 8–30 kJ mol^{-1} so that propagation itself can proceed at a reasonable rate over a broad temperature range. The useful temperature range is much narrower because the other "elementary" steps almost always have a higher activation energy[†]. These circumstances are the reason for the characteristic behaviour of polymerizing systems, viz.

(a) Living polymerizations lose their living character at higher temperatures. Growth centres are destroyed[††] and propagation passes into a kinetically unstable range.

(b) The rates of radical and coordination polymerizations rapidly increase with increasing temperature. This is caused by the high activation energy of initiation (E_i).

(c) The rates of ionic polymerization increases little with increasing temperature. Even an increase of the polymerization rate with decreasing temperature is not exceptional. The reasons are usually the high activation energy of termination and/or participation of reversible reactions with $\Delta H < 0$ in the initiation process.

(d) Isomerization polymerization can occur (providing the structures of monomer and active centre are suitable) only at the lowest temperatures where isomerization is more rapid than propagation.

(e) The highest possible polymerization temperature is limited by one or several of the following factors: the polymerization rate, which may exceed a controllable limit; the appearance of unwanted structures or too low molecular weights of the products (transfer to monomer, polymer or transfer agent); thermal destruction of monomer or polymer; and T_c.

2. Environment of the active centre (external influence)

Propagation cannot be reduced to a simple chemical reaction between the active centre and the monomer. Even if no compounds other than these two partners were present, molecules of the monomer not reacting at the moment

---

[†] The only exceptions are terminations of radical centres by combination, some transfers and isomerizations.
[††] In styrene polymerization on the ion pairs

$\sim CH_2-CH^-, Na^+$ (with phenyl group)

in THF, the centres already begin to decay at 273 K.

would affect addition by their force fields. In addition, the growing macromolecules and even the inactive polymer chains are usually not inert. The presence of solvent and/or a monomer-dispersing phase is likewise reflected in the course of propagation. Two effects which are important in this connection will be discussed.

2.1 The medium

Radical propagations proceed smoothly in non-polar and also in highly polar media. This is understandable as radical solvation is weaker than ion solvation by one or two orders of magnitude. Propagation on coordination polymerization catalysts is only possible in non-polar media which do not interfere with monomer (usually hydrocarbon) coordination on the transition metal atom. Ionic polymerizations also proceed in non-polar media and they are accelerated with increasing medium polarity.

This statement has its limitations. Ionic polymerization in hydrocarbons is always kinetically complicated, it often starts only after the addition of a polar compound (co-initiator), and it is affected by the aggregation of initiating and propagating particles. In strongly polar media, activation of initiator by dissociation of acids and bases is easy. Such solvent is simultaneously a reactive transfer agent. Propagation usually does not occur, and only low molecular products are formed. Exceptions can, of course, be found. During anionic polymerization of lactams in DMF, the solvent only increases the amount of dissociated initiator [27].

Media of moderate polarity, with permittivities in the range 5–15 are therefore the domain of anionic and cationic polymerizations.

Another factor affecting addition is the tendency of monomers to form complexes with other components of the polymerizing system. This was discussed in Chap. 2, Sects. 4.2 and 5.2. The effect of the medium is, of course, also reflected in the mechanisms of termination and transfer. A somewhat unusual but interesting and instructive effect of the environment of an active centre on the course of propagation was observed in polymerization on a liquid—gas phase boundary, and during polymerization of liquid crystals.

A change in rate and/or mode of propagation can be brought about by orientation of monomer molecules in the liquid crystalline state. For vinyloleate in the smectic phase, a higher polymerization rate than in the isotropic phase was observed by Amerik and Krentsel [28]. A significant reduction in the polymerization rate of a relatively complex monomer [N-(p-acryloyloxybenzylidene)-p-methoxyaniline] in the nematic state was described by Perplies et al. [29]. On the other hand, Paleos and Labes observed no change in the polymerization kinetics of a monomer, also of Schiff base character

References pp. 372–382

[*N*-*p*-(methoxy-*o*-hydroxybenzylidene)-*p*-aminostyrene], in the isotropic, nematic and oriented nematic states [30]. In addition, Hsu and Blumstein observed no change in the radical polymerization mechanism of another Schiff base [*N*-(*p*-cyanobenzylidene)-*p*-aminostyrene]

$$CH_2=CH-\langle\bigcirc\rangle-N=CH-\langle\bigcirc\rangle-CN$$

on transition from the nematic to the isotropic phase [31]. Investigation of the effect of monomer ordering in liquid crystals on polymerization in general, and on propagation in particular, will require much more work.

Any kind of monomer orientation prior to addition will affect propagation. During adsorption of an insoluble compound carrying a polar group on a liquid surface, the molecules are oriented but remain mobile (Langmuir–Blodgett films) [32]. The degree of orientation can be widely modified by variations in temperature and surface tension [33].

Nagele and Ringsforf [34] polymerized octadecylmethacrylate at the water—nitrogen interface at a constant surface tension of 0–0.1 Nm^{-1}. In the layer at high surface tensions, polymerization was considerably more rapid than in the expanded state (at low surface tensions). The polymer generated exhibited higher tacticity than that formed by conventional radical polymerization in solution.

More effective monomer orientation than on a liquid surface can be achieved on the surface of a suitable solid phase or by the interaction of monomer and polymer polar groups. Under these conditions, polymerization is characterized by specific features collectively described as the matrix effect. This is treated separately in Sect. 2.2.

At this point it is convenient to mention a special effect of polyelectrolytes. Kabanov et al. were the first to mention the possibility of monomer polymerization on a polyelectrolyte matrix [35], particularly of vinylpyridine on polyacids. Blumstein et al. described the radical polymerization of vinylsulphonic acid on a polydiazabicyclo-[2,2,2-octyl-1-butane matrix [36]

$$\left[-\overset{+}{N}\langle\bigcirc\rangle\overset{+}{N}-(CH_2)_4-\right]_n \longrightarrow \left[-\overset{+}{N}\langle\bigcirc\rangle\overset{+}{N}-(CH_2)_4-\right]_n$$

$$\begin{array}{cc} ^-SO_3 & ^-SO_3 \\ | & | \\ (n-1)\,HC=CH_2 + HC=CH_2 \end{array} \qquad \begin{array}{cc} ^-SO_3 & ^-SO_3 \\ | & | \\ -[CH-CH_2-CH-CH_2]_n \end{array}$$

A polymer with short isotactic sequences was formed. During polymerization of the same monomer in solution, syndiotactic addition is preferred [37]. The counter ions of matrix and monomer are bound by electrostatic forces. These give rise either to mutually relatively firmly bound ion pairs, and/or to a looser "atmospheric attachment" which does not appreciably affect anion mobility allong equipotential surfaces around the polyelectrolyte macromolecule. The authors presented an indirect proof of a larger contribution of relatively free counter anions migrating along the chain matrix [38].

Another type of matrix effect on the properties of the product was described by Doiuchi and Minoura who copolymerized indene with maleic anhydride in the presence of lecithine as a chiral surface active agent [39]. The rate of radical copolymerization in benzene is reduced in the presence of lecithine. An optically active copolymer is formed (the asymmetric carbon is marked by the asterisk,*)

It may be expected that matrix effects can direct polymerizations of complex monomers towards the required structures. Matrix function could even be fulfilled by the catalyst support; combinations of coordination, matrix--controlled polymerizations could be envisaged.

2.2 The consequences of polymer formation

Monomer to polymer transformation must necessarily affect all properties of the polymerizing system. Most changes are back-reflected into the polymerization mechanism and kinetics, either as a result of the direct effects of the generated polymer on some of the basic steps or on all of them simultaneously.

2.2.1 VISCOSITY INCREASE

Monomer and/or solvated polymer chains remain in solution, increasing its viscosity; this effect grows with the strength of solvation, i. e. when the monomer or solvent are a "better" solvent for the polymer. Statistical polymer coils cause a greater increase in medium viscosity when they are loose and better wetted by monomer or solvent [40]. The reacting molecules diffuse in the liquid medium until the reacting groups reach the range of an effective collision. In this position they can remain for a sufficient time to "outlive" 10–100 mutual collisions [22]. When the reaction occurs with high probability at the first collision, its rate will be controlled by the diffusion of both components.

References pp. 372–382

Let us consider the bimolecular reaction yielding the product AB from the colliding pair $\overline{AB}$

$$A + B \underset{k_\omega}{\overset{k_\alpha}{\rightleftharpoons}} \overline{AB} \qquad (16)$$

$$\overline{AB} \underset{k_\psi}{\overset{k_\beta}{\rightleftharpoons}} AB \qquad (17)$$

In the equilibrium or stationary states, the concentration of the colliding pair is given by the relation

$$\frac{d[\overline{AB}]}{dt} = k_\alpha[A][B] + k_\psi[AB] - (k_\beta + k_\omega)[\overline{AB}] = 0 \qquad (18)$$

and thus

$$[\overline{AB}] = \frac{k_\alpha[A][B] + k_\psi[AB]}{k_\beta + k_\omega} \qquad (19)$$

In a kinetically simple medium, the overall rate constant of product formation, k_{as}, can be defined by the relations

$$\frac{d[AB]}{dt} = k_{as}[A][B] \qquad (20)$$

$$k_{as} = \frac{k_\alpha k_\beta}{k_\beta + k_\omega} \qquad (21)$$

The rate constant of AB dissociation is similarly given by

$$-\frac{d[AB]}{dt} = k_{dis}[AB] \qquad (22)$$

$$k_{dis} = \frac{k_\omega k_\psi}{k_\beta + k_\omega} \qquad (23)$$

The equilibrium constant of AB formation from A and B is then defined as

$$K_{AB} = \frac{[AB]}{[A][B]} = \frac{k_{as}}{k_{dis}} = \frac{k_\alpha k_\beta}{k_\omega k_\psi} \qquad (24)$$

K_{AB} must only depend on thermodynamic parameters so that the ratio k_α/k_ω cannot be a function of such kinetic effects as, for example, the rate of diffusion. As k_α includes the diffusion rate effects of both A and B, the same factor must also be included in k_ω. When the reaction between A and B is diffusion-controlled, the dissociation of AB must also be controlled by diffusion [41].

The ratio k_α/k_ω is the equilibrium constant K_{sr}, of collision pair $\overline{AB}$ formation; for colliding uncharged molecules or radicals it can be approximated by the equation

$$K_{sr} = 4\pi N \left(\frac{r_{AB}}{1442}\right)^3 \quad \text{mol}^{-1}\,\text{dm}^3 \tag{25}$$

where r_{AB} is the diameter of the pair in m.

Choosing for r_{AB} of small molecules the probable value of 5×10^{-10} m, K_{sr} becomes about 0.3 mol^{-1} dm^3. Small molecules thus generate a collision pair at a rate characterized by a rate constant often reaching the value 10^9 to 10^{10} mol^{-1} dm^3 s^{-1}. For k_ω this gives a value of 10^{10} s^{-1}. Thus the mean life time of the pair is about 10^{-10} s. Ion association si affected by strong electrostatic interactions. These forces are considered in association constant calculations; they are usually included in the Boltzmann factor.

The rate of the bimolecular formation of the product AB is given by eqns. (20) and (21).

$$\text{rate} = k_\alpha k_\beta (k_\beta + k_\omega)^{-1} [A][B] \tag{26}$$

and at $k_\beta > k_\omega$ its magnitude is predominantly diffusion-controlled. It may be regarded as purely diffusion-controlled for $k_\omega \ll k_\beta$. The quantity k_β is the rate constant of product formation from the collision pair, and according to the collision theory it is given by

$$k_\beta = v_\beta P \exp(-E/RT)\ \text{s}^{-1} \tag{27}$$

where v_β is the frequency of collision of two molecules in the collision pair and P is a steric factor.

The value of k_ω depends on the frequency of diffusion jumps, and thus on the rate of diffusion, similar to k_α. This model was used by Rabinovich [42] as a basis in rate constant calculations of diffusion-controlled reactions; its variant for the reactions of large and small molecules was used by Allen and Patrick [13]. Another very well known, and actually the first approach, is the treatment of this problem by Smoluchowski and his successors [41, 44–48].

The calculation of $k_\alpha k_\beta (k_\beta + k_\omega)^{-1} = k_{as}$ according to these authors is described in many publications [41, 49]. For small molecules the calculation of k_{as} leads to values of 2×10^9 to $5 \times 10^9 \text{ mol}^{-1} \text{ dm}^3 \text{ s}^{-1}$, in good agreement with many observations. For example, the combination of the atoms of iodine dissolved in benzene or carbon tetrachloride at 298 K exhibits rate constants of 1×10^{10} and $7 \times 10^9 \text{ mol}^{-1} \text{ dm}^3 \text{ s}^{-1}$, respectively [50]; k_{as} for the mutual reaction of 2-cyanopropyl or 2-cyanohexyl radicals is also about $2 \times 10^9 \text{ mol}^{-1} \text{ dm}^3 \text{ s}^{-1}$ at 298 K in benzene [51].

The rate constant of diffusion-controlled quenching of photo-excited β-naphthylamine by CCl_4 (in isooctane) is $1.3 \times 10^{10} \text{ mol}^{-1} \text{ dm}^3 \text{ s}^{-1}$, and in cyclohexane $7.2 \times 10^9 \text{ mol}^{-1} \text{ dm}^3 \text{ s}^{-1}$ at 298 K [52]. Another example is a reaction between ions with opposite charge. At 298 K, the rate constant of hydroxyl and ammonium ion association in water is $3 \times 10^{10} \text{ mol}^{-1} \text{ dm}^3 \text{ s}^{-1}$ and that of oxonium and chloroacetate ion association is $1.3 \times 10^{10} \text{ mol}^{-1} \text{ dm}^3 \text{ s}^{-1}$. Values k_{as} that are higher than or roughly equal to $1 \times 10^{10} \text{ mol}^{-1} \text{ dm}^3 \text{ s}^{-1}$ are affected by a contribution of long-range electrostatic interactions [53].

The rate constants of diffusion-controlled reactions are proportional to the diffusion rates of both reaction components. They are therefore a function of solvent viscosity; inversely proportional to it according to Einstein [54]. Except in extreme cases, propagation is not diffusion-controlled. Monomer addition to the radical requires the crossing of an energy barrier of about 10 kJ mol^{-1} and larger; the steric factor of this reaction is 10^{-5} and less. Therefore the rate constant of propagation does not decrease even at medium conversion at considerably increased viscosity (see Chap. 4, Tables 3 and 4). However, at high conversions, k_p falls dramatically (e.g. during radical polymerization of MMA in bulk).

With growing viscosity, the diffusion rate of propagating macromolecules decreases, the probability of their effective collision is diminished, and thus also the rate of bimolecular termination. As the rate of initiation is not appreciably affected by increasing viscosity, in a certain phase some radical polymerizations pass from a stationary to a non-stationary state in which the number of radicals increases[†]. The polymerization is appreciably accelerated. This situation is called the gel effect or the Norrish–Tromsdorff effect [55]. As the change in the rate of termination contributes considerably to the gel effect, it will be discussed in detail in Chap. 6, Sect. 1.3.

2.2.2 SOLID PHASE FORMATION

When the solvating ability of the monomer and/or the solvent is insufficient for the solvation of whole chains of the generated polymer, van der Waals attractive forces between the macromolecules predominate and a solid phase

---

[†] Where viscosity increases strongly with conversion, i.e. mainly in polymerizations in bulk or in good solvents and in systems with limited transfer.

is formed. The separated polymer is source of complications in the process technology. In cases where it separates in the form of individual non-sticking flakes, the problems are less severe, although even in these cases technological problems with the transportation of the suspension, with turbulence variations during mixing, and with flake sedimentation in unstirred spaces, etc. should not be underestimated. When the polymer particles tend to stick together, the situation is more serious. Sediments form on stirrers and reactor walls, hindering heat dissipation. Whole polymer blocks can be formed, hindering stirrer motion and blocking pipes. These are extreme situations but sediment formation is frequent; polymer block formation must be prevented as it completely wrecks polymerization on an industrial scale. This can sometimes be achieved by the choice of polymerization medium, for example by a change in the stirring regime, by substituting emulsion or suspension polymerization for the bulk process, etc.

Polymer separation also changes the physico-chemical conditions, directly or indirectly affecting propagation in the subsequent stages. Probably the most important of these effects is the interaction of the solid polymer with the active centres.

2.2.3 ACTIVE CENTRES AND POLYMER

When the active centre is surrounded by a layer of solid polymer, further propagation will be controlled by the rate of monomer diffusion through the polymer layer. Usually it will be retarded. With a porous polymer layer surrounding the active centres, monomer transport will be easier. These effects must be considered when highly crystalline polymers are formed, especially when the chains grow from a non-transferring monomer as, for example, with coordination polymerizations [56].

Another example of the interference of a propagating active centre with solid polymer is the occlusion of a centre at the surface of a solid particle [57–59]. This is treated in Chap. 6, Sect. 1.3.

Even with reactions of a non-radical active centre, the generated polymer is not always inert. Carbanions react with —C≡N and —COOR substitutents, carboxonium ions produce less acid centres by reaction with an ether-type chain (see Chap. 4, Sect. 2.3), carbocations alkylate aromatic groups, etc. All these reactions affect propagation. Sometimes the physical effect of the generated insoluble polymer is combined with its ability to react chemically in a certain way.

We have observed a reduction in the rate of trioxane polymerization initiated by the siloxonium dication (di-ion pair)

$$HSO_4^- Si^+ \sim\sim Si^+ HSO_4^-$$

References pp. 372–382

The precipitating polyformaldehyde deactivated the centres by formation of the oxonium complex

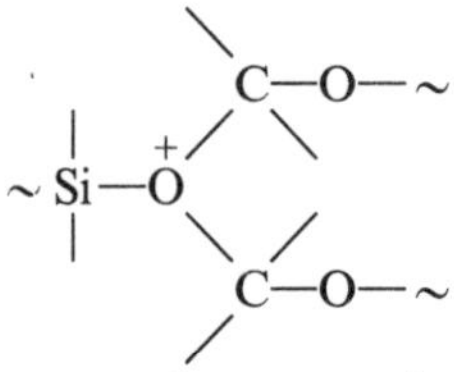

$$\sim Si - \overset{+}{O} \underset{C-O-\sim}{\overset{C-O-\sim}{\diagup\diagdown}}$$

simultaneously preventing monomer access [60].

2.2.4 THE MATRIX (TEMPLATE) EFFECT

Monomer molecules with strongly polar substituents mutually associate and form complexes with the solvent and with their own polymer. All kinds of association cause changes in polymerization kinetics and sometimes even in the properties of the generated polymer.

Kabanov et al. have studied the radical polymerization of acrylic acid in aqueous solutions [61] (see Chap. 4, Sect. 1.4). A crystaline, syndiotactic polymer was formed at pH 10.2–10.8. The kinetics of the photosensitized polymerization of acrylic acid was studied by Galperina et. al. [62]. They observed a strong solvent dependence for the rate constant of propagation. At 293 K, $k_p[\text{mol}^{-1}\,\text{dm}^3\,\text{s}^{-1}] = 22\,500$ in water, 4 200 in formamide, and 500 in dimethylsulphoxide.

Chapiro and Dulieu have established the dominant role of macromolecular associations from the reasons for the effects described [63]. Acrylic acid associates by hydrogen bonds, yielding a cyclic dimer or an oligomer

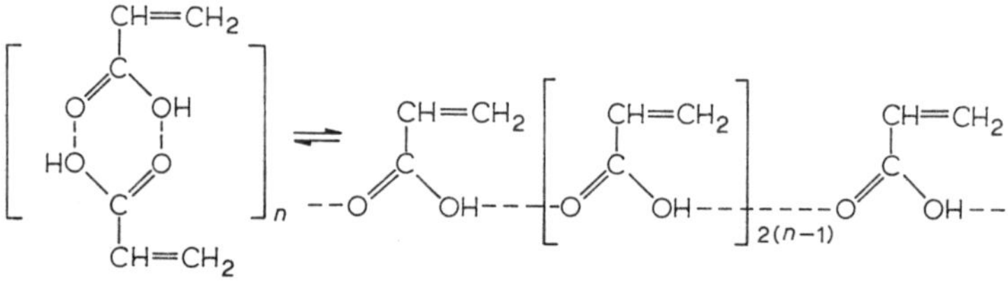

$$\left[\begin{array}{c}CH=CH_2\\ \text{(cyclic dimer)}\\ CH=CH_2\end{array}\right]_n \rightleftharpoons \cdots \left[\begin{array}{c}CH=CH_2\\ \cdots\end{array}\right]_{2(n-1)} \cdots \quad (28)$$

Solvents such as water, methanol, dioxan etc. stabilize the oligomeric associate. Hydrocarbons and their chlorinated derivatives do not associate with acrylic acid, and they shift equilibrium (28) to the left. The polymerization of an equilibrium mixture of oligomers and dimers in polar complexing solvents is almost 20 times more rapid than the polymerization of dimers only [64]. Moreover, even with polymerization in bulk, it proceeds autocatalytically under these conditions. Oligomeric aggregates associate by hydrogen bonds on to the polymer matrix which is formed immediately after the start of the reaction. In this way an organized structure is formed suitable for rapid "zip" propagation along the oriented double bonds.

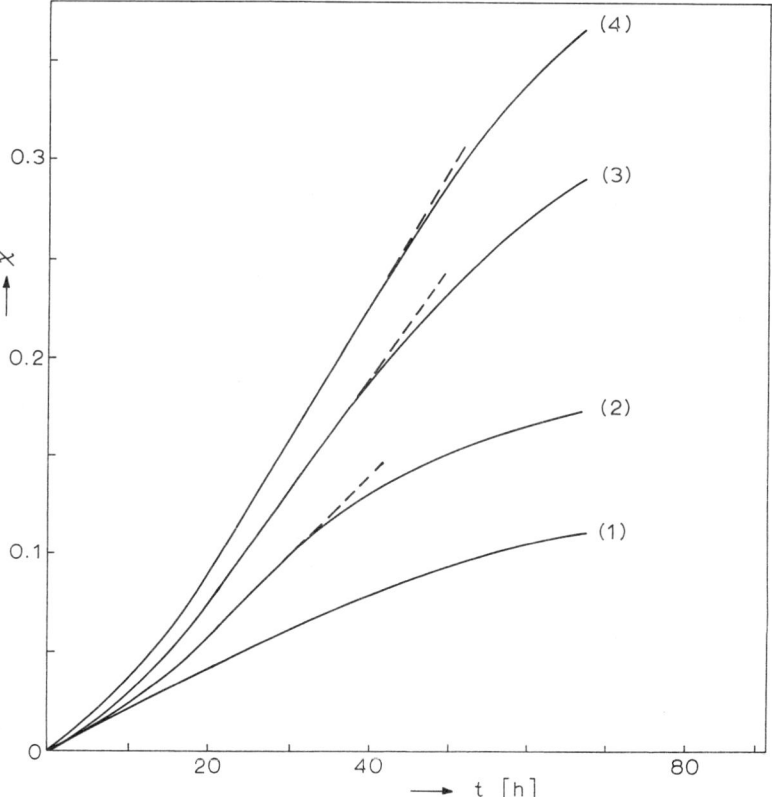

Polymerization on the matrix is 63 times more rapid than polymerization of dimers [64]. The product is a stereo block polymer of the syndiotactic and atactic polyacrylic acid configurations. When hydrocarbons or their chlorinated derivatives are used as solvent, the matrix effect does not occur.

Fig. 3. Polymerization of MMA in acetone in the presence of isotactic PMMA [69]. Temperature 277.5 K; concentration of PMMA (mol dm^{-3}): (1) 0; (2) 0.240; (3) 0.343; (4) 0.446.

References pp. 372–382

Another example of the matrix effect can be demonstrated by methyl methacrylate polymerization. Orlova et al. [65] postulated association of the growing radical with a molecule of the poly(methylmethacrylate) matrix. Propagation continues along the matrix at a reduced rate [66], yielding a

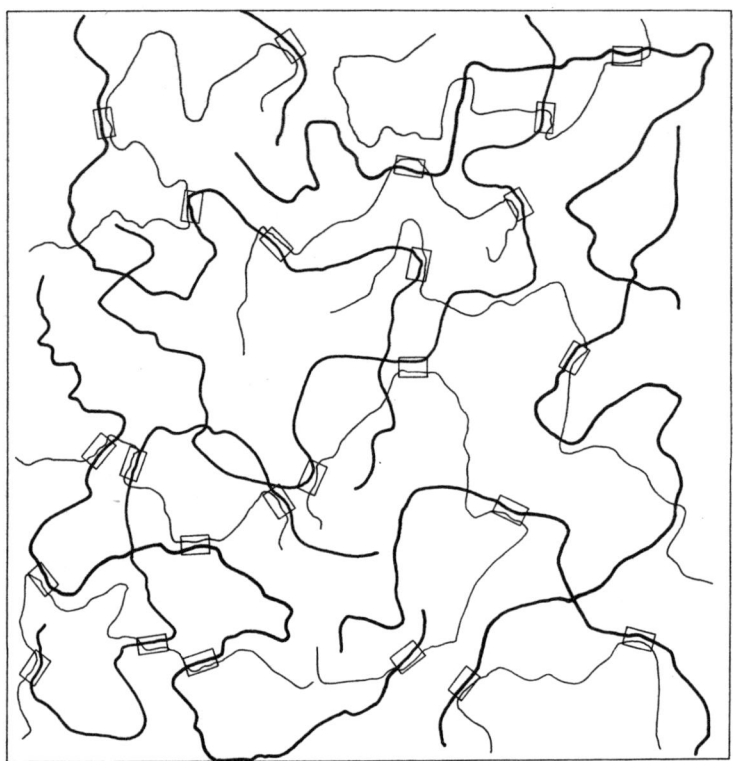

Fig. 4. Schematic representation of the association between growing chains and matrix macromolecules in toluene. The locations of short segment association are marked by rectangles.

product with a higher content of stereoregular polymer [67]. Association of the growing radical with the matrix considerably reduces its segmental mobility, resulting in slower termination [68]. Termination is relatively more retarded than propagation, therefore the overall effect is more rapid polymerization [66] as compared with the reaction where the matrix effect does not operate. At higher conversions, polymerization is even more accelerated (see Fig. 3), especially in weakly complexing solvents, e. g. toluene. This situation is shown schematically in Fig. 4. The polymer chains are relatively little solvated by toluene, and permits association of a part of the growing chain with various matrix macromolecules. A dense network is formed, hindering translational diffusion of polymeric radicals; this reduces the number of their effective collisions per unit time, and thus also the rate of termination. In well-solvating solvents (e. g dimethylformamide) a looser network is formed with fewer branching points (Fig. 5). Translational dif-

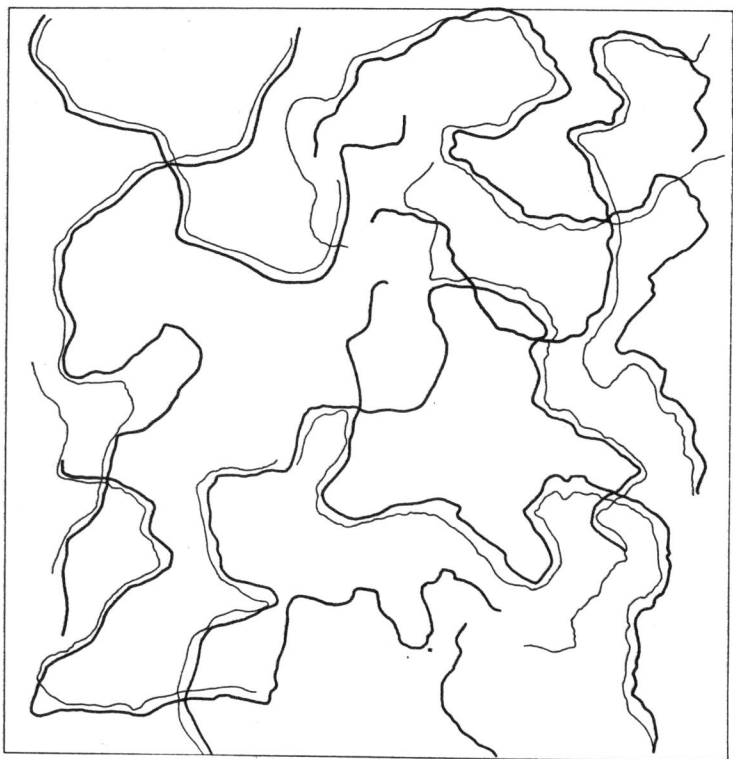

Fig. 5. Schematic representation of the association between growing chains and matrix macromolecules in dimethylformamide

fusion of radicals is less hindered so that, under these conditions, the secondary acceleration by the matrix is smaller; it is better pronounced at low temperatures [69].

In dimethylformamide, in the range of the first acceleration of methyl methacrylate polymerization, changes in the mean lifetime of the radicals were observed [8.4 s for the control and 64 s for polymerization with the matrix effect caused by the presence of isotactic poly(methyl methacrylate)]; k_p fell from 26.6 to 5.9 mol^{-1} dm^3 s^{-1} and k_t from 140×10^4 to 1.7×10^4 mol^{-1} dm^3 s^{-1} [66].

References pp. 372–382

3. General features of propagation

Transformation of monomer to polymer is accompanied by many effects, which can be mostly divided into groups, and the origin of the effect described in general terms. It is this way that all the chapters and sections of this book are actually composed. Some basic manifestations of propagation are of a quite general character and deserve to be specially mentioned.

3.1 Conversion

Conversion is defined as the ratio of the amount of monomer molecules that have been consumed in polymer formation $([M]_0 - [M])$ to the initial amount of monomer $([M]_0)$

$$\varkappa = \frac{([M]_0 - [M])}{[M]_0} = \frac{(g_0 - g)}{g_0} \qquad 0 \leq \varkappa \leq 1 \qquad (29)$$

where g_0 and g are the initial and instantaneous masses of monomer in dm^3 of the batch.

Conversion is a very useful quanity for expressing the momentary state of propagation and of the position of the monomer $\rightleftharpoons$ polymer equilibrium. When the mean number average degree of polymerization $\bar{P}$ is known, then the amount (or concentration) of macromolecules N can be easily calculated from the conversion

$$[N] = \frac{([M]_0 - [M])}{\bar{P}} = \varkappa \frac{[M]_0}{\bar{P}} \qquad (30)$$

The inverse problem is even more important. With living polymerizations, the number of generated macromolecules is given by the number of growth centres, i. e. by the amount of initiator molecules. The value of $[N]$ can also be measured by a suitable physical method (see Chap. 8 Sect. 2.1). Therefore the number average degree of polymerization can be easily calculated[†]

$$\bar{P} = \frac{([M]_0 - [M])}{[N]} = \varkappa \frac{[M]_0}{[N]} = \varkappa \frac{[M]_0}{[I]} \qquad (31)$$

where $[I]$ is the concentration of initiator.

[†] The final term on the right of eqn. (31) is only valid for the simplest cases of living polymerizations with non-combining chains growing at one end only.

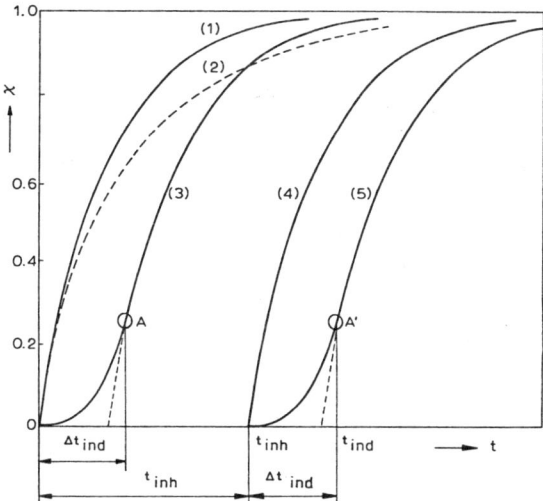

Fig. 6. Types of conversion curves. Conversion curve: 1, 2, 4 polymerization with rapid initiation; rate decreases: (1) only in consequence of monomer consumption (living polymerization); (2) due to the consumption of monomer and of active centres; (3), (5) polymerization with slow initiation; Δt_{ind} is the time interval of the concentration growth of active centres; (4), (5) polymerization with an inhibition period t_{inh} (A and A' are points of inflection).

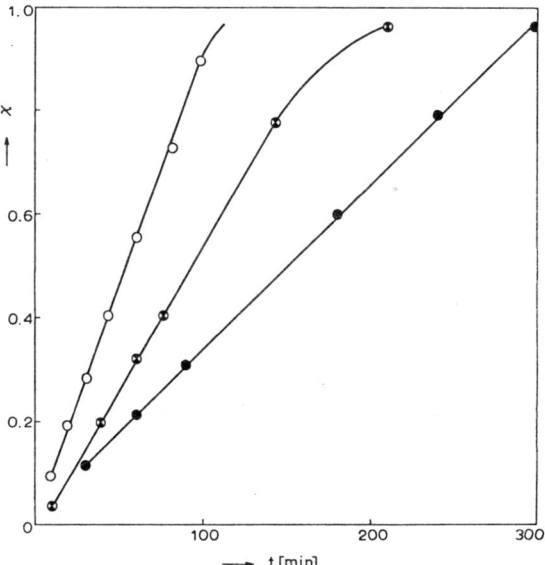

Fig. 7. Example of conversion curves. Emulsion polymerization of VC initiated by the redox system $K_2S_2O_8N_2H_4 \cdot H_2SO_4$ [70]. Temperature: ○, 313 K; ⊗, 303 K; ●, 293 K.

References pp. 372–382

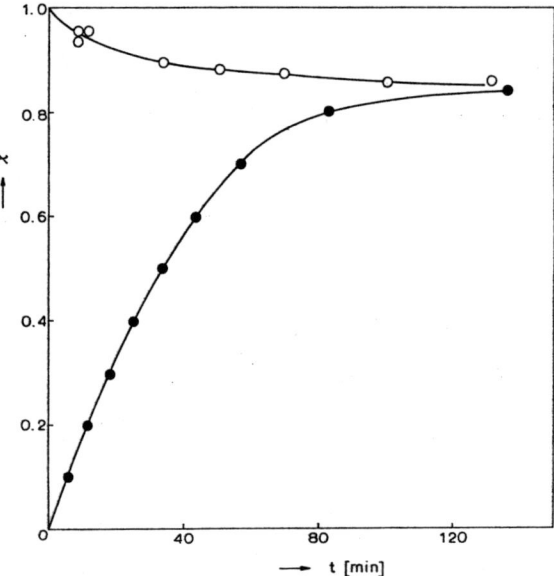

Fig. 8. Example of conversion curve. Reversible anionic polymerization of octamethylcyclotetrasiloxane [71]. Polymerization ● and depolymerization ○ at 423 K in the presence of 1.78×10^{-3} mol kg^{-1} KOH.

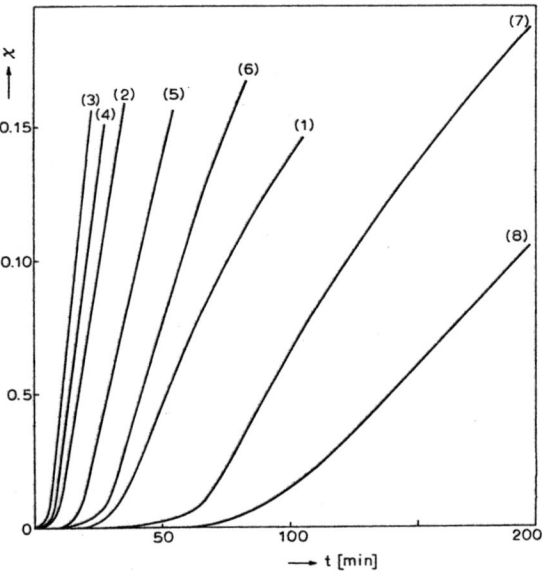

Fig. 9. Examples of conversion curves. Effect of water [at concentrations linearly increasing from case (1) to (8)] on cationic polymerization of trioxane [72].

The plot of conversion vs. time is called the conversion curve. It is of basic importance in kinetic studies, either in the graphic or tabular form. The interval of non-stationary propagation can be derived from a conversion curve by comparison of the external and internal orders[††] of polymerization with respect to monomer. In stationary polymerizations, both orders are identical. A general shape of the conversion curve is shown in Fig 6. Some examples [70–72] are shown in Figs. 3 and 7–9.

3.2 Transfer of heat during polymerization

The enthalpy change during the propagation of most monomers is high enough (see Sect. 1.1) to make the removal of the heat generated a problem in technological processes, adding to the production costs of polymerization. In small laboratory reactors, with large surfaces and small batches, difficulties are not usually observed (although even here isothermal conditions cannot always be maintained, for example during the cationic polymerization of 2-methylpropene or the anionic living polymerization of styrene).

Heat transfer from industrial reactors belongs to the category of engineering problems which are generally solved by other specialists [73] and therefore only a few comments will be made here. The walls of large reactors are usually not sufficient for removing the generated heat. The cooling surface must often be enlarged by cooling coils or some other suitable way. Utilization of the heat of evaporation of the monomer or of the solvent has already been mentioned in Chap. 1. For a long time we have been exploiting the advantages of polymerizations in particles surrounded by a medium of high molar heat and good thermal conductivity, i. e. water. Radical polymerizations in aqueous suspensions and emulsions are a partial solution of the problem. They bring other complications, however, for example contamination of the polymer by tensides.

Utilization of the heat of polymerization as an energy source has so far been sporadic. The quantity of the generated heat is relatively large, for technological reasons, however, it is transferred to other media at relatively low temperatures. Therefore the efficiency of the eventually attached thermal engines is low, and investments in isolation and heat-exchangers are high.

[†] The external reaction order with respect to the substrate M is equal to the experimentally determined exponent n (the power of $[M]$ in the rate equation), as determined from the initial rates at various values of $[M]_0$. The internal order is equal to the exponent n as determined with a single value of $[M]_0$ in the course of M consumption.

References pp. 372–382

3.3 Volume contraction of the polymerizing system

The monomer–polymer transition is accompanied by volume changes, usually by contraction. There are several reasons for the reduction in volume. One of the most important is the change of van der Waals distances between monomer molecules to spacings given by the covalent bond lengths between the monomeric units in a polymer chain [74]. Another factor may be the packing of monomer or polymer molecules in a crystalline phase or the volume changes accompanying the formation of a polymer solution from its components, etc. The volume change of a propagating monomer can be formulated [75] as

$$\Delta V = (R°)V_R° - [(I)_0 - (I)]V_I - [(M)_0 - (M)]V_M \qquad (32)$$

where $(R°)$, $(I)_0$, (I), $(M)_0$, (M) are the mole number (amount of substance) of the propagating chains, initiator and monomer and $(V_R°)$, V_I and V_M are the corresponding molar volumes. The volume contractions accompanying the propagation of some monomers, expressed in vol. %, are summarized in Table 2.

TABLE 2

Calculated contraction values for polymerizations of some monomers [76]

Monomer	Contraction (%)	Monomer	Contraction (%)
Ethylene	66.0	Cyclobutene	18
Propene	39.0	Methyloxirane	17
1,3-Butadiene	36.0	Cyclopentene	15
Vinyl chloride	34.4	Cyclopentane	12
Acrylonitrile	31.0	Tetrahydrofuran	10
Methyl methacrylate	21.2	Cyclohexane	9
Vinyl acetate	20.9	Phenyloxirane	9
Styrene	14.5	Cycloheptane	5
Diallyl phtalate	11.8	Cyclooctene	5
N-vinylcarbazole	7.5	Cyclooctadiene	3
1-Vinylpyrene	6.0	Octamethylcyclotetrasiloxane	2
Oxirane	23		
1,1-Dimethyloxirane	20	Cyclooctane	2

To a first approximation, the contraction of hydrocarbon monomers appears to be related to the number of their atoms (the contraction of styrene with eight carbon atoms is four times less than that of ethylene). Polar substituents cause deviations from this rule. In ring-opening propagation, contraction is the consequence of two mutually opposed effects, and is therefore smaller. The first of these effects is the already mentioned change in van der Waals distances; with oxirane it should cause a contraction of about 40 %. The simultaneous splitting of the covalent CO bond leads to an increase in the spacing of the C and O atoms to almost the van der Waals distance, and this is accompanied by a volume expansion of about 17 % [76]. The larger the ring, the more pronounced is the contribution of the second effect, and the overall contraction is smaller. Sufficiently large rings would polymerize without any contraction.

Monomers opening at least two rings in linear chains formation polymerize either without any volume change, or with a small expansion[†].

Many such monomers have been found [77–79], one of which is shown as an example [80].

$$\begin{array}{c}\text{CH}_2\text{-O} \diagdown \diagup \text{O-CH}_2 \\ \text{CH}_2 \diagup \text{C} \diagdown \text{CH}_2 \\ \text{CH}_2\text{-O} \diagup \diagdown \text{O-CH}_2\end{array} \longrightarrow \text{\{O-CH}_2\text{-CH}_2\text{-CH}_2\text{-O-}\overset{\overset{\text{O}}{\|}}{\text{C}}\text{-O-CH}_2\text{-CH}_2\text{-CH}_2\}_n$$

(33)

1,5,7,11-tetraoxaspiro [5,5] undecane is a crystalline compound of m. p. 514 K. It is polymerized both in the solid and in the liquid phases. By cationic polymerization at 523 K, it yields a macromolecular polycarbonate with ether bonds with a volume expansion of 2 %. Polymerization of the crystalline monomer yields the same polymer with a volume expansion of 17 %.

4. Controlled propagation

Chain propagation can be affected by variations in the mechanism or kinetics of addition. Controlled propagation in the proper sense is usually

---

[†] Volume increase during polymerization is very useful for precise casting, strong adhesives, prestrained plastics, elastomeric packing, dental fillings etc. Considerable progress is to be expected concerning microcrack formation in the polymer matrix and at the component interface, expecially in the field of adhesives and of fibre filled composites. An analogy may be seen in the freezing of water. The volume of ice is larger by 4 % than that of water. When water crystallizes at some surface (including teflon which is not wetted), ice firmly adheres to it. Freezing water expands into the cracks of the imperfectly smooth surface, giving rise to strong micromechanical adhesion. It has been confirmed by experiment that a monomer polymerizing with an increase in volume may also act in a similar way. Most such monomers can be copolymerized with suitable contracting monomers. Volume changes in polymerization can be controlled in this way [76].

References pp. 372–382

connected with the formation of tactic[†] and/or stereoregular[††] polymers, i. e. with a controlled mechanism. However, an intentional change in addition rate and in the number of additions in the generation of one chain (degree of polymerization) also belongs to controlled changes. Propagation can be controlled by the selection of the active centre, by modifications of its environment, and by changes in rate constants, possibly also in the rates of the other basic steps (initation, termination and transfer). Some of these effects have been discussed in Sect. 2 and those conected with the rates of termination and transfer will be treated in detail in Chap. 6. and 7. The very important manner of propagation control by the addition of another monomer (other monomers) will be described in a special part of this chapter dealing with copolymerization (see Sect. 5). The remaining ways will now be treated.

4.1 Stereospecific growth

Completely irregular and atactic, or fully ordered chains only exist as our idealized concepts because we cannot isolate them and prove such completely random or completely regular placement of monomeric units. More or less ordered polymers are formed by all known polymerization techniques.

4.1.1 CHAIN CONFIGURATION AND STATISTICS OF STEREOCHEMICAL PROPAGATION

Let us consider a vinyl polymer from an unsymmetrically substituted monomer in the planar conformation. The neighbouring C—C bonds can form one of the two possible diads, meso (m) and racemic (r),

---

[†] According to the IUPAC definition, a tactic polymer is a regular polymer the molecules of which can be described in terms of only one species of configurational repeating unit in a single sequential arrangement.

[††] A stereoregular polymer is a regular polymer the molecules of which can be described in terms of only one species of stereorepeating unit in a single sequential arrangement. A stereorepeating unit is a configurational repeating unit having defined configuration at all sites of stereoisomerism in the main chain of a polymer molecule. A configurational repeating unit has defined configuration at one, but not all sites of stereoisomerism. Thus a stereoregular polymer is more highly ordered than a tactic polymer

and of the three triads, isotactic (i), syndiotactic (s), and heterotactic (h)

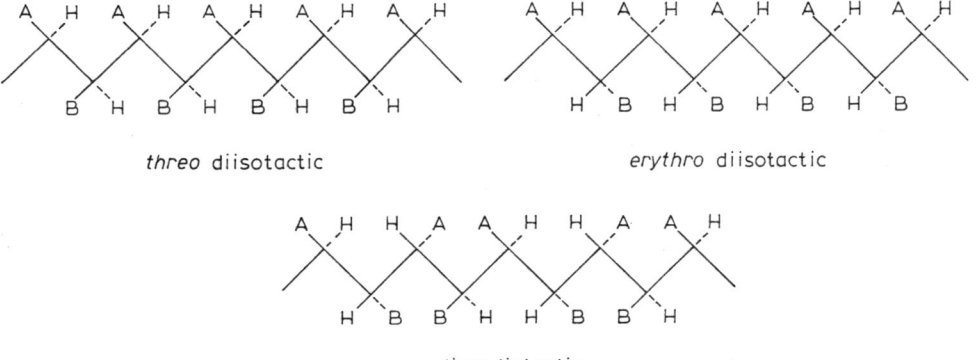

In sterically random propagation, the probability of meso diad generation P_m in each addition is equal to 0.5, and the ratio of triads in the chain is i:h:s = = 1:2:1. In an atactic polymer, the diads m and r are statistically distributed in the chain. Natta [81] called the simplest stereoregular chains isotactic —mmmm— and syndiotactic —rrrr—. Monomers with both α and β carbons unsymmetrically substituted can form diisotactic polymers

In the simplest case, when the structure of the propagating chain does not affect the configuration of the generated diad, the formation probabilities of meso and racemic diads, P_m and P_r, are related as $P_r = (1 - P_m)$. Chain structure obeys Bernoulli statistics as if the added units were selected at random from a "reservoir" in which the fraction P_m of the total amount is m, and the fraction $(1-P_m)$ is r. An isotactic polymer will be formed for $P_m \to 1$, and a syndiotactic polymer for $P_m \to 0$. Within these limits the chains will consists of randomly ordered m and r structures.

References pp. 372–382

When the manner of addition is affected by the growing chain end, the configurations of the added units will not obey Bernoulli statistics. In the simplest case, first-order Markov statistics will operate. Addition will be characterized by two parameters because the probability of r diad generation by monomer addition to an m end unit, P_{mr}, will not be identical with monomer addition to an r end unit, P_{rr}. The probabilities of m or r diad generation by addition to m or r chain ends will be bound by the relations $P_{mm} = (1 - P_{mr})$ and $P_{rm} = (1 - P_{rr})$. According to first-order Markov statistics, an isotactic polymer will be formed for $P_{mr} \to 0$ and $P_{rm} \to 1$, i. e. when the generation of a meso unit remains unaffected, and a tendency to racemic diad formation is prevented by the preference of a meso diad. The condition for the generation of a meso unit remains unaffected, and a tendency to racemic diad formation is prevented by the preference of a meso diad. The condition for the generation of a syndiotactic polymer is $P_{mr} \to 1$, $P_{rm} \to 0$. First-order Markov statistics permits two other non-bernoullian forms of addition: heterotactic for $P_{mr} \to 1$, $P_{rm} \to 1$, and stereoblock for small but finite P_{mr} and P_{rm} values. The smaller the value of P_{mr} or P_{rm}, the larger will be the generated m or r blocks, respectively.

Coleman and Fox published an alternative mechanism [82]. According to these authors, the propagating centres exist in two forms, each of which favours the generation of either the m or r configuration. When both centres are in equilibrium, and when this equilibrium is rapidly established, the chain structure can be described by a modified Bernoulli statistics [83, 84]. The configurations of some polymers agrees better with this model than with first- or even second-order Markov models [84, 85].

4.1.2 GENERAL FACTORS CONTROLLING THE MODE OF ADDITION

It may be expected that the stereochemistry of monomer addition to the centre will depend on the kind of the centre and on its momentary state. Let us call the radical or ionic end of a growing chain a "free" centre and an ion pair or complex a "bound" centre.

(a) Growth on free chain ends

The moment of fixation of a newly added monomer molecule depends on the geometry of the centre [86].

A tetrahedral carbon atom in the active centre usually preserves its configuration even during monomer addition. With monomer addition to a chain end, the configuration of the new end depends on the mode of double bond opening, *cis* [(a) in eqn. (34)] or *trans* [(b) in eqn. (34)]. If the approaching monomer is turned through 180°, the configurations produced by *cis* or *trans* opening are exchanged [(c) and (d) in eqn. (34)].

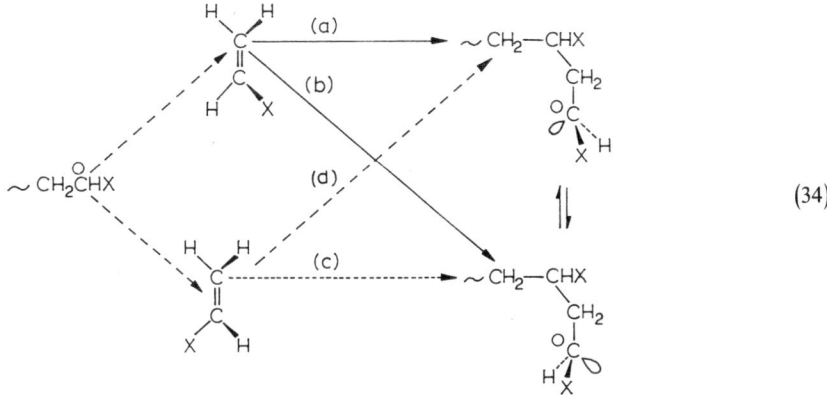

(34)

A planar form of the growing chain end is, of course, much more frequent. Most hydrocarbon radicals, carbenium ions, and carbanions are in the sp^2 state. From a planar chain end, a specific structure is only formed after the addition of a further unit. When the possibility of (momentary) rotation of the chain end is excluded the sp^3 configuration generated from the sp^2 state depends on the direction of monomer approach, which is thus one of the factors determining the stereochemistry of addition

(35)

The stereochemistry of addition to a free centre is mostly determined by interactions between the monomer and active centre during approach to the transition state. In simple cases, represented by equations (34) and (35) only the two primary components will interact, and Bernoulli statistics with a single probability parameter P_m will predominate. For $P_m \neq 0.5$, the propagation rate constants of isotactic and syndiotactic growth, k_{pi} and k_{ps}, will differ

$$\frac{k_{pi}}{k_{ps}} = \frac{P_m}{1 - P_m} \qquad (36)$$

References pp. 372–382

Radical polymerizations usually obey this model. Methyl methacrylate and vinyl chloride have a tendency to syndiotactic propagation $(P_m < 0.5)$ [86]. Low temperatures and bulky substituents cause deviations from Bernoulli statistics.

Polymerizations at lower temperatures (by means of redox systems or radiation) lead to higher contents of sequences with stereoregular order in the polymer. In common commercial polyvinyl chloride, the syndiotactic fraction amounts to about 40 %. Lowering of the polymerization temperature by 80 K increase this fraction to 60 %. The gain is evidently not very large, and to date it is not customary to prepare isotactic polymers by radical polymerization. For practical pruposes, the effect of temperature on the order in chains formed by radical propagation is not without importance. At lower temperatures, copolymerizations of butadiene with styrene (or acrylonitrile) yield synthetic rubber with a larger contens of syndiotactic sequences. It exhibits better mechanical properties than the copolymer prepared at higher temperatures[†]. A quite different situation occurs in radical stereospecific polymerizations when the necessary monomer orientation is provided by a mechanism other than the directing effect of radicals, for example by a matrix.

(b) Growth on bound chain ends

When the active centre has the character of a covalent or electrovalent bond, the number of factors controlling the stereochemistry of addition increases. The probability of random growth is lower. When alkenes are inserted into a metal–polymer covalent bond, the double bond is almost always opened so as to yield the *cis* configuration, especially when a four-center transition complex is formed

$$\begin{array}{ccc} \overset{\delta+}{\bullet} & -- & \overset{\delta-}{C}\sim \\ | & & | \\ C & ==== & C \\ \delta- & & \delta+ \end{array}$$

The possibility of *trans* opening of the double bond is larger in reactions of active centres with ion pair structures, where the four-center transition state is less probable. Counter-ion electrostatic fields strongly affect the space orientation of approaching polar monomers. Thus they contribute to controlled structure formation.

The stereospecificity of poly(methyl methacrylate) formed after initiation by Grignard or organolithium compounds depends on the solvent, temperature, halogen, and the organic initiator component. A change in any of these factors leads to a substantial change in stereostructure [86, 87]. Also

---

[†] Cold rubber contains fewer branched macromolecules formed by transfer to polymer. This also contributes to its better properties.

the presence of polar compounds leads to changes in the ratio of iso and syndio configurations. Water reacts preferentially with the active syndio centres, which are inactivated. The polymer is more slowly produced, and exhibits higher isotacticity [88]. In non-polar toluene, polymers of high isotacticity are produced. A contact ion pair is assumed to be the active centre. In the presence of a small amount of ethers, stereoblocks grow in toluene[†]. The syndiotactic configuration predominates in polar solvents. In tetrahydrofuran, the propagation stereochemistry resembles that of radical propagation ($P_m = 0.12$ at 195 K). It appears that the solvent-separated ion pair with the remote counter-ion cannot sufficiently orient the approaching monomer. Bovey assumes growth of ethyl-*cis*-β-deuteromethacrylate on a planar end group [as eqn. (35)] where the tetrahedral configuration is not fixed until the attachment of the next unit [89]. He proposed the following scheme of addition.

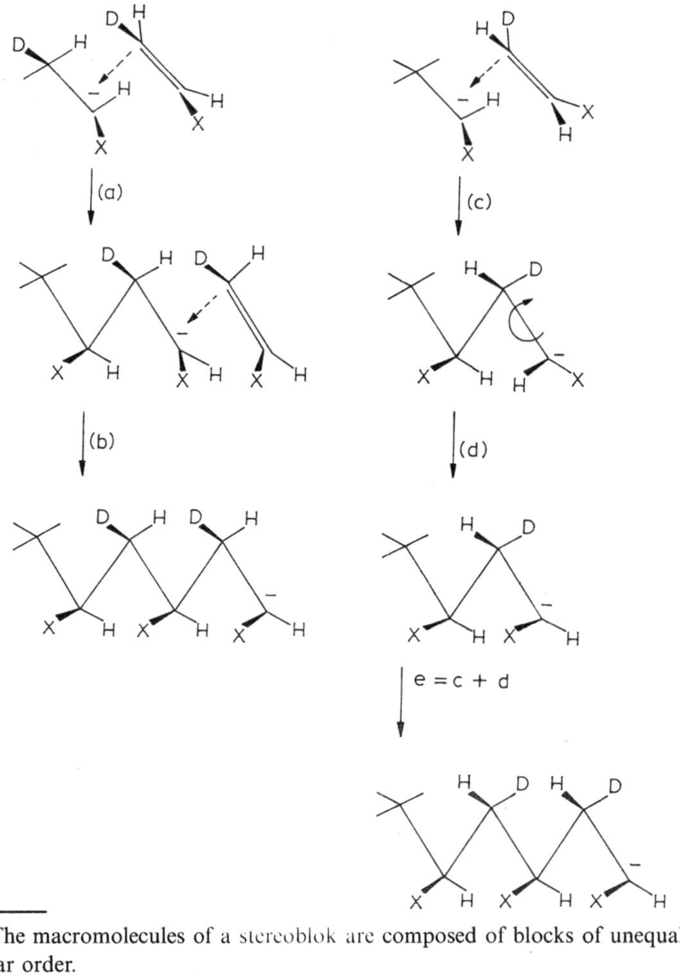

(37)

[†] The macromolecules of a stereoblok are composed of blocks of unequal tactic or stereoregular order.

References pp. 372–382

In toluene, routes (a) and (b) may be expected; the monomer approaches and is added in the isotactic conformation, and this conformation is preserved until the addition of next monomer (after which it cannot change any more). The controlling force, orientating the approaching monomer and maintaining the isotactic conformation, is the Li^+ ion. It is not shown in scheme (37); we assume its coordination to the end atom of the growing chain (carbanion) and to the carbonyl of the penultimate unit.

When the approaching monomer is in syndiotactic conformation and is added as such (c), the β carbon is fixed after addition, but the α carbon can rotate (d) and assume isotactic conformation after the addition of the next unit. The effect of the lithium cation is again assumed to be the driving force for the rotation of the potentially syndiotactic unit to the isotactic conformation. Repeating steps (a) and (b) leads to a threo-diisotactic chain, and repeated steps (c), (d) and (e) to an erythrodiisotactic chain. The addition of tetrahydrofuran reduces the number of threo-diisotactic units because it competes with carbonyl groups for coordination to Li^+. This enhances the amount of monomer approaching the centre by the "syndiotactic" route.

It would be naive to even consider the principles of propagation controlled by ion pairs and free ions as being clarified. Nevertheless, a lot has been done, and the mosaic is being filled in, step by step. Important findings were also recorded with polymerizations of other monomers.

Wicke and Elgert [90] polymerized α-methylstyrene with BuLi in tetrahydrofuran. They observed that the distance between the ions in the pair does not affect polymer configuration. This supports the theory of Coleman and Fox [82] concerning the presence of two types of centre in mutual rapid equilibrium and yielding different tacticities. The overall order can then be described by Bernoulli statistics.

Solvation of the active centre considerably affects even the ratio of 1,2 and 1,4 addition in diene propagation. Solvent can profoundly change the mode of addition in anionic polymerization [91].

Vinyl ethers propagate cationically yielding stereospecific polymers [87, 92]. Itramolecular cyclization is assumed to occur between the growing chain end and the third unit from the end, yielding a six-membered ring. Free rotation of the cationic end is thus prevented, and the orientation of the approching monomer is controlled [93].

4.1.3 ENANTIOELECTIVE (STEREOELECTIVE] AND ENANTIOSELECTIVE (STEREOSELECTIVE) POLYMERIZATION

Enantioelective polymerization [94–98] is a process where, from a mixture of two enantiomeric forms of a monomer, macromolecules are produced containing only one kind of configurational unit. In ideal enantioelective polymerization, exactly one half of the monomer is consumed

$$\text{monomer} \xrightarrow[\text{initiator, selecting R}]{\text{optically active}} \text{polyR + monomer} \quad (38)$$
$$\text{(racemate; R = S)} \qquad\qquad\qquad\qquad (R/S = 0)$$

In polymerizations of cyclic monomers, the product is isotactic. This is not necessarily so when optically active alkenes are polymerized.

By enantioselective polymerization polymer chains, each containing only one configurational kind of monomeric unit, are produced from a mixture of stereoisomeric monomer molecules. The number of kinds of polymer chain generated therefore equals the number of various stereoisomers in the monomer mixture. In the course of propagation, the enantiomeric composition of the polymer and unreacted monomer remains identical to the intial composition. When optically active monosubstituted cyclic monomers are polymerized, stereoregular polymers are formed with both isotactic polyR and polyS chains

$$(S)n \quad \begin{array}{c} CH_3 \\ \diagdown \\ C^* \!-\! CH_2 \\ \diagup \diagdown \\ H \quad S\,\beta \end{array} \longrightarrow \left[\begin{array}{c} CH_3 \\ | \\ S\!-\!C^*\!-\!CH_2 \\ | \\ H \end{array} \right]_{n\,(S)}$$

$$(R)n \quad \begin{array}{c} H \\ \diagdown \\ C^* \!-\! CH_2 \\ \diagup \diagdown \\ CH_3 \quad S\,\beta \end{array} \longrightarrow \left[\begin{array}{c} H \\ | \\ S\!-\!C^*\!-\!CH_2 \\ | \\ CH_3 \end{array} \right]_{n\,(R)} \quad (39)$$

It is important that the ring is opened at the β bond. Attack at the α position leads to irregularities and to asymmetric C inversion.

As initiators usually serve Ziegler–Natta catalysts or Zn, Cd, Mg or Al compounds [for example of the type Et_2Zn + optically active (−)-3,3-dimethyl-1,2-butanediol]. The polymerizations are mostly heterogeneous.

A further type of this class of propagations is the formation of asymmetric polymers from symmetric monomers in the presence of an optically active initiator

$$\begin{array}{c} R \quad\quad R \\ \diagdown \quad \diagup \\ (R)\;C\!-\!-\!-\!C\;(S) \\ \diagup \quad\;\; \diagdown \\ H \;\; X \;\; H \end{array} \xrightarrow[\text{initiator}]{\text{optically active}} \text{optically active polymer}$$

$$X = O, S$$

$$(40)$$

In this case, propagation is controlled by the selection of the site of attack on one the two asymmetric carbons by the optically active initiator. One type of configurational unit predominates in the chains, and the polymer is optically active.

D'Hondt and Lenz polymerized optically active α-phenyl-α-ethyl-β--propiolactone anionically; a racemic and an isotactic product were formed [99]

$$\underset{R_1}{\overset{R_2}{\rule{0pt}{1em}}}\!\!\diagup\!\!\diagdown\overset{O}{\underset{O}{\rule{0pt}{1em}}} \quad \xrightarrow{\text{Ph-C}(=O)\text{-O}^-\overset{+}{\text{N}}\text{Et}_3} \quad \left[\text{-OCH}_2-\underset{R_2}{\overset{R_1}{\underset{|}{\overset{|}{C}}}}-\overset{O}{\underset{}{\overset{\|}{C}}}-\right] \tag{41}$$

Each product exhibits different properties of the crystalline phase. Another example is the enantioelective cationic polymerization of racemic cis-1--methylpropyl-1-propenyl ether initiated by (−)-menthoxyaluminium dichloride in toluene at 351 K. The polymer is optically active (+) and the remaining monomer can be transformed to a polymer with negative optical activity by BF_3OEt_2 [100].

Not all types of propagation described in this paragraph proceed quite specifically so far.

4.1.4 PROPAGATION ON THE TRANSITION METAL—CARBON BOND

The industrially most important method of controlling the propagation of alkenes and conjugated dienes is the forced space orientation of monomers prior to addition. It is realized by monomer coordination to suitable active centres, almost exclusively of the Ziegler–Natta catalyst type [101]. Even though general fatures of 1-alkene and diene propagation are very similar, it will be useful to treat each of these groups separately.

(A) Stereochemical control of the propagation of 1-alkenes

Propene and the higher 1-alkenes can be polymerized to chains with the required degree of tacticity from almost atactic up to very highly tactic structures. However, a syndiotactic polymer can only be obtained from propene, mostly on soluble catalysts. The main factors determining controlled tactic addition are complexation, *cis* or *trans* addition, and "primary" or "secondary" addition. Most authors agree on the point that the interaction of the alkene molecule with the transition metal atom of the active centre leads to complex formation immediately before monomer insertion into the metal—polymer bond. The assumed existence of the complex is based on indirect experimental evidence and on theoretical considerations.

A direct proof has not so far been presented let alone a structural description of the complex.

Evidently 1-alkenes can be inserted into the metal-polymer bond in two ways: by "primary" addition

$$\text{Ti—C} \sim + \text{CH}_2=\overset{\text{Me}}{\underset{|}{\text{CH}}} \rightarrow \text{Ti—CH}_2-\overset{\text{Me}}{\underset{|}{\text{CH}}}-\text{C} \sim \qquad (42)$$

or "secondary" addition

$$\text{Ti—C} \sim + \overset{\text{Me}}{\underset{|}{\text{CH}}}=\text{CH}_2 \rightarrow \text{Ti}-\overset{\text{Me}}{\underset{|}{\text{CH}}}-\text{CH}_2-\text{C} \sim \qquad (43)$$

"Primary" addition is characteristic for isotactic and "secondary" for syndiotactic propagation [102]. Several workers have presented a proof of *cis* addition of the alkene [see eqn. (34), reaction (a)] in both iso and syndiotactic propagation [103–105].

(i) *Iso-specific propagation*. Much attention has been devoted to the search for the cause (driving force) of isotactic propagation. To date this problem has not been clarified; the opinions of various authors can be classified into several groups:

(a) Active participation of the organometallic component of ZN catalysts. Natta assumed the formation of a bimetallic active centre from both catalyst components at the crystal surface of the transition metal compound. This grouping is not symmetrical; at each side, 1-alkene can be complexed in only one of the two possible ways and therefore a chain is formed with predominantly D, or predominantly L configuration of 1-alkene units [106].

(b) Steric repulsion between alkene and the centre ligands. According to this idea, the mode of complexation is determined by steric interactions between the 1-alkene substituent and the ligands on the active centre and neighbouring atoms. The authors of the monometallic [107] and bimetallic [108] centres have pointed out the close relation between isotactic propagation and the geometry of the environment of the Ti atom with a vacancy. The hardly altered ligand field favours complexation, enabling one or the other configurations to be formed. In the complex, the alkene is bound to the vacant site of the active centre octahedron; in the complex, electron shifts resulting in addition are energetically more feasible

(c) Active participation of the helix. Coover proposed the hypothesis that addition which preserves the symmetry of the growing chain helix is more rapid [109]. According to him, the driving force of isotactic propagation is the preferred 1-alkene orientation resulting from the interaction between 1-alkene and the helix (see Fig. 10).

References pp. 372–382

(d) Number of vacancies in the coordination sphere of the transition metal. The effects of internal donors indicate that a transition metal with two vacancies dominates in an atactic centre whereas in an isotactic centre, this metal has a single vacancy in its coordination sphere [110].

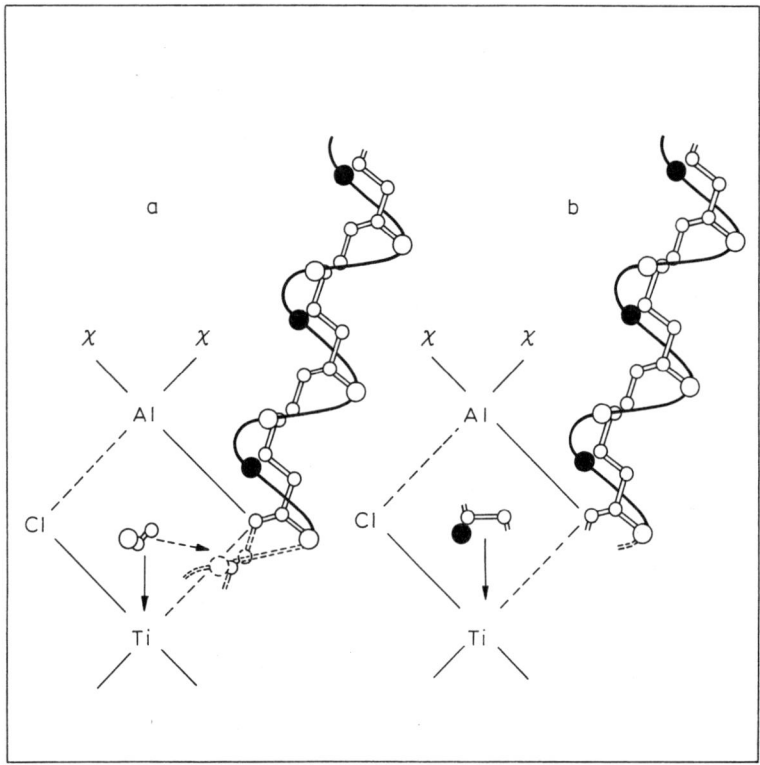

Fig. 10. Model of the interaction of an active centre with the end of an isotactic chain helix [109]. Insertion (a) preserving (b) disturbing helix symmetry.

(ii) *Syndio-specific propagation*. The metal—carbon (polymer) bond is similarly a component of active centres of syndiotactic propagation. Even indirect proof of monomer complexation has not been obtained so far. A single explanation of the cause of syndiotactic chain generation has not been presented. Four explanations may be said to exist.

(a) Alternation of the growth step between two sites of the octahedron. Arlman and Cossee [107] observed that their scheme of isotactic propagation can also lead to an explanation of syndiotactic growth. When the rate of growing chain migration back to its initial site in the octahedron is retarded, a further propene molecule can be added prior to migration. As propene can be complexed at two octahedron sites, it will be added with a reversed configuration. Alternation of the D and L configurations leads to syndiotactic growth. This type of addition is characteristic of centres formed by a transition metal at a crystal surface.

(b) Syndiotacticity is caused by steric repulsion between the methyls of a complexed and of an already added monomer. The scheme of Arlman and Cossee [107] again represents the basic mechanism, this time for soluble catalysts. This model does not require growing chain migration; addition does not proceed at a crystal sufrace. According to Boor and Youngman, propene can form a complex with a vanadium centre only when its methyl is in a position opposite to that of the methyl of the last added unit [111] (□designates a vacancy in the coordination sphere of the transition metal)

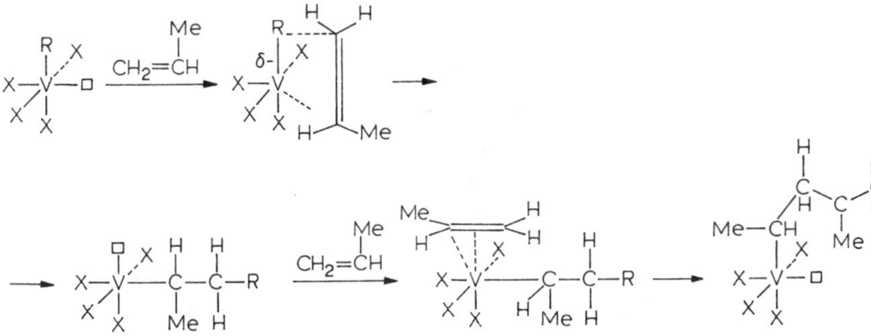

$$(44)$$

The model requires hindered rotation about the V—R bond. Considering the force fields of the Me of the last added unit and of the X ligands, this requirement is fulfilled at low temperatures.

(c) Syndiotacticity is caused by the bimetallic character of the centre. Pasquon assumed that syndiotactic propagation is closely connected with the structure of the bimetallic complex [112]. A more detailed explanation is lacking.

(d) Syndiotacticity is caused by steric repulsions between the alkene and a combination of the steric effects of the last added propene and the centre. Suzuki and Takegami presented a variant of theory (b) [102, 113]. The steric repulsive forces, which determine the manner of attachment of the monomeric units, are enhanced with alkylvanadium.

(B) *Stereochemical control of the propagation of conjugated and non conjugated dienes*

The presence of two double bonds greatly increase the number of the interconnected effects determining the configuration of an added monomeric unit. In this case the stereochemistry may already be established during the transition complex formation between the monomer and the centre or after the insertion of the diene into the metal—carbon bond. The realization of either of these possibilities is determined by the specific driving force of addition, composed of the electronic, steric, and isomeric states of the diene.

Many hypotheses exist at present which attempt to knit all the experimental findings into one concise picture. It appears that the attempts to find a single general mechanism of controlled diene propagation will not lead to a successful solution in the immediate future.

(i) *Controlled addition of conjugated dienes.* According to the point on the reaction path at which the mode of addition is determined, three situations can be distinguished.

Stereochemical control is exerted by the diene–transition metal complex. The structure of the complex is given by a superposition of various forces and effects.

(a) By transition metal diameter. According to Furukawa, butadiene is coordinated to the metal atom predominantly in the 1,4-*cis* mode as long as the transition metal only possesses one coordination site [114].

(b) Coordination by one or by both double bonds. Natta and Porri assume that 1,3-pentadiene is complexed by only one double bond to Ti [in Ti(OR)$_4$—AlR$_3$]

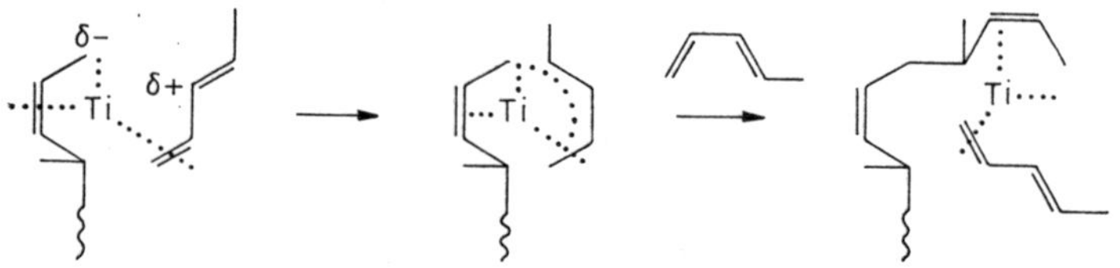

(45)

but by two bonds to Co [in Co(acac)$_3$—Et$_2$AlCl] [115]

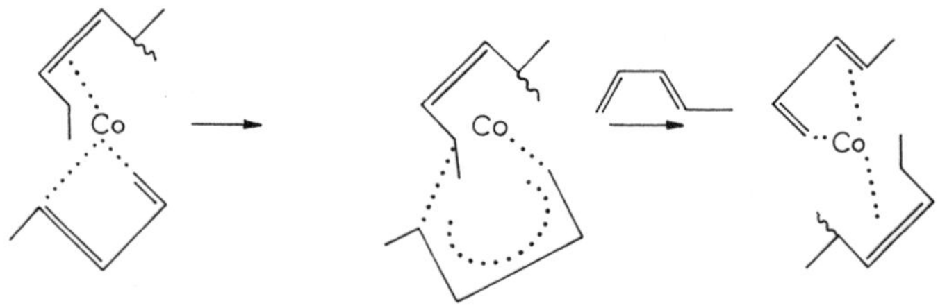

(46)

Cis-1,4 inserted 1,3-pentadiene has the same configuration as the preceding unit [eqn. (45)] and is therefore isotactic. The same monomer inserted by means of Co has the opposite configuration, and the resulting chain is syndiotactic [eqn. (46)]. Before and after isotatic addition the centre remains the same, but its structure changes in syndiotactic addition.

(c) By electron density on the metal of the centre. Matsuzaki and Yasukawa drew attention to the importance of electrostatic interactions between the non-bonding electrons of the transition metal atom and butadiene or of the

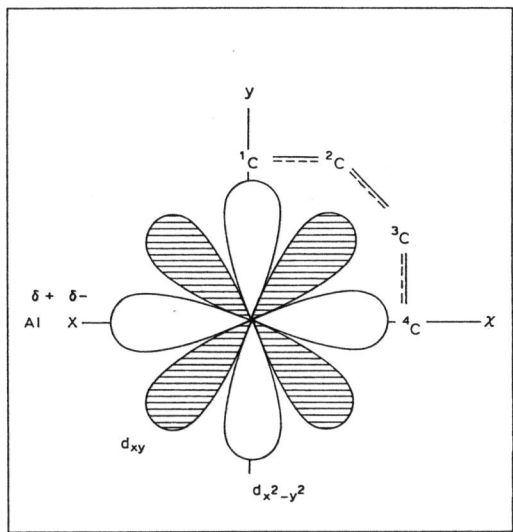

Fig. 11. Schematic representation of the space distribution of d electrons in a transition complex [116].

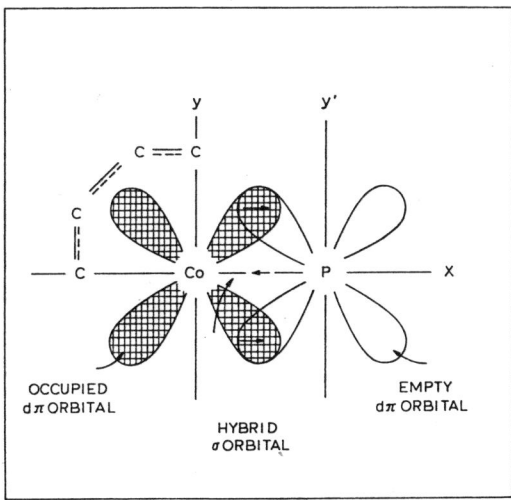

Fig. 12. Schematic representation of donor–acceptor interactions between a Co atom and tetraphenylphosphine [116].

References pp. 372–382

growing chain end [116]. They assume the transition metal to be almost electroneutral. When the electron distribution is spherically symmetric (as in centres with V), then butadiene is predominantly coordinated in its natural configuration *(trans)*. In Ni and Co complexes, electron distribution is non-spherical; they give rise to *cis*-1,4-polybutadiene. This is ascribed to reduced repulsion of the end-standing C_1 and C_4 atoms caused by the interpenetrating shielding effect of d_{xy} electrons (Fig. 11). Electron energy in d_{xy} is lower than in $d_{x^2-y^2}$ orbitals; therefore in most complexes d_{xy} orbitals are preferentially occupied (with the exception of the d^0, d^5 and d^{10} orbitals). Transition metal complexation with the base reduces the fraction of *cis*-1,4 configuration in polybutadiene. The density of $d\pi$ electrons on the transition metal atom is reduced by reverse donation (see Fig. 12); their shielding effect becomes smaller.

(d) By the number of vacancies in the transition metal valence sphere in the crystal lattice. Butadiene is polymerized on α, γ, and δ modifications of $TiCl_3 \cdot AlCl_3$ mixed crystals (containing 0–50 % $AlCl_3$) to 1,4-*trans*-polybutadiene. On the β modification of $TiCl_3$ with the same amount of $AlCl_3$, a

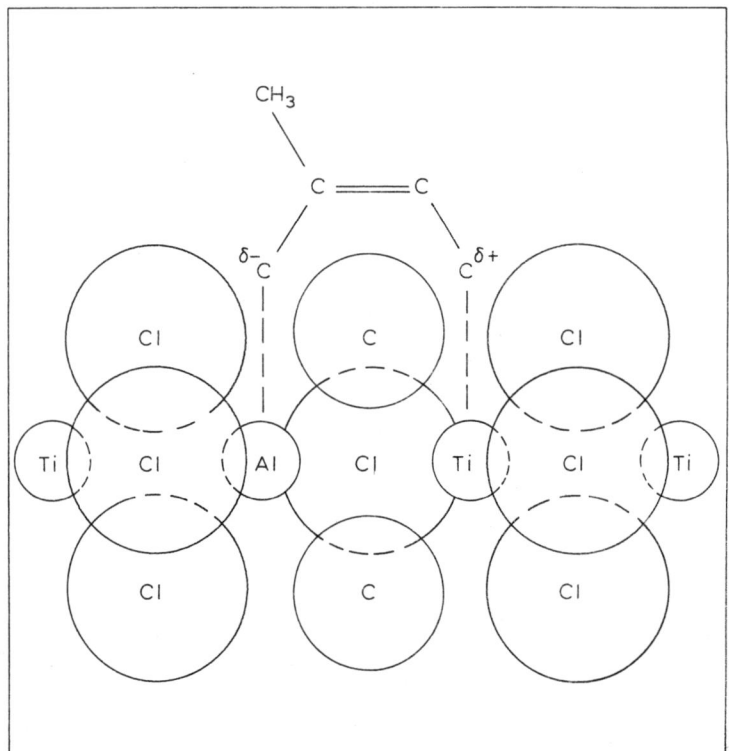

Fig. 13. Schematic representation of the coordination of isoprene molecule at the surface of β $TiCl_3 \cdot AlCl_3$ crystals.

mixture of *trans*-1,4 and *cis*-1,4 configurations is formed. Arlman [117] assumes that only one type of active centre ($\alpha|$) exists on the α, γ, and δ modifications of TiCl$_3$. β-TiCl$_3$ gives rise to two types of centre ($\beta|$ and $\beta\|$) differing in the number of vacancies. $\alpha|$ and $\beta|$ centres complex butadiene in the same manner on a single vacancy. The $\beta\|$ centre has two coordination sites on which butadiene can be chemisorbed in the *cis* configuration as a kind of chelate ligand. This centre thus yields *cis*-1,4-polybutadiene [117].

(e) By atom arrangement on the TiCl$_3$ surface. Saltman assumes *cis*-1,4 coordination of diene on the crystal surface of TiCl$_3$ with accommodation of Ti, Al, C, and Cl atoms in the lattice [118] (see Fig. 13).

(f) By blocking of coordination sites. The homogeneous catalyst LiAlH$_4$ + TiI$_4$ and Et$_2$AlCl + Co salt produce *cis*-1,4-polybutadiene in benzene. The addition of a small amount of Lewis base changes the polymer chain configuration to the *trans*-1,4 derivative [119]. This change is ascribed by Cooper et al. to the blocking of one vacancy by the base. Only one double bond of the diene can then be used for coordination; its natural *trans* configuration remains undisturbed, and *trans*-1,4 addition predominates [120].

(g) By *cis*—*trans* isomerization. *Cis*-1,3-pentadiene is not polymerized in the presence of Et$_2$AlCl—CoCl$_2$ whereas the *trans* isomer is, yielding the syndiotactic *cis*-1,4 polymer. This behaviour is ascribed by Natta et al. to the inability of the *cis* derivatives, to be coordinated in the *cis* mode [121]. In the presence of another catalyst, R$_3$Al—TiCl$_4$, *cis*-1,3-pentadiene is predominantly polymerized to the *trans*-1,4 polymer because it cannot assume the *cis* configuration in the transition complex. On the same catalyst, *cis*-1,4-polyisoprene is produced in high yield.

(h) By steric hindrance. 4-Methyl-1,3-pentadiene is polymerized predominantly by 1,2 addition on the catalysts Et$_3$Al—VCl$_3$, VCl$_4$, TiCl$_4$, or TiCl$_3$. This is explained by steric hindrance caused by the methyl groups [122]. Butadiene is polymerized exclusively to *trans*-1,4 polymers on the same catalysts.

Product stereochemistry is controlled during transition complex generation. Two monomers differing only in the position of the methyl group, isoprene and 1,3-pentadiene, are polymerized in different ways with Et$_3$Al—Ti(OR)$_4$. This is ascribed to the induction effect of the methyl groups, leading to changes in the electron density on the monomer carbons in the transition complex [119]. Isoprene is added by the 3,4 mode (● catalyst)

(47)

1,3-Pentadiene can be coordinated in both ways to the metal. This results in the formation of a mixture of chains with 1,2 and cis-1,4 addition

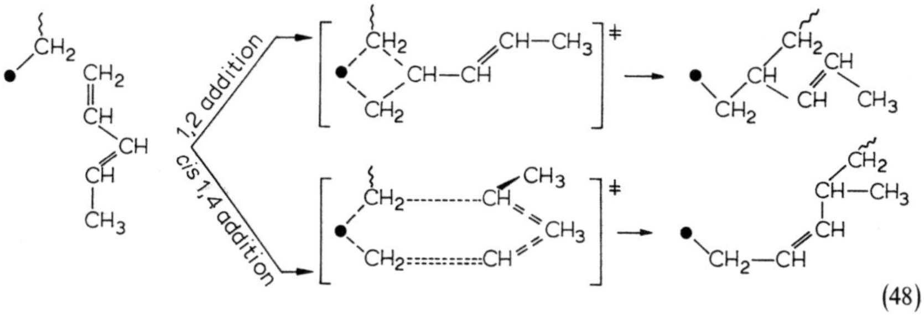

$$(48)$$

A different mode of addition control can predominate on other active centres.

Stereochemical control is exerted by forces operating after diene insertion.

(a) Double bond rotation [101]. The double bond of the last added diene is excited when coordinated to the metal atom. This excitation permits rotation about the bond, and thus *cis* and *trans* isomerization.

(b) Manner of bonding. Natta and Porri [119] assume that the ligands on the transition metal play a determining role in 1,2 or 1,4 addition. With butadiene polymerization on Et_2AlCl—$V(acac)_3$ and on Et_3Al—$V(acac)_3$, 1,4 addtion is typical of the former, containing Cl, whereas 1,2 addition is typical of the latter.

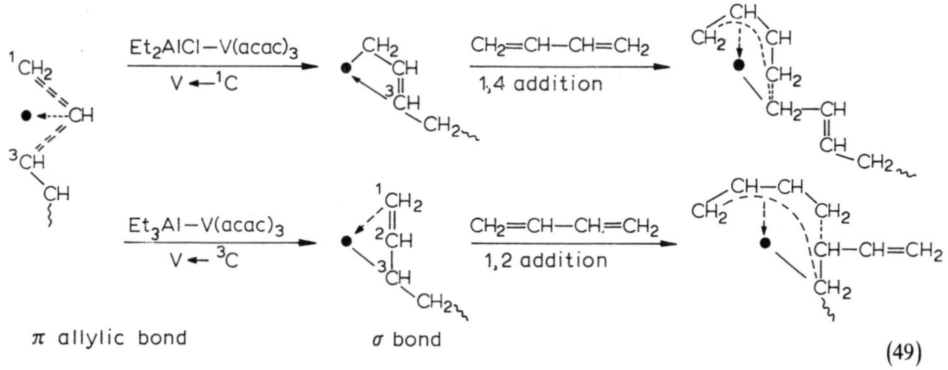

$$(49)$$

During diene coordination to a metal atom with a Cl ligand, the π allylic attachment is transformed into a σ bond between ^1C and the transition metal, •, whereas with the latter catalyst it is transformed into a σ bond between ^3C and •.

(c) The direction of the approaching diene. The approach of the complexing diene is assumed by Marconi [123] to be the deciding factor. When the monomer is coordinated to the metal—$CH_2\sim$ bond with the participation of both ^1C and ^4C, a chain with 1,4 (*cis* or *trans*) configuration is produced

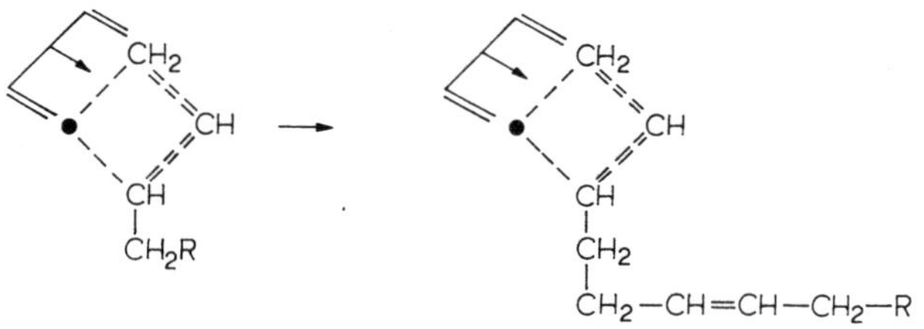

$$\text{(50)}$$

Monomer coordination on the opposite side will lead to 1,2 addition (3,4 with isoprene)

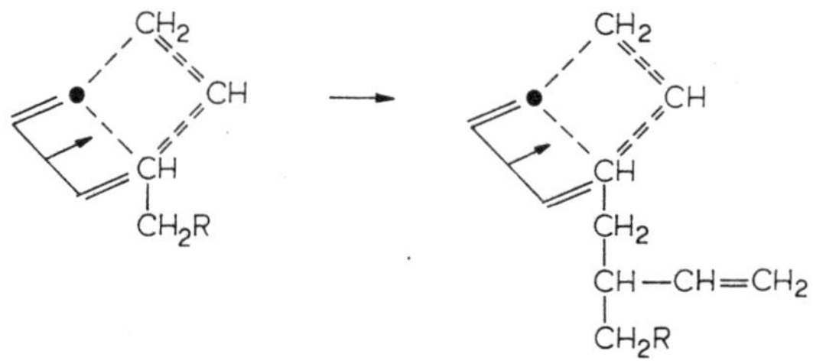

$$\text{(51)}$$

The double bond of the penultimate diene unit appears to be affected by the electric field of the centre with the transition metal [124].

(ii) *Controlled addition of non-conjugated dienes.* Non-conjugated dienes can be polymerized by simple 1,2 addition or by cycloaddition

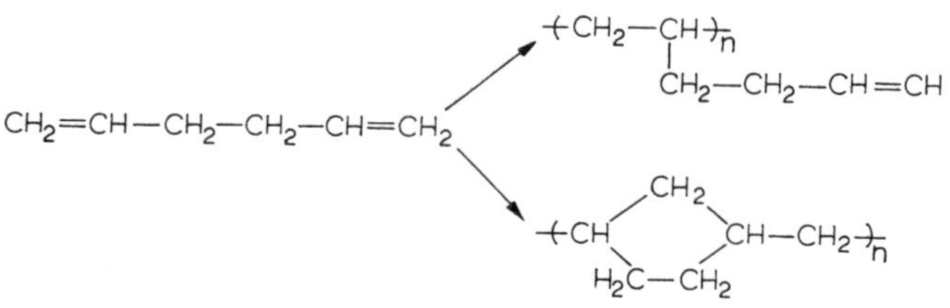

$$\text{(52)}$$

1,2 Addition proceeds predominantly on soluble or colloidal, nonisotactically active catalysts [e. g. $Et_2AlCl + V(acac)_3$, $Et_2AlCl + VO(OEt)_2Cl$] and at low temperatures. Cycloaddition, on the other hand, is favoured by heterogeneous, isotactically operating catalysts (e. g. Et_2AlCl or $Et_3Al + TiCl_3 \cdot xAlCl_3$ or $TiCl_4$), polymerization temperatures in the range 300–400 K, low monomer concentrations, and a certain number of methylene units separat-

References pp. 372–382

ing the double bonds. A two-step mechanism of cycloaddition was proposed by Marvel and Garrison [125]

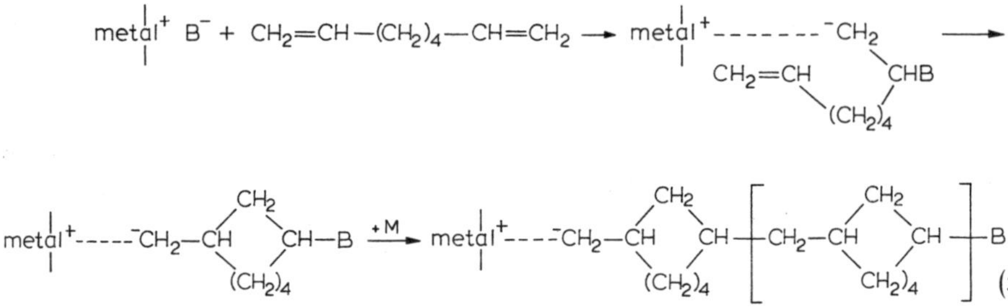

$$\text{metal}^+ \text{B}^- + CH_2=CH-(CH_2)_4-CH=CH_2 \rightarrow \text{metal}^+ \cdots \text{-}CH_2\text{-}CH\underset{(CH_2)_4}{\overset{CH_2=CH}{\diagdown}}CHB \rightarrow$$

$$\text{metal}^+ \cdots \text{-}CH_2\text{-}CH\underset{(CH_2)_4}{\overset{CH_2}{\diagup\diagdown}}CH\text{-}B \xrightarrow{+M} \text{metal}^+ \cdots \text{-}CH_2\text{-}CH\underset{(CH_2)_4}{\overset{CH_2}{\diagup\diagdown}}CH\text{-}\left[CH_2\text{-}CH\underset{CH_2)_4}{\overset{CH_2}{\diagup\diagdown}}CH\right]\text{-}B \quad (53)$$

or a single-step reaction with concerted addition and cyclization.

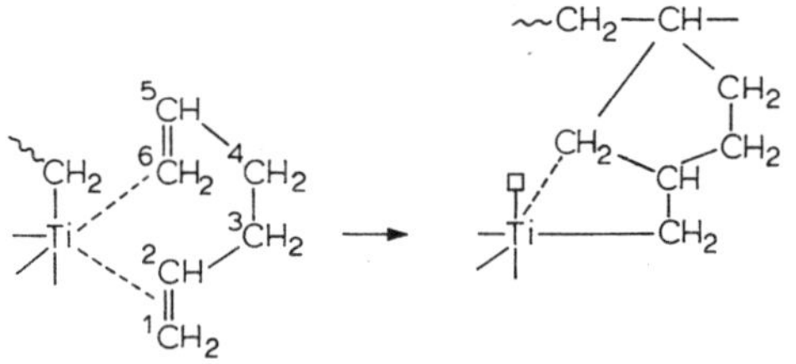

$$\text{(54)}$$

It should be noted that even some radical or ionic polymerizations yield chains composed of cycles (see Sect. 6.4).

4.2 Polymerization in micelles (emulsion polymerization)

In the previous sections, methods of qualitatively controlling the course of propagation were described. Indirect control as well as the quantitative effects caused by intentional control of the other partial processes in polymerization have still to be mentioned. The separation of initiation from propagation alters the kinetic character of the whole reaction. With ionic polymerizations, initiation can be separated from propagation by the selection of conditions suitable for rapid initiation. With radical polymerizations, this is not possible. Therefore both partial processes must be separated in space. Fortunately, radical active centres operate both in polar and in non polar media. Thus it is not difficult to confine initiation and propagation to mutually immiscible components of the medium. Emulsion polymerization remains the most important representative of quantitative control of propagation.

4.2.1 INITIATION IN THE AQUEOUS PHASE

Most monomers polymerizing by the radical mechanism are almost insoluble in water. Intensive stirring of a mixture of such a monomer with water produces an emulsion which remains stable, however, only in the presence of a surface active compound (tenside), e. g. soap. By the addition of a water-soluble initiator to this emulsion, the monomer polymerizes at a rate several times higher than would be observed by any other radical method with an initiator of equal efficiency. At the same time, a higher polymer with a narrower molecular mass distribution is formed. At the initial stages of the reaction, the monomer is present as three types of particle: in tenside-stabilized monomer droplets of diameter 10^{-3} to 10^{-4} cm (about 10^{12} such droplets are present in 1 cm^3 of emulsion of average concentration); in solubilized micelles about 10 nm in size and concentration $\sim 10^{18}$ cm^{-3}; and in the growing, emulsifier-stabilized monomer—polymer particles 50–100 nm in size. This situation is illustrated schematically in Fig. 14(a).

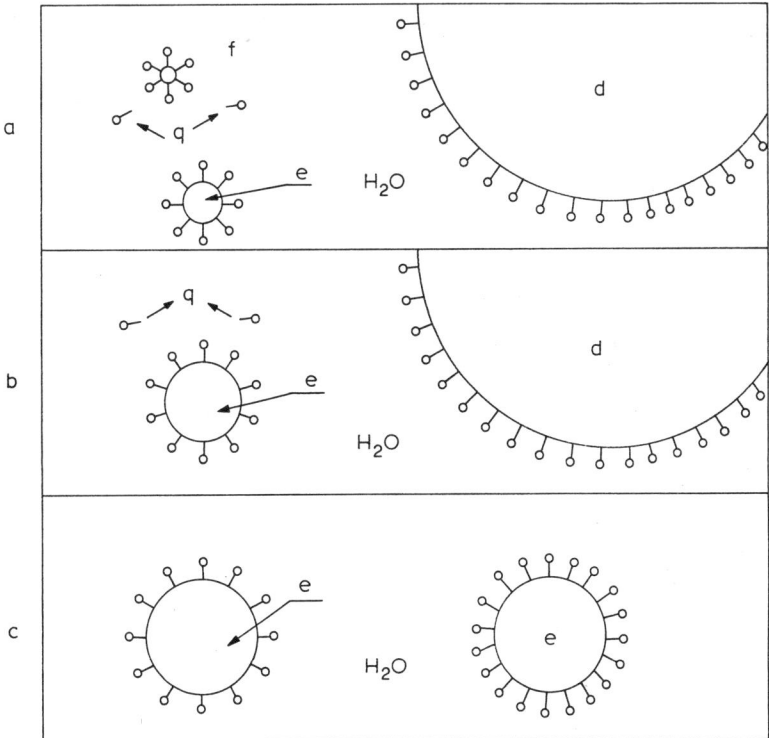

Fig. 14. Monomer placement in particles during emulsion polymerization. a, b, c, Various stages of polymerization (see text); d monomer particle stabilized by emulsifier; diam. = $\sim 10^4$ nm; e monomer-polymer particle; f, monomer solubilized in a micelle; q, emulsifier. See p. 180 of ref. 126.

References pp. 372–382

The proceeding monomer—polymer transformation is accompanied by an increasing need for stabilization of the monomer—polymer particles. The necessary emulsifier is drawn from the micelles so that, at a certain moment, the micelles disppear [at a conversion $\varkappa = 0.1$–0.2; see Fig. 14(b).

Finally, the monomer supply from the droplets is exhausted so that, in the final stage of emulsion polymerization, only the monomer—polymer particles are present in the system in an amount of about 10^{16} cm^{-3} [see Fig. 14(c)].

The first comprehensive qualitative concept of emulsion polymerization was presented by Harkins [127].

The radicals are formed from the initiator in the aqueous phase. They penetrate into the micelles where they initiate polymerization. The generated polymer particles grow at the expense of the monomer dispersed in the aqueous phase. The monomer droplets and micelles decay, supplying the polymer particles. When the micelles disappear, the generation of new particles ceases. The existing particles grow until all monomer has been exhausted. In a polymer particle, a single radical is growing. Both active centres are terminated by introduction of a second radical. When a third radical enters, polymerization in the particle is restarted.

The simple Harkins model is the basis of most quantitative theories of emulsion polymerization.

(a) *The Smith–Ewart theory*

Smith and Ewart in the U.S.A [128], Haward in Britain [129], Yurzhenko and Kolechova in the U.S.S.R. [130], and later in a different way Fikentscher et al. in F.R.G. [131] derived a kinetic scheme of emulsion polymerization explaining most experimental evidence known at that time. Their theory is recognized as valid to this day, and it forms the basis of more modern developments.

A mathematical model of emulsion polymerization as a whole would be too complicated. Therefore Smith and Ewart divided the process into three phases, and appropriately simplified the situation in each of these. The start of polymerization and the monomer—polymer transformation up to the disappearance of micelles were designated as phase I, the subsequent time interval until the complete consumption of monomer droplets as phase II, and the remaining part of polymerization as phase III†.

Phase I: The dependence of the number of polymer particles swollen by monomer at the end of phase I on the concentration of emulsifier is given

$$N = k([\text{E}]p)^{3/5}(\varrho/\mu)^{2/5} \tag{55}$$

---

† Phase I corresponds to scheme (a), Fig. 14, phase II to (b) and phase III to scheme (c).

where N is the total number of particles $\equiv \sum_{i=0}^{\infty} \int_0^{\infty} f_i(V, t) \, dV$ in a volume element (mol cm^{-3}) of emulsion, [E] is the concentration of emulsifier (g cm^{-3}), p is the surface occupied by 1 g of emulsifier, ϱ is the rate of radical generation in the aqueous phase $\equiv fk_d I/V$ (mol h^{-1} cm^{-3}) and μ is the rate of particle growth $\equiv (k_p d_m/N d_p)[\phi_m/(1 - \phi_m)]$ (cm^3 h^{-1}) in which d_m and d_p are the densites of monomer and polymer (g cm^{-3}), ϕ_m is the volume fraction of monomer in the particle (dimensionless). The meanings of the other symbols are evident.

When all generated radicals enter the micelles forming monomer—polymer particles, then $k = 0.53$. When the radicals simultaneously diffuse into micelles and into monomer particles at rates proportional to the surface of each of these formations, then $k = 0.37$ (the experimentally found value of k in styrene polymerization is 0.40).

Phase II: The authors assume a quasistationary state in which the number of particles N_i containing i radicals does not change with time. The concentrations of these particles must obey the relations

$$[N_{i-1} - N_i]\frac{\varrho_A}{N} + \frac{k_u a_p}{V}[(i + 1) N_{i+1} - i N_i] +$$

$$+ \frac{k_t}{VN}[(i + 2)(i + 1) N_{i+2} - i(i - 1) N_i] = \frac{dN_i}{dt} = 0 \qquad (56)$$

where N_i is the number of particles containing i radicals $\equiv \int_0^{\infty} f_i(V, t) \, dV$ (mol cm^{-3}) of emulsion, ϱ_A is the rate of radical entrance into the particles (mol cm^{-3} h^{-1}), k_u is the rate constant of radical desorption from the particle (h^{-1}), a_p is the surface area of a particle of volume V, proportional to $V^{2/3}$ and k_t is the rate constant of termination in the particle (mol^{-1} cm^{-3} h^{-1}). The first term on the left-hand side of eqn. (56) describes the entrance of the radical, the second term its exit, and the third term the mutual termination of the growing radicals in the particle. A general solution of eqn. (56) is difficult. For specific cases it can be greatly simplified, particularly when

(a) The rate of radical entrance into the particle is small compared with their desorption. This situation is considered as very improbable. The diffusion rate of a polymeric radical out of the particle can hardly be greater than that of a small radical into the particle.

(b) The rates of desorption and mutual termination of radicals are much smaller than the rate of their supply. In this case the polymerization rate is given by the relation derived for bulk polymerization

$$v_p = k_p[M] \sum_{i=0}^{\infty} i N_i = k_p[M] \, \overline{i} N = k_p[M]\left(\frac{fk_d[I]}{2k_t}\right)^{1/2} \qquad (57)$$

where $\overline{i}$ is the mean number of radicals in a particle.

References pp. 372–382

(c) The rate of termination is much higher than the rates of radical entrance or escape. This means that at any moment each particle will contain either one growing radical or no radical ($i = 1/2$). In this case the polymerization rate is given by the rate of propagation

$$-\frac{d[M]}{dr} = v_p = k_p[M][R^*] = k_p[M]\frac{N}{2} \qquad (58)$$

The basic prerequisities of the Smith–Ewart theory are equal dimensions of all polymer–monomer particles, and their constant growth rate

$$\frac{dv}{dt} = \frac{k_p d_m}{N d_p} \frac{\phi_m}{1 - \phi_m} i = \mu i \qquad (59)$$

Equation (58) indicates that an increase in initiatior concentration will not enhance the rate of polymerization. It can be used for estimating the molecular mass of the polymer assuming, of course, the absence of transfer. The ratio N/ϱ corresponds to the mean time of polymer growth and molecular mass is equal to the product of the number of additions per unit time and the length of the active life time of the radical, $k_p N/\varrho$. An increase in [I] also means a higher value of ϱ, and thus a shortening of the chains. As in Phase II, the polymerized monomer in the particles is supplemented by monomer diffusion from the droplets across the aqueous phase; a stationary state is rapidly established with constant monomer concentration in the particle. The rate of polymerization is then independent of conversion (see, for example the conversion curves in Fig. 7). We assume that the Smith–Ewart theory does not hold for those polymerizations where the mentioned dependence is not linear [132]. The valdity of the Smith–Ewart theory is limited by many other factors.

(b) *More recent theories*

The more recent theories can be divided into two groups. They either develop, refine or broaden the Smith–Ewart (SE) theory, or they put forward new ideas.

The description of phase I of the SE theory was refined by Gardon [133] and Harada et al. [134]; the effect of particles in which polymerization has been terminated by radical entrance is included. Paris et al. [135] and Sautin et al. [136] calculate the balance of micelles and of the growing and dead particles. Pismen and Kuchanov [137] and Sundberg and Eliassen [138] included the effect of particle size distribution in their calculations. According to Fitch and Tsai [134] and Roe [140], the monomer swollen particles are produced by the polymerization of the monomer which is dissolved in water.

As soon as the growing radicals exceed the solubility limit, the oligomers form a second phase which is stabilized by the emulsifier (see Fig. 15). In contrast to the SE theory, the critical emulsifier concentration is unimportant in the

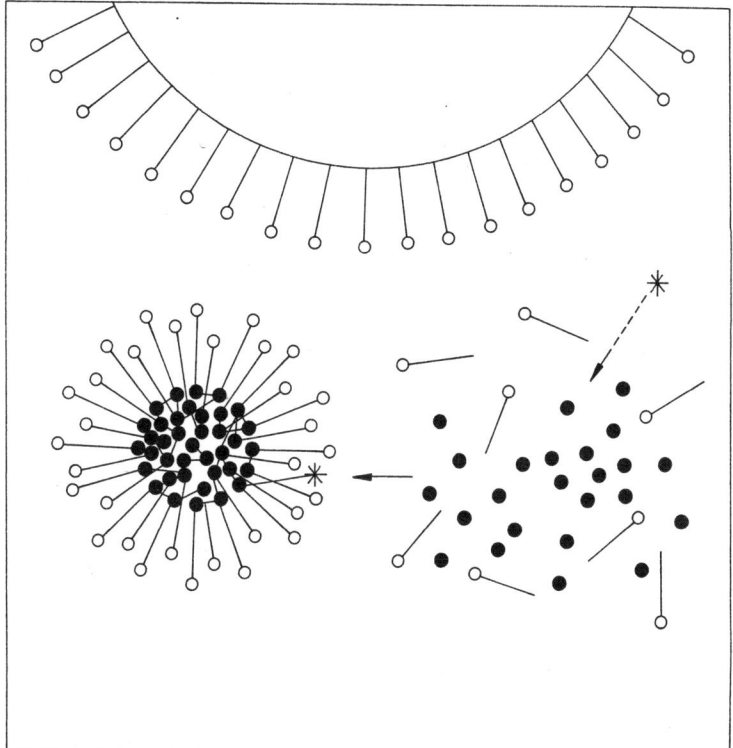

Fig. 15. Formation of a monomer–polymer particle according to and Fitch and Roe.

Fitch–Roe approach. (At lower than critical emulsifier concentrations, micelles are not generated. With increasing amount of emulsifier, the physical properties of aqueous medium discontinuously change in the vicinity of its critical concentration. The critical emulsifier concentration is an important material constant.) Roe [140] proved that the two theories can be described by very similar quantitative relations[†]. This latter theory stress the importance of dissolution equilibria for the equilibrium monomer concentration in the aqueous phase.

[†] The consequences of simplifications introduced into the mathematical formulae are of considerable importance here.

References pp. 372–382

Stockmayer [141] and O'Toole [142] presented an analytical solution of eqn. (56), and thus generatized the formulation of the processes in phase II

$$N_i = a^i 2^{(b-1-3i)} \frac{J_{b+i-1}(a/2^{1/2})}{i! J_{b-1}(a)} \tag{60}$$

where $J_k(x)$ is a modified Bessel function of the first kind, parameter a expresses the relative contributions of initiation and termination, and b represents the relative rates of radical escape and termination

$$a = \left(\frac{8VN\varrho_A}{k_t N}\right)^{1/2}$$

$$b = \frac{k_u a_p N_a}{k_t}$$

For the mean number of radicals $\bar{i}$ in the particle, eqn. (60) yields

$$\bar{i} = \frac{\sum_{i=0}^{\infty} iN_i}{N} = \begin{cases} \dfrac{1-b}{2} + \dfrac{a J_{b-2}(a)}{4 J_{b-1}(a)} & \text{for } b \geq 1 \\[2mm] \dfrac{a}{4} \dfrac{J_b(a)}{J_{b-1}(a)} & \text{for } 0 < b \leq 1 \\[2mm] \dfrac{a}{4} \dfrac{J_0(a)}{J_1(a)} & \text{for } b = 0 \end{cases} \tag{61}$$

By substitution of eqn. (61) into eqn. (57), a more general relation for the rate of propagation is obtained as a function of a and b. It should be noted that the parameter a is a function of time because the particle volume V changes with time. The description of phase II was further refined by Ugelstad et al. [143], by Katz and Saidel [144], and many others.

Phase III was divided into three parts by Alexander and Napper namely into a period of smooth monomer supply of the active centres from aqueous phase and from monomer-rich particle layers, a period of viscosity increase in the particles (by affecting monomer diffusion, viscosity begins to control the reaction rate); and the period of approach to the polymer glass transition [126, 145, 146].

(c) *Theory of emulsion polymerizations considering the internal structure (morphology) of the particles*

Sheinker and Medvedev [147, 148] assume that polymerization proceeds in the micelle layer adsorbed at the particle surface. According to these authors, the interior of the particle consists only of monomer and dead polymer. The reaction rate is thus proportional to the total surface, and is constant for a given concentration of emulsifier.

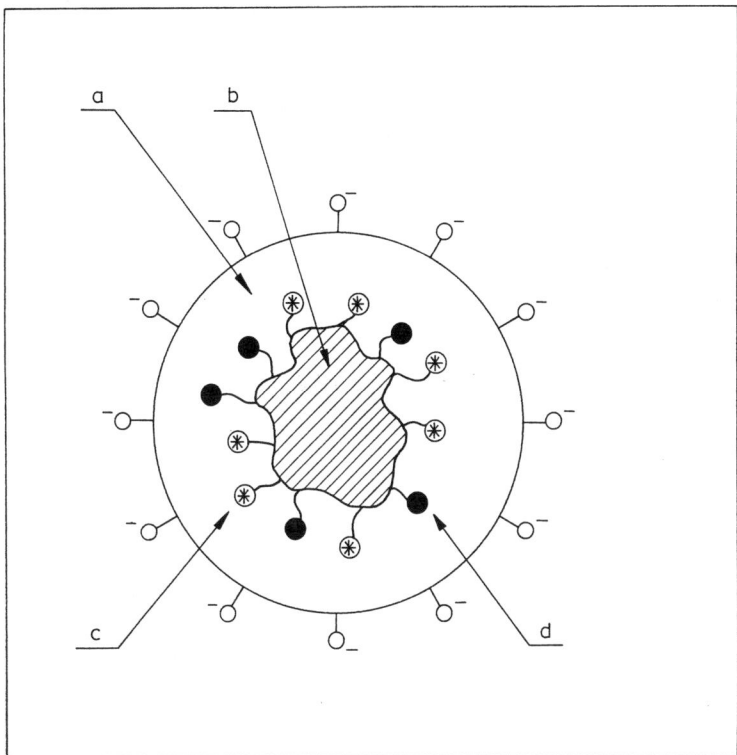

Fig. 16. Structure of a monomer–polymer particle.

The idea of particle inhomogeneity was supported experimentally by Williams [149]. However, his representation of growth is more complicated. In phase II, the monomer concentration in the particle decreases with conversion, while the rate remains constant. The particle has a core with a relatively high polymer content surrounded by a monomer-rich layer (see Fig. 16). Polymerization occurs at the polymer—monomer interface. Under these conditions, monomer concentration at the interface remains constant, even when its amount in the particle decreases. Napper presented the idea of an exactly opposite composition of the monomer—polymer particle [150]. The core should be enriched in monomer and surrounded by a layer with a higher polymer content. Van den Hul and van der Hoff found most growing ends of macromolecules at the particle surface [151], which supports Napper's model.

References pp. 372–382

These examples illustrate the great variety of current ideas. A generally accepted theory of emulsion polymerization, based on a non-uniform composition of the monomer—polymer particles, is presently being developed [152–155].

The presence of the emulsifier, and thus of the micelles, is not a necessary condition for emulsion polymerization. Styrene, the acrylates, vinyl acetate, and vinyl chloride can be polymerized in aqueous medium without an emulsifier [126]. The mechanism of the formation of monomer—polymer particles is in agreement with the ideas of Fitch and Tsai [139] which were refered to above.

4.2.2 POLYMERIZATION IN EMULSION INITIATED BY AN INSOLUBLE INITIATOR

A basic feature of emulsion polymerization is the separation of initiation from propagation. When motion of the initiating radicals is prevented, for example by their fixation at a solid surface, termination by primary radicals is prevented. When propagation proceeds in the emulsion, termination is completely excluded[†]. Mobility of the growing chains is limited by their anchoring at the surface of the solid phase. The emulsifier layer prevents contact between the monomer—polymer particles, and thus termination of the macroradicals. These radical polymerizations proceed by the living mechanism. For their realization, insoluble initiatiors are usually used with functional groups (e. g. hydrogen peroxides) bound on the polymer chain. Graft or block copolymers (or even homopolymers) are produced. This method was first used by Allen et al., grafting poly(methyl methacrylate) chains on a polystyrene containing hydroperoxide groups [156]

Many systems of this kind have been described up to the pressent time. An interesting example is the application of ozonized polypropylene as initiator. By this kind of emulsion polymerization, chains of methyl methacrylate, styrene, and of other monomers [157, 158] can be grafted on polypropylene, thus modifying its properties.

4.2.3 INVERSE EMULSION POLYMERIZATION

When a water-soluble monomer dispersed in a continuous oil phase is polymerized with an initiator soluble in the continuous phase, we speak of inverse emulsion polymerization [159]. This system has all the advantages of emulsion polymerizations (rapid polymerization, high degree of polymeriza-

---

[†] Transfer reactions can weaken or even nullify the validity of this statement.

tion), and moreover it permits easy transfer of the polymeric particles into aqueous phase by simple latex inversion. The hydrophilic phase is usually not formed by the monomer itself but by its aqueous solution [153]. The transfer of the product into the aqueous phase is therefore even easier. The kinetics of inverse emulsion polymerization is, of course, not a "mirror" image of the more common variant; it has its own peculiarities. One of these is the phase-transfer effect of the tenside on the initiator so that the site of its decomposition is not limited to a single phase with both the water-soluble and the water-insoluble initiators. The polymerization of acrylamide dispersed in an aromatic hydrocarbon is the most frequently studied case of this kind of process [160].

4.2.4 IONIC EMULSION POLYMERIZATIONS

Until recently, the term emulsion polymerization was used exclusively for radical heterogeneous polymerizations in aqueous medium. Naturally non-aqueous [161] and non-radical systems are also permitted but only very few have been described so far. Non-polar solvents mostly dissolve monomers, polar solvents strongly solvate the active centres and usually make other than radical polymerizations impossible. In spite of that, such atypical emulsion polymerizations do exist. An example is the ionic polymerization of cyclic siloxanes [162, 163]. These monomers can be emulsified in water by means of a suitable cationic or anionic surface active agent which simultaneously acts as an initiator. In the presence of benzensulphonic acid derivatives, anionic polymerization should occur; with R_4NCl the process should be cationic, according to the authors. Chain propagation consists of ionic opening of the cyclosiloxane yielding silanediol, with subsequent polycondensation (etherification) of the silanol groups.

So far these polymerizations are of no practical importance, and their theoretical studies have hardly begun.

5. Copolymerization

Propagation in a mixture of two or more monomers, as any other chemical reaction, is determined by thermodynamic and kinetic factors. Either a mixture of homopolymers or, more often a copolymer is formed with varying monomer order in the chains. The propagation of two mutually non-complexing monomers $(M_1$ and $M_2)$ can be represented by the scheme

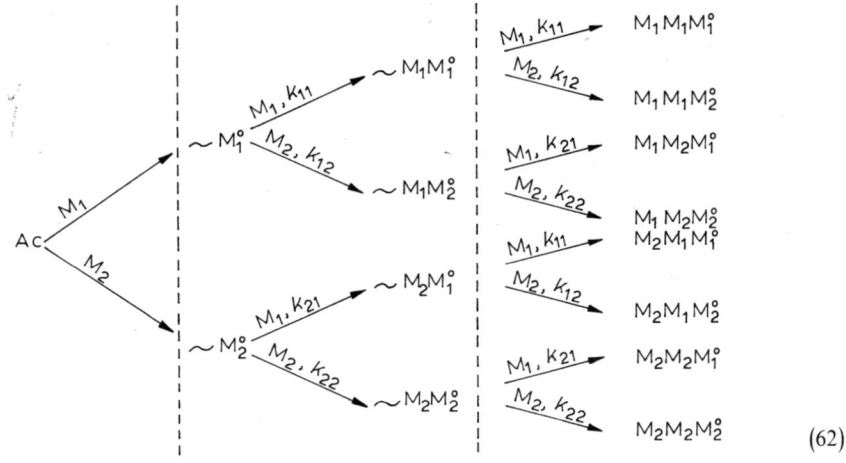

(62)

The simplest copropagation is composed of consecutive additions of one of two monomers to one of two centres of generally different reactivity.

5.1 Copolymerization equation

Let us demonstrate the connection between the compositions of the generated copolymer and the monomer mixture in the simplest example. The monomer addition rate will be assumed to be independent of the penultimate unit type. The rate of monomer consumption will then be given as [164–167]

$$-\frac{d[M_1]}{dt} = k_{11}[\sim M_1^\circ][M_1] + k_{21}[\sim M_2^\circ][M_1] \tag{63}$$

$$-\frac{d[M_2]}{dt} = k_{22}[\sim M_2^\circ][M_2] + k_{12}[\sim M_1^\circ][M_2] \tag{64}$$

The concentrations $[\sim M_1^\circ]$ and $[\sim M_2^\circ]$ (let alone the activities) are almost never known. We know, however, that in all stationary copolymerizations (and such are all living, as wel as a great majority of the radical copolymerizations) the decay of $\sim M_1^\circ$ must be accompanied by the generation of $\sim M_2^\circ$, and vice versa†. Thus

$$k_{12}[\sim M_1^\circ][M_2] = k_{21}[\sim M_2^\circ][M_1] \tag{65}$$

Division of eqn. (63) by eqn. (64), substituting the ratio $k_{12}[\sim M_1^\circ][M_2]/(k_{21}[M_1])$ from eqn. (65) for $[\sim M_2^\circ]$, and rearranging yields the required expression, called the copolymerization equation

$$\frac{d[M_1]}{d[M_2]} = \frac{[M_1]r_1[M_1] + [M_2]}{[M_2]r_2[M_2] + [M_1]} \tag{66}$$

Equation (66) is the most frequently used form of the copolymerization equation. It is also quite often written in the notation

$$n = \frac{r_1 y + 1}{(r_2/y) + 1} \tag{67}$$

This follows from eqn. (66) by substituting n for $d[M_1]/d[M_2]$ and y for $[M_1]/[M_2]$. The copolymerization parameters r_1 and r_2 are defined by the ratio of the rate constants

$$r_1 = \frac{k_{11}}{k_{12}}$$

$$r_2 = \frac{k_{22}}{k_{21}}$$

They are of considerable importance in macromolecular chemistry; they carry valuable information on the reactivities of monomers, transfer agents, and active centres.

The copolymerization equation can also be derived from the mean sequence lengths, $\bar{\lambda}$, of M_1 and M_2 units in the chains. The number of M_1 monomer additions per single M_2 addition will evidently be given by the corresponding rate ratio, and the sequence will be longer by one unit than the rate ratio

$$\bar{\lambda}_1 = \frac{v_{11}}{v_{12}} + 1 = \frac{k_{11}[\sim M_1^\circ][M_1]}{k_{12}[\sim M_1^\circ][M_2]} + 1 = r_1 \frac{[M_1]}{[M_2]} + 1 \tag{68}$$

---

† $\dfrac{d([M_1^\circ] + [M_2^\circ])}{dt} = 0; \quad \dfrac{d[M_1^\circ]}{dt} = -\dfrac{d(M_2^\circ)}{dt}.$

References pp. 372–382

where v is the rate, and the index denotes the reacting active centre and monomer, respectively. Similarly

$$\bar{\lambda}_2 = \frac{v_{22}}{v_{21}} + 1 = \frac{k_{22}[\sim M_2^\circ][M_2]}{k_{21}[\sim M_2^\circ][M_1]} + 1 = r_2\frac{[M_2]}{[M_1]} + 1 \qquad (69)$$

The copolymer composition must be identical with the populations of the two monomers in the chains, i. e. with the ratio $\bar{\lambda}_1/\bar{\lambda}_2$. Dividing the expression (68) by the relation (69) we obtain the copolymerization equation (66) or (67) without any need to assume the existence of a stationary state.

Equations (66) and (67) do not illustrate the behaviour of the monomer pair during copropagation particularly well, and therefore they have been

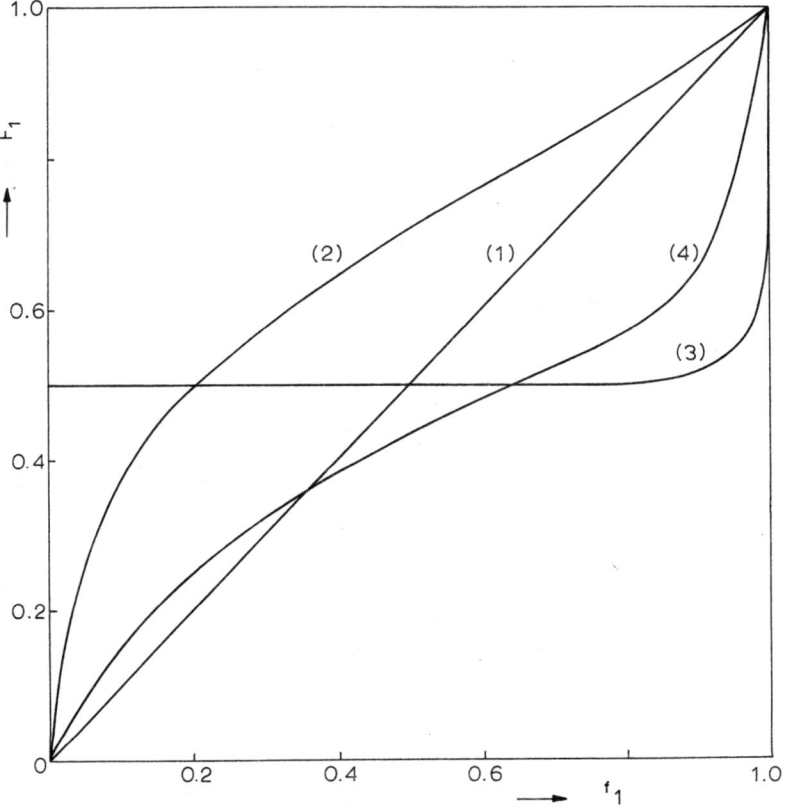

Fig. 17. Copolymerization diagram. Shape of copolymerization curves. (1) $r_1 = r_2 = 1$; (2) $r_1 = 1.25$, $r_2 = 0.10$; (3) $r_1 \approx 0.001$, $r_2 \approx 0$; (4) $r_1 = 0.05$, $r_2 = 0.5$.

rearranged by Skeist [167]. The simple concentrations have been replaced by the mole fractions of monomer M_1 in the copolymer

$$F_1 = \frac{d[M_1]}{(d[M_1] + d[M_2])}$$

and in the monomer mixture

$$f_1 = \frac{[M_1]}{([M_1] + [M_2])}$$

An equation of the copolymerization curve in the form $F_1 = f(f_1)$ has thus been obtained.

$$F_1 = \frac{r_1 f_1^2 + f_1 f_2}{r_1 f_1^2 + 2 f_1 f_2 + r_2 f_2^2} = \frac{f_1^2(r_1 - 1) + f_1}{f_1^2(r_1 + r_2 - 2) + 2 f_1(1 - r_2) + r_2} \quad (70)$$

Equation (70) itself is, of course, no more illustrative than eqn. (66) but its graphical representation provides a clearer picture of the dependence of copolymer composition on the momentary relative contents of the two monomers in the mixture. The copolymerization diagram, as the $F_1 = f(f_1)$ dependence is called, also clearly illustrates the role of the copolymerization parameters in copropagation (see Fig. 17).

5.2 Copolymerization parameters

It has become customary to divide the common copolymerizations into several groups. The grouping depends on the parameter values

(a)
$$r_1 = r_2 = 1 \quad (71)$$

In this case, copolymerization is "ideal"; the relative contents of the two monomers in the monomer mixture and in the product are identical. The copolymerization curve forms a diagonal in the square copolymerization diagram.

(b)
$$r_1 = r_2 = 0 \quad (72)$$

Neither of the monomers is able to homopolymerize. If a polymeric product is formed, it can only be a strictly alternating copolymer. The copolymerization curve is a horizontal straight line dividing the diagram into two equal rectangles ($F_1 = 1/2$, independent of f_1).

(c)
$$r_1 = \frac{1}{r_2} \tag{73}$$

Copolymer composition can be expressed by a single parameter. The copolymerization equation (66) is simplified to a form derived by Wall [168] in the first attempt to characterize copolymerization quantitatively

$$\frac{d[M_1]}{d[M_2]} = r_1 \frac{[M_1]}{[M_2]} \qquad n = r_1 y \tag{74}$$

Also, the copolymerization curve equation is simple

$$F_1 = \frac{r_1 f_1}{1 + f_1(r_1 - 1)}$$

Equation (74) describes copropagation where the rate of M_1 (M_2) addition to the growing end $\sim M_1^\circ$ is equal to the adition rate to $\sim M_2^\circ$. Many radical and most ionic copolymerizations belong to this class [169].

(d)
$$r_1 = 0; \qquad r_2 > 0 \tag{75}$$

M_1 is not able to homopolymerize, the great majority of M_1 units will be contained in alternating sequences, $F_1 = f_1(2f_1 + r_2 f_2)^{-1} \leq 1/2$. Equation (66) is simplified to

$$\frac{d[M_1]}{d[M_2]} = \frac{[M_1]}{r_2[M_2] + [M_1]} \qquad n = \frac{1}{(r_2/y) + 1}$$

(e)
$$r_1 > 1 > r_2 > 0 \tag{76}$$

This relationship describes a frequent case of uncomplicated copolymerizations[†]. Monomer M_1 is consumed at a greater rate than M_2. In batchwise

---

[†] Complications occur due to penultimate unit effects, complexation, the presence of "hot" radicals etc. See further text in Sect. 5.5, 5.6, and 5.8.

copolymerization, the composition of the monomer mixture changes in the course of the copropagation. The greater the numerical difference between the parameters, and the higher the conversion, the more severe is the composition heterogeneity of the product.

(f)
$$0 < r_1 < 1$$
$$0 < r_2 < 1 \qquad (77)$$

From the copolymerization equation, a situation can easily be derived where the compositions of the monomer mixture and the generated polymer will be equal $(F_1 = f_1)$ and the copolymerization curve will intersect the diagonal of the square copolymerization diagram. This will occur, apart from the case of eqn. (71) when $r_1[M_1] + [M_2])/(r_2[M_2] + [M_1]) = 1$, i. e. when

$$\frac{[M_1]}{[M_2]} = \frac{r_2 - 1}{r_1 - 1} \qquad (78)$$

Because $[M_1]/[M_2] > 0$, both parameters must be smaller (or larger) than 1 in this case, which is called azeotropic copolymerization. The situation represented by eqn. (77) is fairly frequent.

(g)
$$r_1 > 1, \qquad r_2 > 1 \qquad (79)$$

Relation (79) is rarely met but more often with ionic than with radical polymerizations. This kind of parameter promotes the formation of long sequences of either monomer, producing block copolymers.

5.2.1 THE EXPERIMENTAL DETERMINATION OF COPOLYMERIZATION PARAMETERS

The rate constants of propagation, k_{11} or k_{22}, can be derived and determined, although the procedure may not be simple. On the other hand k_{12} or k_{21} are not directly measurable at present. They are usually inferred from experience with copolymerizing systems or with transfers. To date, copolymerization parameters much more frequently serve as the basis for the calculation of elementary constants than the reverse. For their determination, according to eqn. (66), we have to know the populations of M_1 and M_2 both in the monomer mixture and in the copolymer. In principle, two approaches are known.

(a) The first is based on the knowledge of the momentary composition of the monomer mixture and product, i. e. it works with the differential form of

References pp. 372–382

the copolymerization equation. The values obtained can, of course only be correct for strictly momentary monomer concentrations. Experimentally this can be relatively easily realized for the monomer mixture, from which a sample can be withdrawn and quickly analyzed (by spectroscopic or chromatographic methods, etc.). However, the problem of separation and analysis of chains formed at a defined moment has not been solved so far. This difficulty is circumvented by terminating copropagation at low conversions ($\varkappa = 0.04-0.06$) where the copolymer composition may be regarded as practically homogeneous.

(b) The other approach makes use of the integrated form of eqn. (66), and thus it avoids the limitation mentioned above. Evaluation of the results is more complicated.

(i) There exist several variants of parameter determination from the differential form of the copolymerization equation (67). Mayo and Lewis [164] used the variant $r_2 = f(f_1, n, y)$

$$r_2 = r_1 \frac{y^2}{n} + \frac{y}{n} - y \tag{80}$$

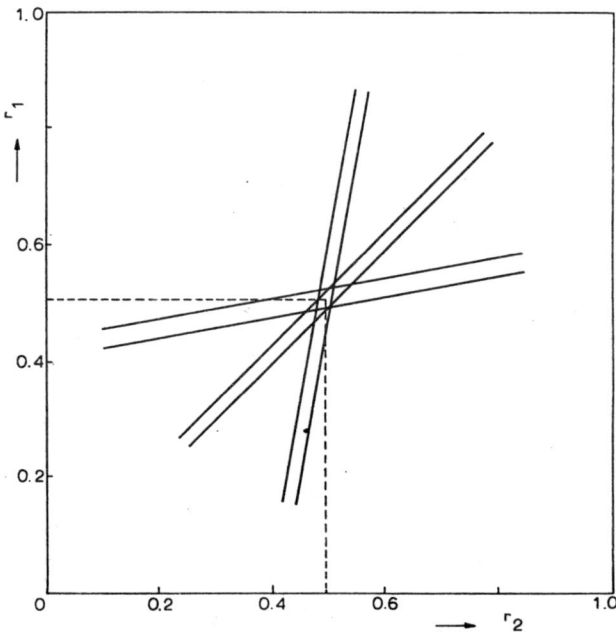

Fig. 18. Determination of copolymerization parameters by the Mayo–Lewis method. See p. 255 of ref. 170.

The graphical method of parameter determination was also found to be useful: For known n and y values a value of r_1 is selected, the corresponding r_2 is calculated and plotted on the graph whose ordinate is r_1 and abscissa r_2. For the same n and y, r_1 is modified in a suitable way (for example by 0.1), r_2 is calculated and plotted on the graph as the second point defining a straight line. The position of this line can be refined by construction of further points in the same way. The construction is then repeated for other experimentally determined n and y values. This results in further straight lines with other slopes. The inersection coordinates correspond to the required values of r_1 and r_2[†] (see Fig. 18).

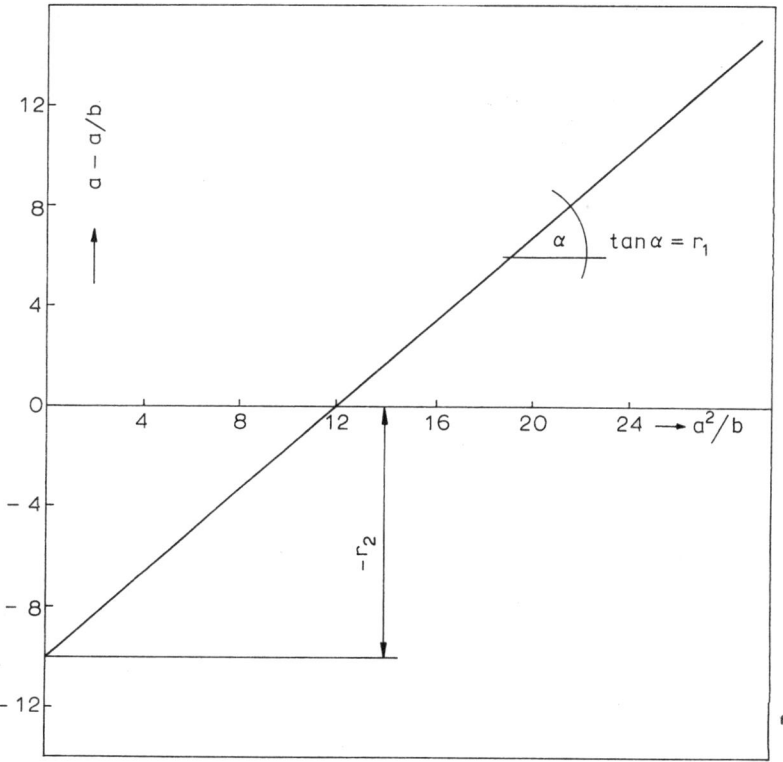

Fig. 19. Determination of copolymerization parameters by the Fineman–Ross method. $a = y$, $b = n$.

[†] It is recommended that three or more n, y pairs are measured. The straight lines derived from these values rarely intersect at one point. The scatter should not be large (it indicates the error). The most probable values of r_1 and r_2 are supposed to correspond to the centre of gravity of the figure formed by the intersections.

References pp. 372–382

Fineman and Ross [171] proposed another form of graphical evaluation of eqn. (80)

$$y - \frac{y}{n} = r_1 \frac{y^2}{n} - r_2 \tag{80a}$$

Equation (80a) is an equation of a straight line with $[y - (y/n)]$ as the ordinate and y^2/n as the abscissa. For several known pairs n and y, a straight line of slope r_1 is constructed whose intercept with the ordinate is equal to $-r_2$ (see Fig. 19).

Tidwell and Mortimer noted that the Fineman and Ross method does not treat experimental data with equal weight, yielding physically unrealistic results in some cases [172]. In order to remove this drawback, Tűdős et al. [173] proposed a modification of eqn. (80) which can also be evaluated graphically

$$\frac{G}{\alpha + H} = \left(r_1 + \frac{r_2}{\alpha}\right)\frac{H}{\alpha + H} - \frac{r_2}{\alpha} \tag{81}$$

where $G = y - (y/n)$, $H = y^2/n$ and $\alpha = (H_{min}H_{max})^{1/2}$. Substituting $\eta = G(\alpha + H)^{-1}$ and $\xi = H(\alpha + H)^{-1}$ in eqn. (81) we obtain a straight line equation in the coordinates η and ξ

$$\eta = \left(r_1 + \frac{r_2}{\alpha}\right)\xi - \frac{r_2}{\alpha} \tag{81a}$$

The main advantage of this equation is that it is possible to check visually assumptions with which the equation has been derived. When these are not fulfilled, the experimental points deviate from a straight line [the slope is a function of both parameters and a compensation of errors is less probable than with the simpler equation (80a)].

(ii) Another method for the determination of copolymerization parameters from integrated copolymerization equation has been also proposed by Mayo and Lewis [164]. Equation (66) is easily integrated, and as such it can be presented in various ways. One of these forms

$$r_2 = \frac{\ln \frac{[M_2]_0}{[M_2]} - \frac{1}{p} \ln \frac{1 - p([M_1]/[M_2])}{1 - p([M_1]_0/[M_2]_0)}}{\ln \frac{[M_1]_0}{[M_1]} + \ln \frac{1 - p([M_1]/[M_2])}{1 - p([M_1]_0/[M_2]_0)}} \tag{82}$$

where $p = (1 - r_1)/(1 - r_2)$

is used for a graphical determination of the numerical values of r_1 and r_2. From the initial and instantaneous concentrations $[M_1]_0$, $[M_1]$, $[M_2]_0$, $[M_2]$, and an arbitrary value of $p^†$, r_2 is calculated and plotted against r_1. The value of p is then changed and the process repeated. The coordinates of the points determined in this way are connected by a straight line. Another straight line with a different slope is obtained for some other ratio of initial and momentary monomer concetrations (determined either by conducting the reaction to higher conversion, whereby only $[M_1]$ and $[M_2]$ are changed, or better by a new copolymerization with different $[M_1]_0$, $[M_2]_0$ and $[M_1]$, $[M_2]$). The intersection coordinates of the two straight lines are the required values. Two straight lines are generally not sufficient; more are usually constructed, and the centre of the figure generated by the points of intersec-

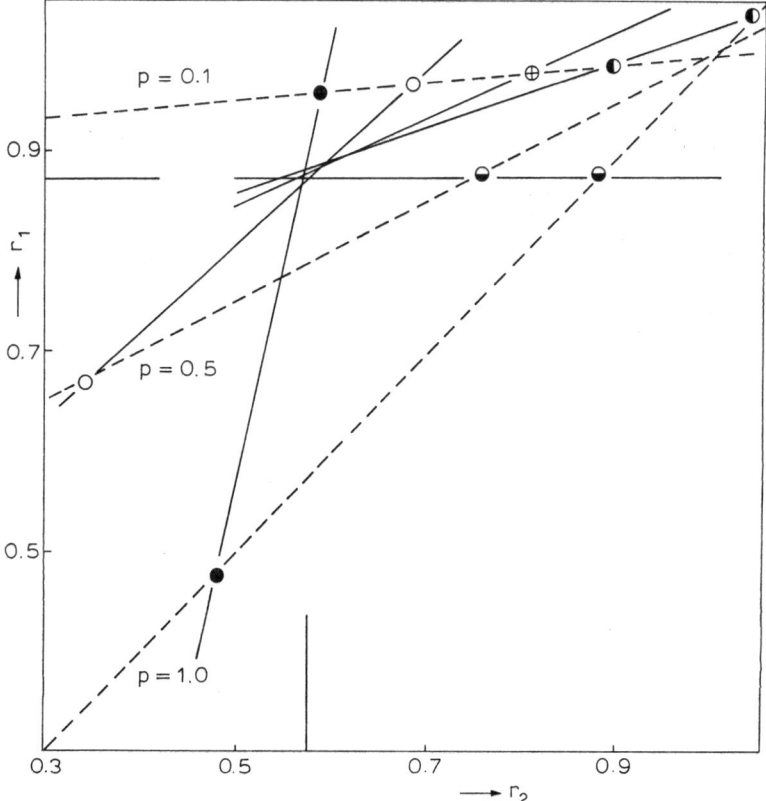

Fig. 20. Determination of copolymerization parameters from the integrated polymerization equation [174]. $r_1 = 0.87 \pm 0.012$; $r_2 = 0.57 \pm 0.025$.

---

† A suitable choice of p considerably reduces the number of calculations necessary. A preliminary calculation of the parameters from the non-integrated equation (66) is therefore recommended.

tion then defines the copolymerization parameters with the required accuracy (see Fig. 20).

The methods of determining copolymerization parameters are continually being improved [175–177].

Kelen et al. published another variant of eqn. (81) suitable for parameter determination in copolymerizations proceeding to high conversion. They use somewhat different definitions of G and H

$$G = \left(\frac{\Delta[M_1]}{\Delta[M_2]} - 1\right) \frac{\log([M_2]/[M_2]_0)}{\log([M_1]/[M_1]_0)}$$

$$H = \frac{\Delta[M_1]}{\Delta[M_2]} \left(\frac{\log([M_2]/[M_2]_0)}{\log([M_1]/[M_1]_0)}\right)^2$$

More detailed computer analyses indicate that methods based on measurements at low conversion are not sufficiently accurate. The method should avoid systematic errors caused by unsubstantiated linearization of the copolymerization equation, and it should respect the structure of experimental errors for which it should be able to compensate.

5.2.2 THEORY OF COPOLYMERIZATION PARAMETERS

Alfrey et al. [178, 179] proposed an empirical rate constant function for the reaction of the growing radical $\sim M_1^*$ with the monomer M_2

$$k_{12} = P_1 Q_2 \exp(-e_1 e_2) \tag{84}$$

where P_1 is a factor characterizing the reactivity of the growing radical formed from M_1, Q_2 is a measure of the resonance stabilization of the monomer M_2, giving rise to the end unit of the growing chain radical $\sim M_2^*$; and e_1 and e_2 are a measure of the radical and monomer polarities ($\sim M_1^*$ and M_1, or $\sim M_2^*$ and M_2, respectively).

Copolymerization parameters can be described, by means of Q and e and the basic features of copropagation can be revealed by observation of objectively defined effects and changes.

$$r_1 = \frac{k_{11}}{k_{12}} = \frac{Q_1}{Q_2} \exp\left[-e_1(e_1 - e_2)\right]$$

$$r_2 = \frac{k_{22}}{k_{21}} = \frac{Q_2}{Q_1} \exp\left[-e_2(e_2 - e_1)\right]$$

(85)

TABLE 3

Copolymerization parameters of selected monomer pairs [180]

M_1	M_2	r_1	r_2	Temperature (K)
Styrene	Vinylidene chloride	1.705	0.108	333
	N-vinylcarbazole	5.7	0.035	348
	Vinyl isocyanate	8.13	0.08	333
	Vinyl chloride	23.92	0.16	323
	Vinyl isobutyl ether	11.58	0.65	323
	Vinyl acetate	24.26	0.02	333
	Methyl methacrylate	0.58	0.56	333
	Methyl acrylate	0.93	0.96	343
	Butyl acrylate	0.836	0.184	353
	Butyl acrylate	0.106	1.233	298
	Acrylonitrile	0.46	0.04	348
	Methacrylonitrile	0.36	0.25	358
	Maleic anhydride	0.02	0.003	333
Methyl methacrylate	Acrylonitrile	1.20	0.15	333
	Methyl acrylate	2.3	0.47	403
	Vinyl chloride	8.976	0.062	318
	Vinyl acetate	2.724	-0.191^a	333
Vinyl acetate	Maleic anhydride	0.019	-0.058^a	348
	2-Chlorobutadiene	-0.02^a	33.52	338
Vinyl chloride	Vinyl acetate	1.034	0.983	313
	2-Methylpropene	2.05	0.055	333
Butadiene	Butyl acrylate	1.048	0.084	278
	3-Chlorostyrene	1.208	1.573	323
	Styrene	1.601	0.332	278
	Vinylidene chloride	1.942	0.043	278
	Vinyl chloride	8.8	0.035	323
	α-Methylstyrene	1.499	0.139	285
	Methyl acrylate	1.107	0.077	323
	Methyl methacrylate	0.526	0.062	278
	Methacrylonitrile	0.058	0.04	278
	Acrylonitrile	0.359	0.046	323
Acrylonitrile	Vinylidene chloride	0.941	0.357	
	Vinyl chloride	2.621	0.023	313
	Vinyl acetate	5.572	0.066	343

[a] Negative value indicates an error in the experimental determination. It lacks physical sense.

References pp. 372–382

TABLE 3A

Q, e, q, ε and p values of selected monomers [179]

Monomer	Q	e	q (kJ mol^{-1})	$10^{20}\,\varepsilon$ (C)	p
α-Methylstyrene	0.98	−0.81[a]	−12.14	−1.23	0.52
Styrene	1.00	−0.80	−12.98	−0.80	0.47
Methacrylic acid	0.98[a]	0.65	−12.56	0.73	0.39
Methyl methacrylate	0.74	0.40	−11.30	0.23	0.39
Methacrylonitrile	0.86[a]	0.68[a]	−12.56	0.77	0.30
Butadiene	1.7[a]	−0.5[a]	−15.07	−0.80	0.29
Acrylonitrile	0.48[a]	1.20	−11.30	1.13	0.28
Methyl acrylate	0.42	0.60	−10.47	0.60	0.26
Vinylidene chloride	0.31[a]	0.36	−8.80	0.37	0.17
Vinyl chloride	0.044	0.20	−4.60	−0.13	0.054
N-Vinylpyrrolidone	0.088[a]	−1.62[a]			0.053
Vinyl acetate	0.026	−0.88[a]	−3.01	−0.37	0.045
Vinyl benzoate	0.030[a]	−0.70[a]	−2.30	−0.20	0.045
Vinyl carbonate	0.000 73	−0.65			0.026
Maleic anhydride	0.23	2.25			
Vinylidene cyanide	20.13	2.58			

[a] Revised values from ref. 181.

The Q and e scheme has been the subject of considerable attention. It promised new possibilities of monomer classification, calculation of copolymerization parameters from known values of Q and e[†] and predictions about the behaviour of copolymerizing systems. However, to such ends the Q and e values should be known, and they are not directly measurable. In order to calculate them from copolymerization parameters, the easily copolymerized styrene has been selected as the reference monomer, with the assigned values $Q = 1$ and $e = -0.8$. Data on the Q and e factors of practically all copolymerizing monomers are now available (see Tables 3 and 3A); some kinetic significance is ascribed to monomer position in the $Q-e$ plane (see Fig. 21). Monomers with a high Q value are expected to form poorly reactive radicals with a low tendency to add further monomer units; monomers with widely differing e values usually copolymerize easily, etc.

Some systems actually behave as the $Q-e$ scheme predicts while other copolymerizations deviate from the proposed pattern. On the whole, the scheme is regarded with some misgivings at present. It is quite clear that a complicated chemical reaction, comprising the mutual interaction of many kinds of radicals and molecules in various media, can hardly be described in its entirety

---

[†] Knowing the copolymerization parameters r_1, r_2 and r_3 of the monomer pairs M_1-M_2 and M_1-M_3, the parameters of the pair M_2-M_3 can be calculated by means of Q_2, e_2 and Q_3, e_3.

by a simple relation with four variable parameters. By deliberate and suitable choice of these parameters, for certain reacting partners and conditions agreement between experiment and calculation can almost always be obtained. On the other hand, even an inaccurate theory is better than none. From this point of view, the $Q-e$ scheme is of great importance. It is constantly being refined and it remains the source for new hypotheses.

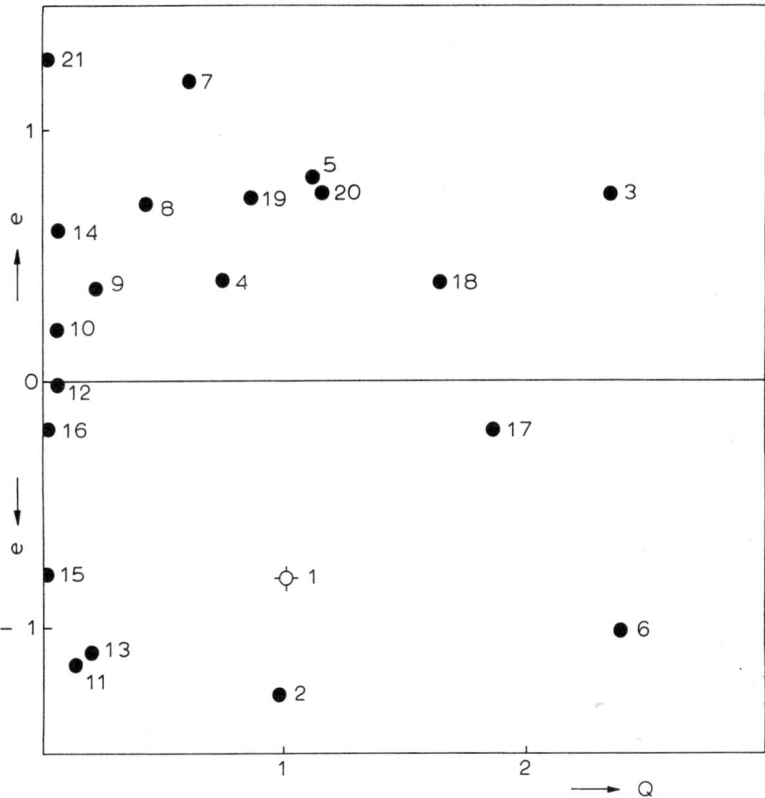

Fig. 21. Q, e plane. 1, S; 2, α-MeS; 3, methacrylic acid; 4, MMA; 5, methacrylonitrile; 6, B; 7, AN; 8, MA; 9, VDC; 10, VC; 11, N-vinylpyrrolidone; 12, VAC; 13, IB; 14, allyl chloride; 15, P; 16, E; 17, p-cyanostyrene; 18, p-nitrostyrene; 19, acrolein; 20, acrylic acid and 21, vinyl fluoride.

While proposing eqn. (84) the authors postulated a more general meaning of the quantities Q and e

$$Q = \exp(-q/RT)$$

$$e = \frac{\varepsilon}{(rDRT)^{1/2}} \quad (86)$$

References pp. 372–382

where q is a function of resonance stabilization of the new radical, ε is the substituent-induced charge on the two carbons participating in the generated covalent bond, i. e. in the growing radical and monomers, r is the distance between these charges in the transition complex, and D is the relative permitivity of the intervening medium [182]. An increase in Q by 0.1 means destabilization by 6.49 kJ mol$_\odot^{-1}$. The lowest measured value is $Q \approx 0.01$; therefore the value $q = 0$, was assigned to $Q = 0.01$, giving $q = -12.98$ kJ mol^{-1} for styrene. For the same monomer $e = -0.8$, $\varepsilon = -8 \times 10^{-19}$ C (coulombs).

Schwan and Price modified eqn. (85) to the form

$$r_1 = \exp\left(-\frac{q_1 - q_2}{RT}\right) \exp\left[\frac{A\varepsilon_1(\varepsilon_1 - \varepsilon_2)}{RT}\right]$$

$$r_2 = \exp\left(-\frac{q_2 - q_1}{RT}\right) \exp\left[\frac{A\varepsilon_2(\varepsilon_2 - \varepsilon_1)}{RT}\right]$$

$$r_1 r_2 = \exp\left[\frac{A(\varepsilon_1 - \varepsilon_2)^2}{RT}\right] \tag{87}$$

where A is a constant $\approx (rD)^{-1}$, and used them to calculate q and ε from experimental data. In this form q and ε are independent of temperature.

The changes in copolymerization parameters with temperature were first studied by Lewis et al. [183] who expressed r_i as a function of the activation entropies $S_{ii}^{\ddagger}$, $S_{ij}^{\ddagger}$ and enthalpies $H_{i2}^{\ddagger}$ and $H_{ij}^{\ddagger\dagger}$

$$r_i = \exp\left(\frac{S_{ii}^{\ddagger} - S_{ij}^{\ddagger}}{R} - \frac{H_{ii}^{\ddagger} - H_{ij}^{\ddagger}}{RT}\right) \tag{88}$$

According to eqn. (88), the product $r_1 r_2$ can be written as

$$r_1 r_2 = \exp\left[\frac{(H_{12}^{\ddagger} + H_{21}^{\ddagger}) - (H_{11}^{\ddagger} + H_{22}^{\ddagger})}{RT} - \frac{(S_{12}^{\ddagger} + S_{21}^{\ddagger}) - (S_{11}^{\ddagger} + S_{22}^{\ddagger})}{R}\right] =$$

$$= \exp(\Omega_{12}^{\ddagger}/RT - \chi_{12}^{\ddagger}/R) = \exp(\Delta G_{12}^{\ddagger}/RT) \tag{89}$$

---

† According to the Eyring theory of chemical reaction rates

$$k_{ii} = \frac{kT}{h} K_c^{\ddagger} = \frac{kT}{h} \exp\left(\frac{\Delta S_{ij}^{\ddagger}}{R} - \frac{\Delta H_{ij}^{\ddagger}}{RT}\right).$$

where $\Omega^{\neq}$ and $\chi^{\neq}$ are the differences in activation enthalpies and entropies, $\Delta G^{\neq}$ is the Gibbs activation energy. From eqn. (89) they derived its temperature change (assuming $d\Omega_{12}^{\neq}/dT = d\chi_{12}^{\neq}/dT = 0$)

$$\frac{\partial \ln r_1 r_2}{\partial T} = -\frac{\Omega_{12}^{\neq}}{RT^2} \qquad (90)$$

Depending on the sign of $\Omega_{12}^{\neq}$, the product of the copolymerization parameters will either increase or decrease with temperature. The values of $\Omega_{12}^{\neq}$ and $\chi_{12}^{\neq}$ can be determined experimentally by calculation or graphically from the logarithmic form of eqn. (89).

$$\ln r_1 r_2 = \frac{\Omega_{12}^{\neq}}{RT} - \frac{\chi_{12}^{\neq}}{R} \qquad (91)$$

A straight line in the coordinates $\ln r_1 r_2$ and T^{-1} has a slope of $\Omega_{12}^{\neq}/R$, and its intercept with the ordinate[†] at $T^{-1} = 0$ has the value of $-\chi_{12}^{\neq}/R$. When the functions $\Omega^{\neq}$ and $\chi^{\neq}$ are both positive or both negative, there must exist a temperature

$$T_n = \frac{\Omega_{12}^{\neq}}{\chi_{12}^{\neq}} \qquad (92)$$

TABLE 4

Values $\Omega_{12}^{\neq}$ and $\chi_{12}^{\neq}$ for three pairs of copolymerizing monomers [184]

System	$\Omega_{12}^{\neq}$ (kJ mol^{-1})	$\chi_{12}^{\neq}$ (J K^{-1} mol^{-1})
Styrene–methyl methacrylate [185]	−4.31	1.38
	−3.31 [186]	
Acrylonitrile–methyl methacrylate [187]	−5.78	1.93
Methacrylonitrile–styrene [188]	−4.52	−3.56

where $\Delta G_{12}^{\neq} = 0$. At this temperature $r_1 r_2 = 1$, and a statistical copolymer is formed. Some $\Omega^{\neq}$ and $\chi^{\neq}$ values are summarized in Table 4 for illustration. Negative $\Omega_{12}^{\neq}$ values mean that $r_1 r_2$ increases with temperature. The values of $\chi_{12}^{\neq}$ shown are very low[††]. Completely random ordering does not,

---

[†] Extrapolated, usually very inaccurate value.
[††] For most polymerizations, the entropy change is very similar, in the narrow range 105–126 J K^{-1} mol^{-1} [184]. With cyclic molecules it strongly depends on ring size, and with increasing n increases from negative to positive values (see Table 1).

References pp. 372–382

therefore, occur in the cited systems but it is approached with increasing temperature.

Similarly to their product, each of the copolymerization parameters can also be studied independently. Only very small or very large parameters will significantly change with temperature. A parameter smaller than 1 will increase with increasing temperature and vice versa. With increasing temperature the parameter value will always approach 1, and thus the tendency to random order will increase, with decreasing temperature the order will tend to alternation.

The Alfrey–Price scheme [178] assumes an influence of medium permittivity on the ratio of the addition rates of the two monomers. A great amount of data have been collected on insignificant solvent-induced changes in the course of radical copolymerizations. The situation is different with strongly basic or acid or ionizable monomers (acrylic and methacrylic acid derivatives, acrylamide, etc.), where the Gibbs energies of the initial and final states of the system are affected by solvent. In these cases the course of polymerization is changed by some solvents, for example those which can associate with the monomers by means of hydrogen bonds, dipole—dipole interactions or other intermolecular forces. The first to note the importance of these effects for the occurrence of copolymerization anomalies was Joshi [189].

5.2.3 THE DISTRIBUTION OF MONOMERIC UNITS IN THE COPOLYMER

The composition of instantaneously generated chains is not sufficiently described by the copolymerization equation which only informs us about the mean populations of monomers in a statistical copolymer. This problem was studied by Goldfinger and Kane [190] based on the following considerations.

Let the probability of monomer M_1 addition to a growing active centre $\sim M_1^\circ$ be P_{11}, and of monomer M_2 addition to the same centre P_{12}. The probatility of further M_1 addition to $\sim M_1 M_1^\circ$ will be a product of the partial probabilities, P_{11}^2, the fourth M_1 will be added with the probability P_{11}^3, the P_nth one with the probability P_{11}^{n-1}. The growth of the sequence λ_1 from the monomer M_1 will be stopped by an addition of M_2, with the probability P_{12}. The degree of polymerization of a mean M_1 sequence will be

$$\overline{\lambda}_1 = P_{12} + 2P_{11}P_{12} + 3P_{12}^2 + \ldots + nP_{11}^{n-1}P_{12} + \ldots$$

$$= \frac{P_{12}}{P_{11}} \sum_{n=1}^{\infty} nP_{11}^n = \frac{1 - P_{11}}{P_{11}} \sum_{n=1}^{\infty} nP_{11}^n = P_{12}^{-1\dagger} \quad (93)$$

and similarly

$$\overline{\lambda}_2 = P_{21}^{-1} \quad (97)$$

The probability of a certain event is given by the ratio of the number of favourable cases to the number of all possible cases. Thus

$$P_{12} = \frac{k_{12}[\sim M_1^\circ][M_2]}{k_{11}[\sim M_1^\circ][M_1] + k_{12}[\sim M_1^\circ][M_2]} = \frac{1}{r_1 y + 1} \tag{98}$$

$$P_{21} = \frac{k_{21}[\sim M_2^\circ][M_1]}{k_{22}[\sim M_2^\circ][M_2] + k_{21}[\sim M_2^\circ][M_1]} = \frac{y}{r_2 y + y} \tag{99}$$

The mean copolymer composition will be

$$n = \frac{\bar{\lambda}_1}{\bar{\lambda}_2} = \frac{P_{21}}{P_{12}} = \frac{r_1 y + 1}{(r_2/y) + 1}$$

It is seen to be given by the copolymerization eqn. (67), derived once more without any assumption of the existence of a stationary state. The whole procedure reminds us of the derivation of eqn. (67) from mean sequence lengths [see eqn. (68) and (69)]. The population of an M_1 sequence of degree of polymerization $\hat{P}_n$ can be determined by means of the relation

$$^1\hat{P}_n = P_{11}^{n-1} P_{12} = (1 - P_{12})^{n-1} P_{12} \tag{100}$$

and similarly for an M_2 sequence

$$^2\hat{P}_n = P_{22}^{n-1} P_{21} = (1 - P_{21})^{n-1} P_{21} \tag{101}$$

◀

---

† The composite series in egn. (93) is of the type $\sum\limits_{n=1} nx^n$. For $|x| < 1$

$$\sum_{n=1}^{\infty} nx^n = \frac{x}{1-1} \left(\frac{1-x^n}{1-x} - nx^n \right), \tag{94}$$

thus

$$\bar{\lambda}_1 = \frac{1-x}{x} \frac{x}{1-x} \left(\frac{1-x^n}{1-x} - nx^n \right) \tag{95}$$

and for $n \to \infty$

$$\lim_{n \to \infty} \bar{\lambda}_1 = (1-x)^{-1} = P_{12}^{-1}. \tag{96}$$

References pp. 372–382

after substitution from eqns. (98) and (99). The calculated data approach reality better the larger the number of sequences in the chain, i. e. the longer the macromolecules generated.

When the copolymerizing monomers have different parameters and when the concentrations of the two monomers are not kept constant (as is customary in continous reactors), the copolymer composition changes with conversion. This problem was treated mathematically by Skeist [167].

From the material balance for the more rapidly polymerizing monomer he derived a differential equation ($[M_1] + [M_2] = M$)

$$f_1 M \quad - (M - dM)(f_1 - df_1) = F_1 dM$$

amount of M_1 amount of reacted   amount of

in the charge   monomer M_1 M_1 in the copolymer

(102)

which, by omitting the second order differential, he transformed to

$$\frac{dM}{M} = \frac{df_1}{F_1 - f_1} \qquad (103)$$

and integrated

$$\ln \frac{M}{M_0} = \int_{(f_1)_0}^{f_1} \frac{df_1}{F_1 - f_1} \qquad (104)$$

either graphically, or analytically [after substituting for F_1 from eqn. (70)] [191][†]. Graphical integration is simpler. In the copolymerization graph we have the necessary data: F_1 (the copolymerization curve) and f_1 (the composition of the monomer mixture). Thus the curve $(F - f_1)^{-1}$ is easily constructed (see Fig. 22); the area below it is divided into the necessary number of vertical strips and their areas are determined. By this integration, the right–hand side of the Skeist equation (104) for a change in monomer mixture composition from $(f_1)_0$ to f_1 is found. From it, M/M_0 and the conversion $\varkappa$ are calculated. As an illustration, a space diagram can be constructed from the determined values with a perpendicular conversion axis attached to the origin of the copolymerization diagram (see Fig. 23).

[†] For azeotropic copolymerizations, $(F_1 - f_1)^{-1}$ is not continuous. Integrations has to be carried out from $f_1 = 0$ to $f_{1(az)}$ (f_1 at the azeotropic point) and from $f_{1(az)}$ to $f_1 = 1$.

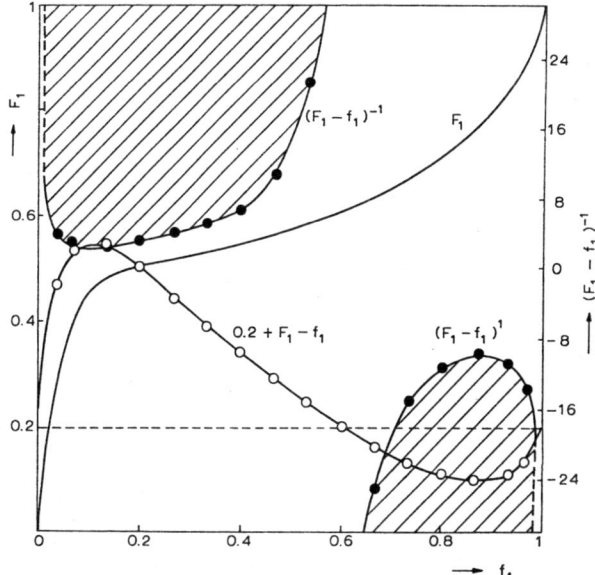

Fig. 22. Construction of the $(F_1 - f_1)^{-1}$ curve from the (S – AN) copolymerization diagram.

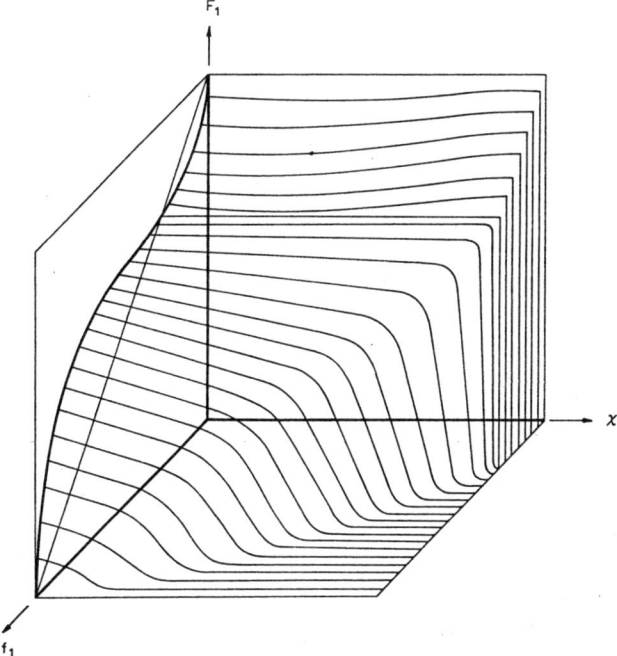

Fig. 23. Space diagram of the copolymerization surface. Copolymer composition as a function of the monomer mixture composition and conversion (S–AN).

References pp. 372–382

5.3 Ionic copolymerization

The theory of ionic copolymerization has been derived from the radical model. Strong electrostatic interactions are a source of deviations and complications. On the other hand, the effect of electrostatic forces in the neighbourhood of the reacting particles is so large that in most cases it overshadows other effects and enables copolymerization to be described by simple relations.

According to O'Driscoll, the rate constants of monomer addition to the end and to the other centre respectively are mutually directly proportional

$$k_{11} = ak_{21}$$

$$k_{12} = ak_{22} \qquad (105)$$

For a large number of ionic copolymerizations $a = 1$, and thus $r_1 r_2 = 1$ [192]. O'Driscoll observed that, for very dissimilar monomers, their ratio in the copolymer is directly proportional to a power of their concentration ratio in the monomer mixture [193]

$$\frac{d[M_1]}{d[M_2]} = \frac{k_1 k_{11}}{k_2 k_{22}} \left(\frac{M_1}{M_2}\right)^m \qquad (106)$$

where k_1 and k_2 are the rate coefficients of reactions

$$Ac^- + M_1 \xrightarrow{k_1} AcM_1^- \quad \text{and} \quad Ac^- + M_2 \xrightarrow{k_2} AcM_2^-,$$

respectively.

The value of the exponent in eqn. (106) lies in the range $1 \leq m \leq 2$. Monomers of widely differing polarity represent the extreme case with $m = 2$. This behaviour is observed, for example, with the pairs styrene—methyl methacrylate and acrylonitrile—methyl methacrylate. The other extreme, represented by the condition (105), was observed in copolymerizing monomers of very similar polarity with $m = 1$ (e. g. styrene—isoprene). The rate of co-addition generally increases with increasing temperature, and m decreases to 1, even for very dissimilar monomers (acrylonitrile—methyl methacrylate [194]).

A special kind of ionic copolymerizations is copropagation connected with ring opening.

Yamashita et al. have shown that the basicity of cyclic ethers, represented by pK_b, plays an important role [195]. The plot of $-\log r_1$ vs. pK_b of monomer M_2 is often linear. Nucleophilic attack of a cyclic ether on the cationic end is thus one of the driving forces of copolymerization. The slope of the straight line in this graph is a function of the centre selectivity. The same author observed a large difference between the kinetic and thermodynamic polymerizability of cyclic ethers. During copolymerization of trioxane with dioxolane, the amount of dioxolane in the product decreased with increasing conversion, temperature, and dilution. Thus dioxolane is kinetically more reactive than trioxane. Its polymerizability is, however, limited by thermodynamic factors (T_c) [196]. Yamashita et al. also studied the copolymerization of tetroxane with dioxolane. Tetroxane is consumed soon after the start of the reaction, and thus it is both kinetically and thermodynamically more reactive than dioxolane [197].

Tanaka [198] made an attempt to determine the contribution of ring strain and basicity to the reactivity of cyclic ethers in copolymerization by means of Gibbs polymerization energy. He derived a relation between the relative reactivities of an m-membered ring with i substituents to that of an n-membered cyclic ether with j substituents. He included a linear combination of their basicity differences, $\Delta(pK_b)_{m,\,i-n,\,j}$ and of Gibbs energy, $\Delta(\Delta G)_{m,\,i-n,\,j}$, in his relation $(a, b,$ and c are constants)

$$-\ln r_n = a\,\Delta(\Delta G)_{m,\,i-n,\,j} + b\,\Delta(pK_b)_{m,\,i-n,\,j} + c \tag{107}$$

Satisfactory agreement was observed.

5.4 Coordination copolymerization

Most kinetic studies of coordination copolymerization are limited to the determination of the copolymerization parameters. The possibility of simplifying assumptions which have been found useful in radical reactions is not self-evident in coordination copolymerizations. It is highly probable that copolymerization proceeds independently on the reaction centres of various activity.

In such cases, the mean copolymerization parameter values would be given by the ratio of the sums

$$\bar{r}_1 = \frac{\sum r_{1j}k_{12j}Z_j}{\sum k_{12j}Z_j},$$

$$\bar{r}_2 = \frac{\sum r_{2j}k_{12j}Z_j}{\sum k_{12j}Z_j} \tag{108}$$

where Z_j is the fraction of the j-type centre from all centre types, and r_{1j}, r_{2j}, and k_{12j} are the copolymerization parameters and coaddition rate constant at the j-type centre [199]. Monomers of structure similar to substituted styrenes deviate but little from the simple copolymerization scheme. In most cases significant, so far unexpained, discrepancies are observed.

5.5 Multicomponent copolymerization

When chains grow from more than two monomers, we speak of multicomponent copolymerization. In a simple case, the number of active centre types in the medium equals the number of copropagating monomers. Many systems are also known where the number of participating centres exceeds the number of monomers. In other cases, the monomers undergo complex formation so that their complexes take part as individual components in copropagation. Let us first pay attention to some cases generated by complications in nominally two-component systems.

5.5.1 THE PENULTIMATE UNIT EFFECT

When the reactivity of the centre is determined not only by the last added unit but also by the last but one unit, we speak of the penultimate effect. Merz et al. treated this problem using eight independent reactions [200, 201].

$$\sim M_1 M_1^\circ + M_1 \xrightarrow{k_{111}} \sim M_1 M_1 M_1^\circ$$

$$\sim M_1 M_1^\circ + M_2 \xrightarrow{k_{112}} \sim M_1 M_1 M_2^\circ$$

$$\sim M_2 M_2^\circ + M_2 \xrightarrow{k_{222}} \sim M_2 M_2 M_2^\circ$$

$$\sim M_2 M_2^\circ + M_1 \xrightarrow{k_{221}} \sim M_2 M_2 M_1^\circ \quad (109)$$

$$\sim M_2 M_1^\circ + M_1 \xrightarrow{k_{211}} \sim M_2 M_1 M_1^\circ$$

$$\sim M_2 M_1^\circ + M_2 \xrightarrow{k_{212}} \sim M_2 M_1 M_2^\circ$$

$$\sim M_1 M_2^\circ + M_2 \xrightarrow{k_{122}} \sim M_1 M_2 M_2^\circ$$

$$\sim M_1 M_2^\circ + M_1 \xrightarrow{k_{121}} \sim M_1 M_2 M_1^\circ$$

With the stationary state assumption, they derived and equation for copolymer composition ($r_1 = k_{111}/k_{112}$, $r_2 = k_{222}/k_{221}$, $r'_1 = k_{211}/k_{212}$, and $r'_2 = k_{122}/k_{121}$)

$$n = \frac{1 + [r_1 y(r_2 y + 1)/(r'_1 y + 1)]}{1 + [r'_2 y(r_2 + y)/(r'_2 + y)]} \tag{110}$$

When one of the copolymerizing monomers is unable to homopolymerize, eqn. (110) is simplified to

$$n - 1 = \frac{r'_1 y(r_1 y + 1)}{r'_1 y + 1} \tag{111}$$

The penultimate effect concept is useful for explaining many deviations from the simpler scheme, for example in the styrene copolymerization with fumaronitrile. Some doubts have been expressed concerning the physical sense of this concept. It is supposed only to increase the number of coefficients, thus rendering the copolymerization equation more fexible so that suitably selected coefficients yield better agreement with experiment. At present, however, the penultimate unit effect is regarded as firmly established. The evidence is largely based on studies of the type of bond between the added monomers and the growing end in coordination polymerizations; even the antepenultimate unit effect need not be negligible [202]. The directing effect of the penultimate unit is particularly prounounced in some types of ionic copolymerizations of polar monomers.

5.5.2 COPOLYMERIZATION WITH MONOMERS AND THEIR COMPLEXES

When donor—acceptor complexes are formed from the monomers, they can take part in copolymerization. When the equilibrium constants of complex formation are not extremely high, both complexes and monomers coexist and compete with active centres in the reaction. In addition, the reverse case may occur when one part of the active centres forms complexes with some component of the medium. the reactivity of the complexed centers is, of course, different from that of the free centres. The situation is formally similar to that of the preceding paragraph.

Let us present an illustrative example. Shaikhutdinov et al. observed that small amounts of naphthalene, anthracene or chrysene in a mixture of butyl vinyl ether and methyl methacrylate significantly increase the contents of the poorly reactive butyl vinyl ether in the product of radical copolymerization [203]. This is explained by the participation of eight additions types

$$
\begin{aligned}
\sim M_1^\circ + M_1 &\xrightarrow{k_{11}} \sim M_1^\circ \\
\sim M_1^\circ + M_2 &\xrightarrow{k_{12}} \sim M_2^\circ \\
\sim M_2^\circ + M_1 &\xrightarrow{k_{21}} \sim M_1^\circ \\
\sim M_2^\circ + M_2 &\xrightarrow{k_{22}} \sim M_2^\circ \\
\sim M_1^\circ\text{---}D + M_1 &\xrightarrow{k_1 D_1} \sim M_1^\circ \\
\sim M_1^\circ\text{---}D + M_2 &\xrightarrow{k_1 D_2} \sim M_2^\circ \\
\sim M_2^\circ\text{---}D + M_1 &\xrightarrow{k_1 D_1} \sim M_1^\circ \\
\sim M_2^\circ\text{---}D + M_2 &\xrightarrow{k_2 D_2} \sim M_2^\circ
\end{aligned}
\qquad (112)
$$

where D represents the electron donor molecule, the aromatic hydrocarbon. The solution is formally similar to the solution of scheme (109) or better eqn. (110) because vinyl ethers are not radically polymerized.

5.5.3 TERNARY COPOLYMERIZATION AND COPOLYMERIZATION OF MORE THAN THREE MONOMERS

The mathematical description of the copropagation of more than two monomers is quite similar to the binary copolymerization scheme. Alfrey and Goldfinger [204] based their analysis on nine addition types

$$
\begin{aligned}
\sim M_1^\circ + M_1 &\xrightarrow{k_{11}} \sim M_1 M_1^\circ \\
\sim M_1^\circ + M_2 &\xrightarrow{k_{12}} \sim M_1 M_2^\circ \\
\sim M_1^\circ + M_3 &\xrightarrow{k_{13}} \sim M_1 M_3^\circ \\
\sim M_2^\circ + M_1 &\xrightarrow{k_{21}} \sim M_2 M_1^\circ
\end{aligned}
$$

$$\sim M_2^\circ + M_2 \xrightarrow{k_{22}} \sim M_2 M_2^\circ$$

$$\sim M_2^\circ + M_3 \xrightarrow{k_{23}} \sim M_2 M_3^\circ$$

$$\sim M_3^\circ + M_1 \xrightarrow{k_{31}} \sim M_3 M_1^\circ \qquad (113)$$

$$\sim M_3^\circ + M_2 \xrightarrow{k_{32}} \sim M_3 M_2^\circ$$

$$\sim M_3^\circ + M_3 \xrightarrow{k_{33}} \sim M_3 M_3^\circ$$

The rate of monomer consumption is expressed by the differential equations

$$-\frac{d[M_1]}{dt} = k_{11}[\sim M_1^\circ][M_1] + k_{21}[\sim M_2^\circ][M_1] + k_{31}[\sim M_3^\circ][M_1]$$

$$-\frac{d[M_2]}{dt} = k_{12}[\sim M_1^\circ][M_2] + k_{22}[\sim M_2^\circ][M_2] + k_{32}[\sim M_3^\circ][M_2]$$

$$-\frac{d[M_3]}{dt} = k_{13}[\sim M_1^\circ][M_3] + k_{23}[\sim M_2^\circ][M_3] + k_{33}[\sim M_3^\circ][M_3]$$

$$(114)$$

An exact solution of scheme (114) would require the knowledge of various active centre concentrations, $[\sim M_1^\circ, \sim M_2^\circ,$ and $\sim M_3^\circ]$. These data are unknown even for the simplest systems. Therefore use must again be made of the stationary state assumption[†] yielding

$$k_{12}[\sim M_1^\circ][M_2] + k_{13}[\sim M_1^\circ][M_3] = k_{21}[\sim M_2^\circ][M_1] + k_{31}[\sim M_3^\circ][M_1]$$

$$k_{21}[\sim M_2^\circ][M_1] + k_{23}[\sim M_2^\circ][M_3] = k_{12}[\sim M_1^\circ][M_2] + k_{32}[\sim M_3^\circ][M_2]$$

$$k_{31}[\sim M_3^\circ][M_1] + k_{32}[\sim M_3^\circ][M_2] = k_{13}[\sim M_1^\circ][M_3] + k_{23}[\sim M_2^\circ][M_3]$$

$$(115)$$

[†] When evaluating the results of ternary copolymerizations by means of the Alfrey–Goldfinger scheme, the stationary character of copropagation should be critically established. With non-stationary processes, the uncertainty of interpretation becomes more serious, even when the experimental results agree with theory.

Substitution from eqns. (115) into eqns. (114) and rearrangement produces a relation between the compositions of the copolymer and of the monomer phase in which $r_{12} = k_{11}/k_{12}$, $r_{13} = k_{11}/k_{13}$, $r_{21} = k_{21} = k_{22}/k_{21}$, $r_{23} = r_{23} = k_{22}/k_{23}$, $r_{31} = k_{33}/k_{31}$, and $r_{32} = k_{33}/k_{32}$.

$$d[M_1]:d[M_2]:d[M_3] = \dot{m}_1:m_2:m_3 =$$

$$= [M_1]\left\{\frac{[M_1]}{r_{31}r_{21}} + \frac{[M_2]}{r_{21}r_{32}} + \frac{[M_3]}{r_{31}r_{23}}\right\}\left\{[M_1] + \frac{[M_2]}{r_{12}} + \frac{[M_3]}{r_{13}}\right\}:$$

$$[M_2]\left\{\frac{[M_1]}{r_{12}r_{31}} + \frac{[M_2]}{r_{12}r_{32}} + \frac{[M_3]}{r_{32}r_{13}}\right\}\left\{[M_2] + \frac{[M_1]}{r_{21}} + \frac{[M_3]}{r_{23}}\right\}: \quad (116)$$

$$[M_3]\left\{\frac{[M_1]}{r_{13}r_{21}} + \frac{[M_2]}{r_{23}r_{12}} + \frac{[M_3]}{r_{13}r_{23}}\right\}\left\{[M_3] + \frac{[M_1]}{r_{31}} + \frac{[M_3]}{r_{32}}\right\}$$

It may be true that many terpolymerizations are well described by these relations, for example the radical copolymerization of styrene, methyl methacrylate and acrylonitrile or vinyl chloride [205, 206] but the uncertainty of the basic assumptions (stationary state, non existence of monomer or active centre complexes, etc.) reduces the meaning of relation (116) to a mere illustration of a complicated copropagation. No way is known to derive the values of the six constants involved independently. Unless they are determined by independent procedures, it is very probable that a good agreement between experiment and the tested relation will be obtained by a suitable choice of these constants. The value of such an agreement should be regarded with caution.

The expression (116) is considerably simplified when one or two of the monomers are not capable of self-propagation. When M_2 and M_3 do not polymerize it holds that

$$\frac{m_1}{m_2} = 1 + \frac{r_2[M_1]}{[M_2]} + \frac{r_{12}[M_3]}{r_{13}[M_2]}$$

$$\frac{m_2}{m_3} = \frac{r_{13}[M_2]}{r_{12}[M_3]}$$

$$\frac{m_1}{m_3} = 1 + \frac{r_{13}[M_1]}{[M_3]} + \frac{r_{13}[M_2]}{r_{12}[M_3]} \quad (117)$$

An independent determination of the constants in the Alfrey–Goldfinger terpolymerization equation was attempted by Ham [207]. His approach is based on an assumption about the mode of sequence ordering in terpolymers and in polymers with a higher number of components.

Ham assumes that the formation probability of the sequence $\sim M_1 M_2^\circ (\sim M_1 M_3^\circ, \sim M_2 M_3^\circ)$ is equal to the formation probability of the opposite sequence $\sim M_2 M_1^\circ (\sim M_3 M_1^\circ, \sim M_3 M_2^\circ)^\dagger$. In a ternary system, the probability of monomer M_2 addition to the active centre $\sim M_1^\circ$ will be given by the ratio of this addition rate to the sum of the rates of all possible addition types

$$P_{12} = \frac{k_{12}[\sim M_1^\circ][M_2]}{k_{11}[\sim M_1^\circ][M_1] + k_{12}[\sim M_1^\circ][M_2] + k_{13}[\sim M_1^\circ][M_3]}$$
$$= \frac{[M_2]/r_{12}}{[M_1] + ([M_2]/r_{12}) + ([M_3]/r_{13})} \qquad (118)$$

and similarly

$$P_{21} = \frac{[M_1]/r_{21}}{([M_1]/r_{21}) + [M_2] + ([M_3]/r_{23})}$$

$$P_{23} = \frac{[M_3]/r_{23}}{[M_1]r_{21} + [M_2] + [M_3]/r_{23}} \quad \text{etc.}$$

The complicated expression (116) can then be simplified to

$$\frac{m_1}{m_2} = \frac{P_{21}P_{31} + P_{32}P_{21} + P_{23}P_{31}}{P_{12}P_{32} + P_{31}P_{12} + P_{13}P_{32}} \qquad (119)$$

$$\frac{m_1}{m_3} = \frac{P_{21}P_{31} + P_{32}P_{21} + P_{23}P_{31}}{P_{13}P_{23} + P_{21}P_{13} + P_{12}P_{23}} \qquad (120)$$

The numerator in both these equations is identical. It expresses the probability of sequence λ_1 generation in the chains. The denominators correspond to

---

$\dagger$ The addition probability of λ_2, λ_3 and λ_1 sequences at $\sim M_1^\circ$, $\sim M_2^\circ$ and $\sim M_3^\circ$ centres is equal to the termination probability of the same sequences by M_1, M_2 and M_3 units.

References pp. 372–382

the generation of sequences λ_2 and λ_3. When the probabilities of sequence generation from left to right are the same as from right to left, it holds that

$$P_{12}P_{23}P_{31} = P_{13}P_{32}P_{21} \tag{121}$$

The expressions (119) and (120) can be greatly simplified by means of eqn. (121). It is sufficient to add the product $P_{21}P_{12}(P_{32} + P_{31}) = P_{12}P_{21}(P_{31} + P_{32})$ to each side of (121). The result and comparison with eqn. (119) yields the relation

$$\frac{m_1}{m_2} = \frac{P_{21}}{P_{12}} \tag{122}$$

Similarly, addition of the product $P_{13}P_{31}(P_{23} + P_{21})$ to both sides of eqn. (121) yields for m_1/m_3

$$\frac{m_1}{m_3} = \frac{P_{31}}{P_{13}} \tag{123}$$

By means of eqn. (122) and (123) terpolymer composition can be very simply predicted, essentially from the binary relations. The latter expressions were tested experimentally and satisfactory agreement was found [169].

Ham also observed that the product

$$P_{12}P_{23}P_{31} = P_{13}P_{32}P_{21} = \mathscr{P} \tag{121a}$$

exhibits a constant value for conjugated monomers (styrene, butadine, methyl methacrylate, etc.) or for a mixture of conjugated with non-conjugated monomers (vinyl chloride, vinyl acetate, etc.). For and equimolar monomer ratio, $\mathscr{P} = 0.037$ in the first case 0.006 in the second. This enables us to calculate the binary copolymerization parameters of a new monomer with any known partner under a single condition: we must know that its bonds are conjugated.

For this calculation we use a relation obtained by transformation of eqn. (121a)

$$\frac{r_{13}}{r_{13} + r_{12}r_{13} + r_{12}} \frac{r_{21}}{r_{21} + r_{21}r_{23} + r_{23}} \frac{r_{32}}{r_{32} + r_{32}r_{31} + r_{31}} =$$

$$= \frac{r_{12}}{r_{12} + r_{12}r_{13} + r_{13}} \frac{r_{31}}{r_{31} + r_{31}r_{32} + r_{32}} \frac{r_{23}}{r_{23} + r_{23}r_{21} + r_{21}} = \tag{124}$$

$$= 0.037 \text{ or } 0.006$$

Equation (121a) can also be regarded as a product of the probabilities, p, characteristic of the individual monomers. For two conjugated and one non-conjugated monomer, it holds that

$$p_1 p_2 p_3 = \mathscr{P} \qquad (125)$$

For two non-conjugated and one conjugated monomer

$$p_1 = \frac{\mathscr{P}}{p_2 p_3} \frac{0.006}{0.037}$$

and for the so far little studied group of three non-conjugated monomers

$$p_1 = \frac{\mathscr{P}}{p_2 p_3} \left(\frac{0.006}{0.037}\right)^3 \qquad (125b)$$

Values of p are summarized in Table 3A; values of p higher than 0.25 are typical of conjugated monomers.

Strictly speaking, the exact theoretical background controlling multicomponent propagation is as yet unknown. We do not therefore know how to estimate the frequency of chance in the more or less empirical relations presented in this paragraph. The agreement with experiment is rather surprising. The existence of some hidden rules directing the composition of multicomponent copolymers is suggested by a further extension of eqn. (121) or of its modification, eqn. (124). It yields the relation

$$r_{13} r_{32} r_{21} = r_{12} r_{23} r_{31} \qquad (126)$$

which can be derived from the Alfrey–Price scheme by substituting Q and e values for every rate constant ($r_{13} = k_{11}/k_{13}$, $r_{32} = k_{33}/k_{32}$, $r_{21} = k_{22}/k_{21}$, etc.)

$$k_{ij} = P_i Q_j \exp(-e_i e_j) \qquad (84a)$$

Demonstration of the agreement between eqn. (121), (126), and an approach based on completely different assumptions prompted Ham to their extension

$$P_{12} P_{23} P_{34} P_{41} = P_{14} P_{43} P_{32} P_{21} \qquad (127)$$

In eqn. (127) each probability P is defined by the ratio of favourable additions to the number of all additions to the active centre

$$P_{12} = \frac{k_{12}[M_1^\circ][M_2]}{k_{12}[M_1^\circ][M_2] + k_{11}[M_1^\circ][M_1] + k_{13}[M_1^\circ][M_3] + k_{14}[M_1^\circ][M_4]}$$

$$= \frac{[M_2]/r_{12}}{([M_2]/r_{12}) + [M_1] + ([M_3]/r_{13}) + ([M_4]/r_{14})} \qquad (128)$$

Agreement was again found in the copolymerization in the styrene—methyl methacrylate—acrylonitrile—vinylidene chloride system. Thus there are no serious objections to the validity of the equation

$$r_{14}r_{43}r_{32}r_{21} = r_{12}r_{23}r_{34}r_{41} \tag{129}$$

This simplifies predictions of product composition in multicomponent copropagations; the results from binary copolymerizations can be used in the calculation

$$\frac{m_1}{m_2} = \frac{P_{21}}{P_{12}} \tag{122}$$

$$\frac{m_1}{m_3} = \frac{P_{31}}{P_{13}} \tag{123}$$

$$\frac{m_i}{m_j} = \frac{P_{ji}}{P_{ij}} \tag{130}$$

Living polymerizations are kinetically much simpler than the cases discussed so far. Szwarc and Perrin derived a general solution for living copolymerizations of two and more monomers provided each addition is reversible [208]. The calculation is very elegant and permits and insight into the course of a complicated reaction; it leads to the determination of the equilibrium monomer concentration, of the molecular mass distribution in the generated chains, of the mean copolymer composition at the degree of polymerization n, and of the placement probability of a certain monomeric unit in the m-th segment of an n-mer. The initial concentrations of the monomers and of the initiating particles, as well as the equilibrium constants of the additions must be known. The approach of Szwarc and Perrin will be described in Chap. 8, Sect. 1.3.

5.6 Thermodynamics of copolymerization

Let us consider the generation of long chains of a statistical binary copolymer at low conversions. The reaction of the active centre with the monomer exhibits the characteristic molar heat of reaction H and molar entropy change S

$$\sim M_i^\circ + M_j \xrightarrow{k_{ij}} \sim M_i M_j^\circ \qquad H_{ij}, S_{ij} \tag{131}$$

The molar heat of copolymerization ΔH will be given by the sum of the heat contributions of the individual aditions[†]

$$-\Delta H = H_{11}F_1 + H_{22}F_2 + F_1F_2[H_{12} + H_{21} - (H_{11} + H_{22})] =$$
$$= H_{11}F_1 + H_{22}F_2 + F_1F_2\Omega \qquad (132)$$

Under these conditions the entropy change will be described by the equation [209]

$$\Delta S = -\{S_{11}F_1 + S_{22}F_2 + F_1F_2[(S_{12} + S_{21}) - (S_{11} + S_{22})]\} =$$
$$= S_{11}F_1 + S_{22}F_2 + F_1F_2\chi \qquad (133)$$

In more general formulations of eqns. (132) and (133) their last term right-hand side is sometimes multiplied by the randomness coefficient ψ, defined as $\psi = P_{21}/F_1 = P_{12}/F_2$. ψ refines the relation between F_1F_2 and the population of the —M_1—M_2— pairs in copolymers of various types.

The entropy change will, of course, also contain the following contributions

(a) the residual entropy at 0 K,
(b) the entropy of configurational changes in the chain, and
(c) the entropy of unit placement in the chain.

According to Orr [210] the residual entropy is negligible as is, according to Temperley [211], the configurational entropy. Sawada [212] described the entropy of copolymerization randomness (unit placement in the chain) by the relation

$$\Delta S_n = -R[P_{11} \ln P_{11} + (1 - P_{11}) \ln (1 - P_{11})] =$$
$$= -R[(1 - \Psi F_2) \ln (1 - \Psi F_2) + \Psi F_2 \ln \Psi F_2] \qquad (134)$$

[†] The content of M_1 and M_2 in the copolymer is equal to F_1 and F_2. Incorporation of the unit ←M→ between M_1 or M_2 units libetares or absorbs the corresponding heat

$M_1\to$	$\leftarrow M_1\to$	$\leftarrow M_1\to$	$\leftarrow M_1\to$	$\leftarrow M_2\to$	$\leftarrow M_2\to$	$\leftarrow M_2\to$	$\leftarrow M_1\to$	$\leftarrow M_1\to$	$\leftarrow M_2\to$
H_{11}	H_{11}	H_{11}	H_{12}	H_{22}	H_{22}	H_{21}	H_{11}	H_{12}	

In simple cases of statistical copolymerizations, F_1F_2 intersequence bonds occur per single intermonomer bond. The difference of the products $H_{11}F_1$ and $H_{11}F_1F_2$ is thus a measure of the heat effect of bond formation within M_1 sequences. Also the enthalpy of internal bond formation in M_2 seguences is smaller than $H_{22}F_2$ by the amount corresponding to the formation of the terminal bonds, $H_{22}F_1F_2$, which have not been formed. They were substituted by intersequence bonds, of enthalpy $F_1F_2(H_{12} + H_{21})$. The overall heat effect of copropagation is thus given by eqn. (132).

References pp. 372–382

According to Orr, $\Delta S_n = 2 - 6.3\ J\ K^{-1}$; it rarely amounts to more than 5 % of the total entropy change in copolymerization which can therefore be expressed by the relation [212]

$$\Delta S = -(S_{11}F_1 + S_{22}F_2 + \chi \Psi F_1 F_2) - R[(1 - \Psi F_2)\ln(1 - \Psi F_2) + \Psi F_2 \ln \Psi F_2] \tag{135}$$

The sum of eqn. (132) and $T\Delta S$ from eqn. (135) represents the Gibbs energy of the reaction as a function of copolymer composition [212]

$$\Delta G = \Delta H - T\Delta S =$$
$$= -(H_{11}F_1 + H_{22}F_2 + F_1 F_2 + F_1 F_2 \Omega \psi + T(S_{11} + F_1 + S_{22}F_2 + \chi \psi F_1 F_2) + RT[(1 - \psi F_2)\ln(1 - \psi F_2) + \psi F_2 \ln \psi F_2] \tag{136}$$

Even though some inaccuracy will be involved, let us assume for the sake of simplicity that ψ is independent of F_2 (ψ is actually independent of F_2 when statistical or alternating copolymers are formed. For $F_2 \to 0$, ψ may be close to 1 even in the remaining cases). Differentiating ΔG with respect to F_2 at constant temperature we obtain

$$\frac{\partial \Delta G}{\partial F_2} = (H_{11} - H_{22}) - T(S_{11} - S_{22}) - \psi(1 - 2F_2)(\Omega - \chi T) + RT\psi \ln\left(\frac{\psi F_2}{1 - \psi F_2}\right) \tag{137}$$

Close to $F_2 \to 0$, we may set $(1 - \psi F_2) \approx 1$. Then

$$\frac{\partial \Delta G}{\partial F_2} = (H_{11} - H_{22}) - T(S_{11} - S_{22}) - \psi(1 - 2F_2)(\Omega - \chi T) + RT \ln \psi F_2 \tag{137a}$$

As

$$\lim_{F_2 \to 0} (\ln \psi F_2) = -\infty,$$

the quantity $\partial \Delta G/\partial F_2$ will be negative at sufficiently small values of F_2. Therefore the introduction of the first comonomer units into the homopolymer is always accompanied by a reduction in the Gibbs energy of the system. In the presence of the comonomer, the homopolymer is always thermodynamically unstable[†].

Let us further consider the general case where Ω is sufficiently large and positive. According to the measurements of Orr, in this case χ may be considered as negligible [210]. At a temperature T_1, sufficiently far below the T_c of any component, the change in ΔG with F_2 will be represented by a curve which is convex towards the F_2 axis. This is shown schematically in Fig. 24. At a temperature T_2, above T_c for both components, the curves will be shifted as shown in Fig. 25. The ceiling temperature exhibits a maximum at a certain copolymer composition. The highest T_c should be observed when alternating copolymers are generated (copolymerization of 2-methylpropene with SO_2, for example [213]). The plot of ΔG vs. F_2 should exhibit a minimum at

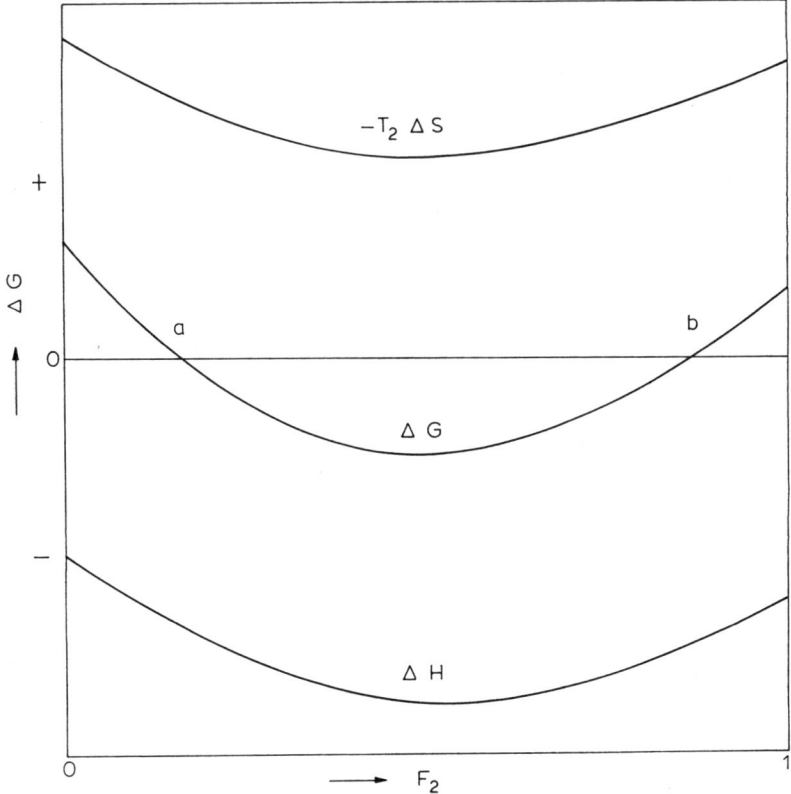

Fig. 24. Dependence of ΔG on F_2 at $T = T_1$. See p. 312 of ref. 212.

[†] The comonomer need not be an intentionally added monomer. A copolymerizing impurity, e. g. oxygen, can have the same effect.

References pp. 372–382

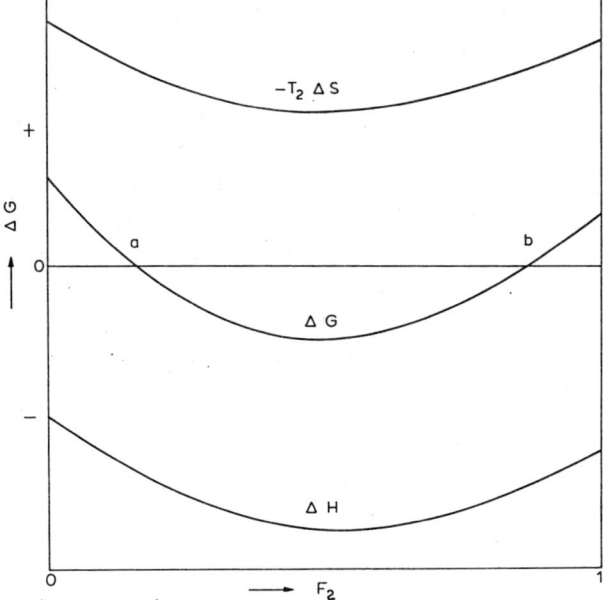

Fig. 25. Dependence of ΔG on F_2 at $T_2 > T_1$. See p. 313 of ref. 212.

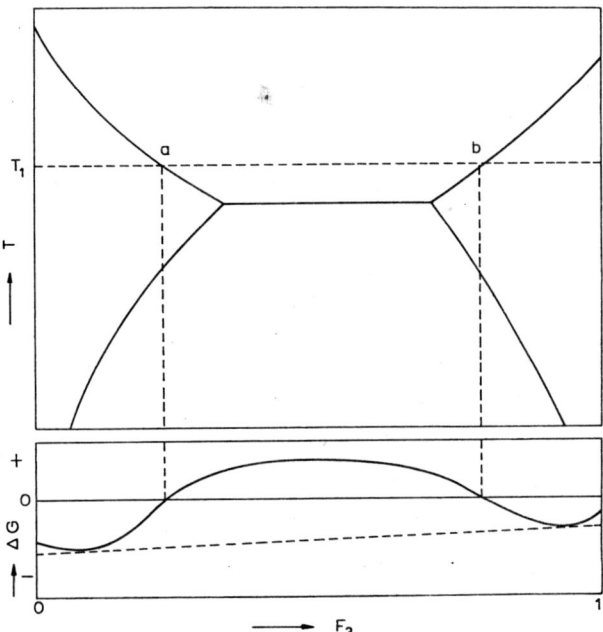

Fig. 26. Dependence of ΔG on F_2. Range of thermodynamic copolymer stability. $\Omega \ll 0$. See p. 315 of ref. 212.

$\Omega \approx 0$. A T_c maximum was actually observed in the copolymerization of acetaldehyde with propionaldehyde [214].

Let us now consider situation with $\Omega \ll 0$. The dependence of Gibbs energy on composition will exhibit two minima (Fig. 26); with increasing F_2 copolymer will pass from the stable range $(0 < F_2 < a)$ into an unstable one $(a < F_2 < b)$ and back into a stable range $(F_2 > b)$. The system will have a T_c minimum. Copolymerizations of vinyl and cyclic monomers present examples of such behaviour. The interactions of these compounds are of repulsive character; coaddition is difficult to achieve. The product is a block copolymer [214] (for example styrene-β-propiolactone [215]). At only weakly negative ω values, the minimum on the T_c–F_2 curve will not occur (see Fig. 27). Two local minima will, however, appear in the ΔG vs. F_2 plot.

A copolymer of composition d and e will be thermodynamically more stable than c. By dissociation of copolymer c to the more stable d and e, the Gibbs energy of the system will be reduced. For $\chi \leqslant 0$ and $\Omega < 0$, the

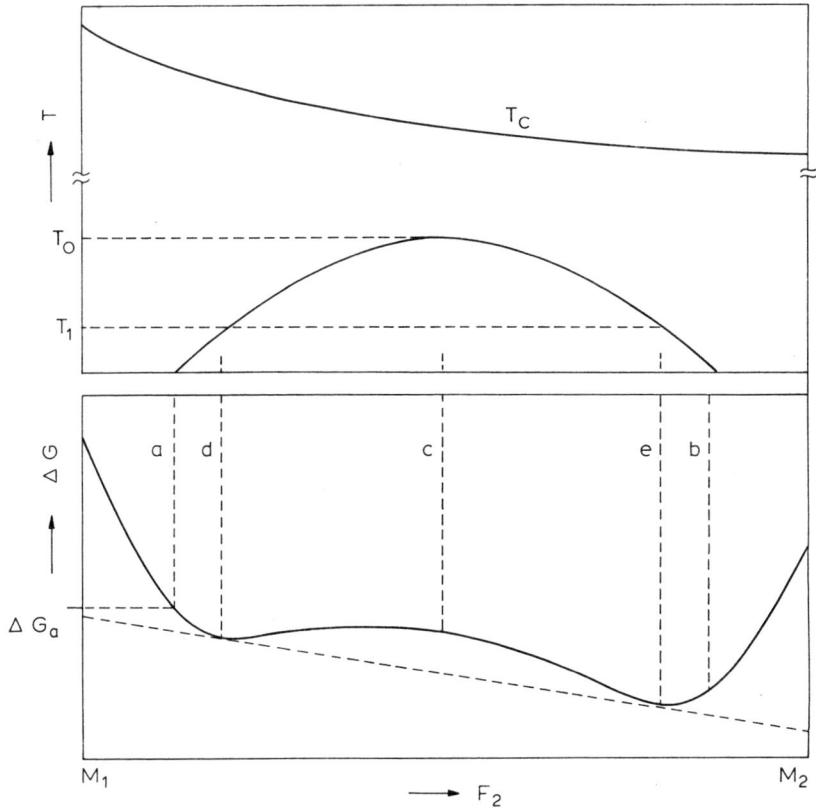

Fig. 27. Dependence of ΔG on F_2. Range of thermodynamic copolymer stability. See p. 316 of ref. 212.

References pp. 372–382

copolymer in the range $d < F_2 < e$ should always be unstable, and two copolymers of different composition should be generated.

This case has not, however, been experimentally observed. This may be because of kinetic factors, particularly slow bond exchange between the two chain types.

5.6.1 COPOLYMERIZATION OF ONE OF THE MONOMERS AT TEMPERATURES ABOVE T_c

Anomalies in the course of copolymerizations which have been ascribed to penultimate unit effects can alternatively be explained by the reversibility of one or several addition types.

TABLE 5

Depropagation during copolymerization. Solved cases

Equilibrium stage		Case					
		1[a]	2[a]	3[a]	4[b]	5[c]	6[d]
$\sim M_1M_1M_1^\circ \underset{}{\overset{K_{111}}{\rightleftharpoons}} \sim M_1M_1^\circ + M_1$							+
$\sim M_1M_2M_1^\circ \underset{}{\overset{K_{121}}{\rightleftharpoons}} \sim M_1M_2^\circ + M_1$						+	
$\sim M_2M_2M_1^\circ \underset{}{\overset{K_{221}}{\rightleftharpoons}} \sim M_2M_2^\circ + M_1$				+		+	
$\sim M_2M_2M_2^\circ \underset{}{\overset{K_{222}}{\rightleftharpoons}} \sim M_2M_2^\circ + M_2$		+	+	+	+	+	+
$\sim M_1M_2M_2^\circ \underset{}{\overset{K_{122}}{\rightleftharpoons}} \sim M_1M_2^\circ + M_2$		+			+	+	
$\sim M_2M_1M_2^\circ \underset{}{\overset{K_{212}}{\rightleftharpoons}} \sim M_2M_1^\circ + M_2$						+	
$\sim M_1M_1M_2^\circ \underset{}{\overset{K_{112}}{\rightleftharpoons}} \sim M_1M_1^\circ + M_2$						+	

[a] See refs. 216 and 220.
[b] See refs. 217 and 220.
[c] See ref. 218.
[d] See ref. 219.

This solution was first suggested by Lowry [216] and later by Razell and Ivin [217], Yamashita et al. [218], and Wittmer [219]. These authors tried to solve the copolymerization equation for a total of six possibilities (see Table 5). For case 1, the copolymer composition is given by the equation

$$\frac{d[M_2]}{d[M_1]} = \frac{[M_2](1-\alpha)^{-1}}{r_1[M_1] + [M_2]} \tag{138}$$

where

$$\alpha = \frac{1}{2}\left\{\left(1 + K_{122}[M_2] + \frac{K_{122}[M_1]}{r_2}\right) - \right.$$

$$\left. - \left[\left(1 + K_{122}[M_2] + \frac{K_{122}[M_1]}{r_2}\right)^2 - 4K_{122}[M_2]\right]^{1/2}\right\}^{\dagger} \quad (139)$$

A copolymerization conforming to eqn. (138) is manifested by several features. Qualitatively

(a) temperature will strongly affect the shape of the copolymerization curve in the vicinity of the ceiling temperature, T_c, of the monomer M_2;

(b) dilution of monomers by an inert solvent will deform the copolymerization curves already at temperatures below T_c; and

(c) in copolymers prepared at high temperatures, F_2 will be limited to 1/2.

For case 2, the copolymerization equation was derived in the form

$$\frac{d[M_2]}{d[M_1]} = \frac{\alpha v - 1 + (1 - \alpha)^{-2}}{(r_1[M_1]/[M_2] + 1)\{\alpha v + [\alpha/(1 - \alpha)]\}}$$

$$v = \frac{K_{1222}[M_2] + (K_{1222}[M_1]/r_2) - \alpha}{K_{1222}[M_2]} \quad (140)$$

Qualitatively it is manifested by the same features as in case 1. F_2 will be limited to 2/3 rather than to 1/2.

Studies by many authors, e. q. on copolymerizations of styrene with α-methylstyrene (characterized by low T_c, 334 K), appear to agree with the ideas of Lowry. Some author claim, however, that even copolymerization of this pair can be described by the simple copolymerization equation [221]. Johnston and Rudin are of the opinion that the depropagation reaction is not as important in this case because only short sequences of α-methylstyrene are produced. The formation of short blocks is accompanied by relatively high polymerization enthalpy. They are thermodynamically more stable than the homopolymer and have a higher T_c. Only at considerably higher copolymerization temperatures (with the pair styrene—α-methylstyrene >420 K) does the depropagation effect become important.

---

† K_{122} is defined by the equilibrium

$$\text{mmM}_1\text{M}_2\text{M}_2^\circ \underset{}{\overset{K_{122}}{\rightleftharpoons}} \text{mmM}_1\text{M}_2^\circ + M_2.$$

K_{1222} in v from eqn. (140) for an λ_2 sequence longer by one unit is defined similarly.

The remaining (and more complicated) cases from Table 6 are described by complicated differential equations [212]. Their complexity rests in the number of parameters, coefficients, and constants rather than in the mathematical expression. They are therefore difficult to verify.

For living equilibrium copolymerizations, Szwarc and Perrin [208] derived a theoretical procedure, enabling them to obtain from the initial state the most extensive information on the generated copolymers so far (see Chap. 8, Sect. 1.3).

5.7 Copolymerization kinetics

Calculations of copolymer composition are based on kinetic considerations and procedures. In spite of this, less attention has been paid to the copropagation rate than to other copolymerization problems. Today a single concise theory is available, solving the rate of the simplest radical binary copolymerization. Other cases described have not been generalized so far; they treat the kinetic behaviour of specific monomer pairs or triplets in specific polymerization circumstances.

Melville et al. [222] started from the copolymerization scheme (62), extended to include initiation (141) and termination by mutual combination (142). $M^*_{1(r)}$, $M^*_{2(r)}$ is a copolymer chain of r monomeric units with an M_1 (or M_2) radical at the end. A chain containing a number of units different from r is designated by the index s.

$$M_1 \rightarrow M^*_{1(1)} \quad \text{initiation of rate } R_1 \tag{141}$$

$$M_2 \rightarrow M^*_{2(1)} \quad \text{initiation of rate } R_2$$

$$\begin{aligned}
M^*_{1(r)} + M_1 &\xrightarrow{k_{11}} M^*_{1(r+1)} \\
M^*_{1(r)} + M_2 &\xrightarrow{k_{12}} M^*_{2(r+1)} \\
M^*_{2(r)} + M_1 &\xrightarrow{k_{21}} M^*_{1(r+1)} \\
M^*_{2(r)} + M_2 &\xrightarrow{k_{22}} M^*_{2(r+1)} \\
M^*_{1(r)} + M^*_{1(s)} &\xrightarrow{k_{(11)t}} \text{non-radical products} \\
M^*_{1(r)} + M^*_{2(s)} &\xrightarrow{k_{(12)t}} \text{non-radical products} \\
M^*_{2(r)} + M^*_{2(s)} &\xrightarrow{k_{(22)t}} \text{non-radical products}
\end{aligned} \tag{62a}$$

The consumption rate of both monomers is given by the sum of the rates in eqn. (63) and (64) (for long chains, monomer consumption by initiation is negligible)

$$-\frac{d([M_1] + [M_2])}{dt} = k_{11}[M_1^*][M_1] + k_{21}[M_2^*][M_1] +$$

$$+ k_{12}[M_1^*][M_2] + k_{22}[M_1^*][M_2] \quad (143)$$

For stationary copolymerizations, this can be transformed by means of eqn. (65) and of the copolymerization parameters to the simpler form

$$-\frac{d([M_1] + [M_2])}{dt} = \frac{k_{12}[M_1^*]}{[M_1]}(r_1[M_1]^2 + 2[M_1][M_2] + r_2[M_2]^2) \quad (144)$$

With simplifying assumptions (stationary state and termination by mutual radical combination) the radical concentration $[M_1^*]$ can be derived. The generation rate of each radical type is equal to its rate of termination

$$R_1 = k_{(11)t}[M_{1(s)}^*][M_{1(r)}^*] + k_{(12)t}[M_{1(r)}^*][M_{2(s)}^*]$$

$$R_2 = k_{(22)t}[M_{2(r)}^*][M_{2(s)}^*] + k_{(21)t}[M_{2(r)}^*][M_{1(s)}^*]$$

As the reactivity of the radicals $M_{1(s)}^*$ and $M_{1(r)}^*$ ($M_{2(s)}^*$, $M_{2(r)}^*$) is equal. and $k_{(12)t} = k_{(21)t}$, the sum of R_1 and R_2 will be given by

$$R_1 + R_2 = k_{(11)t}[M_1^*]^2 + 2k_{(12)t}[M_1^*][M_2^*] + k_{(22)t}[M_2^*]^2 \quad (145)$$

By means of eqn. (65), M_2^* can be expressed as a function of M_1^*; then

$$R_1 + R_2 = [M_1^*]^2 \left(k_{(11)t} + 2k_{(12)t}\frac{k_{12}[M_2]}{k_{21}[M_1]} + k_{(22)t}\frac{k_{12}^2[M_2]^2}{k_{21}^2[M_1]^2} \right) \quad (146)$$

Substituting for $[M_1^*]$ from eqn. (146) into eqn. (144) and setting

$$k_{(11)t}^{1/2} = \delta_1 k_{11}$$

$$k_{(22)t}^{1/2} = \delta_2 k_{22}$$

$$\Phi = \frac{k_{(12)t}}{k_{(11)t}^{1/2} k_{(22)t}^{1/2}} = \frac{k_{(12)t}}{\delta_1 \delta_2 k_{11} k_{22}}$$

References pp. 372–382

we finally obtain the equation which is the most general expression for the rate of radical copolymerization

$$-\frac{d([M_1]+[M_2])}{dt} =$$

$$= \frac{(r_1[M_1]^2 + 2[M_1][M_2] + r_2[M_2]^2)(R_1 + R_2)^{1/2}}{(r_1^2\delta_1^2[M_1]^2 + 2\Phi\delta_1\delta_2 r_1 r_2[M_1][M_2] + r_2^2\delta_2^2[M_2]^2)^{1/2}} \quad (147)$$

When one of the monomers does not polymerize ($r_2 = 0$), eqn. (147) is simplified to [170]

$$-\frac{d([M_1]+[M_2])}{dt} =$$

$$= \frac{[M_1](r_1[M_1] + 2[M_2])(R_1 + R_2)^{1/2}}{(r_1^2\delta_1^2[M_1]^2 + 2\Phi\delta_1 r_1\xi_2[M_1][M_2] + \xi_2^2[M_2]^2)^{1/2}} \quad (148)$$

where $\xi_2^2 = k_{(22)t}/k_{21}^2$ and for $r_1 = r_2 = 0$ to

$$-\frac{d([M_1]+[M_2])}{dt} =$$

$$= \frac{2[M_1][M_2](R_1 + R_2)^{1/2}}{(\xi_1^2[M_1]^2 + 2\Phi\xi_1\xi_2[M_1][M_2] + \xi_2^2[M_2]^2)^{1/2}} \quad (149)$$

where $\xi_1^2 = k_{(11)t}/k_{12}^2$. When the termination rate constant of the radicals M_1^* and M_2^* is given by the geometric mean of the termination rates $M_1^* + M_1^*$ and $M_2^* + M_2^*$ (in other words $\Phi = 1$)†, eqn. (147) is simplified to [170]

$$-\frac{d([M_1]+[M_2])}{dt} =$$

$$= \frac{(r_1[M_1]^2 + 2[M_1][M_2] + r_2[M_2]^2(R_1 + R_2)^{1/2}}{r_1\delta_1[M_1] + r_2\delta_2[M_2]} \quad (150)$$

---

† Cases in which $\Phi = 1$ are exceptions, usually $\Phi \gg 1$.

Both rate equations (147) and (150) were tested by Matsuo and Stockmayer for the copolymerization of vinyl chloride with vinylidene chloride [223]. They found that this case is adequately described not only by the general form (147) but even by the simplified shape (150).

Numerous authors continue to investigate the rates of radical copolymerizations; mostly they study special copolymerization cases, and compare the results with the presented scheme [224, 225]. Agreement is often obtained; possible discrepancies are no proof of the invalidity of the Melwille approach; they may be caused by the wrong choice of constants and coefficients.

The radical model cannot be applied for ionic and coordination polymerizations. With a few exceptions, termination by mutual combination of active centres does not occur. The only possibility is to measure the rate of each copolymerization independently. The situation can be greatly simplified for copolymerizations in living systems. The constants k_{11} and k_{22} can usually be measured easily in homopolymerizations. Also, the coaddition constants k_{12} or k_{21} are often directly accessible when the M_1° and M_2° active centres can be differentiated spectroscopically or when the rate of monomer M_2 (M_1) consumption at $M_1^\circ M_2^\circ$ centres can be measured. Ionic equilibria, association, polarity of medium and solvation must be respected, even when their quantitative effect is not known exactly. The unusual situations confronting macromolecular chemistry will be demonstrated by the example of the anionic copolymerization of styrene with butadiene initiated by lithium alkyls in hydrocarbon medium.

Rakova and Korotkov compared the rates of homopolymerization and copolymerization of styrene and butadiene [226]. Styrene polymerizes very rapidly and butadiene slowly. Their copolymerization is slow at first, with preferential consumption of butadiene. When most of the butadiene is consumed, the reaction gradually accelerates yielding a product with a high styrene content. In the authors' opinion, this is caused by selective solvation of the active centres by butadiene; only after butadiene has polymerized, does styrene gain access to the centres [227]. A similar behaviour was observed by Medvedev and his co-workes [228] and by many others. In our laboratory we observed this kind of behaviour in the cationic polymerization of trioxane with dioxolane. Although trioxane is polymerized much more rapidly than dioxolane, their copolymerization starts slowly, and is accelerated with progressing depletion of dioxolane from the monomer mixture [229].

O'Driscoll and Kuntz [230] interpreted the results of the Korotkov anionic copolymerization by a set of rates and equilibria essentially representing a transcription of the classical copolymerization equation. By a suitable choice of parameters†, with $k_{SB} \gg k_{SS}$ and $k_{BB} \gg k_{BS}$, good agreement with

---

† The subscripts of the rate constants indicate type: SB = butadiene addition to a styrene centre, etc.

experiment was obtained [230]. Direct rate measurements of the coaddition $\sim B^-, Li^+ + S \rightarrow \sim BS^-, Li^+$ also contradict the selective solvation hypothesis because the measured rate was not suppressed in the presence of butadiene (B) [231]. Agreement between experimental findings and calculations with suitable parameters was also observed by Chinese authors [232].

Selective solvation has been proved in many cases [233–235]. On the other hand, the behaviour $r_1 \gg 1$, $r_2 \ll 1$ is a common feature of anionic copolymerization. One monomer is usually much more reactive to either type of active centre in the order acrylonitrile > methyl methacrylate > styrene > butadiene > isoprene, in agreement with its electron affinity [235].

Copolymerization remains at the centre of attention of polymer chemists. The frequency of the appearance of new findings is increasing. Some very good review articles on this topic have already appeared (see, for example, ref. 236).

5.8 Copolymerization types

Statistical and random copolymers are generated by processes described in the preceding text. Copolymerizations leading to other types of polymers have also been mentioned. They deserve, however, a more detailed discussion.

5.8.1 POLYMERIZATIONS PRODUCING PERIODIC COPOLYMERS

Until recently periodic copolymers (consisting of more than two monomers) have been a more or less theoretical domain, treating special kinds of polycondensations rather than polymerizations. Not long ago, donor—acceptor complexes (often zwitterions) composed of two or more compounds (monomers) were discovered, homopolymerizing as a whole to yield periodic polymers.

Saegusa et al. synthesized a number of nucleophilic (M_N) and electrophilic (M_E) monomers (see Chap. 2, Sect. 5.2), yielding zwitterions which can mutually combine by their oppositely charge ends [237].

$$nM_N + nM_E \rightarrow n(^+M_N{-}M_E^-) \rightarrow {}^+M_N{-}(M_E{-}M_N)_{(n-1)}{-}M_E^- \tag{151}$$

By suitable choice of monomers, zwitterions composed of three units can be prepared, either from one M_N and two kinds of M_E

$$M_N \xrightarrow{M_{E_1}} {}^+M_N{-}M_{E_1}^- \xrightarrow{M_{E_2}} {}^+M_N{-}M_{E_1}{-}M_{E_2}^- \tag{152}$$

or from two kinds of M_N and one M_E

$$M_E \xrightarrow{M_{N_1}} {}^+M_N\!-\!M_{E_1}^- \xrightarrow{M_{N_2}} {}^+M_{N_2}\!-\!M_{N_1}\!-\!M_{E_2}^- \qquad (153)$$

Zwitterions formed in this way can mutually combine yielding a ternary, periodic copolymer. An example of this kind of copropagation is the copolymerization of the three units

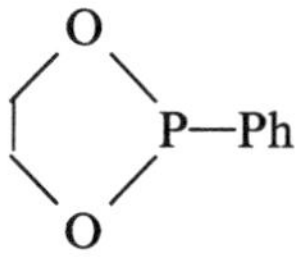

$$\underset{O}{\overset{O}{\diagdown}}\!\!\!\diagup P\!-\!Ph = M_N \text{ (2-phenyl-1,3 dioxaphospholane)}$$

$$CH_2\!=\!CHCN \text{ or} \underset{O=\overset{|}{C}\!-\!OMe}{CH_2\!=\!CH} = M_{E_1}$$

$$CO_2 = M_{E_2}$$

to poly[(ethylene phenylphosphonite)-*per*-(methyl acrylate)-*per*-(carbon dioxide)].

The number of this kind of periodic copolymer is rapidly increasing [238]. The existence of DA complexes naturally offers a much broader choice for the synthesis of alternating copolymers.

Alternating copolymerization

The generation of chains with alternating units has for a long time been the subject of great attention. Investigators are interested in the causes of such behaviour, in the properties of the copolymers, and, last but not least, in the use of non-polymerizing compounds as monomers (CO, SO_2, some molecules with a double bond between two nitrogen atoms, etc.). The first alternating copolymerizations were, of course, of radical type. Originally copolymerization of "free" monomers was assumed [173, 183] with the monomers added only to the "different" radical. This hypothesis was based on the transition complex stabilization by donor—acceptor and electrostatic interactions resulting in an increase of the coaddition rate constant $(k_{ij}, i \neq j)$. Contrasting with this idea is the hypothesis concerning the generation of homopolymerizing donor—acceptor complexes of both monomers [239–242]. The latter is supported by specific changes in copolymerization rates following composition changes in the monomer phase, and by reduced transfer.

Both these hypotheses represent extreme cases. Under special conditions, copropagation can follow one or the other. In most situations, however, both mechanisms are operating. Georgiev and Zubov developed a kinetic method for a quantitative determination of the two contributions [243].

References pp. 372–382

There exist many alternating copolymerizations: ethylene or propene with alkyl acrylates [244], vinyl acetate with maleic anhydride [245], styrene with acrylonitrile [246], styrene with fumaronitrile [247], vinyl carbazol with fumaronitrile, vinyl ferrocenne with diethylfumarate [248], and further pairs or systems of three monomers [238, 249–253]. External conditions can support or hinder alternation. At not too high temperatures, vinyl acetate forms a donor—acceptor complex with maleic anhydride. Under these conditions (and in the presence of a radical initiator), an alternating copolymer is formed. The concentration of the complex decreases with increasing temperature; above 363 K the complex cannot exist. Under these conditions, copolymerization yields a statistical copolymer whose composition depends on the composition of the monomer mixture[†] [245].

Alternating copolymers are, of course, formed also by other than radical copropagations. Soga et al. obtained an alternating copolymer from oxirane and SO_2 in the presence of aromatic amines (quinoline, pyridine) [254] and Koinuma and Hirai did so with CO_2 in the presence of Et_2Zn/H_2O (1 : 1) [255]. Many similar studies have been published [256].

A special kind of alternating copolymer synthesis is the extreme case mentioned in the last lines of the preceding paragraph, namely the formation of donor—acceptor complexes with separated centres of positive and negative charge mutually combining by their oppositely charged groups [see eqn. (151)].

An example is the reaction of 2-phenylimino-1,3-dioxolane with β-propiolactone [257]

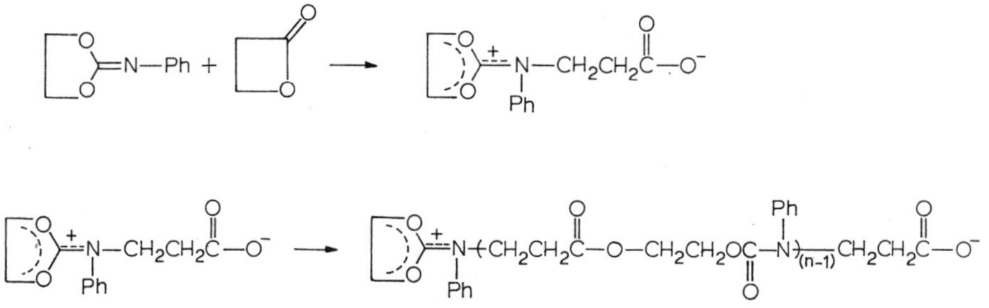

Kobayashi and Saegusa published a review article about this kind of macromolecular synthesis [258]. DA complexes remain at the centre of specialist interest [259] as do the corresponding processes of alternating polymer generation [260, 261].

[†] Thermodynamic effects also help to suppress the tendency to alternation at higher temperatures (see Sect. 5.6).

5.8.2 BLOCK COPOLYMERS

Polymers are not usually mutually miscible. Even polymer mixtures with other substances (fillers, dyes, stabilizers, softeners, etc.) are not always stable. At the same time, materials are often required to have the properties of a mixture of two or more components. Mutually insoluble, incompatible components can be held together by the addition of a compound exhibiting affinity to all components. Block and graft copolymers often possess the required property to affect the van der Waals force distribution at phase boundaries.

Block copolymers themselves are also finding rapidly expanding applications on an industrial scale. A sandwich copolymer (triblock) with an elastomeric core (polybutadiene, polyisoprene, etc.) and plastomeric ends (polystyrene, etc.) represents a physically vulcanizing rubber (plastomeric elastomer). It can be processed above the glass transition temperature of the plastomeric blocks by work-efficient technologies (injection molding, extrusion, etc.). At temperatures below the T_g of the plastic blocks, the copolymer behaves as vulcanized rubber.

The above sentences should indicate the reasons why block copolymers are the subject of increasing attention. They can be prepared by any of the known polymerization techniques. The problems of their synthesis have been treated in several large monographs [262–268].

Mechanochemical preparation of block copolymers is of historic interest. In a polymer blend, mechanical stress causes degradation, producing macroradicals which in turn lead to the formation of a large number of block and graft copolymers. Berlin assumes [269] that the radicals predominantly attack stress-activated backbone C—C bonds. Radical combination is not regarded as important as radical concentration in the system is negligible.

The radical mechanism generates block copolymers mainly by means of multifunctional or polymeric initiators [270] or by the combination of radical chain ends produced by the separation of two propagating monomers into an aqueous and a micellar phase (in emulsion) [271].

Block copolymers are also formed by condensation of two chain types with functional end groups. Some kinds of polyurethanes are examples. Dissimilar and mutually immiscible polymers can be connected in this way. Henderson and Szwarc condensed α, ω-polystyrenedicarboxylic acid with a polyamide carrying terminal amino groups [272], producing the copolymer poly(styrene)-*block*-poly(amide). The condensation was slow; the probability of finding functional groups at an effective distance is small.

The deciding factor which aroused increased interest in block copolymer production was the discovery of living anionic polymerization. New types of materials could be prepared by stepwise polymerization of several monomers.

References pp. 372–382

The synthesis was further simplified by the introduction of difunctional initiators, usually of α-methylstyrene tetramer dianion [271] $(-\alpha-)$.

$$-\alpha- + 2(n+1)\,\text{A} \rightarrow {}^-\text{A}-\text{A}_n-\alpha-\text{A}_n-\text{A}^-$$

$$^-\text{A}-\text{A}_n-\alpha-\text{A}_n-\text{A}^- + 2(m+1)\,\text{B} \rightarrow$$

$$\rightarrow {}^-\text{B}-\text{B}_m-\text{A}_{n+1}-\alpha-\text{A}_{n+1}-\text{B}_m-\text{B}^-$$

(154)

The number of copolymer types accessible in this way is only limited by the number of monomers able to polymerize by the living anionic mechanism. Similarly block copolymers can also be formed by living cationic polymerization [274–276].

An interesting synthesis of block copolymers by cationic polymerization of vinyl compounds was described by Kennedy and Melby [277] who used 2-chloro-6-bromo-2,6-dimethylheptane as coinitiator. Br^- is eliminated by triethylaluminium, and styrene can be polymerized, without transfer, on the generated carbocation. After all the styrene has reacted, diethylaluminium chloride is added to eliminate Cl^- from the coinitiator and thus produce new carbocations on the polymer chain. In the presence of 2-methylpropene, the two-block copolymer poly(styrene)-*block*-poly(2-methylpropene) is formed.

New types of synthesis are based on the combination of macroions (or polymeric di-ions) of opposite charge [278]

(155)

and transformation of centres (see Chap. 4, Sect. 5). In this way, copolymers are formed from a variety of combinations of ionically polymerizing monomers [279–283].

The properties of block copolymers are determined by block composition (a block may be a homopolymer, a statistical or periodic copolymer), block length, and the number and arrangement of the blocks. Block copolymers have the perspective of meeting the demanding requirements of modern plastics applications.

5.8.3 GRAFT COPOLYMERS

Graft copolymers are also formed by radical and ionic processes [262, 263, 284–286]. They are usually difficult to characterize, and their synthesis is poorly reproducible.

As with block copolymers, to describe their synthesis would require an extensive treatise. In principle they can be formed by some variant of the following processes

(a) active centre formation on an existing polymer chain (grafting from), e. g. by decomposition of a hydroperoxide group, elimination of halogen yielding a carbocation, metallation of a multiple bond in a polydiene, etc. In the presence of a suitable monomer, a polymeric branch (graft) grows on the centre (see, for example, refs. 262–265, 287–290).

(b) condensation of a chain with a suitable end group or of a macroion with a suitable group on a polymer chain (grafting on) (see, for example, refs. 262–265, 291–294).

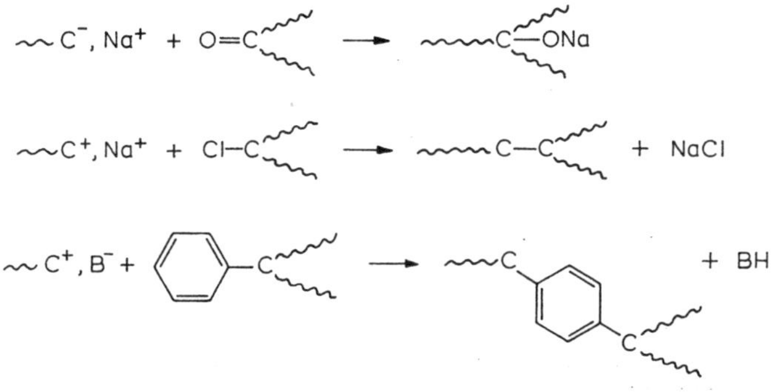

(156)

Preparation of graft copolymers by "chain biting" reactions

Carbanions can react with Cl in PVC macromolecules [295] and with the ester group of PMMA [284]. The rates of the two reactions are probably not very different; by the addition of α-methylstyrene tetramer dianion to a PVC + PMMA solution, the copolymer poly(vinyl chloride)-*graft*-poly-(methyl methacrylate) was obtained [296]. Macrocations formed by the reaction of strong acids with polyalkenes (see Chap. 3, Sect. 3.2) react with polyethers (polysiloxanes) yielding graft and block copolymers, e. g. poly(propylene)-*graft*-poly(oxyethylene) [297], poly(propylene)-*block*-poly(oxyethylene)

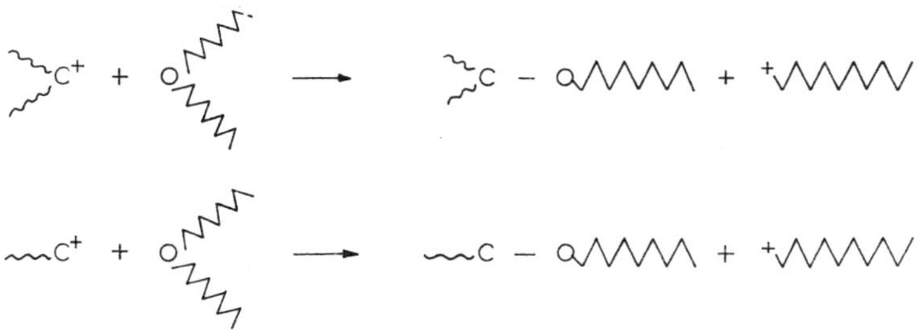

References pp. 372–382

These reactions are just an example of many other possibilities; since a backbone substituent, or the chain backbone directly is attacked, I propose to call these processes *"chain biting"* reactions.

Graft copolymers are also simultaneously branched polymers with all the pertinent consequences. In some of their properties they are similar to block copolymers. Their technical importance is not great so far but interest in their research and development will probably lead to extending applications [298].

6. The mechanism of propagation

Propagation is composed of many consecutive connections of an active centre with monomer molecules. In a broader sense, this connection is called addition. In a narrower sense, addition designates only the simplest monomer connection to a radical centre or to a centre of free ion character. In all other cases, the growth step has another name.

6.1 Addition

Concerning the reactions of a radical or an ion with an unsaturated molecule, general and qualitatively well founded ideas exist (see Chap. 3, Sect. 1). Quantitative data are mostly lacking, especially for reactions in the condensed phase. Even the empirical solution of this problem is not yet satisfactory. We know that every deviation from a unique mode of addition leads to structural defects in the chain. Irregularities usually adversely affect its properties (e. g. they reduce the thermal stability of polymers).

With decreasing temperature of propagation, the number of deviations from the predominant mode of addition decreases. Defects of this kind can be greatly limited by the orientation of monomer molecules prior to addition, for example by sorption on a suitable matrix.

An example of the possiblities in this respect is the polymerization of acetylene and diacetylene derivatives, i. e. of multifunctional monomers, with several degrees of freedom in the addition to the active centre. Generally a poorly defined mixture of mostly insoluble products is formed. On the other hand, polymerization of many substituted diacetylenes in the solid phase is easy. By short-wave irradiation (UV, γ) or simple tempering below the melting point, each monomer molecule is connected with two neighbours, yielding linear chains with conjugated multiple bonds

The polymer formed in this way consists of large, almost perfect crystals [299].

6.2 Insertion

When a monomer molecule is inserted between two or more parts of the same active centre, we speak of insertion (inserting reaction). This kind of propagation step can be very simply described by the scheme

$$\sim\sim^- \cdots\cdots^+ \xrightarrow{M} \sim\sim^-M^+ \longrightarrow \sim\sim^- \cdots\cdots^+ \qquad (157)$$

in which $\sim\sim^- \cdots\cdots^+$ designates a growing chain with a coordination active centre or ion pair. A sharp boundary between simple addition and insertion probably does not exist. Even radical centres can sometimes form complexes with non-polymerizing molecules, and the attachment of a further monomer may be regarded as a kind of insertion. On the other hand, counter-ions can be solvated by monomer or form a complex with it. In this case the growth step may be qualified as simple solvate or complex addition to the free ion. The most important case of insertion is growth at coordination centres.

6.2.1 COORDINATION INSERTION

The discovery of Ziegler–Natta catalysts was accompanied by intense efforts to gain understanding of the propagation mechanism. Many hypotheses were produced which attempted to explain the experimental findings. These include the radical and ionic (anionic and cationic) mechanisms as well as the mechanisms ascribing the main role to the transition metal atom or to the non-transition metal atom or to both at the same time. To date sufficient proof has not been produced in favour of a unique theory of coordination propagation. According to the concept of Cossee [300], the

References pp. 372–382

monometallic growth mechanism, a single growth step, can be represented by the scheme

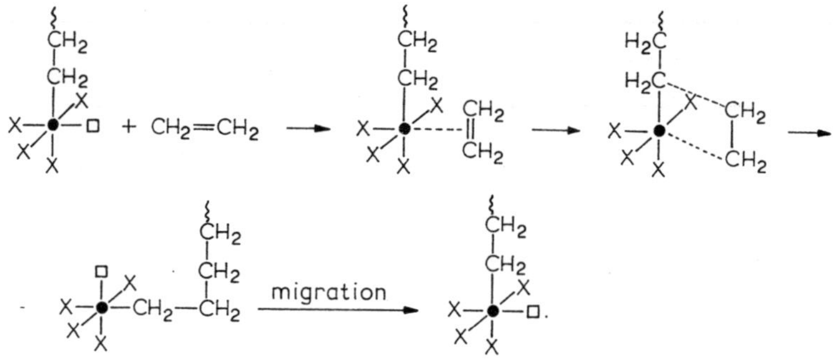

$$(158)$$

where ● is a transition metal atom, X its ligand and □ its unoccupied orbital vacancy.

As the transition metal atom is usually part of the crystal lattice of the insoluble catalyst component, the growing chain must migrate to its original position after each monomer insertion. According to De Bruin, the monomer is coordinated to the transition metal atom, from which it is redirected into the metal—carbon bond of the neighbouring transition metal atom [301]

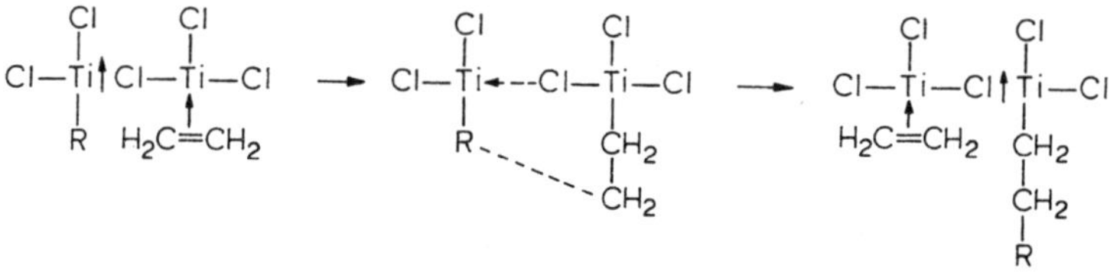

$$(159)$$

The growing alkyl migrates between the neighbouring Ti atoms at the crystal surface. Rodriguez and Van Looy consider the complexed alkylaluminium compound as an integral part of the active centre even though growth only takes place at the transition metal—carbon bond [302].

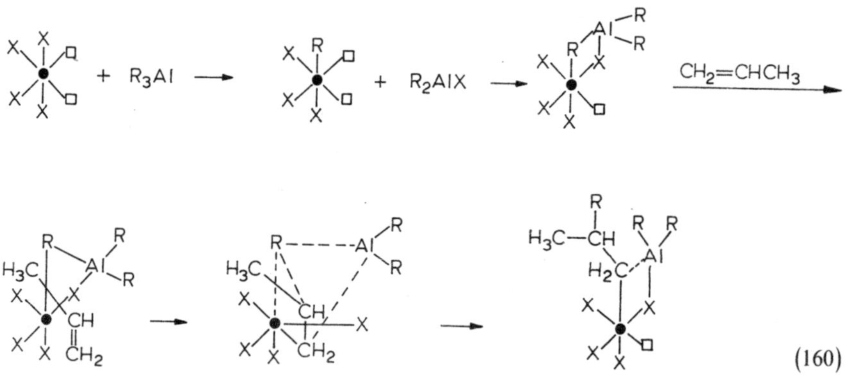

$$(160)$$

Ivin et al. doubt the general possibility of alkene insertion into the transition metal—carbon bond [303]. But insertion into the metal—H bond is regarded as established. They noted the similarity of disproportionation [303] and ZN catalysts, and of the respective reactions. They postulated the following mechanism of homogeneous and heterogeneous alkene polymerization.

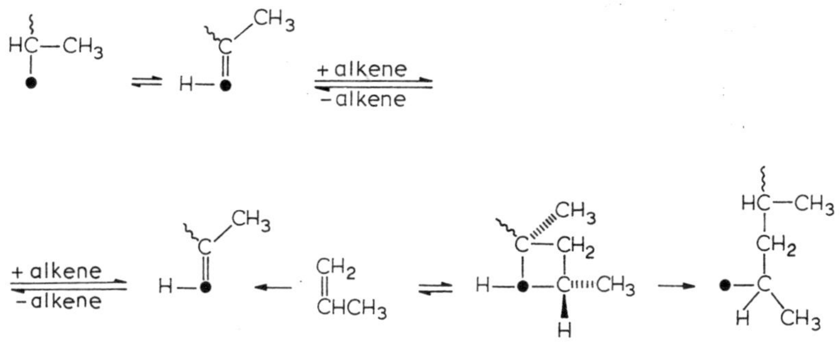

$$(161)$$

A similar mechanism can also be written for alkene oligomerizations on Al centres in the idealized, transition form, because the Al d orbitals can hardly contribute to other bonds in the organoaluminium compounds

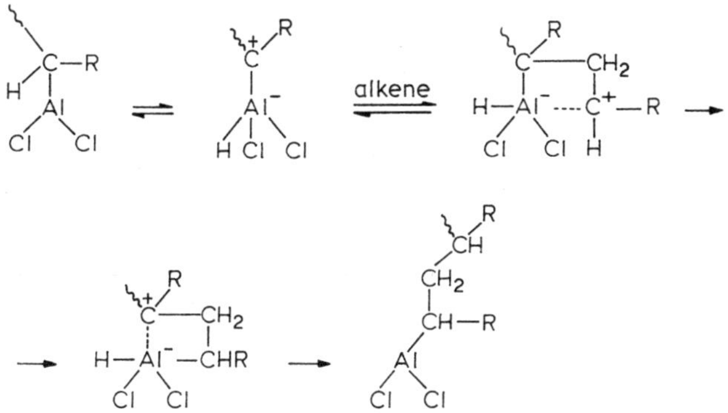

$$(162)$$

The stereospecific activity of the catalysts is given by the possibilities of the space arrangement of the aletane transition form in scheme (162). According to Ivin et al., insertions into the simple metal—carbon bond are an exception rather than a rule [303].

6.3 Ring-opening polymerizations

This is a very important polymerization mode. Usually these polymerizations can be made to proceed by the living mechanism, even cationic. Important polymers are produced in this way (some polyamides, polysiloxanes, polyacetals, polyethers, polyimines, etc.).

References pp. 372–382

Most polymerizations of cyclic monomers are ionic processes. Coordination catalysts are effective only for some heterocycles (oxirane and its derivatives, lactones). Ziegler–Natta catalysts can only be used for cycloalkene polymerization by metathesis; heterocycles act as a catalytic poison. Smooth radical polymerization of hydrocarbon monomers with ring strain is unsuccessful [304]. The deep-rooted faith that ring strain represents a major contribution to the driving force in ring opening (polymerization) has to be revised [305, 306].

One of the so far not very numerous group of cyclic monomers able to polymerize and to copolymerize with styrene, methyl methacrylate, etc. under the effect of peroxides is spiroorthocarbonate [307].

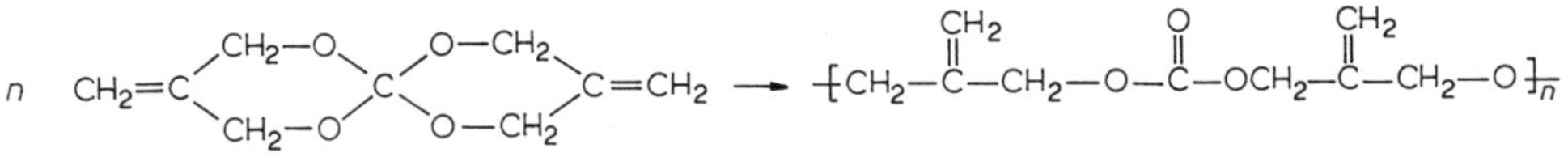

Radical ring-opening polymerization has been intensively studied by Bailey (see for, example, refs. 308–310).

Matsuzaki and Ito polymerized *cis* and *trans* dideuterated oxirane by both ionic and coordination polymerization. They observed that ring opening and chain growth proceeds almost exclusively with configuration inversion [311]

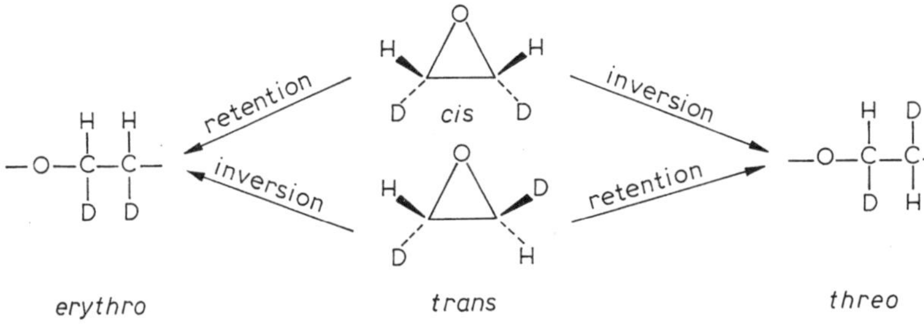

(163)

Tetrahydrofuran is often used as a polymerization model for multimembered cyclic ethers. Its propagation proceeds by nucleophilic attack of the monomer oxygen on the α carbon of the active centre with a polarized bond [312]

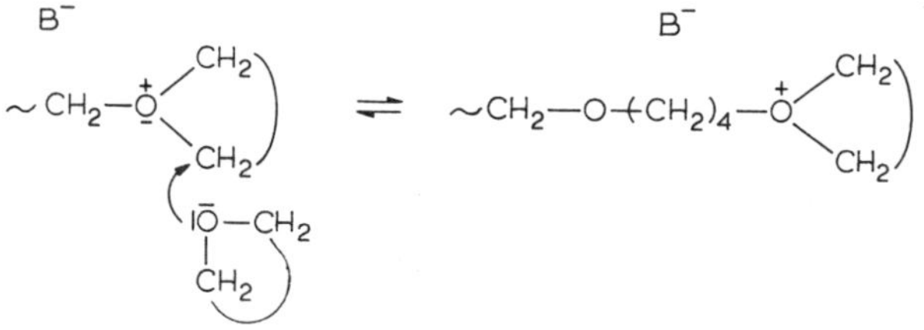

(164)

The polymerization of dioxolane remains the subject of discussion. Most authors believe that its propagation does not differ from the common model, which in a simple form can be written as (without the counter-ion) [313]

$$\sim CH_2\text{-}\overset{+}{O}\underset{}{\frown}O + O\underset{}{\frown}O \rightleftarrows \sim OCH_2CH_2OCH_2\text{-}\overset{+}{O}\underset{}{\frown}O$$

$$\sim CH_2OCH_2CH_2\overset{+}{O}=CH_2 \xrightleftharpoons[-\text{ }\langle\underset{O}{\overset{O}{\frown}}\rangle\text{, polymer}]{+\text{ }\langle\underset{O}{\overset{O}{\frown}}\rangle\text{, polymer}} \sim (CH_2OCH_2CH_2O)_2 C\overset{+}{H_2}\cdots\overset{O\diagdown CH_2CH_2\sim}{\underset{O\diagup CH_2CH_2\sim}{\diagdown CH_2}} \quad (165)$$

Plesch [314] advocates the ring-expansion concept

$$\overset{B}{\underset{}{\overset{+}{\bigsqcup}}} + \langle\underset{O}{\overset{O}{\frown}}\rangle \rightleftarrows \overset{B}{\underset{}{\overset{+}{\bigsqcup\text{-}O\text{-}O}}} \xrightarrow{+n\langle\underset{O}{\overset{O}{\frown}}\rangle} \rightleftarrows \overset{B}{\underset{}{\overset{+}{\bigsqcup\text{-}O\text{-}CH_2\text{-}[O\text{-}CH_2O\text{-}]_{n+1}}}} \quad (166)$$

The mechanism of eqn. (166) is less supported by experiment. It is certain that macrocycles are formed in ring-opening polymerizations. The ratio between linear and cyclic forms may depend on the type of initiator (or more exactly on the position of the equilibria ester $\rightleftarrows$ ion pair $\rightleftarrows$ free ion). Under certain circumstances, relatively many macrocycles are formed, though by a mechanism different from eqn. (166).

Some heterocycles have both nucleophilic and electrophilic atoms in their molecule. Thus they can be opened and polymerized by the anionic, cationic or coordination mechanisms. Examples are lactams, lactones, and cyclic siloxanes. Investigations of the mechanism of lactam propagation are complicated by the occurence of side reactions. In principle, the mechanism described in Chap. 3 by the schemes (55)–(57) and (71) is accepted. Anionic polymerization of cyclic esters consists, in most cases (see Chap. 4, Sect. 2.2) of repeated reversible attacks on the carbonyl carbon by the anion $\sim\overline{O}|^-$. From ε-caprolactone, polyester chains grow according to [315]

$$\sim \overset{O}{\underset{}{\overset{\|}{C}}}(CH_2)_5\text{-}\overline{O}|^- + \overset{O}{\underset{}{\bigcirc}} \rightleftarrows \sim \overset{O}{\underset{}{\overset{\|}{C}}}(CH_2)_5 O\text{-}\overset{O}{\underset{}{\overset{\|}{C}}}(CH_2)_5\text{-}\overline{O}|^- \quad (167)$$

References pp. 372–382

Even carboxylate ions can serve as active centres of β-propiolactone polymerization [316]. Cationic polymerization is characterized by the formation of an oxonium transition salt generated by the reaction of an active centre with an exo- or endocyclic oxygen atom. The reaction mode depends on the kind of initiator and monomer [317]

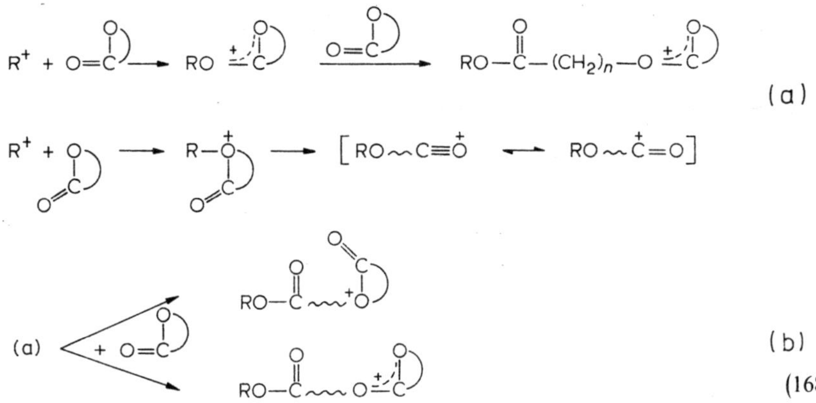

$$\text{(a)}$$

$$\text{(b)}$$

(168)

Propagation according to scheme (168a) is typical of carbocation initiators; both modes were observed with acylium ($RC^+\equiv O$) initiators. During chain growth from β-propiolactone, the concentracion of acylium ions decreases until oxonium ions become the active centres. In ε-caprolactone polymerization, both types of centre continue to operate.

Propagation on coordination centres is formulated as simultaneous insertion with the breaking of a covalent bond (alcoholate) [315, 318] (M_t is the metal atom)

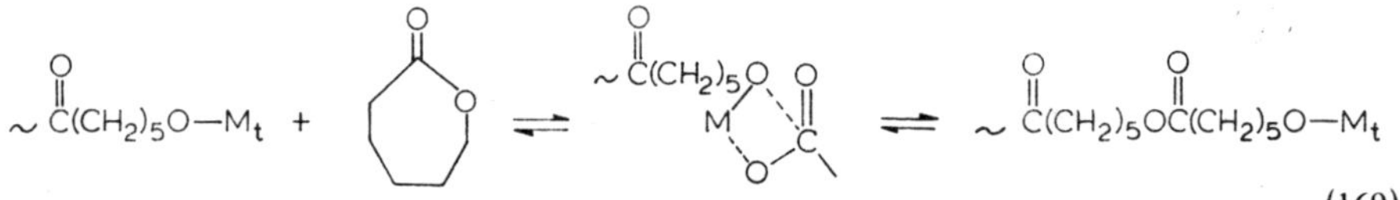

(169)

Siloxane cycles can polymerize by both ionic processes. They are readily attacked by nucleophilic agents, yielding complexes of anionic nature. When the growing siloxane chain acts as the attacking anion, a complex with three practically identical siloxane bonds is generated [319]

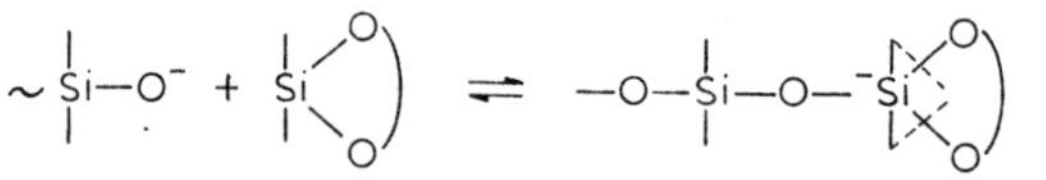

(170)

Spliting in any of the indicated ways leads to chain propagation. In non-polar media, free ions only exist for a short time during bond rearrangement. Thus the silanolate group, rather than a silanolate anion, should be regarded as the active centre [320]. Therefore we have proposed a scheme respecting this circumstance[†]

$$\sim\!\overset{|}{\underset{|}{Si}}\!-\!O^-M_t^+ + \overset{|}{\underset{|}{Si}}\!\!\left(\!\!\begin{array}{c}O\\ \\O\end{array}\!\!\right) \rightleftarrows \sim\!Si\!-\!O\!-\!\overset{|}{\underset{|}{Si}}\!\!\left(\!\!\begin{array}{c}O\\ \\O\end{array}\!\!\right)^{\!-} + M_t^+ \rightleftarrows \sim\!\overset{|}{\underset{|}{Si}}\!-\!O^-M_t^+ \qquad(171)$$

Cationic propagation of siloxane monomers has been known for a long time [322] and has been studied extensively [323, 325]. However, the underlying mechanism is only just emerging. The complex with oxonium oxygen is almost certainly the transition form. Writing the reaction as

$$\sim\!O\!-\!\overset{|}{\underset{|}{Si}}\!^+ + O\!\!\left(\!\!\begin{array}{c}Si\\ \\Si\end{array}\!\!\right) \rightleftarrows \sim\!O\!-\!\overset{|}{\underset{|}{Si}}\!-\!^+O\!\!\left(\!\!\begin{array}{c}Si\\ \\Si\end{array}\!\!\right) \rightleftarrows \sim\!O\overset{|}{\underset{|}{Si}}\!-\!(O\overset{|}{\underset{|}{Si}})\!-\!O\overset{|}{\underset{|}{Si}}^+ \qquad(172)$$

probably is a gross oversimplification (and not only by omitting the counter-ions). Propagation probably proceeds by a reaction analogous to eqn. (164).

Cyclic sulphides and imines are often used as monomers. On the industrial scale they are not so important as to induce the development of a cheap production method. Their propagation is, however, interesting, representing a more complex analogy of the cases described above.

An increasing number of investigators are interested in the propagation of cyclic monomers without carbon. The reasons are theoretical interest and anticipation of the future need of monomer sources other than the fossil fuels. Theoretically siloxane units can be arranged into lamellae or bundles (an analogy of mica or asbestos) yielding thermoplastic materials. Another possibility could be the application of the cyclic esters of phosphoric acid [326], phosphazenes, etc.

[†] The proposed scheme includes liberation of polysiloxanes from the growing chain ends. At the time when this result was published, but more frequently much later, other examples of such behaviour were observed in ring-opening polymerization [321] and the effect was named back-biting reaction.

References pp. 372–382

Whith and Singler polymerized a hexachlorotriphosphazene melt to linear chains

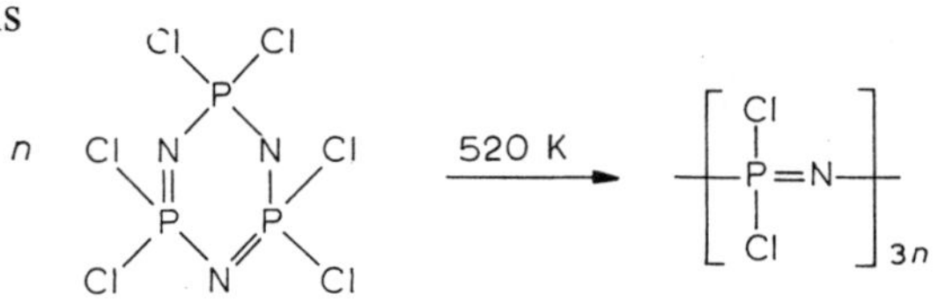

$$n \; \text{N}_3\text{P}_3\text{Cl}_6 \xrightarrow{520 \text{ K}} \left[\begin{array}{c} \text{Cl} \\ | \\ -\text{P}=\text{N}- \\ | \\ \text{Cl} \end{array} \right]_{3n}$$

(173)

which could be modified by a polymer-analogous reaction (exchanging chlorine for more suitable substituents) [327]. A great effort is being made to assemble details concerning the mechanism of these reactions [328].

6.3.1. THE FORMATION OF MACROCYCLES DURING RING-OPENING POLYMERIZATION

The enthalpy change in cyclic monomer propagation is usually small. With the general low entropy change of polymerizations this means, according to eqn. (3), a low ceiling temperature of polymerization. Propagation of cycles is clearly a reversible reaction. The direction from right to left in schemes (165)–(172) need not necessarily mean monomer separation. Quite often larger molecules are eliminated and the product is then a mixture of linear and cyclic fractions. The cycle size depends on a number of factors of which the main ones are steric hindrance, temperature, kind of counter-ion, and solvent. Sometimes defined oligomers are formed in high yield. Eastham and co-workers [321, 329] observed dioxane formation during oxirane polymerization. Tetraoxane and pentoxane were found in trioxane polymerization, [330] cyclic oligomers to macrocycles in polymerizations of tetrahydrofuran [331, 332], oxetane and of larger cyclic ethers, thiirans and aziridines [333].

Cyclic fractions are formed either from monomers by some variant of propagation or, and this is the main source, from the growing chain. Examples of the former could be that already cited [314] or the observation of Weissermel and Nölken [334] that chloromethyloxirane reacts with triethyaluminium at low temperature to give a product which is a mixture of polymer and cyclic dimer

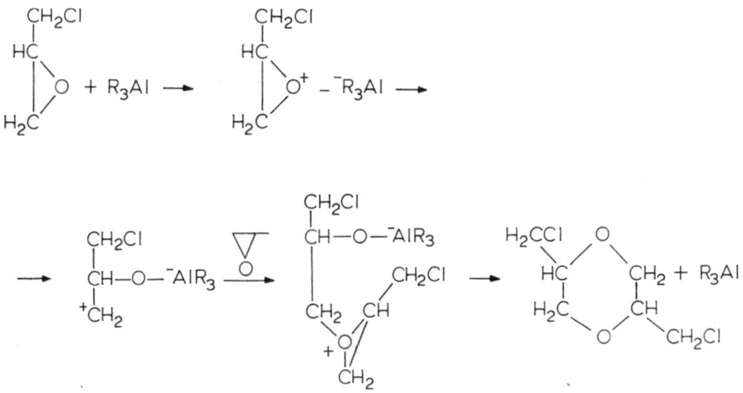

(174)

Oligomers and macrocycles are generated from the growing chain by reaction of the active centre with a suitable place on its own chain, as shown in the following scheme in which X represents the heteroatoms O, N, P ...).

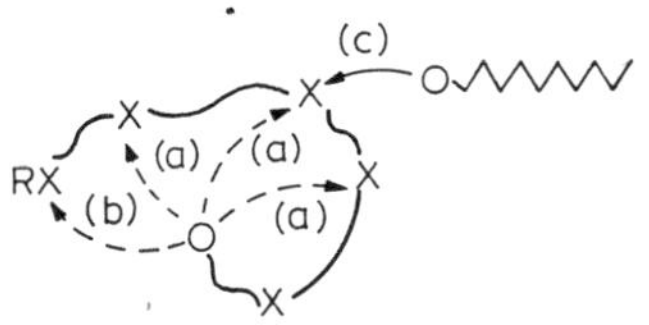

(a) back biting
(b) end biting
(c) chain biting

(175)

The occurrence of these reactions is always determined by thermodynamic factors. Oxirane has a large ring strain. Its polymerization around room temperature exhibits $\Delta G_p < 0$. For 1,4-dioxane under the same conditions, $\Delta G_p > 0$. In other words, polyoxirane will split off 1,4-dioxane because the Gibbs energy of its depolymerization is negative. Actually the polymer should depolymerize completely. That this is not the case, is caused by kinetic factors. Termination of depolymerization need not coincide with termination of polymerization.

The ratio of back- and end-biting reactions depends on the reaction conditions, and may differ considerably in polymerizations of various monomers. In cationic polymerization, methyloxirane produces mainly a mixture of cyclic tetramers [335] and chloromethyloxirane yields both dimers and tetramers [336]. Under similar conditions, 1,3-dioxolane or 1,3-dioxepane yield a number of cyclic derivatives [337], the distribution of cyclics being in agreement with the Jacobson–Stockmayer theory [338].

The occurence of macrocycles in the product is always interesting from the theoretical point of view. From the practical point of view it is sometimes undesirable, e. g. when macromers should be produced by ring-opening polymerization. In some cases it can be utilized; simple methods for preparing crown-ethers in this way have been proposed.

The ratio of linear and cyclic macromolecules will be sharply affected by a change in the kinetics and mechanism of propagation, e. g. by transition from growth on activated chain ends to polymerization with activated monomer. An inactive macromolecule cannot produce macrocycles, neither by back- nor by end-biting reactions [339].

Macrocycles are sometimes only formed in the presence of monomers. When all monomer has been consumed by polymerization, depolymerization also stops [333]. The tendency of common heterocycles to the production of giant molecules, and the possibility of generating cyclic fractions from their chains is illustrated in Table 6.

References pp. 372–382

TABLE 6

Equilibrium concentration of some heterocyclic monomers at 300 K [333].

Monomer	Ring strain[a] (kJ mol^{-1})	ΔG_p° [b] (kJ mol^{-1})	[M] (mol dm^{-3})
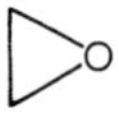 (oxirane)	113	−87.9	6×10^{-16}
 (thiirane)	84	−58.6	7×10^{-11}
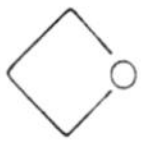 (oxetane)	107	−69.1	1×10^{-12}
 (thietane)	84	−58,6.	$7 . 10^{-11}$
 (tetrahydrofuran)	23	+ 1.7	1.9
 (tetrahydrothiophene)	8	+16.7	8×10^2
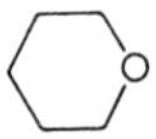 (tetrahydropyran)	−1	+26.4	3.6×10^4

[a] Ref. 340.
[b] Assuming that $-H_p^{\circ}$ = ring strain and $S_p^{\circ} = -83.7$ J K^{-1} mol^{-1}.

6.3.2 METATHESIS

Polymerization of cycloalkenes by metathesis occurs on metal–carbene and metallacyclobutane complexes. The transition metal is usually tungsten or, rarely another metal such as gallium, and the cocatalyst an organometal of aluminium (or tin). Cycloalkenes with higher ring strain usually polymerize more rapidly. This led to the incorrect assumption that the rate is a function of monomer ring strain [341, 342]. More recent studies have proved, however, that monomer reactivity is mostly determined by the geometry of the growing chain in the vicinity of the active centre [343].

Propagation during metathesis polymerization of cycloalkenes can be visualized by a scheme where the addition rate of the next monomer is essentially determined by structure 1.

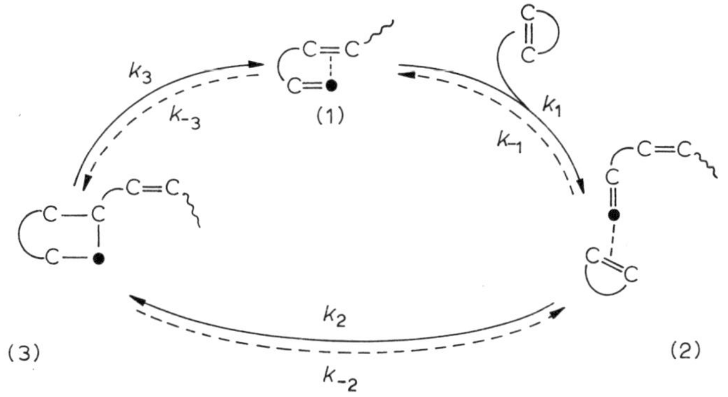

(176)

The metathesis products, the polyalkenamers, exhibit excellent properties as polymeric fillers for elastomers. The interests of the rubber industry stimulate research orientation to this kind of ring-opening polymerization (see, for example, refs. 344–346).

6.3.3 POLYMERIZATION ON ACTIVATED LIGANDS

In studies of cyclic monomer opening on transition metal derivatives, particularly on $TiCl_4$, we have observed a peculiar kind of propagation. It occurs at activated ligands (chloromethyloxirane, acetylchloride) coordinated to Ti by chloride atoms.

We assume that contact of the ligands with the monomer (methyloxirane, tetrahydrofuran) leads to bond redistribution and monomer insertion between the atoms of the growing ligand [347]

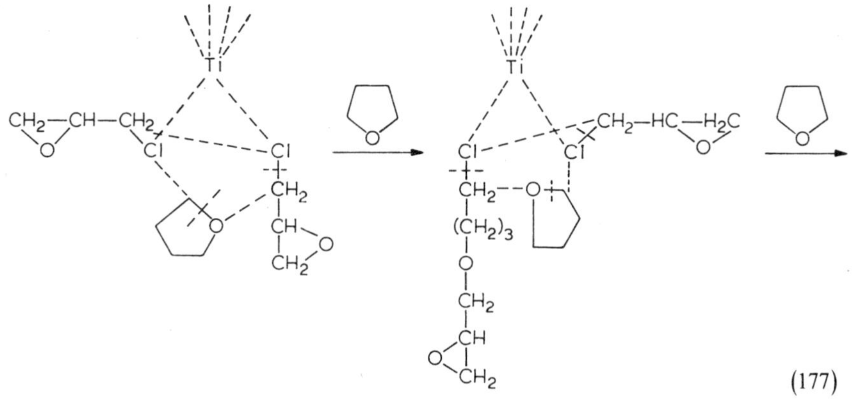

(177)

References pp. 372–382

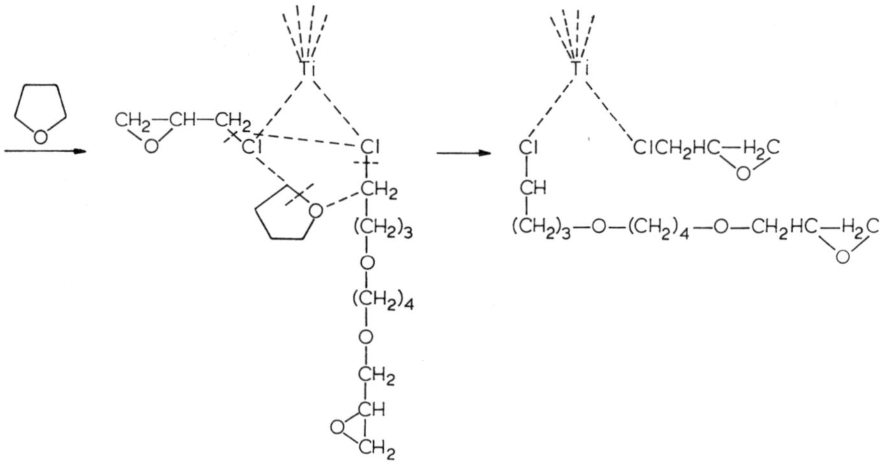

[Scheme showing reaction mechanism with Ti catalyst and epoxide/ether structures]

The proposed mechanism plausibly explains the observed facts. An independent, direct proof of its validity is not yet available.

6.4 Cyclopolymerization

Acyclic monomers, mostly dienes, diynes, and multifunctional monomers with heteroatoms, can under certain circumstances cyclize during propagation. Chains of condensed or isolated ring structures are generated either directly or with isomerization. Many examples of cyclopolymerizations have been observed, but so far they are not technically important. From the theoretical point of view, their importance should not be overstressed; they are nevertheless interesting, illustrating the rich possibilities of organic syntheses. With several mechanisms usually overlapping, they are hard to control at present.

The cyclopolymerization process will be demonstrated by several examples.

Costa et. al. [348] polymerized *o*-divinylbenzene by a radical mechanism

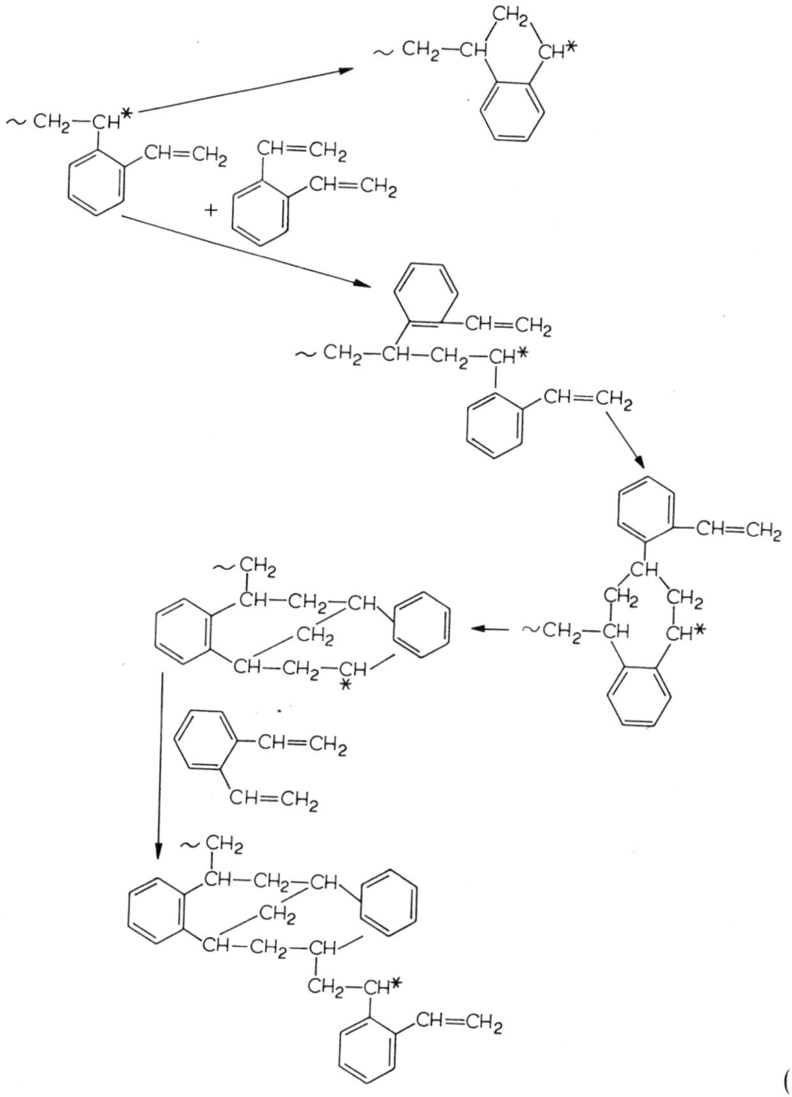

(178)

Kössler et al. [349] described the cyclopolymerization of isoprene in the presence of EtAlCl$_2$. They postulate simple diene addition to the propagating ladder chain

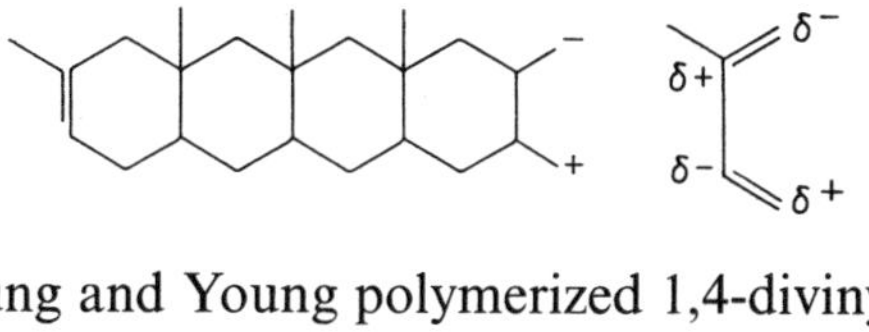

(179)

Seung and Young polymerized 1,4-divinyloxybutane and diethyleneglycol

divinyl ether by means of elementary iodine [350]. Although both monomers are tetrafunctional, the generated product was a soluble, saturated polymer

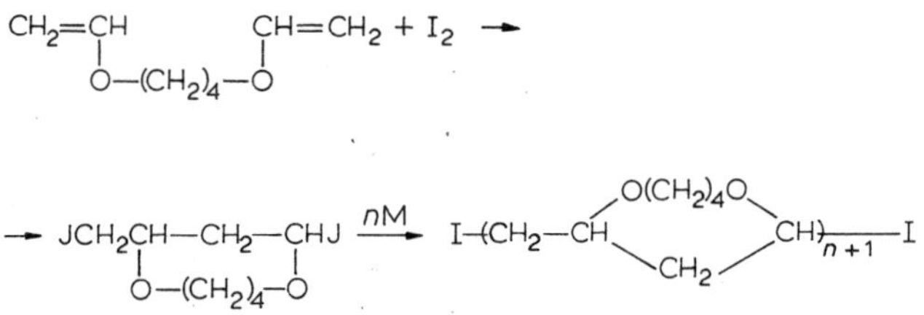

$$\begin{array}{c} CH_2=CH \quad CH=CH_2 + I_2 \longrightarrow \\ | \qquad\qquad | \\ O-(CH_2)_4-O \end{array}$$

$$\longrightarrow JCH_2CH-CH_2-CHJ \xrightarrow{nM} I-(CH_2-CH\underset{CH_2}{\overset{O(CH_2)_4O}{\diagup\diagdown}}CH)_{n+1}-I \qquad (180)$$
$$| \qquad\qquad |$$
$$O-(CH_2)_4-O$$

6.5 Polymerization with activated monomer

This type of polymerization is specific for ring-opening polymerization, and it has already been mentioned several times (in Chap. 1, Sect. 9.3 and Chap. 4, Sects. 2.2 and 2.3). Propagation of lactams mostly proceeds by a system of consecutive reversible reactions

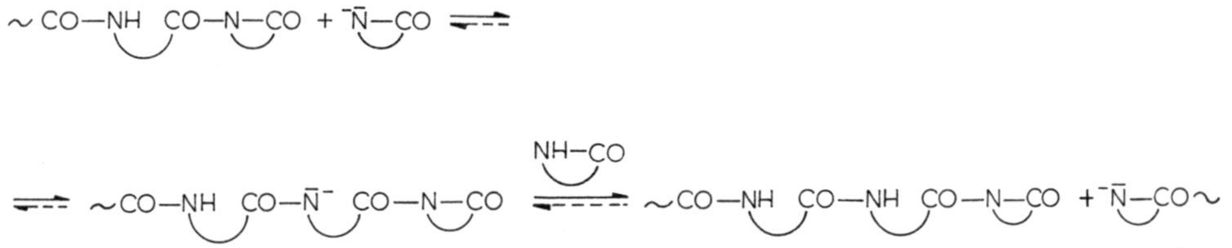

$$\sim CO-NH \quad CO-N-CO + {}^-\bar{N}-CO \rightleftharpoons$$

$$\rightleftharpoons \sim CO-NH \quad CO-\bar{N}^- \quad CO-N-CO \rightleftharpoons \sim CO-NH \quad CO-NH \quad CO-N-CO + {}^-\bar{N}-CO\sim \qquad (181)$$

$$\sim NH_2 + NH-CO---H^+ \rightleftharpoons \sim\overset{H}{\underset{H}{N^+}}-\overset{OH}{\underset{}{C}}-NH \rightleftharpoons \sim NH-CO \quad NH_2 + NH-CO---H^+ \qquad (182)$$

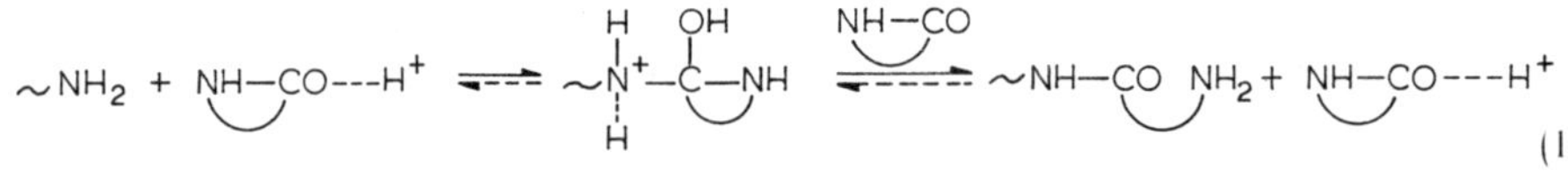

Polymerization of oxirane (and of its derivatives) by the mechanism of activated monomer is so far exclusively cationic and can be represented by schemes (27) and (28) of Chap. 4. In contrast to the ring-opening polymerization of lactams, both the classical and the activated monomer mechanisms are operating in this case. Conditions can be found where one or the other mechanism predominates [339].

6.6 Isomerization polymerization

Instability (reactivity) of an active centre may cause rearrangements of atoms or of whole substituents in the centre prior to or during the addition of further monomer. The structure of the generated chain in such a case is different from that corresponding to simple propagation. The practical importance of isomerization propagations is not as yet great. From the theoretical point of view, interesting information can be obtained about the behaviour of the active centre in the course of propagation. Any kind of centre (radical, ionic or coordination) can undergo isomerization. The driving force is usually the higher thermodynamic stability of the isomerized form of the active centre (or structural unit).

Orr derived high $-\Delta H$ and ΔS values for an imaginary polymerization of cyclooctatetraene, not stabilized by resonance, into resonance-stabilized polyene chains [9, 85]. The heat of polymerization would be even higher for the transformation of cyclooctatetraene to polystyrene (~ 460 kJ mol^{-1}). The nature of isomerization reaction will be better illustrated by several examples:

Cho and Ahn studied the radical polymerization of 1,1-disubstituted 2-vinylcyclopropanes (X = COOEt, or CN)

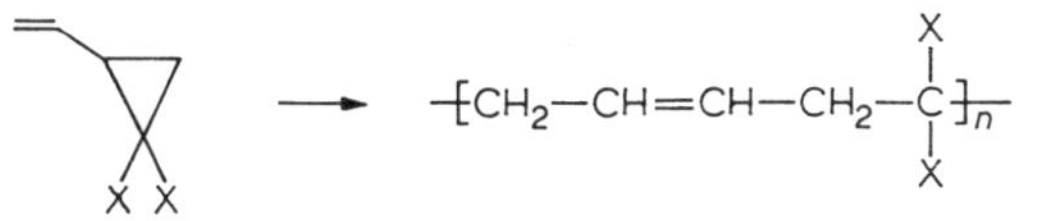

$$\underset{X\ X}{\overset{=}{\bigtriangledown}} \longrightarrow -\!\!\!\!+\!\!CH_2-CH=CH-CH_2-\underset{X}{\overset{X}{C}}\!\!+\!\!\!\!-_n$$ (183)

The authors explain the easy 1,5-polymerization of these monomers by the substituent stabilization of the growing radical [351].

Saegusa et al. observed a proton transfer during anionic polymerization of hydroxyalkylacrylates [352]

$$\sim CH_2CH_2\overset{O}{\overset{\|}{C}}-OCH_2CH_2O^-\ Li^+ + CH_2=\underset{\underset{OCH_2CH_2OH}{|}}{\overset{|}{CH}}_{C=O} \longrightarrow$$

$$\longrightarrow \{CH_2CH_2\overset{O}{\overset{\|}{C}}-OCH_2CH_2-OCH_2-\underset{H}{\overset{-}{\overset{|}{C}}}-\overset{O}{\overset{\|}{C}}-OCH_2CH_2OH\ Li^+] \longrightarrow$$

$$\longrightarrow \sim CH_2CH_2\overset{O}{\overset{\|}{C}}-OCH_2CH_2O-CH_2CH_2\overset{O}{\overset{\|}{C}}-OCH_2CH_2OLi \quad (184)$$

References pp. 372–382

In his review, Reimschuessel described [353] isomerization polymerizations of lactams containing a carboxyl group as a substituent or part of a substituent. Whereas normal lactam ring opening leads to polyamides, isomerization polymerization produces polyimides in a complicated way, with proton transfer and through bicyclic intermediates

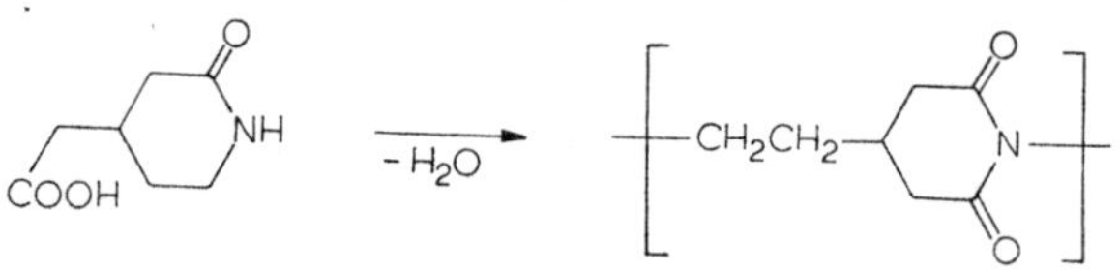

$$\text{(lactam with COOH)} \xrightarrow{-H_2O} \text{[-CH}_2\text{CH}_2\text{-(imide)-]} \qquad (185)$$

Ferraris et al. contributed to the interpretation of cationic isomerization polymerization of 4-methyl-1-pentene [354], proposing the scheme, in which $M = -CH_2-CH-CH_2-CH(CH_3)_2$

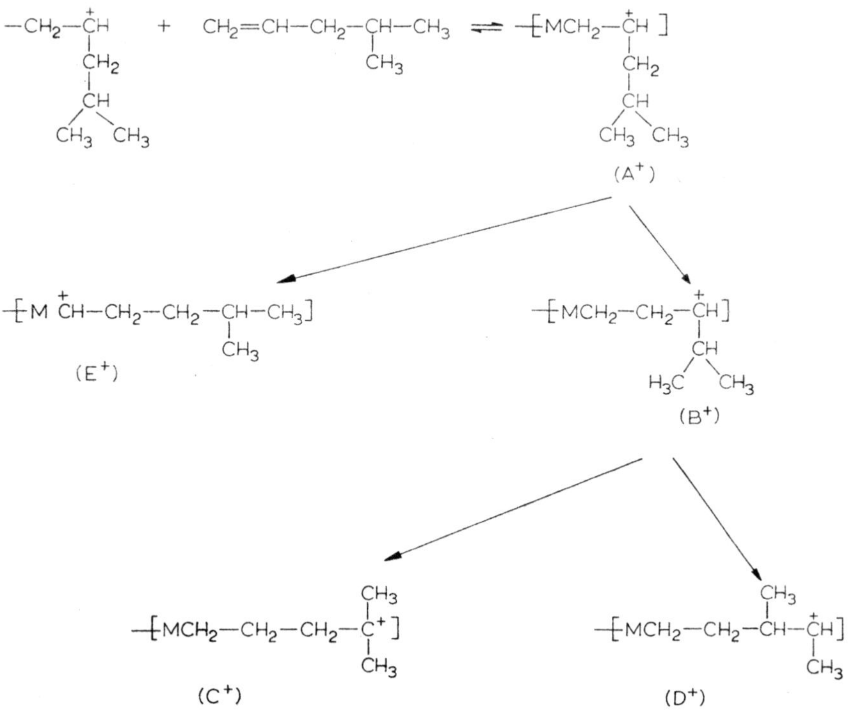

(186)

The existence of all cationic forms A^+ to E^+ was proved by NMR spectroscopy. Their relative abundances change with the polymerization temperature. The proton or methylium cation transfer has a low activation energy which, together with the energy state of the respective cation, determines the structure of the generated polymer.

Coordination polymerizations are often accompanied by isomerization. By means of the ternary catalytic system VCl_4, $(acac)_3Fe$, and Et_3Al, propene can yield crystalline polyethylene and the amorphous ethylene—propene copolymer. Many more such cases have, of course, been observed. Probably of greatest importance are those where a non-polymerizing 2-alkene is isomerized to 1-alkene prior to propagation [355].

6.7 Group transfer polymerization

Methacrylate monomers able to generate the reactive silyl ketene acetal group add further monomer molecules with the silyl group shifting to the growing end after each addition [356]

$$\underset{1}{\underset{Me}{\overset{Me}{|}}\underset{|}{C}=\underset{|}{\overset{OR}{|}}C-OSi(Me)_3} \xrightarrow{\underset{Me}{\overset{COOMe}{|}}CH_2=C} \underset{Me}{\overset{Me}{|}}C-CH_2-\underset{Me}{\overset{COOR}{|}}C=\underset{}{\overset{OMe}{|}}C-OSi(Me)_3 \rightarrow$$

$$\xrightarrow{n\underset{Me}{\overset{COOMe}{|}}CH_2=C} \underset{Me}{\overset{Me}{|}}C-(CH_2-\underset{Me}{\overset{COOR}{|}}C)_n-CH_2\underset{Me}{\overset{COOMe}{|}}C=\underset{}{\overset{OMe}{|}}C-OSi(Me)_3 \qquad (187)$$

Group transfer polymerization (GTP) must be co-initiated by bases or acids. Nucleophilic compounds are preferred because only a small amount is needed, only about 0.1 % with respect to the initiator which most often is 1-alkoxy-1-(trimethylsiloxy)-2-methyl-1-alkene [1 in scheme (187)]. Acid co-initiators must be added in much larger amounts, up to 10% of monomer. GTP only proceeds in media without active hydrogen donors (H^+, H^-), and the methacrylate polymerization is of living character. Acrylates can also be polymerized, at a greater rate than methacrylates, but the polymerization is not living. This circumstance can be utilized; the possibilities of polymer architecture are amplified by the use of monomers with both acrylate and methacrylate groups. An uncross-linked poly(acrylate) is generated with intact pendant methacrylate groups in the chains [356]. When a strong base is used as co-initiator, it can itself initiate anionic monomer propagation so that the polymer is generated both by GTP and by propagation on the anions; the resulting polymer exhibits a bimodal molecular mass distribution curve [357].

References pp. 372–382

6.8 Branching and cross-linking

Propagation of difunctional and of many multifunctional monomers does not always lead to branching or even cross-linking of macromolecules. The generation of non-linear and three-dimensional structures is, in these cases caused by other processes, usually by transfer. Polymer branching can, in its turn, affect propagation. The resulting effects are little known both when caused by transfer, where the branching produced is less dense, or by propagation itself with the participation of monomers or comonomers of multifunctional behaviour. This is due to the insufficient accuracy of analytical methods determining the number, length, and distribution of branches. Insoluble cross-linked polymers are even less amenable to analysis.

A clever way to evade the difficulties mentioned was presented by two groups of workers. Essentially they polymerized or copolymerized a tetrafunctional monomer in which the double bonds were separated by an ester group. After polymerization, which produced an insoluble three-dimensional network, the linear hydrocarbon network fragments (polymer or copolymer macromolecules) could be separated and analyzed after hydrolysis. For the copolymerization of triethyleneglycol dimethacrylate with styrene, Bolbit and Frenkel found that the course of the reaction was at first identical with "linear" copolymerization in solution. Copolymer composition undergoes a linear change with conversion, following the deviation of the feed composition from the azeotrope. The degree of polymerization of the copolymer after hydrolysis and the reaction rate are complicated functions of conversion and of the ratio of the monomer and polymer phases in the system [258].

Braun et al. studied the polymerization kinetics of symmetrical alkenes, particularly of 4-vinylbenzoic acid anhydride. The cross-linked polymer could be transformed to linear macromolecules by hydrolysis [359].

Polymerization of a monomer with two double bonds is essentially a copolymerization. When the monomer is symmetrical, with mutually independent bonds, equal reactivity of the two bonds may be assumed (independent of whether they are placed in the diene or in the chain). In this way the reaction scheme is greatly simplified.

≡ M₁, symmetrical diene;

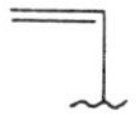

≡ M₃, a double bond in a chain;

≡ P₁*, a growing radical with a neighbouring double bond;

P₁* and M₃ yield ≡ P₃*, thus

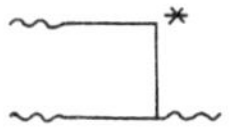

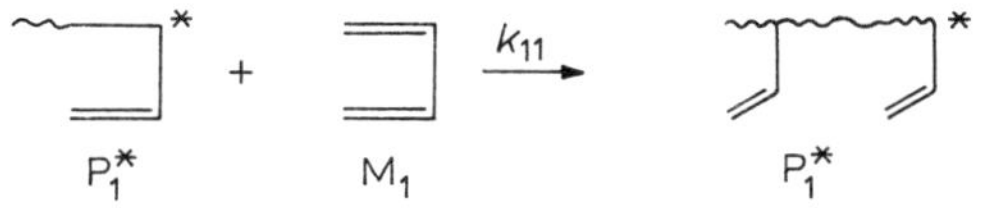

$$P_1^* + M_1 \xrightarrow{k_{11}} P_1^* \qquad r_{11} = 2k_{11}[P_1^*][M_1] \tag{189a}$$

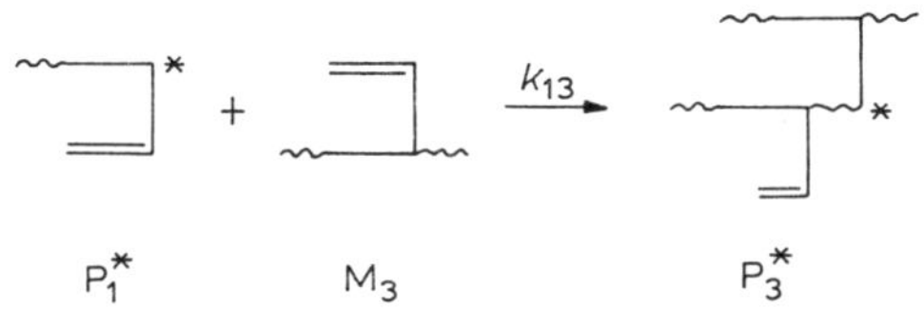

$$P_1^* + M_3 \xrightarrow{k_{13}} P_3^* \qquad r_{13} = k_{13}[P_1^*][M_3] \tag{189b}$$

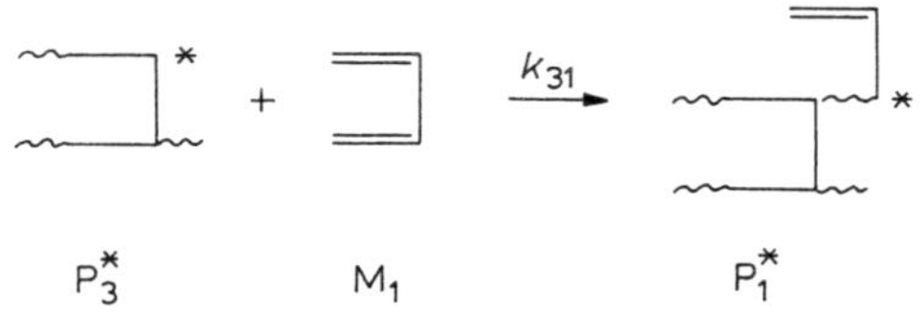

$$P_3^* + M_1 \xrightarrow{k_{31}} P_1^* \qquad r_{31} = 2k_{31}[P_3^*][M_1] \tag{189c}$$

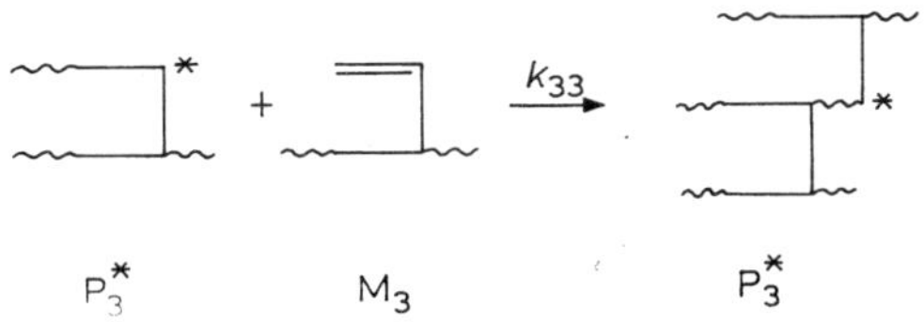

$$P_3^* + M_3 \xrightarrow{k_{33}} P_3^* \qquad r_{33} = k_{33}[P_3^*][M_3] \tag{189d}$$

A diene has two double bonds, hence the factor 2 in eqn. (189a) and (189c). As $k_{11} = k_{13} = k_{31} = k_{33} = k$, the solution of eqns. (189) following the stationary state copolymerization approach is simple, and the concentration of cross-linking or cyclizing points [N], formed only when M_3 is consumed $\{d[N]/dt = v_{13} + v_{33} = k[M]_3([P_1] + [P_3])\}$, is given as

$$\frac{d[N]}{d[M_1]} = -\frac{[M_3]}{2[M_1]} \quad \text{and} \quad \frac{d[N]}{d[M_3]} = \frac{[M_3]}{[M_3] - 2[M_1]}$$

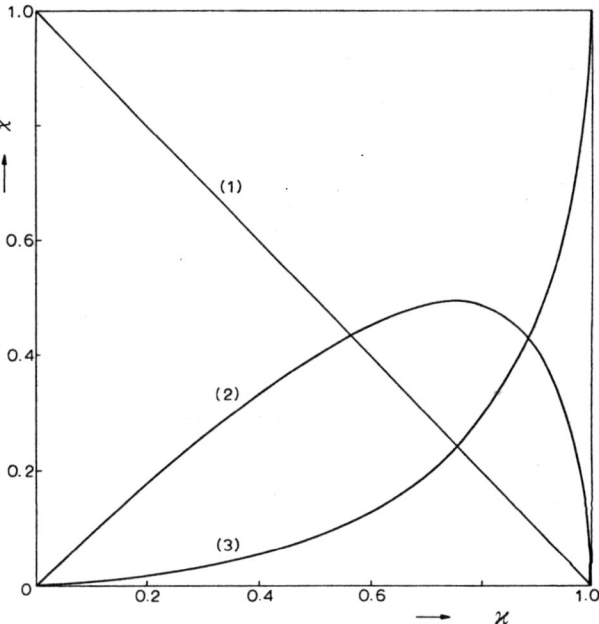

Fig. 28. Change of relative concentrations x with conversion $\varkappa$ for (1) monomer $[M]_1/[M]_0$; (2) incorporated double bonds $[M]_3/[M]_0$; and (3) cross link points $[N]/[M_1]_0$.

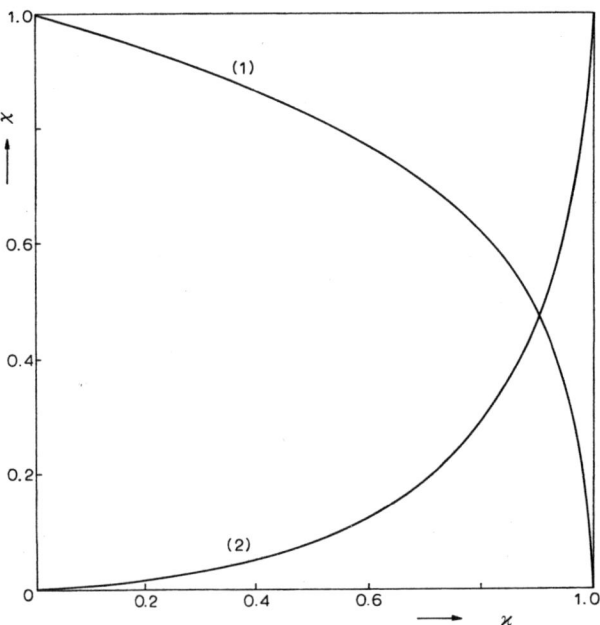

Fig. 29. Decay of the relative contents of double bonds and gain in the contents of cross link points during polymerization of a tetrafunctional monomer. (1) $[M]_3/[M]_0$; (2) $[N]/[M]_0$.

These relations can be verified and compared with experiment. The calculated dependence of the relative concentrations M_1, M_3, and N on conversion is shown in Fig. 28, and the loss of double bonds and gain in cross-link (cyclization) points in Fig. 29. The results are in qualitative agreement with theory; quantitative agreement is not complete [359].

In cross-linking studies it is sometimes useful to adopt an exactly opposite approach: to cross-link ("vulcanize") previously prepared linear chains. This process is applied in practice many different ways, mostly in the rubber industry. For study purposes, reversible cross-linking is most suitable because the network can subsequently be destroyed, for example by hydrolysis. Reactions of suitable macromolecules with chelating compounds are interesting from this point of view.

Brito et al. cross-linked polymers and copolymers of 5-methyl-5-hexene--2,4-dione(methacryloylacetone)

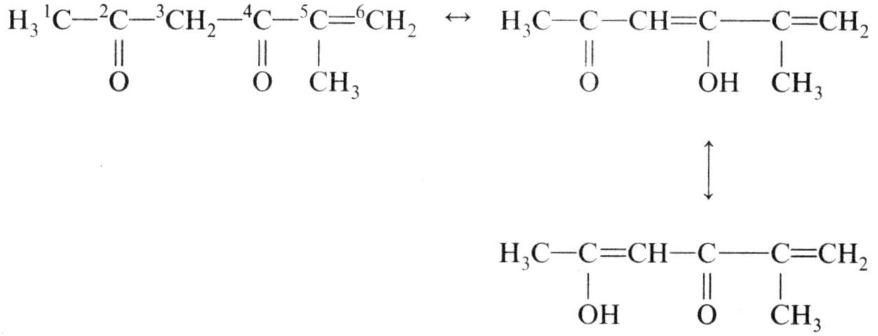

$$H_3{}^1C-{}^2C-{}^3CH_2-{}^4C-{}^5C={}^6CH_2 \leftrightarrow H_3C-C-CH=C-C=CH_2$$

with structural representations showing keto-enol tautomerism, and

with transition metals and their compounds [360]

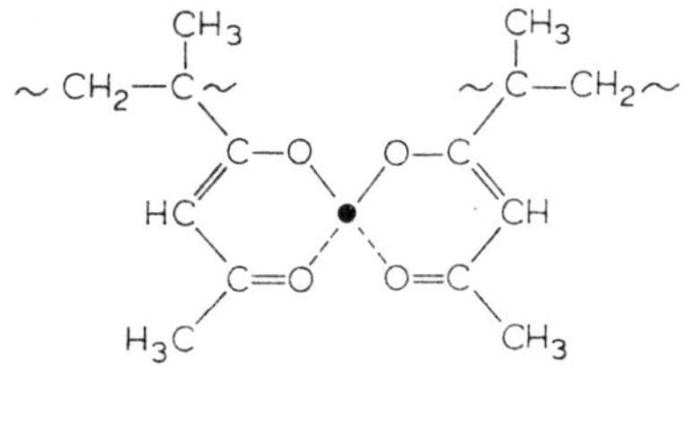

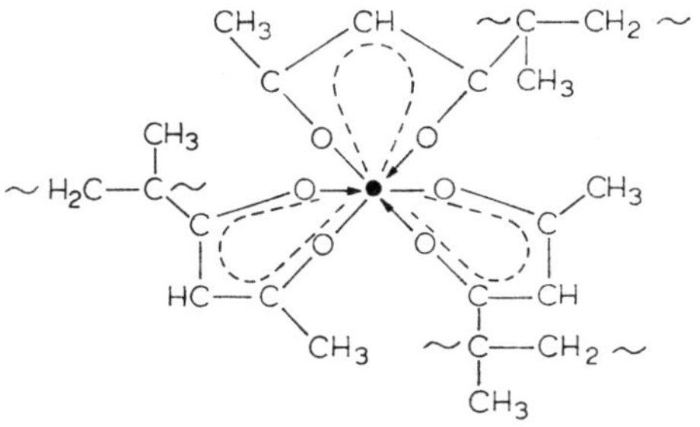

The bond energy in the chelate is comparable with the energy of a covalent bond [361]. More general consequences of this observation have not yet been formulated.

7. Depropagation

From the simple thermodynamic relations of Sect. 1.1, depropagation would appear to be a simple matter. Closer examination indicates that this is not the case [362]. Depropagation is strongly affected by kinetic factors. Its occurrence during addition may considerably alter the polymer structure. Depropagation need not always produce the monomer (this is well documented in Sect. 6.3), and product composition may then be determined by factors other than initiation, propagation, termination, and growth. This is particularly true in the polymerizations of monomers with relatively low polymerization ceiling temperatures. Let us demonstrate the importance of depropagation by several examples.

Wicke and Elgert studied the mode of α-methylstyrene placement in chains polymerized with butyllithium in tetrahydrofuran. They called attention to the possibility of various addition and depropagation rates for isotactic and syndiotactic monomer placement. Isotactic attachment is harder to form, and easier to decompose. For reversible processes this circumstance must be manifested at higher temperatures by thermodynamic, and at lower temperatures by kinetic, control of product stereoregularity [363].

Copolymerization at temperatures above T_c of some of the monomers has already been mentioned. Depropagation of the growing chain end can significantly affect product composition. Cais and Stuk copolymerized vinyl monomers $CH_2=CHX$ with SO_2. The generated polysulphones were not always alternating copolymers. Vinyl chloride, styrene, and acrylamide could successfully compete with SO_2 for addition to the $\sim C^*$ radical. The sulphonyl radical readily depropagates when the penultimate chain unit is also a sulpho group. Depropagation relieves the strain caused by steric and electrostatic repulsion between the oxygen atoms of the adjacent sulphone groups [364].

The mutual relation between thermodynamics and kinetics is nicely demonstrated by some examples of cyclic monomer polymerizations. Jelínek et al. [365] polymerized dodecamethylcyclohexasiloxane. The conversion vs. time plot exhibited an unusual maximum (see Fig. 30). The twelve-membered ring of this monomer is polymerized more rapidly than the eight-membered octamethylcyclotetrasiloxane. Depropagation of polysiloxane chains yields cyclosiloxanes with relatively low ring strain. The equilibrium concentration of the twelve-membered cyclosiloxane is low, and its generation by depropagation is slow. Therefore almost all twelve-membered siloxane rings are rapidly incorporated in the polymer chain. Elimination of the more stable cycles reduces the maximum attainable conversion (hypothetic equilibrium twelve-membered rings ⇌ polymer), and at longer times the measurable

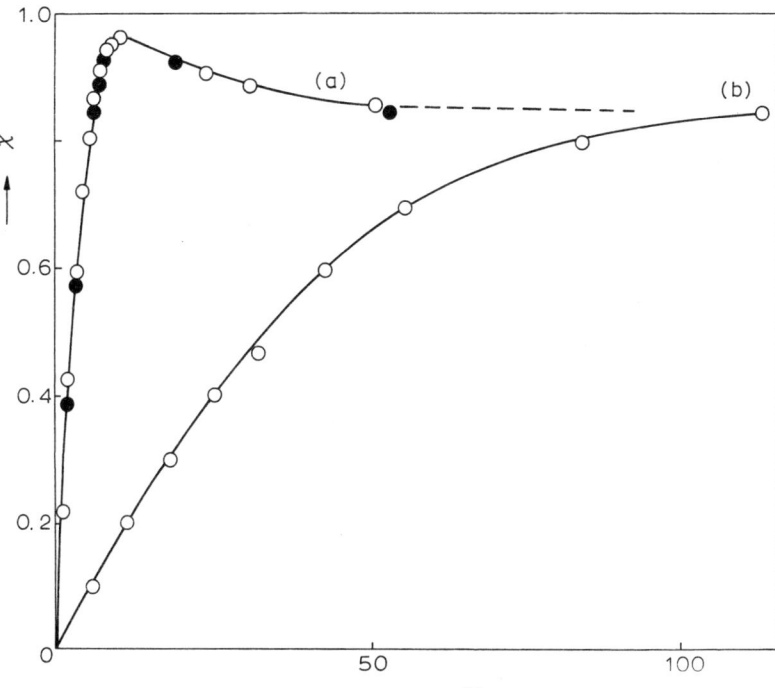

Fig. 30. Conversion curves for the anionic polymerization of (a) dodecamethylcyclotetrasiloxane and (b) octamethylcyclotetrasiloxane at 523 K.

conversion also decreases to a final value which is equal for all cyclic siloxane polymerizations.

The ring size of cycles eliminated from linear siloxane chains is mostly determined by steric effects. From macromolecules containing suitable functional groups only those cycles that are closed with the participation of the respective functional group can be separated. A suitable example are the polyesters from ε-lactones. Only the corresponding ε-lactones or, theoretically, their integral multiples are produced. Ito and Yamashita measured the propagation and depropagation rates of ε-lactones of various sizes. The effective propagation rate constant increased with ring size: the depropagation constant remained almost unchanged. For ε-caprolactone at 273 K, the monomer ⇌ polymer equilibrium is shifted far to the right and monomer amounts are hardly measurable. The equilibrium concentrations of smaller lactones are larger [366].

References pp. 372–382

To date, depropagation has been regarded as a necessary evil complicating polymerization kinetics, reducing polymer yields, and threatening the stability of polymer products. In the future, however, depropagation may attain great technical importance for the regeneration of monomers from used products by controlled depolymerization, (e. g. of polycaprolactam) and for waste removal. For the latter cause, photo- and thermo-oxidative or biodegradative reactions are considered at present.

Even non-construction applications of polymers may gain importance in the future. If a cheap polymer could be synthesized which liberated gaseous products by depolymerization at the required rate at the proper time, it might serve as an energy accumulator. A sufficient number of other possible non--construction applications of polymers, made possible by their controlled depropagation, exist that it cannot be excluded that some of them will be soon realized.

8. Propagation kinetics

In its basic form, propagation, as any elementary reaction, is kinetically very simple

$$v_{\text{prop}} = k_p[\text{M}^\circ][\text{M}] \tag{190}$$

In fact, we know only a few propagations which involve only a single elementary reaction. Usually we must consider two or more types of active centre, each of which can add monomer at a rate characterized by its own k_p

$$v_{\text{prop}} = [\text{M}] \left(\sum_i k_{p_i}[\text{M}_i^\circ] \right) \tag{190a}$$

When the concentration and relative amounts of centres of various reactivity do not change, the product sum

$$\sum_i k_{p_i}[\text{M}_i^\circ]$$

remains constant. In these cases relation (190) for the propagation rate formally remains valid. The elementary k_{p_i} are, of course, replaced by another, compound rate constant, usually termed as overall, effective (apparent, etc.)

$$v_{\text{prop}} = k_{\text{pef}}[\text{M}^\circ][\text{M}] \tag{190b}$$

Enough cases are known where the propagating monomers behave in this formally simple way. For studies of the principles of propagation, living polymerizations are the most important. They enable us to understand mutual relations between additions on various centres, and thus to unravel even more complicated processes in the kinetics of non-stationary polymerizations.

8.1 Propagation in living systems

In the simplest case, with rapid initiation and participation of a single type of active centre, the rate of propagation is equal to the polymerization rate, and k_p is the overall polymerization rate constant. Rapid initiation can be established in ionic processes; the presence of several kinds of centres means unequal numbers of monomer molecule additions to different centres. Long macromolecules will be formed on "rapid" centres, shorter ones on "slow" centres[†]. A practical example of this situation is anionic living polymerization with the participation of contact and solvent-separated ion pairs, and of free ions.

8.1.1 DETERMINATION OF PROPAGATION RATE ON FREE IONS AND ION–PAIRS IN LIVING POLYMERIZATIONS

Styrene polymerization by means of the sodium salt of polystyrene oligomer in dioxan is kinetically the simplest possible polymerization. The low permittivity of the medium should make dissociation of living centres to free ions negligibly small (in dioxan solutions of living polystyrene, no measurable conductivity could actually be detected). Aggregation of living centres does not occur. The rate constant does not change its value with the concentration of living centres or of monomer. It is regarded as the elementary propagation constant on ion pairs $k_p = \bar{k}_{+-}$ [367] ($\bar{k}_{+-}$ designates the rate constant on contact and solvent-separated ion pairs).

In studies of styrene polymerization in tetrahydrofuran, Geacintov et al. observed significant deviations from the simple circumstances described above. The effective rate constant decreased with increasing concentration of living centres [LC] [368]. This was explained by ion-pair dissociation to free ions and proved by the linear dependence of $k_{p_{ef}}$ on $[LC]^{1/2}$ [369, 370].

In the equilibrium

$$\sim CH_2-\underset{Ph}{CH}|^-, Na^+ \;\overset{K_d}{\rightleftharpoons}\; \sim CH_2-\underset{Ph}{CH}|^- + Na^+ \tag{191}$$

[†] As long as the centres are not in dynamic equilibrium, there are oscillations between "slow" and "rapid" form several times during growth.

References pp. 372–382

let us represent the fraction of $\sim$CH$_2$—$^-$CHPh (Na$^+$) liberated by dissociation as x. Then evidently

$$\frac{x^2}{1-x} = \frac{K_d}{[LC]} \tag{192}$$

The dissociation equlibrium is not disturbed by propagation. The effect of chain length is considered to be negligible (with the lower oligomers as a possible exception), and [LC] is not changed by propagation. The observed propagation rate constant k_{Pef} is thus given by the sum of the rate constant products on ion pairs and free ions and their relative amounts

$$k_{Pef} = (1-x)k_{+-} + xk_- \tag{193}$$

For $x \ll 1$ (Worsfold and Bywater [371] found under these conditions at 298 K that $K_d = 1.5 \times 10^{-7}$ mol dm^{-3} and thus $x \sim 1\%$), eqn. (192) can be approximated by $x = K_d^{1/2}[LC]^{-1/2}$, and thus

$$k_{Pef} \cong \bar{k}_{+-} + (k_- - \bar{k}_{+-})K_d^{1/2}[LC]^{-1/2} \tag{194}$$

The value of K_d can be obtained from conductivity or kinetic measurements (see Table 7). The calculation of k_- and $\bar{k}_{+-}$ from experimental data is easy. The correctness of the assumptions and calculations can be verified by the

TABLE 7

Dissociation constant of the equilibrium

$$\sim CH_2\text{---}^-CHPh, \ M_t^+ \ \overset{K_d}{\rightleftharpoons} \ \sim CH_2\text{---}^-CHPh + \dot{M}_t^+$$

in tetrahydrofuran at 298 K

Counter-ion	$K_d \times 10^7$ (mol dm^{-3})		Reference
	From conductivity measurements	From kinetic measurements	
Li$^+$ [a]	1.9	2.2	369
Na$^+$ [a]	1.5		402
Na$^+$ [b]	1.5	1.5 (0.8)	369, 370, 372, 373
K$^+$ [b]	0.7	0.8	369
Rb$^+$ [b]		0.11	369
Cs$^+$ [a]	0.028	0.021	369

[a] Monoanion.
[b] Dianion.

addition of a salt with the same type of cation as that of the living centres. The addition of salt suppresses centre dissociation, and polymerization is slowed down. With an excess of salt such that practically all cations present were formed by its dissociation, the concentration of free anions x is inversely proportional to cation concentration

$$x \cong \frac{K_d}{[M_t^+]} \tag{195}$$

For this case, eqn. (193) yields

$$k_{p_{ef}} \cong \bar{k}_{+-} + (k_- - \bar{k}_{+-})\frac{K_d}{[M_t^+]} \tag{196}$$

$[M_t^+]$ can be calculated from the known dissociation constant of the added

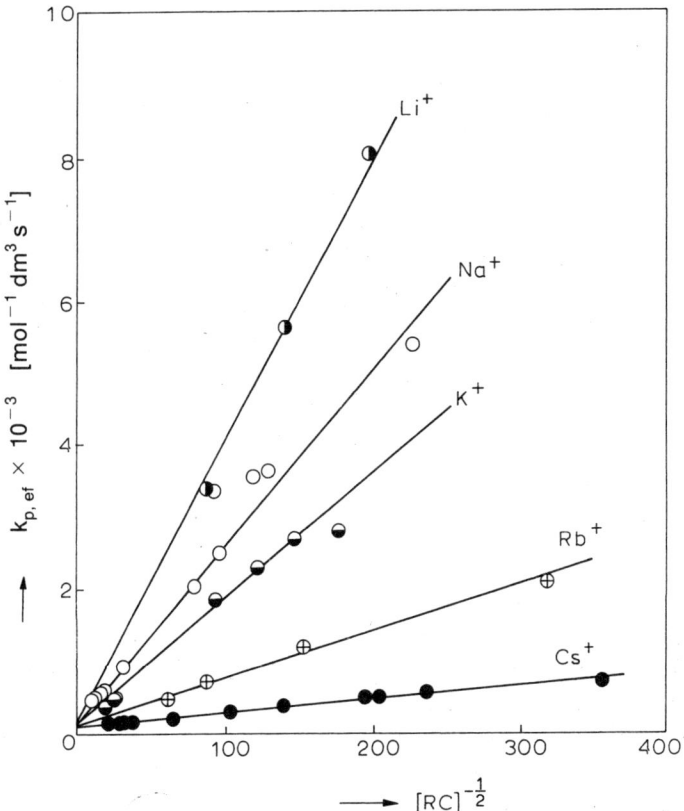

Fig. 31. Dependence of the effective rate constant of propagation on the square root of the growth centre concentration. Polymerization of S in THF at 298 K [369].

References pp. 372–382

salt and from its concentration. An alkali tetraphenylborate salt was found to be suitable in the case discussed. It is easibly soluble, and the position of its dissociation constant in tetrahydrofuran is shifted farther towards free ions than the dissociation equilibrium of living centres. We assume that the

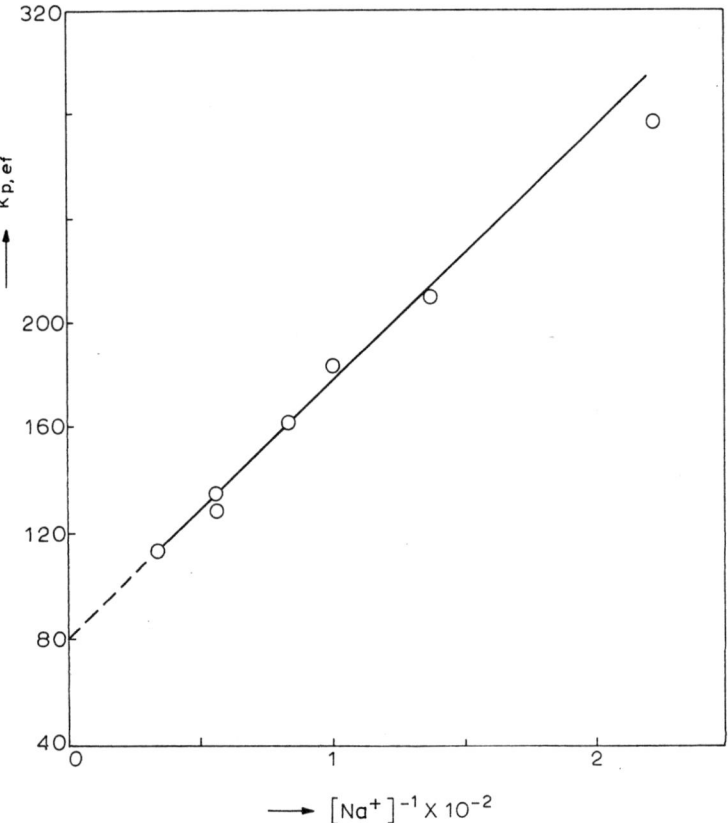

Fig. 32. Dependence of the effective rate constant of propagation on the inverse concentration value of the alkali metal cation. Polymerization of S in THF at 298 K in the pressence of $Ph_4B^-Na^+$ [374].

tetraphenylborate ion does not initiate or otherwise affect polymerization. An example of the evaluation of kinetic measurements by means of eqns. (194) and (196) is shown in Figs. 31 and 32. K_d, k_- and $\bar{k}_{+-}$ can be calculated from the slope of the straight lines constructed for polymerizations in the absence and presence of the salt. Evidently this procedure can also be applied for determining k_+ and $\bar{k}_{+-}$ in living cationic polymerizations. The added salt must, of course, have a common anion.

The value of k_- is independent of the type of solvent and of the counter-ion [374–376]; k_{+-} is generally much smaller and depend on these factors.

Based on this dependence and on the temperature course, the concept of propagation on contact and solvent separated ion pairs can be quantified (see Fig. 33). In solvents that are not too polar the plot of $\ln \bar{k}_{+-}$ vs. T^{-1} is non-linear, revealing the complexity of the constant. The reactivity of solvent-

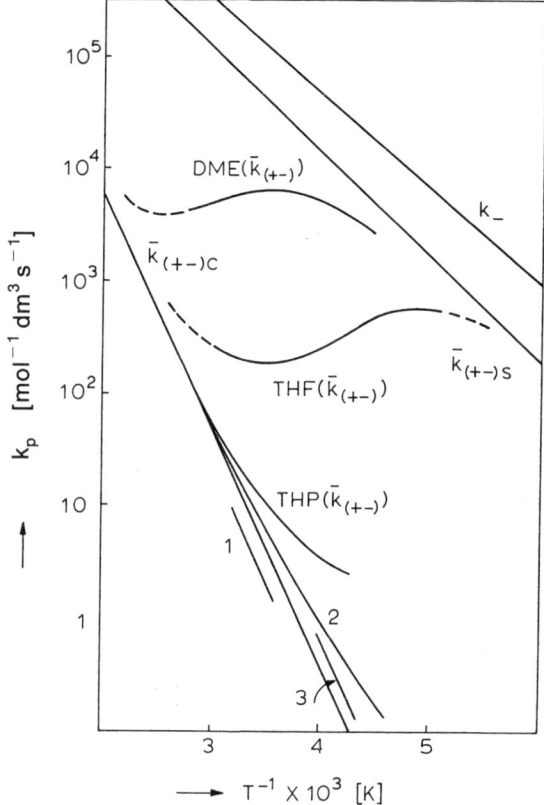

Fig. 33. Arrhenius plot of k_p vs. T^{-1}. The k_- values are the highest in the whole temperature range, $\bar{k}_{(+-)_s}$ are slightly lower, and $\bar{k}_{(+-)_c}$ are the lowest. The dependence is linear. When one ion pair type is transformed to another at the changing temperature $[k_{(+-)_c} \rightleftharpoons k_{(+-)_s}]$ the dependence ceases to be linear. (1) Dioxane, (2) oxepane, (3) cumyl methyl ether. See p. 524 of ref. 19 and ref. 375.

separated ion pairs is about one third [377][†] of that of free ions: the reactivity of contact pairs is lower by three to five orders of magnitude.

[†] This statement is not absolute. Under certain circumstances, the strong dipole of ion pair can suitably orient the approaching monomer molecule and thus facilitate addition. Addition to the ion pair may then be more rapid than to a free ion [374].

References pp. 372–382

We assume that even reactions of contact ion pairs are unaffected by solvent [378].

Williams et al. studied the radiation-initiated cationic polymerization of styrene. By conductivity measurements they were able to determine very small amounts of ions and thus directly determine the propagation rate constant [381] under otherwise comparable conditions, k_+ is about 30 times larger than k_-.

Cationic polymerizations are less well understood than their anionic counterparts, particularly concerning the participation of various ionic forms of active centres in propagation. The values k_+, $k_{(+-)_c}$, and $k_{(+-)_s}$ have mostly not been safely determined (with the exception of some heterocycles, see below). The main reason is probably "contamination" of centres by solvating molecules, and the instability of various centre types caused by the simultaneous solvating and polymerizing ability of the monomers.

Our picture of the transitions between centres is very incomplete so far, based on studies of distribution curve shapes in the products. When a monomer is polymerized by a living mechanism on two or more centres of widely differing reactivity, chains of characteristic legth are produced on each centre type. In a strictly living medium where centres of one type are not transformed to another, a product with a bi- or multimodal distribution curve of degrees of polymerization is formed. When the various centre types are in a dynamic equilibrium where the centre type changes in the course of propagation, the distribution curve of the product will be broader than the width of either of the peaks in the previous case, but narrower than the overall

TABLE 8

Approximate rate constant values for the transition of living centres between contact (c) and solvent-separated (s) ion pairs and free ions (v) at ~255 K
From data in ref. 375.

Constant	Value of constant $k_{i,j}$ in solvent (s^{-1})	
	Tetrahydrofuran	Tetrahydropyran
$k_{c,s}$	110	1
$k_{s,c}$	8 000	4 000
$k_{s,v}$		300
$k_{v,s}$		5×10^9

TABLE 9

Rate constants of propagation[a]

Monomer	Medium[b]	T (K)	Type k_p	k_p (mol^{-1} dm^3 s^{-1})	$\ln \dfrac{A_p}{(\text{mol}^{-1}\,\text{dm}^3\,\text{s}^{-1})}$	E_p (kJ mol^{-1})	Ref.
Styrene	Bulk	303	k_*	55	15.31	30.5	382
	Emulsion				14.60	24.7	382
	THF, Li$^+$	298	k_-	13×10^4	19.8–20.5	41–42	382
	Na$^+$			6.5×10^4			383
	Cs$^+$			13×10^4	20.8	24.7	384
	THP, Li$^+$	303		16×10^4			383
	Na$^+$			13×10^4			384
	K$^+$			14×10^4			376
	Rb$^+$			10×10^4			376
	THF, Na$^+$	298	$k_{(+-)s}$	6×10^4	19.1	20	376
	THF, Na$^+$		$k_{(+-)c}$	15	18.0	36	375
	Cs$^+$			22			375
	DX[c], Li$^+$			0.9			384
	Na$^+$			5			384
	K$^+$			20			375
	DX[c], Cs$^+$			25			384
	Bulk, radiolysis	288	k_+	350×10^4	13.8–18.4	0–8	384
Methyl acrylate	Bulk		k_*		15.7	23	379
Methyl methacrylate	Bulk, benzene				14.2	20.1	382
n-Butyl methacrylate	Bulk				12.7	17	382
Tert. butyl methacrylate					13.3	18	385
2-Vinylpyridine	THF, Na$^+$	298	k_-	10^5	18.4	34	386
			$\overline{k}_{(+-)}$	2 100	24.4		388

TABLE 9 (continued)

Monomer	Medium[b]	T (K)	Type k_p	k_p (mol^{-1} dm^3 s^{-1})	ln $\dfrac{A_p}{\text{(mol}^{-1}\text{ dm}^3\text{ s}^{-1})}$	E_p (kJ mol^{-1})	Ref.
Methacrylonitrile	Bulk		k_*		23.2	48	389
Acrylonitrile	DMF[d]				12.36	27	390
Methacrylamide	H$_2$O				12.87	15.5	391
N-Vinyl carbazole	THF	273			17.68	29	392
	CH$_2$Cl$_2$, C$_7$H$_7^+$ClO$_4^-$		k_+	22×10^4	(26.5)	(27)	393
	C$_7$H$_7^+$SbCl$_6^-$			46×10^4	(25.8)	(24)	393
	(CN)$_2$C=C(CN)$_2$	307		13			394
Vinyl chloride	Emulsion	323	k_*	4.68×10^3			395
Vinyl bromide					14.0	30	392
α-Methylstyrene	THF, NA$^+$	298		830	18.90	30	396
			$k_{-(+-)s}$			25	396
			$k_{(+-)c}$			33–54	396
Oxirane	Bulk, radiolysis	273	k_+	400×10^4	13.8–18.4	0–8	379
	DMSO[e], K$^+$	298	k_-	0.1			397
Methylthiirane	THF, Na$^+$	243	k_-	41	8.75		398
			$\bar{k}_{(+-)}$	0.003			398
Cyclopentadiene	Bulk radiolysis	198	k_+	6×10^8	18.4–23	0–8	379
Isobutyl vinyl ether	Bulk, radiolysis	303	k_+	30×10^4		(28)	379
	CH$_2$Cl$_2$, PH$_3$C$^+$SbCl$_6^-$	273		0.4×10^4			394
Tetrahydrofuran	CH$_2$Cl$_2$, Et$_3$O$^+$BF$_4^-$		k_+	0.01			399
			$\bar{k}_{(+-)}$	0.0014			399
	HOSO$_2$F			0.0016			400
3,3-Dimethylthiethane	Et$_3$O$^+$BF$_4^-$	293	$\bar{k}_{(+-)}$	0.0065	16.35	52.3	401
			k_+	0.067			401
			$\bar{k}_+$	0.0017			401
3,3-Bischloromethyloxetane	PhCl, iso-Bu$_3$Al/H$_2$O	343	$\bar{k}_{(+-)}$	8.5			399
Dioxolane	CH$_2$Cl$_2$, HClO$_4$	273	$\bar{k}_{(+-)}$	10			402
Dioxepan			$\bar{k}_{(+-)}$	3 000			402

Propene	$TiCl_3$–Et_2AlCl[g]	323	k_p	1.2		380
	$TiCl_3$–Et_3Al			0.426	381	
	$TiCl_4$, $MgCl_2$, Et_3Al[g]	333		800		403
	{$TiCl_4$, EB[i], $MgCl_2$, Et_3Al, EB[i,g]}			160		403
Ethylene	$TiCl_4$—iso-Bu_3Al[g]	328		127		21
	cp_2TiCl_2—Me_2AlCl[h]	303		13.6		381
	$\delta TiCl_3 1/3 AlCl_3$, $(^iBu)_3Al$[g]	333		14 000		404
	$TiCl_4$, $MgCl_2$, Et_3Al[g]	338		13 000		403

[a] Condition of k_p determination can only be indicated. Details can be found in the literature cited.
[b] Solvent, in ionic polymerizations counter-ion or initiator.
[c] 1,3-Dioxane.
[d] Dimethylformamide.
[e] Dimethylsulfoxide.
[g] Coordination heterogeneous initiating system.
[h] Coordination homogeneous initiating system.
[i] Ethyl benzoate.

distribution curve of the product from stable centres of various reactivity. An estimate of the transition rates for polystyryl$^-$ Na$^+$ centres

$$\sim CH_2\text{---}^-CHPh, Na^+ \underset{k_{s,c}}{\overset{k_{c,s}}{\rightleftarrows}} \sim CH_2\text{---}^-CHPh \parallel Na^+ \underset{k_{v,s}}{\overset{k_{s,v}}{\rightleftarrows}}$$

$$\underset{k_{v,s}}{\overset{k_{s,v}}{\rightleftarrows}} \sim CH_2\text{---}^-CHPh + Na^+$$

(197)

in tetrahydrofuran and tetrahydropyran at 225 K is summarized in Table 8.

8.2 The rate constant of propagation

The determination of the elementary propagation rate constant is connected with considerable difficulties. Measurements can rarely be made in the gaseous phase, and in the condensed phase the consequences of attractive intermolecular forces cannot be excluded. The reaction studied depends on the medium. Therefore the micromechanism of propagation must be known in detail, and suitable kinetic methods must be applied. Even then, the "elementarity" of the measured constant will not be guaranteed at the present state of the art.

On the other hand, a certain dose of creative spirit is appropriate. When the requirements of modern research methods are respected, good reproducibility can be achieved, disturbing effects can be limited, and our knowledge can be promoted by a further step. The measured constants, even though only defined for a certain system, form an excellent basis for further discoveries. The values of propagation rate constants for some monomers in radical, ionic and coordination polymerizations are summarized in Table 9.

References

1 F. S. Dainton and K. J. Ivin, Q. Rev., 12 (1958) 61.
2 R. D. Snow and F. E. Frey, Ind. Eng. Chem., 30 (1938) 176.
3 A. Eisenberg and A. V. Tobolsky, J. Polym. Sci., 46 (1960) 19.
4 J. C. Wheeler and P. Pfeuty, J. Chem. Phys., 74 (1981) 6415.
5 S. J. Kennedy and J. C. Wheeler, J. Chem. Phys., 78 (1983) 953.
6 M. Szwarc and C. L. Perrin, Macromolecules, 18 (1985) 528.
7 R. M. Joshi and B. J. Zwolinski, in G. E. Ham. (Ed.), Vinyl Polymerization, Part I, Dekker, New York, 1967, pp. 479–490.
8 G. S. Parks and H. P. Mosher, J. Polym. Sci. Part A–1, (1) (1963) 1979.
9 R. J. Orr, Polymer, 4 (1963) 187.
10 C. T. Mortimer, Reaction Heats and Bond Strengths, Pergamon Press, London 1962.

11 T. L. Flitcroft and H. A. Skinner, Trans. Faraday Soc., 54 (1958) 47
12 L. L. Ferstandig, J. Am. Chem. Soc., 84 (1962) 1323, 3553.
13 L. L. Ferstandig and F. C. Goodrich, J. Polym. Sci., 43 (1960) 373.
14 C. E. H. Bawn, W. A. Janes and A. M. North, J. Polym. Sci., Part C, (1963) 427.
15 W. G. Barb, J. Polym. Sci., 11 (1953) 117.
16 R. H. Wiley and D. J. Parish, J. Polym. Sci., 45 (1960) 503.
17 M. L. Miller and J. Skogman, J. Polym. Sci. Part A-1, (2) (1964) 4551.
18 H. C. Brown, R. S. Fletcher and R. B. Johanesen, J. Am. Chem. Soc., 73 (1951) 212; 74 (1952) 1896; H. C. Brown, J. H. Brewster, H. Schechter, 76 (1954) 467.
19 P. E. M. Allen and C. R. Patrick, Kinetics and Mechanisms of Polymerization Reactions, Horwood, Chichester, 1974, pp. 210–220, 230.
20 M. Bodenstein, Z. Phys. Chem., 82 (1913) 329.
21 J. A. Christiansen and H. A. Kramers, Z. Phys. Chem., 104 (1923) 451.
22 P. E. M. Allen and C. R. Patrick, Kinetics and Mechanisms of Polymerization Reactions, Horwood, Chichester, 1974, (a) p. 135; (b) p. 83.
23 G. M. Burnett, Trans Faraday Soc., 46 (1950) 772.
24 W. I. Bengough and H. W. Melville, Proc. R. Soc. (London) Ser. A, 225 (1954) 30; 230 (1955) 429.
25 S. W. Benson and A. M. North, J. Am. Chem. Soc., 81 (1959) 1339.
26 M. J. Hayes and D. C. Pepper, Proc. R. Soc. (London) Ser. A, 263 (1961) 63.
27 J. Šebenda private communication, 1983.
28 Y. B. Amerik and B. A. Krencel, J. Polym. Sci. Part C 16 (1967) 1383.
29 L. Perplies, H. Ringsdorf and J. H. Wendorff, Makromol. Chem., 175 (1974) 553.
30 C. M. Paleos and M. M. Labes, Mol. Cryst. Liq. Cryst., 11 (1970) 385.
31 E. C. Hsu and A. Blumstein, J. Polym. Sci. Polym. Lett. Ed., 15 (1977) 129.
32 G. L. Gaines, Jr. International Reviews of Science Physical Chemistry Series, Vol. 7, Butterworths, London, 1972, p. 1.
33 M. Hatada, M. Nishii and H. Hirota, J. Colloid Interface Sci., 43 (1973) 502; Macromolecules, 8 (1975) 19.
34 D. Nagele and H. Ringsdorf, J. Polym. Sci. Polym. Chem. Ed., 15 (1977) 2821.
35 V. A. Kabanov, T. I. Patrikeeva, K. V. Aliev and O. V. Kargina, J. Polym. Sci. Part C. 16 (1967) 1079.
36 A. Blumstein, S. R. Kakivaya and J. C. Salomone, J. Polym. Sci. Part. B 12 (1974) 651.
37 A. Blumstein, S. R. Kakivaya, R. Blumstein and T. Suzuki, Macromolecules, 8 (1975) 435.
38 G. Weill and A. Blumstein, Polym. Prepr. Am. Chem. Div. Polym. Chem., 17 (2) (1976) 479.
39 T. Doiuchi and Y. Minoura, Macromolecules, 11 (1978) 270.
40 C. Tanford, Physical Chemistry of Macromolecules, Wiley, New York, 1961.
41 P. E. M. Allen and C. R. Patrick, Kinetics and Mechanisms of Polymerization Reactions, Horwood, Chichester, 1974, pp. 85–107.
42 E. Rabinowich, Trans. Faraday Soc., 33 (1937) 1225.
43 P. E. M. Allen and C. R. Patrick, Makromol. Chem. 72 (1964) 106.
44 H. von Smoluchowski, Z. Phys. Chem., 92 (1919) 129.
45 R. M. Noyes, Progress in Reaction Kinetics, Vol. 1, Pergamon Press, London, 1961.
46 F. C. Collins and G. E. Kimball, J. Colloid Sci., 4 (1949) 425.
47 M. Berkowitz, J. D. Morgan, J. A. McCammon and S. H. Northrup, J. Chem. Phys., 79 (11) (1983) 5563.
48 J. F. Garst, J. Chem. Soc. Chem. Commun., (1987) 589.
49 P. Debye, Trans. Electrochem. Soc. 82 (1942) 265.
50 R. M. Noyes, J. Am. Chem. Soc., 76 (1954) 2140; 86 (1964) 4529.
51 R. D. Burkhardt, J. Am. Chem. Soc., 90 (1968) 273.

52 H. G. Carme and G. K. Rollefson, J. Am. Chem. Soc., 74 (1952) 3766.
53 M. Z. Eigen, Z. Electrochem., 64 (1960) 115.
54 A. Einstein, Ann. Phys., 17 (1905) 549.
55 E. Tromsdorff, H. Kohle and P. Lagally, Makromol. Chem., 1 (1948) 169.
56 See for example J. Boor, Jr. Ziegler–Natta Catalysts and Polymerizations, Academic Press, New York, 1979, p. 210.
57 C. H. Bamford, W. G. Barb, A. D. Jenkins and P. F. Onyon, Kinetics of Vinyl Polymerization by Radical Mechanisms, Academic Press, New York, 1958, p. 103.
58 W. I. Bengough and R. G. W. Norrish, Proc. R. Soc. (London) Ser. A., 200 (1950) 310.
59 H. S. Mickley, A. S. Michaels and A. L. Moore, J. Polym. Sci., 60 (1962) 121.
60 M. Kučera, D.Sc. thesis, Research Institute of Macromolecular Chemistry, Brno, 1966.
61 V. A. Kabanov, D. A. Topchiev and T. M. Karapuladze, J. Polym. Sci. Part. C, 42 (1973) 173.
62 N. I. Galperina, T. A. Gugunava, V. F. Gromov, P. M. Komikovskii and A. D. Abkin, Vysokomol. Soedin. Sec. A 17 (1975) 1455; Vysokomol. Soedin. Ser. B, 18 (1976) 384
63 A. Chapiro and J. Dulieu, Eur. Polym. J., 13 (1977) 563.
64 F. Laborie, J. Polym. Sci. Polym. Chem. Ed., 15 (1977) 1255.
65 O. V. Orlova, J. B. Amerik, B. A. Krencel and V. A. Kargin, Dokl. Akad. Nauk SSSR., 178 (1968) 889.
66 J. Gons, E. J. Vorenkamp and G. Challa, J. Polym. Sci. Polym. Chem. Ed., 13 (1975) 1699; 15 (1977) 3031.
67 R. Buter and Y. Y. Tan and G. Challa, J. Polym. Sci. Polym. Chem. Ed., 10 (1972) 1031.
68 A. M. North and G. A. Reed, Trans. Faraday Soc., 57 (1961) 859.
69 J. Gons, W. O. Slagter and G. Challa, J. Polym. Sci. Polym. Chem. Ed., 15 (1977) 771.
70 M. Kučera, Chem. Prum. 7/32 (1957) 443.
71 M. Kučera, J. Lániková and M. Jelínek, J. Polym. Sci., 53 (1961) 301.
72 M. Kučera and E. Spousta, Makromol. Chem., 82 (1965) 60.
73 H. Kramers and K. Westerterp, Elements of Chemical Reactor Design and Operation, Netherland Univ. Press, Amsterdam 1963; Khimicheskie Reaktory. Khimia, Moscow, 1967, pp. 113–143.
74 A. V. Tobolsky, F. Leonard and G. P. Roeser, J. Polym. Sci., 3 (1948) 604.
75 G. C. East, D. Margerison and E. Pulat, Trans. Faraday Soc., 62 (1966) 1301.
76 W. J. Bailey, Polym. Prepr. Am. Chem. Soc. Div. Polym. Chem., 18 (1) (1977) 17.
77 T. Endo, Kobunshi Kako, 33 (5) (1984) 222; Chem. Abstr., 101 (1984) 192501n.
78 T. Endo and T. Ogasawara, Netsu Kokasei Jushi, 5 (2) (1984) 98; Chem. Abstr., 102 (1985) 46258t.
79 M. S. Cohen, C. Bluestein and M. Dunkel, Proc. 8th Int. Radiat. Curing 1984, pp. 11/1–11//12. Chem. Abstr. 104 (1986) 110192 n.
80 W. J. Bailey, H. Katsuki and T. Endo, Polym. Prepr. Am. Chem. Soc. Div. Polym. Chem. 14 (1973) 1976.
81 G. Natta, J. Polym. Sci., 16 (1955) 143.
82 B. D. Coleman and T. G. Fox, J. Am. Chem. Soc. 85 (1963) 1241
83 F. A. Bovey, Polymer Conformation and Configuration, Academic Press, New York, 1969.
84 B. D. Coleman, T. G. Fox and M. Reinmöller, J. Polym. Sci. Part. B, 4 (1966) 1029.
85 H. L. Frisch, C. L. Mallows, F. Heatley and F. A. Bovey, Macromolecules, 1 (1968) 533.
86 P. E. M. Allen and C. R. Patrick, Kinetics and Mechanisms of Polymerization Reactions, Horwood, Chichester, 1974, pp. 502–509.
87 M. Goodman, Top. Stereochem. 2 (1967) 73.
88 K. Hatada, Y. Umemura, M. Furomoto, S. Kokan, K. Ohta and H. Yuki, Makromol. Chem., 178 (1977) 1215.
89 F. A. Bovey, Polymer Conformation and Configuration, Academic Press, New York, 1969; Acc. Chem. Res., 1 (1968) 175.

90 R. Wicke and K.-H. Elgert, Makromol. Chem., 178 (1977) 3075.
91 V. P. Judin, Vysokomol. Soedin. Ser. A, 20 (1978) 1001.
92 M. Biswas and G. M. A. Kabir, Polymer, 19 (1978) 375.
93 C. E. Schildknecht, A. O. Zoss and C. McKinley, Ind. Eng. Chem., 39 (1947) 180.
94 See for example T. Tsuruta, J. Polym. Sci. Part D., 6 (1972) 179.
95 N. Spassky, P. Dumas, M. Sepulchre and P. Sigwalt, J. Polym. Sci. Part C, 52 (1975) 327.
96 N. Spassky, Polym. Prepr. An. Chem. Soc. Div. Polym. Chem., 18 (1) (1977) 76.
97 Y. Izumi and A. Tai, Stereodifferentiating Reactions, Academic Press, New York 1977, Chap. 6.
98 Stereochemical Definitions and Notations Relating to Polymers (Recommendations 1980), Pure Appl. Chem., 53 (1981) 733.
99 Ch. G. D'Hondt and R. W. Lenz, J. Polym. Sci. Polym. Chem. Ed., 16 (1978) 261.
100 T. Higashimura and Y. Hirokawa J. Polym. Sci. Polym. Chem. Ed., 15 (1977) 1137.
101 J. Boor Jr., Ziegler–Natta Catalysts and Polymerizations, Academic Press, New York, 1979, pp. 382–440.
102 T. Suzuki and Y. Takegami, Bull. Chem. Soc. Jpn., 43 (1970) 1484.
103 G. Natta, M. Farina and M. Peraldo, Chim. Ind. (Milan), 42 (1960) 255.
104 T. Miyazawa and T. Ideguchi, Makromol. Chem., 79 (1964) 89
105 A. Zambelli, M. G. Giongo and G. Natta, Makromol. Chem., 112 (1968) 183.
106 G. Natta, J. Inorg. Nucl. Chem., 8 (1958) 589.
107 E. J. Arlman and P. Cossee, J. Catal., 3 (1964) 99.
108 J. Boor, J. Polym. Sci. Part C, 1 (1963) 237.
109 H. W. Coover, Jr., J. Polym. Sci. Part C, 4 (1963) 1511.
110 P. Pino, B. Rotzinger and E. von Achenbach, Makromol. Chem. Suppl., 13 (1985) 105.
111 J. Boor and E. A. Youngman, J. Polym. Sci. Part A–1, (4) (1966) 1861.
112 I. Pasquon, Pure Appl. Chem., 15 (1967) 465.
113 Y. Takegami and T. Suzuki, Bull. Chem. Soc. Jpn., 42 (3) (1969) 848.
114 J. Furukawa, Bull. Inst. Chem. Res. Kyoto Univ., 40 (1962) 130.
115 G. Natta and L. Porri,, Polym. Prepr. Am. Chem. Soc. Div. Polym. Chem., 5 (2) (1964) 1163.
116 K. Matsuzaki and T. Yasukawa, J. Polym. Sci. Part A–1, (5) (1967) 511, 521.
117 E. J. Arlman, J. Catal., 5 (1966) 178.
118 W. M. Saltman, J. Polym. Sci., 46 (1960) 375.
119 G. Natta and L. Porri, in J. P. Kennedy and E. G. M. Törnqvist, (Eds.), Polymer Chemistry of Synthetic Elastomers, Part 2, Chap. 7A, Wiley, Interscience, New York, 1969, p. 600.
120 W. Cooper, G. Degler, D. E. Eaves, R. Hank and G. Vaughan, Adv. Chem. Ser., 52 (1966) 46.
121 G. Natta, L. Porri, A. Carbonaro, F. Ciampelli and G. Allegra, Makromol. Chem., 51 (1962) 229.
122 L. Porri and M. C. Gallazzi, Eur. Polym. J., 2 (1966) 189.
123 W. Marconi, in A. D. Ketley (Ed.), The Stereochemistry of Macromolecules, Dekker, New York, 1967, pp. 239–307.
124 V. I. Klepikova, G. P. Kondratenkov, V. A. Kormer, M. I. Lobachi and L. A. Khurlyaeva, J. Polym. Sci. Polym. Lett. Ed., 11 (1973) 193.
125 C. S. Marvel and W. E. Garrison, J. Am. Chem. Soc., 81 (1959) 4737.
126 K. W. Min and W. H. Ray, J. Macromol. Sci. Rev. Macromol. Chem., C11 (1974) 177.
127 W. D. Harkins, J. Am. Chem. Soc., 69 (1947) 1428; J. Polym. Sci., 5 (1950) 217.
128 W. V. Smith and R. H. Ewart, J. Chem. Phys., 16 (1948) 592.
129 R. N. Haward, J. Polym. Sci., 4 (1949) 273
130 A. T. Yurzhenko and M. Kolechova, Dokl. Akad. Nauk SSSR, 47 (1945) 354.
131 H. Fikentscher, H. Gerrens and H. Schuller, Angew. Chem., 72 (1960) 856.

132 Lindemann, in G. E. Ham (Ed.), Vinyl Polymerization, Dekker, New York, 1967, p. 263
133 J. L. Gardon, J. Polym. Sci. Part A–1, (9) (1971) 2763.
134 M. Harada, M. Nomura, H. Kojima, W. Eguchi and S. Nagata, J. Appl. Polym. Sci., 16 (1972) 811.
135 A. G. Parts, D. E. Moore and J. G. Watterson, Makromol. Chem., 89 (1965) 156.
136 S. N. Sautin, P. A. Kulle and N. I. Smirnov, Zh. Prikl. Khim., 42 (1969) 1812.
137 L. M. Pis'men and S. I. Kuchanov, Vysokomol. Soedin. Ser. A, 13 (1971) 1055.
138 D. C. Sundberg and J. D. Eliassen, in R. M. Fitch (Ed.), Polymer Colloids, Plenum Press, New York, 1971, p. 153.
139 R. M. Fitch and C. H. Tsai, J. Polym. Sci. Part B, 8 (1970) 703.
140 C. P. Roe, Ind. Eng. Chem., 60 (9) (1968) 20.
141 W. H. Stockmayer, J. Polym. Sci., 24 (1957) 314.
142 J. T. O'Toole, J. Appl. Polym. Sci., 9 (1965) 1291.
143 J. Ugelstad, P. C. Mörk and J. O. Aasen, J. Polym. Sci. Part A–1, (5) (1967) 2281.
144 S. Katz and G. M. Saidel, Polym, Prepr. Am. Chem. Soc. Div. Polym. Chem., 7 (1966) 737; J. Polym. Sci. Part C, 27 (1969) 149.
145 A. E. Alexander and D. H. Napper, Prog. Polym. Sci., 3 (1971) 145.
146 M. J. Ballard, R. G. Gilbert, D. H. Napper, P. J. Pomery, P. W. O'Sullivan and J. H. O'Donnell, Macromolecules, 19 (5) (1986) 1303.
147 A. P. Sheinker and S. S. Medvedev, Dokl. Akad. Nauk SSSR, 97 (1954) 111.
148 S. S. Medvedev, IUPAC International Symposium on Macromolecular Chemistry, Plenary Lecture, 1969, Academiai Kiadó, Budapest 1971, p. 39.
149 D. J. Williams, J. Polym. Sci. Polym. Chem. Ed., 11 (1973) 301.
150 D. H. Napper, J. Polym. Sci., Part A–1 (9) (1971) 2089.
151 H. J. van den Hul and J. W. Vanderhoff, Br. Polym. J., 2 (1970) 121.
152 G. I. Litvienko, V. A. Kaminski and M. G. Slinko, Khim. Prom., Khimia (Moscow), (8) (1984) 463.
153 J. W. Vanderhoff, J. Polym. Sci. Polym. Symp. 72 (1985) 161.
154 P. J. Feeney, R. G. Gilbert and D. H. Napper, J. Coll. Interface Sci., 107 (1) (1985) 159, D. R. Stuyman, A. Klein, M. S. El-Aasser and J. W. Vanderhoff, Ind. Eng. Chem. Prod. Res. Rev., 24 (3) (1985) 404.
156 P. E. M. Allen, J. M. Downer, G. W. Hastings, H. W. Melville, P. Molyneux and J. R. Urwin, Nature (London), 177 (1956) 910.
157 D. Mikulášová, V. Chrástová and P. Citovický, Eur. Polym. J., 10 (1974) 551.
158 P. Citovický, D. Mikulášová, V. Chrástová and G. Beňo, Eur. Polym. J., 13, (1977) 655.
159 J. W. Vanderhoff, H. L. Tarkowski, J. B. Shaffer, E. B. Bradford and R. M. Wiley, Adv. Chem. Sci., 34 (1962) 32.
160 D. L. Visioli, M. S. El-Aasser and J. W. Vanderhoff, Polym. Mater. Sci. Eng., 51 (1984) 258.
161 K. Barret, Dispersion Polymerization in Organic Media, Wiley, London, 1975.
162 J. F. Hyde and J. R. Wehrly, U.S. Pat. 2, 891, 920 (1959).
163 D. R. Weyenberg, D. E. Finlay, J. Cekada, Jr. and A. E. Bey, J. Polym. Sci. Part C, 27 (1969) 27.
164 F. R. Mayo and F. M. Lewis, J. Am. Chem. Soc., 66 (1944) 1594.
165 T. Alfrey Jr. and G. Goldfinger G., J. Chem. Phys., 12 (1944) 205.
166 F. T. Wall, J. Am. Chem. Soc., 66 (1944) 2050.
167 I. Skeist, J. Am. Chem. Soc., 68 (1946) 1781.
168 F. T. Wall, J. Am. Chem. Soc., 63 (1941) 1862.
169 G. E. Ham in G. E. Ham (Ed.) Vinyl Polymerization, Part I, Dekker, New York, 1967, pp. 25–46.
170 G. M. Burnett, in Mechanism of Polymer Reactions, Interscience, London, 1954, p. 121.

171 M. Fineman and S. D. Ross, J. Polym. Sci., 5 (1950) 259.
172 P. V. Tidwell and G. A. Mortimer, J. Polym. Sci. Part A–1 (3) (1965) 369; J. Macromol. Sci. Rev. Macromol. Chem., C4 (1970) 281.
173 F. Tüdös, T. Kelen and T. F. Berezhnikh, J. Polym. Sci. Symp. 50 (1075) 109.
174 Z. Macháček, Chem. Listy, 48 (1954) 477.
175 T. Kelen, F. Tűdős and B. Turczányi, Polym. Bull., 2 (1) (1980) 71.
176 H. N. Linssen, Proceedings 1st European Meeting of Statisticians, Grenoble, 1976, North-Holland, Amsterdam, 1977.
177 F. L. M. Hautus, H. N. Linssen and A. L. German, J. Polym. Sci. Polym. Chem. Ed., 22 (1984) 3487.
178 T. Alfrey, Jr. and C. L. Price, J. Polym. Sci., 2 (1947) 101.
179 T. Alfrey and L. J. Young in G. E. Ham (Ed.), Copolymerization, Wiley-Interscience, New York, 1964, Chap. 11.
180 R. Z. Greenley, J. Macromol. Sci. Chem., A14 (4) (1980) 445.
181 R. Z. Greenley, J. Macromol. Sci. Chem., A14 (4) (1980) 427.
182 T. C. Schwan and C. C. Price, J. Polym. Sci., 40 (1959) 457.
183 F. M. Lewis, C. Walling, W. Cummings, E. R. Briggs and F. R. Mayo, J. Am. Chem. Soc., 70 (1948) 1519.
184 H. Sawada, J. Macromol. Sci. Rev. Macromol. Chem., C11 (1974) 257.
185 W. E. Meyer, J. Polym. Sci. Part A–1, (5) (1967) 1289.
186 S. Russo, B. M. Gallo and G. Bonta, Chim. Ind., 54 (1972) 521.
187 R. M. Joshi and S. L. Kapur, J. Sci. Ind Res. Sect. B, 16 (1957) 379.
188 A. Rudin and R. G. Yule, J. Polym. Sci. Part A–1 (9) (1971) 3009.
189 R. M. Joshi, J. Polym. Sci., 60 (1962) 556.
190 G. Goldfinger and T. Kane, J. Polym. Sci., 3 (1948) 462.
191 A. W. Snow, Macromolecules, 10 (1977) 1371.
192 J. P. Kennedy, T. Kellen and F. Tűdős, J. Polym. Sci. Polym. Chem. Ed., 15 (1977) 2277.
193 K. F. O'Driscoll, J. Polym. Sci., 57 (1962) 721.
194 F. Dawans and G. Smets, Makromol. Chem., 59 (1963) 163.
195 Y. Yamashita, T. Tsuda, M. Okada and S. Iwatsuki, J. Polym. Sci. Part A–1, (4) (1966) 2121.
196 Y. Yamashita, T. Asakura, M. Okada and K. Ito, Makromol. Chem., 129 (1969) 1.
197 Y. Yamashita, T. Inoue, G. Hattori and K. Ito, Makromol. Chem., 151 (1972) 91.
198 Y. Tanaka, J. Macromol. Sci. Chem., A1 (1967) 1059.
199 C. Cozewith and G. ver Strate, Macromolecules, 4 (1971) 482.
200 E. Merz, T. Alfrey Jr. and G. Goldfinger, J. Polym. Sci., 1 (1946) 75.
201 W. G. Barb, J. Polym. Sci., 11 (1953) 117.
202 A. Guyot and J. Guillot, J. Macromol. Sci. Chem., A1 (1967) 793.
203 E. M. Shaikhutdinov, B. A. Zhubanov, S. R. Rafikov and S. Kh. Khusainova, Vysokomol. Soedin., 19 (1977) 1961.
204 T. Alfrey, Jr. and G. Goldfinger, J. Chem. Phys., 12 (1944) 322; 14 (1946) 115.
205 C. Walling and E. R. Briggs, J. Am. Chem. Soc., 67 (1945) 1774.
206 R. G. Fordyce, E. C. Chapin, G. E. Ham, J. Am. Chem. Soc., 70 (1948) 2489.
207 G. E. Ham, J. Polym. Sci. Part A, 2 (1964) 2735, 4181.
208 M. Szwarc and C. L. Perrin, Macromolecules, 18 (1965) 528.
209 H. Sawada, J. Polym. Sci., Part A–1, (1965) 2483.
210 R. Orr, Polymer, 2 (1961) 74.
211 H. N. V. Temperley, J. Res. Natl. Bur. Stand., 56 (1956) 55.
212 H. Sawada, J. Macromol. Sci. Rev. Macromol. Chem., C10 (1974) 293.
213 R. E. Cook, K. J. Ivin and J. H. O'Donnell, Trans. Faraday Soc., 61 (1965) 1888.
214 A. M. North and D. Richardson, Polymer, 6 (1965) 333.

215 T. Tsuda and Y. Yamashita, Makromol. Chem., 86 (1965) 304.
216 G. G. Lowry, J. Polym. Sci., 45 (1960) 463.
217 J. E. Hazell and K. J. Ivin, Trans. Faraday Soc., 58 (1962) 176.
218 Y. Yamashita, H. Kasahara, K. Suyama and M. Okada, Makromol. Chem., 117 (1968) 242.
219 P. Wittmer, in N. A. J. Platzer (Ed.), Multi-Component Polymer Systems, American Chemical Society, Washington, DC, 1970, p. 140.
220 K. J. Ivin, Pure Appl. Chem., 4 (1963) 271.
221 H. K. Johnston and A. Rudin, Macromolecules, 4 (1971) 661.
222 H. W. Melville, B. Noble and V. F. Watson, J. Polym. Sci., 2 (1947) 229.
223 K. Matsuo and W. H. Stockmayer, Macromolecules, 10 (3) (1977) 658.
224 P. Wittmer, Makromol. Chem. Suppl, 3 (1979) 129.
225 D. Bralin, G. Disselhoff and F. Quella, Makromol. Chem., 182 (1981) 2951.
226 G. V. Rakova and A. A. Korotkov, Dokl. Akad. Nauk SSSR, 119 (1958) 982.
227 A. A. Korotkov and G. V. Rakova, Vysokomol. Soedin., 3 (1961) 1482.
228 Y. L. Spirin, D. K. Polyakov, A. R. Gantmakher and S. S. Medvedev, J. Polym. Sci., 53 (1961) 233.
229 M. Kučera and J. Pichler, Polymer, 5 (1964) 371.
230 K. F. O'Driscoll and I. Kuntz, J. Polym. Sci., 61 (1962) 19.
231 M. Szwarc, Carbanions, Living Polymers and Electron Transfer Processes, Wiley–Interscience, New York, 1968, p. 529.
232 Z. Feng, F. Yu. and S. Ying, Huadong Huagong Xueyuan Xuebao, 12 (2) (1986) 219; Chem. Abstr., 105 (1986) 209423w.
233 K. Remerie and J. Engberts, J. Phys. Chem., 87 (26) (1983) 5449.
234 M. Apostolopoulos, M. Morcellet and C. Loucheux, Makromol. Chem., 184 (12) (1983) 2519.
235 S. Bywater, in C. H. Bamford and C. F. H. Tipper (Eds.), Comprehensive Chemical Kinetics Vol. 15, Elsevier, Amsterdam, 1976, p. 56.
236 D. A. Tirrel, in J. I. Kroschwitz (Ed.), Encyclopedia of Polymer Science Engineering, Vol. 4, Wiley, New York, 1985, pp. 1920–233.
237 T. Saegusa, S. Kobayashi and Y. Kimura, Macromolecules, 10 (1977) 68.
238 S. Iwatsuki, T. Itoh, M. Shimizu and S. Ishikawa, Macromolecules, 16 (1983) 1407.
239 P. D. Bartlett and K. Nozaki, J. Am. Chem. Soc., 68 (1946) 1495.
240 W. G. Barb, J. Polym. Sci., 16 (1955) 315.
241 F. S. Dainton and G. B. Bristow, Proc. R. Soc. (London) Ser. A, 229 (1955) 509, 525.
242 I. Ito, T. Saegusa and J. Furukawa, J. Chem. Soc. Jpn., 65 (1962) 1878.
243 G. S. Georgiev and V. P. Zubov, Eur. Polym. J., 14 (1978) 93.
244 A. L. Logothetis and J. M. McKenna, Polym. Prepr. Am. Chem. Soc. Div. Polym. Chem., 17 (2) (1976) 642.
245 R. B. Seymour, D. P. Garner, G. A. Stahl and L. J. Sanders, Polym. Prepr. Am. Chem. Soc. Div. Polym. Chem., 17 (2) (1976) 660.
246 N. G. Gaylord and A. Takahashi, Adv. Chem. Sci., 91 (1969) 94.
247 M. H. Litt and J. Radovic-Wellinghoff, Prepr. Polym. Am. Chem. Soc. Div. Polym. Chem., 17 (2) (1976) 596.
248 M. Yoshimura, Y. Shirota and H. Mikawa, Polym. Prepr. Am. Chem. Soc. Div. Polym. Chem., 17 (2) (1976) 590; J. Polym. Sci. Polym. Lett. Ed., 11 (1973) 457.
249 T. L. Petrova, A. I. Smirnov, V. B. Golubev and V. P. Zubov, Deposited Docum. 1981, VINITI 575–82, pp. 17–21. Chem. Abstr. 98 (1983) 198788c.
250 S. Jiao, B. Huang and Y. Qi, Gaodeng Xuexiao Huaxue Xuebao, 5 (3) (1984) 415; Chem. Abstr., 101 (1984) 1984131174m.
251 H. Hirai, M. Komiyama and T. Mori, Kobunshi Ronbunshu, 41 (7) (1984) 421.

252 X. Han, F. Guo and D. Wang, Yingyong Huaxue, 1 (3) (1984) 41; Chem. Abstr., 101 (1984) 171827h.
253 B. Tizianel, C. Caze and C. Loucheux, J. Macromol. Sci. Chem., A22 (11) (1985) 1477.
254 K. Soga, I. Hattori, J. Kinoshita and S. Ikeda, J. Polym. Sci. Polym. Chem. Ed., 15 (1977) 745.
255 K. Koinuma and H. Hirai, Makromol. Chem., 178 (1977) 1283.
256 Y. Hino, Y. Yoshida and S. Inoue, Polym. J. (Tokyo), 16 (2) (1984) 159.
257 T. Saegusa, S. Kobayashi and K. Hayashi, Macromolecules, 11 (1978) 360.
258 S. Kobayashi and T. Saegusa, in J. M. G. Cowie (Ed.), Alternating Copolymers, Plenum Press, New York, 1985, pp. 189–238.
259 Y. Shirota, in J. I. Kroschwitz, Encyclopedia of Polymer Science Engineering, Vol. 3, Wiley, New York, 1985, pp. 327–363.
260 J. Fukurawa, in J. I. Kroschwitz (Ed.) Encyclopedia of Polymer Science Engineering, Vol. 4, Wiley, New York, 1985, pp. 233–261.
261 R. F. Righettini, Diss. Abstr. Int., 46 (12) (1986) 4268.
262 W. J. Burlant and A. S. Hoffman, Block and Graft Copolymers, Reinhold, New York, 1960.
263 J. R. Ceresa, Block and Graft Copolymers, Butterworths, London, 1962.
264 D. C. Allport and W. H. Janes, Block Copolymers, Applied Science Publishers, London, 1973.
265 A. Noshay and J. E. McGrath, Block Copolymers, Academic Press, New York, 1977.
266 L. J. Fetters, Block Copolymers, MMI Press Symposium Series, Vol. 3, 1983, pp 17–38.
267 R. Jerome, R. Fayt and T. Ouhapi, Prog. Polym. Sci., 10 (2–3), 1984, 87.
268 G. Riess, G. Hurtrez and P. Bahadur, in J. I. Kroschwitz (Ed.), Encyclopedia of Polymer Science Engineering, Vol. 2, Wiley, New York, 1985, pp. 324–434.
269 A. A. Berlin, Usp. Khim., 27 (1958) 94.
270 I. Piirma and B. Gunesin, Polym. Prepr. Am. Chem. Soc. Div. Polym. Chem., 18 (1977) 687.
271 M. Lambla, B. Valentin, S. Guerrero and A. Banderet, J. Macromol. Sci. Chem., A11 (8) (1977) 1439.
272 J. F. Henderson and M. Szwarc, Macromol. Rev., 3 (1968) 372.
273 C. L. Lee, J. Smid and M. Szwarc, J. Phys. Chem., 66 (1962) 904.
274 J. L. Lambert and E. J. Goethals, Makromol. Chem., 133 (1970) 289.
275 T. Saegusa, S. Matsumoto and Y. Hashimoto, Macromolecules, 3 (1970) 377.
276 Y. Yamashita, M. Okada and M. Hirota, Angew. Makromol. Chem., 9 (1969) 136.
277 J. P. Kennedy and E. B. Melby, J. Polym. Sci., Polym. Chem. Ed., 13 (1975) 29.
278 M. Kučera, CS 94, 883 (1958).
279 G. Berger, M. Levy and D. Vofsi, J. Polym. Sci. Part B4 (1966) 183.
280 Y. Yamashita, K. Nobutoki, Y. Nakamura and M. Hirota, Macromolecules, 4 (5) (1971) 548.
281 F. J. Burgess, A. V. Cunliffe, D. H. Richards and D. C. Sherrington, J. Polym. Sci. Polym. Lett. Ed., 14 (1976) 471, 483.
282 D. H. Richard, S. B. Kingston and T. Souel, Polymer, 19 (1978) 806.
283 M. Kučera, F. Božek and K. Majerová, Polymer, 20 (1979) 1013.
284 S. P. Mitsengendler, G. A. Andreeva, K. I. Sokolova and A. A. Korotkov, Vysokomol. Soedin., 4 (1962) 1366.
285 L. Reibel, J. C. Schweickert and E. Franta, Proc. IUPAC, 28th IUPAC Macromol. Symp. 1982, p. 114.
286 Y. Kawakami, R. A. N. Murthy and Y. Yamashita, Makromol. Chem., 9 (1984) 9.
287 B. M. Grieveson, Polymer, 1 (1960) 499.
288 M. R. Muidinov, D. P. Kiryukhin, M. K. Asamov and I. M. Barkalov, Vysokomol. Soedin., 20 (1968) 360.

289 K. I. Lee and P. Dreyfuss, Polym. Prepr. Am. Chem. Soc. Div. Polym. Chem., 18 (1) (1977) 11.
290 A. S. Teblina, J. G. Freidlin and V. V. Korshak, Vysokomol. Soedin., 19 (1977) 1482.
291 See M. Szwarc, Carbanions, Living Polymers and Electron Transfer Processes, Wiley-Interscience, New York, 1968, pp. 90–94.
292 M. Kučera, CS 205, 887 (1980).
293 M. Kučera, M. Kunz and K. Majerová, CS 203, 510 (1980).
294 M. Kučera and M. Kunz, CS 212, 627 (1981).
295 S. P. Mitsengendler, K. I. Sokolova, G. A. Andreeva, A. A. Korotkov, T. Kadyrov, S. I. Klenin and S. J. Magarik, Vysokomol. Soedin. Ser. A, 9 (1967) 1133.
296 M. Kučera, Z. Salajka and K. Majerová, Polymer, 26 (1985) 1575.
297 M. Kučera, D. Kimmer, K. Majerová, Paper presented at IUPAC Symposium on Macromolecular Chemistry, Merseburg, D.D.R., 1987.
298 P. Dreyfuss and R. P. Quirk, in J. I. Kroschwitz (Ed.), Encyclopedia of Polymer Science and Engineering, Vol. 7, Wiley, New York, 1987, pp. 551–579.
299 V. Enkelmann and G. Wenger, Makromol. Chem., 178 (1977) 635.
300 P. Cossee, J. Catal., 3 (1964) 80; Rec. Trav. Chim. Pays Bas, 85 (9–10) (1966) 1152.
301 P. H. De Bruin, Chem. Weekbl., 56 (1960) 161.
302 L. A. M. Rodriguez and H. M. van Looy, J. Polym. Sci. Part A–1 (4) (1966) 1971.
303 K. J. Ivin, J. J. Rooney, C. D. Stewart, M. L. H. Green and R. Mahtab, J. Chem. Soc. Chem. Commun., (1978) 604.
304 H. K. Hall, H. Tsuchiya, P. Ykman, J. Otton, V. Papanu, S. C. Snider and A. Deutschman Jr., Polym. Prepr. Am. Chem. Soc. Div. Polym. Chem., 18 (1) (1978) 104.
305 M. L. di Vona, G. Illuminati and C. Lillocci, J. Chem. Soc. Chem. Commun., (1985) 380.
306 K. Matyjaszewski, J. Macromol. Sci. Rev. Macromol. Chem. Phys., C26 (1) (1986) 1.
307 W. J. Bailey, Polym. Prepr. Am. Chem. Soc. Div. Polym. Chem., 18 (1), (1978) 1.
308 W. J. Bailey, P. Y. Chen, S. C. Chen, W. B. Chiao, T. Endo, B. Gapud, V. Kuruganti, Y. N. Lin, Z. Ni, C.-Y. Pan, S. E. Shaffer, L. Sidney, S.-R. Wu, N. Yamamoto, N. Yamazaki, K. Yonezava and L.-L. Zhou, Makromol. Chem. Macromol. Symp., 6, (1986) 81.
309 W. J. Bailey, Makromol. Chem. Suppl., 13 (1985) 171.
310 W. J. Bailey, Ring Opening Polymerization, ACS Symposium Series, Vol. 286, American Chemical Society, Washington, DC, 1985, pp. 47–65.
311 K. Matsuzaki and M. Ito, J. Polym. Sci. Polym. Chem. Ed., 15 (1977) 647.
312 B. A. Rosenberg, E. B. Lyudvig, A. R. Gantmakher and S. S. Medvedev, J. Polym. Sci. Part C, 16 (1967) 1917.
313 S. Penczek, Polym. Prepr. Am. Chem. Soc. Div. Polym. Chem., 18 (1) (1977) 23.
314 P. H. Plesch, in Kinetics and Mechanisms of Polyreactions, Akadémiai Kiadó, Budapest, 1971, p. 213.
315 G. L. Brode and J. V. Koleska, J. Macromol. Sci. Chem., A6 (1972) 1109.
316 A. Hofman, S. Slomkowski and S. Penczek, Makromol. Chem., 185 (1) (1984) 91.
317 A. Hofman, R. Szymanski, S. Slomkowski and S. Penczek, Makromol. Chem., 185 (4) (1984) 655.
318 R. H. Yong, M. Matzner and L. A. Pilato, Polym. Prepr. Am. Chem. Soc. Div. Polym. Chem., 18 (1) (1977) 57.
319 W. T. Grubb and R. C. Osthoff, J. Am. Chem. Soc., 77 (1955) 1405
320 M. Kučera, Collect. Czech. Chem. Commun., 25 (1960) 547.
321 G. A. Latremouille, G. A. Merall and A. M. Eastham, J. Am. Chem. Soc., 82 (1960) 120.
322 Br. Pat. 594, 481 (1945).
323 E. E. Bostick, in K. C. Frisch and S. L. Reegen (Eds.), Ring Opening Polymerization, Dekker, London, 1969, pp. 337–338.

324 J. Chojnowski, S. Rubinsztajn and L. Wilczek, Actual Chim., (3) (1986) 56.
325 G. Sauvet, J. J. Lebrun, P. Sigwalt and E. S. Coethals (Ed.), Proc. 6th Int. Symp. Cationic Polym. Relat. Processes, 1983, Academic, London (1984) 273.
326 W. Vogt, Makromol. Chem., 178 (1977) 3179.
327 J. E. White and R. E. Singler, J. Polym. Sci. Polym. Chem. Ed., 15 (1977) 1169.
328 H. R. Allcock, Makromol. Chem. Macromol. Symp., 6 (1986) 101.
329 D. J. Worsfold and A. M. Eastham, J. Am. Chem. Soc., 79 (1957) 897, 900.
330 S. Okamura, T. Higashimura and T. Miki, Progr. Polym. Sci. Jpn., 3 (1972) 97.
331 G. Pruckmayr and T. K. Wu, Macromolecules, 11 (1978) 265.
332 M. Kučera, F. Božek, K. Majerová and L. Kahle, Polymer, 24 (1983) 217.
333 E. J. Goethals, Adv. Polym. Sci., 23 (1977) 104.
334 K. Weissermel and E. Nölken, Makromol. Chem., 68 (1963) 140.
335 R. J. Kern, J. Org. Chem., 33 (1968) 384.
336 K. Ito, N. Usami and Y. Yamashita, Polym. J., 11 (1979) 171.
337 J. M. Andrews and J. A. Semlyen, Polymer, 13 (1972) 141.
338 H. Jacobson and W. H. Stockmayer, J. Chem. Phys., 18 (1950) 1600.
339 M. Wojtania, P. Kubisa and S. Penczek, Makromol. Chem. Macromol. Symp., 6 (1986) 201.
340 A. S. Pell and G. Pilcher, Trans. Faraday Soc., 61 (1965) 71.
341 K. J. Ivin, Olefin Metathesis, Academic Press, London, 1983, Chap. 11.
342 R. H. Grubbs, in G. Wilkinson, F. G. A. Stone and E. W. Asbel (Eds.), Compendium of Organometallic chemistry, Vol. 8, Pergamon Press, Oxford, 1982, p. 499.
343 P. A. Patton and T. J. McCarthy, Polym. Prepr. Am. Chem. Soc. Div. Polym. Chem., 26 (1) (1985) 66.
344 M. Doherty, A. Stove, A. Parlier, H. Rudler and M. Fontanille, Makromol. Chem. Macromol. Symp., 6 (1986) 33.
345 K. Weiss and R. Goller, J. Mol. Catal., 36 (1-2) (1986) 39.
346 J. Kress, J. A. Osborn, R. M. E. Greene, K. J. Ivin and J. J. Rooney, J. Am. Chem. Soc., 109 (3) (1987) 899.
347 M. Kučera, A. Zahradníčková and K. Majerová, Polymer, 17 (1976) 519, 528, 535.
348 L. Costa, O. Chiantore and M. Guaita, Polymer, 19 (1978) 202.
349 I. Kössler, M. Štolka and K. Mach, J. Polym. Sci. Part C, 4 (1963) 977.
350 S. L. N. Seung and R. N. Young, J. Polym. Sci. Polym. Lett. Ed., 16 (1978) 367.
351 I. Cho and K. D. Ahn, J. Polym. Sci. Polym. Lett. Ed., 15 (1977) 751.
352 T. Saegusa, S. Kobayashi and Y. Kimura, Macromolecules, 8 (1975) 950.
353 H. K. Reimschuessel, Polym. Prepr. Am. Chem. Div. Polym. Chem., 18 (1) (1977) 91.
354 G. Ferraris, C. Corno, A. Priola and S. Cesca, Macromolecules, 10 (1) (1977) 188.
355 A. Shimizu, T. Otsu and M. Imoto, J. Polym. Sci. Part B, 3 (1965) 449.
356 O. W. Webster, Polym. Prepr. Am. Chem. Soc. Div. Polym. Chem., 27 (1) (1986) 161.
357 F. Bandermann and R. Witkowski, Makromol. Chem., 187 (11) (1986) 2691.
358 N. M. Bolbit and S. J. Frenkel, Vysokomol. Soedin., 20 (1978) 294
359 D. Braun, W. Brendlein and Jin Chul Jung, Makromol. Chem., 178 (1977) 2515.
360 H. Brito, V. Brito and J. Springer, Makromol. Chem., 178 (1977) 2507.
361 E. B. Trostyanskaya and P. G. Badaevskii, Usp. Khim., 40 (1) (1971) 117.
362 H. Sawada, in J. I. Kroschwitz, Encyclopedia of Polymer Science Engineering, Vol. 4, Wiley, New York, 1985, pp. 719-745.
363 R. Wicke and K.-F. Elgert, Makromol. Chem., 178 (1977) 3085.
364 R. E. Cais, G. J. Stuk, Polymer, 19 (1978) 179.
365 M. Jelínek, Z. Laita and M. Kučera, J. Polym. Sci. Part C, 16 (1967) 431.
366 K. Ito and Y. Yamashita, Macromolecules, 11 (1978) 68.
367 D. N. Bhattacharyya, J. Smid and M. Szwarc, J. Phys. Chem., 69 (1965) 624.

368 C. Geacintov, J. Smid and M. Szwarc, J. Am. Chem. Soc., 83 (1961) 1253; 84 (1962) 2508.
369 D. N. Bhattacharyya, C. L. Lee, J. Smid and M. Szwarc, Polymer, 5 (1964) 54; J. Phys. Chem., 69 (1965) 612.
370 H. Hostalka, R. V. Figini and G. V. Schulz, Makromol. Chem., 71 (1964) 198.
371 D. J. Worsfold and S. Bywater, J. Chem. Soc., (1960) 5234.
372 T. Shimomura, K. J. Tölle, J. Smid and M. Szwarc, J. Am. Chem. Soc., 89 (1967) 796.
373 H. Hostalka and G. V. Schulz, Z. Phys. Chem. (Frankfurt), 45 (1965) 286.
374 See M. Szwarc, Carbanions, Living Polymers and Electron Transfer Processes, Wiley–Interscience, New York, 1968, pp. 405–424.
375 L. L. Böhm, M. Chmelíř, G. Löhr, B. J. Schmitt and G. V. Schulz, Fortschr. Hochopolym. Forsch., 9, (1972) 1.
376 A. Parry, J. E. L. Roovers and S. Bywater, Macromolecules, 3 (1970) 355.
377 P. Sigwalt, J. Polym. Sci., Symp. 50 (1975) 95
378 See P. E. M. Allen and C. R. Patrick, Kinetics and Mechanisms of Polymerization Reactions, Horwood, Chichester, 1974, p. 527.
379 F. Williams, K. Hayashi, K. Ueno and S. Okamura, Trans. Faraday Soc., 63 (1967) 1501.
380 G. Bier, W. Hoffman, G. Lehmann and G. Seydel, Makromol. Chem., 58 (1962) 1.
381 J. C. W. Chien, J. Polym. Sci. Part A, 1 (1963) 425.
382 See P. E. M. Allen and C. R. Patrick, Kinetics and Mechanisms of Polymerization Reactions, Horwood, Chichester, 1974, pp. 562–569.
383 N. Ise, H. Hirohara and K. Takaya, Macromolecules, 4 (1971) 288.
384 M. van Beylen, M. Fischer, J. Smid and M. Szwarc, Macromolecules, 2 (1969) 575.
385 G. M. Burnett, P. Evans and H. W. Melville, Trans. Faraday Soc., 49 (1953) 1096.
386 D. H. Grant and N. Grassie, Trans. Faraday Soc., 55 (1959) 1042
387 W. I. Bengough and W. Henderson, Trans. Faraday Soc., 61 (1965) 141.
388 M. Fischer and M. Szwarc, Macromolecules, 3 (1970) 23.
389 N. Grassie and N. Vance, Trans. Faraday Soc., 52 (1956) 727.
390 A. F. Revzin, Usp. Khim., 35 (1966) 172.
391 F. S. Dainton and W. D. Sisley, Trans. Faraday Soc., 59 (1963) 1369.
392 J. Hughes and A. M. North, Trans. Faraday Soc., 62 (1966) 1866.
393 P. M. Bowyer, A. Ledwith and D. C. Sherington, Polymer, 12 (1971) 509.
394 C. E. H. Bawn, A. Ledwith and M. Sambi, Polymer, 12 (1971) 209.
395 J. M. Liegeois, J. Macromol. Sci. Chem., A11 (1977) 1379
396 F. S. Dainton, K. M. Hui, and K. J. Ivin, Eur. Polym. J., 5 (1969) 387.
397 C. E. H. Bawn, A. Ledwith and N. R. McFarlane, Polymer, 10 (1969) 653.
398 P. Guerin, P. Hemery, S. Boileau and P. Sigwalt, Eur. Polym. J., 7 (1971) 953.
399 J. M. Sangster and D. J. Worsfold, Macromolecules, 5 (1972) 229.
400 M. U. Mahmud, G. Wenger, W. Kern, J. B. Lando and Y. Osada, J. Macromol. Sci. Chem., A11 (12) (1977) 2233.
401 E. J. Goethals and W. Drijvers, Makromol. Chem., 136 (1970) 73; Polym. Prepr. Am. Chem. Soc. Div. Polym. Chem., 12 (1971) 663.
402 P. H. Plesch and P. H. Westermann, J. Polym. Sci. Part C, 16 (1968) 3837; Polymer, 10 (1969) 105; Adv. Polym. Sci., 8 (1971) 137.
403 Y. I. Ermakov, in F. Ciardelli and P. Giusti (Eds.), IUPAC, Structural Order in Polymers, Pergamon Press, Oxford, New York, 1981, p. 37.
404 N. B. Khumaevskii, V. A. Zakharov, G. D. Bukatov, G. I. Kuznetsova and Y. I. Ermakov, Makromol. Chem., 177 (1976) 747.

Chapter 6

Termination

> ... modern theory can explain more things, and enables us to pose simpler — that is more difficult — questions with the hope for a satisfactory answer
>
> Le Sutton, Oxford

Our knowledge of termination (and also of transfer) is based on relatively little information. Termination usually occurs in the presence of polymer. Analysis of intermediates is therefore often impossible, and analysis of products is extremely difficult. The mechanism of termination reactions is usually extrapolated from similar reactions of organic chemistry.

Of equal theoretical importance as the termination mechanism itself are the consequences of termination rate changes on the overall course of polymerization. They will therefore also be discussed in this chapter.

Further progress in the theory and practice of polymerization processes can be achieved by a transformation of most of the present procedures into stationary or living processes. This, of course, requires the control of termination and its suppression. Thus the situation calls for intensive studies of termination.

To date, by far the greatest experience has been accumulated on the termination of radical polymerizations.

1. Decay of radical activity

There is no reason to doubt that a mutual reaction of two radicals, many times observed in the chemistry of low-molecular-weight compounds and verified by quantum calculations, should also proceed when the atoms with the unpaired electron are part of macromolecules. The behaviour of polymeric radicals is not, however, quite identical to the behaviour of small radicals.

1.1 Combination, disproportionation of radicals

The mutual reaction of two radicals can lead to their simple connection (combination)

$$^1R^* + {}^2R^* \xrightarrow{k_{t,c}} {}^1R-{}^2R \qquad (1)$$

or to the transfer of a hydrogen atom from one radical to the other (disproportionation)

$$^1RCH_2-{}^*CH-{}^2R + {}^*CH^3R-CH_2{}^4R \xrightarrow{k_{t,d}}$$
$$\xrightarrow{k_{t,d}} {}^1RCH_2,CH_2{}^2R + {}^4RCH={}CH^3R \quad (2)$$

The pre-exponential Arrhenius parameter, A of the rate constants $k_{t,c}$ and $k_{t,d}$ is unusually high, of the order of 10^{10} mol^{-1} dm^3 s^{-1} (for small alkyls C_2 to C_4)[†], and the activation energy is very low, practically zero [1]. A high A value and low activation energy of the rate constant are characteristic of a reaction occurring with high probability at every collision of the reacting components. The reaction rate is determined by a parameter controlling molecular motion. We speak of diffusion-controlled reactions.

Combination and disproportionation are, of course, preceded by the occurrence of a different transition complex. It seems that the transition complex leading to the disproportionation of small radicals (ethyl) is more compact than that from which the combined complex is generated. Nevertheless, they both have a loosened structure. The mechanism of disproportionation is not known in detail. The high A factor indicates that a simple abstraction of the β hydrogen by one radical from the other is not probable. The transition complex may be polar [2]; a direct proof is not accessible. The rate is determined by diffusion [3]; the effect of solvent polarity on the activation parameters cannot be measured. The value of k_t (comprising combination and disproportionation) of two small radicals is of the order of 10^9 mol^{-1} dm^3 s^{-1} at temperatures around 273 K.

Experiments have shown that the termination rate constant for two macroradicals is lower by several orders. This finding has, of course, far reaching consequences. It signifies that the rate of decay of a radical pair by mutual interaction is a function of the radical lengths. The basic postulate is breaking down, radical reactivity can no longer be regarded as independent of the degree of polymerization. Derivations of rate equations, kinetic chain lengths, transfer constants or molecular mass distribution curves are inaccurate when changes in k_t during the course of polymerization are neglected.

Fortunately the observed deviations are not usually very large. When differences between the results of calculations and experimental measurements appear, it can not at present always be safely stated to what extent they are caused by variations in k_t or by termination of, or by, primary radicals (non-ideality of polymerization).

[†] Higher by two or more orders than is usual for a bimolecular reaction of the two simple molecules (species).

North confirmed the controlling effect of diffusion in the bimolecular termination of two macroradicals [4]. It is useful to remember that diffusion is manifested by several overall effects, each of which can be considered independently; the main of these are

(a) translational diffusion of gravity centres of macromolecules,

(b) translational and rotational diffusion of segments (micro-Brownian motion, self-diffusion) carrying the active centres [5],

(c) reptations of macromolecules, transfer of their loops or entanglements in a hypothetical tube with an axis coinciding with the bending long axis of the macromolecule (reptation diffusion [6]).

The rate and mode of diffusion are a function of polymer–polymer and polymer–solvent interactions. In a dilute solution where the individual macromolecular coils are separated, hydrodynamic (solvodynamic) and frictional interactions of the deformed coil with the solvent play the predominant role (a). At higher polymer concentrations, the importance of polymer–polymer interactions increases: (i) intermolecular hydrodynamics (better solvodynamics) overshadows the intramolecular hydrodynamic interactions; (ii) long chains can locally hinder the mobility of neighbouring macromolecules or of their parts (entanglements) (a), (b); (iii) at very high polymer concentrations even direct friction of neighbouring chains can occur, the system behaves as a swollen microgel. It contains temporary network cross-links, of finite life–time. Reptation diffusion dominates (b), (c) [6].

Some authors consider diffusion (a), (b) as consecutive processes, and assume the existence of colliding pairs [7–9]. Other models stress the importance of segmental diffusion of the active ends in a common volume of the two colliding macromolecules [10–12]. A common drawback of the mathematical models is the lack of a generally formulated expression for the effective diffusion coefficient of the active end in a coiling chain. Most models try to solve this difficulty by introducing suitable parameters with some physical meaning.

Mahabadi and O'Driscoll [5] derived the termination rate constant for two flexible polymeric radicals, R_m^* and R_n^* in solution, based on the chain lengths m and n, intramolecular coefficients of linear expansion α_A and α_B, and on the excluded volume [9].

Different kinds of approach have recently appeared in a number of important studies, incorporating into the kinetic relations changes in k_t with chain length. They start from the empirical functional dependence of the rate constant of bimolecular termination on the degree of polymerization of the reacting radicals

$$R_m^* + R_n^* \xrightarrow{k_{t(m,n)}} N$$

$$k_{t(m,n)} = k_{t(1,1)} F_{(m,n)}$$

Olaj et al. [13] and Mahabadi [14] set $F(m, n) = (m, n)^{-a}$ (a is different with the two groups of workers). Using the harmonic mean:

$$F(m, n) = [2mn/(m + n)]^{-a}$$

Olaj et al. obtained a better agreement with experiment; however, the geometric mean $F(m, n) = (mn)^{-b/2}$ was only a little less accurate, but led to a significant simplification of the calculation. The approach used by the two groups in deriving their expressions was very similar. Olaj et al. produced a more general solution. Their procedure is shown here and in Chapter 8, Sect. 2.1 [12].

The initiator, I, is decomposed into two primary radicals[†] at the rate v_i

$$I \rightarrow 2 R_0^* \tag{3}$$

The chains grow

$$R_m^* + M \rightarrow R_{m+1}^*$$

$$v_p = k_p[M] \sum_m [R_m^*] \tag{4}$$

and the radical ends of the macromolecules decay at a rate which is dependent on their length

$$R_m^* + R_n^* \rightarrow N$$

$$v_t = \sum_m \sum_n k_{t(m, n)} [R_m^*] [R_n^*]$$

$$= k_{t(1, 1)} F(m, n) [R_m^*] [R_n^*] \tag{5}$$

Transfer is caused by the transfer agent XT (which may even be the monomer)

$$R_m^* + XT \rightarrow N + T^*$$

$$v_{tr} = k_{tr}[XT] \sum_m [R_m^*] \tag{6}$$

Reaction (6) produces the radical T^* with properties identical to those of the primary radicals R_0^* ($T^* = R_0^*$). Both particles R_0^* and T^* react quantitatively with the monomer at a rate controlled by k_p, and they do not terminate (thus they do not contribute to changes in k_t with macroradical length).

[†] The primary radical is the initiating radical fragment of the initiator.

Let us now introduce the quantity of propagation probability P_m, defined by the concentration ratio of the radicals of length $(m - 1)$ (R^*_{m-1}), and those longer by one unit, of length $m(R^*_m)$, generated from R_{m-1} by propagation. P_m is also a measure of the stationary concentration of growing radicals, differing by one unit in length.

$$P_m = \frac{[R^*]}{[R^*_{m-1}]} = \left(1 + \frac{k_{tr}[XT]}{k_p[M]} + \frac{k_t(1,1)}{k_p[M]} \sum_m F(m,n)[R^*_n]\right)^{-1} \quad (7)$$

In this way the stationary concentrations of the radicals $[R^*_m]$ and $[R^*_n]$, in relation to $[R^*_0]$, can be determined

$$[R^*_m] = [R^*_0] \prod_{i=1}^{m} P_i \quad (8)$$

$$[R^*_n] = [R^*_0] \prod_{j=1}^{n} P_j$$

The expressions for the rates of propagation v_p, termination v_t, and transfer are then better written in the form

$$v_p = k_p[M][R^*_0] \sum_m \prod_{i=1}^{m} P_i \quad (4a)$$

$$v_t = k_{t(1,1)}[R^*_0]^2 \sum_m \sum_n F(m,n) \prod_{i=1}^{m} P_i \prod_{j=1}^{n} P_j \quad (5a)$$

$$v_{tr} = k_{tr}[XT][R^*_0] \sum_m \prod_{i=1}^{m} P_i \quad (6a)$$

In eqn. (5a), the dependence of termination on chain length $F(m, n)$, is explicitly included. In the remaining two equations, $F(m, n)$ is only contained through P_i.

In eqn. (5a), the double summation is set equal to the square of a single summation in order to simplify the formulation of the further steps

$$S_1^2 \equiv \sum_m \sum_n F(m,n) \prod_{i=1}^{m} P_i \prod_{j=1}^{n} P_j \quad (9)$$

and the simple summations from eqns. (4a) and (6a) are designated as S_2

$$S_2 \equiv \sum_m \prod_{i=1}^{m} P_i \quad (10)$$

The kinetic chain length v is then given by the expression

$$v = \frac{v_p}{v_t} = \frac{k_p[R_0^*][M] S_2}{k_t(1,1)[R_0^*]^2 S_1^2} = \frac{k_p[M]}{k_{t(1,1)}[R_0^*] S_1} \frac{S_2}{S_1} \qquad (11)$$

The member $k_p[M]/k_{t(1,1)}[R_0^*] S_1$ has the meaning of the kinetic length of chains $v_0 \equiv C^{-1}$ which would be formed under identical conditions, but in a system where termination is independent of macroradical length.

$$v_0 = \frac{k_p[M]}{k_t(1,1)[R_0^*] S_1} = \frac{1}{C} \qquad (12)$$

The expressions derived can be further formally simplified by setting $k_{tr}[XT]/k_p[M] = D$. Using eqns. (8) and (12), eqn. (7) can then be transcribed in the form

$$P_m = \left[1 + D + \frac{C}{S_1} \sum_n F(m,n) \prod_{j=1}^n P_j \right]^{-1} \qquad (7a)$$

Calculating $\sum_n F(m,n) \sum_{j=1}^n P_j$ from eqn. (9) and substituting into eqn. (7a), we obtain a formal equation for S_1

$$S_1 = C^{-1} \sum_{m=1}^\infty (P_m^{-1} - D - 1) \prod_{i=1}^m P_i \qquad (13)$$

S_1 and S_2 have the meaning of (zeroth) distribution moments

$$f_1(m) = C^{-1}(P_m^{-1} - D - 1) \prod_{i=1}^m P_i \qquad (14)$$

and

$$f_2(m) = \prod_{i=1}^m P_i \qquad (15)$$

Thus $f_1(m)$ is the (unnormalized) length distribution of inactive chains formed by disproportionation, particularly in systems where disproportionation represents an exclusive or predominating termination mechanism. $f_2(m)$ corresponds to the (unnormalized) length distribution of macroradicals.

The general moments $S_1^{(k)}$ and $S_2^{(k)}$ are expressed by the equations

$$S_1^{(k)} = C^{-1} \sum_{m=1}^\infty m^k (P_m^{-1} - D - 1) \prod_{i=1}^m P_i \qquad (16a)$$

or, in the sense of the definition of S_1 according to eqn. (9)

$$S_1^{(k)} = \frac{1}{S_1^{(0)}} \sum_m m^k \sum_n F(m, n) \prod_{i=1}^m P_i \prod_{j=1}^n P_j \qquad (16b)$$

and

$$S_2^{(k)} = \sum_{m=1}^{\infty} m^k \prod_{i=1}^m P_i \qquad (17)$$

With the moments $S_1^{(k)}$ and $S_2^{(k)}$, $S_3^{(k)}$ can also be defined

$$S_3^{(k)} = CS_1^{(k)} + DS_2^{(k)} \qquad (18)$$

By means of the latter relations, kinetic schemes can be completely solved, mean degrees of polymerization can be derived, as well as the polydispersity coefficients of polymers terminated by disproportionation $(= S_1^{(k+1)}/S_1^{(k)})$ and of macroradicals or inactive macroradicals after transfer $(= S_2^{(k+1)}/S_2^{(k)})$. For the number, weight, and z average, $k = 0, 1$, and 2, respectively.

The problem of a reliable determination of the ratio of combining and disproportionating radicals remains unsolved.

The chains of the produced polymer are lengthened by combination, but not by disproportionation. This affects the molecular mass distribution but the differences are not very large, differing by a factor of 2 at most. Due to the inaccuracies in molecular mass determinations, it is almost impossible to make estimates of the relation between termination and disproportionation from the distributions. Even labelling of the initiator and determination of the average number of its fragments in a macromolecule (one for disproportionation and two for combination) is usually unsafe because of transfer.

The opinions of various authors on the mode of termination therefore do not always agree. Bamford et al. have found that polymethacrylate radicals mostly disproportionate while those of polyacrylonitrile mostly combine at 363 K [15]. More detailed measurements by Schulz et al., who also considered transfer in methyl methacrylate polymerization, enabled them to express the ratio of the two termination modes quantitatively [16]

$$\log_{10}\left(\frac{k_{t,d}}{k_{t,c}}\right) = 2.89 - \frac{1\,000}{T} \qquad (19)$$

Bevington et al. [17, 18] published the equation

$$\log_{10}\left(\frac{k_{t,d}}{k_{t,c}}\right) = 3.34 - \frac{880}{T} \qquad (20)$$

According to these authors, disproportionation of polystyryl radicals is unimportant. Also, poly(methyl acrylate) radicals almost exclusively combine around 300 K [19].

1.1.1 THE EFFECT OF PRESSURE ON TERMINATION

According to general ideas on polymerization, the rate of a termination with a diffusion-controlled rate constant should decrease with increasing viscosity caused by growing pressure.

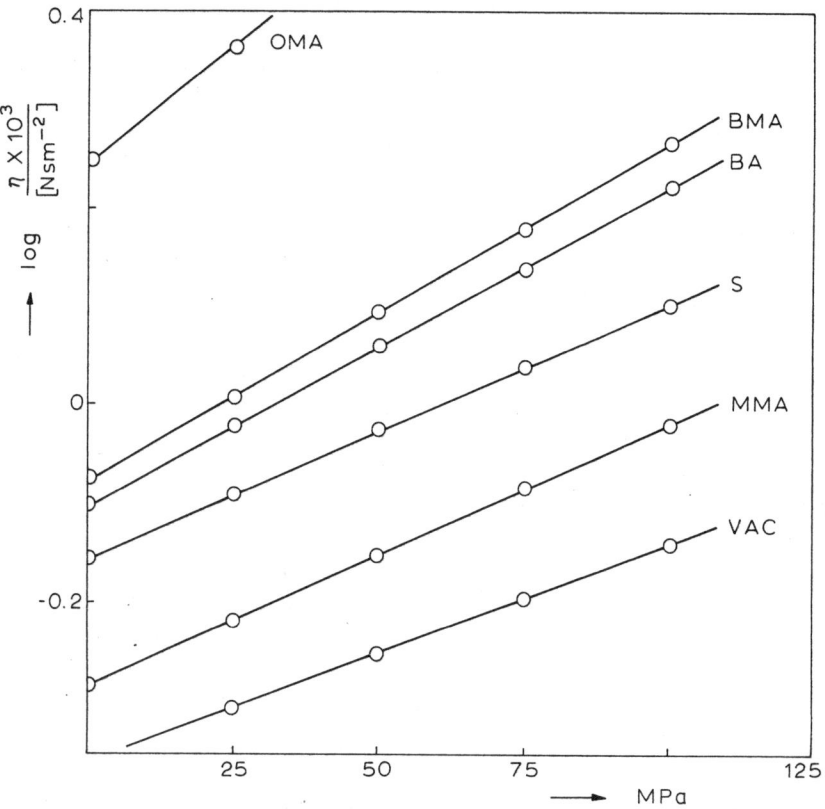

Fig. 1. Dependence of viscosity on pressure for selected monomers. BA = butyl acrylate, BMA = butyl methacrylate, OMA = octyl methacrylate. S, MMA, and VAC have their usual meanings.

Ogo and Kyotani measured the viscosity dependence on pressure for some monomers [20]. The results are shown in Fig. 1. Their mathematical expression has the form

$$\eta_p = \eta_0 \exp(\varrho p) \qquad (21)$$

where p is the pressure and ϱ a constant. The activation volume of viscous flow $\Delta V_{visc}^{\ddagger}$ is a measure of volume expansion during viscous flow of the liquid. It corresponds to the volume of cavities accessible to the molecules and can be derived from the viscosity of the liquid at high pressure

$$\Delta V_{visc}^{\ddagger} = RT \left(\frac{\partial \ln \eta}{\partial p}\right)_T \tag{22}$$

Differentiating eqn. (21) and substituting the result in eqn. (22), we obtain the simple dependence

$$\Delta V_{visc}^{\ddagger} = \varrho RT \tag{23}$$

Thus the slopes of the straight lines in Fig. 1 are directly proportional to $\Delta V_{visc}^{\ddagger}$ [†]. Ogo and Kyotani [20] have used these data for calculating k_t at high pressures. They started from the equation

$$k_t = F_1(T,\zeta_0) \, F_2(\alpha, n) \tag{24}$$

where ζ_0 is the friction coefficient of the polymer segment, α the linear expansion coefficient of the statistical coil, and n the number of monomer units in a segment [5], with some simplifying assumptions; first of all they selected the relative constants and functions defined by the equations

$$k_{t(rel)} = \frac{k_{t(p)}}{k_{t(0)}} \tag{25}$$

where $k_{t(rel)}$, $k_{t(p)}$ and $k_{t(0)}$ are termination rate constants (relative, at presure p, and at atmospheric pressure, respectively). Further, they used the relations

$$F_{1(rel)} = \frac{\eta(0)}{\eta(p)} \tag{26}$$

$$F_{2(rel)} = \frac{k_{t(rel)}}{F_{1(rel)}} \tag{27}$$

which are valid when k_t is inversely proportional (i.e. through F_1) to viscosity. Equation (27) describes the dependence of F_2 on pressure. The effect of pressure on $k_{t(rel)}$, $F_{1(rel)}$, and $F_{2(rel)}$ in polymerizations of the monomers from

---

[†] $\Delta V_{visc}^{\ddagger}$ is also a function of molecular mass.

References pp. 438–442

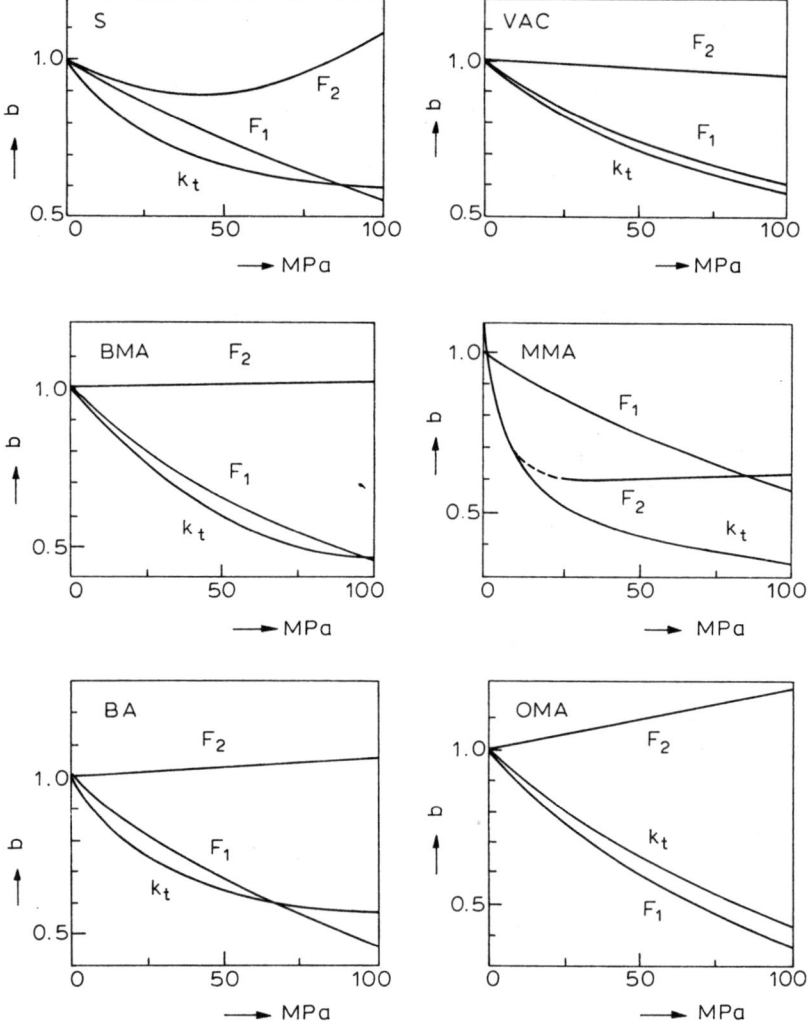

Fig. 2. Effect of pressure on the relative factors of termination (b) arbitraty units.

Fig. 1 is shown in Fig. 2. When the cases of styrene and methyl methacrylate below 25 MPa are excluded because of non-linear $F_{2(rel)}$ dependence[†], the following relations (in which β is the slope of the $F_{2(rel)}$ straight lines in Fig. 2) can be formulated.

[†] According to O'Driscoll, $F_{2(rel)}$ should increase with pressure; in the cases discussed, the counteracting effects of viscosity changes in monomer viscosity and polymeric coil dimensions are very pronounced.

$$F_{1(\text{rel})} = \frac{\eta_{(0)}}{\eta_{(p)}} = \exp(-\varrho p)$$

$$F_{2(\text{rel})} = \frac{F_{2(p)}}{F_{2(0)}} = 1.00 + \beta p$$

$$k_{t(\text{rel})} = \frac{k_{t(p)}}{k_{t(0)}} = F_{2(\text{rel})} F_{1(\text{rel})} = (1.00 + \beta p) \exp(-\varrho p)$$

$$k_{t(p)} = k_{t(0)}(1.00 + \beta p) \exp(-\varrho p)$$

thus

$$\ln \frac{k_{t(p)}}{1.00 + \beta p} = C - \varrho p \tag{28}$$

where $C = \ln k_{t(0)}$. The situation is further simplified, for negligible values of β, to

$$\ln k_{t(p)} = C - \varrho p \tag{28a}$$

The authors [20] regard eqn. (28) as a semiempirical relation which is generally useful for predicting the effect of pressure on k_t in all radical polymerizations.

1.1.2 THE EFFECT OF POLYMER CONCENTRATION AND CHAIN LENGTHS

Radical termination can naturally only occur when both reaction participants have mutually approached to within reaction distance. Diffusion of all macroradicals is strongly affected by the medium for the whole polymerization time. The dependence of k_t on the quantities that vary in the course of this process, i.e. on concentrations, dimensions of radicals and of inactive molecules, temperature or solvent is not known with sufficient accuracy. Only with some systems it has been possible to isolate the remaining variables from the measured variable, and in this way important data have been obtained.

North and Reed [21] have shown that in a system with a small poly(methyl methacrylate) content (up to ca. 10 wt. %)

(i) the diffusion coefficient of PMMA is inversely proportional to η_0, the viscosity of the system prior to the start of polymerization;

References pp. 438–442

(ii) in a good solvent, k_t increases with conversion, but decreases in a poor solvent;

(iii) for PMMA in a good solvent in the same concentration range, the mutual diffusion coefficient (D_{mut}) increases with concentration whereas the self-diffusion coefficient of reptation (D_s) decreases;

(iv) neither k_t, nor D_s or D_{mut} are (inversely) proportional to solution viscosity at finite polymer concentrations;

(v) the value of k_t exhibits a maximum in its dependence on both the polymer concentration and its molecular mass.

Mahabadi and O'Driscoll derived equations forming the theoretical background of the observed effect [22]. Good agreement between the calculated and measured values was obtained. They ascribe the growth of k_t in the initial phases of polymerization to the effect of intermolecular interactions on the self-diffusion rate of the end segments in growing chains, which controls termination.

An increase in k_t at low conversions was also observed for polymerizations of styrene and of other monomers.

Ludwico and Rosen [23] polymerized styrene in solution in the presence of a known amount of polystyrene of known molecular mass. The value of k_t again reached a maximum; its decrease was more or less linear. The effect of the molar mass of macroradicals and of added polystyrene was small in the stage of decreasing k_t.

The diffusion mode of macroradicals is predominantly determined by the molecular mass distribution both of the radicals and of the inactive polymer. The self-diffusion coefficient of a discrete species depends, of course, on the molecular mass of that particle itself [24].

1.2 Termination of and by primary radicals

For "ideal" radical polymerization to occur, three prerequisites must be fulfilled: for both macro- and primary radicals, a stationary state must exist; primary radicals have to be for initiation only and termination of macroradicals only occur by their mutual combination or disproportionation. The rate equation for an "ideal" polymerization is simple (see Chap. 8, Sect. 1.2); it reflects the simple course of this chain reaction. When the primary radicals are deactivated either mutually or with macroradicals, kinetic complications arise. Deviations from ideality are logically expected to be larger the higher the concentration of initiator and the lower the concentration of monomer. Today termination by primary radicals is an exclusively kinetic problem. Almost nothing has been published on the mechanism of radical liberation from the aggregation of other initiator fragments and from the cage of the surrounding molecules. Our knowledge of the chemistry of termination by primary radicals is no better.

$$\sim\sim M^* + R_0^* \xrightarrow{k_{t_{pr}}} \sim\sim MR \tag{29}$$

Participation of termination according to eqn. (29) during polymerization is reflected in the reaction orders with respect to [I] and [M], in changes of $v_p/([I]^{1/2}[M])$ with concentration of monomer or initiator, etc. A modification of the rate equation with participation of termination by primary radicals was derived by Olaj [12] and by Deb and Meyerhoff [25] (see Chap. 8, Sect. 1.2). Many authors have started to recognize the existence of reaction (29), and to measure at least the value of the group of rate constants comprising $k_{t_{pr}}$, for example $k_{t_{pr}}/(k_i k_p)$, and their activation energy. Unfortunately not all deviations from "ideality" are caused by termination by primary radicals. Induced initiator decomposition can also be a complicating factor. Many authors have noted that the kinetics of mutual termination of polymer radicals is a function of their dimensions [27, 28] (see Sect. 1.1). Mahabadi and O'Driscoll recommend measurements of $k_{t_{pr}}$ at constant $\bar{P}$ (the effect of macroradical length on $k_{t_{pr}}$ is thus eliminated) and at low conversion (while the effect of concentration on $k_{t_{pr}}$ is negligible). These conditions can be fulfilled in polymerizations with a transfer agent. The authors call attention to the existence of a functional dependence between $\bar{k}_t$ and chain length, which is manifested in the same way as termination by primary radicals. In literature data neglecting this fact, the value of $k_{t_{pr}}$ is overestimated. An opposite error, the underestimation of $k_{t_{pr}}$ or of the ratio $k_{t_{pr}}/(k_i k_p)$, may be caused by neglect of a possible thermal initiation [28].

1.3 Systems with retarded termination

The ratio of the rates of initiation and termination is an important factor determining the character of a polymerization. When termination is more rapid than initiation, propagation is limited and a high polymer is not formed. Such a situation is sometimes created intentionally for study purposes. For practical use it is of no great importance. Real systems have two extremes: either initiation is equally rapid as termination or termination does not occur in the respective medium. In the first case, polymerization is stationary; in the other, it is living (proceeding without transfer). In between these limits we find systems where termination is slower than initiation. They all exhibit an autoaccelerated course. The polymerizations are usually stationary at first with termination limited by the effect of the generated polymer.

References pp. 438–442

1.3.1 THE GEL EFFECT

Norrish and Smith [29] and later Tromsdorff et al. [30] described a polymerization of methyl methacrylate, the rate of which increased from a certain conversion. The number of monomers of similar behaviour was extended by methyl acrylate [31], butyl acrylate [32] and other acrylates [33] and methacrylates [34], and vinyl acetate. The effect was explained by the reduction of the termination rate caused by hindered macroradical mobility[†] in viscous medium; it was called the gel effect, or the Norrish-Tromsdorff effect. The gel effect is clearly manifested in radical polymerizations of weakly transferring monomers in bulk. It is significant also in the presence of a good solvent. The gel effect is suppressed by the presence of poor solvents[††] and by

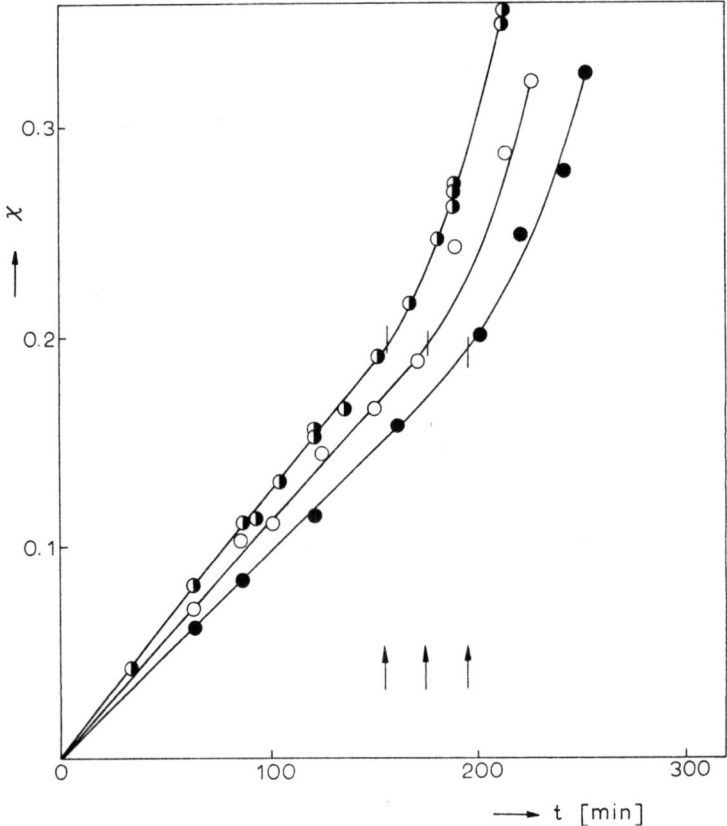

Fig. 3. Conversion curves of methyl methacrylate polymerization. Temperature 323 K, initiator AIBN at concentrations (mass %) of ●, 0.3; ○, 0.4; ◐, 0.5. Original results [35, 41] redrawn according to ref. 4.

[†] The mobility of the small molecules (monomer, primary radicals and transfer agents) is hardly affected.
[††] The statistical coils of macromolecular chains in poor solvents are closely packed.

transfer reactions. Externally this effect is manifested by a sudden and steep increase of molecular mass and by an autoacceleration character of the conversion vs. time dependence, until monomer depletion or transition of the system from a liquid into a viscoelastic phase. Both these manifestations are illustrated in Figs. 3 and 4.

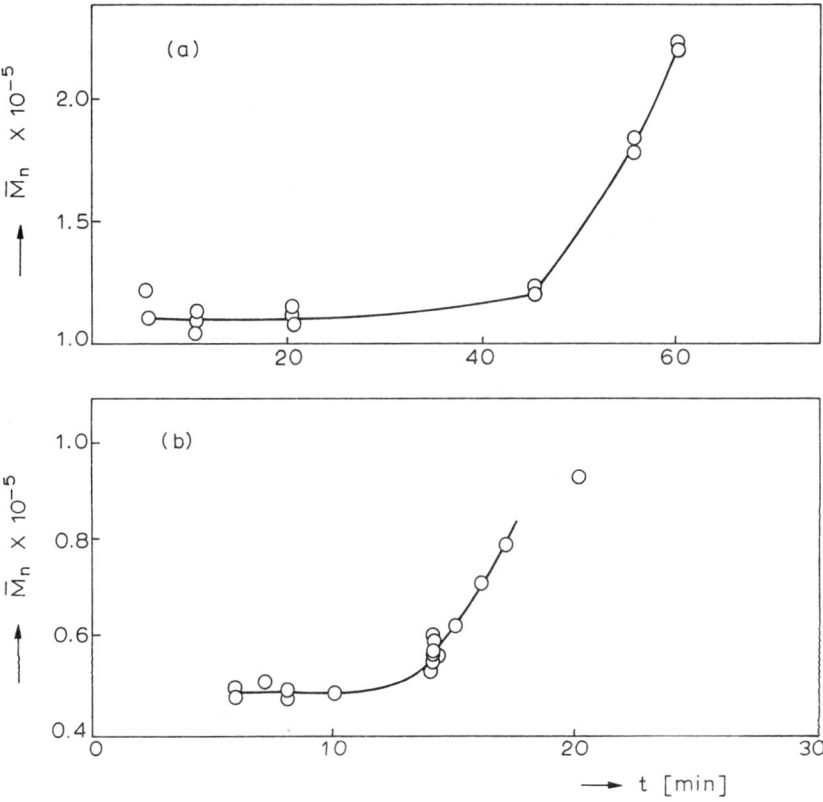

Fig. 4. Dependence of the molecular mass of PMMA on time at polymerization temperatures of (a) 343 K and (b) 363 K [35, 41].

Diffusion control of termination is presently beyond any doubt. Various authors differ, however, in their opinion concerning the type of diffusion and the quantitative contribution of diffusion in the collection of processes manifested as the gel effect. Attempts to explain the Norrish-Tromsdorff effect on a molecular level are still of empirical character [8, 12, 36–38].

According to Cardenas and O'Driscoll [39], there exist two populations of macroradicals, of a polymerization degree greater and smaller than a certain critical size $\bar{P}_c$. $\bar{P}_c$ is defined by a bend in the plot of log η (system viscosity) vs. log DP (degree of polymerization of the dissolved polymer). With increasing concentration and chain length of the polymer, the importance of poly-

References pp. 438–442

mer–polymer interactions increases until, at the critical point, the statistical coils begin to interpenetrate and entangle; the mobility of the clinched molecule decreases abruptly. The termination of radicals R_m^* and R_n^* of $m < \bar{P}_c$ and $n < \bar{P}_c$ is given by $k_{t,0}$ corresponding to the reaction in a dilute solution. When $m > \bar{P}_c$ and $n > \bar{P}_c$, the termination rate constant of these entangled radicals is $k_{t,e}$, whereas the radical pair with $m > \bar{P}_c$ and $n < \bar{P}_c$ is terminated with the rate constant $k_{t,c}$, which is a geometric mean of $k_{t,0}$ and $k_{t,e}$: $k_{t,c} = (k_{t,0} k_{t,e})^{1/2}$.

To a first approximation, $k_{t,c}$ is inversely proportional to the density of entangled coils, resulting in the functional dependence

$$k_{t,e} \sim (\Phi_p \bar{P})^{-1}$$

where Φ_p is the volume fraction of polymer, and $\bar{P}$ is the degree of polymerization of inactive macromolecules.

A variant of the concept of Cardenas and O'Driscoll is the approach of Marten and Hamielec [40]. According to these authors, k_t is proportional to the difusion coefficients, which are a function of the viscosity of the medium. This viscosity depends on the weight average mean molecular mass and thus k_t can be written as

$$k_t \approx \bar{M}_w^{-b} e^{-a/V_f}$$

where a is a constant, b is a parameter with an assigned value of 1.75, and V_f is the free volume.

Turner et al. [41–43] do not consider reduced coil mobility with increasing polymer concentration as decisive. Under these conditions, macromolecules behave as rigid balls, mutually hardly penetrable; the probability of radical encounter decreases. The onset of autoacceleration is also ascribed to mutual coil entanglement which only occurs at and above the critical polymer concentration C_c and a certain minimum mean molecular mass $\bar{M}_c^a (0.5 \lesssim a \lesssim 1)$. The product of these critical quantities is constant

$$K = C_c \bar{M}_c^a \tag{30}$$

The onset of the gel effect is identified either by the beginning of coil overlap ($a = 0.5$) or by the onset of entanglement ($a = 1$).

All the described procedures can reproduce surprisingly well the autoaccelerating course of the rate and the degrees of polymerization for the products of radical polymerizations in bulk or in concentrated solutions. From the point of view of the technological control of these systems it is a success. On the other hand, this very agreement hinders our approach to a real understanding of the basis of the Norrish-Tromsdorff effect because it is difficult to abandon incorrect ideas that have, nevertheless, been successful.

A further gain is undoubtedly the inclusion of the concept of reptation diffusion into the gel effect model, largely due to Tulig and Tirrell and their co-workers [24, 44, 45]. Solving the kinetic scheme (similar to that in Sect. 1.1) with the use of the Smoluchowski equation [46], they derived a relationship analogous to eqn. (30) (based on rheologic concepts)

$$K = c^{**} \bar{P}_n^{0.24} \tag{31}$$

where c^{**} is the concentration of the onset of entanglement and of reptation diffusion, and $\bar{P}_n$ is the number average degree of polymerization of the inactive polymer. The termination rate constants, derived for various diffusion regimes ($k_{t,\,trans}$ corresponding to termination controlled by translational diffusion in dilute solution, $k_{t,\,seg}$ to termination controlled by segmental diffusion, and $k_{t,\,rep}$ to termination controlled by reptation diffusion at concentrations $> c^{**}$), when substituted into the respective kinetic relations, again reproduce very well the effects actually observed.

The studies of Olaj et al. [13] also undoubtedly contribute to our understanding of the basis of the autoaccelerating effects. At the same time, however, they confirm the empirical state of our present knowledge of the gel effect. We propose models which are then tested by confrontation with reality. So far there exist several suitable models, but we lack a test for discarding or modifying the less correct ones.

1.3.2 TRAPPED RADICALS

The rate of precipitating radical polymerizations usually increases during the course of the polymerization process from the beginning up to high conversions. This behaviour is typical for halogenated vinyl monomers, acrylonitrile, etc. The apparent autocatalytic behaviour of these polymerizations in bulk and in most common solvents contrasts with the behaviour of homogeneous systems where the rate decreases with decreasing monomer concentration. The viscosity of the liquid phase does not increase substantially; the acceleration is not caused by the gel effect. The precipitation polymer has a higher molecular mass than the polymer formed under otherwise similar conditions in homogeneous medium.

Bengough and Norrish observed this behaviour during vinyl chloride polymerization. They explained it by transfer to polymer chains on which immobile, long-lived and propagating radicals are formed. These centres decay by transfer to monomer or by termination with untrapped radicals from the liquid phase [47]. According to these two authors, the acceleration is proportional to the surface area of the solid particles. A similar acceleration of polymerization was observed by Bamford et al. [18] with acrylonitrile

References pp. 438–442

which is known to have a small tendency to transfer reactions. In this case, the explanation of Bengough and Norrish could not be applied. Therefore the authors proposed the occlusion theory, according to which the growing radical are sorbed at the surface of solid particles. The occluded radicals can propagate, but their termination by mutual collisions is effectively suppressed. The resulting increased radical concentration is the reason for the acceleration of polymerization [48].

Mickley et al. have proved that immobilization of radicals by occlusion predominates over transfer to polymer even in the polymerization of the strongly transferring vinyl chloride [49]. Currently, the occlusion theory is preferred for explaining autoacceleration in precipitating radical polymerizations. Transfer to polymer is of little importance for autoacceleration.

For vinyl chloride polymerization in bulk, the rate equation

$$-\frac{d[M]}{dt} = k_p \left(\frac{fk_d[I]}{2k_t}\right)^{1/2} ([M] + v) \qquad (32)$$

was derived [49] where v is the variable contribution of autoacceleration, comprising the rate of occlusion, and migration of radicals to and from the surface of solid particles. The concentration of growing immobilized radicals is a function of the rate constant of transfer to monomer (or solvent). A calculation of changes in radical concentration with time is generally difficult, and cannot be made without simplifying assumptions.

The presence of the solid phase is not the only possible cause for radical immobilization. The matrix effect caused by retarded termination due to reduced radical mobility, is another example of polymerization acceleration (see Chap. 5, Sect. 2.2, Fig. 3). Under suitable conditions, the matrix effect can even be caused by solvents.

Bolshakov et al. described low-temperature (100–130 K) polymerizations of acrylic acid, acrylamide, and methyl acrylate, in which radical mobility was strongly hindered by their specific interactions with the molecules of a melting alcohol matrix [50].

Radicals trapped at a solid surface or in a matrix lattice (network) can terminate with untrapped radicals or, though only rarely, by mutual collision, when they can migrate over the surface or in the lattice. When a suitable polymeric polyfunctional initiator is used for initiation, free radicals cannot occur in the system[†]. The growing chains are attached to the polymer surface by covalent bonds. Consequently they cannot migrate, and termination hardly occurs. When the residual, and in any case very small possibility of

---

[†] When transfer does not occur.

contact between the growing chains is futher limited, e.g. by their confinement in soap micelles, a living radical polymerization results [51–53]. Observations of living radical polymerizations are becoming more numerous [54–56]. With their increasing number and the quality of our knowledge of termination and transfers, the number of known living systems will increase, and their practical applications can be anticipated.

1.4 Inhibition and inhibitors

Polymerizations sometimes exhibit just the opposite behaviour to that described in the previous paragraph, i.e. rapid termination. In the extreme where the termination rate exceeds that of initiation or propagation, we speak of inhibition. Such a situation is usually caused by the presence of compounds eagerly reacting with primary radicals to yield non-initiating substances. Such compounds are called inhibitors.

A radical reacting with a molecule must produce a particle with an unpaired electron. For the reacting molecule to be called an inhibitor, these secondary radicals should have negligible tendency to propagation. They can have various fates. They either dimerize or react with a further radical that is able to propagate. The initiation rate can be determined by means of an "ideal" inhibitor. The most important inhibitors are some quinones, nitro and nitroso aromatics, polycyclic aromatic hydrocarbons, some metal chlorides, and stable radicals (e.g. 1,1-diphenyl-2-picrylhydrazyl-DPPH, etc.).

The inhibition reaction itself is kinetically simple

$$R^* + Q \xrightarrow{k_q} \text{products} \tag{33}$$

When Q is an ideal inhibitor[†], and when it only reacts with one radical, then the rate of its consumption can be expressed as

$$-\frac{d[Q]}{dt} = k_q[R^*][Q] \tag{34}$$

and the concentration of radicals in its presence as

$$\frac{d[R^*]}{dt} = v_{\text{init}} - k_q[R^*][Q] - 2k_t[R^*]^2 \tag{35}$$

[†] An ideal inhibitor Q is a substance that stoichiometrically reacts in a known way with radicals to give reaction products which are completely inert to initiation, propagation, termination and transfer. The ratio $d[Q]/d[M] = k_q[Q]/(k_p[M])$ must be sufficiently large.

The course of the inhibited polymerization can be analyzed by means of eqns. (34) and (35). When the product $k_q[Q]$ is sufficiently large, each radical produced by initiation is immediately inactivated by the inhibitor. Polymerization does not occur, and the duration of such situation is called the inhibition period. The further course of the beginning of the inhibited polymerization will be divided into two phases. In the first of these, $k_q[R^*][Q] \gg 2k_t[R^*]^2$, and the development of radical concentration may be approximated by

$$\frac{d[R^*]}{dt} = v_{init} - k_q[R^*][Q] \tag{36}$$

Integration of eqn. (36) gives an expression for the radical concentration. In the period of negligible Q consumption

$$[R^*] = [R^*]_{s,i} \{1 - \exp[-k_q[Q](t - t_{inh})]\} \tag{37}$$

where

$$[R^*]_{s,i} = \frac{v_{init}}{k_q[Q]} \tag{38}$$

for $[Q(t - t_{inh})] = 0$. After the start, $[Q]$ decreases and t increases. Thus in the first phase, $[R^*]$ also increases at a varying rate. In the second phase, where the values of $k_q[R^*][Q]$ and $2k_t[R^*]^2$ are comparable, $[R^*]$ increases more rapidly until the disappearance of Q. The concentration of radicals reaches the stationary value $[R^*]_s = (v_{init}/2k_t)^{1/2}$ for $[Q] = 0$.

The boundaries between the stages are not sharp, and are essentially a matter of convention. The transition between the first and second stages can be defined by a certain value of the ratio $k_q[R^*][Q]/2k_t[R^*]^2)$. The length of the first stage is usually proportional to the ratio $[Q]_0/v_{init}$. The duration of the second stage is independent of $[Q]_0$ for each set of measurements with constant v_{init}. Thus in the presence of inhibitor, the monomer only begins to be consumed after a certain time, called the inhibition period, and the polymerization rate increases during the induction period. The shape of a conversion curve deformed by inhibition and induction is shown schematically in Fig. 6 Chap. 5. The length of the induction period is a function of both phases

$$\Delta t_{ind} = t_{ind} - t_{inh} = \frac{[Q]_0}{v_{init}} + \text{residual contribution of phase 1}$$
$$+ \text{contribution of phase 2} \tag{39}$$

The contribution of the second phase is constant; the residual contribution of the first phase is small (and roughly proportional to $[Q]_0$). Thus t_{inh} is directly proportional to $[Q]_0$, with the proportionality constant v_{init}^{-1} [53].

An exact solution of eqn. (35) describing the inhibition kinetics is not available. Some authors simplified the situation by assuming the existence of a stationary state [59]. This is, of course, an unacceptable assuption; as long as $[Q] \neq 0$, $d[R^*]/dt \neq 0$. Bamford et al. [60] settled on a compromise: they also assume that $d[R^*]/dt = 0$, but only up to the ratio of radical concentrations $[R^*]/[R^*]_s = 0.684$. This is more or less acceptable because in the first phase, the change of radical concentrations can be small.

The reactions of inhibitors with radicals have not been sufficiently studied so far. The structures and reactivities of the reaction products are largely unknown. This is particularly odd when the initiation rate is to be measured from the length of the inhibition period. In an ideal case, an inhibitor molecule inactivates two radicals

$$\sim\sim\sim^* + Q \longrightarrow \sim\sim\sim Q^* \tag{40}$$

$$\sim\sim Q^* + {}^*\sim\sim \longrightarrow \sim\sim Q \sim\sim \tag{41}$$

so that the stoichiometric coefficient of the inhibitor is 2. Actually this is rarely the case [65]; for most inhibitors the stoichiometric coefficient is less. It is usually assumed that the reactivity of the radical $\sim\sim Q^*$ is sufficient to regenerate a propagating radical by copolymerization reaction with monomer[†] [62]

$$\sim\sim Q^* + M \longrightarrow \sim\sim QM^* \tag{42}$$

Tüdös [61] regards reaction (42) as a reaction of a "hot" radical. The heat-excited radical $\sim\sim Q^\star$, "hot" after the exothermic reaction (40), can attack the monomer. However, after losing its excess energy, the radical can no longer react according eqn. to (42). The validity of the hypothesis about hot radical reactions in the liquid phase depends on the relaxation time of the thermally excited radical. When its excited states are efficiently quenched by collisions with neighbouring molecules, it will probably lose the excitation energy prior to reacting. When quenching is slow, the reaction can occur before inactivation. It appears that the reactions of "hot" radicals are possible not only in the gaseous phase, but even in an undiluted liquid substrate. In dilute solutions, the chances of excess energy loss by collisions with the

---

[†] The concentration ($[M] \gg [\sim Q^*]$) is also important. The reaction $\sim Q^* + {}^*Q\sim$ is rather improbable.

"inert" solvent are greater, and the probability to react while still "hot" is reduced. Direct proof of the kinetic arguments is not yet available. Generally there exists little information on the reactions of thermally excited particles in solution. The reactions of photochemically excited "hot" particles have been much more studied, and some of the results support the "hot" radical hypotheses [61].

The practical importance of radical inhibitors in macromolecular chemistry is of a double kind. They are used deliberately only for the stabilization of stored monomers, in determinations of initiation rates, and for fast control of polymerization accidents: an uncontrollable radical chain reaction can be stopped by inhibitor addition. The unwanted presence of inhibiting impurities in polymerization media leads to adverse effects. Impurities can lead to the appearance of inhibition and induction periods which cannot be tolerated on an industrial scale. They decrease equipment productivity, and increase the consumption of initiator. Probably the most dangerous inhibitor is oxygen, either as such or more often in hydroperoxides formed by its reaction with monomers. It belongs to the copolymerizing inhibitors [reaction (42)]. The generated unstable polyperoxides $+(CH_2-CXY-O-O+)_n$ greatly deteriorate the service properties of the products. This kind of behaviour was observed during the polymerization of styrene both in bulk [63] and in emulsion [64], acrylonitrile [65], methyl acrylate [66], vinyl acetate [67], vinyl chloride [68], and of other monomers [69, 70].

Oxygen reacts very eagerly with radicals but it is not a good inhibitor, mainly because it leads to the production of labile peroxides. In addition, its concentration in the stored monomer is usually low. Antioxidants of the phenol or aniline type are themselves very inefficient inhibitors. However, they significantly reduce the rate of O_2 consumption when the latter is present [71]. This probably occurs by way of ROO decomposition to non-radical products so that ROO propagation is prevented [72]. More recently, therefore, the kinetics and mechanism of the inhibition effects of antioxidants (hydroquinone, pyrocatechol derivatives, etc.) in radical polymerizations and copolymerizations have been studied mainly in the presence of oxygen [73, 74].

The rate constant of the reaction of most inhibitors with radicals is much larger than the rate constant of the reaction of the same radical with monomer $[k_q = (10^3 - 10^4) k_p]$. Thus polymerization can be suppressed by very small amounts of inhibitor.

1.4.1 AUTOINHIBITION

Electron delocalization in a conjugated system is generally accompanied by a decrease in radical reactivity. When the extent of conjugation increases with each monomer adition, the active centre soon loses its ability to attack the monomer, and polymerization stops. The weakly reactive radicals decay by some termination reaction, often yielding a cyclic conjugated product. Such behaviour is observed in the polymerizations of acetylenes, and to a lesser degree also with their alkyl-substituted derivatives: this is ascribed to H atom migration from the side groups to the backbone. In this way the content of alkylidene substituents increases, and backbone conjugation is interrupted [75]. Aryl-substituted acetylenes yield polymers with a polyene chain.

The above-mentioned reactions have not been sufficiently studied so far, although they are of considerable interest, as they could yield thermostable polymers or organic semiconductors. With radical initiators, only oligomers are always formed, with a large content of cycles[†]. Most authors assume that the reluctance of acetylenes to yield long chains is connected with the attenuation of centre reactivity, i.e. with inter- and intramolecular delocalization of the unpaired electron in the growing macroradical [79].

Criticisms have been voiced [80]. Nevertheless, termination of polyphenylacetylene growth is of the first order, as proved by Amdur et al. Most probably it is also connected with intramolecular electron transfer or structural change, leading to a less active centre [81].

1.5 Non-stationary states of radical polymerizations

It is a gross error to apply stationary state theory to non-stationary polymerizations, of which many exist. In a great majority of cases, the stationarity conditions are only fulfilled after a certain non-stationary phase in which the concentration of active centres increases. When this occurs in a kinetically pure medium, i.e. in the absence of inhibitors or other intervening compounds, it usually signifies slow initiation and is called the pre-effect. The general shape and meaning of the pre-effect is represented graphically in Fig. 5. For very small conversions ($[M] \approx [M]_0$) we can write [57]

$$\frac{d \ln [M]}{dt} = \frac{U}{t_n} \tag{43}$$

When monomer consumption in the initiation reaction is neglected (this is only permitted when long chains are generated), both sides of eqn. (43) are directly proportional to the concentration of growing radicals $[M^*]$.

[†] A relatively high-molecular-weight, crystalline polyphenylacetylene of molecular mass of the order of 10^4 is formed by ionic or coordination polymerization (with WCl_6 or $(acac)_3Fe$) [77]. In these polymerizations, termination is also a function of chain length [78].

References pp. 438–442

The intercept U is defined by

$$U = \int_0^\infty \left[\left(\frac{d \ln [M]}{dt}\right)_s - \left(\frac{d \ln [M]}{dt}\right)\right] dt \qquad (44)$$

where the index s designates a stationary quantity. By means of eqns. (43) and (44) some rate constant combinations can be determined which are different from those that are accessible in the stationary range. In this way, the otherwise undesirable non-stationary states can be utilized for the determination of elementary constants.

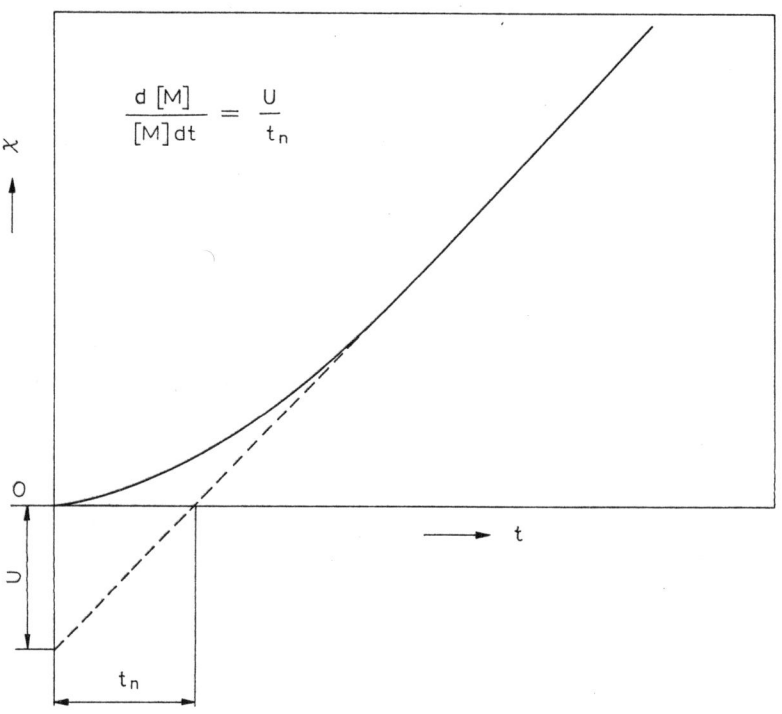

Fig. 5. Non-stationary phase caused by slow initiation; pre-effect.

1.5.1 THE PRE-EFFECT OF RADICAL POLYMERIZATION ACCOMPANIED BY MUTUAL RADICAL TERMINATION

In order to measure U and t_n with the maximum possible accuracy, the moment the polymerization starts must be known [57]. This condition is best fulfilled with photoinitiated polymerizations which start immediately after illumination.

Under these conditions, the concentration of radicals is given by

$$\frac{d[R^*]}{dt} = v_{init} - 2k_t[R^*]^2 \tag{45}$$

At constant light intensity, the rate of photochemical initiation, v_{init}, is constant. Therefore integration yields

$$[R^*] = [R^*]_s \tanh\left[(2k_t v_{init})^{1/2} t\right] \tag{46}$$

At a stationary radical concentration $[R^*]_s$ it holds, from eqn. (45) for cases where $d[R^*]/dt = 0$, that

$$[R^*]_s = \left(\frac{v_{init}}{2k_t}\right)^{1/2} \tag{47}$$

When monomer consumption in initiation is neglected, then its decay is given by propagation, i.e. by the equation (see Chap. 8, Sect. 1)

$$\frac{d \ln[M]}{dt} = k_p [R^*] \tag{48}$$

Substituting $[R^*]$ from eqn. (46) into eqn. (48), and the obtained relation for monomer decay into eqn. (44), we can write

$$U = k_p [R^*]_s \int_0^\infty \left(1 - \tanh\left[(2k_t v_{init})^{1/2} t\right]\right) dt \tag{49}$$

Integration of the right-hand side of eqn. (49) and presentation of the result in exponential form yields

$$U = \frac{k_p}{2k_t} \ln 2 \tag{50}$$

Finally, substituting eqn. (47) into eqn. (48), we obtain

$$\left(\frac{d \ln[M]}{dt}\right)_s = k_p \left(\frac{v_{init}}{2k_t}\right)^{1/2} \tag{51}$$

and from eqn. (43)

$$t_n = \frac{\ln 2}{(2k_t v_{init})^{1/2}} \tag{52}$$

References pp. 438–442

When v_{init} is known, for example from measurements with inhibitors, the value of $k_p/k_t^{1/2}$ can be derived from the stationary polymerization rate (see Chap. 8, Sect. 1. 2). The ratio k_p/k_t can be calculated from eqn. (50), and by comparing the two rations, k_p and k_t can be separated.

1.5.2 THE DECAY OF RADICAL POLYMERIZATION DUE TO MUTUAL TERMINATION OF RADICALS

When radical generation is stopped abruptly, polymerization dies out at a rate proportional to the decreasing concentration of radicals. This is called the post or after effect. In a defined manner it can be brought about only in photochemically initiated polymerizations simply by switching off the light. In the course of polymerization decay, radicals are consumed by mutual termination

$$-\frac{d[R^*]}{dt} = 2k_t[R^*]^2 \tag{53}$$

and thus

$$[R^*]^{-1} - [R^*]_s^{-1} = 2k_t t \tag{54}$$

At the moment when the light is switched off, the system still contains the stationary concentration of radicals $[R^*]_s$, $t = 0$. At this moment the polymerization rate will be [by substiting eqn. (54) into eqn. (48)]

$$-\frac{d \ln[M]}{dt} = \frac{k_p}{2k_t t + [R^*]_s^{-1}} \tag{55}$$

The ratio k_t/k_p can also be obtained, from relation (55). This is a somewhat difficult, but relatively accurate method [34]. Reliable determination of the instantaneous rate during polymerization decay is not simple. The measurements are much easier when the pre- and post-effects are alternated in close order. An elegant way of producing such conditions is illumination of photoinitiated polymerization either with intermittent switching or by the motion of the polymerizing mixture in bands of light and darkness.

1.5.3 ROTATING SECTORS

The transitions between light and darkness should be as sharp as possible in the whole polymerizing medium. Partial illumination should be avoided. So far these conditions are best realized by means of an opaque disc with one or more transparent segments. The disc rotates between the light source and the reactor, usually as near as possible to the source in order that the boundary between light and shade passes through the polymerizing medium in a negligible time. During the light phase, radicals are generated in the

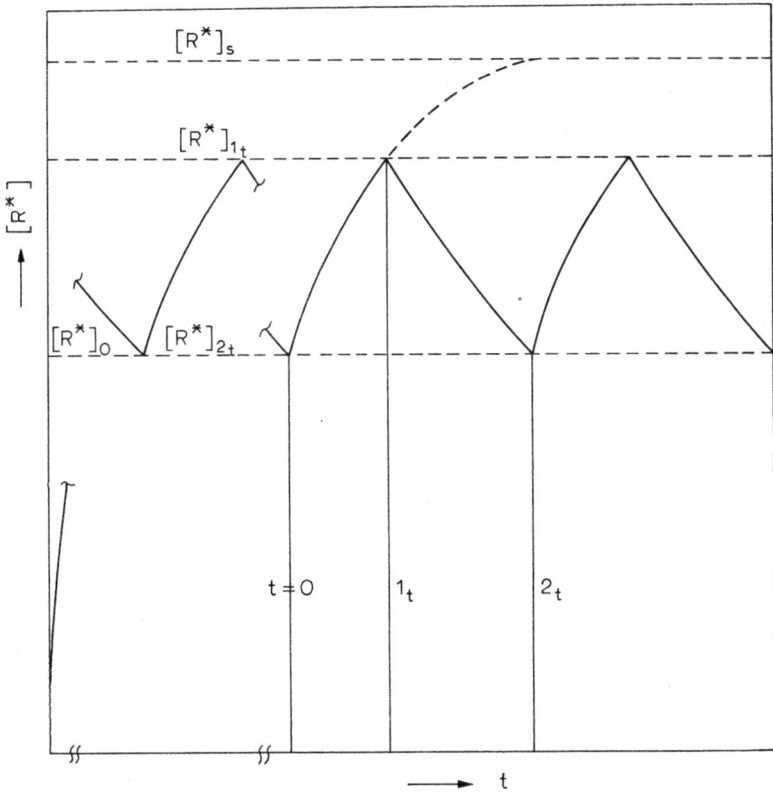

Fig. 6. Dependence of radical concentration on time with periodic illumination

reactor by photoreaction from a suitable initiator; in the dark phase, their concentration decreases. The method of rotating sectors was proposed by Berthoud and Bellenot [82] and by Chapman et al. [83]. In polymerization studies it was first applied by Melville [84].

By selecting a suitable width for the segments and the number of disc revolutions, conditions can be found where the loss of radicals in the dark is just compensated by their generation in the light. Under these conditions, the mean lifetime of radicals [defined by the ratio of radical concentration and their termination rate, $\tau = [R^*]/2k_t[R^*]^2 = (2k_t[R^*])^{-1}$] is constant and can be measured. Let us designate the ratio of dark and light periods in the reactor as $z[z = (^2t - {}^1t)/{}^1t$, see Fig. 6]. When ${}^1t < \tau$, the situation is

References pp. 438–442

equivalent to a reduction of light intensity by the factor $(z + 1)^{-1}$, so that at the initiation rate $I\phi_{(t)}$ the rate of the photoreaction is [85, 86]

$$v_1 = k \left(\frac{I\phi_{(t)}}{z + 1} \right)^{1/2} \tag{56}$$

where $\phi_{(t)}$ is a parameter describing the efficiency of radical generation in light and in darkness. In an ideal case, $\phi_{(t)} = 1$ (during illumination) and $\phi_{(t)} = 0$ (in darkness)[†], and k is a proportionality constant. On the other hand, when the illumination period is longer than the mean lifetime of the radicals, $^1t > \tau$, the situation is equivalent to full illumination for a time shorter by the factor $(z + 1)^{-1}$. In this case, the rate of the photoreaction is

$$v_2 = \frac{k(I\phi_{(t)})^{1/2}}{z + 1} \tag{57}$$

When the polymerization rate at constant illumination is $v_s = k(I\phi_{(t)})^{1/2}$[††], the relative rates are given by the fractions

$$\frac{v_1}{v_s} = (z + 1)^{-1/2} \tag{58}$$

$$\frac{v_2}{v_s} = (z + 1)^{-1} \tag{59}$$

Thus the rate ratios are different for illumination times $^1t < \tau$ and $^1t > \tau$. When ^1t approaches τ, the relative rates pass from one limiting value to the other. In this way, τ and k_p/k_t can be determined by comparison of the experimental and calculated values. For a practical description of this situation, the arrangement $(z + 1)^{1/2} v/v_s$ vs. $^1t/\tau$ was found as useful (see Fig. 7); this can be derived theoretically by the following analysis.

The change of radical concentration during alternating illumination and darkness is given by the equation [87]

$$\frac{d[R^*]}{dt} = I\phi_{(t)} - 2k_t[R^*]^2 \tag{60}$$

[†] When the initiator is also thermally decomposed in the dark, $\phi_{(t)} \neq 0$ and the rate of thermal initiation I' must be added to the product $I\phi_{(t)}$. The higher I', the relatively lower the contribution of $I\phi_{(t)}$, and the overall error increases.

[††] Monomer concentration, which is constant within acceptable limits, is included in k.

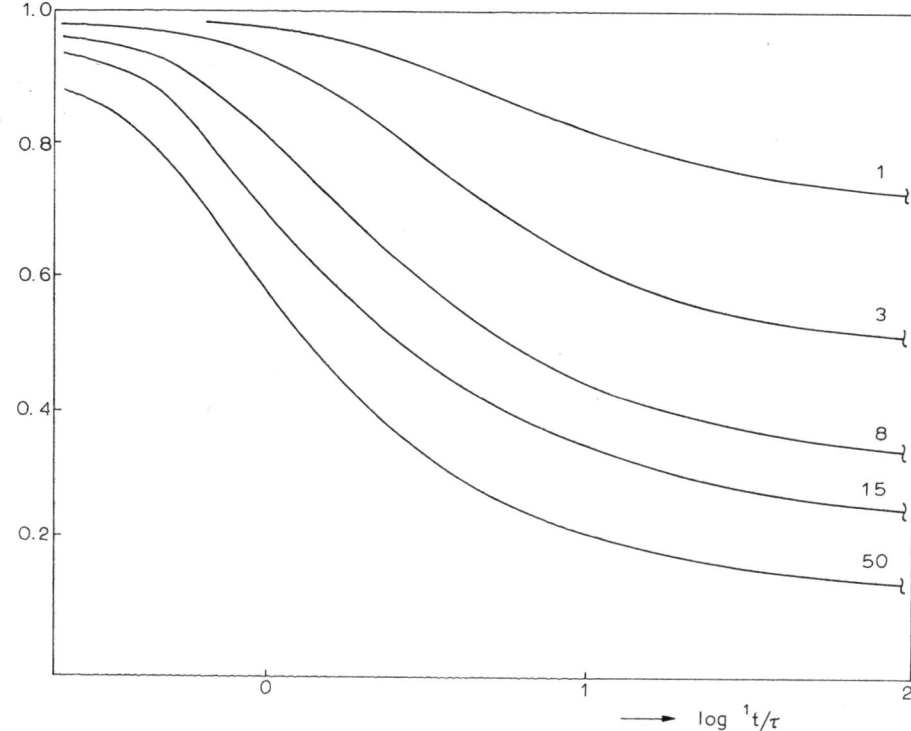

Fig. 7. Relative rate of photoinitiated polymerization as a function of log $[^1t(2I\phi_{(t)}k_t)^{1/2}]$ for various z values. For constructing the graph, data were taken from ref. 86 pp. 192 and 193.

General integration of this equation is difficult, and therefore it is usually solved separately for the light and dark periods. For the light period (interval $0 - {}^1t$, see Fig. 6), integration of eqn. (60) yields

$$[R^*] = [R^*]_s \tanh \left[(2I\phi_{(t)}k_t)^{1/2} \, t + c \right] \tag{61}$$

where $[R^*]_s$ is defined by the expression $[R^*]_0 = [R^*]_s \tanh [c]$. $[R^*]_0$ and $[R^*]_s$ are radical concentrations at the beginning of illumination and in the stationary state. Radical concentration in the dark is given by eqn. (53) which yields relation (54) by integration. Let us reformulate it as

$$[R^*]^{-1} - [R^*]_{1_t}^{-1} = 2k_t({}^2t - t) \tag{54a}$$

$[R^*]_{1_t}$ is the radical concentration at the beginning of the dark period and

${}^1t \leqq t \leqq {}^2t$. When, during alternating illumination, the system reaches the situation where $[R^*]_{1_t}$ and $[R]_{2_t}$ are both constant, the radical concentra-

References pp. 438–442

tions at the beginning of the light period and at the end of the dark period must be identical

$$[R^*]_0 = [R^*]_{2t} \qquad (62)$$

and their concentrations at the end of the light period and at the beginning of the dark period ^1t will be

$$[R^*]_{1t} = [R^*]_s \tanh\left[(2I\phi_{(t)}k_t)^{1/2}\,{}^1t + c\right] \qquad (61a)$$

These last two equations define the boundary conditions for both periods. For a large number of light-dark cycles, the observed mean rate of polymerization will be

$$\bar{v}_p = k_p[M]\,[\overline{R^*}] \qquad (63)$$

with the mean radical concentration $[\overline{R^*}]$

$$[\overline{R^*}] = ({}^1t)^{-1} \int_0^{{}^1t} [R^*]\,dt + ({}^2t - {}^1t)^{-1} \int_{{}^1t}^{{}^2t} [R^*]\,dt \qquad (64)$$

The solution of eqn. (64)† is usually presented in the form of a dependence of $[R^*]/[R^*]_s = \bar{v}_p/v_s$ on ${}^1t(2I\phi_{(t)}k_t)^{1/2}$ ($= {}^1t/\tau$ at stationary state conditions) for various values of z. The necessary values have already been published [85]. Thus the determination of τ is based on experimental values of $\bar{v}_p/v_s$ for known values of z and illumination time ^1t. The required ratio $^1t/\tau$ is found (interpolated) from Table 1.

According to eqn. (58), $(z + 1)^{1/2}\,\bar{v}_p/v_s$ should be equal to 1. With increasing illumination time (at constant z), the validity of eqn. (58) breaks down, as exhibited by a sharp fall in $\bar{v}_p/v_s$. This is exploited for determining $^1t/\tau$ from a graph by comparing the experimentally found values of $\bar{v}_p/v_s$ at the values of z and ^1t used, with the calculated curves as shown in Fig. 7. From the mean lifetime of the radicals and the rates of initiation (determined, for example, by means of inhibitors) and polymerization, the values of the elementary constants k_p and k_t can be derived††.

With correspondingly more complicated mathematical techniques, the method of rotating sectors can even be applied to the analysis of situations with non-zero polymerization rate in the dark, with first- and second-order

---

† Consists of substitution for $[R^*]$ from eqn. (50) and (54a) into eqn. (64) and calculation using eqn. (61a) and (62).
†† $\tau = (2k_t[R^*])^{-1} = k_p[M]/2k_t\bar{v}_p$ (see eqn. (63)). From this $k_p/2k_t = a$ can be determined. According to eqn. (23), $k_p^2/k_t = v_p^2/([M]^2 v_{init}) = b$. Then $b/a = 2k_p$, $k_p/a = 2k_t$.

TABLE 1

Values of the ratio $[\overline{R^*}]/[R^*]_s$ $(=\bar{v}_p/v_s)$ as a function of $^1t/\tau$ for various values of z [85]

$^1t/\tau$	z				
	1	2	3	4	5
≪0.1	0.707	0.577	0.500	0.447	0.408
0.1	0.707	0.577	0.499	0.446	0.407
0.2	0.707	0.576	0.498	0.445	0.406
0.3	0.706	0.574	0.496	0.442	0.403
0.4	0.705	0.572	0.493	0.438	0.399
0.5	0.704	0.569	0.489	0.433	0.393
0.6	0.702	0.566	0.484	0.429	0.387
0.7	0.701	0.562	0.480	0.423	0.381
0.8	0.699	0.558	0.476	0.418	0.375
0.9	0.697	0.555	0.471	0.412	0.369
1.0	0.695	0.552	0.467	0.407	0.363
2.0	0.672	0.515	0.422	0.360	0.315
3.0	0.653	0.487	0.392	0.330	0.286
4.0	0.637	0.467	0.372	0.310	0.267
5.0	0.625	0.453	0.357	0.296	0.253
6.0	0.616	0.442	0.346	0.286	0.244
7.0	0.607	0.432	0.337	0.277	0.236
8.0	0.601	0.425	0.330	0.271	0.230
9.0	0.595	0.419	0.324	0.265	0.225
10.0	0.590	0.413	0.319	0.261	0.221
20	0.560	0.384	0.293	0.237	0.199
30	0.546	0.372	0.282	0.227	0.191
40	0.538	0.364	0.276	0.222	0.186
50	0.533	0.360	0.272	0.218	0.183
60	0.529	0.356	0.269	0.216	0.181
70	0.526	0.354	0.267	0.214	0.179
80	0.523	0.352	0.265	0.213	0.178
90	0.521	0.350	0.264	0.212	0.177
100	0.520	0.349	0.263	0.211	0.176
200	0.512	0.342	0.257	0.206	0.172
300	0.508	0.340	0.255	0.204	0.170
400	0.507	0.338	0.254	0.203 3	0.169 5
500	0.506 5	0.337 5	0.253 3	0.202 8	0.169 0
600	0.504 8	0.336 9	0.252 8	0.202 4	0.168 7
700	0.504 2	0.336 4	0.252 5	0.202 1	0.168 4
800	0.503 7	0.336 1	0.252 2	0.201 8	0.168 2
900	0.503 4	0.335 8	0.252 0	0.201 6	0.168 0
1 000	0.503 1	0.335 6	0.251 8	0.201 5	0.167 9
≫1 000	0.500 0	0.333 3	0.250 0	0.200 0	0.166 7

References pp. 438–442

termination, with varying light absorption and non-negligible interval between illumination and darkening [85, 86].

1.5.4 SPATIALLY INTERMITTENT POLYMERIZATION (SIP)

O'Driscoll and Mahabadi described a variant of the rotating sector method [88], permitting calculation of k_p and k_t from a single set of experimental measurements. The amount of product and its molecular mass are determined in the course of photo-initiated polymerization. A small contribution of themal initiation can be tolerated. The method is based on kinetic analysis of the polymerization of a monomer flowing through a tube reactor, with alternating sections of light and darkness[†]. During passage through the light zones, decomposition of the photosensitive initiator takes place, radicals are generated at a rate varying with time, in dependence on the width of the light slits, on their distance, and on the velocity of monomer flow (see Fig. 8).

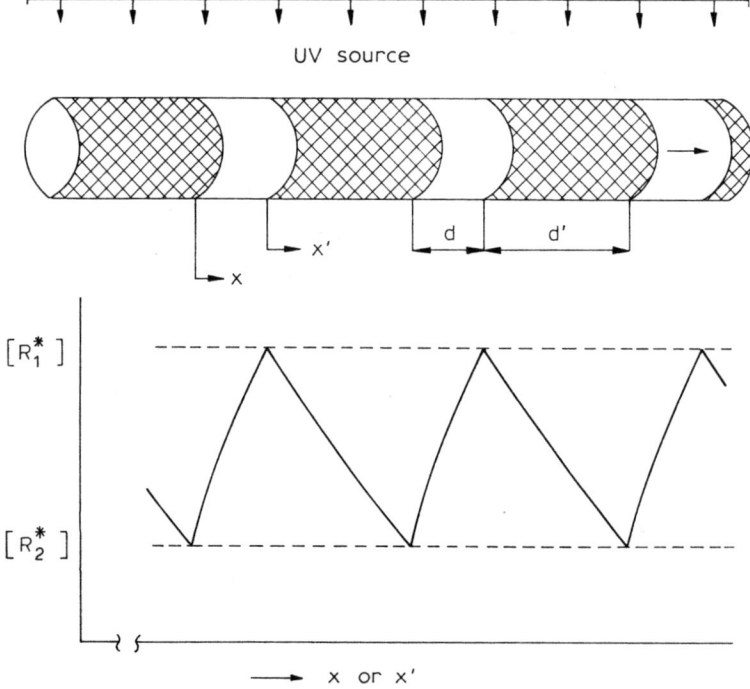

Fig. 8. Schematic representation and function of a reactor for spatially intermittent polymerization.

[†] A glass tube through which the polymerizing mixture is flowing, is inserted into a metal cylinder with many narrow, regularly spaced slits permitting illumination. Formally this is a reactor with piston flow of the reacting system.

At zero flow rate of the mixture, a stationary radical concentration $[R^*]_s$ is established in the illuminated sections, with a lifetime, τ, inversely proportional to the initiation rate, v_{init}

$$\tau = \frac{[R^*]_s}{v_{init}} \tag{65}$$

In the stationary state, the rate of initiation is equal to that of termination, $v_{init} = 2k_t[R^*]_s^2$. The monomer will polymerize at the rate

$$v_s = -\frac{d[M]}{dt} = k_p[R^*]_s[M] \tag{63a}$$

and therefore

$$\tau = \frac{k_p[M]}{2k_t v_s} \tag{66}$$

During flow of the polymerizing mixture, the radical concentration changes with the coordinate of the respective elementary volume dV. This situation is schematically represented in Fig. 8. The mean polymerization rate, $\bar{v}_p$, is a function of the mean concentration of radicals $\overline{[R^*]}$

$$\bar{v}_p = k_p\overline{[R^*]}[M] \tag{63}$$

and the number average degree of polymerization $\bar{P}$ will be

$$\bar{P} = \frac{k_p\overline{[R^*]}[M]}{k_t\overline{[R^*]}^2 + k_{tr,M}\overline{[R^*]}[M] + k_{tr,Y}\overline{[R^*]}[Y]} \tag{67}$$

Assuming termination by combination, the factor $1/2$ must be applied to $2k_t$; $k_{tr,M}$ and $k_{tr,Y}$ are the rate constants of transfer to monomer and to the transfer agent Y, respectively. Substituting eqns. (63) and (66) into eqn. (67), we obtain expressions for k_p or k_t in which all quantities can be measured by independent methods

$$k_p = \frac{\bar{P}}{[M]}\left(\frac{\bar{v}_p}{2\tau v_s} + k_{tr,M}[M] + k_{tr,Y}[Y]\right) \tag{68}$$

$$k_t = \frac{\bar{P}}{2\tau v_s}\left(\frac{\bar{v}_p}{2\tau v_s} + k_{tr,M}[M] + k_{tr,Y}[Y]\right) \tag{69}$$

References pp. 438–442

while τ is calculated from equations which are derived in the same way as for the light-dark situation with the method of rotating sectors. For the illuminated space, the authors of SIP derived the equation (the meaning of the symbols is evident from Fig. 8).

$$\tanh^{-1} \frac{[R^*]_1}{[R^*]_s} - \tanh^{-1} \frac{[R^*]_2}{[R^*]_s} = \frac{d}{\bar{v}_p \tau} \tag{70}$$

and for the dark space

$$\tanh^{-1} \frac{[R^*]_2}{Z[R^*]_s} - \tanh^{-1} \frac{[R^*]_1}{Z[R^*]_s} = \frac{Zd'}{\bar{v}_p \tau} \tag{71}$$

where Z is the contribution of thermal polymerization in darkness[†], defined by the ratio $Z = \bar{v}_p \text{(dark)}/v_s$). Integration of the equation (72)

$$\frac{d[R^*]}{d(\bar{u}/X)} = v_{\text{init}} - v_t \tag{72}$$

from which eqns. (70) and (71) have also been calculated[††], over illuminated and dark sections leads to an expression for the ratio of the mean radical concentration in the reactor under non-stationary conditions to that at constant illumination, $[\overline{R^*}]/[R^*]_s$. Substituting this ratio by the ratio of the experimentally measured rates, and d'/d by the symbol $\bar{d}$ (the ratio of the dark and light zone lengths), $d/\bar{v}_p \tau = m$ (the ratio of the time spent in the illuminated volume dV to the mean life time of radicals under stationary conditions), and setting $[R^*]_1/[R^*]_s = a_1$, $[R^*]_2/[R^*]_s = a_2$, we obtain the equation

$$\frac{\bar{v}_p}{v_s} = (1 + \bar{d})^{-1} \left\{ 1 + \bar{d}Z + m^{-1} \ln \left[\frac{(1 + a_2)(1 + a_1/Z)}{(1 + a_1)(1 + a_2/Z)} \right] \right\} \tag{73}$$

Equations (70)–(73) can be solved numerically using the experimental values of $\bar{v}_p$ and v_s, Z, m, d, d' and $\bar{d}$, and in this way a_1, a_2 and τ can be determined.

[†] The authors mutely assume that this is negligible compared with the contribution of photopolymerization in the light zone. They therefore neglect it in deriving eqn. (70).
[††] v_t and $\bar{u}$ are the rates of termination and of motion of the elementary volume dV of the polymerizing mixture in the direction of the longitudinal axis X, or X' of the reactor; $\bar{u} = dX/dt = dX'/dt$.

By means of the spatially intermittent reactor, Ito and O'Driscoll determined, for example, the absolute rate constant of termination, k_t, for methyl methacrylate copolymerization with butyl or dodecyl methacrylate. The value of k_t is a function of the monomer mixture composition [89].

2. Termination of ionic polymerizations

Under certain stringent conditions, all ion reactivity is concentrated on propagation. The resulting polymerizations are living. In other cases, the instability of the ions is manifested by a complex of poorly defined reactions leading to transfers, retardation, and decay of the polymerizing activity of the centres, i.e. termination. Termination is either an inherent feature of the respective polymerizing system or it may be caused by accidental impurities or, finally, it may be a consequence of deliberately added compounds.

2.1 Termination involving counter-ions

Polymerizations are generally carried out in externally electroneutral media[†]. The presence of counter-ions always involves the possibility of neutralization or of some other reaction leading to a temporary or permanent activity loss of the centre.

2.1.1 SHIFT OF EQUILIBRIA BETWEEN IONS, ION PAIRS, AND ESTERS

Equilibria between various forms of living centres were treated in Chap. 5, Sect. 8.1. Equilibria of similar character control the arrangement and reactivity of all ionic centres. When polymerization-inactive structures participate in the equilibria, the number of active centres is reduced by the equilibrium amount of inactive forms. This phenomenon is usually not considered as termination; the unreactive particles are treated as dormant. In the course of polymerization, however, the physico-chemical parameters of the system change as a function of the monomer–polymer transformation. Changes in permittivity, viscosity and the amount of polymer can cause shifts in ionization and dissociation equilibria. The kinetic manifestations of such changes are identical with the occurrence of termination.

These phenomena have not been studied quantitatively. The interest of most authors studying dissociation equilibria is limited to polymerizations

---

[†] Except for typically theoretical cases, where polymerizations are initiated by low energy electron or proton beams. Such polymerizations are of no practical importance at present.

References pp. 438–442

which are not appreciably affected by the presence of polymer. In the kinetics of high-conversion polymerizations, termination is involved to such an extent that the contribution of possible equilibrium shifts cannot be separated.

Exchanges of counter-ions

Ionic equilibria can be shifted by the addition of salts containing a common ion with the active centre. These shifts have been treated in Chap. 5, Sect. 8.1; they cannot, of course, be designated as termination even though they are accompanied by a considerable lowering of the polymerization rate.

An addition of a suitable electrolyte to an ionically polymerizing system can, however, lead to almost instantaneous and quantitative termination of centres when the ionic equilibria are shifted completely towards the inactive species. Saegusa and Matsumoto stopped the polymerization of tetrahydrofuran by sodium phenolate [90]

$$\sim\!\!\stackrel{+}{\underset{B^-}{O}}\!\!\big] + NaOPh \rightleftharpoons \sim\!\!\sim O(CH_2)_4 OPh + BNa \tag{74}$$

The equilibrium is shifted completely to the right. Thanks to the presence of the phenolate group at the chain ends, the macromolecules could be counted by UV spectrophotometry (polytetramethylene oxide is transparent in the UV range). With the counter-ion $B^- \equiv BF_4^-$ termination was slow; with $B^- \equiv AlCl_4^-$ it was rapid. The active centres can be counted using the Saegusa and Matsumoto method and, together with a determination of the degree of polymerization and the concentration of macromolecules, the elementary constants can be determined (the dead-stop method).

2.1.2 COMBINATION OF IONS

The situation has already been described in the preceding paragraph where one of the combining ions comes from the added electrolyte. When both combining ions are natural components of the polymerizing system we have a case of pure termination. Combination of the counter-ion with the active centre is the reason why many sufficiently strong acids and bases cannot be used for the initiation of ionic polymerizations.

The interaction of HCl with an alkene in a non-polar medium results in a simple addition of the hydrogen halide. The centre generated from the proton and monomer is immediately deactivated by the strongly basic and mobile Cl^-. In a solvating medium, e.g. nitromethane, styrene oligomerizes in the presence of HCl [91]. Thus the combination of the solvated ions is relatively slow so that propagation can compete with it.

By the introduction of styrene into trifluoroacetic acid, polystyrene with a molecular mass of the order of 10^4 is formed. The trifluoroacetate counterions are stabilized by molecules of the acid by means of hydrogen bonds. The rate of their combination with carbenium chain ends is strongly reduced[†] [92]. This enables several hundred styrene molecules to be added to each centre prior to its termination. When the concentration of styrene is much larger than that of the initiator, which can be realized by the addition of trifluoroacetic acid to styrene, polymerization does not take place. In a non-solvating medium, the combination rate is sufficiently high to prevent propagation.

Penczek et al. polymerized oxetane in nitrobenzene with trinitromethane. They proved the termination to be a bimolecular reaction

$$\sim CH_2\overset{+}{O}\begin{array}{c}CH_2\\ \diagup \diagdown \\ CH_2 \quad CH_2 \\ \diagdown \diagup \\ CH_2\end{array} + {}^-C(NO_2)_3 \longrightarrow \sim CH_2O(CH_2)_2CH_2C(NO_2)_3 \qquad (75)$$

Combination is slow; the authors assume this to be due to the solvation sphere of the anion $^-C(NO_2)_3$ and to the opening of the oxonium oxygen--containing ring [93].

Thus there exist experimentally accessible situations that are intermediate between immediate termination by a counter-ion and living polymerizations. Combination of a counter-ion with a growing anion (cation) can be prevented, or at least retarded, by its coordination with a suitable particle. Some of the activators discussed in Chap. 3. Sect. 2.5 operate in this way.

Polymerization of 2-methylpropene is not initiated by hydrofluoric acid alone. In the presence of $TiCl_4$, polymerization is very rapid even at low temperatures [94]. Termination by the F^- counter-ion is prevented by its complexation with $TiCl_4$. The basicity of the $TiCl_4F^-$ anion is low, and this anion as such does not combine with the growing cation.

Complex counter-ions are, of course, not stable under all practical circumstances. They can cause termination by mechanisms other than simple combination.

2.1.3 TERMINATION BY UNSTABLE COUNTER-ION

In anionic polymerization, the counter-ions are usually simple, derived mainly from a single element. as a whole they are therefore highly stable. One of the most important termination mechanisms is absent, and the polymeriza-

---

[†] This idea can also be expressed so that in a polar acid, reactive separated ion pairs and (solvated) free ions may be present (see Chap. 3, Sect. 3.1).

References pp. 438–442

tions are often living. In cationic polymerizations the counter-ions are composed of two, and frequently of even more, atoms. They can therefore decompose, and the decomposition products can combine with the growing cation.

Even at low temperatures, the ions BF_4^- and BF_3OH^- are decomposed to BF_3 and a combining anion [95]

$$\sim CH_2-\underset{Me}{\overset{Me}{\underset{|}{\overset{|}{C^+}}}}BF_3OH^- \rightarrow \sim CH_2-\underset{Me}{\overset{Me}{\underset{|}{\overset{|}{C}}}}-OH + BF_3 \qquad (76)$$

A similar decomposition is also assumed to occur with the counter-ions $AlCl_4^-$, $AlBr_4^-$, $AlCl_3\bar{O}H$, $AlCl_3\bar{O}R$, $TiCl_5^-$, $TiCl_4\bar{O}H$, $SnCl_5^-$, $SbCl_6^-$, and with many other halogenides, mixed halogenides, hydroxy-(alkoxy)-halogenides, etc. from Friedel–Crafts catalysts. In all cases, termination is accelerated with increasing temperature. This is one of the main reasons why cationic polymerizations are non-stationary and lead to chains of high molecular mass only at low temperatures.

Several compound counter-ions are known which are stable even at temperatures around 273 K. They are SbF_6^-, AsF_6^-, PF_6^-, $CF_3SO_3^-$, FSO_3^-, and ClO_4^-. Initiators producing these counter-ions can start living (when transfer is prevented) or at least stationary polymerizations.

2.2. Transformation of an active centre to an inactive species or to an ion of low activity

Some reactions can cause active centres to lose part of their reactivity. When such a transformation results in an inactive species, we speak of termination. When less active centres are formed which are still able to propagate, we speak of retardation. Naturally there is no sharp boundary between these two phenomena. Sometimes it is even hard to decide if termination of a part of the centres or a reduction in the reactivity of all centres is taking place.

2.2.1 AGGREGATION (ASSOCIATION) OF CENTRES

In our studies of the polymerization of cyclic siloxane monomers we observed anomalies in the kinetics of the process and in the viscosity changes of the medium. The growing polydimethylsiloxane chain ends were found to associate

$$\sim \overset{|}{\underset{|}{Si}}-OK + KO-\overset{|}{\underset{|}{Si}}\sim \rightleftharpoons \sim Si\underset{\underset{K^+}{O^-}}{\overset{\overset{K^+}{O^-}}{\diamond}}Si\sim \rightleftharpoons \sim \overset{|}{\underset{|}{Si}}-O\underset{\underset{K^+}{}}{\overset{\overset{K^+}{}}{\diamond}}O-\overset{|}{\underset{|}{Si}}\sim \qquad (77)$$

The aggregates are polymerization inactive; their formation is accompanied by a corresponding decrease in the polymerization rate [96]. Aggregate formation is not limited to siloxane monomers and polymers. Independently, and in the same year, associate formation during anionic styrene polymerization was described by Worsfold and Bywater

$$\sim CH_2-\underset{Ph}{CH^-}K^+ + {}^+K\underset{Ph}{\bar{C}H}-CH_2\sim \rightleftharpoons \sim(CH_2-\underset{Ph}{\bar{C}HK^+})_2\sim \qquad (78)$$

The same authors also found that the aggregates are polymerization inactive [97]. In benzene, the degree of association of the living ends increases with decreasing cation size [98] and with increasing length of the hydrocarbon polyisoprenyllithium chains. It also depends on the kind of solvent and on the dilution [99].

The formation of aggregates with low to zero polymerization activity is quite general in ionic polymerizations. This statement can be further documented by the observation of centre aggregation during anionic polymerization of oxirane [100–103]

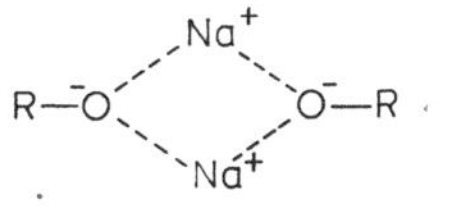

$$R-\bar{O}\underset{Na^+}{\overset{Na^+}{\cdots}}\bar{O}-R \qquad (79)$$

Evidence on association in cationic polymerizations is less frequent. Nevertheless it does exist, and the occurrence of the effect is beyond any doubt. Belenkaya et al. described ion aggregates in cationic lactone polymerization

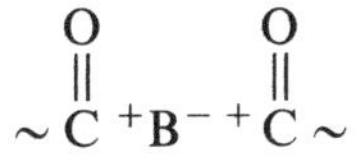

$$\sim \overset{O}{\overset{\|}{C}}{}^+B^-{}^+\overset{O}{\overset{\|}{C}}\sim$$

and

$$B^-\overset{O}{\overset{\|}{C}}{}^+B^-$$

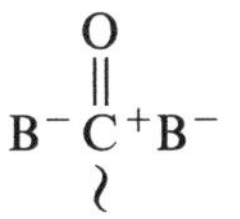

Only the first of these is an active centre [104].

Aggregation of centres is a thermodynamic category manifested in the process kinetics. By changes of temperature, component concentration, and medium (polarity and solvent power of the solvent), equilibria are shifted to new values. The number of inactive centres is generally changed. Therefore termination by aggregation of centres does not necessarily mean final and total loss of polymerizing activity. This usually occurs by other, irreversible processes.

References pp. 438–442

2.2.2 POLYMERIZATION-INACTIVE COMPLEXES OF IONS

Coordination of molecules to some part of an ionic active centre may increase its activity or it may lead to its complete deactivation. The detailed structure and properties of these $\sim C^+$ and $\sim C^-$ complexes are not suf-

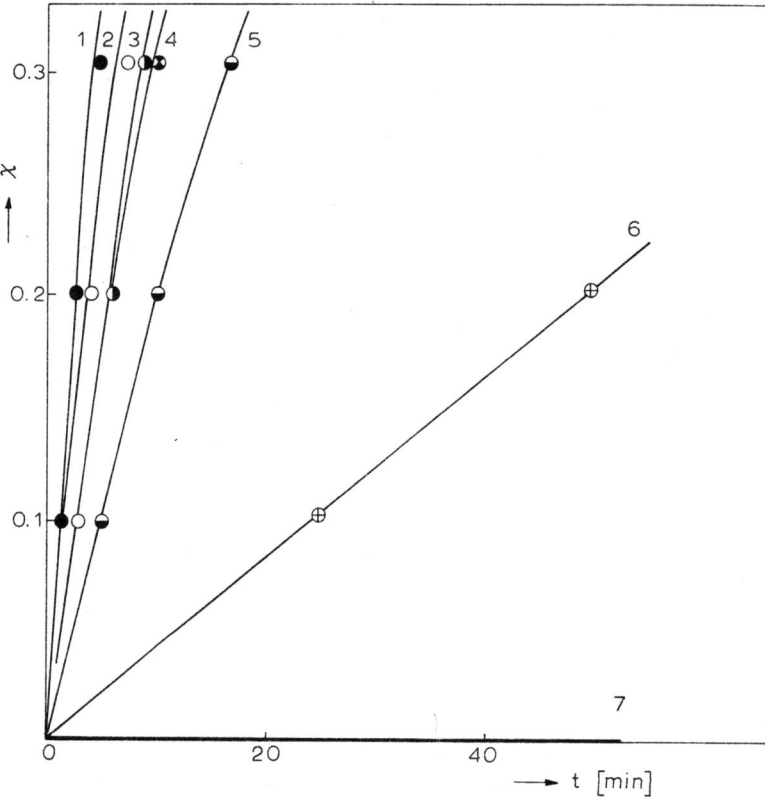

Fig. 9. Effect of LiOH and NaOH on the polymerization of octamethylcyclotetrasiloxane initiated by KOH. Temperature 423 K; [NaOH]/[KOH]: (1) 0, (2) 0.7, (3) 1, (6) 10; [LiOH]/[KOH]: (4) 0.7, (5) 1, (7) 10.

ficiently known. Formally it is easier to derive complexes of active centres containing an atom with a coordination number higher than 4, such as the complexes of silicon.

Already the equilibrium (77) has described the loss of the polymerization activity of a certain part of the centres by interaction of growing ends of equal basicity. Therefore we have called this reaction "*isobasic termination*". An addition of alkali metal hydroxides[†] to the initiator (potassium hydroxide)

[†] Which themselves initiate polymerization of cyclic siloxanes.

leads to a rate decrease down to the complete cessation of polymerization [105] (see Fig. 9). We assume this to be caused by the species

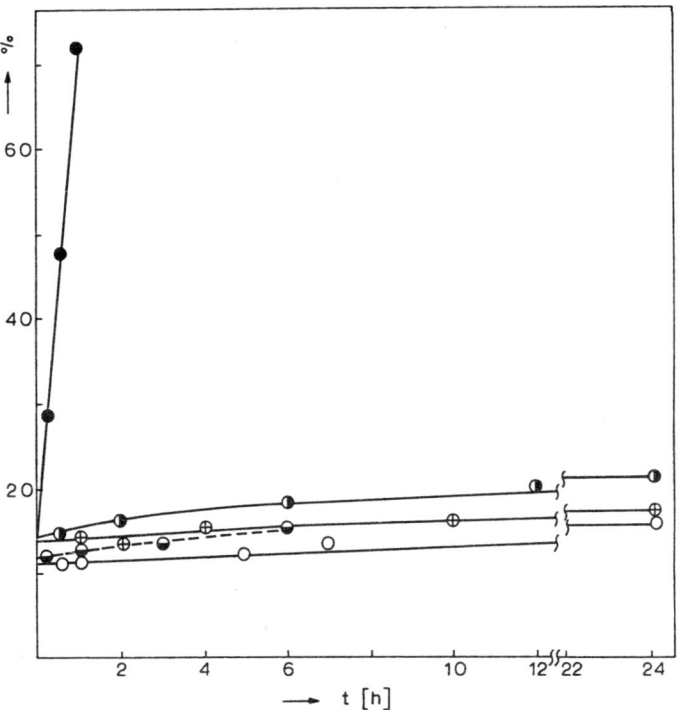

Much more stable complexes are formed from silanolate active centres and oxides, hydroxides or partly dehydrated hydroxides of amphoteric metals [106]. A polymer whose centres have been complexed in this way is completely resistant to depolymerization even at 523 K†. This situation is illustrated in Fig. 10.

Fig. 10. Dependence of mass loss on time for polydimethylsiloxane in the presence of metal hydroxides. Temperature 543 K; ●, unstabilized polymer; polymer with 5 % $Be(OH_2)$ (◐), 5 % $Al(OH)_3$ (⊕), 5 % $Fe(OH)_3$ (○), and 5 % $Zn(OH)_3$ (◉).

---

† Also the thermo-oxidative stability of the deactivated polymer is excellent. At the present time, hydroxides and partially dehydrated hydroxides of amphoteric metals, especially of aluminium, are applied as additives suppressing the flammability of polymers.

References pp. 438–442

The transformation of reactive centres to stable complexes may be of considerable practical importance. The expensive washing out of initiator residues can be substituted by their complexation. Suitable procedures and agents will also be sought for other ionic and coordination polymerizations.

2.2.3 AUTOINHIBITION (AUTOTERMINATION)

A growing chain end can lose its polymerizing ability without any external influence. The reason usually is the formation of a conjugated system where the ion loses part of its reactivity by charge delocalization. Such a system is logically formed by additions of suitable monomers (usually of the alkyne type; this case is similar to the gradual reactivity loss of a growing radical as described in Chap. 6, Sect. 1.4) or by isomerization through hydride anion or proton transfer. An example of the loss of the addition ability of a growing anion is the abstraction of the allylic proton from the unsaturated end group of polystyrene, yielding the unreactive diphenylallyl ion [107]

$$\sim CH_2-\bar{C}H + CH=CH-\underset{Ph}{\overset{H}{\underset{|}{C}}}\sim \longrightarrow \sim CH_2-CH_2Ph + CH\overset{-}{=\!=}CH\overset{-}{=\!=}C\sim \quad (80)$$
$$\qquad Ph \qquad Ph \qquad\qquad\qquad\qquad\qquad Ph \qquad Ph$$

Chains with a double bond in their end group are formed by a reaction generating an alkali metal hydride (see Chap. 7, Sect. 3.1).

The formation of the unreactive allylic ion causes growth termination of cationically polymerizing 2-methylpropene chains [108]

$$\sim CH_2-\underset{Me}{\overset{Me}{\underset{|}{C^+}}} + CH_2=\underset{Me}{\overset{Me}{\underset{|}{C}}} \longrightarrow \sim CH_2-\underset{Me}{\overset{Me}{\underset{|}{CH}}} + CH_2\overset{-}{=\!=}\underset{Me}{\overset{}{\underset{|}{\bar{C}}}}{=\!=}CH_2 \quad (81)$$

The loss of the addition ability of an ion by delocalization of the electron defect in a conjugated polyene chain is documented by the observation of Subramanian et al. They polymerized monomers with C≡C and C≡N bonds by electroinitiation. Anionic polymerization of phenylacetylene started very easily but it rapidly died out without any external cause [109]. The polymerization of n-alkyl-2-cyanoacrylate[†] in the presence of tetracyanoethylene yields only low-molecular-weight products. This is due to the formation

---

[†] These monomers are unusually reactive and polymerize at a high rate at almost any surface. They are therefore used as adhesives both in industry and in medicine. Polymerization is anionic and initiated by traces of bases, including water [110].

of an end-group anion stabilized by charge delocalization to such an extent that it cannot propagate further [111].

When the anionically produced polymer is more acid than the monomer, then the active centres are neutralized by protons dissociating from the chains.

This kind of autotermination was described in the polymerization of m, β- and p, β-dinitrostyrenes

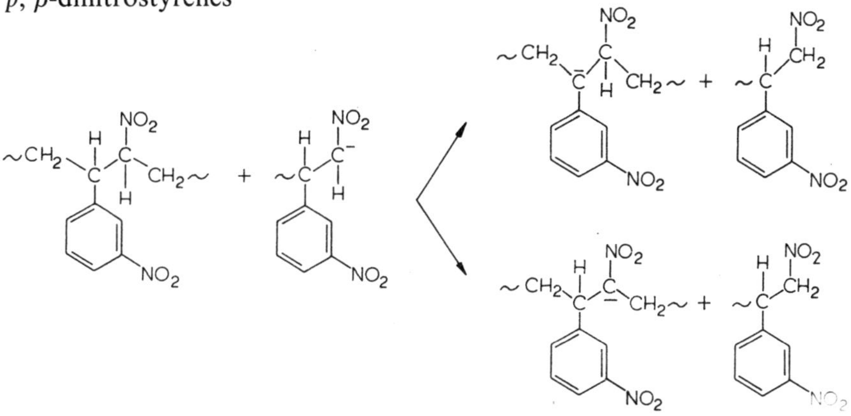

(82)

The generated carbanion is inactive, either for steric reasons or because of its weaker nucleophilic character [112].

Similarly, when the polymer produced by cationic polymerization is more basic than the monomer, the centres can lose their reactivity by interaction with the nucleophilic sites on the chains. This situation can be demonstrated by the observation of Penczek and Kubisa during polymerization of 3,3-bis-(chloromethyl)-oxetane [113]

(83)

The trialkyloxonium form is much more stable than the original active centre.

There probably exist more reactions leading to autotermination than we are aware of at present. These also include processes listed with other effects, for example pseudotermination [114]

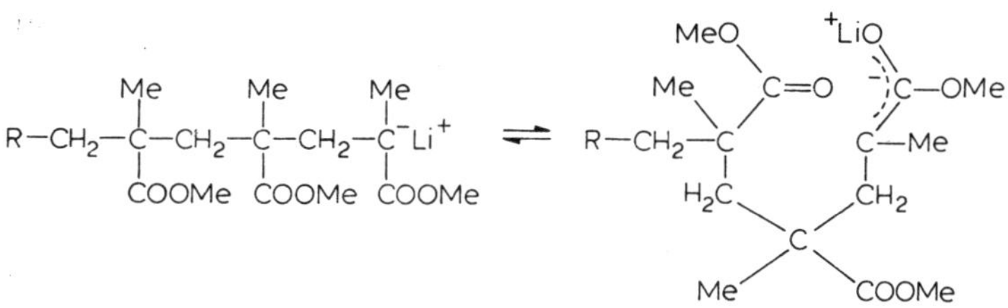

(84)

References pp. 438–442

or isomerization [115]

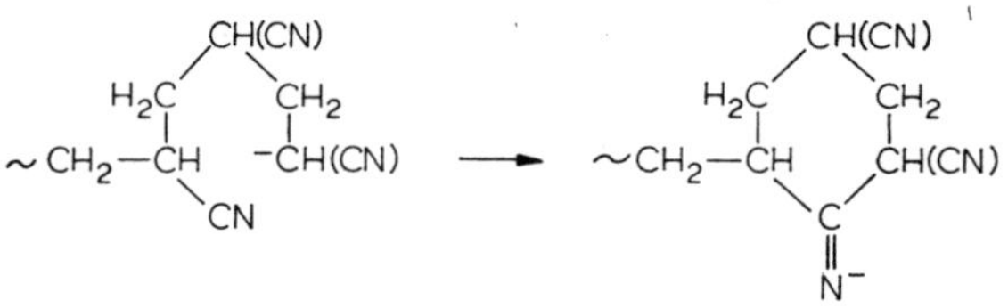

$$\sim CH_2-CH(CN)-CH(CN)-CH_2-C(=N^-)-CH(CN) \quad (85)$$

Research in this direction should elucidate the causes of prematurely extinguishing polymerizations and might even lead to the control of unwanted termination.

2.2.4 TERMINATION DUE TO IMPROPER ADDITION OF MONOMERS

When the monomer is attached to the active centre in a manner leading to the formation of a weakly reactive ion, growth is terminated. A typical example of this kind of termination is the formation of an unreactive ion during cationic polymerization of vinyl carbazole. Instead of propagation

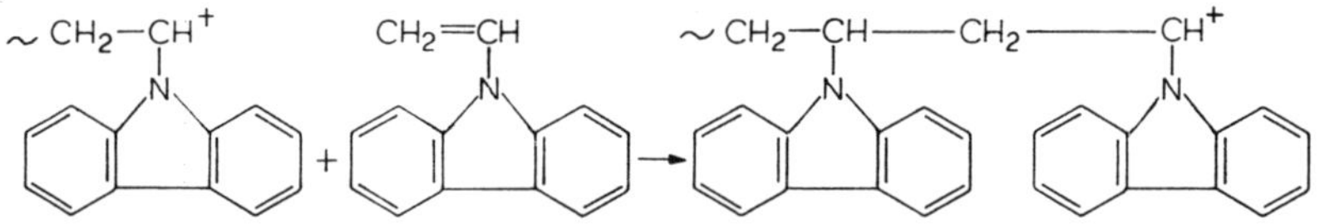

$$\sim CH_2-CH^+(Cbz) + CH_2=CH(Cbz) \rightarrow \sim CH_2-CH(Cbz)-CH_2-CH^+(Cbz) \quad (86)$$

head-to-head addition probably occurs [116]

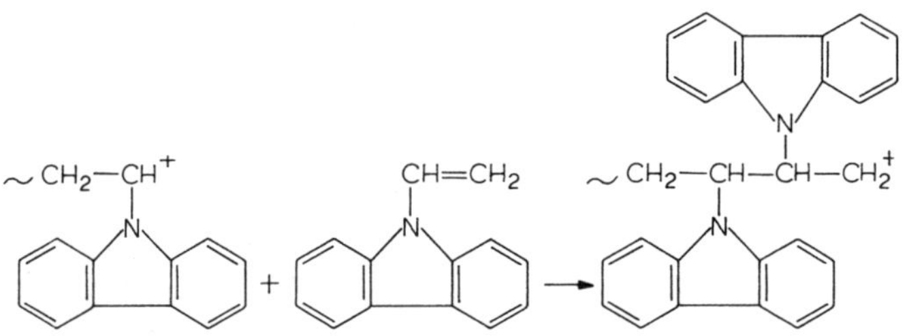

$$\sim CH_2-CH^+(Cbz) + CH=CH_2(Cbz) \rightarrow \sim CH_2-CH(Cbz)-CH(Cbz)-CH_2^+ \quad (87)$$

The generated ion does not react with monomer. It easily combines with the counter-ion.

2.2.5 TERMINATION BY IMPURITIES

When non-reproducible inhibition or induction periods are observed in polymerizations, when polymerization dies out at various degrees of conversion, or when its rate is poorly reproducible, terminating impurities should be suspected. Very little is known about this kind of reaction. Most of them can be equally well classified as degradative transfer.

Living anionic polymerizations are terminated by oxygen-yielding chains of double length [117]

$$\sim CH_2-\underset{Ph}{CH^-},Li^+ + {}^+Li,{}^-\underset{Ph}{HC}-CH_2\sim + O_2 \rightarrow$$

$$\rightarrow \sim CH_2-\underset{Ph}{CH}-\underset{Ph}{CH}-CH_2\sim + Li_2O_2 \qquad (88)$$

Water reacts with anionic and cationic centres yielding ^-OH or H_3O^+, respectively, and according to the conditions these can either initiate further polymerization, or combine with the counter-ion to an inactive product. Alcohols and ethers can react similarly and, in an analogous way, so do ammonia, amines, etc. Tetrahydrofuran, an often-used solvent for anionically polymerizing monomers, is opened by carbanions [118]. The generated alkoxide ion does not usually initiate polymerization. Highly effective terminating agents of polymerizations on carbanion centres are the halogenated hydrocarbons

$$\sim CH_2-\underset{X}{C^-}A^+ + RCl \rightarrow \sim CH_2-\underset{X}{CH}-R + ACl \qquad (89)$$

Quite generally, and logically, anionic polymerizations are terminated by acids, and cationic polymerizations by bases.

2.3 Combination of macroions

A special kind of termination in ionic polymerizations is the mutual combination of anionic and cationic living chains (see Chap. 5, Sect. 5.8). When the two polymers consist of different monomers, block copolymers are formed. The two macroions can also consist of the same monomer.

Kučera et al. combined anionic and cationic polydimethylsiloxane. With the ratio of active centres 1:1, a perfectly stable polymer was produced which did not depolymerize even under conditions where a trace of acid or base would lead to a rapid decomposition of all polymer chains [105]. This was the first combination of macroions described in the literature.

As far as termination is concerned, combination of polymeric ions is unimportant because most other termination methods are cheaper. Rather, it represents an elegant way to utilize termination for producing otherwise inaccessible block or graft copolymers.

References pp. 438–442

3. Termination of coordination polymerizations

The understading and control of termination in coordination polymerizations would represent an important technological advance. The productivity of catalysts would be enhanced [as expressed by the ratio of polymer mass (kg) produced per 1 kg of transition metal, or sometimes per 1 kg of catalyst] and the content of dangerous catalyst residues in the product would be reduced.

3.1 Termination by donors

When monomer coordination to the active centre is prevented, polymerization cannot occur. Coordination is a reversible reaction; strongly solvating agents deactivate centres in a ratio very close to 1:1 (acetylenes, allene, ketene, tetrahydrofuran, ROH, H_2O, COS, CO, CO_2, R_3N) but weaker donors must be present in some excess in order to cause total inhibition. It appears that isotactic and atactic centres of Ziegler–Natta polymerizations, particularly centres with $TiCl_3$ and Et_2AlCl, exhibit a different ability to coordinate donors (different acidity).

3.2 Termination by breaking of the metal–polymer bond in an active centre

The metal—carbon (metal—polymer) bond is strongly polar, and thus easily attacked by solvolyzing agents. Water and alcohols are the most efficient terminators of this kind.

Labelled alcohols, particularly *tert.* BuOT, are used to determine the number of active centres. Polymerization is terminated by a mixture of *tert.* BuOH and *tert.* BuOT

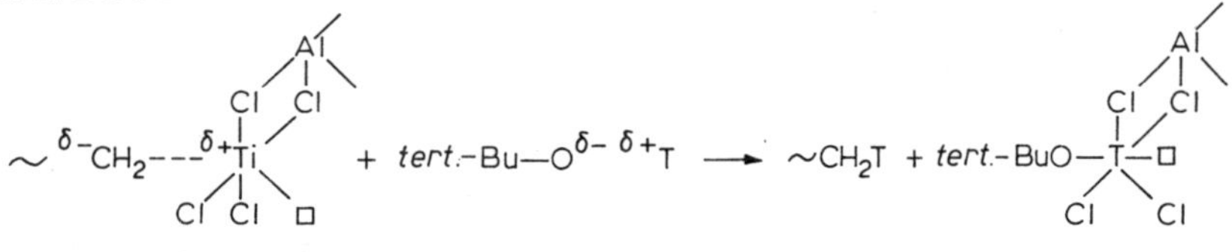

$$\sim\!{}^{\delta-}CH_2\text{---}{}^{\delta+}Ti(Cl)(Cl)(Cl)(Cl)(Al)(\square) + \textit{tert.}\text{-Bu}\text{---}O^{\delta-}\ {}^{\delta+}T \longrightarrow \sim\!CH_2T + \textit{tert.}\text{-BuO}\text{---}Ti(Cl)(Cl)(Cl)(Cl)(Al)(\square)$$

(90)

The reaction products are separated and ignited; the tritium content of the polymer, as determined by radiation trace analysis, is proportional to the number of metal—polymer bonds. A part of these bonds was generated by transfer to organometal (see Chap. 7, Sect. 5.1). The measurement is therefore repeated several times at various conversions. The number of active centres is found by extrapolation of the measured dependence to zero conversion (zero time), where the number of metal—polymer bonds generated by transfer is negligible.

The breaking of the metal—polymer bond in the active centre is a fairly general termination reaction. It can be induced by various agents, for example RX, RSH, H_2S, $SnCl_4$, $SiCl_4$, CCl_4, NaX, BCl_3, etc. [119] where X represents a halogen.

$$\sim\!{}^{\delta-}CH_2\text{---}{}^{\delta+}Ti + {}^{\delta+}BuCl^{\delta-} \rightarrow Ti\text{---}Cl + \sim\!CH_2Bu \qquad (91)$$

3.3 Termination by deactivation of the organometallic component

Solvolyzing compounds do not, of course, react selectively only with the transition metal compounds. They also attack the organometal by breaking the alkyl—metal bond. Thus their effect is similar to that of the former case, and the effects are additive.

3.4 "Spontaneous" termination

All the processes terminating coordination polymerizations discussed above are caused by the presence of impurities or of intentionally introduced additives. We know, however, that in the polymerizing medium termination occurs even without the participation of the these substances. Coordination polymerizations are notoriously non-stationary. Some kinds of centre have long lives, and others only short ones. So far we cannot intentionally change the ratio of the number of short-living to long-living centres, neither can we affect their lifetimes. To this end the basic prerequisite, a knowledge of the terminating mechanism, is lacking.

A first step towards acquiring such knowledge is the work of Chien who polymerized ethylene with homogeneous bis cyclopentadienyltitanium dichloride (cp_2TiCl_2) and Et_2AlCl. He assumes termination to be a bimolecular reaction of the active centres

$$2 \sim\!CH_2\text{---}{}^{\delta-}CH_2\text{---}\underset{cp}{\overset{cp}{Ti^{4+}}}\text{---}X \rightarrow \qquad (92)$$

$$\rightarrow \sim\!CH=CH_2 + \sim\!CH_2\text{---}CH_3 + 2\,\underset{cp}{\overset{cp}{Ti^{3+}}}\text{---}X$$

It involves disproportionation of the chain end groups, and reduction of two Ti^{4+} to Ti^{3+} [120].

It appears that the reduction of the transition metal oxidation state is a more frequent cause of activity loss of the centre than is usually acknowledged. Such reduction can occur as a consequence of a bimolecular reaction according to eqn. (92) but it may also be caused by the organoaluminium (organomagnesium) catalyst component, by H_2 intentionally added in order to reduce the molecular mass of the product, and perhaps also by other, so far unknown, processes.

4. Use of termination

Uncontrolled terminations are undesirable; they prevent polymerizations being carried out by a living mechanism. On the other hand, a suitable method of intentional termination is necessary for the deactivation of centres which would otherwise induce unwanted reactions during processing and application of the polymer.

4.1 Macromers (Macromonomers)

Deactivation of an active centre of ionic and coordination polymerizations is always connected with the formation of a certain end group at the originally living chain end. By the use of a suitable terminating agent at the right moment, polymers of a desired chain length and with a desired end group can be prepared

$$\sim\oplus, A^+ + Br_2 \rightarrow \sim Br + ABr$$

$$\sim\ominus, A^+ + RCOCl \rightarrow \sim\underset{O}{\overset{\|}{C}}-R + ACl$$

$$\sim\oplus, B^- + HOH \rightarrow \sim OH + HB$$

$$\sim\oplus, B^- + ROH \rightarrow \sim OR + HB$$

$$\sim\oplus, B^- + RC\overset{\overset{O}{\|}}{\underset{OR'}{\diagdown}} \rightarrow \sim O\underset{O}{\overset{\|}{C}}R + R'B$$

when living di-ions are terminated by suitable agents, chains with functional groups at both ends can be prepared†. Such polymers are called macromers (telechelic polymers) [121–123]. They are used as "monomers" for the preparation of liquid rubbers, copolymers [124, 125], special types of polyurethanes, polyesters, polyamides, etc. by polycondensation reactions.

4.2 Deactivation of centres

Active centres and their residues (in non-stationary polymerizations) are always a source of unwanted reactions of macromolecules during further operations. On an industrial scale, radical centres are terminated by inhibitors. Even the inhibiting effect of atmospheric oxygen is exploited.

The effectiveness of such cheap termination by oxygen is, of course, low. In high-pressure polyethylene, polyvinyl chloride, polytetrafluoroethylene, etc., a large amount of "frozen" radicals was detected by ESR [126]. Actually, even after oxygen addition to a radical, the resulting ROO* is sufficiently unstable. Research in this direction could lead to an acceptably cheap and efficient deactivation.

Decay of ionic and coordination centres always leads to the formation of some end groups and centre residues. The centres usually lose their polymerizing activity on contact with atmospheric humidity. A residue of very active centres, which are rare, is usually not removed from the polymer (e.g. of the order of one ppm of the transition metal in low-pressure polyethylene). Larger residues have to be washed out (some types of polypropylene are still washed at the present time).

4.2.1 DETERMINATION OF THE NUMBER OF CENTRES

Otsu et al. published a series of studies [127–132] in which they detected and identified primary and propagating radicals by the "spin-trapping" technique. The generation of initiating radicals, represented by the scheme

$$I \xrightarrow{k_d} R^*$$

or $\quad$ (93)

$$n M \longrightarrow R^*$$

and initiation

$$R^* + M \xrightarrow{k_i} M^* \quad\quad\quad (94)$$

---

† For synthesis it is often useful to use transfer instead of termination (see Chap. 7, Sect. 8. 3).

could be quantified if the momentary concentrations [R*] and [M*] were known. The lifetimes of these radicals are usually short, so that they are not easily detected by resonance methods. When the reactions (93) proceed in the presence of a spin trapping agent, Z, the unstable species R* and M* react with Z yielding stable radicals RZ* and MZ*

$$R^* + Z \rightarrow RZ^* \tag{95}$$

$$M^* + Z \rightarrow MZ^* \tag{96}$$

When Z selectively reacts with either R*, or M*, the ESR spectrum of the reaction mixture can reveal the structure of the originally generated R* and/or M*. 2-Methyl-2-nitrosopropane, $Me_3C-N=O$ [131–133], and derivatives of nitrosocompounds [133] are used as spin-trapping agents.

A practical examle of this kind of measurement is the study by Sato and Otsu of di-*tert.*-butyl peroxalate decomposition in *p*-xylene as solvent in the presence of vinyl monomer and 2-methyl-2-nitrosopropane (at room temperature the initiator is decomposed into Me_3CO^* and CO_2). Due to repulsive forces, the spin-trapping agents reluctantly react with oxyl radicals. Therefore the following reactions mutually compete

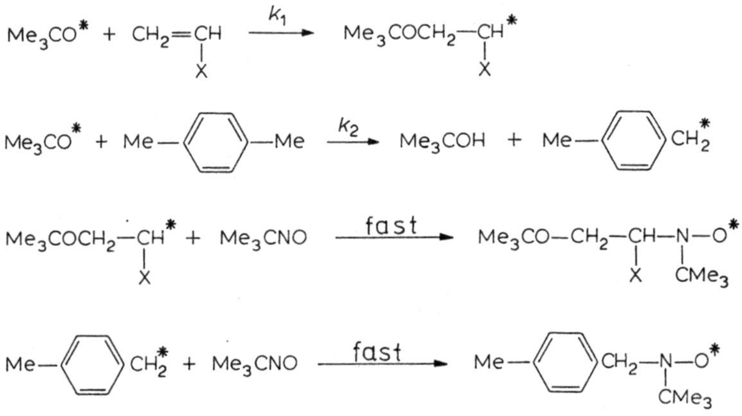

$$Me_3CO^* + CH_2=CHX \xrightarrow{k_1} Me_3COCH_2-CHX^* \tag{97}$$

$$Me_3CO^* + Me-C_6H_4-Me \xrightarrow{k_2} Me_3COH + Me-C_6H_4-CH_2^* \tag{98}$$

$$Me_3COCH_2-CHX^* + Me_3CNO \xrightarrow{fast} Me_3CO-CH_2-CHX-N(CMe_3)-O^* \tag{99}$$

$$Me-C_6H_4-CH_2^* + Me_3CNO \xrightarrow{fast} Me-C_6H_4-CH_2-N(CMe_3)-O^* \tag{100}$$

From ESR spectra, the relative concentration of the radical products of reactions (99) and (100) was determined. When both reactions are sufficiently rapid, the concentration ratio is equal to k_1/k_2, i.e. to the relative reactivity of the monomer $CH_2=CHX$ with Me_3CO^* radicals. In the same solvent, measurements with various monomers yield relative initiation rates [132] according to reaction (94). By a combination of this procedure with an absolute method (e.g. with inhibitors), for which the most favourable conditions can be selected, the accuracy of the determination of the kinetics and mechanism of initiation can be significantly enhanced.

Recent progress in microelectronics has led to the construction of very sensitive instruments so that ever the concentration of active centres can be measured directly with increasing frequency. Lau et al. determined the stationary radical concentration in emulsion polymerizations of MMA, methacrylic acid, and butyl acrylate by ESR. Thus k_p could be derived directly [134].

An ionic analogy of the spin-trapping measurements are the ion-trapping methods. 2-Methyl-2-nitrosopropane also reacts with carbanions [133]

$$n\ CH_2=\underset{\underset{OMe}{\overset{|}{O=C}}}{\overset{\overset{Me}{|}}{C}} \xrightarrow{Bu^-Li^+} \sim CH_2-\underset{\underset{OMe}{\overset{|}{O=C}}}{\overset{\overset{Me}{|}}{C^-Li^+}} \xrightarrow{Me_3CNO}$$

$$\longrightarrow \sim CH_2-\underset{\underset{O=COMe}{|}}{\overset{\overset{Me}{|}}{C}}-N\underset{O^-Li^+}{\overset{CMe_3}{\diagup}} \xrightarrow[-e]{MeOH} \quad (101)$$

$$\longrightarrow \sim CH_2\underset{\underset{O=CMe}{|}}{\overset{\overset{Me}{|}}{C}}-N\underset{O^*}{\overset{CMe_3}{\diagup}}$$

Brzezinska et al. described a procedure converting an unstable cation into a stable one containing a phosphorus atom [135]. Phosphorus can be easily determined by ^{31}P NMR

$$\sim + \ + PR_3 \xrightarrow{k_z} \sim P^+R_3 \quad (102)$$

The rate constant of the ion-trapping reaction, k_z, is much larger than the propagation rate constant, at least in cyclic ether and acetal polymerizations. The ratio $k_z[PR_3]/k_p[M]$ is large and termination is instantaneous compared with growth [136]. In this way, the number of centres can be determined at any moment during polymerization. The chemical shifts of some quaternary phosphonium ions (of structure corresponding to terminated polyheterocyclic chains) are summarized in Table 2.

Among the methods discussed for detecting active centres, some successful applications of spectroscopic and resonance methods should be mentioned. These include the spectrophotometry of carbanions and polystyryl cations in the UV and visible ranges [137, 138], and ^1H NMR of sulphonium [139] and oxonium [140] centres in polymerizations of the corresponding heterocycles. The use of sodium phenolate has already been mentioned. Tetraethylammonium picrate was also used to the same end [141], i.e. counting of centres during tetrahydrofuran polymerization. However, each of the mentioned

TABLE 2

$^{31}P\{^1H\}$ NMR Chemical shifts of tetraalkylphosphonium ions formed by traping of model cations or macrocations[135]

Parent cation	Quarternary phosphonium ion	Chemical shift[b]
"H$^+$"	H—$\overset{+}{P}$(n-C$_4$H$_9$)$_3$SbF$_6^-$	−11,9[c]
C$_6$H$_5\overset{+}{\equiv}$O	C$_6$H$_5$C(O)$\overset{+}{P}$(C$_6$H$_5$)$_3$SbF$_6^-$	−11,6[c]
CH$_3$O$\overset{+}{C}$H$_2$	CH$_3$OCH$_2\overset{+}{P}$(C$_6$H$_5$)$_3$Cl$^-$	−17,3[c]
CH$_3^+$	CH$_3\overset{+}{P}$(C$_6$H$_5$)$_3$CF$_3$SO$_3^-$	−21,2[c]
CH$_3$CH$_2^+$	CH$_3$CH$_2\overset{+}{P}$(C$_6$H$_5$)$_3$CF$_3$SO$_3^-$	−25,1[c]
...$\overset{+}{C}$H$_2$—O ⟨△⟩	...—O—(CH$_2$)$_3$—$\overset{+}{P}$(C$_6$H$_5$)$_3$SbF$_6^-$	−23,8[c]
...—$\overset{+}{C}$H$_2$—O ⟨⬠⟩	...—O—(CH$_2$)$_4$—$\overset{+}{P}$(C$_6$H$_5$)$_3$SbF$_6^-$	−23,4[c]
	...—O—(CH$_2$)$_4$—$\overset{+}{P}$(n-C$_4$H$_9$)$_3$SbF$_6^-$	−32,6[d]
...—$\overset{+}{C}$H$_2$—O ⟨⬡⟩	...—O—(CH$_2$)$_6$—$\overset{+}{P}$(C$_6$H$_5$)$_3$SbF$_6^-$	−23,0[c]
	...—O—(CH$_2$)$_6$—$\overset{+}{P}$(n-C$_4$H$_9$)$_3$SbF$_6^-$	−32,1[d]
...—OCH$_2$CH$_2\overset{+}{O}$CH$_2$	...OCH$_2$CH$_2$OCH$_2\overset{+}{P}$(C$_6$H$_5$)$_3$SbF$_6^-$	−16,7[d]
...—OCH$_2\overset{+}{O}$CH$_2$	...OCH$_2$OCH$_2\overset{+}{P}$(C$_6$H$_5$)$_3$SbF$_6^-$	−17,5[d]

[a] (C$_6$H$_5$)$_3$P: $\delta = +5.96$(CH$_2$Cl$_2$), $\delta = +6,70$(CH$_3$NO$_2$); (n-C$_4$H$_9$)$_3$P: $\delta = +33,1$(CH$_2$Cl$_2$).
[b] Chemical shifts δ in p.p.m. from 85 % H$_3$PO$_4$.
[c] In CH$_2$Cl$_2$.
[d] In CH$_3$NO$_2$.

methods has its specific limitations. Spectrophotometry can be applied only to polymerizations of a small number of monomers (most active centres do not absorb in a suitable range), and ^{1}H NMR only for measuring high concentrations of centres ($\gtrsim 10^{-3}$ mol dm^{-3}). For unknown reasons, labelling of macrocations by phenolate anions can only be used with a limited number of monomers [142].

Much effort has been devoted to the development of methods for determining (counting) centres in coordination polymerizations. If these polymerizations were living it would be sufficient to count the macromolecules in the final product, and this number would also be equal to the number of growth centres. There exist, however, only a few living coordination polymerizations, and these do not include the technologically important ones. At the present time, a method suitable for accurately counding the active centres and also eliminating the consequences of termination and transfer, has been developed and tested.

In principle, two methods are known: either we label the macromolecules produced by the centres or we measure the consumption of an inhibitor safely inactivating the centres. Several variants of each of these methods exist with specific advantages and limitations.

(a) The macromolecules are marked either by means of an organometallic co-catalyst, i.e. an active centre is constructed so as to provide each chain growing on it with a label (atom group) which can be detected in the polymer, or more often by determining the number of metal-polymer bonds (see Sect. 3.2) or by some other selective marking (by CS_2, CO, CO_2, COS, SO_2 etc.) of the growing macromolecule.

(b) The methods of selective marking are usually interconnected with determinations of polymerization inhibitor consumption (CS_2, CO, CO_2, COS, SO_2, acetylenes, dienes with cumulated bonds are catalytic poisons).

The determination of the number of centres from the known number average polymerization degree of the product is also of some importance [143]

$$\frac{1}{\bar{P}_n} = \frac{1}{k_p[M]}\left(\frac{R_p}{Q} + \sum k_{tr}[XT]^n\right)$$

where Q is the amount of product in mass units.

When the constants, the concentration, and R_p are independent of time, the plot of $\bar{P}_n^{-1}$ vs. R_p/Q yields a straight line of slope $(k_p[M])^{-1}$. The amount of active centres $[C^a] = R_p/k_p[M]$.

The above-described procedures are very sensitive to external conditions. The application of the same methods at various laboratories often leads to conflicting results. A study of the original literature is therefore necessary for the specialist. Fortunately a number of expert review articles have been recently published [119, 144–146].

5. Kinetics of termination

The difficulties involved in the direct determination of the momentary concentration of active centres are the most serious shortcoming in studies of termination itself. With radical polymerizations we at least know the most probable method of centre decay, and thus the molecular scheme of the termination reaction. In ionic and coordination polymerizations, the termination mechanism is mostly unknown. Quite generally we can write

$$\sim M^\circ \xrightarrow{k_t} \text{inactive polymer} \quad (103)$$

or

$$\sim M^\circ + {}^\circ M \sim \xrightarrow{k_t} \text{inactive polymer} \tag{104}$$

The kinetics of termination is usually approximated by simple relations where a speculative reaction order is inserted into the rate equation[†]

$$-\frac{d[Ac]}{dt} = k_t \tag{105}$$

$$-\frac{d[Ac]}{dt} = \bar{k}_t[Ac] \tag{106}$$

$$-\frac{d[Ac]}{dt} = \bar{k}_t[Ac]^2 \tag{107}$$

According to eqns. (105)–(107), the dependence of the decay of active centres on time is linear, exponential, and hyperbolic, respectively. The best approximation is obtained by a comparison of the observed course with the rate of monomer decay calculated for the assumed termination mode. This is, of course, a very rough method, which often does not reveal the effect of other components on termination, for example that of the monomer (its "wrong" addition etc.)

$$\sim M^\circ + M \xrightarrow{k_t} \text{inactive polymer} \tag{108}$$

thus

$$-\frac{d[Ac]}{dt} = k_t[Ac][M] \tag{109}$$

or of the generated polymer

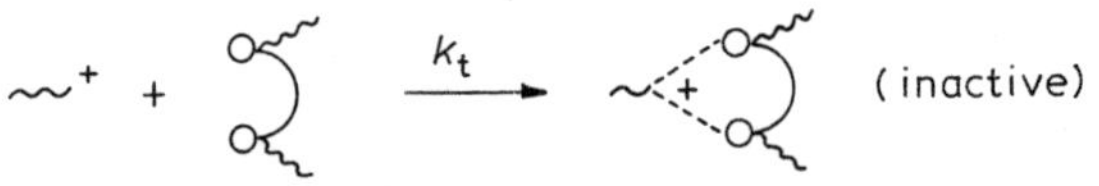

$$\tag{110}$$

[†] The possibilities are not very numerous. The decay of centres is either independent of their concentration, or is directly proportional to it. Dependence on the square of their concentration is typical of radical and some coordination processes, otherwise it is rare. Thus the order of termination is 0, 1, or 2. Fractional orders are not considered without serious independent evidence.

so that

$$-\frac{d[Ac]}{dt} = k_t[Ac]([M]_0 - [M]) \tag{111}$$

These examples represent only a part of the possible situations.

The calculated functional dependence for [Ac] is inserted into the rate equation. An agreement between calculation and experiment should not be overestimated. An exact kinetic analysis of polymerization requires proof of the partial steps, and is very difficult.

The preceding statement should not invoke in the reader an idea about the intractability of the problem. In studies of polymerizations, even of the complicated ones, the necessary direct and indirect evidence is almost always accessible, although it may not be easy to obtain.

One example of a direct determination of the time dependence of active centres is the already cited work of Saegusa and Matsumoto [90] (see Fig. 11).

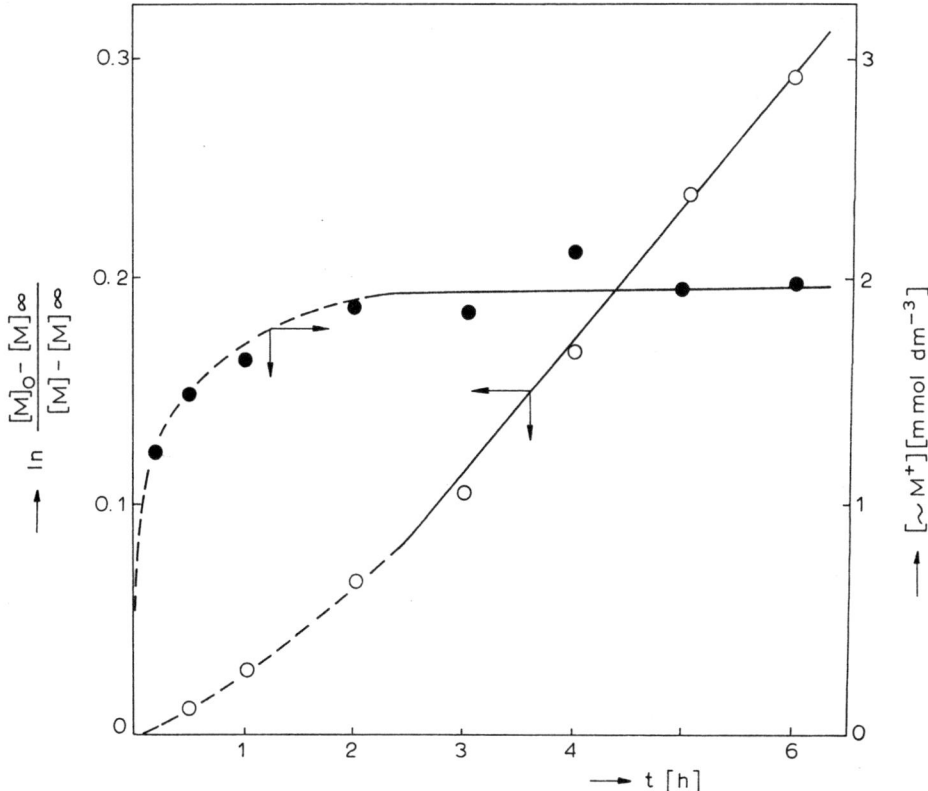

Fig. 11. Dependence of active centre concentration on time during THF polymerization. Temperature 273 K, initiator Et_3Al/H_2O (2:1) with chloromethyloxirane as co-initiator.

Information of this kind opens new exact access to the kinetic analysis of further polymerization steps.

Rapid accumulation of the missing data on termination is documented by well-founded theoretical studies, leading to the solution of the complicated problems connected with the consequences of termination and transfer on the molecular mass distribution by means of graphical analysis [147, 148]. Equilibrium polymerization with reversible transfer has also aroused the interest of theorists [149] even though it appears that this kind of transfer in particular has only a small efect on the resulting polymer parameters.

References

1. P. E. M. Allen and C. R. Patrick, Kinetics and Mechanisms of Polymerization Reactions, Horwood, Chichester, 1974, pp. 289–292, 377.
2. S. W. Benson, Adv. Photochem., 2 (1964) 1.
3. G. V. Schulz, Z. Phys. Chem. (Frankfurt), 8 (1956) 284.
4. A. M. North, in J. C. Robb and F. W. Peaker (Eds.), Progress in High Polymers, Pergamon Press London, Vol. 2, 1968.
5. H. K. Mahabadi and K. F. O'Driscoll, J. Polym. Sci. Polym. Chem. Ed., 15 (1977) 283.
6. P. G. de Gennes, Macromolecules, 9 (1976) 587, 594.
7. S. W. Benson and A. M. North, J. Am. Chem. Soc., 84 (1962) 935.
8. R. D. Burkhart, J. Polym. Sci. Part A-1, (3) (1965) 883.
9. Kh. S. Bagdasaryan, Vysokomol. Soedin. Ser. B, 9 (1967) 10.
10. P. E. M. Allen and C. R. Patrick, Makromol. Chem., 72 (1964) 106.
11. A. F. Moroni and G. V. Schulz, Makromol. Chem., 118 (1969) 313.
12. O. F. Olaj, Makromol. Chem., 147 (1971) 235.
13. O. F. Olaj, G. Zifferer and G. Gleixner, Macromolecules, 20 (1987) 839, 850, and previous papers.
14. H. Kh. Mahabadi, Macromolecules, 18 (1985) 1319.
15. C. H. Bamford, A. D. Jenkins and R. Johnston, Trans. Faraday Soc., 55 (1959) 179.
16. G. V. Schulz, G. Henrici-Olivé and S. Olivé, Z. Phys. Chem. (Frankfurt), 27 (1961) 1.
17. J. C. Bevington, H. W. Melville and R. P. Taylor, J. Polym. Sci., 14 (1954) 463.
18. J. C. Bevington, Radical Polymerization, Academic Press, London, 1961.
19. G. Ayrey, M. J. Humphrey and R. C. Poller, Polymer, 18 (1977) 840.
20. Y. Ogo and T. Kyotani, Makromol. Chem., 179 (1978) 2407.
21. A. M. North and G. A. Reed, Trans. Faraday Soc., 57 (1961) 859.
22. H. K. Mahabadi and K. F. O'Driscoll, Macromolecules, 10(1) (1977) 55; J. Polym. Sci. Polym. Lett. Ed., 16 (1978) 351.
23. W. A. Ludwico and S. L. Rosen, J. Polym. Sci. Polym. Chem. Ed., 14 (1976) 2121.
24. T. Tulig and M. Tirrell, Macromolecules, 14 (1981) 1501.
25. P. C. Deb and G. Meyerhoff, Eur. Polym. J., 10 (1974) 709.
26. P. C. Deb and I. D. Gaba, Makromol. Chem., 179 (1978) 1549, 1559.
27. S. W. Benson and A. M. North, J. Chem. Soc., 81 (1959) 1339.
28. H. K. Mahabadi and K. F. O'Driscoll, Makromol. Chem., 178 (1977) 2629.
29. R. G. Norrish and R. R. Smith, Nature (London), 150 (1942) 366.
30. E. Tromsdorff, H. Kohle and P. Lagally, Makromol. Chem., 1 (1948) 169.

31 M. S. Matheson, E. E. Auer, E. B. Bevilacqua and E. J. Hart, J. Am. Chem. Soc., 73 (1951) 5395.
32 H. W. Melville, Z. Electrochem., 60 (1956) 276.
33 D. Mangaraj and S. K. Patra, Makromol. Chem., 104 (1967) 125.
34 G. M. Burnett and G. L. Duncan, Makromol. Chem., 51 (1962) 154, 171, 177.
35 S. T. Balke and A. E. Hamielec, J. Appl. Polym. Sci., 17 (1973) 905.
36 P. J. Flory, J. Chem. Phys., 17 (1949) 303.
37 W. Kuhn and H. Kuhn, Helv. Chim. Acta, 29 (1945) 1533.
38 G. V. Schulz and J. P. Fischer, Makromol. Chem., 107 (1967) 253.
39 J. N. Cardenas and K. F. O'Driscoll, J. Polym. Sci. Polym. Chem. Ed., 14 (1976) 883. J. Polym. Sci. Polym. Chem. Ed., 15 (1977) 1883.
40 F. L. Marten, A. E. Hamielec: Am. Chem. Soc., Symp. ser. N°104, 1979, 43.
41 D. T. Turner, Macromolecules, 10(2) (1977) 221.
42 H. B. Lee and D. T. Turner, Macromolecules, 10(2) (1977) 226, 231.
43 H. B. Lee and D. T. Turner, Polym. Prepr. Am. Chem. Soc. Div. Polym. Chem., 18(2) (1977) 539; 19 (1978) 603.
44 T. J. Tulig and M. Tirrell, Macromolecules, 15 (1982) 459.
45 M. Tirrell, B. Hanley, S. Balloge and T. Tulig, Polym. Prepr. Am. Chem. Soc. Div. Polym. Chem., 26(1) (1985) 299.
46 M. Smoluchowski, Z. Phys. Chem., 92 (1918) 129.
47 W. I. Bengough and R. G. W. Norrish, Proc. R. Soc. (London), Ser. A, 200 (1950) 310.
48 C. H. Bamford, W. G. Barb, A. D. Jenkins and P. F. Onyon, Kinetics of Vinyl Polymerization by Radical Mechanisms, Academic Press, New York, 1958, p. 103.
49 H.S. Mickley, A. S. Michaels and A. L. Moore, J. Polym. Sci., 60 (1962) 121.
50 A. I. Bolshakov, A. I. Mikhailov and I. M. Barkalov, Vysokomol. Soedin. Ser. A, 20 (1978) 1820.
51 P. E. M. Allen, J. M. Downer, G. W. Hastings, H. W. Melville, F. Molyneaux and J. R. Urwin, Nature (London), 177 (1956) 910.
52 D. Mikulášová, V. Chrástová and P. Citovický, Eur. Polym. J., 10 (1974) 551.
53 E. Bigdeli, R. W. Lenz, B. Oster and R. D. Lundberg, J. Polym. Sci. Polym. Chem. Ed., 16 (1978) 469.
54 M. Ueda, S. Shouji, T. Ogata, M. Kamachi and C. U. Pittman, Jr., Macromolecules, 17(12) (1984) 2800.
55 C. I. Simionescu and B. C. Simionescu, Pure Appl. Chem., 56(3) (1984) 427.
56 T. Sato, J. Miyamoto and T. Otsu, J. Polym. Sci. Polym. Chem. Ed., 22 (1984) 3921.
57 See P. E. M. Allen and C. R. Patrick, Kinetics and Mechanisms of Polymerization Reactions, Horwood, Chichester, 1974, pp. 136–138, 159.
58 W. I. Bengough and W. H. Fairservice, Trans. Faraday Soc., 61 (1965) 1206; 63 (1967) 382; 67 (1971) 414.
59 G. M. Burnett and P. R. E. J. Cowley, Trans Faraday Soc., 49 (1953) 1490.
60 C. H. Bamford, A. D. Jenkins and R. Johnston, Proc. R. Soc. Ser. A, 239 (1957) 214.
61 F. Tüdös, Makromol. Chem., 79 (1974) 8; Acta Chim. Hung., 43 (1965) 397; J. Polym. Sci. Part C, 17 (1968) 3461.
62 J. C. Bevington, Radical Polymerization, Academic Press, London, 1961.
63 A. A. Miller and F. R. Mayo, J. Am. Chem. Soc., 78 (1956) 1017.
64 I. M. Kolthoff and W. J. Dale, J. Am. Chem. Soc., 69 (1947) 441.
65 E. Dyer, S. C. Brown and R. W. Medeiros, J. Am. Chem. Soc., 81 (1959) 4243.
66 G. V. Schulz and G. Henrici, Makromol. Chem., 18/19 (1956) 437.
67 D. Stein and G. V. Schulz, Makromol. Chem., 38 (1960) 248.
68 Z. Macháček and F. Čermák, Chem. Prum., 15(40) (1965) 484.

69 E. Dyer, O. A. Pickett, S. F. Strause and H. E. Worrel, J. Am. Chem. Soc., 78 (1956) 3384.
70 G. A. Razuvaev and K. S. Minsker, Vysokomol. Soedin., 2 (1960) 1239.
71 R. G. Caldwell and J. L. Ihrig, J. Am. Chem. Soc., 84 (1962) 2878.
72 N. N. Tvorogov, I. A. Matreyeva, A. A. Volodkin and A. G. Kondrateva, Polym. Sci. USSR, A18(2) (1976) 399.
73 S. A. Chen and L. C. Tsai, Makromol. Chem., 187 (1986) 633.
74 R. K. Samal, P. K. Sahoo, S. P. Bhattacharjee and M. S. Samantaray, J. Macromol. Sci. Chem., A23(2) (1986) 203.
75 M. G. Chauser, J. M. Rodionov, V. M. Misin and M. I. Cherkasin, Usp. Khim. 45 (1976) 695.
76 T. Masuda, K. Hasegawa, T. Higashimura, Macromolecules, 7 (1974) 728; 8 (1975) 225.
77 H. X. Nguyen, S. Amdur and P. Ehrlich, Polym. Prepr. Am. Chem. Soc. Div. Polym. Chem., 18 (1977) 200.
78 B. Biyani, A. J. Campagna, D. Daruwalla C. M. Srivastava and P. Ehrlich, J. Macromol. Sci. Chem., A9 (1975) 327.
79 J. Manassen and R. Rein, J. Polym. Sci. Part A-1, (8) (1970) 1403.
80 M. G. Chauser, J. M. Rodionov and M. I. Cherkasin, J. Macromol. Sci. Chem., A11 (1977) 1113.
81 S. Amdur, A. T. Y. Cheng, C. J. Wong, P. Ehrlich and R. D. Allendoerfer, J. Polym. Sci. Polym. Chem. Ed., 16 (1978) 407.
82 A. Berthoud and H. Bellenot, Helv. Chim. Acta, 7 (1923) 303.
83 D. L. Chapman, F. Briers and E. Walters, J. Chem. Soc., (1926) 562.
84 H. W. Melville, Proc. R. Soc. (London) Ser. A, 163 (1937) 511.
85 G. M. Burnett and H. W. Melville, in A. Weissberger, (Ed.) Technique of Organic Chemistry, Vol. VIII, Part II, Interscience, New York, 1963, pp. 1111–1126.
86 G. M. Burnett, in H. Mark, H. W. Melville, C. S. Marvel and G. S Whitby (Eds.) Mechanisms of Polymer Reactions, Vol. III, Interscience, London, 1954, pp. 190–203.
87 See P. E. M. Allen and C. R. Patrick, Kinetics and Mechanisms of Polymerization Reactions, Horwood, Chichester, 1974, pp. 156–158.
88 K. F. O'Driscoll and H. K. Mahabadi, J. Polym. Sci. Polym. Chem. Ed., 14 (1976) 869.
89 K. Ito and K. F. O'Driscoll, J. Polym. Sci. Polym. Chem. Ed., 17 (1979) 3913.
90 T. Saegusa and S. Matsumoto, J. Polym. Sci. Part. A-1, (6) (1968) 1559; J. Macromol. Sci. Chem., A4 (1970) 873.
91 D. C. Pepper, Q. Rev. (London), 8 (1954) 88.
92 J. J. Trossell, S. P. Sood, M. Szwarc and V. Stannett, J. Am. Chem. Soc., 78 (1956) 1122.
93 S. Penczek, J. Jagur-Grodzinski and M. Szwarc, J. Am. Chem. Soc., 90 (1968) 2174.
94 M. Marek, L. Toman, V. Halaška and P. Lopour, Polyisobutylene 4-80, Research Report, Institute of Macromolecular Chemistry, Czechoslovk Academy of Sciences, Prague, 1981.
95 P. H. Plesch, Progr. High Polym. 2 (1968) 139.
96 M. Kučera and M. Jelínek, Vysokomol. Soedin., 2 (1960) 1860.
97 D. J. Worsfold and S. Bywater, Can. J. Chem., 38 (1960) 1891.
98 J. E. C. Roovers and S. Bywater, Trans. Faraday Soc., 62 (1966) 701.
99 S. Bywater and D. J. Worsfold, Polym. Prepr. Am. Chem. Soc. Div. Polym. Chem., 27(1) (1986) 140.
100 J. E. Figueruelo and D. J. Worsfold, Eur. Polym. J., 4 (1968) 439.
101 J. E. Figueruelo and A. J. Bello, J. Macromol. Sci., A3 (1969) 311.
102 K. S. Kazansij, A. A. Solovyanov and A. N. Tarasov, Prepr. IUPAC Int. Symp. Macromol. Helsinki, 2 (1972) 535.
103 A. A. Solovyanov and K. S. Kazansij, Vysokomol. Soedin. Ser. A, 14 (1972) 1063, 1071.
104 B. G. Belenkaya, A. I. Lebenko and E. B. Lyudvig, Vysokomol. Soedin. Ser. A, 20 (1978) 559.

105 M. Kučera, M. Jelínek, J. Láníková and K. Veselý, J. Polym. Sci., 53 (1961) 311.
106 M. Kučera and J. Láníková, J. Polym. Sci., 59 (1962) 79.
107 M. Szwarc, Carbanions, Living Polymers and Electron Transfer Processes, Interscience, New York, 1968, p. 649.
108 J. P. Kennedy and R. G. Squires, J. Macromol. Sci. Chem., A1 (1967) 861.
109 R. V. Subramanian, J. Jakubovski a nd B. K. Garg, Polym. Repr. Am. Chem. Soc. Div. Polym. Chem., 19(1) (1978) 517.
110 F. Leonard, J. A. Collins and H. A. Porter, J. Appl. Polym. Sci., 10 (1966) 1617.
111 E. F. Donnelly, D. S. Johnston, D. C. Pepper and D. J. Dunn, J. Polym. Sci. Polym. Lett. Ed., 15 (1977) 399.
112 M. E. Carter, J. L. Nash, Jr., J. W. Drueke, Jr., J. W. Schwietert and G. B. Butler, J. Polym. Sci. Polym. Chem. Ed., 16 (1978) 937.
113 S. Penczek and P. Kubisa, Makromol. Chem., 130 (1969) 186.
114 D. L. Glusker, E. Stiles and B. Yoncoskie, J. Polym. Sci., 49 (1961) 297.
115 See M. Szwarc, Carbanions, Living Polymers and Electron Transfer Processes, Interscience, New York, 1968, p. 656.
116 R. Gumbs, S. Penczek, J. Jagur-Grodzinsky and M. Szwarc, Macromolecules, 2 (1969) 77.
117 J. M. Catala, G. Riess and J. Brossas, Makromol. Chem., 178 (1977) 1249.
118 G. E. Coates and K. Wade, Organometallic Compounds, Vol. 1, Methuen, London, 3rd edn., 1967.
119 J. Boor, Jr., Ziegler–Natta Catalysts and Polymerizations, Academic Press, New York, 1979, pp. 245, 246.
120 J. C. W. Chien, J. Am. Chem. Soc., 81 (1959) 86.
121 R. Asami and M. Takaki, Makromol. Chem. Suppl., 12 (1985) 163.
122 G. G. Cameron and M. S. Chisholm, Polymer, 26 (1985) 437.
123 M. Sawamoto, T. Enoki and T. Higashimura, Macromolecules, 20(1) (1987) 1.
124 Y. Yamashita, Polym. Bull., 5 (1981) 361.
125 M. Niwa, T. Hayashi and T. Matsumoto, J. Macromol. Sci. Chem., A23(4) (1986) 433.
126 A. Saika, H. Ohya-Nishiguchi, T. Satokawa and A. Ohmori, J. Polym. Sci. Polym. Chem. Ed., 15 (1977) 1073.
127 T. Otsu, T. Sato and M. Abe, Polym. Prepr. Am. Chem. Soc. Div. Polym. Chem. 17(2) (1976) 615.
128 T. Sato and T. Otsu, Polymer, 16 (1975) 389.
129 M. Ko, T. Sato and T. Otsu, Makromol. Chem., 176 (1975) 643; J. Macromol. Sci. Chem., A9 (1975) 199.
130 T. Sato, S. Kita and T. Otsu, Makromol. Chem., 176 (1975) 561.
131 T. Sato, K. Hibino and T. Otsu, Chem. Ind., (1973) 745.
132 T. Sato and T. Otsu, Makromol. Chem., 178 (1977) 1941.
133 A. T. Bullock, G. G. Cameron and J. M. Elsom, Eur. Polym. J., 13 (1977) 751.
134 W. Lau, D. Westmoreland and R. W. Novak, Macromolecules, 20(2) (1987) 457.
135 K. Brzezinska, W. Chwialkowska, P. Kubisa, K. Matyjaszewski and S. Penczek, Makromol. Chem., 178 (1977) 2491.
136 Y. Maeda, P. Schmid, D. Griller and K. O. Ingold, J. Chem. Soc. Chem. Commun., 13 (1978) 525.
137 M. Szwarc, Carbanions, Living Polymers and Electron Transfer Processes, Interscience, New York, 1968, Chap. VII.
138 M. de Sorgo, D. C. Pepper and M. Szwarc, J. Chem. Soc. Chem. Commun., (1973) 419.
139 E. J. Goethals and E. H. Schacht, J. Polym. Sci. Polym. Lett. Ed., 11 (1973) 497.
140 K. Matyjaszewski and S. Penczek, J. Polym. Sci. Polym. Chem. Ed., 12 (1974) 1905.
141 R. A. Barzykina, G. N. Komratov, T. V. Korovina and S. G. Entelis, Vysokomol. Soedin. Ser. A, A16 (1974) 906.

142 P. E. Black and D. J. Worsfold J. Macromol. Sci. Chem., 9 (1975) 1533.
143 Yu. I. Ermakov and V. A. Zakharov, Usp. Khim., 41 (1972) 401.
144 W. Cooper, in H. C. Bamford and C. F. H. Tipper (Eds.), Comprehensive Chemical Kinetics, Vol. 15, Elsevier, Amsterdam, 1976, Chap. 3.
145 V. A. Zakharov, G. D. Bukatov and Yu. I. Ermakov, Usp. Khim., 49 (1980) 2213.
146 J. Mejzlík, M. Lesná and J. Kratochvíla, Advances in Polymer Science, Vol. 81, Springer-Verlag, Berlin, Heidelberg, 1987, pp. 83–120.
147 Shen Jiacong, Yang Meilin, Yian Deyue, Guan Shilin, Jilin Daxue Ziran Kexue Xuebao, (3) (1983) 75; Chem. Abstr., 100 (1984) 52042 t.
148 Zhou Pu, Sun Zaijian, Liu Quirong, Polymer, 27 (1986) 275.
149 G. Cai and D. Yan, Makromol. Chem., 187(3) (1986) 667.

Chapter 7

Transfer

> ...processes evidently occur leading to
> the generation of several molecules
> from the initiating radical
>
> Paul J. Flory

The discrepancy between the mean kinetic chain length and the observed degrees of polymerization of some polymerization products was overcome by Flory's discovery of transfer [1]. By transfer the active centre on the growing chain is lost, but at the same time a species is generated, initiating the growth of a new macromolecule

$$\sim M° + XT \xrightarrow{k_{tr}} \sim MX + T° \quad (1)$$

$$T° + M \xrightarrow{k_{i,tr}} TM° \quad (2)$$

An accidentally present impurity, the solvent, but also the initiator, monomer or polymer may act as transfer agents, XT. As a rule, transfer is undesirable. It leads to lower molecular weight of the product, and sometimes to branching. However, with increasing frequency it is being used to advantage. An intentionally added transfer agent is called a (molecular mass) regulator or telogen.

1. General characteristics of transfer reactions

The activation energy of the rate constant k_{tr} of reaction (1) is generally different (usually higher) from that of k_p[†]. By lowering the polymerization temperature, transfer is relatively suppressed. Lowering of the polymerization rate is caused by the re-initiation rate constant, $k_{i,tr}$ [reaction (2)], when the latter is lower than k_p. When it is much lower, we speak of degradative transfer. On the other hand, polymerization is not accelerated in practice when $k_{i,tr} > k_p$. The more rapid addition of one monomer per several

---

[†] Some very efficient transfer agents, e. g. mercaptans, have activation energies $E_{tr} < E_p$ for most radically polymerization of vinyl monomers [2].

hundred "normal" additions must, in principle, lead to an increased rate, but the difference is small and has not been experimentally proved so far.

In most cases it is much easier to determine the ratio of the transfer and propagation rate constants, $k_{tr}/k_p = C_{tr}$, than the value of k_{tr}. The ratio C_{tr} is called the apparent transfer constant. Formally, transfer may be regarded as copolymerization with the transfer agent XT acting as a non-polymerizing "monomer".

1.1 Retardation

In the presence of a transfer agent, the relations between the values of k_p, k_{tr} and $k_{i,tr}$ substantially affect the kinetics of polymerizations. Let us consider several extreme cases

$$k_{tr} \gg k_p \qquad k_{i,tr} = 0 \tag{3}$$

$$k_{tr} \gg k_p \qquad k_{i,tr} \geq k_p \tag{4}$$

$$k_{tr} > 0 \qquad 0 \leq k_{i,tr} < k_p \tag{5}$$

$$k_{tr} > 0 \qquad k_{i,tr} \geq k_p \tag{6}$$

Under conditions where eqn. (3) holds, inhibition occurs as described in Chap. 6, Sect. 1.4. In cases described by eqn. (4), only low polymers will be formed at an undiminished rate (when very short chains are formed, the rate of oligomerization might even increase). Chains shortened by transfer will also be formed at an undiminished rate in case of eqn. (6). Perhaps the most interesting is the situation described by eqn. (5). Transfer is accompanied by slow re-initiation. A part of the active centres is present in an inactivated form. Thus, at a given moment, fewer chains are growing and monomer consumption is slower. With stationary polymerizations, the external manifestations of this situation are called retardation. For various reasons, a stationary polymerization may change to a non-stationary one due to transfer. Such cases are described as degradative transfer (see Fig. 1).

A general mathematical solution of transfer and the corresponding retardation is not available. Many years ago, a solution was published for stationary radical polymerizations (with mutual termination of radicals). The derived relations remain valid. Other cases of transfer and of degradative transfer are solved individually.

A change in radical concentration R_n^* of degree of polymerization n is described by the general equation

$$\frac{d[R_n^*]}{dt} = k_p[R_{n-1}^*][M] - k_p[R_n^*][M] - k_{tr}[R_n^*][XT] - Z_{t,n} \tag{7}$$

where, for the sake of simplicity, k_p is equal for monomer addition to radicals with $n \geq 1$ and $Z_{t,n}$ symbolizes the rate of all kinds of termination of the radical R_n. As R_1 is formed by reaction (2) and initiation, the rate of its concentration change is given by

$$\frac{d[R_1^*]}{dt} = k_i[R_0^*][M] + k_{i,tr}[T^*][M] - k_p[R_1^*] - Z_{t,1} \tag{8}$$

where R_0^* is a radical from the initiator and $Z_{t,1}$ represents the rate of R_1^* termination.

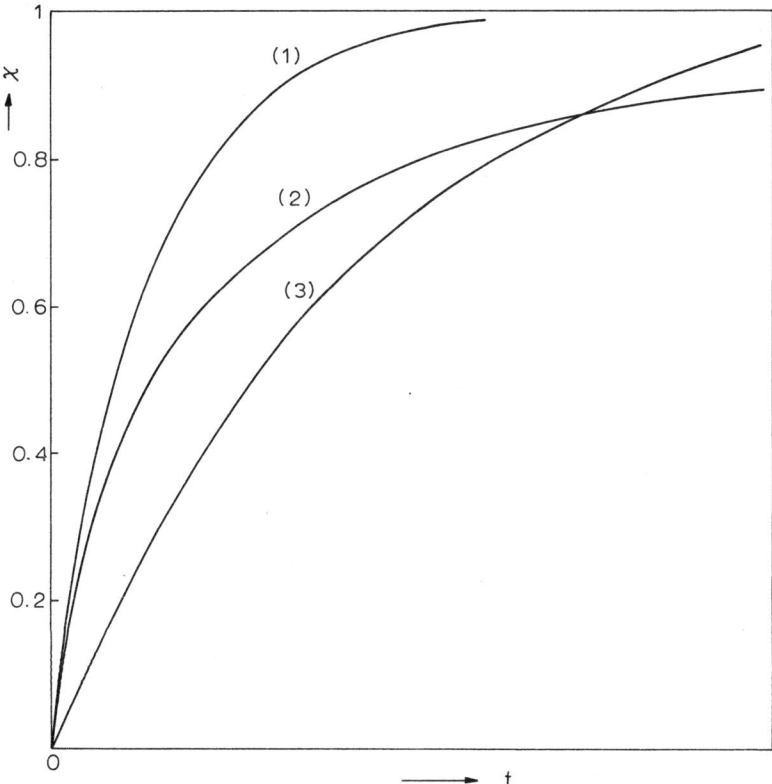

Fig. 1 Polymerization of a monomer with high T_c. (1) Stationary, (2) with degradative transfer, (3) retarded.

The change in concentration of T* is formulated as

$$\frac{d[T^*]}{dt} = k_{tr} \sum_{n=1}^{\infty} [R_n^*][XT] - k_{i,tr}[T^*][M] - Z_{t,T} \tag{9}$$

References pp. 476–479

where $Z_{t,T}$ represents the destruction rate of T*. In eqn. (9), some simplification is evidently involved; R_0^* can also react with XT. This simplification does not significantly modify the final conclusions [3].

In a stationary state, eqn. (9) assumes the form

$$k_{tr}[XT] \sum_{n=1}^{\infty} [R_n^*] - k_{i,tr}[T^*][M] - Z_{t,T} = 0 \tag{10}$$

When the decay rate of the radicals T* by $Z_{t,T}$ processes is small compared with the other terms of eqn. (10), $Z_{t,T}$ can be neglected and we can write approximately

$$k_{tr}[XT] \sum_{n=1}^{\infty} [R_n^*] = k_{i,tr}[T^*][M] \tag{11}$$

The condition of eqn. (11) must be fulfilled when the overall polymerization rate remains unaffected by the presence of the transfer agent XT. In other words, the difference in the decay rate of active centres in the presence or absence of XT must be negligible.

Setting $\sum_{n=1}^{\infty} [R_n^*] = [R^*]$ and $k_i = k_p$, the polymerization rate will be given by

$$v_p = -\frac{d[M]}{dt} = k_p[R^*][M] + k_{i,tr}[T^*][M] \tag{12}$$

and the consumption rate of the transfer agent will be

$$-\frac{d[XT]}{dt} = k_{tr}[R^*][XT] \tag{13}$$

Substituting for [T*] from the approximate relation (11) into (12) and dividing by (13) we obtain

$$\frac{d[M]}{d[XT]} = \frac{d[M]/dt}{d[XT]/dt} = \frac{k_p[R^*][M] + k_{tr}[R^*][XT]}{k_{tr}[R^*][XT]} \tag{14}$$

Long chains will only be formed when $k_p[M] \gg k_{tr}[XT]$. Under these conditions, a further simplification is feasible

$$\frac{d \ln[M]}{d \ln[XT]} = \frac{k_p}{k_{tr}} \tag{15}$$

Equation (14) is generally valid, even though it is mostly applied to radical polymerizations [3, 4].

When the growing chains are inactivated by mutual collisions in the presence of a transfer agent they can be terminated in three ways

(a) $\quad$ R* + R* $\rightarrow$ inactive polymer

$$-\frac{d[R^*]}{dt} = 2k_{tRR}[R^*]^2$$

(b) $\quad$ T* + R* $\rightarrow$ inactive polymer

$$-\frac{d[R^*]}{dt} = -\frac{d[T^*]}{dt} = -k_{tTR}[T^*][R^*]$$

(c) $\quad$ T* + T* $\rightarrow$ inactive polymer

$$-\frac{d[T^*]}{dt} = 2k_{tTT}[T^*]^2$$

thus

$$Z_{t,T} = 2k_{tTT}[T^*]^2 + k_{tTR}[T^*][R^*] \tag{16}$$

$$\sum_{n=0}^{\infty} Z_{t,n} = 2k_{tRR}[R^*]^2 + k_{tTR}[T^*][R^*] \tag{17}$$

In a stationary state, at equal initiation and termination rates, it holds that

$$v_{init} = 2\{k_{tRR}[R^*]^2 + k_{tTR}[T^*][R^*] + k_{tTT}[T^*]^2\} \tag{18}$$

When long polymer chains are formed, monomer consumption by transfer reactions is negligible and the rate of monomer consumption is simply expressed as [see Chapt. 8, Sect. 1.1, eqn. (21)]

$$-\frac{d[M]}{dt} = k_p[R^*][M] \tag{19}$$

Substituting for T* from eqn. (11) into eqn. (18), we can calculate [R*] and substitute this into eqn. (19). In the presence of a transfer agent, the overall

References pp. 476–479

polymerization rate will depend, among other factors, on the rate constants of termination and transfer and on the concentration of the transfer agent

$$-\frac{d[M]}{dt} = \frac{(v_{init}/2)^{1/2}k_p[M]^2}{(k_{tRR}[M]^2 + k_{tTR}k_{tr}k_p^{-1}[XT][M] + k_{tTT}k_{tr}^2k_p^{-2}[XT]^2)^{1/2}} \quad (20)$$

By means of eqn. (20), it has been possible to explain the unusual dependence of the initiation and polymerization rate of methyl methacrylate and vinyl acetate in benzene and in other aromatics [5].

Relation (18) can be used for analyzing retardation and inhibition processes. Their kinetics is independent on the mechanism of transfer. Both disproportionation in the sense of eqn. (1) or electron transfer in ionic reactions may be involved.

When T° cannot be inactivated by mutual termination (i. e. when T° is a stable free radical or ion), $k_{tTT} = 0$, and the last term on the right-hand side of eqn. (18) will be absent[†] [6]. In this case, the stationary concentration [R*] is calculated by means of the expression

$$v_{init} - k_{tr}[R^*][XT] + k_{i,tr}[T^*][M] - k_{tRT}[R^*][T^*] - 2k_{tRR}[R^*]^2 = 0 \quad (21)$$

from which a functional dependence can be derived for [T*]

$$[T^*] = \frac{v_{init} - k_{tr}[R^*][XT] - 2k_{tRR}[R^*]^2}{k_{tRT}[R^*] - k_{i,tr}[M]} \quad (22)$$

Substituting for [T*] from eqn. (22) back into eqn. (18), a quadratic equation with resp to [R*] is obtained; at least one of its roots is physically significant.

When the retarded polymerization proceeds at a measurable rate (v_p) and long chains are formed, it is better to use the approximations of Kice [7] or Jenkins [8].

From eqn. (19) it holds that

$$[R^*] = -k_p^{-1}\frac{d\ln[M]}{dt} = k_p^{-1}[M]^{-1}v_p \quad (23)$$

For the polymerization rate in the absence of transfer agent, $v_{p,0}$, we have

$$v_{init} = 2k_{tRR}[R^*]^2 = 2k_{tRR}k_p^{-2}[M]^{-2}v_{p,0}^2 \quad (24)$$

[†] Complications are likely to occur when XT reacts with more than two radicals.

Introducing v_p and $v_{p,0}$ into eqn. (18) we obtain $(c = 4k_{tRR}k_{tTT}/k_{tTR}^2)$

$$[T^*] = \frac{k_{tTR}}{2k_p k_{tTT}} \{[M]^{-2}v_p^2 + c[M]^{-1}(v_{p,0}^2 - v_p^2)^{1/2} - [M]^{-1}v_p\} \quad (25)$$

Expression (25) can be simplified by defining the ratio

$$\frac{v_p}{v_{p,0}} = W \quad (26)$$

so that

$$[T^*] = \frac{k_{tTR} v_p}{(2k_p k_{tTT})[M]} \left\{ \left[1 + \frac{c(1-W)^2}{W^2}\right]^{1/2} - 1 \right\} \quad (27)$$

Of all the generated radicals T*, a part decays by termination with the radicals R* (γ_1) and another part by mutual termination (γ_2). The fraction of radicals inactivated by attack of the radical R* on transfer agent XT is thus ($\gamma_1 + \gamma_2$).

$$\gamma_1 = \frac{k_{tTR}[R^*]}{k_{i,tr}[M] + k_{tTR}[R^*] + 2k_{tTT}[T^*]} \quad (28)$$

$$2\gamma_2 = \frac{2k_{tTT}[T^*]}{k_{i,tr}[M] + k_{tTR}[R^*] + 2k_{tTT}[T^*]} \quad (29)$$

In the stationary state the relation

$$v_{init} = k_{tr}[R^*][XT](\gamma_1 + \gamma_2) + 2k_{tRR}[R^*]^2 \quad (30)$$

is valid.

Expressions for γ_1 and γ_2 can be substituted from eqn. (28) and (29), for R* from eqn. (23), for T* from eqn. (25), and for W from eqn. (26). The most general available form of a mathematical description of retardation is obtained

$$\frac{W^2[XT]}{1-W^2} \left(1 + \left[1 + \frac{c(1-W^2)}{W^2}\right]^{1/2}\right) =$$

$$= \frac{2k_{tRR} v_p}{k_p k_{tr}[M]} \left[1 + \frac{c(1-W^2)}{W^2}\right]^{1/2} + \frac{2k_{tRR} k_{i,tr}[M]}{k_{tTR} k_{tr}} \quad (31)$$

References pp. 476–479

This differs from an exact solution only by the neglect of the second term on the right-hand side of eqn. (12) and of the change in k_t with the length of the terminating radicals. Its verification in real systems led to qualitatively satisfactory results (effects of benzoquinone [7, 9] and aromatics [8] in methyl methacrylate polymerization). Many authors introduced further simplifications [5, 10, 11]. Their approximations are always special cases of eqn. (31). Unfortunately, directions for the simple application of this equation have not been presented so far. Rigorous analysis of the problems of retardation therefore remains in the hands of specialists.

1.1.1 DEGRADATIVE TRANSFER

In Chap. 8, Sect. 1.2, the derivation will be given of the equation for an "ideal" radical polymerization

$$-\frac{d[M]}{dt} = k_p \left(\frac{v_{init}}{2k_t}\right)^{1/2} [M] \tag{32}$$

Precise studies of the kinetics of many polymerizations have revealed deviations from this formula both with respect to initiator and to monomer. The causes of the observed anomalies have been gradully elucidated. They include the already discussed change in diffusion coefficients with increasing chain length, termination by primary radicals, retardation, the occurrence of "hot" radicals, and complexes of growing radicals with solvents and monomers. One of the reasons of non-ideal behaviour is degradative transfer. By this we understand transfer forming a weakly active centre unable to re-initiate chain growth. Originally, a limiting case of situation (5) was evidently concerned, with $k_{i,tr} = 0$, so that the generated radical could only combine with the growing chain or with its own kind

$$\sim\sim^* + T^* \xrightarrow{k_t} \sim\sim T \tag{33}$$

$$2\, T^* \xrightarrow{k_t'} T_2 \tag{34}$$

This situation is a special case of retardation.

Degradative transfer cannot be differentiated by simple kinetic analysis from a copolymerization producing a weakly reactive centre [12]

$$\sim\sim M_1^* + M_2 \xrightarrow{k_{12}} \sim\sim M_1 M_2^*$$

$$\sim\sim M_1^* + {}^*M_2\sim\sim \xrightarrow{k_t} \text{inactive polymer}$$

$$\sim\sim M_2^* + {}^*M_2\sim\sim \xrightarrow{k_t'} \text{inactive polymer} \tag{35}$$

A comparison of the degrees of polymerization and of the molecular mass distribution curves of the products may help to reveal the nature of the reactions which occur. Another kind of degradative transfer was described by Scott and Senogles [12]. It involves intramolecular transfer with the formation of a non-propagating centre and occurs during the generation of chains with a tendency to form five- and six-membered cyclic structures as a transition stage

$$\sim\!\!-\!CH\!-\!CH_2\!-\!CH\!-\!CH_2\!-\!\overset{*}{CH} \longrightarrow \sim\!\!-\!CH\underset{X}{\overset{CH_2-CHX}{\underset{*CH-CH_2}{\big\langle}}} \longrightarrow \sim\!\!-\!\overset{X}{\underset{*}{C}}\!-\!CH_2CHXCH_2CH_2X \qquad (36)$$

(with X substituents as shown)

A similar reaction was observed during the polymerization of ethylene [13] and probably also of vinyl acetate [14].

The original concept of degradative transfer was limited to a few cases where the solvent acted as transfer agent. More detailed studies revealed that even monomers and fragments of initiators, (and also impurities, of course) may cause degradative transfer. From this point of view alkenes are particularly active, easily eliminating hydrogen, with allylic structure formation. Therefore, i. e. propene, butenes, etc., do not undergo radical polymerization. In copolymerizations, for example with vinyl chloride [15, 16] and with alkyl acrylates [17, 18], they strongly reduce the overall reaction rate and stop polymerization even before the more rapidly reacting monomer is exhausted [19]. Copolymerization here evidently occurs in the non-stationary range. Other monomers where degradative transfer occurs during polymerization are vinyl chloride [20], vinyl-1,3-dioxolane [21], etc. Naturally, among the transfer agents we can find those which also cause degradative transfer in addition to "normal" transfer. An example are the chlorophoshines [22].

Degradative transfer is not limited to radical processes. It can occur in many ionic polymerizations which are not exactly living. Degradative transfer here is, of course, harder to prove; an "ideal" standard is missing. Nevertheless, degradative transfer has been proved, for example, during the cationic polymerization of oxetane [23].

1.2 Branching (cross-linking) by transfer

Thermoplastics, i. e. polymers with linear chains, are mainly prepared by polymerization. Branched or cross-linked macromolecules are only needed in special cases. When a polymer is produced for further processing, a cross-

References pp. 476–479

linked product must never, without any exception, be formed during polymerization[†]. Therefore considerable attention is being paid to processes leading to branching or even crosslinking of macromolecules.

1.2.1 BRANCHING BY THE "WRONG" ADDITION OF MONOMER OR INITIATOR

"Wrong" monomer addition to the active centre can lead both to termination (see Chap. 6, Sect. 2.2) and to transfer

$$\sim\sim^+ + \text{PhC(=CH}_2\text{)CH} \xrightarrow{k_{tr}} \sim\sim\text{PhC(=CH}_2\text{)CH} + H^+$$

because the eliminated proton will initiate the growth of a new macromolecule. The double bond at the end of the polymeric chain can polymerize even when the substituent on the α carbon is very large. A branched polymer is formed by its incorporation into other chains.

Trossarelli et al. studied the formation of poly-β-alanine by the polymerization of acrylamide with *tert.*-BuNa. The initiating anion is in equilibrium with the monomer [24]

$$CH_2=CH\text{-}C(=O)NH_2 + tert.\text{-}BuO^- \rightleftharpoons CH_2=CH\text{-}C(=O)\bar{N}H + tert.\text{-}BuOH \quad (37)$$

With chain growth, equilibrium (37) is shifted to the right so that all the initiator is consumed at the initial stages of polymerization. The growing anions can react even with incorporated double bonds. Simultaneously, the generated carbanion is isomerized by intra- or intermolecular transfer of H^+ to an nitrogen anion

$$H_2C=CHC(=O)NH^- + CH_2=CH\text{-}C(=O)NH_2 \rightarrow CH_2=CHCONHCH_2\text{-}\bar{C}H\text{-}C(=O)NH_2 \rightleftharpoons CH_2=CHCONHCH_2CH_2CONH^-$$

$$\sim\sim CONH^- + CH_2=CH\sim\sim \rightarrow \sim\sim CO\underline{N}CH_2CH_2\sim\sim$$

$$\sim\sim C(=O)\text{-}N^- + CH_2=CH\text{-}C(=O)\text{-}NH\sim\sim \rightarrow \sim\sim CON\sim\sim\text{-}CH_2\text{-}H_2C\text{-}C(=O)\text{-}N^-\sim\sim \quad (38)$$

[†] On the other hand, final products made of cross-linked macromolecules are often desirable. For reason, rubber is vulcanized, and processes are sought for the cheap cross-linking of products made from thermoplastics.

The number of branching sites rapidly increases at first, but the rate decreases with conversion.

1.2.2 MULTIFUNCTIONAL TRANSFER AGENTS

The most efficient transfer agents are the mercaptans. Practially every —SH group is used for transfer, and does not retard polymerization [4]. Molecules with several mercaptan groups are copolymerizing cross-linking agents of a well-defined reaction mode [25]

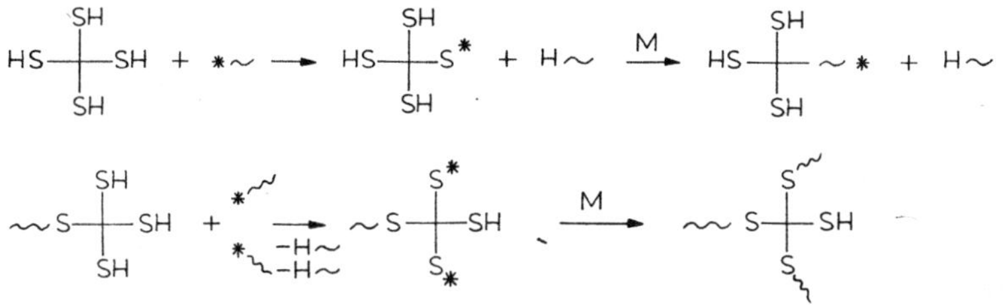

$$\text{(39)}$$

Multifunctional transfer agents are of no practical importance. They can, however, contribute significantly to our knowledge of transfer to polymer.

1.2.3 TRANSFER TO POLYMER

Each polymer is really a potential multifunctional transfer agent. It depends only on the reactivity of the growing radicals or ions under the given conditions[†] as to how often it will be attacked. Transfer to polymer has many variants. In principle, the reaction is always of the type[††]

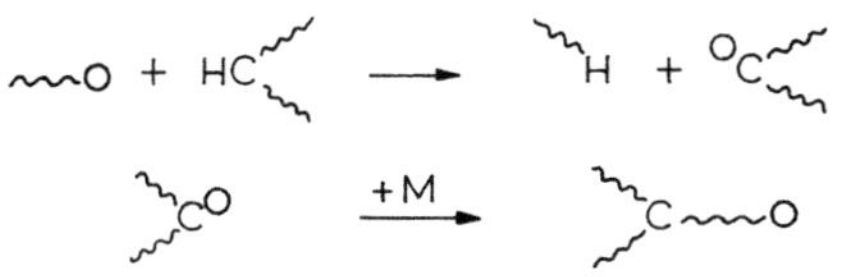

$$\text{(40)}$$

Teblina et al. [26] studied the kinetics of radical polymerization of methacrylic acid in the presence of a hexafluoropropene–vinylidene fluoride copolymer. The polymerization rate increased with conversion. A grafted copolymer was formed by a reaction analogous to eqn. (40) (with F instead of

---

[†] These include temperature, concentration (pressure), the quality of the medium, as well as the composition and ordering of polymer chains.

[††] Not all transfers to polymer give rise to branched polymers. Many lead to a redistribution of chain lengths.

References pp. 476–479

H). Transfer to polyalkenes is similar when these are present during radical polymerization of vinyl monomers. Cross-linking transfer was also observed in ionic polymerizations [27].

The value of the radio $k_{tr,P}/k_p$, where $k_{tr,P}$ is the rate constant of transfer to polymer, inereases with temperature. Unwanted branching can therefore be suppressed by polymerization at the lowest possible temperature and by limited conversion.

1.3 Reversible transfer

Kennedy and Kelen studied a cationic polymerization with transfer proceeding by reversible reactions [28]. The process can be represented by the scheme

$$HA + P_n \underset{k_{tr}}{\overset{k_a}{\rightleftharpoons}} HR_n^+ A^- \qquad (41)$$

$$HR_{n-1}^+ A^- + M \xrightarrow{k_p} HR_n^+ A^- \qquad (42)$$

in which HA is the initiator, $HR_n^+ A^-$ the irreversibly propagating chain, and P_n an inactive macromolecule of degree of polymerization n.

Szwarc and Zimm described the consequences of transfer in the sense of eqn. (41) by means of differential equations [29], and O'Driscoll simulated the system corresponding to the above scheme by the Monte Carlo method [30]. Both kinds of approach produced polydispersity coefficient $(\bar{P}_w/\bar{P}_n)$ vs. time plots which were practically identical at longer times.

2. Transfer in radical polymerizations

A certain kind of radical transfer can be modelled by the transfer of a hydrogen atom from an alkane molecule to a small alkyl radical. This reaction was studied in detail in the gas phase. With hydrocarbon partners, heats of reaction are a fairly safe measure of the relative rate of transfer, as the pre-exponential Arrhenius factors remain approximately constant for a series of transfers to a given radical. Tabulated thermodynamic data indicate, however, [31, 32] that the correlation between the heat of reaction and the transfer rate is not valid for reactions of a radical with polar substrates [32, 33]. In condensed phases, transfer reactions have not been sufficiently studied. Polymerizations themselves are the source of the most valuable, though incomplete, information.

2.1 Mechanism of some transfers

A very efficient transfer agent in polymerizations of styrene, vinyl acetate and methyl methacrylate is CBr_4.

Barson and Ensor assume that transfer is stimulated by electron migration from an active centre to the transfer agent, i. e. by a contribution of non-bonded resonance of the type [34]

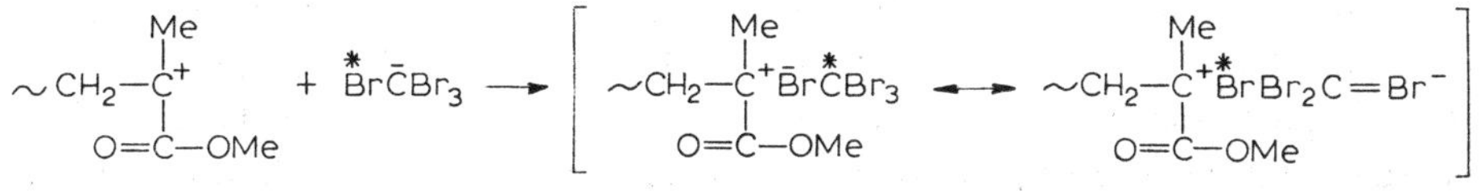

$$\sim CH_2-\underset{O=C-OMe}{\overset{Me}{\underset{|}{C^+}}} + Br\overset{*}{\bar{C}}Br_3 \longrightarrow \left[\sim CH_2-\underset{O=C-OMe}{\overset{Me}{\underset{|}{\bar{C}^+}}}Br\overset{*}{C}Br_3 \longleftrightarrow \sim CH_2-\underset{O=C-OMe}{\overset{Me}{\underset{|}{C^+}}}\overset{*}{Br}Br_2C=Br^- \right]$$

Not only carbon–halogen bonds but also halogen with other atoms, for example with P, are efficient sources of transfer, especially when a resonance-stabilized radical is generated by the transfer.

The value of C_{tr} decreases in the chlorophosphine series $Ph_2PCl > PhPCl_2 > PCl_3$. The differences are thought to be due to the ability of the phenyl ring to stabilize the transition complex [22]

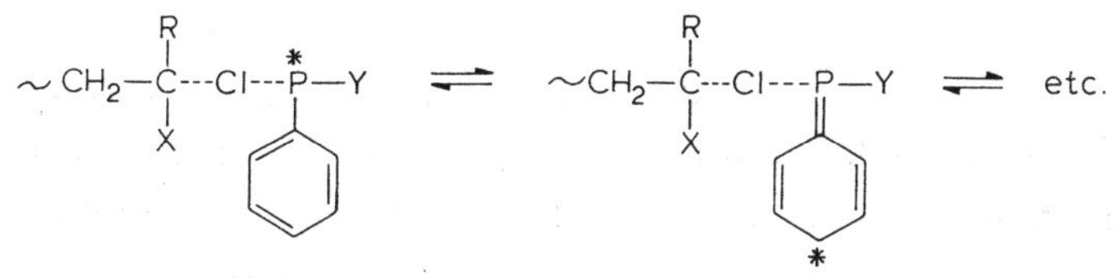

$$\sim CH_2-\underset{X}{\overset{R}{\underset{|}{\overset{|}{C}}}}---Cl---\overset{*}{P}-Y \; \rightleftharpoons \; \sim CH_2-\underset{X}{\overset{R}{\underset{|}{\overset{|}{C}}}}---Cl---P-Y \; \rightleftharpoons \; etc.$$

(43)

(R = H and X = Ph for styrene; R = Me and X = —CO—OMe for methyl methacrylate, Y = Cl or Ph)

A carbon–halogen bond in the monomer, polymer or solvent is always a centre of transfer reactions. Transfer to monomer controls chain length in polyvinyl chloride so efficiently that the molecular mass of the products is independent of the amount of initiator over a rather wide concentration range [35]. Also, vinylidene chloride lowers the polymerization degree of its own polymer by transfer [36].

The presence of a carbon–halogen bond is not absolutely essential for the occurrence of tranfer to monomer. Moore et al. [37] studied styrene polymerization with γ-irradiation. They measured thermodynamic quantities, particularly the Gibbs energy, enthalpy, entropy and volume changes by the method of rotating sectors and found that transfer to monomer is negligible at temperatures < 298 K and pressures < 220 MPa. Its contribution rapidly increases with growing temperature and pressure [37]. Transfer to monomer usually results in the formation of a polymer end group with a double bond

$$\sim CH_2-\underset{X}{CH^*} + CH_2=\underset{X}{CH} \xrightarrow{k_{tr,M}} \sim CH=\underset{X}{CH} + CH_3-\underset{X}{CH^*} \quad (44)$$

Transfer to monomer is often less efficient than to a monomeric unit incorporated in the polymer chain and therefore, usually, $k_{tr,P}/k_{tr,M} > 1$. This is particularly true of styrene and methyl methacrylate. Poly(methyl methacrylate) end groups have a greater tendency to transfer than the chain interior and therefore $C_{tr,P}$ considerably depends on the molecular mass of the polymer [38].

Many authors observed transfer in polymerizations of styrene [39] and methyl methacrylate [40] initiated by 2,2'-azobisisobutyronitrile. Athey proposed a mechanism explaining the observed generation of acrylonitrile [41]

$$\sim * + \underset{Me}{\overset{Me}{C}}-N=N-\underset{CN}{\overset{Me}{C}}\underset{Me}{\overset{Me}{}} \xrightarrow{-\sim H} \left[*CH_2-\underset{CN}{\overset{Me}{C}}-N=N-\underset{CN}{\overset{Me}{C}}-Me \right] \rightarrow CH_2=\underset{CN}{CH} + N_2 + Me-\underset{CN}{\overset{Me}{C}}-CH_2^* \quad (45)$$

Transfer need not always be undesirable. Compounds of high transfer activity are used industrially as molecular mass regulators. The C_{tr} values of some transfer agents are summarized in Table 1. We see that sulphur com-

TABLE 1

Apparent transfer constants for selected compounds in the polymerization of styrene [42, 43]. Spontaneous initiation at 333 K.

Compound	$C_{tr} \times 10^4$	Compound	$C_{tr} \times 10^4$
Benzene	0.018	CH_2Cl_2	0.15
Cyclohexane	0.024	Acetone	0.5
Toluene	0.125	$CHCl_3$	0.5
Decalin	0.4	Tert.-Butyl disulphide	1.4
n-Heptane	0.42	Butyldisulphide	24
Ethylbenzene	0.67	Trichloroacetic acid	66
Cumene	0.82	CCl_4	92
Isopropylbenzene	0.82	CH_2Br_2	110
Diphenylmethane	2.3	CH_2I_2	710
Triphenylmethane	3.5	Phenyl disulphide	1470
Fluorene	75	1-Butanethiol	120 000
Pentaphenylethane	20 000	Dodecanethiol	148 000

pounds are very efficient telogens. Polysulphides with —S—S— bridges act as multifunctional polymeric transfer agents.

Homopolymerization of a vinylic monomer (styrene, methyl methacrylate) in the presence of polysulphide yields a block copolymer [39]

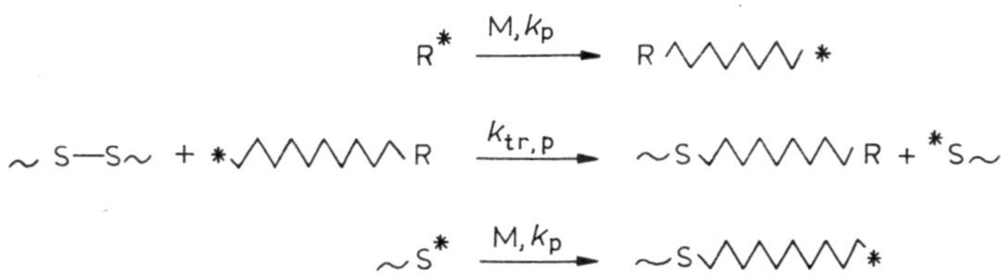

$$R^* \xrightarrow{M, k_p} R\sim\sim\sim *$$

$$\sim S-S\sim + *\sim\sim\sim R \xrightarrow{k_{tr,p}} \sim S\sim\sim\sim R + {}^*S\sim$$

$$\sim S^* \xrightarrow{M, k_p} \sim S\sim\sim\sim * \qquad (46)$$

where $\sim\sim *$ designates a growing polystyrene or poly(methyl methacrylate) chain and $\sim$ a polysulphide chain.

The values of transfer constants directly indicate the danger of unfounded estimates of radical reactivities with respect to various substrates. The ratio of transfer rates of polystyrene and poly(methyl methacrylate) radicals to various substrates assumes a range of values [38] which are dependent on the substrate properties. The former radical is more reactive towards mercaptans, CBr_4 or CCl_4, and the latter towards hydrocarbon transfer agents and trialkylamines which assume donor character in the transition complex. The interpretation of polar effects in macroradical reactivities is not yet satisfactory.

3. Transfers in anionic polymerizations

For more than twenty years the attention of investigators studying anionic polymerizations has been concentrated on living systems. Observations of anionic transfer are mostly the byproducts in studies of other processes. Transfer of a hydride ion, or of a proton, is mostly encountered.

3.1 Transfers of H^- and H^+

Anionic polymerizations can be terminated by acids, but when the conjugated base is sufficiently reactive, transfer may occur [44]

$$\sim\underset{\underset{Ph}{|}}{CH_2-CH^-} + NH_3 \rightarrow \sim\underset{\underset{Ph}{|}}{CH_2-CH_2} + NH_2^- \qquad (47)$$

This case occurs in the polymerization of styrene initiated by $NaNH_2$ in liquid ammonia. The transfer is very rapid, resulting in low molecular mass of the produced polystyrene (about 3×10^3, independent of initiator concentration). The rate of proton transfer is a function of anion basicity. Methacrylonitrile polymerized under comparable conditions [45] yields chains of molecular mass of the order of 10^5. When α-methylstyrene or butadiene are added to styrene polymerizing in liquid ammonia, the degree of polymerization of the product is further decreased [44]. The more basic carbanion (of these monomers) evidently accelerates transfer [46]. A similar case of transfer to solvent was observed during the polymerization of styrene in dimethylsulphoxide, initiated by the anion $MeSOCH_2^- Na^+$ [47].

$$\sim CH_2-\underset{Ph}{CH^-} + MeSOMe \rightarrow \sim CH_2-\underset{Ph}{CH_2} + MeSOCH_2^- \qquad (48)$$

Transfer to monomer is considered to be very probable in the polymerization of isoprene [46]. Vinylmesitylene and other monomers eliminate a proton yielding a more stable anion [48].

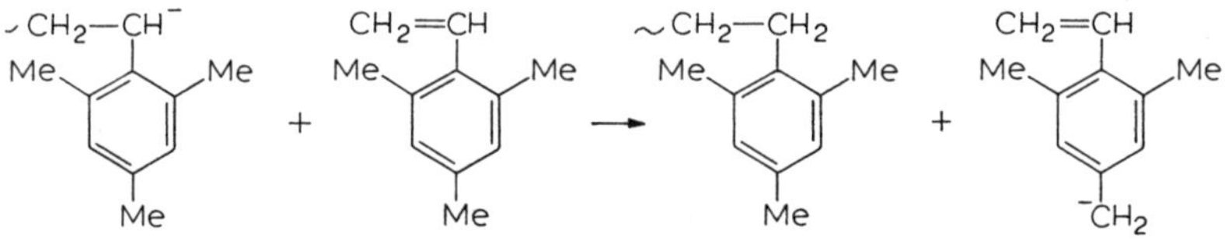

$$\text{(49)}$$

In this case, transfer to monomer leads to retardation [49].

Transfer to polymer was observed, for example, in the polymerization of p-isopropyl-α-methylstyrene initiated by sodium naphthalene. The macromolecule can eliminate a proton (in the presence of cationic initiators a hydride ion, in radical polymerizations of this monomer a hydrogen atom). An active centre is generated

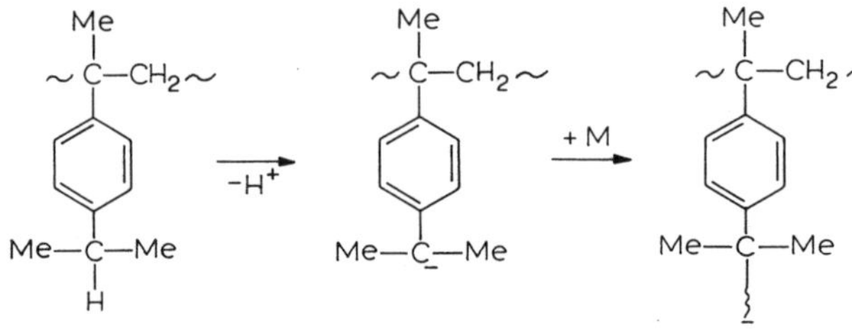

$$\text{(50)}$$

The eliminated H^+ terminates the attacking centre. In addition to the indicated reaction, the macroanions also attack completed chains so that isomerization and cross-linking result [50].

An example of anionic transfer to initiator is the formation of NaH in solutions of living polystyrene or poly-α-methylstyrene in inert solvents. The process is started by the elimination of a hydride anion

$$\sim CH_2-\underset{Ph}{CH^-Na^+} \rightarrow \sim CH=\underset{Ph}{CH} + Na^+H^- \qquad (51)$$

Hydrides are reactive nucleophilic agents able to initiate growth slowly. The whole process will have the character of degradative transfer; the polymerization of α-methylstyrene will be stopped at a certain conversion. The double bond at the chain end will increase the acidity of the penultimate styrene unit and the corresponding C—H bond will be weakened by resonance stabilization of the anion

$$\sim \underset{Ph}{\bar{C}}-\underset{Ph}{CH}=CH$$

This enables one active centre to be terminated

$$\sim CH_2-\underset{Ph}{CH^-} + \sim \underset{Ph}{CH}-CH=\underset{Ph}{CH} \rightarrow \sim CH_2-\underset{Ph}{CH_2} + \sim \underset{Ph}{C}\!=\!=\!=\!\underset{}{\bar{C}H}\!=\!=\!=\!\underset{Ph}{CH}$$
(52)

The indicated mode of transfer and termination is very slow [46, 51].

A typical transferring impurity is water. The resulting HO⁻ anion is mostly not sufficiently active to initiate further polymerization. Nevertheless, in many cases transfer does occur [52]

$$\sim O-\underset{Me}{\overset{Me}{Si}}-O^-K^+ + HOH \rightleftharpoons \sim O-\underset{Me}{\overset{Me}{Si}}-OH + KOH \qquad (53)$$

The behaviour of alcohols is similar to that of water.

Living polymerizations continue to attract the attention of theorists studying transfer. Chinese authors have analyzed the penultimate unit effect on transfer to a monomer mathematically [53]. According to their conclusions, the penultimate effect is important when the activities of the growth centres on the polymer chain and on the monomer (after transfer) are widely different, otherwise it can be neglected.

4. Transfers in cationic polymerizations

Cationic polymerizations are notorious by their transfers which are more intensive than in other polymerizations. This is caused by the high reactivity

References pp. 476–479

of the carbocation, mostly solvated by nucleophilic components of the polymerizing medium. Most transfers correspond to the migration of protons (exceptionally of positively charged atom groups) to the transfer agent or of a hydride ion from the agent to the propagating cation. Monomer, polymer, initiator, solvent or accidentally present impurities can again serve as transfer agents.

Most transfers are caused by a proton shift from the macrocation to the counter-ion. This reaction is generally formulated as [54]

$$\sim CH_2-\underset{X}{CH^+} \; B^- \xrightarrow{k_{tr}} \sim \underset{X}{CH=CH} + BH$$

$$BH + CH_2=\underset{X}{CH} \xrightarrow{k_{i,tr}} CH_3-\underset{X}{CH^+}B^- \quad (55)$$

Williams et al. [55] and Pfeiffer and Jewett [56] independently studied the structure of the ethenium cation which they assume to model the behaviour of carbenium ions reliably. They found equilibrium to exist between the classical and non-classical forms, the former being more stable by about 40 kJ mol^{-1} than the species on the right-hand side of the equation.

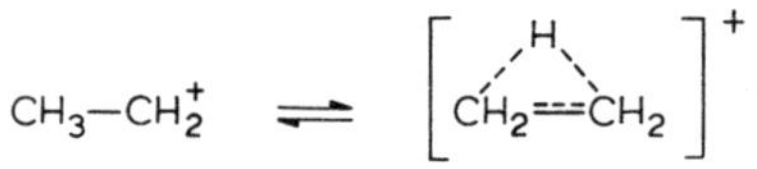

$$CH_3-CH_2^+ \rightleftharpoons \left[\underset{CH_2 = = CH_2}{\overset{H}{\diagup \diagdown}} \right]^+ \quad (56)$$

A connection between eqn. (56) and transfer in cationic polymerization has not yet been observed. It is assumed that the great intensity of proton transfer is caused by the existence of the equilibrium (57), comprising a non-classical ion with a loose, more mobile proton

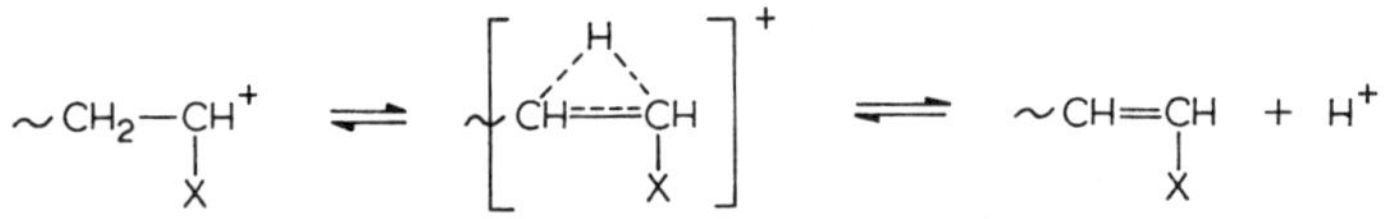

$$\sim CH_2-\underset{X}{CH^+} \rightleftharpoons \left[\sim \underset{X}{\overset{H}{\diagup \diagdown}}_{CH===CH} \right]^+ \rightleftharpoons \sim \underset{X}{CH=CH} + H^+ \quad (57)$$

Another general mode of transfer is represented by the reaction of a cation with a base having a sufficiently strong conjugated acid to initiate growth

$$\sim + \; + \; HOH \rightarrow \sim OH + H^+ \quad (58)$$

4.1 Transfer during polymerization of hydrocarbon monomers and of monomers containing heteroatoms in side chains

Cationic polymerizations of styrene, 2-methylpropene, vinylnaphthalene, indene, etc. at temperatures about 273 K yield only low polymers. Polymerization of styrene with $HClO_4$ in chlorinated solvents at room temperature is accompanied by extremely rapid transfer. Even at about 240 K, several tens of chains are generated for every initiator molecule when the reaction is allowed to proceed to high conversion [57].

The contribution of the various kinds of transfer is not known. Transfer to monomer is considered as a general reaction, occurring with almost all cationic polymerizations [58]. When a complex of Brönsted and Lewis acids serves as the initiating species, the initiator is regenerated by proton transfer to the counter-ion [59]

$$\sim CH_2-\overset{Me}{\underset{Me}{C}}{}^+ BF_3OH^- \rightarrow \sim CH=\overset{Me}{\underset{Me}{C}} + HBF_3OH \qquad (59)$$

The comparable polymerization of 2-methylpropene is always more rapid and leads to a higher molecular mass when initiated by γ-irradiation [60] than when carried out in $MeCl_2$ with the initiators BF_3, $AlCl_3$ or $EtAlCl_2$ [61].

With monomers and solvents containing aromatic groups, alkylation can compete with proton transfer [58] according to

$$\sim^+ + \langle\bigcirc\rangle R \longrightarrow \sim\langle\bigcirc\rangle R + H^+ \xrightarrow{+M} \sim\langle\bigcirc\rangle R + HM^+$$

4.2 Transfers during polymerization of heterocyclic monomers

These are much slower than to the preceding group of monomers, evidently because of the lower reactivity of oxonium, sulphonium, ammonium, phosphonium and siloxonium, ions. Moreover, monomers with these heteroatoms are strongly basic, and therefore cations are preferentially solvated by the monomers. This reduces the probability of other kinds of transfer: to solvent, impurities, etc. Many heterocycles, e. g. N-substituted aziridines, thiethanes [62], tetrahydrofuran [63], under suitable conditions polymerize by a living mechanism, i. e. without transfer. In situations where transfer does occur, it is assumed to proceed by the mechanism disscussed previously, for example by transfer to the counter-ion. With regard to transfer intensity, vinyl ethers can be ordered between the hydrocarbon monomers and the heterocycles. The mechanism of transfer in their polymerization has yet to be studied.

References pp. 476–479

5. Transfers in coordination polymerizations

Briefly, these can be characterized as being of low intensity. This is probably caused by the close connection between termination and transfer in coordination polymerizations. In recent years, the connection between Lewis acidity of centres and their ability to participate in reactions leading to transfer is becoming ever more evident [64]. Only two types of transfer have been proved, and may be considered as roughly known: to organometals and to hydrogen. In industrial production, the former must be taken into account and the latter is being exploited.

5.1 Transfers by organometallics

Natta and Pasquon were the first to publish a study on the reduction of the molecular mass of polypropylene caused by an increasing Et_3Al concentration. According to these authors, the inverse value of the degree of polymerization is directly proportional to the square root of the organometal concentration [65]. Later, Natta et al. demonstrated the high transfer efficiency of Et_2Zn added to the system $TiCl_3$—Et_3Al in propene polymerization [66]. They explained the mechanism of transfer to organometallics by the exchange of the alkyls in Et_2Zn (Et_3Al) for the growing chain in the active centre.

The transfer can be described by the scheme

$$\square \sim\sim\sim\sim + Et_2Zn \longrightarrow \square\text{-Et} + \sim\sim\sim\sim ZnEt \quad (60)$$

$\square$ represents the active centre of coordination polymerization on which the macromolecule propagates. The exchange rate is of the first order with respect to $[Et_2Zn]$ and is independent of monomer concentration. The generated $\sim\sim$—ZnEt does not initiate further growth; Et_2Zn is consumed by the transfer reaction. The preparation of a polymer of constant molecular mass requires a supply of new Et_2Zn replacing the consumed part. According to Natta, the lower transfer efficiency of Et_3Al compared with Et_2Zn is caused by its occurrence in dimeric form; Et_2Zn does not associate [67]. Et_2Zn is an efficient transfer agent, even in polymerizations of butadiene initiated by $CoCl_2$—Et_2AlCl complexes [68]. Et_2Cd and Et_2Hg also serve as transfer agents in polymerizations of propene with α-$TiCl_3$—Et_3Al [68]. Borisova et al. [70] proved the triple role of $EtAlCl_2$ which serves as a cocatalyst, inhibitor and transfer agent. This conclusion followed from their studies of ethylene polymerization with $cp_2TiEtCl + EtAlCl_2$. They found a larger number of macromolecules than the introduced number of Ti atoms [70].

5.2 Transfers by molecular H_2

It was again Natta who discovered transfers to hydrogen in coordination polymerizations [71]. He assumed breakage of the metal–polymer bond in the active centre

$$\square\text{---polymer} + H_2 \longrightarrow \square\text{---H} + H\text{---polymer} \qquad (61)$$

The centre is regenerated by monomer addition to $\square$—H

$$\square\text{---H} + CH_2=CHMe \longrightarrow \square\text{---}CH_2CH_2CH_3 \qquad (62)$$

Scheme (62) is based on measurements where the rates of propene polymerization were compared in the absence and presence of hydrogen (see Fig. 2). According to Natta, the re-initiation rate of the centres of alkene addition to $\square$–H is slow. This leads to a reduced rate.

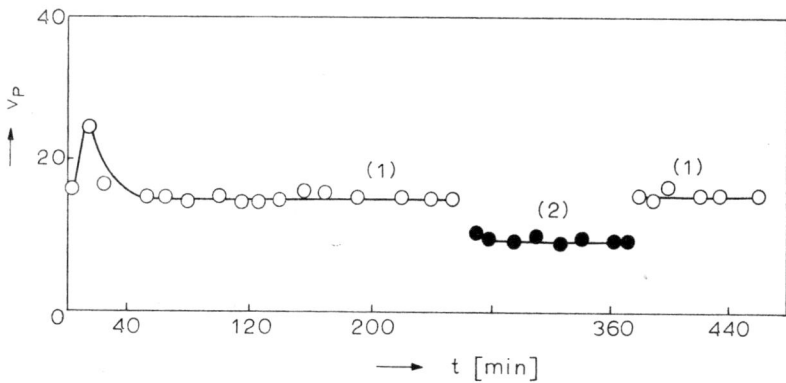

Fig. 2 Polymerization rate of propene (1) in absence and (2) in the presence of H_2. Temperature 348 K, $TiCl_3$–Et_3Al.

Many apparently contradicting observations can be found in the literature. According to some authors, hydrogen is not a transfer agent but an inhibitor [72]. Instead of retardation, other authors observed more rapid polymerization in the presense of hydrogen [73]. The behaviour of the system is evidently affected by the quality of the heterogeneous catalyst ($TiCl_3$), of the activating organometal and of the solvent [74]. Even the hydrogen side reactions might be important. Triethyluminium can be partially hydrogenated to Et_2AlH [71] and hydrogenation of monomer has also been observed [75].

Vandenberg has shown that H_2 is an efficient transfer agent in polymerizations of ethylene, propene and styrene with $TiCl_4$ + iso-Bu_3Al (Et_2AlCl, Et_3Al) [76].

References pp. 476–479

Most catalytic systems used in industrial production yield polyalkenes with very long chains which are unsuitable for current processing procedures and applications. For regulating molecular mass, H_2 is preferred to organometallics. Hydrogen is not a suitable transfer agent in diene polymerizations on cobalt complexes [67] because it reduces the Co (II) π-allylic centre to inactive Co (I) particles.

5.3 Other transfers of coordination polymerizations

Almost nothing is known as yet about these transfers. Rabovskaya et al. considered the possibility of transfer to monomer in butadiene polymerization on π-alkenyl NiCl–acceptor complexes [77].

6. Molecular mass of polymerization products

The molecular mass of the generated macromolecules is most easily predicted for living polymerizations with rapid initiation (see Chap. 5, Sect. 8.1). Somewhat more complicated is the case of systems that, although are not living, at least polymerize without transfer. The number average polymerization degree, $\bar{P}$, of the product is a simple function of the mean kinetic chain length, v

$$v = k_p[M]\pi \tag{63}$$

The product $k_p[M]$ is equivalent to the number of monomer molecules added to the active centre per unit time (s) and π is the mean life time of the centres. In radical polymerization, which are the most frequent representatives of the systems discussed, the relation

$$v \leq \bar{P} \leq 2v \tag{64}$$

holds, depending on the kind of termination. With a mole fraction x_{disp} of chains terminated by disproportionation, we have [78]

$$\bar{P} = \frac{2v}{1 + x_{\text{disp}}} \tag{65}$$

The situation is considerably complicated by transfer in the polymerizing medium. The mean degree of polymerization is reduced

$$\bar{P} = \frac{\text{total number of monomer molecules consumed}}{\text{total number of generated macromoleecules}} =$$

$$= \frac{\text{summed rates of all monomer-consuming reactions}}{\begin{pmatrix}\text{summed rates of all reactions leading to}\\ \text{macromolecular growth termination}\end{pmatrix}} =$$

$$= \frac{v_{\text{prop}}}{v_t + v_{\text{tr}}} =$$

$$= -\frac{\int_0^t (d[M]/dt)\, dt}{\int_0^t (d[P]/dt)\, dt} \tag{66}$$

where v_{prop}, v_t and v_{tr} are the rates of propagation, termination and transfer, respectively. The definition (66) is quite general. It is valid for stationary and non-stationary polymerizations. The degree of polymerization of the products of stationary systems can, of course, be expressed in a simpler form

$$\bar{P} = \frac{(d[M]/dt)_s}{d[P]/dt} =$$

$$= \frac{(d[M]/dt)_s}{(d[P]/dt)_t + k_{\text{tr}}[R^\circ][XT]} \tag{67}$$

where $(d[P]/dt)_t$ is the rate of the termination reactions.

When termination occurs by a bimolecular reaction, relation (67) is further simplified

$$\bar{P}^{-1} = \frac{2k_t[R^\circ]^2 + k_{\text{tr}}[R^\circ][XT]}{k_p[R^\circ][M]} =$$

$$= \bar{P}_0^{-1} + \frac{k_{\text{tr}}}{k_p} \frac{[XT]}{[M]} \tag{68}$$

Relation (68) is called the Mayo equation after the author by whom it was first derived [79]. Its graphical representation, in the from of a $\bar{P}^{-1}$ vs.

References pp. 476–479

[XT]/[M] plot is called the Mayo diagram. In it $\bar{P}_0$ is defined by the relation†

$$\bar{P}_0^{-1} = (1 + x_{\text{disp}}) \frac{(2k_{\text{tRR}} v_{\text{init}})^{1/2}}{2k_p[M]} \tag{69}$$

Terms expressing transfer by several transfer agents, including monomer, initiator, etc., can be incorporated in eqn. (68). Transfer to monomer extends eqn. (68) by the term $k_{\text{tr,M}}/k_p$; thus it is independent of concentration. It may be concealed in the term $\bar{P}_0^{-1}$. It is usually very difficult to prove or to exclude the existence of slow transfer to monomer.

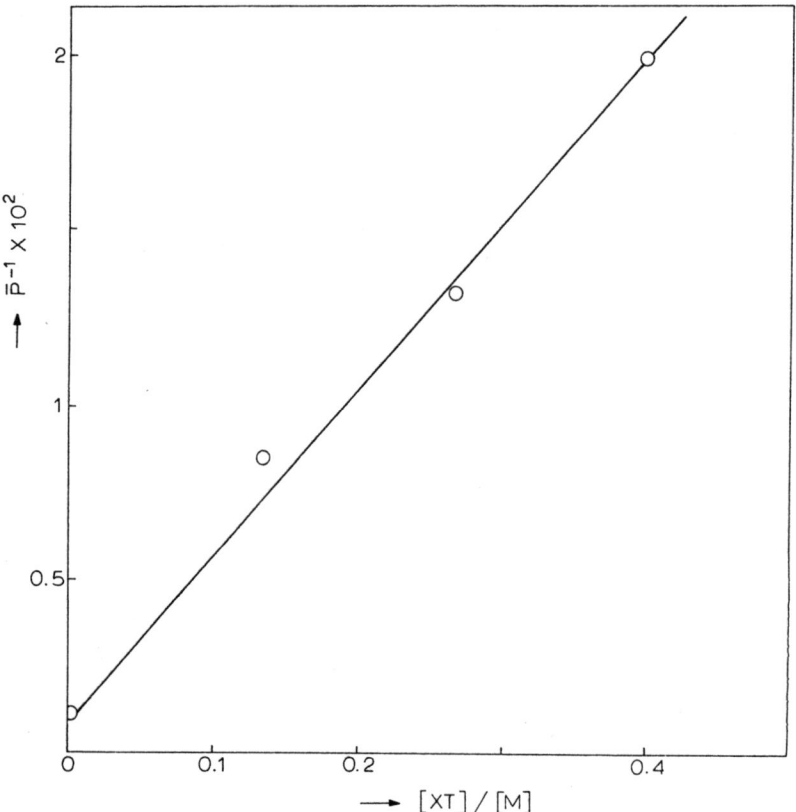

Fig. 3 Mayo graph for the polymerization of S in the presence of PCl$_3$ [22].

---

† In the stationary state, $v_{\text{init}} = 2k_t[R^*]^2$ and $\tau = (k_t[R^*])^{-1}$ thus $\tau = (2k_t v_{\text{init}})^{-1/2}$. Substituting for τ in eqn. (63), we obtain $v = k_p[M]/(2k_t v_{\text{init}})^{1/2}$. Expressing v in this form in eqn. (65), we obtain eqn. (69).

Examples of the application of the Mayo equation are shown in Figs. 3–5. It should be noted that eqn. (68) is valid only when the termination rate is not altered by transfer. More general equations have been derived by Kice for retarded and partly inhibited polymerizations. They provide a satisfactory description of experimental findings [7, 9].

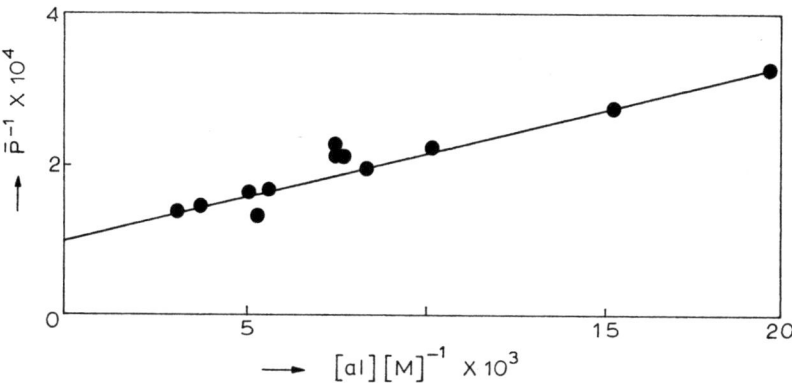

Fig. 4 Transfer to organometal in ethylene polymerization. Dependence of $\bar{P}^{-1}$ on the ratio $[Et_3Al]/[M]$ [80].

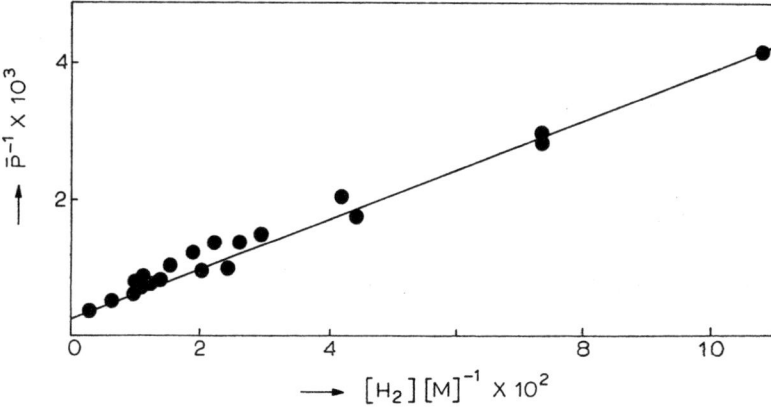

Fig. 5 Transfer to hydrogen in ethylene polymerization. Dependence of $\bar{P}^{-1}$ on the ratio $[H_2]/[M]$ [80].

Even when all prerequisites are fulfilled, solution of eqn. (68) only yields the degree of polymerization of the instantaneously produced polymer. In the stationary state, of course, the ratio of the growth and termination rates does not change. If the ratio $[XT]/[M]$ also remains essentially unchanged, a polymer with equal mean chain length is formed in the course of the whole stationary period. When $[XT]/[M]$ increases, the degree of polymerization

References pp. 476–479

decreases with conversion, and vice versa. Qualitatively this is not difficult to explain; the concentration of the transfer agent increases (decreases) relatively and transfers are more efficient (suppressed). This situation was described quantitatively by Mark and Tobolsky [81].

When sufficiently long chains are formed, the rate of monomer consumption is described by the well-known relation [see Chap. 8, eqn. (21)]

$$-\frac{d[M]}{dt} = k_p[M^\circ][M] \qquad (70)$$

For a transfer agent XT, transferring only once, the rate of concentration change (consumption by transfer) is

$$-\frac{d[XT]}{dt} = k_{tr}[M^\circ][XT] \qquad (71)$$

Mutual division of the two equations and integration yields

$$[XT] = [XT]_0 \left(\frac{[M]}{[M]_0}\right)^{C_{tr}} \qquad (72)$$

in which $C_{tr} = k_{tr}/k_p$. When chain growth is only terminated by transfer[†] the mean degree of polymerization can be calculated from the ratio of the consumed molecules [see eqn. (66)]

$$\bar{P} = \frac{[M]_0 - [M]}{[XT]_0 - [XT]} \qquad (73)$$

Substituting [XT] from eqn. (72) into eqn. (73) and defining the conversion $\varkappa$ as $([M]_0 - [M])/[M]_0$, we obtain an expression relating $\bar{P}$ to conversion

$$\bar{P} = \frac{[M]_0}{[XT]_0} \frac{\varkappa}{1 - (1 - \varkappa)^{C_{tr}}} \qquad (74)$$

Even though eqn. (74) has been derived using simplifying assumptions, it reflects fairly well the actual situation in some cases (see Fig. 6).

The solutions of some more complicated relations between transfer and degree of polymerization will be mentioned in Sect. 8. Examples of theoretical

---

[†] This assumption means that the decay rate of centres and other transfers are negligible compared with transfer to compound XT. Each polymer chain is terminated by an XT fragment.

solutions of the termination and transfer effects (spontaneous and to impurities) on molecular mass distribution in products of ionic polymerizations can be found, for example, in the work of De–Yue Yan and Cui–Ming Yuan [83, 84].

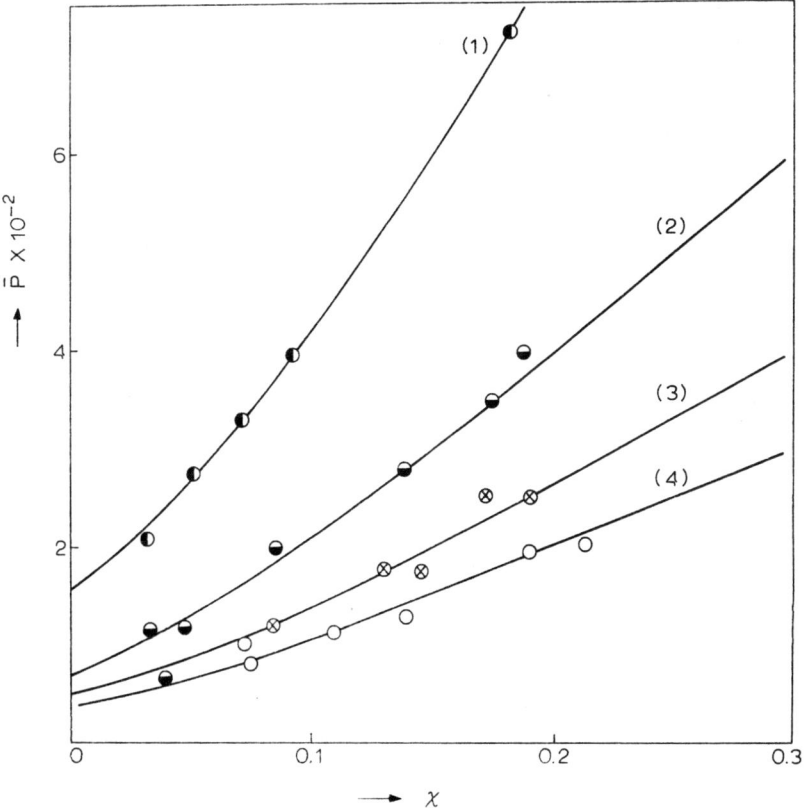

Fig. 6 Dependence of the degree of polymerization of polyoxymethylene on conversion during trioxane polymerization. Contents of water as transfer agent [10^{-3} mol kg^{-1}]: (1) 2.8, (2) 5.6, (3) 8.3, (4) 11.1. The points correspond to experimentally measured values; the full curves were calculated by means of eqn. (74) [82].

7. Copolymerizing transfer agents

The scheme of propagation and transfer is strikingly similar to the scheme of copolymerization with a non-polymerizing comonomer [compare eqn. (1) with Chap. 5, eqn. (62)]. Our experience with copolymerization can therefore be applied to transfer. An undoubtedly favourable circumstance is the availability of the corresponding values of the apparent transfer constant C_{tr}. Using

References pp. 476–479

the terminology of Chap. 5, Sect. 5.2 and eqn. (1), $k_{11} = k_p$, $k_{12} = k_{tr}$, $k_{22} = 0$. Then

$$C_{tr} = \frac{k_{tr}}{k_p} = \frac{k_{12}}{k_{11}} = r_1^{-1} \qquad (75)$$

Substituting from eqn. (75) into Chap. 5, eqn. (85) and rearranging, we obtain [85]

$$\ln Q_1 C_{tr} - e^2 = \ln Q_{tr} - e_1 e_{tr} \qquad (76)$$

For vinyl monomers polymerizing by a radical mechanism, Q_1 and e_1 are known (see Chap. 5, Table 3A). Therefore Q_{tr} end e_{tr} can be determined by

TABLE 2

C_{tr} and e_{tr} values for some transfer agents [86]

Transfer agent	Polymerization at			
	333 K		353 K	
	$Q_{tr} \times 10^4$	e_{tr}	$Q_{tr} \times 10^4$	e_{tr}
Acetone	0.11	0.38	0.30	0.63
Benzene	0.05	−1.41	0.07	−0.70
Toluene	0.21	−0.56	0.35	−0.42
Dibenzoylperoxide	45.0	−1.41		
Butanol	0.53	−0.57		
Butanone	0.83	0.54	1.30	1.18
Cyclohexane	0.08	−0.98	0.11	−0.91
1,2-Dichloroethane	0.19	0.04	0.61	0.31
Ethyl acetate	0.08	−1.03		
Alkyl mercaptans	15 000		2.31	
CH_2Cl_2	0.009	−0.70		
$CHCl_3$	0.35	−0.31	0.54	−0.14
CCl_4	3.61	3.62	4.09	3.48
CBr_4	3300	1.51		
Methanol	0.13	−0.87		
PCl_3^a	5.8[a]	4.71[a]		
$PhPCl_2^a$	78.9[a]	2.56[a]		
Ph_2PCl^a	114[a]	2.45[a]		

[a] Polymerization at 323 K [22].

a simple calculation, when C_{tr} values are measured for each of two selected monomers. In polymerizations of other monomers, the effect of the transfer agent can be predicted from eqn. (76). The accuracy of such calculations is, of course, low. It suffers from all the drawbacks of the Q, e scheme, and moreover it is difficult to eliminate other kinds of transfer during measurement of C_{tr}. Nevertheless, the derived Q_{tr} and e_{tr} values can illustrate the transfer efficiency of some transfer agents (see Table 2).

According to the recommendation of Ham [87], the calculation can be improved by the use of the principle of chain reversibility as treated in Chapter 5, Sect. 5.5. Let us consider the copolymerization of two monomers, 1 and 2, in the presence of the transfer agent 3. According to the reversibility principle, the probability of termination of some sequence order by the transfer agent and initiation of new chain growth is equal to the probability of occurrence of all partial reactions in reverse order

$$P_M P_{11}^{m-1} P_{22}^{n-1} P_{23} P_{31} = P_{13} P_{32} P_{22}^{n-1} P_{21} P_{11}^{m-1} P_M$$

$$P_{12} P_{23} P_{31} = P_{13} P_{32} P_{21} \tag{77}$$

where m and n are the number of monomers 1 and 2, respectively, in the sequences and M is either M_1 or M_2, but always only one of the two. Then, for example, P_{12} is the probability of addition of monomer 2 to centre 1 in the presence of 1 and 3 and P_{31} is the probability of addition of 1 to the transfer agent fragment 3. By the procedure described in Chap. 5, Sect. 5.5, a relation valid for equimolar concentrations of all three components can be derived from eqn. (77)

$$\frac{1}{1+r_1+C_{tr1}r_1} \frac{C_{tr2}r_2}{1+r_2+C_{tr2}r_2} \frac{1}{1+(k_{32}/k_{31})} =$$
$$= \frac{C_{tr}r_1}{1+r_1+C_{tr1}} \frac{k_{32}/k_{31}}{1+(k_{32}/k_{31})} \frac{1}{1+r_2+C_{tr2}r_2} = \mathscr{P} \tag{78}$$

r_1 and r_2 are the copolymerization parameters, C_{tr1} and C_{tr2} the apparent transfer constants of the agent with monomers 1 and 2, and k_{ij} represent the bimolecular rate constants of the components defined by the index.

The denominators both sides of eqn. (78) are equal, therefore

$$\frac{k_{32}}{k_{31}} = \frac{°C_{tr2}r_2}{°C_{tr1}r_1} \tag{79}$$

The activity of the transfer agent can be predicted from eqn. (78) for homopolymerizations of a further monomer, without any direct measurement of transfer. To this end, the binary copolymerization parameters of the pair 1 and 2 must be known, as well as the apparent transfer constants, $°C_{tr}$ of the given agent with each of the monomers. Substituting these values into eqn. (78), we can determine the value characteristic $\mathscr{P}$ for the given system. The tested transfer agent, when present during the polymerization of any monomer, will cause chain shortening at a rate determined by the ratio k_{tr}/k_p. Hypothetically, it is thus sufficient to "exchange" the original parameters for the parameters of the new pair, r_1, r'_2. From eqns. (78) and (79) holds that

$$1 + \frac{k_{32}}{k_{31}} = \mathscr{P}^{-1}(1 + r_1 + C_{tr1}r_1)^{-1} C_{tr2}r'_2(1 + r'_2 + C_{tr2}r'_2)^{-1} = \\ = 1 + C_{tr2}r'_2(C_{tr1}r_1)^{-1} \quad (80)$$

The values of $\mathscr{P}$ and $°C_{tr1}$ are known from the original system and r_1 and r'_2 are the known copolymerization parameters of the monomer pair after "exchange" of the latter. The required value of C_{tr2} is obtained by calculation using eqn. (80) after rearrangement

$$C_{tr2} = \frac{1}{2r'_2}\left\{b\left(\frac{a}{\mathscr{P}} - 1\right) - c \pm \left(\left[b\left(1 - \frac{a}{\mathscr{P}}\right) + c\right]^2 - 4bc\right)^{1/2}\right\} \quad (81)$$

where $a = (1 + r_1 + r_1°C_{tr1})^{-1}$, $b = r_1°C_{tr1}$, and $c = 1 + r'_2$. The application of Ham's method [87] is unique. The author himself tested it by means of the copolymerization parameters of styrene with methyl methacrylate and of their transfer constants to n-butylmercaptan. He calculated C_{tr} of this transfer agent with methyl acrylate (more precisely with its macroradical) and made a similar calculation with the constants for the telogen CBr_4. Good agreement of the calculated and experimental values was found in both cases. According to the autor, the approach described can yield much useful information from the values of P for monomers and transfer agents, e. g. estimates of the resonance stabilization and polar character of telogens. By these criteria, CBr_4 is only slightly less reactive than methyl acrylate. The reactivity of monomers with the telogen radical can also be estimated by this method.

Ham's method of calculating k_{tr}/k_p is certainly interesting. So far, however, the principle of chain reversibility is not supported by any rational justification, and the calculation of the probability P is subject to some error; the error in $\mathscr{P}$, a product of three P values, is correspondingly larger.

8. The use of transfers

Transfer reactions are utilized to regulate molecular mass of polymerizing macromolecules, for grafting, and to introduce functional groups into the chains of polymeric "monomers", the macromers.

8.1. Control of molecular mass

For various reasons, the production of either low or high polymers is required. These requirements can usually be fulfilled by suitable selection of polymerization conditions. Almost always $E_{tr} - E_p > 0$ and transfer is promoted by increasing temperature. Cationic polymerizations of hydrocarbon monomers yield a high-molecular product only at low temperatures[†]. With most radical polymerizations, the length of the macromolecules produced (e. g. PVC, PS, PMMA, PVAc, PE) can be controlled by temperature. From a technological standpoint, a higher polymerization temperature is of advantage because the heat of reaction can be more easily removed. However, the rate of transfer to polymer also increases with increasing temperature, and this is almost always undesirable. Even chain configuration is a function of temperature. Because of these circumstances, relatively low polymerization temperatures must sometimes be used. In such cases, chain growth to undesired length is prevented by the addition of a transfer agent as regulator. An example of this situation is the synthesis of "cold" rubber. The molecular mass of polyalkenes prepared by coordination polymerizations must, in most cases (but not always), be reduced by the addition of hydrogen. Regulators are almost always added in telomer syntheses.

A suitable regulator (transfer agent) is selected according to its C_{tr} (see Table 1) and conditions of use. Regulators with a high C_{tr} are very efficient, but they are rapidly consumed [see eqn. (72)], yielding polydisperse products. Continuous processes can be arranged so that the concentration changes of the regulator remain small, and the enormous growth of the polydispersity coefficient need not occur.

8.2 Grafting by transfer process

When growing polyvinyl chloride radicals react with polyalkene macromolecules, transfer to polymer occurs according to scheme (40) (see Sect. 1.2). A polyalkene-*graft*-polyvinyl chloride results. This polymer is already being

---

[†] 2-Methylpropene must be polymerized at temperatures below 240 K, depending on the quality of initiator, in order to give rise to a polymer of molecular mass 3×10^5.

References pp. 476–479

produced on a small scale. Another example is the grafting of methacrylic acid on to the hexafluoropropene-vinylidene-fluoride copolymer [25]. Macrocations can be grafted to polymers with aromatic groups [88]. Grafting by transfer and branching, particularly transfer to polymer, are very similar reactions. They have been roughly treated in Sect. 1.2.

8.3 Formation of macromers by transfer

The reaction of an active centre with a transfer agent produces chains in which fragments of the transfer agent are incorporated as chain ends.

Ikada et al. [89] attached acetyl chloride to the chain ends of polystyrene oligomers by means of trichloroacetyl chloride as transfer agent, and amino groups to poly (methyl methacrylate) by means of 2-aminothioethanol, always in high yield. The preparation of a difunctional macromer by transfer in cationic polymerization of 4-methyl-1-pentene was described by Ver-Strate and Baldwin [90]. Chain growth was interrupted by reaction with a difunctional transfer agent having one active and one less active Br atom. Difunctional oligomers were produced

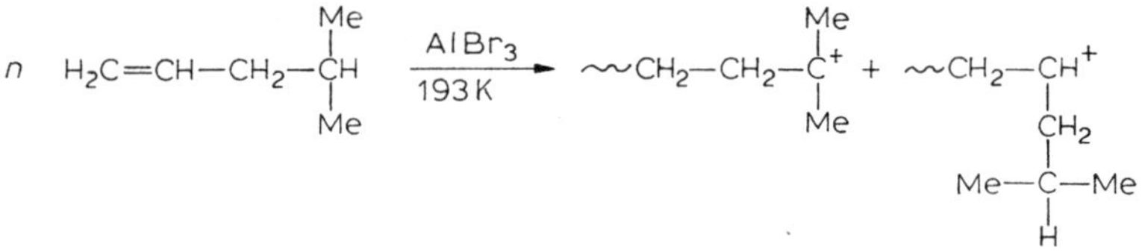

$$n\ H_2C=CH-CH_2-\underset{Me}{\underset{|}{CH}}-\overset{Me}{\overset{|}{CH}} \xrightarrow[193K]{AlBr_3} \sim CH_2-CH_2-\overset{Me}{\underset{Me}{\overset{|}{C}}}^+ + \sim CH_2-\underset{\underset{Me-\underset{H}{\overset{|}{C}}-Me}{\overset{|}{CH_2}}}{CH^+} \qquad (82)$$

$$\sim CH_2-\overset{+}{CH}-CH_2-\underset{Me}{\underset{|}{CH}}-\overset{Me}{\overset{|}{CH}} + Br-CH_2-CH_2-\underset{Me}{\underset{|}{C}}-\overset{Me}{\overset{|}{C}}-Br \xrightarrow{k_{tr}}$$

$$\xrightarrow{k_{tr}} \sim CH_2-\underset{Br}{\underset{|}{CH}}-CH_2-\underset{Me}{\underset{|}{CH}}-\overset{Me}{\overset{|}{CH}} + Br-CH_2-CH_2-\underset{Me}{\underset{|}{C}}-\overset{Me}{\overset{|}{C}}^+ \qquad (83)$$

$$BrCH_2-CH_2-\underset{Me}{\underset{|}{C}}-\overset{Me}{\overset{|}{C}}^+ + CH_2=\underset{\underset{Me}{\underset{|}{CH_2-CH}}}{\overset{|}{CH}} \longrightarrow Br\sim\underset{Me}{\underset{|}{CH}}-CH_2-\overset{Me}{\overset{|}{\overset{+}{CH}}}-CH_2-\underset{Me}{\underset{|}{CH}}-\overset{Me}{\overset{|}{CH}} \qquad (84)$$

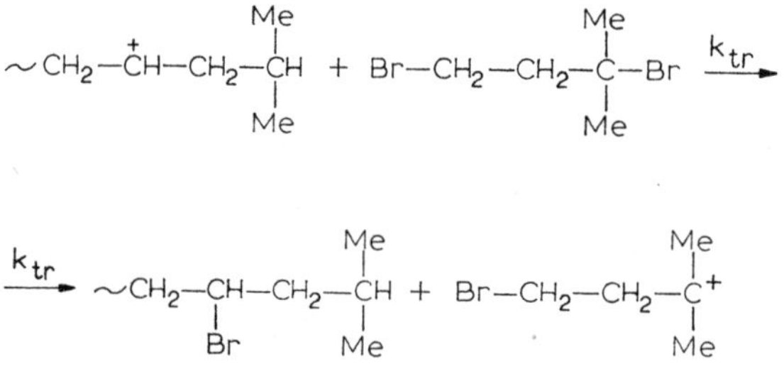

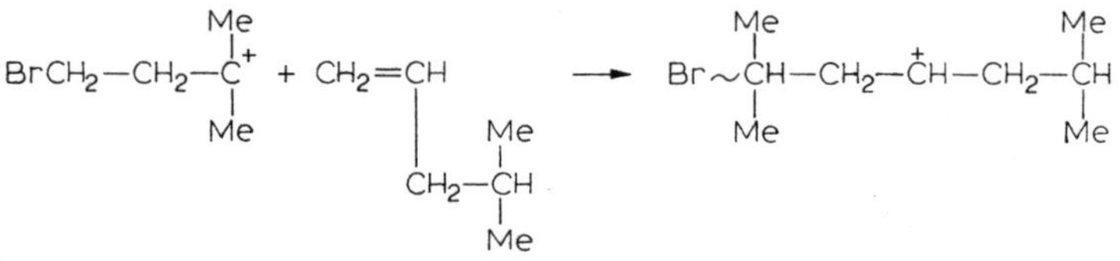

By reactions (82)–(84), 8 wt. % Br was introduced into the oligomers, an average of 1.95 Br atoms per chain. This work was extended by Kennedy and Smith [91] who selected the co-initiator from a group of transfer agents, which thus had a double function. This new technique they called the inifer method (from co-*ini*tiator–trans*fer* agent). When the rate of transfer to monomer is much less than to the inifer, the formation of the reactive macromer end groups can be better controlled. Macromer formation can be represented by the scheme

$$\begin{align}
\text{Hlg R Hlg} + M_t\text{Hlg}_x &\longrightarrow \text{Hlg R}^+ + M_t\text{Hlg}^-_{(x+1)} \\
\text{Hlg}_tR^+ + M &\longrightarrow \text{Hlg R M}^+ \\
\text{Hlg R M}^+ + n\,M &\longrightarrow \text{Hlg R M}\sim M^+ \\
\text{Hlg R M}\sim M^+ + \text{Hlg R Hlg} &\longrightarrow \text{Hlg R M}\sim M\text{Hlg} + \text{Hlg R}^+
\end{align} \quad (85)$$

where $M_t\text{Hlg}_x$ is an initiator of the Friedel–Crafts catalyst type and Hlg R Hlg is a co-initiator + transfer agent, the inifer. The method of Kennedy and Smith is very promising, and it is also being developed by other authors.

8.3.1 INIFER, INIFERTER TECHNIQUES

The inifer technique is specific for cationic and radical polymerizations. The most frequently used inifers in the first case are 1-chloro-1-methylethylbenzene (*a*); 1,4-bis(1-chloro-1-methylethyl)benzene (*b*); and 1,3,5-tris(1-chloro-1-methyl)benzene (*c*)

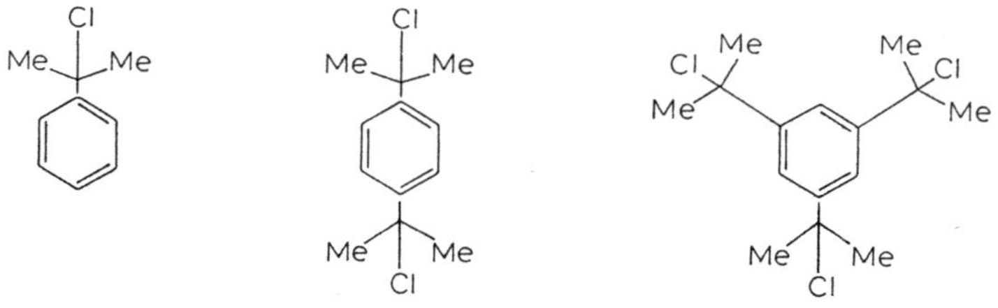

a
inifer

b
binifer

c
trinifer

which produce macromers by the general scheme (85). For example as:

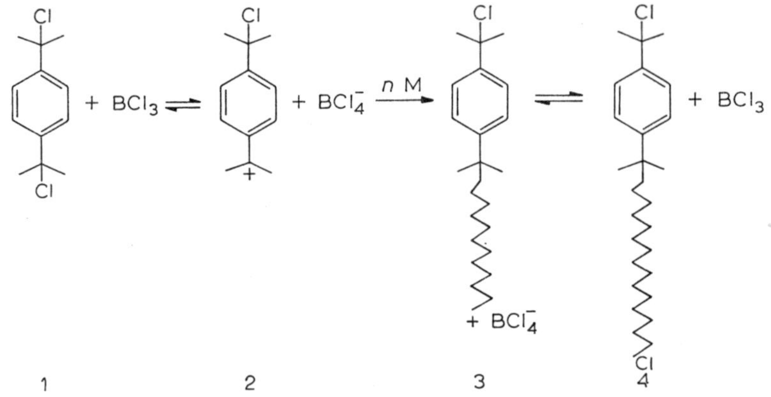

(86)

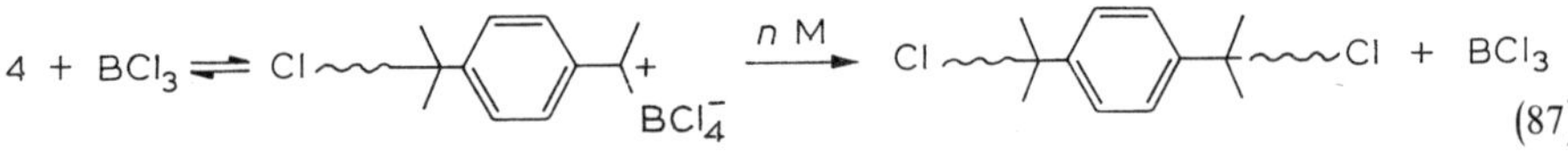

$$\text{1} \quad \text{2} \quad \text{3} \quad \text{4}$$

$$4 + BCl_3 \rightleftharpoons Cl\sim\!\!\!\!\!-\!\!\!\!\bigcirc\!\!\!\!-\!\!\!\!\stackrel{+}{\underset{BCl_4^-}{}} \xrightarrow{n\,M} Cl\sim\!\!\!\!\!-\!\!\!\!\bigcirc\!\!\!\!-\!\!\!\!\sim Cl + BCl_3 \quad (87)$$

With 2-methylpropene as M, both linear and star macromers have been prepared [92–94]. Many kinds of inifers may, of course, be used. For example Kress and Heitz prepared macromers from poly(oxytetramethylene) chains with acrylate or methacrylate end groups, by THF polymerization initiated by superacids with anhydrides as co-initiators – transfer agents [95].

The Inifer technique enables us to fulfil some requirements of polymer architecture even in some radical processes. An amplified form may be applied, the Iniferter variant, where the radical initiator simultaneously acts as a transfer and terminating agent. Otsu et al. used sulphides and disulphides (tetraethylthiuram disulphide, PhSSPh, Ph_2S, $PhCH_2SSCH_2Ph$) [96] and carbamates (benzyl-N,N-diethyldithiocarbamate, p-xylylene-N,N-diethyl-dithiocarbamate) [97] in the photopolymerization of methyl methacrylate and styrene, and phenylazotriphenylmethane in the polymerization of methyl methacrylate [98]. Living radical polymerizations yield polymers with defined end groups or the required block copolymers.

References

1. P. J. Flory, J. Am. Chem. Soc., 59 (1937) 241.
2. G. E. Ham, Vinyl Polymerization, Dekker, New York, 1967, p. 49.
3. P. E. M. Allen and C. R. Patrick, Kinetics and Mechanisms of Polymerization Reactions, Horwood, Chichester, 1974, pp. 144, 145.
4. C. H. Bamford, W. G. Barb, A, D. Jenkins and P. F. Onyon, The Kinetics of Vinyl Polymerization by Radical Mechanisms, Butterworths, London, 1958.
5. G. M. Burnett and L. D. Loan, Trans. Faraday Soc., 51 (1955) pp. 214, 219.
6. See P. E. M. Allen and C. R. Patrick, Kinetics and Mechanisms of Polymerization Reactions, Horwood, Chichester, 1974, pp. 186–188.

7   J. L. Kice, J. Am. Chem. Soc., 76 (1954) 6274.
8   A. D. Jenkins, Trans. Faraday Soc., 54 (1958) 1885, 1895.
9   J. L. Kice, J. Polym. Sci., 19 (1956) 123.
10 W. H. Atkinson, C. H. Bamford and G. C. Eastmond, Trans. Faraday Soc., 66 (1970) 1446.
11 P. W. Allen, F. M. Merrett and J. Scanlan, Trans. Faraday Soc., 51 (1955) 95.
12 G. E. Scott and E. Senogles, J. Macromol. Sci. Rev. Macromol. Chem., C9 (1973) 49.
13 M. J. Roedel, J. Am. Chem. Soc., 75 (1953) 6110.
14 H. W. Melville and P. R. Sewell, Makromol. Chem., 32 (1959) 139.
15 M. J. R. Canton, C. W. Cline, C. A. Heilberger, D. T. Huibers and R. Phillips, Mod. Plast. 46(6) (1963) 128.
16 M. Ravey and J. A. Waterman, J. Polym. Sci. Polym. Chem. Ed., 15 (1977) 2521.
17 K. Nozaki, Discuss. Faraday Soc., 2 (1947) 337.
18 A. L. Logothetis and J. M. McKenna, Polym. Prepr. Am. Chem. Soc. Div. Polym. Chem., 19(1) (1978) 528.
19 Z. Mrázek, Thesis, Institute of Macromolecular Chemistry, Czechoslovak. Academy of Sciences, Prague, 1980.
20 J. Ugelstad, J. Macromol. Sci. Chem., A11(7) (1977) 1281.
21 T. Ouchi and Y. Komatsu, J. Macromol. Sci. Chem., A11(3) (1977) 487.
22 H. Uemura, T. Taninaka and Y. Minoura, J. Polym. Sci. Polym. Lett. Ed., 15 (1977) 493.
23 S. Penczek and P. Kubisa, Makromol. Chem., 130 (1969) 186.
24 L. Trossareli, M. Guaita and G. Camino, Makromol. Chem., 105 (1967) 285.
25 B. Ullisch and W. Burchard, Makromol. Chem., 178 (1977) 1403.
26 A. S. Teblina, J. G. Freidlin and V. V. Korshak, Vysokomol. Soedin., Ser. A19 (1977) 1482.
27 See, for example, K. A. Andrianov, I. I. Tverdokhlebova, C. C. A. Pavlova, I. V. Perchova and I. L. Orlovskaya, Vysokomol. Soedin. Ser., A19 (1977) 19.
28 J. P. Kennedy and T. Kelen, J. Macromol. Sci. Chem., A18 (1983) 1189.
29 M. Szwarc and B. H. Zimm, Macromolecules, 16 (1984) 1918.
30 K. F. O'Driscoll, Macromolecules, 18 (1985) 1508.
31 E. Ratajczak and A. F. Trotman-Dickenson, Suplementary Tables of Bimolecular Gas Reactions, University of Wales, Institute of Science and Technology, Cardiff, 1970.
32 A. F. Trotman-Dickenson and E. Ratajczak, Table of Bimolecular Gas Reactions, National Standard Reference Data Series, Vol. 9, National Bureau of Standards, Washington, 1967.
33 J. A. Kerr, in J. K. Kochi (Ed.), Free Radicals, Vol. 1, Wiley, New York, 1972.
34 C. A. Barson and R. Ensor, Eur. Polym. J., 13 (1977) 113.
35 J. W. Breitenbach, Makromol. Chem., 8 (1952) 147.
36 K. Matsuo, G. W. Nelb, R. G. Nelb and W. H. Stockmayer, Macromolecules, 10 (1977) 654.
37 P. W. Moore, F. W. Ayscough and J. G. Clouston, J. Polym. Sci. Polym. Chem. Ed., 15 (1977) 1291.
38 G. Henrici-Olivé and S. Olivé, Fortschr. Hochpolym. Forsch., 2 (1961) 496.
39 W. A. Pryor and T. R. Fiske, Macromolecules, 2 (1969) 62.
40 G. Ayrey and A. C. Haynes, Makromol. Chem., 175 (1974) 1463.
41 R. D. Athey, Jr., J. Polym. Sci. Polym. Chem. Ed., 15 (1977) 1517.
42 M. L. Hallensleben, Eur. Polym. J., 13 (1977) 437.
43 J. Brandrup and E. H. Immergut (Eds.), Polymer Handbook, Interscience, New York, 1966.
44 J. J. Sanderson and C. R. Hauser, J. Am. Chem. Soc., 71 (1949) 1595.
45 C. B. Wooster and J. F. Ryan, J. Am. Chem. Soc., 56 (1934) 1133.
46 M. Szwarc, Carbanions, Living Polymers and Electron Transfer Processes, Interscience, New York, 1968, pp. 650–652.
47 G. E. Molan and J. E. Mason, J. Polym. Sci. Part A-1(4) (1966) 2236.
48 H. K. Hall, Jr. and R. C. Daly, Macromolecules, 8 (1975) 22.

49 D. N. Bhattacharyya, J. Smid and M. Szwarc, J. Polym. Sci. Part A–1, (3) (1965) 3099.
50 J. Léonard and S. L. Malholtra, J. Macromol. Sci. Chem., A11 (1977) 2087.
51 G. Sprach, M. Levy and M. Szwarc. J. Chem. Soc., (1962) 355.
52 M. Kučera, Thesis, Research Institute of Macromolecular Chemistry, Brno, 1959.
53 Pu Zhoui, Yan De-Yue, Tang We-Fung: Makromol. Chem., 186(1) (1985) 159.
54 T. Higashimura and H. Nishi, J. Polym. Sci. Polym. Chem. Ed., 15 (1977) 329.
55 J. E. Williams, V. Buss, L. C. Allen and P. R. Schleyer, J. Am. Chem. Soc., 92 (1970) 2141.
56 G. V. Pfeiffer and J. G. Jewett, J. Am. Chem. Soc., 92 (1970) 2143.
57 D. C. Pepper, J. Polym. Sci. Polym. Symp., 50 (1975) 51.
58 P. H. Plesch, Prog. High Polym., 2 (1968) 139.
59 See P. E. M. Allen and C. R. Partick, Kinetics and Mechanisms of Polymerization Reactions, Horwood, Chichester, 1974, p. 555.
60 F. Williams, A. Shinkawa and J. P. Kennedy, J. Polym. Sci. Symp. 56 (1976) 421.
61 J. P. Kennedy and R. G. Squires, Polymer, 6 (1965) 579.
62 P. K. Bossaer, E. J. Goethals, P. J. Hackett and D. C. Pepper, Eur. Polym. J., 13 (1977) 489.
63 T. G. Croucher and R. E. Wetton, Polymer, 17 (1976) 205).
64 P. Pino, B. Rotzinger and E. v. Achenbach, Makromol. Chem. Suppl., 13 (1985) 105.
65 G. Natta and I. Pasquon, Adv. Catal., 11 (1959) 1.
66 G. Natta, I. Pasquon and L. Giuffre, Chim. Ind. (Milan), 43(8) (1961) 871.
67 J. Boor, Jr., Ziegler–Natta Catalysts and Polymerizations, Academic Press, New York, 1979, pp. 245–260.
68 E. A. Yongman, K. Nozaki and J. Boor, (Shell Oil Company) US Pat. 3,084,148 (1963).
69 R. Vilím, Chem. Prum., 12 (1962) 102.
70 L. F. Borisová, E. A. Fushman, E. I. Vizen and N. M. Chirkov, Eur. Polym. J., 9 (1973) 953.
71 G. Natta, Chim. Ind. (Milan), 41(6) (1959) 519.
72 G. Bourat, J. Ferrier and A. Perez, J. Polym. Sci. Part. C, 4 (1963) 103.
73 See, for example, E. M. J. Pijpers and B. C. Roest, Eur. Polym. J., 8 (1972) 1151.
74 I. Okura, K. Soga, A. Kojima and T. Keii, J. Polym. Sci. Part A–1, (8) (1970) 2717.
75 O. N. Pirogov and N. M. Chirkov, Polym. Sci. USSR, 8 (1966) 1985.
76 E. J. Vandenberg, (Hercules Powder Co.): US Pat. 3,051,690 (1962).
77 R. V. Rabovskaya, E. I. Tinyakova and G. A. Parfenova, Vysokomol. Soedin. Ser. A, 19 (1977) 2352.
78 See P. E. M. Allen and C. R. Patrick, Kinetics and Mechanisms of Polymerization Reactions, Horwood, Chichester, 1974, p. 143.
79 F. A. Mayo, J. Am. Chem. Soc., 65 (1943) 2324.
80 L. L. Böhm, Polymer, 19 (1978) 562.
81 H. Mark and A. V. Tobolsky, Physical Chemistry of High Polymeric Systems, Vol. II, New York, 1950, p. 416.
82 M. Kučera and E.. Spousta, Makromol. Chem., 76 (1964) 183.
83 Cui-Ming Yuan and De-Yue-Yan: Makromol. Chem., 188 (1987) 341.
84 De-Yue Yan and Cui-Ming Yuan: Makromol. Chem., 188 (1987) 333 and previous papers.
85 N. Fuhrman and R. B. Mesrobian, J. Am. Chem. Soc., 76 (1954) 3281.
86 R. Z. Greenley, J. Macromol. Sci. Chem., A11 (1977) 933.
87 G. E. Ham, J. Polym. Sci. Part. B, 3 (1965) 459.
88 M. Kučera, CS 205,887 (1980).
89 Y. Ikada, H. Iwata and S. Nagaoka, Macromolecules, 10 (1977) 1364.
90 G. Ver-Strate and F. P. Baldwin, Polym. Prepr. Am. Chem. Soc. Div. Polym. Chem. 17(2) (1976) 808.
91 J. P. Kennedy and R. A. Smith, J. Polym. Sci. Polym. Chem. Ed., 18 (1980) 1523.
92 R. Santos, A. Femervari and J. P. Kennedy, J. Polym. Sci. Polym. Chem. Ed., 22 (1984) 2685.

93 O. Nuyken, S. D. Pask, A. Vischer and M. Walter, Polym. Prepr. Am. Chem. Soc. Div. Polym. Chem., 26(1) (1985) 44.
94 O. Nuyken, S. D. Pask, A. Vischer and M. Walter, Makromol. Chem. Macromol. Symp., 3 (1986) 129.
95 H. J. Kress and W. Heitz, Makromol, Chem. Rapid Commun., 2 (1981) 427.
96 T. Otsu, A. Kuriyama and M. Yoshida, Kobunshi Ronbunshu, 40(10) (1983) 583; Chem. Abstr., 100 (1984) 86176e.
97 T. Otsu and A. Kuriyama, Polym. Bull. (Berlin), 11(2) (1984) 135.
98 T. Otsu and T. Tazaki, Polym. Bull. (Berlin), 16(4) (1986) 277.

Chapter 8

Kinetics

> *The mother tongue of Nature*
> *is mathematics*
>
> A free interpretation of a statement by G. Galilei

The accumulation of data needed to predict the course of chemical reactions is often hindered by the presence of small concentrations of very unstable intermediates. It is just these highly reactive species, difficult to determine analytically, that are the indispensable link in the chain of causes and consequences in all processes where the substrate is transformed to the product by a chemical reaction.

Thus chemical kinetics of polymerizations is often expected to describe the consequences of the existence of species whose concentration and structure remain unknown. On the other hand, kinetic studies reveal the existence of these species and thus provide a basis for their future analysis. The mastering of macromolecular syntheses is unthinkable without a knowledge of the corresponding process kinetics. Interest is centred on the determination of the overall polymerization rate and of the rates of individual steps, of the product molecular mass, and of the changes in these quantities induced by external conditions. Kinetic data also form the basis of the chemical engineering of polymer production.

1. Polymerization rate

Cationic polymerization of 2-methylpropene at temperatures about 170 K may be almost flash-like; the transformation of tetrahydrofuran to an equilibrium polymer–monomer mixture may last tens to hundreds of hours at 260 K. Evidently the overall polymerization rate is a function of many factors which may be interconnected or may act separately. The aim of kinetic measurements is to describe the polymerization, and to find conditions under which it would proceed in the desired manner. This is usually only possible after the various factors and their consequences have been isolated and investigated. The rate of monomer consumption during polymerization mostly depends on the generation rate of active centres, and on their concentration and reactivity.

References pp. 544–545

1.1 Rate of initiation

In general the rate of initiation is determined by two reactions: the generation of the initiating particle R° from the initiator I, and of the active centre from R° and monomer

$$I \xrightarrow{k_d} R° \quad (1)$$

$$R° + M \xrightarrow{k_i} RM° \quad (2)$$

In simpler cases, one of these reactions is significantly slower than the other, thus representing the controlling step.

Measurement of the initiation rate has been best developed for radical polymerizations.

The generation of primary radicals can generally be represented by eqn. (1) which for an ideal case assumes the rate form[†]

$$\frac{d[R^*]}{dt} = -\frac{d[I]}{dt} = k_d[I] \quad (3)$$

A part of the radicals just generated decays by mutual collisions. The creation rate of primary initiating particles will therefore be lower than indicated by eqn. (3). The lowering of the rate is quantitatively described by the coefficient of initiation efficiency, f

$$\frac{d[R^*]}{dt} = f k_d[I] \quad (4)$$

Initiator consumption during the whole course of polymerization is usually small (2–5%). Therefore the initiation rate may be regarded as practically constant in a broad conversion range. When reaction (1) is the controlling step, v_{init} is determined by the rate of initiator consumption

$$v_{init} = -\frac{d[I]}{dt} = f k_d[I] \quad (5)$$

[†] When more radicals are formed from one initiator molecule, eqn. (3) is modified
  a) I → R* ⇒ eqn. (3)
  b) I → 2R* ⇒ $-\frac{d[I]}{dt} = \frac{2d[R^*]}{dt} = k_d[I]$.

This is irrelevant for practical purposes. It is, however, important when absolute rate constant values are to be determined. In the case referred to, it modifies k_d by a multiplicative factor of 2.

According to some observations, situations do exist where reaction (2) is slower. Than

$$v_{init} = k_i[R^*][M] \tag{6}$$

In the stationary state, $v_{init} = v_{term} = 2k_t[R^*]^2$; substituting for $[R^*]$ into the rate equation for monomer consumption [see eqn. (21)] we obtain

$$v_p = k_p\left(\frac{v_{init}}{2k_t}\right)^{1/2}[M] \tag{7}$$

For calculating v_{init} from eqn. (7), k_p^2/k_t must be determined independently; this can be done, for example, from the degree of polymerization of the product by means of the Mayo equation

$$\bar{P}^{-1} = (1+x)\frac{k_t v_p}{k_p^2[M]^2} + \frac{k_{trM}}{k_p} + \frac{k_{tr}[XT]}{k_p[M]} \tag{8}$$

Probably the most frequently used method of measuring initiation rates is the determination with inhibitors. With an "ideal" inhibitor, polymerization starts only after the inhibitor has been completely consumed. The easily measurable length of the inhibition period, t_{inh}, is a function of the initiation rate†

$$\frac{[Q]_0}{t_{inh}} = \frac{v'_{init}}{f} \tag{9}$$

Under suitable conditions, when the inhibitor exhibits a characteristic spectrum, the rate of inhibitor consumption can be directly measured spectrophotometrically. After the initial phase we can write

$$v'_{init} - k_{t,Q}[R^*][Q] = 0 \tag{10}$$

$$v'_{init} = -\frac{d[Q]}{dt} \tag{11}$$

For relation (11) to be useful, the number of radicals destroyed by the inhibitor molecule must be known (usually 1 or 2).

---

† It is evident that $v_{init} = v'_{init}$ only in ideal situations where $f = 1$. In general, the values of f from eqn. (5) and (9) are not identical but the differences are mostly regarded as negligible and f is supposed to be equal to 0.66; or, more often but less correctly, to 1 (see Table 2 in Chap. 3).

References pp. 544–545

The most direct method of measuring the initiation rate is the determination of the incorporation rate of initiator fragments into the polymer. Experimentally this is a complicated method. The evaluation is usually based on the assumption that, in the stationary state at slow initiator consumption, $\bar{v}_p$, v_{init} and $\bar{v}$ (mean kinetic chain length) remain almost constant.

$$\bar{v} = \frac{v_p}{v_{init}} = \frac{\text{number of monomeric units in the polymer}}{\text{number of initiating fragments in the polymer}} \qquad (12)$$

Various types of transfer have to be corrected for. The number of initiating fragments is usually measured by radiochemical tracer methods [1, 2].

Each of the mentioned methods has its own limitations. A reliable determination of initiation rate is difficult both theoretically and experimentally. A more detailed review of the possibilities and pitfalls of the various methods is given, for example, by Allen and Patrick [3].

1.2 Radical polymerization

Almost all radical polymerizations follow the simple scheme

$$I \underset{k'_d}{\overset{k_d}{\rightleftharpoons}} R^* \qquad (13)$$

$$R^* + M \xrightarrow{k_i} RM_1^* \qquad (14)$$

$$RM_n^* + M \xrightarrow{k_p} RM_{n+1}^* \qquad (15)$$

$$M_n^* + XT \xrightarrow{k_{tr}} M_nX + T^* \qquad (16)$$

$$T^* + M \xrightarrow{k_{i,tr}} TM_1^* \qquad (17)$$

$$M_n^* + M_m^* \begin{array}{c} \xrightarrow{k_{t,c}} M_{m+n} \\ \xrightarrow{k_{t,d}} M_m + M_n \end{array} \qquad (18)$$

$$M_n^* + R^* \xrightarrow{k_{t,pr}} RM_n \qquad (19)$$

Each of the elementary reactions of any polymerization can exhibit its own specific features. For example, the decomposition of initiator I [reaction (13)] can proceed as written in the scheme or by a more complicated induced decomposition. Transfer [reactions (16) and (17)] need not occur or it may involve several types of transfer agents each with its characteristic rate.

Sometimes termination by primary radicals is negligible [reaction (19)]. Polymerizations where termination occurs by primary radicals and/or by degradative transfer are called „non-ideal". A number of other causes may lead to deviation from "ideality".

1.2.1 OVERALL REACTION RATE OF IDEAL POLYMERIZATIONS

The rate of monomer consumption in polymerizations proceeding according to reactions (13)–(19) is given by

$$v_p = -\frac{d[M]}{dt} = k_i[R^*][M] + k_{i,tr}[T^*][M] + k_p[M_n^*][M] \quad (20)$$

In writing eqn. (20), the assumption is made that the reactivities of radical M_n^* of various degrees of polymerization n is equal. Let us introduce the further hypothesis that the reactivities of the radicals R^* and T^* are equal to that of M_n^*, i. e. $k_i = k_{i,tr} = k_p$. Equation (20) is then simplified to as

$$\sum_{n=1}^{\infty}[M_n^*] + [R^*] + [T^*] = [M^*]$$

and

$$v_p = -\frac{d[M]}{dt} = k_p[M^*][M] \quad (21)$$

The concentration $[M^*]$ is, of course, unknown but for stationary conditions it can be calculated (remembering that an error is introduced by neglect of the changes in termination rate with the length of the interacting radicals) with $k_d' = 0$ as

$$\frac{d[M^*]}{dt} = 0$$

$$v_{init} = v_{term} \quad (22)$$

$$fk_d[I] = 2k_t[M^*]^2$$

Substituting for $[M^*]$ from eqn. (22) into eqn. (21) we finally obtain

$$-\frac{d[M]}{dt} = k_p\left[\frac{fk_d[I]}{2k_t}\right]^{1/2}[M] \quad (23)$$

$$\ln\frac{[M]_0}{[M]} = k_p\left[\frac{fk_d[I]}{2k_t}\right]^{1/2}t \quad (24)$$

References pp. 544–545

The assumptions under which eqn. (23) has been derived are very stringent, and they are rarely fulfilled. Nevertheless, the situation described does exist. An example is the polymerization of butyl-1-chloroacrylate, which is first order with respect to monomer (see Fig. 1).

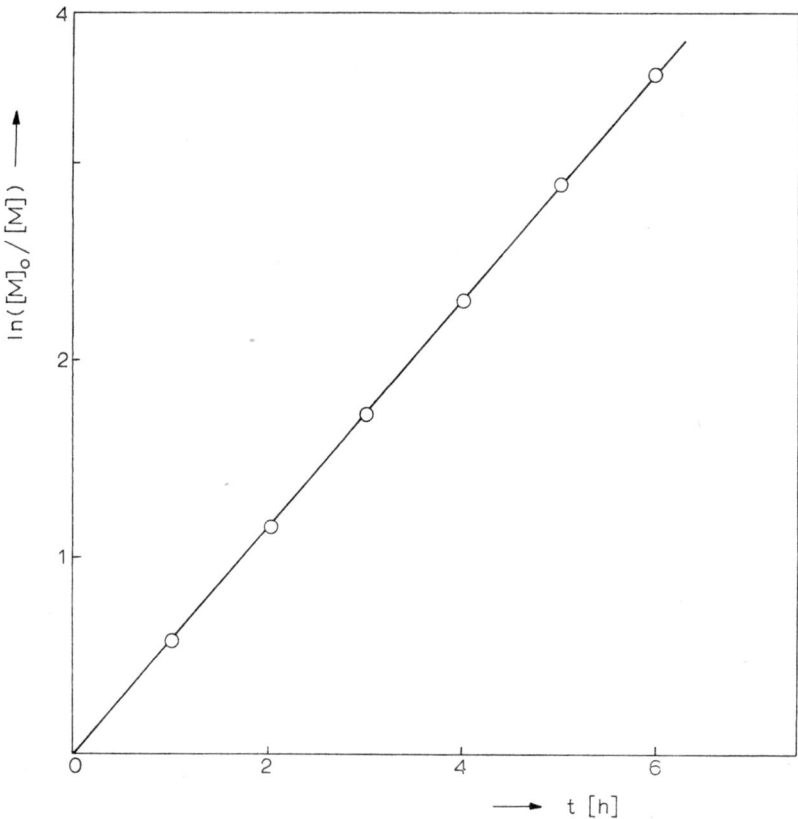

Fig. 1. Plot of the variables of eqn. (24). Polymerization of butyl 1-chloroacrylate [4].

Experimental results of the studies of polymerization kinetics can often be explained by the reversible character of eqn. (13) and by the relatively slow generation rate of the radicals M*. Again assuming equal reactivity of radicals of various length we have

$$v_p = -\frac{d[M]}{dt} = \\ = k_{p_i}[R^*][M] + k_p[M^*][M] \approx \\ \approx k_p[M^*][M] \tag{25}$$

When long chains are generated, monomer consumption for initiation is negligible†. Only the simplified form of eqn. (25) is usually considered, without substantial loss of accuracy. [M*] is calculated for stationary state conditions and substituted into eqn. (25); an expression is obtained according to which the polymerization order is 1.5 with respect to the monomer $(k_d/k_d' = K)$

$$k_i[R^*][M] = fk_iK[I][M] = \\ = 2k_t[M^*]^2 \qquad (26)$$

yielding

$$[M^*] = \left[\frac{fk_iK[I][M]}{2k_t}\right]^{1/2} \qquad (27)$$

$$v_p = -\frac{d[M]}{dt} = \\ = k_p\left[\frac{fk_iK[I]}{2k_t}\right]^{1/2}[M]^{3/2} \qquad (28)$$

By integration

$$[M]^{-1/2} - [M]_0^{-1/2} = k_p\left[\frac{fk_iK[I]}{2k_t}\right]^{1/2} t \qquad (29)$$

An example of this type of polymerization is demonstrated in Fig. 2. It represents the linear dependence between $[M]^{-1/2}$ and time in vinyl chloride polymerization [5].

Expressions formally quite similar to eqns. (28) and (29) were derived for (sensitized) photoinitiated polymerization [6]. The initiator concentration, [I], is replaced by the intensity of absorbed light (J)

$$v_{init} = k_i f(J)[M] \qquad (30)$$

Equation (28) is transformed to

$$-\frac{d[M]}{dt} = k_p\left[\frac{k_i f(J)}{2k_t}\right]^{1/2}[M]^{3/2} \qquad (31)$$

---

† When chains with a polymerization degree of 100 are formed, 1% at most of monomer is consumed by initiation. Macromolecules produced in practice have $\bar{P} = 500-1000$ or more.

References pp. 544–545

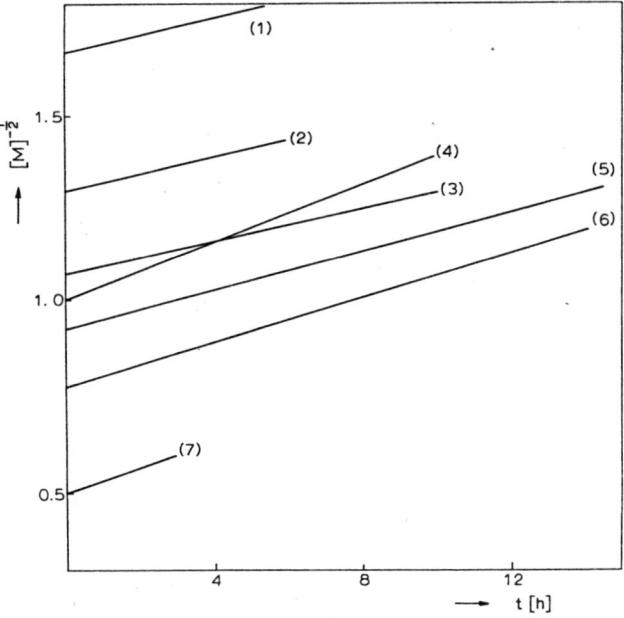

Fig. 2. Plot of the variables of eqn. (29) [5]. Polymerization of VC in solution at 323 K. The concentration of VC increases with curve number from 0,35 to 4.00 mol dm^{-3}.

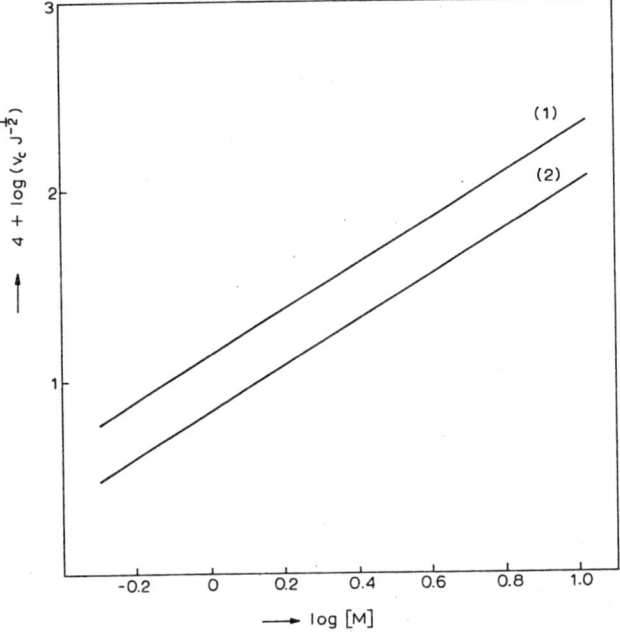

Fig. 3. Plot of the variables of eqn. (32). Photoinitiated polymerization of VC [7]. Temperature (K): (1) 323, (2) 298.

Its integrated form

$$[M]^{-1/2} - [M]_0^{-1/2} = k_p \left[\frac{k_i f(J)}{2k_t} \right]^{1/2} t \qquad (32)$$

has again been verified with vinyl chloride polymerization (see Fig. 3).

When the radicals are thermally generated, initiation for a bimolecular reaction is described by the scheme (see Chap. 3, Sect. 1.1)

$$2M \xrightarrow{k_i} M^* \qquad (33)$$

Therefore in the stationary state it holds that

$$k_i[M]^2 = 2k_t[M^*]^2 \qquad (34)$$

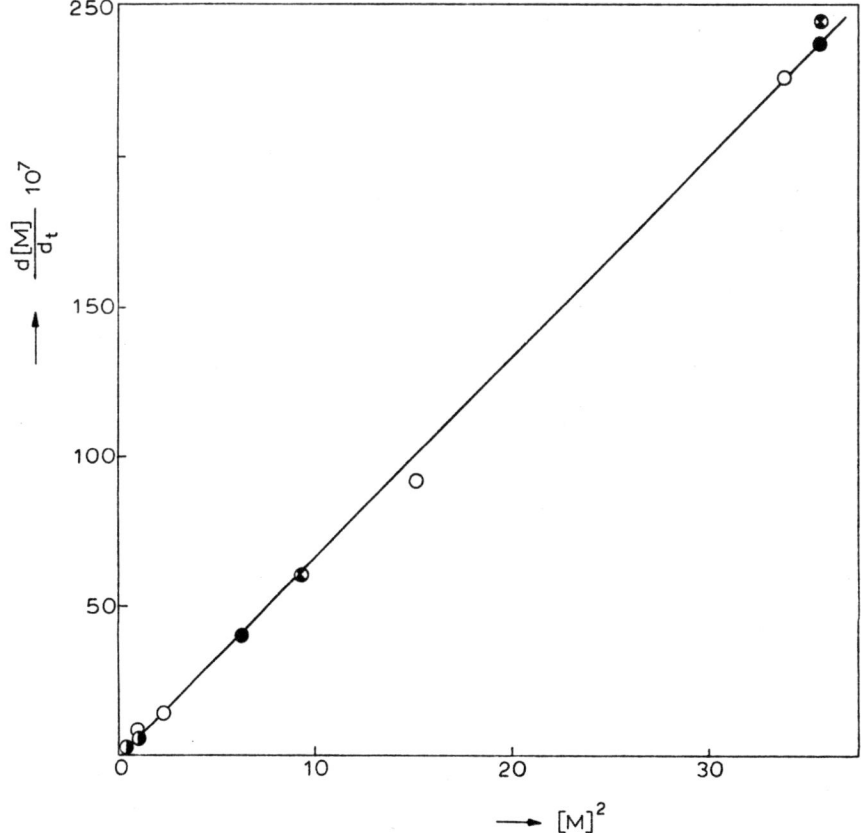

Fig. 4. Dependence of the rate of thermally initiated polymerization of S on the square of monomer concentration [8]. Temperature 373 K, solvent: ○, cyclohexane; ●, benzene; ⊗, ethylbenzene.

and after substitution for $[M^*]$ into eqn. (25)

$$v_p = -\frac{d[M]}{dt} = k_p \frac{k_i}{2k_t}[M]^2 \tag{35}$$

After integration

$$[M]^{-1} - [M]_0^{-1} = \frac{k_i k_p}{2k_t} t \tag{36}$$

Equation adequately describes the course of the thermally initiated polymerization of styrene (see Fig. 4).

The cases described represent the simplest examples of polymerization kinetics. The assumptions mentioned above are often not fulfilled and complications occur.

1.2.2 REACTIONS OF RADICALS OF VARIOUS LENGTHS

Some considerable time called attention to the functional dependence between the length of the growing radical and the rate of propagation [9]. Ito also observed that termination rate may depend on the dimensions of the macroradical [10, 11], even in media of low viscosity. His communication was based on measurements where other reasons for non-ideal polymerization behaviour were suppressed by the appropriate selection of experimental conditions. The deviations from the theoretical course could thus be ascribed to reactivity changes of the growing radicals.

He polymerized methacrylonitrile by means of 2,2'-azobisisobutyronitrile. The isobutyronitrile radical generated by initiator decomposition is structurally almost identical with the macroradical end group. Under these conditions, the rate constant of primary radical termination should be very near to the rate constant of termination between oligomeric and polymeric radicals. The studied polymerizations were carried out in dimethylformamide, which is a poor solvent for polymethacrylonitrile. In poor solvents, the change in termination rate with the length of the growing chain should not depend on the excluded volume [10].

Ito observed a change in the value of the termination rate constant between radicals having degrees of polymerization n and s (n and s also included primary radicals) which could be expressed by the relation

$$k_{t,ns} = \text{const.} \, (ns)^{-a}$$

with $a = 0.091$ at 333 K[†]. In similar experiments with the polymerization of styrene by means of 1-azobis-2-phenylethane in bulk, $a = 0.167$ at the same temperature. Such high values of a cannot be caused by excluded volume effects. The author tried to explain the rate constant changes by a relation between segmental diffusion and chain length.

When the growing radicals are small, their termination constant can be calculated by means of the theory of diffusion-controlled reactions [12]. The mean termination rate constant is then given by the relation

$$\bar{k}_t = \frac{4\pi\varrho k_0 D}{k_0 + D} \tag{37}$$

where k_0 is a constant, D the relative diffusion coefficient, and ϱ the radius of the radical end sweeps. k_0 and the diffusion coefficient of the primary radicals, D_{11} are related by the expression [13]

$$0.5 \approx \frac{D_{1,1}}{k_0 + D_{1,1}}$$

As the diffusion coefficient is directly proportional to the square root of molecular mass, we can write [11]

$$D = \frac{D_{1,1}}{(1 + \bar{P}/2)^{1/2}}$$

and thus

$$\bar{k}_t = \frac{4\pi\varrho D_{1,1}}{1 + (\bar{P}/2 + 1)^{1/2}} \tag{38}$$

The experimentally observed relationship between $k_p \bar{k}_t$ and $\bar{P}$ is shown in Fig. 5. The ratio of the constants decreases with increasing values of $\bar{P}$. According to eqn. (38), for radicals of $\bar{P}/2 = 11$, the value of $\bar{k}_t$ is about half of that for dimeric or trimeric radicals. Thus the propagation constant of an oligomeric radical is at least three times higher than the k_p value of a macroradical.

It appears that oligomeric radicals are more reactive than polymeric radicals, and this should be accounted for in rigorous theoretical considerations. In particular an uncritical application of kinetic quantities from the gaseous phase (which are more easily accessible, better defined and often even tab-

---

[†] For an ideal case, $a = 0$.

References pp. 544–545

ulated) to polymerizing systems should be avoided. From the practical point of view, the unequal reactivity of radicals of various length does not cause serious trouble.

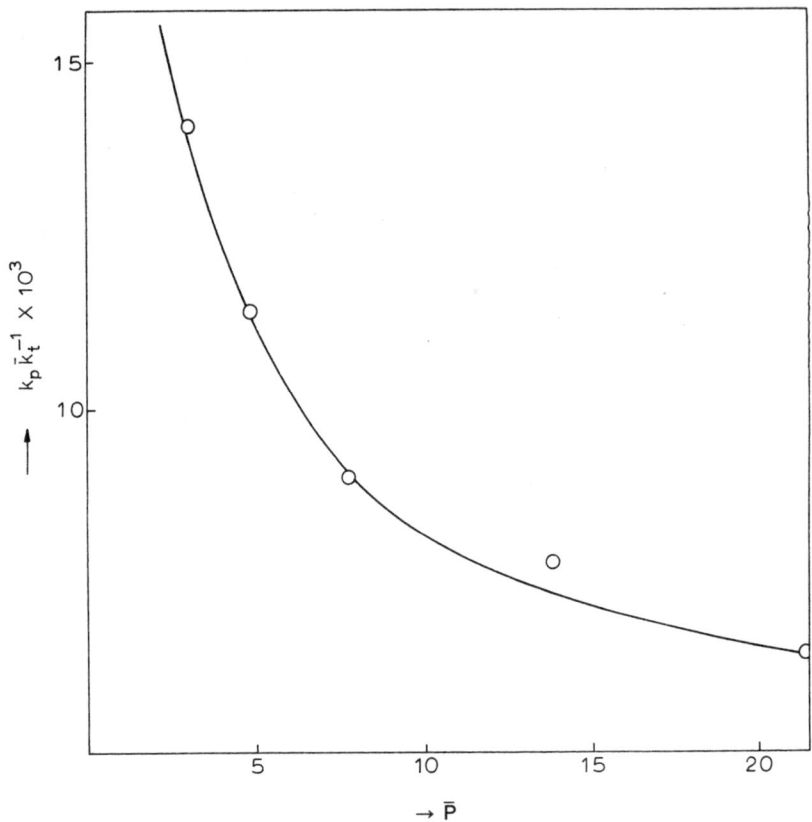

Fig. 5. Plot of $k_p/\bar{k}_t^{-1/2}$ vs. $\bar{P}$ of the growing radical for the polymerization of methacrylonitrile in DMF. Temperature 333 K, [AIBN] = 0.1 mol dm^{-3}.

1.2.3 NON—IDEAL RADICAL POLYMERIZATION

When the assumptions under which relations (23), (28), (31) and (35) have been derived are not fulfilled, these relations are no longer valid. This is manifested by a change in the reaction orders with respect to initiator and to monomer with changing concentration of these components. One of the most general causes of deviations from ideality is termination by primary radicals.

Deb and Gaba [14] studied the polymerization of styrene with dibenzoylperoxide and 2,2'-azobisisobutyronitrile. When deriving the kinetic relations, they considered mutual combination of primary radicals (and of their decomposition products), and also transfer by primary radicals to be negligible. They started from the general scheme

$$I \xrightarrow{k_d} 2R^* \qquad (39)$$

$$R^* + M \xrightarrow{k_i} RM_1^* \qquad (40)$$

$$RM_n^* + M \xrightarrow{k_p} RM_{n+1}^* \qquad (41)$$

$$R^* \xrightarrow{^1k_d} {}^1R^* \qquad (42)$$

$$^1R^* + M \xrightarrow{^1k_i} {}^1RM_1^* \qquad (43)$$

$$M_s^* + M_s^* \xrightarrow{k_t} \text{inactive polymer} \qquad (44)$$

$$R^* + M_n^* \xrightarrow{k_{t,pr}} \text{inactive polymer} \qquad (45)$$

$$^1R^* + M_n^* \xrightarrow{^1k_{t,pr}} \text{inactive polymer} \qquad (46)$$

where reaction (42) describes decomposition of PhCOO* to Ph* + CO_2(Ph≡$^1R^*$). Under the stationary state assumption, the concentrations of the various radicals are given by

$$[R^*] = \frac{2fk_d[I]}{{}^1k_d + k_i[M] + k_{t,pr}[M^*]} \qquad (47)$$

$$[^1R^*] = \frac{2f^1 k_d k_d[I]}{({}^1k_i[M] + {}^1k_{t,pr}[M^*])/({}^1k_d + k_i[M] + k_{t,pr}[M^*])} \qquad (48)$$

$$2k_t[M^*] = k_i[R^*][M] + {}^1k_i[^1R^*][M] - k_{t,pr}[R^*][M] - {}^1k_{t,pr}[^1R^*][M] \qquad (49)$$

Substituting for $[R^*]$ and $[^1R^*]$ from eqns. (47) and (48) into eqn. (49), they obtained the equation

$$2k_t[M] = \frac{2fk_d[I](k_i[M] - k_{t,pr}[M^*])}{k_d + k_i[M] + k_{t,pr}[M^*]} + $$

$$+ \frac{2f^1 k_d k_d[I]({}^1k_i[M] - {}^1k_{t,pr}[M^*])}{({}^1k_d + k_i[M] + k_{t,pr}[M^*])({}^1k_i[M] + {}^1k_{t,pr}[M^*])} \qquad (50)$$

Equation (21) holds for long chains and thus

$$\frac{v_p^2}{[\mathrm{I}][\mathrm{M}]^2} = fk_dk_p^2 / \left[k_t \left(1 + \frac{{}^1k_d}{k_i[\mathrm{M}]} + \frac{k_{t,\mathrm{pr}}v_p}{k_ik_p[\mathrm{M}]^2} \right) \right] \left\{ 1 - \frac{k_{t,\mathrm{pr}}v_p}{k_ik_p[\mathrm{M}]^2} + \right.$$

$$\left. + \left[\frac{{}^1k_d}{k_i[\mathrm{M}]} \left(1 - \frac{{}^1k_{t,\mathrm{pr}}v_p}{{}^1k_ik_p[\mathrm{M}]^2} \right) \right] / \left[1 + \frac{{}^1k_{t,\mathrm{pr}}v_p}{{}^1k_ik_p[\mathrm{M}]^2} \right] \right\} \quad (51)$$

This equation is too complicated to be confronted with experimental results. It can be greatly simplified when reaction (46) is neglected. The assumption about the small importance of macroradical termination by phenyl radicals appears acceptable. First of all, the concentration of phenyl radicals is much smaller than that of benzoyloxyl radicals, $k_{t,\mathrm{pr}}$ should not substantially differ from ${}^1k_{t,\mathrm{pr}}$, and finally the reactive phenyl radical should be immediately consumed by initiation. Equation (51) then assumes the form

$$\frac{v_p^2}{[\mathrm{I}][\mathrm{M}]^2} = \frac{fk_dk_p^2}{k_t[(1-us)/(1+us)]} \quad (52)$$

where

$$u = \frac{k_{t,\mathrm{pr}}/(k_ik_p)}{1 + \{{}^1k_d/k_i[\mathrm{M}]\}} \quad (53)$$

and

$$s = \frac{v_p}{[\mathrm{M}]^2} \quad (54)$$

As $us \ll 1$, the logarithmic form of eqn. (52) can be approximated as

$$\ln \frac{v_p^2}{[\mathrm{I}][\mathrm{M}]^2} = \ln \frac{fk_dk_p^2}{k_t} - 2us \quad (55)$$

The expression for u [eqn. (53)] indicates how important it is to consider primary radical dissociation [the term ${}^1k_d/(k_i[\mathrm{M}])$] when correct values of $k_{t,\mathrm{pr}}/(k_ik_p)$ are to be obtained, even when termination of macroradicals by secondary phenyl radicals is neglected. In Fig. 6, a graphical representation of eqn. (55) for styrene polymerization with dibenzoylperoxide in benzene is shown. When this dependence is measured for two monomer concentrations, $k_{t,\mathrm{pr}}/(k_ik_p)$ and ${}^1k_d/k_i$ can be calculated from the slope u. To reduce error,

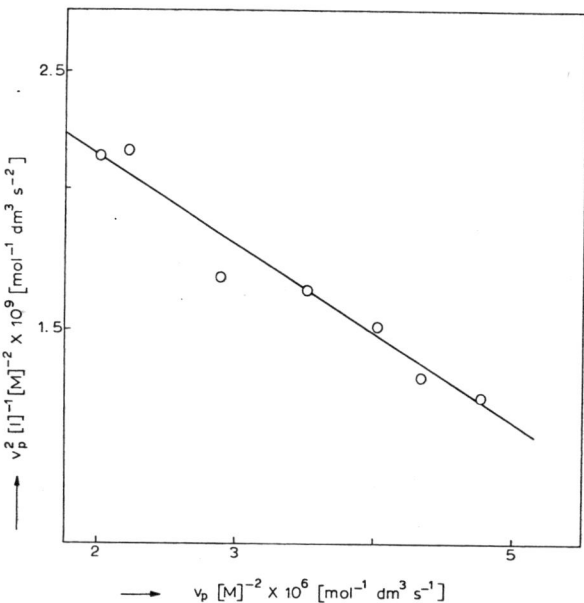

Fig. 6. Polymerization rate as a function of monomer and initiator concentrations according to eqn. (55) [14]. $[M]_0 = 2.58$ mol dm^{-3}, temperature 333 K.

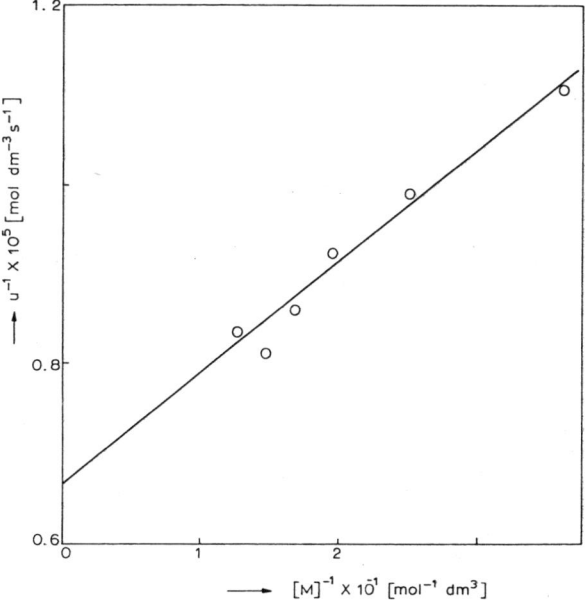

Fig. 7. Dependence of u^{-1} on $[M]^{-1}$ according to eqn. (56).

References pp. 544–545

the measurement is usually repeated with several monomer concentrations and the results are evaluated graphically. From eqn. (53)

$$u^{-1} = \frac{k_i k_p}{k_{t,pr}} + \frac{k_i k_p {}^1k_d}{k_{t,pr} k_i [M]} \tag{56}$$

thus the plot of u^{-1} vs. $[M]^{-1}$ is a straight line whose intercept with the ordinate and slope are equal to the required quantities (see Fig. 7 and Table 1).

TABLE 1

Sets of rate constants for non-ideal styrene polymerization in benzene with dibenzoylperoxide initiation at 333 K [14].

[M] (mol dm^{-3})	$10^9 f k_d k_p^2 / k_t$ (mol^{-1} dm^3 s^{-1})	$10^{-5} \mu$ (mol dm^{-3} s)	$10^{-5} k_{t,pr}/(k_i k_p)$ (mol dm^{-3} s)	${}^1k_d/k_i$ (mol dm^{-3})
2.58	3.0	0.90		
3.86	2.9	1.01		
5.17	2.9	1.10	1.5	1.73
6.03	3.0	1.20		
6.89	2.8	1.27		
7.75	2.5	1.21		

The primary radical from 2,2'-azobisisobutyronitrile is very rapidly decomposed to

$$\begin{array}{c} Me \\ \diagdown \\ \overset{*}{C}-CN \\ \diagup \\ Me \end{array} \longleftrightarrow \begin{array}{c} Me \\ \diagdown \\ C=C=N^* \\ \diagup \\ Me \end{array}$$

Therefore only the secondary radical functions as initiator. This situation is kinetically equivalent to the condition ${}^1k_d = 0$; the radicals do not compete in initiation. Equation (51) is also considerably simplified in this case

$$\frac{v_p^2}{[I][M]^2} = \frac{f k_d k_p^2}{k_t} \frac{1 - (k_{t,pr} v_p)}{k_i k_p [M]^2} \left[\frac{1 + (k_{t,pr} v_p)}{(k_i k_p [M]^2)} \right]^{-1} \tag{57}$$

Deviations from ideality will increase with increasing initiator concentration and decreasing concentration of monomer. Simplifying the logarithmic form of eqn. (57), we obtain the approximate relation

$$\ln \frac{v_p^2}{[I][M]^2} = \ln \frac{f k_d k_p^2}{k_t} - \frac{k_{t,pr}}{k_i k_p} \frac{v_p}{[M]^2} \tag{58}$$

In the way that has been described, Deb and Gaba derived the values of the rate constant sets for the case discussed (see Table 2). The trend towards growth of $fk_d k_p^2/k_t$ with monomer concentration is ascribed by the authors to the effect of monomer on k_d [15].

TABLE 2

Sets of rate constants for non-ideal styrene polymerization in benzene, with 2,2'-azobisisobutyronitrile initiation at 333 K [14].

[M] (mol dm^{-3})	$10^9 fk_d k_p^2/k_t$ (mol^{-1} dm^3 s^{-1})	$10^{-4} k_{t,pr}/(k_i k_p)$ (mol dm^{-3} s)
1.72	4.00	3.35
2.58	4.72	3.40
3.44	5.40	3.26
5.17	5.90	3.30
6.03	6.00	3.42
6.89	6.00	3.50
7.75	5.80	2.80

By the described kinetic analysis, the activation energy of initiation [reaction (40)], which is otherwise accessible only with difficulty, could also be estimated. A value about twice that of the activation energy of propagation was found ($E_i \approx 74$ kJ mol^{-1}, or 66 kJ mol^{-1}, $E_p = 32.6$ kJ mol^{-1} [16]).

1.2.4 EXAMPLE: KINETIC ANALYSIS OF VINYL CHLORIDE POLYMERIZATION

Each year, hundres of thousands of tons of vinyl chloride are polymerized in the world. Commensurate attention is thus paid to studies of its polymerization. Vinyl chloride is one of those monomers that are transformed to polymer by a complicated mechanism. Poly(vinyl chloride) is soluble neither in its own monomer nor in the common solvents. Its formation is therefore connected with the appearance of a solid phase; the process has the character of precipitation polymerization. This greatly complicates the kinetics of "solution" and bulk (suspension) polymerization.

Various groups of investigators so far do not agree in their ideas concerning the mechanism by which the polymer particles, swollen by monomer, are formed. To illustrate how the kinetics of a complicated macromolecular synthesis is being analyzed, the approach of the representatives of several laboratories will be presented.

The polymerization of vinyl chloride evidently proceeds in the monomer phase and in monomer–polymer particles. Crosato-Arnoldi et al. [17] and

Abdel-Alim and Hamielec [18] assume that the radicals do not diffuse between the various phases. Radicals are generated in monomer medium, and here they also decay, either still in solution or already in the solid phase[†]. The active centres, generated in the polymer-phase, decay in the same phase; their desorption does not occur [17]. When the partition coefficient of the initiator between the monomer and polymer-rich phases is equal to unity we can write

$$\frac{d[R_M^*]}{dt} = fk_i[I] - 2\,^M k_t[R_M^*]^2 = 0 \tag{59}$$

$$\frac{d[R_P^*]}{dt} = fk_i[I] - 2\,^P k_t[R_P^*]^2 = 0 \tag{60}$$

thus

$$[R_M^*] = \left(\frac{fk_i[I]}{2\,^M k_t}\right)^{1/2} \tag{61}$$

$$[R_P^*] = \left(\frac{fk_i[I]}{2\,^P k_t}\right)^{1/2} = \left(\frac{^M k_t}{^P k_t}\right)^{1/2} [R_M^*] = Q[R_M^*] \tag{62}$$

Then, for chains of sufficient length

$$v_p = -\frac{d[M]}{dt} = k_p \left(\frac{fk_i[I]}{2\,^M k_t}\right)^{1/2} ([^M M] + Q[^P M]) \tag{63}$$

When expressed by means of changes in conversion, the rate equation (63) will be of the form[††]

$$v_p = \frac{d\varkappa}{dt} = \left(\frac{fk_i[I]}{2\,^M k_t}\right)^{1/2} k_p[1 - \varkappa(1 + a - aQ)] \tag{64}$$

[†] A radical generated in the monomer grows and soon gives rise to a particle which is "precipitated". The degree of polymerization of such a chain is about 25. It is a matter of convention up to which stage this monoradical coil will be called dissolved radical, or precipitated primary particle [19].
[††] $[^M M]/[M]_0$ = total amount of monomer minus the monomer in the monomer–polymer particles is equal to $1 - \varkappa - a\varkappa$. $[^P M]/[M]_0 = a\varkappa$.

where a is the mass ratio of monomer to polymer in the particle. Equation (64) can even be of the form

$$v_p = \frac{d\varkappa}{dt} = (fk_i[I])^{1/2} k_p \left[\frac{1 - \varkappa(1 + a)}{(2^M k_t)^{1/2}} + \frac{a\varkappa}{(2^P k_t)^{1/2}} \right] \quad (65)$$

The authors considered their description of a series of bulk and suspension polymerizations of vinyl chloride with various initiators as very satisfactory [17] with $Q \approx 15$. In spite of that, their approach has been criticized by Ugelstad who considers termination of growing radicals exclusively by mutual collisions to be very improbable [19]. A large number of primary particles are formed in a short time interval, and many of these are then easily absorbed by the already solid flakes. Desorption of sorbed radicals has also to be considered. Radical distribution among the two phases should be controlled by these processes, especially at low conversion.

Ugelstad et al. proposed a model in which the possibility of radical exchange is respected [20]. Under stationary conditions it must hold that

$$^M v_{init} - k_a[R^*_M] + k_d[R^*_P] - 2^M k_t[R^*_M]^2 {^M V} = 0 \quad (66)$$

$$^P v_{init} + k_a[R^*_M] - k_d[R^*_P] - 2^P k_t[R^*_P]^2 {^P V} = 0 \quad (67)$$

In eqns. (66) and (67), $^M v_{init}$ and $^P v_{init}$ designate initiation rates in the monomer and polymer-rich phases, respectively, k_a and k_d are the rate constants of radical absorption and desorption in solid phase particles, and $^M V$ and $^P V$ are the total volumes of the monomer-rich phase and of the precipitated polymer particles, respectively.

When the products $k_a[R^*_M]$ and $k_d[R^*_P]$ play a dominant role in eqns. (66) and (67), it holds approximately [cf. eqn. (62)] that

$$\frac{[R^*_P]}{[R^*_M]} = \frac{k_a}{k_d} = Q \quad (68)$$

Introducing eqn. (68) into eqns. (66) and (67), we obtain

$$[R^*_M] = \left[\frac{fk_i I}{^M V 2^M k_t + Q^2 {^P V} 2^P k_t} \right]^{1/2} \quad (69)$$

$$[R^*_P] = \left[\frac{fk_i I}{^M V 2^M k_t + Q^2 {^P V} 2^P k_t} \right]^{1/2} Q \quad (70)$$

References pp. 544–545

where I is the total number of moles of initiator, and for the rate

$$-\frac{dM}{dt} = \left[\frac{fk_iI}{{}^MV2{}^Mk_t + Q^{2P}V2{}^Pk_t}\right]^{1/2} k_p({}^MM + Q^PM) \qquad (71)$$

Therefore also

$$\frac{d\varkappa}{dt} = \left[\frac{fk_iI}{{}^MV2{}^Mk_t + Q^{2P}V2{}^Pk_t}\right]^{1/2} k_p(1 - \varkappa - a\varkappa + Qa\varkappa) \qquad (72)$$

For describing the rate of conversion changes in a volume unit the author simplified eqn. (72) (substituting volume units for the total volumes ${}^MV$ and ${}^PV$) to[†]

$$\frac{d\varkappa}{dt} = \left[\frac{fk_i[I]_0}{(1 - 1.47\varkappa)2{}^Mk_t + 1.07\varkappa 2{}^Pk_tQ^2}\right]^{1/2} k_p(1 - 1.47\varkappa + 0.47\varkappa Q) \qquad (73)$$

Ugelstad sets ${}^Mk_t = 2.5 \times 10^9 \text{ mol}^{-1} \text{ dm}^3 \text{ s}^{-1}$ and ${}^Pk_t = 5 \times 10^7 \text{ mol}^{-1} \text{ dm}^3 \text{ s}^{-1}$; $Q \approx 200$. According to his ideas, termination occurs mainly in polymer particles, even immediately after the start of the reaction. At a conversion of 0.05, 84% of all polymer was generated in the particles, and also 98% of all termination occurred there. With increasing values of Q, the proportion of these reactions occurring in the particles further increases.

The proposed model explains the rate reduction under conditions favouring polymer precipitation, e. g. in the presence of poor solvents. It also explains the effect of CBr$_4$, leading to a lowering of Q (and thus to an increased rate at low conversions) and to loss of autoacceleration

$$\frac{d\varkappa}{dt} = k_p \left[\frac{fk_i[I]_0}{2{}^Mk_t + 1.07\varkappa Q^2 2{}^Pk_t}\right]^{1/2} \qquad (74)$$

[†] Monomer solubility in polymer at 323 K and saturated vapour pressure is 6 mol dm^{-3}. Assuming additivity of polymeric phase volume, i. e. ${}^PV = {}^Mq/{}^M\varrho + {}^Pq/{}^P\varrho$, it holds

$a = 0.47$,
${}^MV = V_0(1 - \varkappa - a\varkappa) = V_0(1 - 1.47\varkappa)$,
${}^PV = V_0({}^M\varrho/{}^P\varrho + a)\varkappa = V_0 \times 1.07\varkappa$.

${}^Mq$ and ${}^Pq$ are the masses of monomer and polymer in the polymeric phase and ${}^M\varrho$ and ${}^P\varrho$ are monomer and polymer densities ($= 0.85$ and 1.4 g cm^{-3}, respectively).

At conversions above 10% with the above-mentioned values of Q and $^M k_t$, eqn. (73) is reduced to the expression

$$\frac{d\varkappa}{dt} = k_p \left[\frac{fk_i[I]_0}{2^P k_t}\right]^{1/2} 0.47\varkappa^{1/2} \tag{75}$$

indicating once more that polymerization already proceeds in the polymer particles at relatively low conversions. For the initial rate, eqn. (73) yields the dependence

$$\frac{d\varkappa}{dt} = k_p \left[\frac{fk_i[I]_0}{2^M k_t}\right]^{1/2} \tag{76}$$

Olaj proposed a model for the bulk polymerization of vinyl chloride involving solid phase generation from the separating primary radicals [21]. According to his ideas, the radicals are generated both in solution and in the precipitated polymer. Those generated in the monomer-rich phase add monomer and separate when the necessary degree of polymerization is reached. Significant termination does not occur in the liquid phase. The separating radicals produce particles by aggregation, without terminating in this step. Therefore in the stationary state

$$\frac{d[^1R^*]}{dt} = {^M}v_{init} - k_a[^1R^*] = 0 \tag{77}$$

$$\frac{d[R^*]}{dt} = {^P}v_{init} + k_a[^1R^*] - 2^P k_t[^P R^*]^{2}\, ^P V = 0 \tag{78}$$

where $^1R^*$ is a primary radical and k_a is the rate constant of primary radical aggregation. The rate equation is a sum of the contributions of primary radicals separating from the monomer-rich phase, and of the radicals from the polymer-rich phase

$$\frac{d\varkappa}{dt} = k_p {^M}\varPhi \left[\frac{2fk_i[I]}{k_a}\left(1 - \frac{r\varkappa}{{^P}\varPhi}\right) + \left(\frac{fk_i[I]}{2k_t}\frac{r\varkappa}{{^P}\varPhi}\right)^{1/2}\right] \tag{79}$$

where $^M\varPhi$ and $^P\varPhi$ are the volume fractions of monomer and polymer in the polymer-rich phase and $r = {^M}\varrho/{^P}\varrho$. Olaj assumes that the second term in the bracket at the right-hand side of eqn. (79) predominates at low conversions.

At some conversions, eqns. (79) and (73) mutually correspond. The description differs in other polymerization stages. Evidently each author ascribes

References pp. 544–545

different values to the contributions of various partial processes in the overall polymerization rate. Probably it is quite impossible to apply the stationary state theorem from the very start of the polymerization. In a more exact description, the non-stationary conditions must be respected in the stage when growing radicals separate from the liquid phase [19].

1.3. Ionic polymerizations

Active centres of ionic polymerizations do not usually decay by mutual collisions as the radical centres. The stationary state, when it exists at all, results from quite different causes, mostly specific to the given system. Therefore the kinetics of ionic polymerizations is more complicated and its analysis more difficult. The concentration of centres cannot usually be calculated. On the other hand, ionic systems with rapid initiation give rise to the kinetically very simple living polymerizations (see Chap. 5, Sect. 8.1).

1.3.1 LIVING POLYMERIZATIONS WITH SLOW INITIATION

When living centres are slowly generated, the polymerization accelerates with time. "Older" centres have time to grow to larger dimensions than the fresh centres. Several authors [22–24] have paid attention to kinetic analysis of living polymerizations with slow initiation, the most recent of these studies being that of Pepper [25].

Let us consider the simplest case where initiation and propagation are of the same order

$$I + M \xrightarrow{k_i} M° \tag{80}$$

$$M_n° + M \xrightarrow{k_p} M_{n+1}° \tag{81}$$

Termination does not occur in living systems, therefore $[M°] = [I]_0 - [I]$, and the rate of monomer consumption is described by the equations

$$-\frac{d[M]}{dt} = k_i[I][M] + k_p[M°][M] \tag{82}$$

$$-\frac{d \ln[M]}{dt} = (k_i - k_p)[I] + k_p[I]_0 \tag{83}$$

The polymerization will accelerate with time until all initiator molecules are transformed to living centres or until complete monomer consumption. The acceleration is described by a second-order linear differential equation [22, 23]†

$$\frac{d^2 \ln[M]}{dt^2} + k_i[M]\frac{d \ln[M]}{dt} + k_i k_p [I]_0 [M] = 0 \tag{84}$$

When all initiator is consumed in the course of the polymerization, the reaction rate becomes dependent only on the monomer concentration, and $\ln[M]$ will be directly proportional to t. Under these conditions, k_p can be calculated from eqn. (84) because

$$[M^\circ] \to [I]_0 \tag{85}$$

$$\frac{d^2 \ln[M]}{dt^2} \to 0 \tag{86}$$

and

$$\left(-\frac{d \ln[M]}{dt}\right)_{max} \to k_p[I]_0 \tag{87}$$

---

† From eqn. (80)

$$-\frac{d[I]}{dt} = k_i[I][M]. \tag{80a}$$

since, occording to eqn. (83)

$$-[I] = (k_i - k_p)^{-1}\left(k_p[I]_0 + \frac{d \ln[M]}{dt}\right), \tag{83a}$$

$$-\frac{d[I]}{dt}(k_i - k_p)^{-1}\left(\frac{d^2 \ln[M]}{dt^2}\right) \tag{83b}$$

by substitution from eqn. (83a) and (83b) into eqn. (80a) we obtain

$$(k_i - k_p)^{-1}\frac{d^2 \ln[M]}{dt^2} + k_i(k_i - k_p)^{-1}\left(k_p[I]_0 + \frac{d \ln[M]}{dt}\right)[M] = 0$$

yielding by rearrangement eqn. (84).

References pp. 544–545

In addition, k_i can be derived. Beste and Hall expressed the rate of initiator consumption according to eqn. (80) by the equation [22]

$$-\frac{d[I]}{dt} = k_i[I][M]$$

$$d\ln[I] = -k_i[M]\,dt \tag{88}$$

$$\ln\frac{[I]}{[I]_0} = -k_i\int_0^t [M]\,dt$$

Rearrangement of eqn. (83) and substitution from eqn. (88) yields the expression

$$\ln\left(\frac{d\ln[M]}{[I]_0\,dt} + k_p\right) = \ln(k_p - k_i) - k_i\int_0^t [M]\,dt \tag{89}$$

For selected times, the value of the integral on the right-hand side of eqn. (89) can be derived from conversion curves, and the values found can be plotted against the corresponding values of the left-hand side of eqn. (89) calculated by means of trial k_p values. The obtained dependence will be linear only for approximately correct k_p values, and its slope will be equal to the required k_i. The slope, k_i, should be consistent with the intercept on the ordinate which is a logarithmic function of the difference $k_p - k_i$, and this might be used for checking the correctness of the procedure. Trial selection of k_p is unnecessary when the value of $-d\ln[M]/dt$ can be determined after complete transformation of the initiator into living centres.

The transformation of initiator into living centres, referred to monomer conversion, is an important kinetic criterion of the type of reaction discussed. Litt has shown that, at the moment of complete monomer transformation to polymer, the fraction of consumed initiator, $\varkappa_i = ([I]_0 - [I])/[I]_0$, is given by [24]

$$\frac{[M]_0}{[I]_0} = -\frac{k_p}{k_i}[\ln - \varkappa_i) + \varkappa_i] + \varkappa_i \tag{90}$$

and for any conversion [25]

$$\frac{[M]_0 - [M]}{[I]_0} = -\frac{k_p}{k_i}[\ln(1 - \varkappa_{i,\varkappa} + \varkappa_{i,\varkappa}] + \varkappa_{i,\varkappa} \tag{91}$$

where $\varkappa_{i,\varkappa}$ is the conversion of initiator into growth centres at monomer conversion $\varkappa$.

In Table 3, data on initiator fractions transformed into living centres at various conversions and various k_i/k_p and $[M]_0/[I]_0$ values are presented. Several useful conclusions follow from this table.

TABLE 3

Transformation of initiator into living centres [25].

$(k_i[M]_0)/(k_p[I]_0)$	$\varkappa_i$ at		
	$\varkappa = 0.25$	$\varkappa = 0.5$	$\varkappa = 1.0$
0.5	0.42	0.55	0.70
1.0	0.55	0.70	0.84
2.0	0.70	0.84	0.95
4.0	0.84	0.95	0.99
8.0	0.95	0.99	0.999

(a) When $k_i[M]_0/(k_p[I]_0) \leq 1$, the initiator is not completely consumed. The polymerization will be non-stationary throughout. The rate of the imaginary reaction with constant monomer concentration would increase continuously.

(b) At ratio values > 5, the initiator will be completely consumed. Experimental data arranged in a graph for reactions of the first order with respect to monomer will lie on a straight line even for low conversions. The kinetic behaviour of such a system will approach that of an ideal (rapidly initiated) living polymerization.

(c) For ratio values between the above limits, it should be possible to derive k_i and k_p by kinetic analysis using eqn. (89).

(d) With very rapid polymerizations, $[M]_0/[I]_0$ may be $> 10^4$. Ideal living polymerization will be sufficiently approximated at $k_i/k_p > 10^{-4}$. Thus the rate constant of initiation may be smaller by several orders of magnitude compared with k_p.

In more complicated situations, the order of the initiation rate with respect to monomer concentration is higher than the order of propagation [25]. In this case, with decreasing monomer content, the rate of initiation decreases relatively more rapidly than that of propagation. The drop may be so sharp that the number of living centres will stop increasing. The first-order graph for monomer conversion will approach a straight line, but its slope will not be a simple function of $k_p[I]_0$. Cases of this type have actually been observed.

1.3.2 LIVING EQUILIBRIUM COPOLYMERIZATION

The theoretical derivation of equilibrium comonomer concentrations, of molecular mass distribution in the copolymers, of the mean composition of n-mers as a function of n, and of the placement probability of a certain monomer as the mth segment in an n-mer from known data on the initial state and the values of equilibrium constants was published by Szwarc and Perrin [26].

(a) *Basic data*

On the initiating particles IM_1° and IM_2°, copolymer chains grow from the monomers M_1 and M_2. IM_1° and IM_2° cannot undergo depolymerization and termination and transfer do not occur in the system. The propagation constant, k_p, depends only on the kind of monomer and of the growing end of the macromolecule (irrespective of its length). The depropagation constant is determined by the type of the last two monomers

$$I \sim\sim M_1^\circ + M_1 \underset{}{\overset{K_{11}}{\rightleftharpoons}} I \sim\sim M_1 M_1^\circ$$

$$I \sim\sim M_1^\circ + M_2 \underset{}{\overset{K_{12}}{\rightleftharpoons}} I \sim\sim M_1 M_2^\circ$$

$$I \sim\sim M_2^\circ + M_1 \underset{}{\overset{K_{21}}{\rightleftharpoons}} I \sim\sim M_2 M_1^\circ$$

$$I \sim\sim M_2^\circ + M_2 \underset{}{\overset{K_{22}}{\rightleftharpoons}} I \sim\sim M_2 M_2^\circ \qquad (92)$$

After sufficient time, the system reaches an equilibrium where the monomer concentrations remain constant at $M_{1,\infty}$ and $M_{2,\infty}$. The initial concentrations, $IM_{1,0}$ and $IM_{2,0}$, are selected at will for each kind of system, but the corresponding equilibrium concentrations $IM_{1,\infty}$ and $IM_{2,\infty}$ are determined by the pertinent equilibrium constants and by $M_{1,0}$ and $M_{2,0}$ i. e. by the initial monomer concentrations, including those monomeric units which have been incorporated in $IM_{1,0}^\circ$ and $IM_{2,0}^\circ$.

The final state of the system is given by four unknowns: the equilibrium concentrations of the monomers, $M_{1,\infty}$ and $M_{2,\infty}$, and by the equilibrium concentrations of the initiating particles IM_1° and IM_2°. All these unknowns can be calculated from the known concentrations $IM_{1,0}^\circ$, $IM_{2,0}^\circ$ and $M_{1,0}$, $M_{2,0}$.

(b) *Determination of the equilibrium concentrations*

Let $M_{1,n}^\circ$, $M_{2,n}^\circ$ designate the equilibrium concentrations of n-mers terminated by M_1° or M_2°, respectively, irrespective of the remaining chain composition; $IM_{1,n}$ and $IM_{2,n}$ shall designate the equilibrium concentrations of n-mers starting with IM_1 or IM_2, again irrespective of the order and kind of monomer in the rest of the chains. The equilibrium concentrations $IM_{1,\infty}^\circ$ and $IM_{2,\infty}^\circ$ are evidently equal to $M_{1,1}^\circ$ and $M_{2,1}^\circ$, therefore

$$\sum_1^\infty IM_{1,n} = IM_{1,0}^\circ \quad \text{and} \quad \sum_1^\infty IM_{2,n} = IM_{2,0}^\circ \tag{93}$$

$$\sum_1^\infty M_{1,n}^\circ = M_{1,\infty}^\circ \quad \text{and} \quad \sum_1^\infty M_{2,n}^\circ = M_{2,\infty}^\circ \tag{94}$$

After equilibrium has been reached, eqn. (92) yields

$$M_{1,2}^\circ = K_{11} M_{1,\infty} M_{1,1}^\circ + K_{21} M_{1,\infty} M_{2,1}^\circ \tag{95}$$

and

$$M_{2,2}^\circ = K_{12} M_{2,\infty} M_{1,1}^\circ + K_{22} M_{2,\infty} M_{2,1}^\circ \tag{96}$$

or more generally

$$M_{1,n+1}^\circ = K_{11} M_{1,\infty} M_{1,n}^\circ + K_{21} M_{1,\infty} M_{2,n}^\circ \tag{97}$$

$$M_{2,n+1}^\circ = K_{12} M_{2,\infty} M_{2,n}^\circ + K_{22} M_{2,\infty} M_{2,n}^\circ \tag{98}$$

The two equations can be written in the matrix notation

$$\begin{pmatrix} M_{1,n+1}^\circ \\ M_{2,n+1}^\circ \end{pmatrix} = \begin{pmatrix} K_{11} M_{1,\infty} & K_{21} M_{1,\infty} \\ K_{12} M_{2,\infty} & K_{22} M_{2,\infty} \end{pmatrix} \begin{pmatrix} M_{1,n}^\circ \\ M_{2,n}^\circ \end{pmatrix} \tag{99}$$

Designating the first matrix on the right hand side of eqn. (99) as A, the notation is formally simplified to

$$\begin{pmatrix} M_{1,n}^\circ \\ M_{2,n}^\circ \end{pmatrix} = A^{n-1} \begin{pmatrix} M_{1,1}^\circ \\ M_{2,1}^\circ \end{pmatrix} = A^{n-1} \begin{pmatrix} IM_{1,\infty}^\circ \\ IM_{2,\infty}^\circ \end{pmatrix} \tag{100}$$

The equilibrium concentrations of n-mers ending in M_1 ($M_{1,n}^\circ$) or M_2 ($M_{2,n}^\circ$) are given by $M_{1,\infty}$ and $M_{2,\infty}$. $M_{1,1}^\circ = IM_{1,\infty}^\circ$ and $M_{2,1}^\circ = IM_{2,\infty}^\circ$, therefore

$$\sum_1^\infty \begin{pmatrix} M_{1,n}^\circ \\ M_{2,n}^\circ \end{pmatrix} = \begin{pmatrix} M_{1,\infty}^\circ \\ M_{2,\infty}^\circ \end{pmatrix} = (1-A)^{-1} \begin{pmatrix} M_{1,1}^\circ \\ M_{2,1}^\circ \end{pmatrix} \tag{101}$$

It holds that $M_{1,\infty}^\circ + M_{2,\infty}^\circ = IM_{1,0}^\circ + IM_{2,0}^\circ$, but $M_{1,\infty}^\circ \neq IM_{1,0}^\circ$ and $M_{2,\infty}^\circ \neq IM_{2,0}^\circ$.

References pp. 544–545

In a further step, the authors derived the matrix of conditional probabilities, transforming the placement of units M_1 and M_2 in position m into their placement in position $m + 1$ in a living n-mer

$$N_{n,m} = \begin{pmatrix} r_{n,m} & 1 - s_{n,m} \\ 1 - r_{n,m} & s_{n,m} \end{pmatrix}$$

In the original communication, the matrix N has been derived in detail. In this paragraph I merely wish to demonstrate the reasoning of the authors rather than to present detailed instructions for the calculation (a system suitable for such a calculation is being sought at present). The derivation is so compact that hardly anything can added to it, and merely copying it does not seem to be worthwhile. Therefore the derivation of matrix N has been omitted[†].

By means of the matrix N, eqn. (100) can be transformed to

$$\begin{pmatrix} M^\circ_{1,n} \\ M^\circ_{2,n} \end{pmatrix} = \prod_{m=1}^{n-1} N_{n,m} \begin{pmatrix} IM_{1,n} \\ IM_{2,n} \end{pmatrix} \qquad (102)$$

[†] $r_{n,m}/(1 - r_{n,m})$ is the ratio of the conditional probabilies of finding M_1 in position m in an n-mer, followed by M_1 or M_2:

$$\frac{r_{n,m}}{1 - r_{n,m}} = \frac{(1\ \ 1)\, A^{n-m-1} \begin{pmatrix} K_{11} M_{1,\infty} \\ 0 \end{pmatrix}}{(1\ \ 1)\, A^{n-m-1} \begin{pmatrix} 0 \\ K_{12} M_{2,\infty} \end{pmatrix}} ;$$

$$r_{n,m} = \frac{(1\ \ 1)\, A^{n-m-1} \begin{pmatrix} K_{11} M_{1,\infty} \\ 0 \end{pmatrix}}{(1\ \ 1)\, A^{n-m-1} \begin{pmatrix} K_{11} M_{1,\infty} \\ K_{12} M_{2,\infty} \end{pmatrix}}$$

and

$$s_{n,m} = \frac{(1\ \ 1)\, A^{n-m-1} \begin{pmatrix} 0 \\ K_{22} M_{2,\infty} \end{pmatrix}}{(1\ \ 1)\, A^{n-m-1} \begin{pmatrix} K_{21} M_{1,\infty} \\ K_{22} M_{2,\infty} \end{pmatrix}}$$

Again $IM_{1,n} \neq M^\circ_{1,n}$ and $IM_{2,n} \neq M^\circ_{2,n}$, although $IM_{1,n} + IM_{2,n} = M^\circ_{1,n} + M^\circ_{2,n}$. A product of non commuting matrices always corresponds to an order of decreasing indices, here m, from left to right. Matrices $N_{n,0}$ and $N_{n,n}$ are defined as unit matrices. Inversion of eqn. (102) yields

$$\begin{pmatrix} IM_{1,n} \\ IM_{2,n} \end{pmatrix} = \left(\prod_{m=1}^{n-1} N_{n,m} \right)^{-1} \begin{pmatrix} M^\circ_{1,n} \\ M^\circ_{2,n} \end{pmatrix} \tag{103}$$

Considering eqn. (100)

$$\begin{pmatrix} IM_{1,n} \\ IM_{2,n} \end{pmatrix} = \left(\prod_{m=1}^{n-1} N_{n,m} \right)^{-1} A^{n-1} \begin{pmatrix} M^\circ_{1,1} \\ M^\circ_{2,1} \end{pmatrix} \tag{104}$$

As the sum

$$\sum_{n=1}^{\infty} \begin{pmatrix} IM_{1,n} \\ IM_{2,n} \end{pmatrix} = \begin{pmatrix} IM_{1,0} \\ IM_{2,0} \end{pmatrix} \tag{105}$$

we obtain

$$\sum_{n=1}^{\infty} \left(\prod_{m=1}^{n-1} N_{n,m} \right)^{-1} A^{n-1} \begin{pmatrix} IM^\circ_{1,\infty} \\ IM^\circ_{2,\infty} \end{pmatrix} = \begin{pmatrix} IM^\circ_{1,0} \\ IM^\circ_{2,0} \end{pmatrix} \tag{106}$$

Equation (106) solves the relations between the unknowns $M_{1,\infty}$, $M_{2,\infty}$ and $IM^\circ_{1,\infty}$, $IM^\circ_{2,\infty}$ and the known values of $IM^\circ_{1,0}$ and $IM^\circ_{2,0}$ by means of the relevant values of K.

By the other two equations, those four unknowns are derived by means of the initial monomer concentrations, $M_{1,0}$ and $M_{2,0}$. The amounts of the monomeric units M_1 and M_2 in the equilibrium n-mers $M_{1,n}$ or $M_{2,n}$ are

$$\begin{pmatrix} M_{1,n} \\ M_{2,n} \end{pmatrix} = \sum_{i=1}^{n} \left(\prod_{m=0}^{i-1} N_{n,m} \right) \begin{pmatrix} IM_{1,n} \\ IM_{2,n} \end{pmatrix} \tag{107}$$

For monomer balance, the relation

$$\begin{pmatrix} M_{1,0} - M_{1,\infty} \\ M_{2,0} - M_{2,\infty} \end{pmatrix} = \sum_{n=1}^{\infty} \begin{pmatrix} M_{1,n} \\ M_{2,n} \end{pmatrix}$$

must be obeyed; therefore, considering eqn. (103) and (107)

$$\begin{pmatrix} M_{1,0} - M_{1,\infty} \\ M_{2,0} - M_{2,\infty} \end{pmatrix} = \sum_{n=1}^{\infty} \sum_{i=1}^{n} \left(\prod_{m=i}^{n-1} N_{n,m} \right) \begin{pmatrix} IM_{1,\infty} \\ IM_{2,\infty} \end{pmatrix} \tag{108}$$

References pp. 544–545

Equations (106) and (108) enable those four unknowns ($M_{1,\infty}$, $M_{2,\infty}$, $IM_{1,\infty}^\circ$ and $IM_{2,\infty}^\circ$) to be calculated. When these values are known, it is further possible to determine the following.

(i) The distribution of n-mers.

$$\begin{pmatrix} x_{M_{1,n}} \\ x_{M_{2,n}} \end{pmatrix} = A^{n-1} \begin{pmatrix} IM_{1,\infty}^\circ \\ IM_{2,\infty}^\circ \end{pmatrix} \left(IM_{1,0}^\circ + IM_{2,0}^\circ \right)^{-1} \tag{109}$$

where x_{M_1} and x_{M_2} are the mole fractions of n-mers ending in M_1 and M_2, respectively. When x_n is the mole fraction of n-mers irrespective of their end units

$$x_n = (1 \quad 1) A^{n-1} \begin{pmatrix} IM_{1,0}^\circ \\ IM_{2,0}^\circ \end{pmatrix} \left(IM_1^\circ + IM_2^\circ \right)$$

In this way, any moment of the distribution can be calculated and mean values like $\overline{DP}_n$, $\overline{DP}_w$ etc. can be determined. $\overline{DP}_n$ is simply given by

$$\overline{DP}_n = \left(M_{1,0} + M_{2,0} - M_{1,\infty} - M_{2,\infty} \right) \left(IM_{1,0}^\circ + IM_{2,0}^\circ \right)^{-1}$$

(ii) The probabilities of finding M_1 or M_2 in position m in an n-mer. These probabilities, designated as $P_{M_{1,n,m}}$ and $P_{M_{2,n,m}}$ are

$$\begin{pmatrix} P_{M_{1,n,m}} \\ P_{M_{2,n,m}} \end{pmatrix} = \left(\prod_{i=0}^{m-1} N_{n,i} \right) \begin{pmatrix} IM_{1,n}/(IM_{1,n} + IM_{2,n}) \\ IM_{2,n}/(IM_{1,n} + IM_{2,n}) \end{pmatrix} \tag{110}$$

(iii) The mean composition of an n-mer. The mole fraction $f_{M_{1,n}}$ and $f_{M_{2,n}}$ in living n-mers are

$$\begin{pmatrix} f_{M_{1,n}} \\ f_{M_{2,n}} \end{pmatrix} = \sum_{i=1}^{n} \left(\prod_{m=0}^{i-1} N_{n,m} \right) \begin{pmatrix} IM_{1,n}/(IM_{1,n} + IM_{2,n}) \\ IM_{2,n}/(IM_{1,n} + IM_{2,n}) \end{pmatrix} \tag{111}$$

(iv) The frequency of various sequences of monomeric units in equilibrium polymers. The probability of finding a sequence of monomeric units, say $M_1M_2M_2$ starting at position m in an n-mer is given by the product of the placement probability of the first unit in this position and of the corresponding conditional probabilities. For $M_1M_2M_2$

$$P_{M_{1,n,m}}(1 - r_{n,m}) s_{n,m+1}$$

The total number of $M_1M_2M_2$ sequences in all polymers is

$$\sum_{n=3}^{\infty} \sum_{m=1}^{n-2} P_{1,n,m}(1 - r_{n,m})(s_{n,m+1}) A^{n-1} \begin{pmatrix} IM_{1,\infty}^\circ \\ IM_{2,\infty}^\circ \end{pmatrix} \tag{112}$$

and their populations are given by the value of eqn. (112) divided by the total number of triads

$$\sum_{n=3} (n-2) \begin{pmatrix} 1 & 1 \end{pmatrix} A^{n-1} \begin{pmatrix} IM^\circ_{1,\infty} \\ IM^\circ_{2,\infty} \end{pmatrix} \tag{113}$$

The authors of the presently described study have extended the possibilities of these calculations to analysis of systems with reversible initiation, and especially to the copolymerization of a larger number of monomers.

1.3.3 NON-STATIONARY POLYMERIZATION

Radical polymerizations are almost always considered as kinetically stationary. However, the stationarity conditions are not always fulfilled. Living polymerizations with rapid initiation are stationary, but the character of the medium should not significantly change during polymerization in order to prevent shifts in the equilibria between ion pairs and free ions. All other polymerizations are non-stationary even, to some extent, living polymerizations with slow initiation. It is usually very difficult to define initiation and termination rates so as to permit exact kinetic analysis. When the concentration of active centres cannot be directly determined, indirect methods must be applied, and sometimes even just a trial search for best agreement with experiment.

About thirty years ago, all cases of polymerization kinetics used to be solved as statinary reactions[†]. Hayes and Pepper [27] were the first to call attention to the non-stationary character of ionic polymerizations. They noticed the premature decay of styrene polymerization initiated by H_2SO_4 (see Fig. 8). This was a simple case of non-stationarity caused by the slow decay of rapidly generated active centres [27, 28]. They assumed that the polymerization proceeds according to a rather conventional scheme represented in simplified form (without transfer) by the reactions

$$I + M \xrightarrow{k_i} M^+$$
$$M_n^+ + M \xrightarrow{k_p} M_{n+1}^+$$
$$M_n^+ \xrightarrow{k_t} \text{inactive polymer}$$

When the initiation rate is high compared with other partial processes, all the initiator is transformed to active centres in the initial phases of poly-

---

[†] The stationary state principle is being misused to this day. The results cannot, of course, withstand due criticism.

References pp. 544–545

merization, therefore $[I]_0 = [M^+]_0$. The centres decay by termination; their concentration continually decreases. In the simple case discussed, this decay is monomolecular

$$\frac{d[M^+]}{dt} = -k_t[M^+] \tag{114}$$

$$[M^+] = [M^+]_0 \exp(-k_t t) = [I]_0 \exp(-k_t t) \tag{115}$$

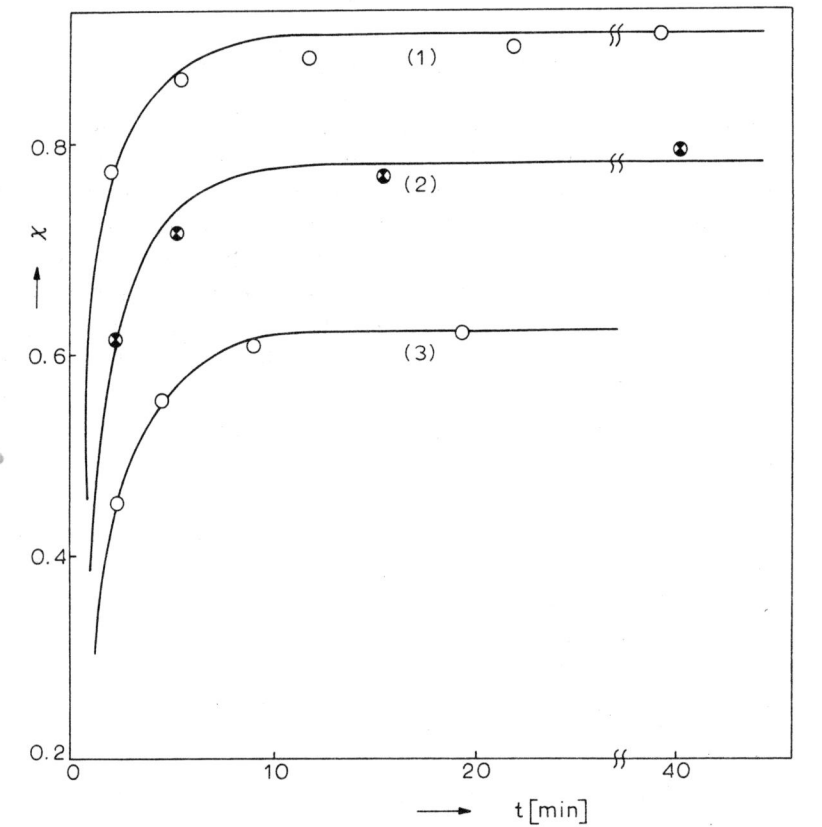

Fig. 8. Polymerization of styrene with H_2SO_4 in dichloroethane [29]. Temperature 298 K, $[M]_0$ = 0.35 mol dm^{-3}. $10^3[I]_0$ (mol dm^{-3}): (1) 3.4, (2) 1.7, (3) 0.81.

The corresponding rate equation for monomer consumption, obtained by substitution of eqn. (115) into the well-known expression (when chains of sufficient length are formed) is of the form

$$-\frac{d[M]}{dt} = k_p[M^+][M] = k_p[I]_0[M]\exp(-k_t t) \tag{116}$$

By its integration we obtain

$$\ln \frac{[M]_0}{[M]} = \frac{k_p}{k_t} [I]_0 [1 - \exp(-k_t t)] \tag{117}$$

For very long polymerization times

$$\ln \frac{[M]_0}{[M]_\infty} = \frac{k_p}{k_t} [I]_0 \tag{118}$$

For $k_t > 0$ and $[M]_\infty > 0$, and for kinetic reasons, the polymerization must die out prior to complete monomer consumption. Maximum conversion will be a function of k_p, k_t and $[I]_0$. It can be derived from eqn. (118) that

$$\varkappa_\infty = 1 - \frac{[M]_\infty}{[M]_0} = 1 - \exp\left[-\frac{k_p}{k_t} [I]_0\right] \tag{119}$$

In a qualitative sense, eqn. (119) provides a satisfactory description of the results of direct measurements (Fig. 8). The authors extended their analysis to situations with transfer, and to calculations of the degree of polymerization of the product.

The indicated procedure for analyzing the kinetics of non-stationary polymerizations is evidently unacceptably simplified, and may serve only as a very rough model for the analysis of actual situations. An exact general procedure should include thermodynamic principles (floor and ceiling polymerization temperatures), detailed concepts on the generation and decay of active centres and on transfer (especially degradative).

Most of the difficulties mentioned would be solved if the instantaneous concentrations of active centres could be determined. The sensitivity of present instrumentation (NMR, ESR, spectrophotometry, etc.) roughly just reaches the applicability limit; in some cases the centres can already be couted at present. In the near future it will undoubtedly be possible to analyze individually cases of non-stationary polymerizations with sufficient accuracy.

References pp. 544–545

1.3.4 EXAMPLE. PROPAGATION OF HETEROCYCLES. PERIODIC CHANGES OF GROWING CENTRE REACTIVITY

The living character of some cationic polymerizations greatly reduces the number of overlapping effects. Even transitions between forms of growth centres of various activity become accessible to reliable measurements. The best studied reaction from this point of view is the polymerization of tetrahydrofuran by centres with stable counter-ions.

In tetrahydrofuran polymerization, chain propagation occurs at the ion pairs

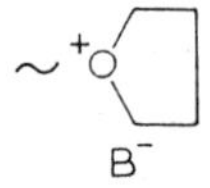

and ester groups $\sim O(CH_2)_4B$, where B^- is a counter-ion (e. g. FSO_3^-). The amounts of free ions and ion aggregates present are negligible. Between the two types of centre an equilibrium is established

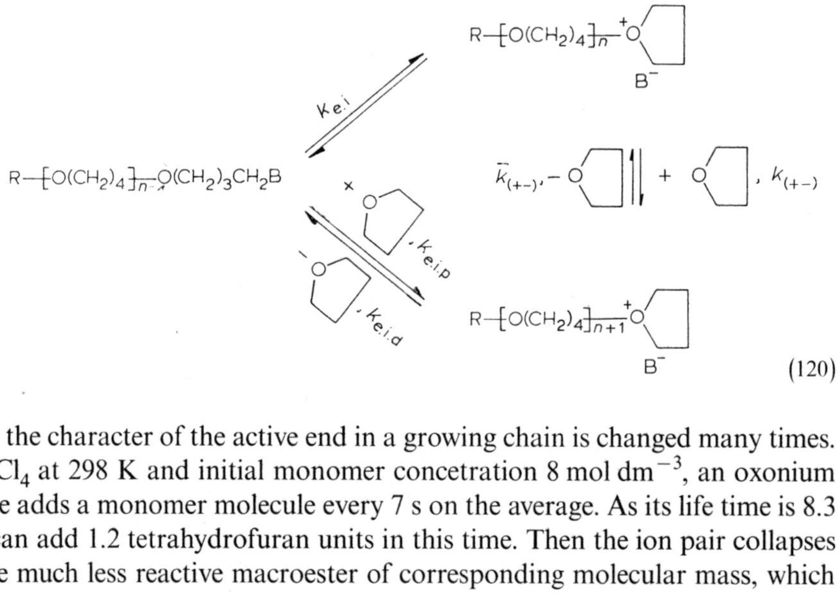

(120)

Thus the character of the active end in a growing chain is changed many times. In CCl_4 at 298 K and initial monomer concetration 8 mol dm^{-3}, an oxonium centre adds a monomer molecule every 7 s on the average. As its life time is 8.3 s, it can add 1.2 tetrahydrofuran units in this time. Then the ion pair collapses to the much less reactive macroester of corresponding molecular mass, which exists for 125 s. After that, the cycle is repeated. Transformation of the macroester to an ion pair by a bimolecular reaction with the rate constant $k_{e,i.p}$ (propagation on ester) only occurs once per 32 monomolecular conversions (with the constant $k_{e,i}$). Thus it can be neglected in practice [30].

It appears that the mean lifetime of various types of centre depends on solvent, temperature and conversion. Propagation predominantly occurs at oxonium ion pairs, mainly after the start of polymerization. Their concentration rapidly decreases with growing conversion and temperature, while the amount of ester increases [31].

Details in our picture of the heterocycles propagation are obscured by conflicting reports. This proves that, in spite of the many simplifications, we still have to deal with very complicated and sensitive reactions. Even related initiators can cause complications. In a medium with the counter-ions FSO_3^-, $CF_3SO_3^-$ aggregation of centres does not occur [30], but it has been observed with SO_3H^- [32].

In polymerizations of cyclic ethers with rings smaller than tetrahydrofuran, ring strain affects the position of the equilibria discussed above. Small cycles (oxiranes, oxetanes) cannot be formed from esters with ion pair formation, and the equilibrium is shifted in favour of the macroesters [33, 34].

1.4 Coordination polymerization

A quarter of a century after the discovery of coordination polymerizations, in spite of tens of highly qualified and hundreds of supplementary studies, and millions of tons of polyalkenes produced each year, we still must confess that we possess only very general ideas about the basic kinetic features of these reactions.

1.4.1 KINETIC MODEL OF ZIEGLER–NATTA POLYMERIZATIONS

A model including all the presently known processes occurring in Ziegler–Natta coordination polymerizations has been published by Böhm [35]. It is a special case of the Rideal mechanism [36], and it can be applied to both

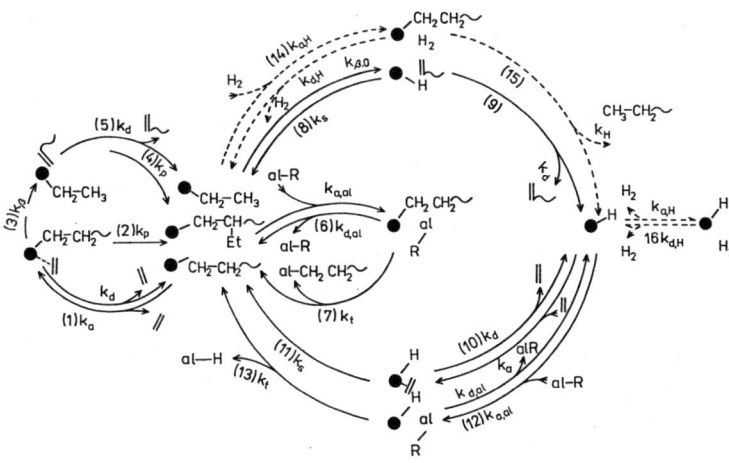

References pp. 544–545

homogeneous and heterogeneous processes [see scheme (121)]. It refers particularly to the polymerization of ethylene, but it is also valid for other 1-alkenes. It consists of two cycles. The first, composed of reactions (1)–(5), describes the insertion of ethylene into the metal–polymer bond. The remaining reactions, forming the second cycle, are transfer processes, proceeding spontaneously or with transfer agents (reactions with hydrogen are indicated by broken lines). The author provides the following explanation to the scheme:

(a) Prior to insertion, the monomer molecule is reversibly coordinated to the active centre by reaction (1).

(b) According to Arlman and Cossee, propagation (2) corresponds to the insertion with a four-centred transition complex [37].

(c) As an alternative to (2), a reaction with a six-centred transition complex (3) may occur, in which H is eliminated from the β-position in the chain and transferred to the monomer. A vinyl group is generated on the polymer chain, continuing to form a complex with the active centre. In the following step, it can be inserted into the metal–polymer bond according to reaction (4), producing a branched polymer, or it can be desorbed by reaction (5).

(d) The ligand of the organoaluminium compound is bound to the active centre vacancy according to reaction (6). Ligand exchange occurs in the generated complex [38, 39]. Complex decomposition liberates the organoaluminium component or, after ligand exchange, the polymer is bound to the dissociating organometal according to reaction (7) with subsequent transfer.

(e) Chain growth is spontaneously terminated (8), yielding a metal–H bond and vinyl groups at the chain end [40]. As in (c), this group remains bound to the centre but it can be desorbed according to reactions (9) or (5).

(f) The active centre with the metal–H bond forms a complex with ethylene by reaction (10), and by subsequent insertion (11) it is transformed to a metal–carbon centre.

(g) Another variant of the transformation of a metal–H to a metal–C bond is described by reactions (12) and (13). In the first step a complex is formed which can liberate either the original organoaluminium compound or aluminium hydride after ligand exchange.

(h) In the presence of molecular hydrogen, reactions (14)–(16) become important, leading either to transfer (15) or to inhibition (16).

A mathematical description of several mutually connected chemical reactions usually requires some simplification. According to Böhm, all reactive intermediates are assumed to exist in a stationary state, and all rate constants are assumed to be independent of the length of the growing macromolecules. As the derived expressions should also describe heterogeneous polymerizations, the author used the numbers of particles in unit volume of reacting medium instead of concentrations. Some expressions can be simplified in this way. The rate constant are defined in scheme (121). All reactions except bimolecular adsorption are monomolecular.

The symbols used in the equations, correspond to the following quantities:

$n°$ = number of all centres; $n_0°$ = number of •—CH_2—CH_2~;

$n_a°$ = number of •$\diagup\!\!\!\diagdown^{CH_2-CH_2\sim}_{/\!/}$; $n_b°$ = number of •$\diagup\!\!\!\diagdown^{\sim}_{CH_2-CH_3}$;

$n_c°$ = number of •$\diagup\!\!\!\diagdown^{\sim}_{H}$; $n_d°$ = number of •—H; $n_{d,H}°$ = •$\diagdown_{H2}^{H}$;

$n_e°$ = number of •$\diagup\!\!\!\diagdown^{H}_{/\!/}$; $n_H°$ = number of •$\diagup\!\!\!\diagdown^{CH_2-CH_2\sim}_{H_2}$;

$n_{al,1}°$ = number of •$\diagup\!\!\!\diagdown^{CH_2-CH_2\sim}_{\diagdown\text{al}}_{R}$; $n_{al,2}°$ = number of •$\diagup\!\!\!\diagdown^{H}_{\diagdown\text{al}}_{R}$;

n_k = number of moles of catalyst; m_P = number of macromolecules; n_M = = number of polymerized monomer molecules; n = number of macromolecules dissociated from active centres ($= 1/2$ of the end groups); [al] = = concentration of monomeric organoaluminium compound at the surface of catalyst; and $/\!/$ = monomer.

The total number of active centres is given by the sum

$$n° = n_0° + n_a° + n_b° + n_c° + n_d° + n_e° + n_{al,1}° + n_{al,2}° \qquad (122)$$

The generation and decay of a discrete type of centre is equally rapid, thus

$$\frac{dn_a°}{dt} = k_a[M]_0 n_0° - (k_p + k_d + k_\beta)n_a° = 0 \qquad (123)$$

$$\frac{dn_b°}{dt} = k_\beta n_a° - (k_p + k_d)n_b° = 0 \qquad (124)$$

References pp. 544–545

$$\frac{dn_c^\circ}{dt} = k_{\beta,0} n_0^\circ - (k_s + k_d) n_c^\circ = 0 \tag{125}$$

$$\frac{dn_d^\circ}{dt} = k_d n_c^\circ + k_d n_e^\circ + k_{d,al} n_{al,2}^\circ - (k_a [M]_0 + k_{a,al} [al]_0) n_d^\circ = 0 \tag{126}$$

$$\frac{dn_e^\circ}{dt} = k_a [M]_0 n_d^\circ - (k_s + k_d) n_e^\circ = 0 \tag{127}$$

$$\frac{dn_{al,1}^\circ}{dt} = k_{d,al} [al]_0 n_0^\circ - (k_{d,al} + k_t) n_{al,1}^\circ = 0 \tag{128}$$

$$\frac{dn_{al,2}^\circ}{dt} = k_{a,al} [al]_0 n_d^\circ - (k_{d,al} + k_t) n_{al,2}^\circ = 0 \tag{129}$$

The amounts of various complexes are described by the following equations

$$n_a^\circ = n_0^\circ \left(\frac{k_a [M]_0}{k_p + k_d + k_\beta} \right) \tag{130}$$

$$n_b^\circ = n_0^\circ \left[\frac{k_\beta k_d [M]_0}{(k_p + k_d)(k_p + k_d + k_\beta)} \right] \tag{131}$$

$$n_c^\circ = n_0^\circ \left(\frac{k_{\beta,0}}{k_s + k_d} \right) \tag{132}$$

$$n_{al,1}^\circ = n_0^\circ \left(\frac{k_{a,al} [al]_0}{k_{d,al} + k_t} \right) \tag{133}$$

where

$$n_0^\circ = \frac{n^\circ}{a + b + c} \tag{134}$$

$$a = \frac{k_p + k_d + k_a[M]_0}{k_p + k_d} \qquad (135)$$

$$b = \frac{k_{\beta,0}}{k_s + k_d} = \left\{ 1 + \frac{\dfrac{k_d + k_d k_a[M]_0}{k_s + k_d} + \dfrac{k_d k_{a,al}[al]_0}{k_{d,al} + k_t}}{\dfrac{k_s k_a[M]_0}{k_s + k_d} \dfrac{k_t k_{a,al}[al]_0}{k_{d,al} + k_t}} \right\} \qquad (136)$$

$$c = \frac{k_{a,al}[al]_0}{k_t + k_{d,al}} \qquad (137)$$

The rate of monomer consumption is then expressed by

$$-\frac{dn_M}{dt} = k_p(n_a^\circ + n_b^\circ) + k_s n_e^\circ \qquad (138)$$

In situations where $k_p(n_a^\circ + n_b^\circ) \gg k_s n_e^\circ$, the second term on the right-hand side of eqn. (138) and (134) can be neglected. Substitution from eqns. (130) and (134) into eqn. (138) yields the equation

$$-\frac{dn_M}{dt} = \frac{dm_p}{Mdt} = \frac{k_p k_a[M]_0}{k_p + k_d} \frac{n^\circ}{a + b + c} \qquad (139)$$

M is the molecular mass of the monomer.

Dividing eqn. (139) by the total amount of catalyst, n_K, we obtain an expression for the polymerization rate referred to unit surface of catalyst[†]

$$(v_{pol})_s = \frac{dm_p}{Mn_K dt} = \frac{k_p k_a[M]_0}{k_p + k_d + k_a[M]_0} \left(1 + \frac{b}{a} + \frac{c}{a}\right)^{-1} \frac{n^\circ}{n_K} \qquad (140)$$

The term $[1 + (b/a) + (c/a)]^{-1}$ in eqn. (140) describes the degree of inhibition by the organoaluminium compound. The special case of eqn. (140) has been derived previously [41]. The agreement may be regarded as confirming the correctness of the procedure.

[†] The number of moles of catalyst is proportional to the active surface of the catalyst.

The presence of hydrogen, controlling the mean length of macromolecules, amplifies the kinetic scheme by reactions (14)–(16). Equation (122) assumes the form

$$n° = n°_0 + n°_a + n°_b + n°_c + n°_d + n°_e + n°_H + n°_{d,H} + n°_{al,1} + n°_{al,2} \quad (141)$$

$$\frac{dn°_d}{dt} = k_d n°_c + k_d n°_e + k_{d,al} n°_{al,2} + k_H n°_H + k_{d,H} n°_{d,H} -$$
$$- (k_a[M]_0 + k_{a,al}[al]_0 + k_{a,H}[H_2]_0) = 0 \quad (142)$$

The stationary amount of complexes with hydrogen is expressed by the relations

$$\frac{dn°_H}{dt} = k_{a,H}[H_2]_0 n°_0 - (k_{d,H} + k_H) n°_H = 0 \quad (143)$$

$$\frac{dn°_{d,H}}{dt} = k_{a,H}[H_2]_0 n°_d - k_{d,H} n°_{d,H} = 0 \quad (144)$$

In the presence of hydrogen, eqn. (134) is transformed to the form

$$n°_0 = \frac{n°}{a + b + c + d} \quad (145)$$

where

$$d = \frac{k_{a,H}[H_2]_0}{k_{d,H} + k_H}\left\{1 + \left[k_H + \frac{k_H k_a[M]_0}{k_s + k_d} + \frac{k_d k_{\beta,0}(k_{d,H} + k_H)}{k_{d,H}(k_s + k_d)} + \right.\right.$$
$$\left.\left. + \frac{k_H k_{a,H}[H_2]_0}{k_{d,H}} + \frac{k_H k_{a,al}[al]_0}{k_{d,al} + k_t}\right]\left[\frac{k_s k_a[M]_0}{k_s + k_d} + \frac{k_t k_{a,al}[al]_0}{k_{d,al} + k_t}\right]^{-1}\right\} \quad (146)$$

The polymerization rate in the presence of hydrogen is given by eqn. (138) which, after substitution from eqns. (143)–(145), can be written as

$$(v_{pol})_s = \frac{k_p k_a[M]_0}{k_p + k_d + k_a[M]_0}\left(1 + \frac{a}{b} + \frac{c}{a} + \frac{d}{a}\right)^{-1}\frac{n°}{n_K} \quad (147)$$

The symbol d in eqn. (147) describes the degree of inhibition by hydrogen.

The number average degree of polymerization of a polyalkene is indirectly proportional to the total number of macromolecules. Thus

$$\bar{P} = \frac{m_{\text{P}}}{M(n° + n)} \tag{148}$$

In the initial polymerization phases transfer may be regarded as negligible, therefore $n = 0$ and

$$\lim_{t \to 0} \bar{P} = \bar{P}_0 \approx \frac{m_{\text{P}}}{M n°} \tag{149}$$

Since, from eqn. (139)

$$(M n°)^{-1} \int_0^{m_{\text{P}}} dm_{\text{P}} = (k_{\text{p}} k_{\text{a}} [\text{M}]_0)(k_{\text{p}} + k_{\text{d}})^{-1} (a + b + c)^{-1} \int_0^t dt \tag{150}$$

integration and substitution from eqn. (135) yields

$$\bar{P}_0 = \frac{m_{\text{P}}}{M n°} = \frac{k_{\text{p}} k_{\text{a}} [\text{M}]_0}{k_{\text{p}} + k_{\text{d}} + k_{\text{a}}[\text{M}]_0} \left(1 + \frac{b}{a} + \frac{c}{a}\right)^{-1} t \tag{151}$$

In the first polymerization phases the degree of polymerization is independent of the amount of catalyst and is therefore proportional to the polymerization time t.

After long polymerization times, on the other hand, the number of macromolecules affected by transfer is such that the number of chains with a bound active centre is negligible and

$$\lim_{t \to \infty} \bar{P} = \bar{P}_\infty \approx \frac{m_{\text{P}}}{M n} \tag{152}$$

For calculating $\bar{P}_\infty$, we use the rearranged equation (150)

$$M^{-1} \int_0^{m_{\text{P}}} dm_{\text{P}} = (k_{\text{p}} k_{\text{a}} [\text{M}]_0)(k_{\text{p}} + k_{\text{d}})^{-1} (a + b + c)^{-1} \int_0^t n°_{(t)} dt \tag{153}$$

According to scheme (121), the growing chain is separated from the active centre by thereee processes. Spontaneous inactivation is neglected; it must be

References pp. 544–545

slow compared with transfer. The change in the number of macromolecules with time is then described by the equation

$$\frac{dn}{dt} = k_p n_b^\circ + k_d n_c^\circ + k_t n_{al,1}^\circ \tag{154}$$

Substitution in eqn. (154) for n_b°, n_c° and $n_{al,1}^\circ$ from eqns. (131)–(133), and for n_0° from eqn. (134) leads to eqn. (155) which can be integrated

$$\int_0^n dn = n = \left[\frac{k_d k_\beta k_a [M]_0}{(k_p + k_d)(k_p + k_d + k_\beta)} + \frac{k_d k_{\beta,0}}{k_s + k_d} + \frac{k_t k_{a,al}[al]_0}{k_{d,al} + k_t} \right] (a + b + c)^{-1} \cdot \int_0^t n_{(t)}^\circ dt \tag{155}$$

Relations (152), (153) and (155) finally yield

$$\frac{1}{\bar{P}_\infty} = A + \frac{B}{[M]_0} + \frac{C[al]_0}{[M]_0} \tag{156}$$

where

$$A = (k_d k_\beta) k_p^{-1} (k_p + k_d + k_\beta)^{-1}$$

$$B = k_d k_{\beta,0}(k_p + k_d)(k_d + k_s)^{-1} k_p^{-1} k_a^{-1}$$

$$C = k_t k_{a,al}(k_p + k_d)(k_{d,al} + k_t)^{-1} k_p^{-1} k_a^{-1}$$

From eqn. (148) an important relation can be derived between the number average molecular mass of the product and catalyst yield, Θ, equal to m_P/n_K ($\bar{M}_n = \bar{P}M$)

$$\frac{\Theta}{M_n} = \frac{n^\circ}{n_K} + \frac{n}{m_P} \Theta \tag{157}$$

After substitution for m_P and n from eqn. (152) and rearrangement, we obtain the equation

$$\frac{\Theta}{M_n} = \frac{n^\circ}{n_K} + \frac{\Theta}{M_{n,\infty}} \tag{158}$$

from which the number n° of active centres can be calculated from the known values of molecular mass and Θ. A similar possibility of determining the number of centres has previously been pointed out by Natta and Pasquon [42].

In the presence of hydrogen, eqn. (154) is extended by the member involving transfer to H_2 according to reaction (15) of scheme (121)

$$\frac{dn}{dt} = k_d n_b^\circ + k_d n_c^\circ + k_t n^\circ_{al,1} + k_H n_H^\circ \tag{159}$$

For $\bar{P}_\infty$ it then holds that

$$\frac{1}{\bar{P}_\infty} = A + \frac{B}{[M]_0} + C \frac{[al]_0}{[M]_0} + D \frac{[H_2]_0}{[M]_0} \tag{160}$$

where

$$D = k_H K_{a,H}(k_p + k_d)(k_{d,H} + k_H)^{-1} k_p^{-1} k_a^{-1} \tag{161}$$

Böhm's scheme summarizes the results of authors whose work on coordination polymerizations is of basic importance [43–47]. The author himself warns that the derived equations are only valid when three conditions are fulfilled.

(a) The rate constants must be independent of the length of the growing chain;

(b) the concentrations of the reacting substrates at the surface of the catalyst or in homogeneous medium are constant and independent of time in the integrated intervals; and

(c) only the number of active centers, n°, depends on time, decreasing slowly.

Homogeneous coordination polymerizations are also well described by the above equations. Instead of the amounts of various components, it is then better to use their concentrations. According to the author, the derived equations are valid even in the case when particles of one kind of various reactivity are present. However, in this case the constants are of a complex character [35].

1.4.2 EXAMPLE. POLYMERIZATION OF ETHYLENE

Böhm polymerized ethylene [48] on a highly active catalyst, formed by the interaction of $Mg(OEt)_2$ with $TiCl_4$. Et_3Al was used as cocatalyst. The activity of the catalytic system depended on the ratio between catalyst and cocatalyst (see Fig. 9)†. After a certain mixing rate (> 500 rev. min^{-1}) has

---

† The maximum of the dependence from Fig. 9 is usually placed to much lower values of the ratio n_{Al}/n_K.

References pp. 544–545

been exceeded, the polymerization rate did not depend on a further increase in the number of mixer revolutions, and the rate of polymer generation was directly proportional to the amount of catalyst, n_K. Also, the molecular mass

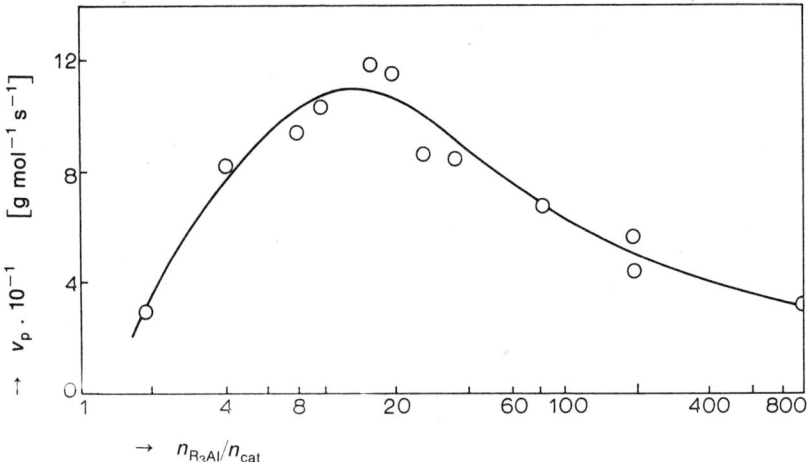

Fig. 9. Initial rate of ethylene polymerization as a function of the cocatalyst/catalyst ratio [48]. Temperature 358 K, $n_{cat} = 2.5 \times 10^{-5}$ mol TiCl$_4$, solvent 0.6 dm^3 of Diesel oil, ethylene pressure 0.1 MPa.

of the product remained constant under the described conditions. These measurements were indispensable. They proved that at $n_{R_3Al}/n_K \geqq 20$, at 750 rev. min^{-1} and n_K up to 5×10^{-5} mol in 1 dm^3 of hydrocarbon solvent, concentration and temperature are constant in the suspended catalyst–polymer particles, or at least the gradient of these quantities remains constant.

The yield of the catalyst, Θ, was measured at various ethylene concentrations (see Fig. 10). According to the results, initiation is rapid and the catalytic system maintains full capacity for a long time, for at least 1 h. In this interval, the polymeric particles increase their size 5–10 fold. Thus the monomer supply into the pores of the particles by diffusion cannot be hindered. In the subsequent phase, activity already decreases. Either the conditions for monomer transport to the centres by diffusion are deteriorating, and/or the centres are slowly decaying. The polymerization rate, v_{pol}, can be determined from the slopes of the curves in Fig. 10. The determined values of the initial rates are directly proportional to monomer concentration (except for the lowest values of [M]), as shown in Fig. 11.

The linear relation between v_{pol} and [M] is in agreement with eqn. (140) for $(k_p + k_d) \gg k_d[M]_0$, $[M]_0 \approx [M]$ and a, b, c and n°/n_K either constant or small. The validity of these conditions, mainly of the relation $(k_p + k_d)$

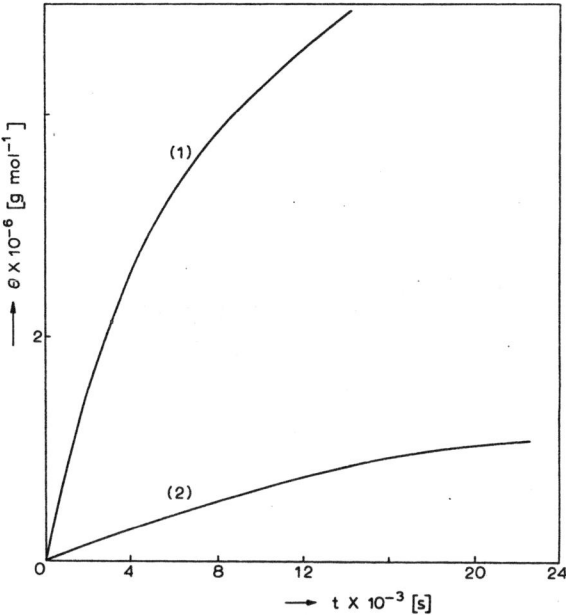

Fig. 10. Time dependence of catalyst yield Θ [48]. [M] (mol dm^{-3}): (1) 0.52, (2) 0.086. Other conditions as for Fig. 9.

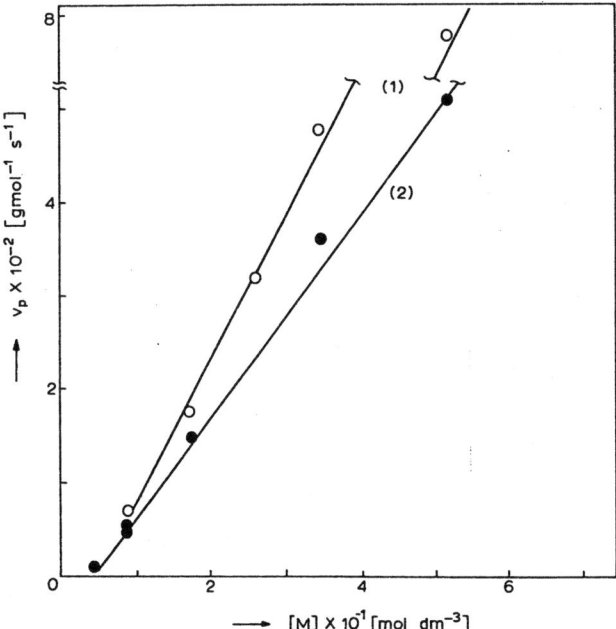

Fig. 11. Initial polymerization rate as a function of monomer concentration [48]. Conditions as in Figs. 9 and 10. n_{R_3Al}/n_{cat}: (1) 80, (2) 200.

References pp. 544–545

$\gg k_a$ was proved by the author by direct and indirect measurements which need not be described here. He could then obtain further information on the constants controlling the rate of polymerization.

When $k_p \gg k_d$, v_{pol} can be approximated by

$$v_{pol} \approx k_a[M]\left(1 + \frac{b}{a} + \frac{c}{a}\right)^{-1} \frac{n°}{n_K} \tag{162}$$

The polymerization rate, v_{pol}, will predominantly depend on k_a. The rate-determining step in this case is the formation of the complex between the monomer and the active centre. When $k_d \gg k_p$, the approximation

$$v_{pol} \approx k_p \frac{k_a}{k_d}[M]\left(1 + \frac{b}{a} + \frac{c}{a}\right)^{-1} \frac{n°}{n_K} \tag{163}$$

holds and the polymerization rate will be a function of the group of constants $k_p k_a k_d^{-1}$.

An important polymerization characteristic is the ratio $n°/n_K$ which can be derived from eqn. (158). The experimental values plotted in a graph of (Θ/M_n) vs. Θ actually yielded a straight line. Its intercept with the ordinate is the required quantity, though subject to large error. In the case discussed, the author estimated the value of $n°/n_K$ to lie in the range 0.4–1. This means that, on average, about 70% of all Ti atoms are used for active centre formation. This is surprisingly high. In catalysts of the activated $TiCl_3$ with Et_2AlCl type, the utilization of Ti atoms is less by about an order of magnitude, according to other sources.

2. The degree of polymerization of the product

Physical properties of polymers and of their solutions generally depend on chain length and on the molecular mass ditribution of the macromolecules. As the interest of consumers, and thus indirectly also of investigators, is mainly determined by the mechanical properties of polymers, knowledge of the conditions for the generation of the required degree of polymerization and of its population in the product is just as important as the control of polymerization rate. The theory of the generation of certain chain lengths operates with the kinetic constants used in the theory of reaction rates. One branch of macromolecular chemistry thus verifies the validity of assumptions and the correctness of the quantities derived in a different discipline, and they mutually complement each other.

2.1 Molecular mass distribution of the product

Macromolecules of unequal length are usually generated by polymerization[†]. Some idea of the relative contents of chains of various size in the studied polymer sample can be obtained by fractionation[††]. The results are represented graphically as distribution curves of molecular mass (or degree

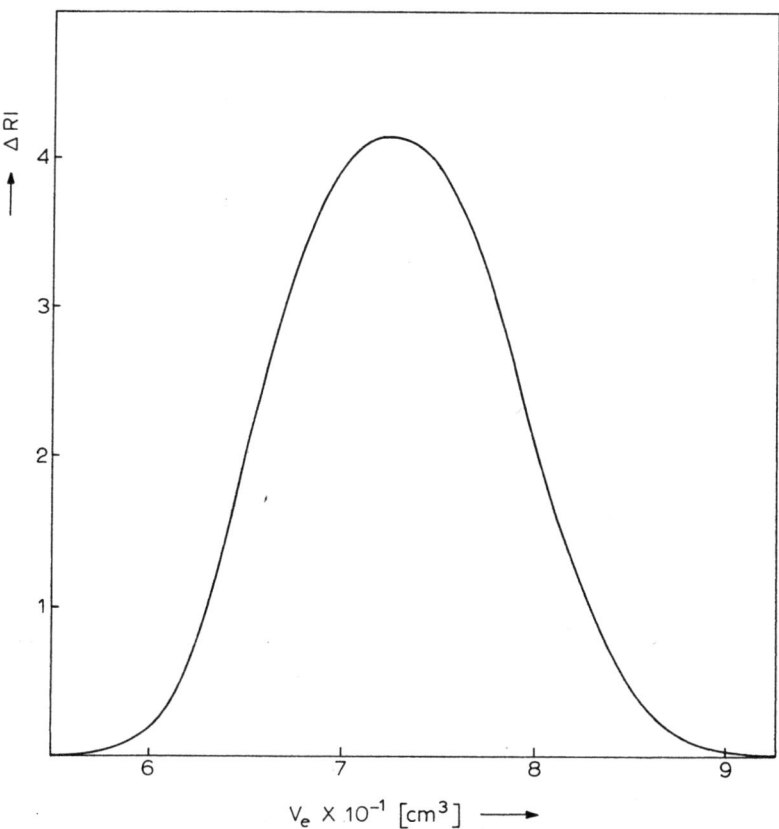

Fig. 12. Example of a GPC curve for a solution of radical polystyrene. Response of a sensor measuring the refraction index (RI) for an elution volume (V_e). $\bar{P}$ and $\bar{P}_w$ can be derived from the curve by means of calibration graphs. $\bar{M} = 66000$, $\bar{M}_w = 368000$.

[†] Except for the polymerization of multifunctional monomers, where a single three-dimensional macromolecule is formed. Chains of exactly equal length are not formed even by strictly living polymerizations because mechanical problems connected with the concentration gradient of rapidly reacting living centres during their mixing.
[††] Chromatography, in the form of gel permeation chromatography (GPC), can also be used.

References pp. 544–545

of polymerization†) (see Fig. 12). Distribution functions can be calculated by the methods of mathematical statistics. The theory of this kind of statistical calculation has bee described in many publications [49].

The fraction of molecules, X_n, of degree of polymerization n (or some other chain length characteristics) can be defined by some function n

$$X_n = F(n) \tag{164}$$

positive for all n and normalized

$$\int_0^\infty F(n)\, dx = 1 \tag{165}$$

The number of monomeric units in one average macromolecule of a polymer sample in which the chain length distribution is represented by the function $F(n)$, is expressed by

$$\bar{n} = \sum_{n=1}^\infty F(n)\, n \tag{166}$$

or, respecting condition (165)

$$\bar{n} = \int_0^\infty n\, F(n)\, dn \tag{167}$$

According to the statistical nomenclature, $\bar{n}$ is regarded as the moment μ'_1, a quantity from a more general set of parameters which can be assigned to the given distribution. Other moments of this kind of distribution (moments with respect to 0) can be defined as

$$\mu'_2 = \sum_{n=1}^\infty n^2\, F(n) \quad \text{or} \quad \mu'_2 = \int_0^\infty n^2\, F(n)\, dn \tag{168}$$

and generally

$$\mu'_r = \sum_{n=1}^\infty n^r\, F(n) \quad \text{or} \quad \mu'_r = \int_0^\infty n^r\, F(n)\, dn \tag{169}$$

---

† Distribution curves of this type are discontinuous in principle. The error caused by approximation to a continuous function is quite negligible.

Another kind of important moment, represented by dispersion variance, are moments referred to the centre. They are defined by

$$\mu_r'' = \sigma_r^2 = \sum_{n=1}^{\infty} (n - \bar{n})^r F(n) \, dn \quad \text{or} \quad \mu_r'' = \sigma_r^2 =$$
$$= \int_0^{\infty} (n - \bar{n})^r F(n) \, dn \tag{170}$$

For dispersion variance, defined as the mean value of the square of the difference between the variable n and its mean value $\bar{n}$, $r = 2$.

In statistics, the square root of the dispersion variance is called the standard, or root mean square, deviation. In holds that [50]

$$\sigma_r^2 = \mu_2' - \bar{n}^2 = \mu_2' - \mu_1' \tag{171}$$

For characterizing a distribution function, we need to know a sufficient number of moments and their mutual relations [51, 52]. The accuracy of the characteristics depends on the number of moments.

2.1.1 NUMBER AND WEIGHT (MASS) AVERAGE DEGREES OF POLYMERIZATION

The quantity $\bar{n}$ of eqns. (166) or (167) is identical with the number average degree of polymerization $\bar{P}$. The expression for $\bar{P}$ is usually written in the form

$$\bar{P} = \frac{\sum_i P_i N_i}{\sum_i N_i} = \sum_i P_i x_i = \int_0^{\infty} P \, F(n) \, dP \tag{172}$$

where x_i is the mole fraction of macromolecules of degree of polymerization i. An analogous equation is used for calculating the number average molecular mass, respecting the relation $\bar{M} = \bar{P} M$.

$$\bar{M} = \frac{\sum_i m_{P,i}}{\sum_i N_i} = \frac{\sum_i M_i N_i}{\sum_i N_i} = \sum_i M_i x_i$$

where $m_{P,i}$ is the mass of macromolecules of degree of polymerization i.

References pp. 544–545

The mass average degree of polymerization, $\bar{P}_w$, is defined as the first moment with respect to zero of the differential mass distribution

$$\bar{P}_w = \frac{\sum\limits_i N_i P_i^2}{\sum\limits_i N_i P_i} = \sum_i P_i w_i = \int_0^\infty P\, F_{(w)}\, dP \qquad (173)$$

and similarly

$$\bar{M}_w = \bar{P}_w M = \frac{\sum\limits_i m_{P,i} P_i}{\sum\limits_i m_{P,i}} = \frac{\sum\limits_i M_i^2 N_i}{\sum\limits_i M_i N_i} = \sum_i M_i w_i$$

where w_i is the weight fraction of the ith polymer fraction. Thus the mass average is the statistically weighted mean degree of polymerization. The statistical weights are equal to the weight fractions of the corresponding fractions of polymer.

The number average degree of polymerization is a very frequently used quantity in polymerization kinetics. It can be determined experimentally by osmometric, cryoscopic, ebullioscopic and end group methods. Viscometry yields values intermediate between the number and weight average values but is nearer to the latter. $\bar{P}_w$ or $\bar{M}_w$ values are determined by light scattering, and in some cases by diffusion and sedimentation methods. The latter are, however, rather used for the determination of higher statistical moments, the so-called z average, or very rarely $(z+1)$ average.

Various average quantities are related as $\bar{P} \leq \bar{P}_w \leq \bar{P}_z \leq \bar{P}_{z+1}$. The equality sign is only valid for strictly monodisperse polymers. The number average values $\bar{P}$ or $\bar{M}$ are very sensitive to the presence of low molecular components, even when these are present in very small amounts. The higher averages, $\bar{P}_z$ or $\bar{M}_z$, are strongly affected by the presence of high molecular components, even in trace concentrations.

The width of the distribution function can be roughly characterized by the polydispersity coefficient, ν, which is equal to the ratio of the weight and number averages

$$\nu = \frac{\bar{P}_w}{\bar{P}} = \frac{\bar{M}_w}{\bar{M}} \qquad (174)$$

The values of ν are only slightly higher than 1 for products of living polymerizations with rapid initiation. They are around 2 for radical polymers, and in extreme cases values of several tens for sets of macromolecules generated by some coordination polymerizations.

2.1.2 RATE OF LIVING POLYMERIZATIONS AND MOLECULAR MASS DISTRIBUTION CURVES OF PRODUCTS

Living polymerizations are charactetized by rapid initiation $(k_i \gg k_p)$, all initiator molecules are transformed to living centres, and a macromolecule grows from each centre. The reaction rate is given by the product of the concentrations of the centres and monomer, and of k_p [see (Chap. 5, eqn. (190)] and the number average degree of polymerization by the ratio of consumed monomer molecules and concentration of living centres. The resulting polymer is monodisperse; all chains are of equal length. When the chains grow at both ends, their number average degree of polymerization will, of course, be doubled[†].

When initiation is not sufficiently rapid, living centres will be generated even during macromolecular growth. The propagation times of various chains will be different, and the molecular mass distribution curve of the product will be broadened.

For cases where initiation is so slow that all the initiator cannot be transformed to living centres in the course of the whole polymerization, the following model is sometimes used [53]. The initiator, I, of a single kinetic type is uniformly distributed [††] in the polymerizing medium, and termination does not occur. The reactivity of the living centres, LC, of all lengths (but not of the initiating particles, I) is considered equal. The rate of concentration change of the growing chains of degree of polymerization $P = 0, 1, 2, ..., n$, must be expressed by the set of equations

$$-\frac{d[I]}{dt} = k_i[I][M]$$

$$\frac{d[RC]_1}{dt} = [M](k_i[I] - k_p[RC]_1)$$

$$\vdots$$

$$\frac{d[RC]_n}{dt} = k_p[M]([RC]_{n-1} - [RC]_n)$$

(175)

[†] With an initiator binding n living centres, an n-fold polymer will theoretically be formed

$$\bar{P} = \frac{n([M]_0 - [M])}{[LC]}.$$

[††] The mixing rate of initiator with monomer must be large, in order to make the range of concentration changes small compared with the product $k_i[M]_0$.

References pp. 544–545

Consumption of a particle I and growth of any $[LC]_n$ by one unit is equivalent to the consumption of a monomer molecule. Thus, for the overall polymerization rate it holds that

$$-\frac{d[M]}{dt} = k_i[I][M] + k_p[M]\left(\sum_{n=1}^{\infty}[RC]_n\right) \quad (176)$$

By means of the substitution

$$dm = [M]dt \quad \text{or} \quad m = \int_0^t [M]dt \quad (177)$$

eqn. (175) can be written as

$$\frac{d[I]}{dm} = k_i[I]$$

$$\frac{d[RC]_1}{dm} = k_i[I] - k_p[RC]_1 \quad (178)$$

$$\vdots$$

$$\frac{d[RC]_n}{dm} = k_p([RC]_{n-1} - [RC]_n)$$

In the absence of termination, the sum of the initiating and living centres is constant

$$[I]_0 = [I] + \sum_{n=1}^{\infty}[RC]_n \quad (179)$$

Integration of the first member of the set of equations (178) yields a relation for the instantaneous concentration of I

$$[I] = [I]_0 \exp[-k_i m] \quad (180)$$

so that

$$\sum_{n=1}^{\infty}[RC]_n = [I]_0(1 - \exp[-k_i m]) \quad (181)$$

After substitution from eqns. (180) and (181) into eqn. (176), the polymerization rate is described by the equation

$$-\frac{d[M]}{dt} = k_i[M][I]_0 \exp[-k_i m] + k_p[M][I]_0 (1 - \exp[-k_i m]) \tag{182}$$

or by substitution according to eqn. (177)

$$-\frac{d[M]}{dm} = k_i[I]_0 \exp[-k_i m] + k_p[I]_0 (1 - \exp[-k_i m]) \tag{183}$$

yielding by integration

$$\frac{[M]_0 - [M]}{[I]_0} = k_p m - \left(\frac{k_p}{k_i} - 1\right)[1 - \exp(-k_i m)] \tag{184}$$

When the relative number of living centres is designated by $X\{= ([I]_0 - [I])/[I]_0 = [LC]/[I]_0\}$, then according to eqn. (181)

$$1 - X = \exp[-k_i m] \tag{185}$$

and expression (184) is simplified to

$$\frac{[M]_0 - [M]}{[I]_0} = -\frac{k_p}{k_i}[X - \ln(1 - X)] + X \tag{186}$$

Within the conditions under which it has been derived, eqn. (186) is of general validity.

At any moment, the number average degree of polymerization is given by the ratio of consumed monomer and concentration of growth centres (macromolecules)

$$\bar{P} = \frac{[M]_0 - [M]}{[I]_0}\frac{1}{X} = -\frac{k_p}{k_i}\frac{X + \ln(1 - X)}{X} + 1 \tag{187}$$

For $k_i = k_p$

$$\bar{P} = -X^{-1}\ln(1 - X) = k_p m X^{-1} \tag{188}$$

References pp. 544–545

For $k_i \gg k_p$ the expression for the amount of consumed monomer is of the shape

$$[M]_0 - [M] = [I]_0 \left(1 + k_p \int_0^t [M]\, dt\right) = \qquad (189)$$
$$= [I]_0 (1 + k_p m)$$

and for the average degree of polymerization

$$\bar{P} = 1 + k_p m \qquad (190)$$

When initiation is slow, i. e. $k_i \ll k_p$ (and generally for $k_i \neq k_p$), summation over $[LC]_n$ is complicated. As

$$-d[M] = d \sum_{n=0}^{\infty} n[RC]_n$$

substitution of this relation into eqn. (184) and integration yields

$$\sum_{n=0}^{\infty} n[RC]_n = [I]_0 \left\{k_p m - \left(\frac{k_p}{k_i} - 1\right)[1 - \exp(-k_i m)]\right\} \qquad (191)$$

The second moment is obtained by integration of the expression

$$\frac{d \sum_{n=0}^{\infty} n^2[RC]}{dm} = k_p \sum_{n=0}^{\infty} (2n + 1)[RC]_n + k_i[I] =$$
$$= [I]_0 \left\{2k_p^2 m + 2k_p\left(\frac{k_p}{k_i} - 1\right)[1 - \exp(-k_i m)] + \right.$$
$$\left. + k_p[1 - \exp(-k_i m)] + k_i \exp(-k_i m)\right\} \qquad (192)$$

thus

$$\sum_{n=0}^{\infty} n^2[RC]_0 = [I]_0 \left\{k_p^2 m^2 + k_p m\left(3 - 2\frac{k_p}{k_i}\right) - \right.$$
$$\left. - \left[3\frac{k_p}{k_i} - 2\left(\frac{k_p}{k_i}\right)^2 - 1\right][1 - \exp(-k_i m)]\right\} \qquad (193)$$

Again, since

$$\sum_{n=0}^{\infty} n[RC]_n = [I]_0 - [I]$$

relations for both $\bar{P}$ and $\bar{P}_w$ can be derived from eqns. (191) and (193).

The problem of molecular mass distribution has been studied by many authors [24, 54–56]. Their solutions resemble the case discussed above and are sometimes more complicated, depending on whether the author considered the incorporated initiator fragment as a monomeric unit or whether he neglected it.

Any deviation from the living mechanism broadens the distribution. Transfers and termination caused by residual impurities and insufficient mixing are the main causes of "improvable" broadening of distribution curves. Internal reasons are inherent to two factors, namely to reversibility of propagation and to propagation on centres with various reactivity.

Equation (175) can be amplified by a term involving the termination reactions. In this way they acquire a much more general character, permitting analysis even of complicated cases of non-stationary ionic polymerizations. Their solution is not simple, of course.

2.1.3 MOLECULAR MASS DISTRIBUTION IN PRODUCTS OF RADICAL POLYMERIZATIONS

The basic kinetics of radical polymerizations were discussed in Sect. 1.2. This should now be extended to calculations of distribution.

For radicals of a certain degree of polymerization in the stationary state, the generation rates are given as

$$\frac{d[M_1^*]}{dt} = v_{init} - k_p[M_1^*][M] - 2k_t[M_1^*]\sum_{j=1}^{\infty}[M_j^*] = 0 \quad (194)$$

$$\frac{d[M_2^*]}{dt} = k_p[M_1^*][M] - k_p[M_2^*][M] - 2k_t[M_2^*]\sum_{j=1}^{\infty}[M_j^*] = 0 \quad (195)$$

$$\vdots$$

$$\frac{d[M_{k-1}^*]}{dt} = k_p[M_{k-2}^*][M] - k_p[M_{k-1}^*][M] - 2k_t[M_{k-1}^*]\sum_{j=1}^{\infty}[M_j^*] = 0 \quad (196)$$

$$\frac{d[M_k^*]}{dt} = k_p[M_{k-1}^*][M] - k_p[M_k^*][M] - 2k_t[M_k^*]\sum_{j=1}^{\infty}[M_j^*] = 0 \quad (197)$$

The ratio of radical concentrations is given by the relation [after substitution for their stationary concentration from eqn. (22)]

$$\frac{[M_k^*]}{[M_{k-1}^*]} = \alpha = \frac{k_p[M]}{k_p[M] + (2k_t v_{init})^{1/2}} \qquad (198)$$

The concentration of a discrete radical $[M_P^*] = [M_1^*]\alpha^{P-1}$. This follows from the relation

$$\sum_{P=1}^{\infty} [M_P^*] = [M_1^*] + [M_2^*] + [M_3^*] + \ldots$$

$$= [M_1^*] + [M_1^*]\frac{[M_2^*]}{[M_1^*]} + [M_1]\left(\frac{[M_2^*]}{[M_1^*]}\frac{[M_3^*]}{[M_2^*]}\right) + \ldots$$

$$= [M_1^*] + [M_1^*]\alpha + [M_1^*]\alpha^2 + \ldots$$

$$= [M_1^*]\sum_{P=1}^{\infty}\alpha^{P-1} = \frac{[M_1^*]}{1-\alpha} \qquad (199)$$

The calculations are simplified when the monomer and initiator concentrations are constant. To a first approximation, this condition may be regarded as fulfilled at low conversions†. At any instant, macromolecules of various length are produced by termination. Let us first consider termination exclusively by disproportionation [49]. The fraction of chains with degree of polymerization P is a function of their generation rate, and this in turn is proportional to the decay rate of a radical of equal length. Thus, according to eqn. (199) the mole fraction of macromolecules, x_P, is

$$x_P = B[M_P] = B\alpha^{P-1}$$

$$\sum_{P=1}^{\infty} x_P = 1 = B\sum_{P=1}^{\infty} \alpha^{P-1} = \frac{B}{1-\alpha} \qquad (200)$$

where B is a constant. Thus

$$B = (1-\alpha)$$

$$x_P = (1-\alpha)\alpha^{P-1} \qquad (201)$$

---

† Even situation in which α varies can be solved, but in a more complicated way.

The distribution (201) is called the "most probable" distribution. In this case, the number average degree of polymerization, $\bar{P}$, is given by

$$\bar{P} = \sum_{P=1}^{\infty} (1 - \alpha) P\alpha^{P-1} = (1 - \alpha) \sum_{P=1}^{\infty} P\alpha^{P-1} = \frac{1}{1 - \alpha} \qquad (202)$$

The weight fraction w_p, is determined by the product

$$w_p = \frac{P}{\bar{P}} x_P = (1 - \alpha)^2 P\alpha^{P-1} \qquad (203)$$

the weight average degree of polymerization is equal to

$$\bar{P}_w = (1 - \alpha)^2 \sum_{P=1}^{\infty} P^2 \alpha^{P-1} = \frac{1 + \alpha}{1 - \alpha} \qquad (204)$$

and the polydispersity coefficient, v, to

$$v = \frac{\bar{P}_w}{\bar{P}} = 1 + \alpha \qquad (205)$$

A high polymer is produced only when α very nearly approaches one, so that for the discussed case of radical polymerization, $v \approx 2$.

When chain growth is terminated only by combination, macromolecules of degree of polymerization P will be generated at the rate

$$v_{t,P} = 2k_t \sum_{i=1}^{P-1} [M_i^*][M_{P-i}^*] \qquad (206)$$

Substituting for $[M_i^*]$ and $[M_{P-i}^*]$ from eqn. (199), we obtain for $v_{t,P}$

$$v_{t,P} = 2k_t \sum_{i=1}^{P-1} \alpha^{i-1}[M_1^*] \alpha^{P-i-1}[M_1^*] = [M_1^*]^2 2k_t \sum_{i=1}^{P-1} \alpha^{P-2} = \qquad (207)$$
$$= [M_1^*]^2 2k_t \alpha^{P-2}(P - 1)$$

The mole fraction of macromolecules with P monomeric units in their chains is proportional to the rate of their generation. Thus

$$x_P = C\alpha^{P-2}(P - 1) \qquad (208)$$

The proportionality constant C can again be derived from the condition

$$\sum_{P=1}^{\infty} x_P = 1$$

thus $C = (1 - \alpha)^2$ and for x_P it holds that

$$x_P = (1 - \alpha)^2 \alpha^{P-2}(P - 1) \tag{209}$$

The number average degree of polymerization is a sum of products

$$\bar{P} = \sum_{P=1}^{\infty} P x_P = \frac{2}{1 - \alpha} \tag{210}$$

After substitution for $\bar{P}$ and x_P, the weight fraction w_p is

$$w_p = \frac{P}{\bar{P}} x_P = \frac{1}{2} (1 - \alpha)^3 P(P - 1) \alpha^{P-2} \tag{211}$$

and the weight average degree of polymerization is

$$\bar{P}_w = \frac{2 + \alpha}{1 - \alpha} \tag{212}$$

so that

$$\nu = \frac{\bar{P}_w}{\bar{P}} = \frac{2 + \alpha}{2}$$

Polymerizations terminating by combination produce polymers with a distribution narrower than the most probable one. As α again very nearly approaches one, $\nu \approx 3/2$.

An alternative derivation of the molecular mass distribution for products of radical polymerizations, including also changes in k_t with the length of macroradicals, was published by Olaj and Zifferer [57]. By means of the general moments $S_1^{(k)}$ and $S_2^{(k)}$ [see Chap. 6, eqns. (16) and (17)], they expressed various kinds of degrees of polymerization by the relations

$$\bar{P}_n = 1 + S_2^{(0)} \approx S_2^{(0)} \tag{213}$$

$$\bar{P}_w = \frac{2 S_2^{(1)}}{S_2^{(0)} + 1} + 1 \approx \frac{2 S_2^{(1)}}{S_2^{(0)}} \tag{214}$$

$$\bar{P}_z = \frac{3 S_2^{(2)} + S_2^{(1)}}{2 S_2^{(1)} + S_1^{(0)} + 1} + 1 \approx \frac{3}{2} \frac{S_2^{(2)}}{S_2^{(1)}} \tag{215}$$

The approximate relations are almost exactly valid for high polymers.

2.1.4 EXAMPLE. KINETICS OF THE CHANGES IN MOLECULAR MASS DISTRIBUTION DURING THE POLYMERIZATION OF HETEROCYCLES

Davtyan et al. analyzed the problem of changes mass distribution with conversion during the polymerization of heterocycles [58]. This case is interesting; the question of the coexistence of one or two active centres on a single growing macromolecule is solved by means of rate equations. The results can also be applied to polymerizations of other monomers, when the conditions

(a) initiation is rapid (instantaneous),
(b) termination does not occur and
(c) transfer occurs to agent XT and to polymer

are fulfilled. Transfer to polymer leads to a redistribution of chain lengths. Propagations of such systems are usually reversible. The authors assumed little affect of the reverse reaction on the distribution width.

They therefore considered only a simple reaction course, in a single direction, as represented by the scheme

$$
\begin{aligned}
M_j^\circ + M &\xrightarrow{k_p} M_{j+1}^\circ & \text{growth} \\
M_j^\circ + XT &\xrightarrow{k_{tr}} M_{j,x} + T^\circ & \text{transfer} \\
T^\circ + M &\xrightarrow{k_{i,tr}} M_1^\circ & \text{re-initiation} \\
T^\circ + M_j &\xrightarrow{k_{tr,P}} M_{j-k}T + M_k^\circ & \text{redistributing transfer to polymer}
\end{aligned}
\quad (216)
$$

Transfer to polymer leads to the appearance of three chain types: (a) chains without an active centre (dead) M_j, (b) chains with one active end M_j° and (c) chains with active centres at both ends $^\circ M_j^\circ$.

The concentration changes of various particle types can be described by the differential equations[†]

$$\frac{d[M_j]}{dt} = \frac{1}{2} k_{tr,P}[T^\circ] \sum_{i=j+1}^{\infty} ([M_i^\circ] + 2[M_i]) + k_{tr}[M_j^\circ][XT] - $$

$$- k_{tr,P}[T^\circ](j-1)[M_j]$$

[†] A polyheterocyclic chain of degree of polymerization j can be attacked by the active centre at a heteroatom of any originally monomeric unit. The members describing this type reaction must therefore be multiplied by the factor $(j-1)$, or $(j-i)$, when the monomer contains a single susceptible heteroatom. The described approach is permissible only when all heteroatoms in the chain are sterically equally accessible.

References pp. 544–545

$$\frac{d[M_j^\circ]}{dt} = k_{tr,P}[T^\circ] \sum_{i=j+1}^{\infty} (M_i] + [M_i^\circ] + [^\circ M_i^\circ]) +$$
$$+ k_{tr}[XT](2[^\circ M_j^\circ] - [M_j^\circ]) + k_p[M]([M_{j-1}^\circ] -$$
$$- [M_j^\circ]) - k_{tr,P}[T^\circ](j-1)[M_j^\circ]$$

$$\frac{d[^\circ M_j^\circ]}{dt} = \frac{1}{2} k_{tr,P}[T^\circ] \sum_{i=j+1}^{\infty} ([M_i^\circ] + 2[^\circ M_i^\circ]) +$$
$$+ 2k_p[M]([^\circ M_{j-1}^\circ] - [^\circ M_j^\circ]) - k_{tr,P}[T^\circ](j-1) \cdot$$
$$\cdot [^\circ M_j^\circ] - 2k_{tr}[^\circ M_j^\circ][XT]$$

$$\frac{d[T^\circ]}{dt} = k_{tr}([I]_0 - [T^\circ])[XT] - k_{tr,P}[T^\circ]\{([M]_0 -$$
$$- [M]) \sum_{i=1}^{\infty} ([M_i] + [M_i^\circ] + [^\circ M_i^\circ])\} - k_{i,tr}[T^\circ][M]$$

$$\frac{d[XT]}{dt} = -k_{tr}[XT] \sum_{i=1}^{\infty} ([M_i^\circ] + 2[^\circ M_i^\circ]) \tag{217}$$

By summation of the first three equations in the system (217) we obtain

$$\frac{d \sum_{j=1}^{\infty} [M_j]}{dt} = \frac{1}{2} k_{tr,P}[T^\circ] \sum_{i=j+1}^{\infty} [(i-j)([M_i^\circ] + 2[M_i])] +$$
$$+ k_{tr}[XT] \sum_{j=1}^{\infty} [M_j^\circ] - k_{tr,P}[T^\circ] \sum_{j=1}^{\infty} [(j-1)[M_j]]$$

$$\frac{d \sum_{j=1}^{\infty} [M_j^\circ]}{dt} = k_{tr,P}[T^\circ] \sum_{i=j+1}^{\infty} [(i-j)([M_i] + [M_i^\circ] + [^\circ M_i^\circ])] +$$
$$+ k_{tr}[XT] \sum_{j=1}^{\infty} (2[^\circ M_j^\circ] - [M_j^\circ]) -$$
$$- k_{tr,P}[T^\circ] \sum_{j=1}^{\infty} [(j-1)[M_j^\circ]]$$

$$\frac{d\sum_{j=1}^{\infty}[^{\circ}M_{j}^{\circ}]}{dt} = \frac{1}{2}k_{tr,P}[T^{\circ}]\sum_{i=j+1}^{\infty}[(i-j)([M_{i}^{\circ}]+2[^{\circ}M_{i}^{\circ}])] -$$

$$- k_{tr,P}[T^{\circ}]\sum_{j=1}^{\infty}[(j-1)[^{\circ}M_{j}^{\circ}]] -$$

$$- 2k_{tr}[XT]\sum_{j=1}^{\infty}[^{\circ}M_{j}^{\circ}] \qquad (218)$$

From eqns. (214) and (215), it follows that

$$d\left\{\sum_{j=1}^{\infty}([M_{j}]+[M_{j}^{\circ}]+[^{\circ}M_{j}^{\circ}]) - [T^{\circ}] - [XT]\right\}dt = 0 \qquad (219)$$

and therefore

$$\sum_{j=1}^{\infty}([M_{j}]+[M_{j}^{\circ}]+[^{\circ}M_{j}^{\circ}]) = [I]_{0} + [XT]_{0} - [XT] - [T^{\circ}] \qquad (220)$$

The sum of all end groups is given by the sum of the amount of initiator and the consumed transfer agent. When the rate of re-initiation after transfer is large, the rate equation assumes the simple shape

$$-\frac{d[M]}{dt} = k_{p}[I]_{0}[M]$$

As the conversion $\varkappa = ([M]_{0} - [M])/[M]_{0}$, the number average degree of polymerization, P, is given by [58]

$$\bar{P} = \frac{a\varkappa[M]_{0}}{[XT]_{0}[1-(1-\varkappa)C^{tr}]+[I]_{0}}$$

where

$$a = \left[1 - \left(1 + \frac{[M]_{0}\left(\varkappa\frac{k_{tr,P}}{k_{p}} + \frac{(1-\varkappa)}{C^{tr}}\right)}{[XT]_{0}\left(1 - \frac{\varkappa}{\varkappa_{\infty}}\right)C^{tr}}\right)^{-1}\right] \qquad (221)$$

and $C^{tr} = k_{tr}/k_{p}$.

References pp. 545–545

When $C^{tr} \ll 1$, $[XT]_0 \approx [I]_0$, then eqn. (221) is simplified to

$$\bar{P} = \varkappa \frac{[M]_0}{[I]_0} \tag{222}$$

and the Poisson distribution of molecular mass is maintained up to high conversions. At $C^{tr} \gg 1$, $(k_{tr,P}/k_p) \approx 1$ and $[XT]_0 \approx [I]_0$, starting from some low conversion the equation assumes the form

$$\bar{P} = \varkappa \frac{[M]_0}{[I]_0 + [XT]_0} \tag{223}$$

From the start of the polymerization, the molecular mass distribution curve widens up to the total consumption of the transfer agent. The rate of the distribution changes depends on the intensity of the transfers and the starting conditions. The populations of various lengths in dead chains and in

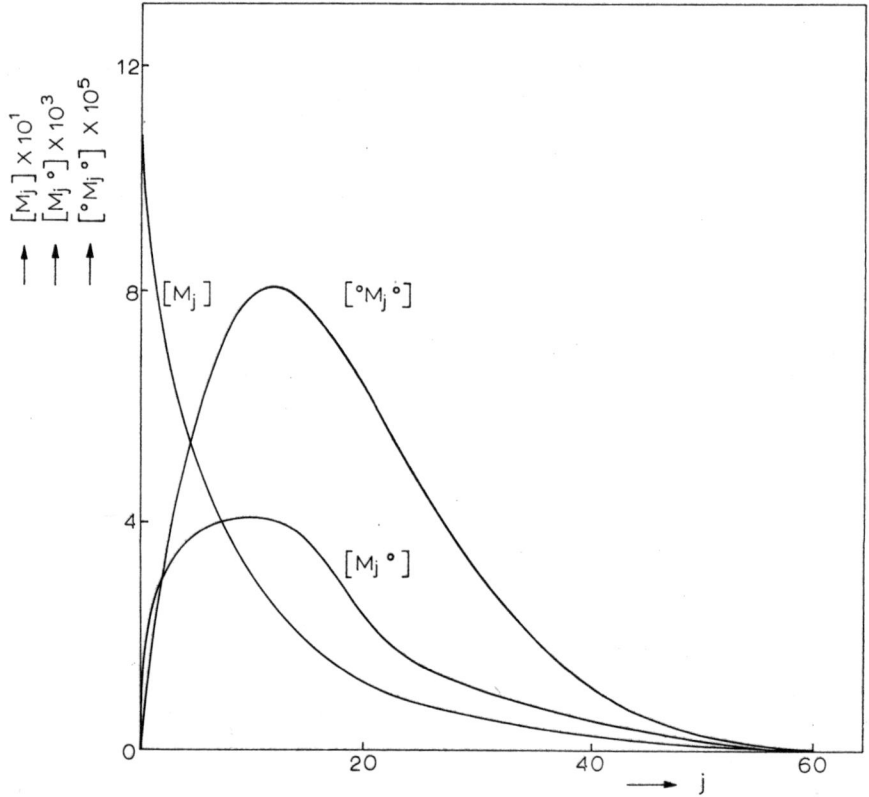

Fig. 13. Dependence of $[M_j]$, $[M_j^\circ]$ and $[^\circ M_j^\circ]$ on the number of units j (for equal conversions) [58].

chains with one and two active ends (with $C^{tr} \gtreqqless 1$, $(k_{tr,P}/k_p) \gtreqqless 1$ and $[XT]_0 \gg [I]_0$) are qualitatively represented in Figs. 13 and 14. Figure 14 also exhibits changes in the polydispersity coefficient with conversion. The effect

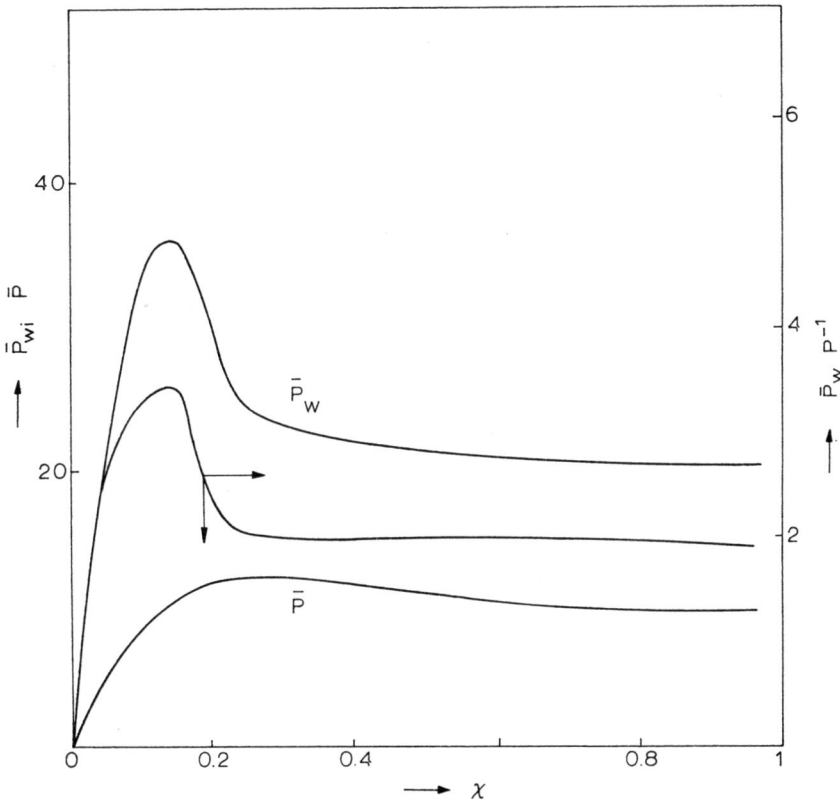

Fig. 14. Dependence of $\bar{P}$ and $\bar{P}_w$ on the polydispersity coefficient and on conversion [58].

TABLE 4

Effect of transfer constant values C_{tr} and $k_{tr,P}/k_p$ on the character of the molecular mass distribution in the product

	$C_{tr} \ll 1$	$C_{tr} \gg 1$	$C_{tr} \ll 1$	$C_{tr} \approx 1$
	$0 < \dfrac{k_{tr,P}}{k_p} < \infty$	$\dfrac{k_{tr,P}}{k_p} \approx 1$	$\dfrac{k_{tr,P}}{k_p} \approx 1$	$\dfrac{k_{tr,P}}{k_p} \approx 1$
	$[I]_0 \approx [XT]_0$	$[I]_0 \approx [XT]_0$	$[I]_0 \ll [XT]_0$	$[I]_0 \ll [XT]_0$
During polymerization	$v = 1$	$v \geqq 1$	$1 < v < 2$	$v = 2$
At end of polymerization	$1 < v \leqq 2$	$v = 1$	$v = 2$	$v = 2$

References pp. 544–545

of transfer constants and of the concentration of transfer agent on the molecular mass distribution in the product is summarized in Table 4.

Other values of the constants and of the initial conditions can lead to broader distributions with $(\bar{P}_w/\bar{P}) > 2$. Analysis of such cases represents a special problem.

References

1. L. M. Arnett and J. H. Peterson, J. Am. Chem. Soc., 74 (1952) 2031.
2. C. J. Bevington, Fortschr. Hochpolym. Forsch., 2 (1960) 1.
3. P. E. M. Allen, C. R. Patrick, Kinetics and Mechanisms of Polymerisation Reactions, Horwood, Chichester, 1974, pp. 415–498.
4. C. S. Marvel, J. Dec and H. G. Cooke, Jr., J. Am. Chem. Soc., 62 (1940) 3499.
5. G. V. Tkachenko, P. M. Khomikovskü and S. S. Medvedev, Zh. Fiz. Khim., 25 (1951) 823.
6. K. Veselý, Polyreakce, SNTL, Prague, 1955, p. 65.
7. P. S. Shantarovich and S. S. Medvedev, Zh. Fiz. Khim., 23 (1949) 1426; 24 (1950) 10.
8. G. V. Schulz, A. Dinglinger and E. Husemann, Z. Phys. Chem. Abt. B, 43 (1939) 305.
9. P. J. Flory, Principles of Polymer Chemistry, Cornell University Press, Ithaca, New York, 1953.
10. K. Ito, J. Polym. Sci. Polym. Chem. Ed., 12 (1974) 1991.
11. K. Ito, J. Polym. Sci. Polym. Chem. Ed., 15 (1977) 1759.
12. F. Collins, J. Colloid Sci., 4 (1949) 202.
13. K. Ito, J. Polym. Sci. Part A-1, (10) (1972) 57.
14. P. C. Deb and I. D. Gaba, Makromol. Chem., 179 (1978) 1549, 1559.
15. K. C. Berger and G. Mayerhoff, Makromol. Chem., 176 (1975) 1983.
16. M. S. Matheson, E. E. Auer, E. B. Bevilacqua and E. J. Hart, J. Am. Chem. Soc., 73 (1951) 1700.
17. A. Crosato-Arnoldi, P. Gasparini and G. Talamini, Makromol. Chem., 117 (1968) 140.
18. A. H. Abdel-Alim and A. E. Hamielec, J. Appl. Polym. Sci., 16 (1972) 783.
19. J. Ugelstad, J. Macromol. Sci. Chem., A11 (1977) 1281.
20. J. Ugelstad, H. Flogstad, T. Hertzberg and E. Sund, Makromol. Chem., 164 (1973) 171.
21. O. F. Olaj, Angew. Makromol. Chem., 47 (1975) 1; J. Macromol. Sci. Chem., A11 (1977) 1307.
22. L. F. Beste and H. K. Hall, J. Phys. Chem., 68 (1964) 264.
23. C. E. H. Bawn, C. Fitzsimmons, A. Ledwith, J. Penfold, D. C. Sherrington and J. A. Weightman, Polymer, 12 (1971) 119.
24. M. Litt, J. Polym. Sci., 58 (1962) 429.
25. D. C. Pepper, Eur. Polym. J., 16 (1980) 407.
26. M. Szwarc and C. L. Perrin, Macromolecules, 18 (1985) 528.
27. M. J. Hayes and D. C. Pepper, Proc. Chem. Soc., (August) (1958) 228.
28. R. E. Burton and D. C. Pepper, Proc. R. Soc. (London), Ser. A, 263 (1961) 58.
29. M. J. Hayes and D. C. Pepper, Proc. R. Soc. (London), Ser. A, 263 (1961) 63.
30. A. M. Buyle, K. Matyjaszewski and S. Penczek, Macromolecules, 10 (1977) 269.
31. M. U. Mahmud, G. Wenger, W. Kern, J. B. Lando and Y. Osada, J. Macromol. Sci. Chem., A11 (1977) 2233.
32. K. Matyjaszewski and S. Penczek, J. Polym. Sci. Polym. Chem. Ed., 15 (1977) 247.
33. S. Kobayashi, H. Danda and T. Saegusa, Bull. Chem. Soc. Jpn., 47 (1974) 2699.

34 S. Kobayashi, K. Morikawa and T. Saegusa, Macromolecules 8 (1975) 386.
35 L. L. Böhm, Poolymer, 19 (1978) 545.
36 E. K. Rideal, Proc. Cambridge Philos. Soc., 35 (1939) 130.
37 E. J. Arlman and P. Cossee, J. Catal., 3 (1964) 99.
38 G. Natta, J. Polym. Sci., 34 (1959) 21.
39 G. Fink, R. Rottler, D. Schnell and W. Zoller, J. Appl. Polym. Sci., 20 (1976) 2779.
40 K. Ziegler, H. G. Gellert, E. Holzkamp, J. Schneider, M. Söll and W. R. Kroll, Annalen, 629 (1960) 121.
41 D. R. Burfield, I. D. McKenzie and P. J. T. Tait, Polymer, 13 (1972) 302.
42 G. Natta and I. Pasquon, Adv. Catal., 11 (1959) 1.
43 L. Reich and A. Schindler, Polymerization by Organometallic Compounds, Interscience, New York, 1966.
44 J. Boor, Macromol. Rev., 2 (1967) 115.
45 J. I. Ermakov and V. A. Zakharov, Russ, Chem. Rev., 41 (1972) 203.
46 T. Keii, Kinetics of Ziegler–Natta Polymerization, Kodansko, Tokyo; Chapman and Hall, London, 1972.
47 J. C. W. Chien, (Ed.), Coordination Polymerization, Academic Press, New York, 1975.
48 L. L. Böhm, Polymer, 19 (1978) 553.
49 M. Kubín, Small Macromolecular Monographs (Malé Makromolekulární Monografie), Vol. 7, Institute of Macromolecular Chemistry, Czechoslovak Academy Sciences, Prague, 1970, pp. 1–38 (in Czech.).
50 See P. E. M. Allen and C. R. Patrick, Kinetics and Mechanisms of Polymerization Reactions, Horwood, Chichester, 1974, pp. 168–170.
51 M. Szwarc, in Carbanions Living Polymers and Electron Transfer Processes, Interscience, New York, 1968, pp. 27–72.
52 C. H. Bamford and H. Tompa, Trans. Faraday Soc., 50 (1954) 1097.
53 See P. E. M. Allen and C. R. Patrick, Kinetics and Mechanisms of Polymerization Reactions, Horwood, Chichester, 1974, pp. 162–167, 181–183.
54 V. S. Nanda and R. K. Jain, J. Polym. Sci. Part. A-1 (2) (1964) 4583.
55 L. Gold, J. Chem. Phys., 28 (1958) 91.
56 W. T. Kyner, J. R. M. Radok and M. Wales, J. Chem. Phys., 30 (1959) 363.
57 O. F. Olaj and G. Zifferer, Macromolecules, 20 (1987) 850.
58 S. P. Davtyan, B. A. Rozenberg, V. I. Irzhak, A. I. Prikhozhenko and N. S. Enikolopyan, Vysokomol. Soedin. Ser. A, 13 (1971) 1630.

Chapter 9

Conclusion

In the sixties, plastics were being produced in amounts of hundreds of millions of tons, the trends of world production were increasing and the prospects of future development were very optimistic. This era, roughly up to the first oil crisis, is sometimes regarded as the golden age of polymer chemistry.

The era of intense development of polymer production has passed. According to present estimates, the production of plastics will continue to increase, but the quality of the polymers produced is continually being stressed. It may be expected that in the specifications of the properties of macromolecular compounds, quality parameters will in the future be dominant, and that exacting requirements will promote the production of speciality polymers.

Some features of this development may already be observed, as new trends are appearing in the development [1].

(i) The importance of radical polymerizations may be expected to recede, as the products are insufficiently defined polymers with branched macromolecules of irregular structure.

(ii) The importance of ionic and mainly of coordination polymerizations is increasing.

Coordination polymerizations are becoming an inspiration source for further methodical development in addition polymerizations. The nearest aim could possibly be the insertion of polar monomers (as indicated by group transfer polymerization) and a deepening of our understanding of catalysis. Research in this field should lead to partial or even total replacement of catalysts by other means. I shall try to indicate one of the possibilities.

1. Activation energy of polymerization and thermodynamics

A chemical reaction proceeds spontaneously when the corresponding Gibbs energy is negative, $\Delta G = \Delta H - T\Delta S < 0$. Under normal conditions, ethylene and other monomers can exist only because the relevant chemical reactions (including polymerization) cannot occur unless the monomer molecules have acquired a certain energy, the activation energy. As soon as this barrier has been overcome, the electrons are spontaneously (following the laws of thermodynamics) rearranged so as to produce a system of valence electrons and atom residues of minimum energy (for the particular situation).

To systems which have previously been prepared for the reaction, the activation energy is often supplied in the form of heat. Thermal energy is suitable for overcoming the activation barrier, but it is non-specific, it also enables the excited molecules to react in undesirable ways. Therefore the effect of heat is usually combined with the addition of compounds which, after overcoming a low activation barrier can form complexes with the substrate and can react further, again after overcoming some activation barrier, to the required product. In this process, the molecules of the added compound (the catalyst) are liberated and the process is repeated. Thus the reaction coordinate is modified (Fig. 1).

In alkene polymerizations, π electrons are transformed to σ electrons, and transition metals with their surroundings act as catalyst. This process can be simply expressed by the scheme

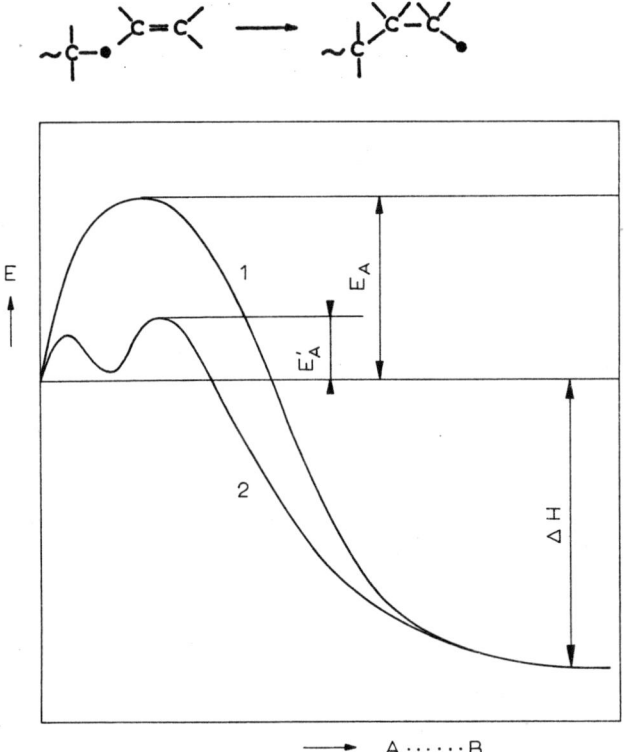

Fig. 1. Reaction coordinate of uncatalyzed (1) and catalyzed (2) reactions.

Such a process cannot occur without temporary electron delocalization followed by rearrangement. Schematically [2-4] this is represented by

[chemical scheme showing electron delocalization and rearrangement steps with transition metal M]

The necessary electron delocalization is possible by the existence of free d orbitals on the transition metals, with small energy level differences; it is aided by a suitable modification of the electron configuration on the transition metal atom by its oxidation state and the presence of electron-attracting or electron-donating ligands. An excess of electrons is manifested by a reduced tendency of the active centre to interact with the π electrons of the monomer, the transition complex is formed only with difficulty or not at all; a lack of electrons results in the formation of a relatively stable complex of the active centre with the monomer with little tendency to decompose at the necessary rate in the required way.

Delocalization of bonding electrons in the transition complex is an important event. In the absence of this far-reaching delocalization, even the most acid alkenes, ethylene and propene only undergo both radical and ionic polymerization reluctantly.

2. Replacement of active centres by a high frequency magnetic field

Delocalization of electrons, especially in the outer layer, i. e. valence electrons, must also occur in a magnetic field. Evidence does exist that a stationary magnetic field actually affects some chemical reactions. Turro

References pp. 550–551

himself and with his co-workers have observed a considerable enhancement of the rate and of the resulting molecular mass in photoinitiated emulsion polymerizations of styrene, methyl methacrylate and acrylic acid under the influence of a magnetic field [5, 6]. Imoto and Nomoto also observed a non-zero magnetic field effect in methyl methacrylate polymerization [7]. Ouchi et al. describe an uncatalyzed radical polymerization of vinyl monomers where, under the effect of a magnetic field, some kinds of hydrophilic macromolecules become initiators [8].

A larger effect may, of course, be expected from a pulsed magnetic field, especially at high frequency, which could induce the valence electrons to vibrate. An indication of an effect generated by a pulsed field was described by Lambrev, who measured the catalytic efficiency of metal oxides (CdO, HgO, Tl_2O_3, PbO, Br_2O_3) under the effect of such a field in methyl methacrylate polymerization [9]. He observed activation and deactivation of the oxides which depended on the intensity of the pulsing field.

There exist communications on chemical reactions catalyzed by magnetic fields in connection with electromagnetic radiation. A photoresist was cross--linked by Panico under such conditions [10]. Teffal and Gourdenne polymerized (without initiator) 2-hydroxyethyl methacrylate by means of microwaves in a waveguide [11]. Actually this should be regarded as thermal polymerization, with the heat generated by dipole vibrations of the polar groups the monomer.

A high-frequancy magnetic field, suitable for displacing the bonding electrons either by a single pulse or by a packet of pulses resonating with electron motions, should by its effects be able to overcome the activation energy barrier. Electron oscillations leading to valence shifts can be induced by energies $\leq 1/3$ of the energy of a covalent C–C bond. The delocalized electrons spontaneously assume (according to the laws of thermodynamics) a position of minimum energy, i. e. the configuration of single (σ) bonds in the polymer. The electromagnetic component of such a field, probably in the range of infrared radiation, does not possess sufficient energy to split covalent bonds; plasma would not be formed.

It should be possible selectively to affect and control the desired reaction by selecting the frequency and intensity of the magnetic field. The equipment for the generation of the necessary field would not be simple; frequencies of the order of $> 10^2$ GHz should probably be considered.

References

1 H. Mark, Paper presented at the Institute of Macromolecular Chemistry, Czechoslovak Academy of Sciences, Prague, 1988.
2 K. J. Ivin, J. Rooney. Jr. and C. D. Stewart, J. Chem. Soc. Chem. Commun., (1978) 604.

3   R. J. McKinney, J. Chem. Soc. Chem. Commun., (1980) 490.
4   M. Kučera, Chem. Listy, 83 (1989) 730; 829; 936. 84 (1990) 69.
5   N. J. Turro, Ind. Eng. Chem. Prod. Res. Dev., 22(2) (1983) 272.
6   N. J. Turro, M. F. Chow, Ch. J. Chung and Ch. H. Tung, J. Am. Chem. Soc., 105(6) (1983) 1572.
7   M. Imoto and K. Nomoto, Makromol. Chem. Rapid Commun., 2 (1981) 703.
8   T. Ouchi, H. Sakamoto, M. Kubo, Y. Hosaka and M. Imoto, J. Macromol. Sci. Chem., A24(2) (1987) 111.
9   D. Lambrev, Scientifie Works (Nauchni Tr.) Plovdivski Univ. Mat. Fiz. Khim. Biol., 20(2) (1982) 247. Chem. Abstr. 101 (1984) 55590b.
10 C. R. Panico, (Xenon Corp.), WO 84,00,506 (1984).
11 M. Teffal and A. Gourdenne, Eur. Polym. J., 19(6) (1983) 543.

Index

abstraction of hydrogen 384
acceptor 42, 43, 68, 114, 115, 143, 144, 145, 147, 200, 202
acenaphthylene 57
acetaldehyde 39
acetone 39, 456, 470
acetonitrile 29, 63, 84, 100, 102, 147, 171, 219
acetophenone 164
acetylacetonate 208
acetyl chloride 349, 474
acetylene(s) 31, 46, 114, 143, 173, 236, 338, 405, 428, 435
acid(s) 40, 102, 124, 418, 419, 457
–, acrylic 44, 46, 94, 177, 180, 199, 252, 303, 306, 400, 550
–, ascorbic 64
–, Brönsted 120, 125
–, carboxylic 43, 85
–, conjugated 460
–, fluorosulphonic 125, 130
–, fumaric 47
–, halogenoacetic 125, 419, 456
–, Lewis 44, 63, 86, **126,** 127, 128, 207, 212, 461
–, maleic 47
–, methacrylic 44, 47, 94, 237, 238, 302, 303, 306, 433, 474
–, vinylsuphonic 246
acrolein 39, 303
acrylamide 43, 46, 85, 89, 114, 199, 238, 306, 360, 400, 452
acrylamine 56
acrylates 30, 91, 93, 185, 288, 355, 396
acrylonitrile 20, 43, 46, 53, 62, 63, 64, 85, 90, 91, 109, 187, 214, 238, 260, 301, 302, 303, 305, 309, 310, 316, 320, 332, 334, 399, 404, 456
activation 128
–, thermal 145
activator 113, 122, 123
active centre(s) 11, 13, 14, 16, 19, 22, 25, 27, 40, 45, 55, 57, 62, 75, 84, 101, 105, 107, 109, 111, 114, 119, 121, 123, 126, 133, 136, 141, 153, 154, **163,** 165, 171, 172, 173, 177, 178, 179, 184, 186, 187, 188, 189, 192, 193, 196, 198, 199, 203, 205, 206, 207, 208, 209, 210, 212, 214, 219, 220, 221, 231, 237, 240, 242, 243, 244, **251,** 266, 268, 270, 278, 282, 286, 289, 290, 291, 292, 306, 311, 312, 313, 315, 319, 338, 347, 352, 368, 404, 405, **420,** 429, 435, 436, 437, 443, 450, 463, 481, 482, 498, 513, 516, 517, 521, 522, 523, 539, 549
– –, anionic 188, **190,** 216, 217, 427
– –, atactic 139, 140, 272, 428
– –, bimetallic 140, 206, 271, 273
– –, "bound" 264, 266
– –, carbanionic 427
– –, carbenium 132, 193
– –, carboxonium 14, **193,** 194
– –, cationic 130, 131, 192, 216, 427
– –, coordination 41, 203, 214, 339, 344, 353, 431, 434, 462
– –, covalent **195,** 197, 266
– –, dead 221
– –, dormant 417
– –, "free" 264, 313
– –, growth 365, 504
active centre(s), homogenous 208
– –, inactive 421, 444
– –, in ZN polymerizations 41, 140, 141, **203,** 210, 211, 215, 217, 428

– –, ionic 180, 190, 195, 338, 353, 417, 422, 431, 502
– –, isotactic 139, 140, 212, 272, 428
– –, living 109, 212, 365, 366, 417, 502, 504, 505, 527, 531, 532, 533
– –, long living 429
– –, monometallic 140, 206, 271
– –, multifunctional 179
– –, nonpropagating 451
– –, oxonium 14, **194**, 195, 216, 344, 433, 514
– –, propagating 264
– –, radical 75, **164**, **172**, 177, 180, 217, 280, 338, 353, 431
– –, "rapid" 212, 363
– –, short living 429
– –, silanolate 188, 423
– –, "slow" 212, 363
– –, sulphonium 433
activity of radicals **383**
acyllactam 122
addition 231, 290, 338, 363, 463
–, 1,2; 1,4 109, 110, 111, 268, 277, 278, 279
–, cis, trans 270
addition, controlled **261**, 264, 274
–, head to head 426
–, head to tail 170
–, isotactic 246
–, monomer to radical 172
–, oxidative 141
–, "primary", "secondary" 270, 271
–, syndiotactic 246
–, "wrong" 436, **452**
adhesives 424
after effect 408
agent, electrophilic 124
–, Grignard 266
–, nucleophilic 37, 121, 344, 459
–, solvating 428
–, solvolyzing 428
–, spin trapping 219, 432
–, surface active 281
–, terminating 221, 427, 431, 476
–, transfer 220, 221, 245, 291, 386, 395, 396, 415, 443, 446, 447, 448, 449, 453, 460, 476
–, with chiral activity 247
aggregate(s) 65, 105, 108, 125, 221, 252, 421
–, of initiator with growing particles 245
aggregation of centres 133, 183, 190, 363, **420**, 515

AIBN 85, 86, 87, 90, 100, 396
alcohol(s) 427, 428, 459
alkenes **28**, 135, 141, 142, 270, 274, 355, 418
–, optically active 269
alkoxides (alcoholates) 107, **112**, 113, 120, 188, 189, 213, 220, 427
alkylacrylates 451
alkylaluminium 340
–, chlorides 63, 64
alkylation 461
alkyleneimines 36
alkyl radical 93, 171
allene 428
allylchloride 53
aluminoxanes 210, 211, 212
alumina **138**, 210, 211, 212
amalgam 200
amides of alkali metals 112
amines 96, 104, 108, 188, 427, 428
–, aliphatic 93, 94
–, aromatic 94
anion(s) 112, 118, 120, 122, 186, 187, 188, 189, 190, 198, 216, 218, 365, 420, 425, 459
–, allyl 424
–, delocalized 185
anions, dienyl 186
–, initiating **102**, 112, **118**, 452, 459
anthracene 116, 313
aromatic amides 123
aromatics 116
associates 61, 106, 163, 188, 191, 421
–, ionic 180
association 105, 184, 213, 220, 238, 252, 254, 331
– number 108
– of ions **180**, 200
autoacceleration 397, 398, 399, 400, 500
autocatalytic course 107
autodissociation 126
autoinhibition **404**, 424
autotermination **424**, 425
aziridine(s) 41, 43, 346, 461
2,2'-azobisisobutyronitrile 63, 85, 86, 87, 100, 456, 490, 492, 496
azo initiators **85**, 86, 96

barrier to internal rotation 186
basicity of heterocycles 59, 310
benzene 32, 104, 105, 106, 250, 421, 456, 470, 489, 494, 496

benzonitrile 202
benzophenone 90, 117
benzoquinone 450
benzyllithium 105, 108
benzyl vinyl ether 50
BF_3 127
3,3'-bis-chloromethyloxetane 238, 371
block 95, 427
boiling point 46, 47, 48, 49
bond, acetylenic 119
–, aldehydic 119
–, π, σ-allylic 140, 141, 142, 209
–, amide 60, 119
– angle (valence) 31, 33, 326
–, carboxylic 119
–, C-halogen 455
–, conjugated 56, 172, 173, 318, 339
–, coordination 213
–, covalent 103, 126, 180, 181, 191, 195, 266, 304, 344, 360, 550
– –, double 173, 187, 236, 237, 264, 266, 273, 274, 277, 279, 333, 356, 357, 358, 359, 424, 452, 459
– –, multiple 337, 339
–, donor-acceptor 143, 145
–, electrovalent 266
–, ester 119
–, ether 261
–, ionic 63, 103, 104
– length 33, 236, 260
–, metal-carbon **103**, 138, 140, 141, 208, 270, 272, 273, 340, 341, 428
–, metal-hydrogen 138, 341, 516
–, metal-polymer 139, 266, 270, **428**, 429, 435, 463, 516
bond, polarized 342, 428
–, vinylic 119, 187
branching 241, **356**, 452
– by transfer 443, **451**, 474
– of chains 45
– of macromolecules 452
bromine 91
butadiene 30, 46, 53, 101, 115, 135, 142, 143, 171, 186, 236, 260, 274, 276, 277, 278, 301, 302, 303, 318, 331, 332, 464
butane 164
butene(s) 20, 46, 101, 134, 139, 140, 143, 179
butyl acrylate 53, 301, 390, 396, 433
butyl hydroperoxide 82
butyl lithium 106, 107, 109, 111, 124, 360

butyl mercaptane 472
butyl methacrylate 370, 390
butyl vinylether 47, 313, 314
butyl vinylketone 109

cage, radical 62, 98, 394
calibration graph 527
calomel electrode 42
ε-caprolactam 35, 49, 60, 61, 64, 111, 122
ε, (γ)-caprolactone 59, 189, 194, 343, 344
carbanion(s) 52, 112, **118**, 119, 120, **184**, 190, **191**, 193, 216, 217, 218, 251, 265, 268, 337, 425, 427, 433, 452, 458
–, coloured 184
carbocations 124, 132, 192, 195, 218, 251, 336, 337, 460
–, initiating 131
carbon dioxide 39, 213
– monoxide 39
N-carboxy-α-aminoacids 38
catalyst(s) 133, 139, 428, 431, 517, 519, 521, 522, 523, 524, 526, 548
–, colloid 279
–, donor modified 139
–, Friedel-Crafts 126, 420, 475
–, heterogeneous 138, 463
–, MAF 141, 142
–, of coordination polymerizations 22, 139, 342
–, Phillips 138, 141
–, Solvay 137, 203
–, soluble 270, 279
–, support 136, 206, 247
–, UCC 137
–, Ziegler-Natta 22, 44, 51, 133, 134, 135, 140, **203**, **206**, **208**, 212, 218, 339, 341, 342
– – –, soluble **208**, 273
catalytic poison **29**, 342
cation(s) 118, 122, 123, 129, 131, 132, 133, 150, 191, 194, 196, 210, 220, 268, 354, 365, 420, 421, 433, 461
–, ethenium 460
–, initiating **124**, 126, 128
–, metal 183
–, methylium 354
– of nitrogen 196, 197
–, radical 202
–, tetraalkylammonium 188
centres, active 101, 137, 180, 186, 188, 189, 190, 210, 264, 451

–, anionic **183**, 184, 201, 216, 217,
–, bound 264, 266
–, cationic **191**, 215, 216, 243
–, coordination **203**, 211, 213, 217, 218
–, dead 221
–, dormant 195, 212, **221**
–, free 264
–, isotactic 139, 140, 212
–, living 163, 184, **221**, 243, 363
–, radical 201, 217, 218, 244
chain biting **337**, 338
–, end 347
– –, carbenium 419
– –, propagating 195, 263
chain(s), atactic 21, 262
–, irregular 21, 262
–, isotactic 267, 268
–, ladder 351
–, linear 451
–, living 427, 431
–, polyene 424
–, reversibility 173
–, syndiotactic 266
– with conjugated multiple bonds 173, 353, 424
change in enthalpy with pressure 239, 346
– in entropy 239, 346
– in volume during polymerization 240
character, autoacceleration 395, 396
–, autocatalytic 107
–, donor 457
–, electrophilic 43
–, ionic 210
–, living 355
–, non ionic 195
–, nucleophilic 43, 129, 183, 425
– –, ionic 103, 104, 112
– –, repulsion 325
characteristics of transfer reactions **443**
chelates 124
chemical shift 50, 434
chloral 39
chloroform 93
chloromethyloxirane 59, 135, 196, 198, 214, 346, 347, 349, 437
chlorophosphine 455
chloroprene (2-chloro-1,3-butadiene) 30, 46, 53, 108, 180
chlorostyrene 53, 301
chromatography, gel permeation 427

chrysene 313
classification of monomers 54, 55
cleavage, heterolytic 39, 180
–, homolytic 90
–, α, β 112, 269
co-catalyst 205, 208, 210, 348, 435, 462, 523, 524
– of coordination polymerizations 134, **137**, 138, 205, 435
coefficient, chance (randomness) 321
coefficient, -diffusion 385, 393, 394, 398, 450, 491
–, expansion 239, 391
– –, friction 391
– of initiator efficiency 482
–, partition 498
–, polydispersity 389, 454, 473, 530, 537, 543
–, stoichiometric 403
coil, polymer 247, 392, 498
–, statistical 391, 396, 398
– –, freely drained by monomer 247
co-initiation 126, 127, 128, 355
– by protogenic molecule 126
– by solvent 126
co-initiator 13, 124, 125, 128, 129, 437, 475, 476
combination 86, 147, 164, 217, 331, 384, 419
–, intramolecular 152
– of ions **418**
– of macroions 216, **427**
– of radicals 199, **383**, 389, 390, 492
– of radical ions 145, 199
– of zwitterions 331, 332
compatibility 16
complex(es) 61, 64, 65, 66, 79, 84, 85, 91, 125, 163, 177, 209, 212, 214, 252, 339, 344, 422, 424, 518, 548
– activated 186
complex(es), catalytic 137, 138, 274,
–, crystalline 123
–, donor-acceptor 64, **65**, 66, 67, 68, 85, 91, 92, 93, 103, 126, 129, 143, 144, 145, 147, 148, 177, 197, 199, 202, 313, 332
– –, collision 145
– –, initiating **133**, **143**
– –, photoexcited 91, 144
–, internal 111
–, of ions **422**

– of monomers **63**, 64, 129, 145, 177, 312, 316, 450, 526, 549
– – with active centres 177, 316, 516, 526, 549
– – with initiator 86
– of pyridine, quinoline 39
–, oxonium 252
–, transition 12, 55, 56, 58, 120, 121, 140, 165, 166, 169, 170, 171, 178, 273, 275, 277, 304, 333, 384, 455, 457
– –, four-centre 266, 516
– –, six-centre 516
– with hydrogen 520
– with solvents 450
composites 16
compound(s), alkylaluminium 104, 105, 134, 341
–, alkyllithium 105, 106, 267
–, alkylmagnesium 134
–, alkylsilicon 104
–, alkyltin 104
–, aryllithium 104, 108
–, solvolyzing 429
concentration, equilibrium 360, 506, 507
– – of monomers 360, 506
– of micelles, critical 285
– of radicals, stationary 387, 408, 433
condensation 35, 38
conductivity, electrical 125, 220, 364, 368
–, thermal 259
configuration, atactic 253
–, cis, trans 266, 277, 279
–, D, L 271, 272
–, iso(tactic) 267
– of chains **262**
–, syndio(tactic) 267
conformation, chair 31
–, planar 262
conjugation 165, 235, 236, 405
constant, dissociation 200, 201, 220, 366
–, equilibrium 115, 200, 232, 248, 249, 313, 320, 327, 364, 506
–, Hammet, Taft 51
–, material 285
–, rate 63, 153, 165, 166, 170, 174, 248, 249, 250, 300, 319, 384, 390, 395, 404, 406, 443, 491, 497, 499, 514, 517, 523
– –, elementary 295, 406
– – of addition 102, 183, 310, 312
constant, rate, of depropagation 232, 361, 506
– –, of initiation 102, 153, 505
– –, of initiator decomposition 153
– –, of propagation 51, 172, 174, 176, 232, 252, 265, 295, 363, 364, 365, 366, 367, 368, 369, 370, 371, 372, 433, 444, 491, 506
– –, of radical absorption, desorption 283
– –, of termination 174, 242, 283, 330, 384, 385, 391, 398, 399, 417, 448, 490, 491
– –, of the ion trapping reaction 433
– –, of transfer 172, 384, 400, 415, 448
– – –, apparent 173
– – –, to polymer 454
– –, overall, apparent, effective 52, 176, 363, 444
contraction 261
–, volume 239, **260**
control of molecular mass **473**
conversion 125, 127, 154, 173, 195, 256, 284, 356, 358, 359, 361, 405, 426, 428, 429, 459, 461, 468, 482, 498, 501, 505, 515, 541, 543
–, maximum 513
coordination 17, 133, 203, 209, 210, 268
– number 37, 188, 422
– of monomers 153, 206, 212, 340, 428
– of solvent molecules with ion 181
– site 213, 277
copolymer(s) 91, 136, 138, 178, 290, 292, 308, 313
–, alternating 15, 64, 66, 67, 147, 199, 294, 322, 323, 333, 334, 360
–, block 15, 95, 97, 108, 119, 132, 150, 211, 216, 218, 288, 292, 325, **336**, 337, 338, 427, 457, 476
–, ethylene-propene 355
crossing, intersystem 78, 89, 90
copolymer(s), graft 16, 97, 119, 132, 150, 288, 335, **337**, 338, 427, 453
–, hexafluoropropene-vinylidenefluoride 453
–, multiblock 216
–, optically active 247
–, statistical 15, 16, 67, 306, 322, 332, 334, 336
copolymerization 14, 51, 56, 62, 66, 67, 112, 139, 140, 147, 171, 199, 213, 219, 235, 237, **289**, 299, 302, 306, 320, 326, 327, 356, 357, 360, 403, 444, 450, 469, 471, 511
–, alternating 65, 66, **333**, 334
–, azeotropic 295, 308
–, binary 320, 328

–, block   15
–, coordination   138, 139, **311**
– curve   292, 293, 294, 295, 327
– diagram   292, 293, 295, 309
– equation   **290**, 291, 292, 294, 295, 296, 298, 299, 306, 307, 313, 326, 327, 331
–, graft   15
–, ideal   293
–, ionic   294, 310
–, living   290, 320, 328, 331, **505**
–, multicomponent   **312**
– of CO_2 with oxiranes   **213**
–, oxidation reduction   199
–, radical   247, 290, 306, 314, 316, 328
–, statistical   15
–, stereospecific   51
–, ternary   314, 316, 333
copropagation   290, 292, 293, 294, 295, 296, 300, 310, 312, 314, 316, 328, 333, 334
–, multicomponent   314, 320
counter-ions   119, 120, 148, 183, 184, 186, 187, 188, 190, 191, 192, 194, 195, 197, 220, 266, 339, 343, 345, 346, 366, **417**, 418, 419, 420, 426, 427, 460, 461, 514, 515
crossing, intersystem   78, 89, 90
cross-linking   **356**, 357, 358, 359, 458
– agents, copolymerizing   453
–, reversible   359
crown ethers   123, 124, 186, 187, 190, 220
cryoscopy   530
cryptands   123, 190, 220
cumyl hydroperoxide   83
curve, conversion   154, 184, 257, 258, 259, 361, 504
–, copolymerization   292, 293, 295
–, distribution   369, 528
– –, molecular mass   368, 527, **531**
– –, multimodal   368
cyanoacrylates   57, 198, 424
cyclic sulphides   345
cycloaddition   279, 280
cycloalkanes   31, 32
cycloalkenes   101, 143, 342, 348
cyclobutane   32, 238
cyclobutene   32, 135, 143, 260
cycloheptane   31
cyclohexane   260, 456, 470, 489
cyclooctadiene   260
cyclooctane   260
cyclooctatetraene   353

cyclooctene   260
cyclopentadienyl-   208
cyclopentane   31, 238, 260
cyclopentene   260
cyclopolymerization   24, **350**, 351
cyclopropane   32, 238

decomposition
–, induced   86, 126
– –, of initiator   80, 395, 484, 490
– –, of peroxide   79, 80, 83, 84, 136
–, monomolecular   79
– of initiator, secondary   86
–, secondary   79, **80**, 153
–, thermal   93
deformation of bond angle   31
delocalization of charge   424, 425
– of electron   36, 105, 173, 404, 405, 549
– of electrons defect   163, 191, 424
density, electron   34, 50, 51, 94, 210
–, optical   68
depolymerization   347, 362
depropagation   180, 232, 234, 326, 327, **360**, 362
determining step   482
deviation, mean   529
diacyl peroxide   92, 95
diads, erythro, meso   262, 264
–, racemic   262, 264
dialkylperoxide   83
dianion   115, 119, 184, 185, 201, 336, 337
diazonium salts   85
dibenzoylperoxide   79, 470, 492, 494, 496
dication   130, 131, 145, 202
–, siloxonium   130, 251
1,2-dichloroethane   150, 512
dienes   114, 115, 135, 138, 141, 142, 143, 186, 212, 268, 277, 278, 356, 357, 435, 464
dienes, conjugated   30, 134, 208, 236, 273, 274
–, nonconjugated   30, 273, 279
–, substituted   114
diethyl aluminium chloride   134, 135, 137, 141, 205, 212, 336
diethylzinc   215
diffusion   254, 255, 284, 285, 393, 397, 399, 450, 491
– controlled reaction   44, 62, 86, 99, 173, 249, 250, 384, 390
– of monomer to centres   212, 243, 251, 524,

diynes 31, 350
diions 115, 199, 336, 431
dimerization 139, 151, 152, 201
dimethoxyethane 114, 115, 117,
dimethyldichlorosilane 130, 241, 242
dimethyl ether 117
dimethylformamide 63, 93, 100, 255, 490
dimethyloxirane 260
dimethylsulphoxide 93, 372, 458
dimethylthiethane 371
dinitrostyrene(s) 425
dioxane 59, 346, 347, 363, 367, 372
dioxepane 347, 371
dioxolane 59, 193, 238, 311, 331, 343, 347, 371
diphenyl 116
diphenylethylene 29, 50, 68, 69, 106, 108
diphenylpicrylhydrazyl 76
diradical 75, 76, 78, 95
disproportionation 24, 121, 164, 211, 217, 341, **383**, 388, 389, 390, 431, 448, 464, 536
dissociation 180, 182, 210, 220, 249
– of acids 245
– of covalent bond 181
– of electrolyte 183
– of initiator, thermal 128
– of ion pairs 363
–, secondary 80
distribution
– function 530
–, molecular mass 116, 235, 281, 384
– – –, of radical polymerization products 527
– of particle size 284
–, spherically symmetric 276
disulphides, cyclic 36
divinylbenzene 30, 58, 350
divinyloxybutane 351
donor 29, 42, 43, 108, 114, 138, **139**, 140, 143, 144, 145, 147, 202, 208, 210, 219, 272, 314, 355, 428

ebulioscopy 530
effect, alternation 51
–, cage 19, **98**, 99
–, +I, −I 61
–, induction 165
–, inductometric 165
–, matrix 246, 247, **252**, 253, 254, 400
effect, Norrish-Trömsdorf 250, 396, 398

– of penultimate member in copolymerization **312**, 459
– of pressure in propagation **239**
–, polar **457**
–, steric 165, 361
efficiency of initiation 62, 86, **99**, 410
electroinitiation 94, **149**, 424
–, direct 149
–, indirect 149
electrolyte 418
–, supporting 150
electromagnetic radiation 550
electron affinity 42, 84, 103, 143, 144, 199, 200, 332
electronegativity of bond 37, 103
electron(s) 114, 183, 275, 455, 548
–, capture 200
–, nonbonding 236
–, π, n, 209, 548
– pair 39, 213
– –, delocalized 60, 61, 173, 404, 405, 549, 550
– –, free 60, 63, 94, 114, 184
–, solvated 200
–, unpaired 116, 164, 173, 179, 180, 383, 401, 405
–, valence 55, 548, 549, 550
electrophilic agent 33
– attack 120
electrostatic field of ions 181, 266
elementary process 62, 68
– transfer process 516
elimination of hydride anion 459
emulgator 281, 282
emulsion 281, 282, 283, 288, 335, 404, 433
energy, activation 76, 122, 165, 166, 168, 170, 171, 172, 173, 175, 232, 244, 354, 384, 395, 443, 549
– –, apparent 52
– –, of initiation 497
– –, of propagation 12, 497
– –, of termination 244
– barrier 237, 250, 550
–, bond 360
–, electrostatic interaction 181
–, Gibbs 25, 182, 183, 200, 231, 239, 305, 306, 311, 322, 323, 324, 325, 347, 455, 547
–, ionization 143, 144, 150
–, localization 132, 167, 168
– of exciplex 144

-, of excitation   89, 144
- of interaction   237
- of plasma   151
-, perturbation   169
-, potential   75, 166, 167, 168
-, resonance   179, 235, 236
-, solvation   114, 200
-, stabilization   164, 235
enthalpy   239, 240, 305, 327, 346, 455
-, activation   304
entropy   37, 183, 234, 235, 238, 239, 240, 455
-, activation   171, 304, 305
- change in polymerization   171, 320, 321
- - - propagation   239, 240
-, dissociation   182
-, molar   320
- of configurational change   321
-, residual   321
-, standard   171
epichlorohydrin   59, 135, 196, 198, 214, 346, 347, 349, 437
epoxybutane   59
equation, copolymerization   **290**
- -, integrated   298
-, rate   331, 394, 395, 498, 501, 512, 539, 541
equilibrium   129, 460
-, deaggregation   106
-, dissociation   364, 366, 417
- ester ⇌ ion pair ⇌ free ions   417, 511, 514
-, ionic   44, 62, 128, 331, 418
- monomer ⇌ polymer   238, 256, 361, 481
- position   190
-, tautomeric   133
equipotential surface   247
ESR   180, 201, 432, 433, 513
ester(s)   40, 195, 221
ethane   104, 164
ethanolysis of oxiranes   60
ethers   104, 108, 134, 137, 182, 186, 188, 193, 194, 220, 427
-, cyclic   **34**, 121, 146, 310, 311, 346, 515
ethyl benzene   100, 456, 489
ethylene   20, 28, 44, 46, 53, 63, 101, 135, 137, 138, 139, 140, 142, 164, 180, 203, 207, 208, 211, 236, 238, 260, 303, 334, 372, 430, 451, 462, 467, 516, **523**, 524, 547, 549
ethylenediamine   178
ethyleneimine   36, 48
ethyleneoxide   48
ethylene sulphide   48

ethyllithium   104
ethyl vinyl ether   53
ethyl vinyl sulphide   55
exciplex   91, 143, 144, 202
excitation by photons   89, 143
expansion, volume   391
external standard   50

factor
-, pre-exponential   171, 172, 232, 454
-, steric   249
fluorene   106, 456
forces, electrostatic   180, 182, 247, 310
-, intermolecular   306
- -, attractive   129, 181, 369
-, repulsive   181, 237, 273
-, van der Waals   129, 235, 238, 250, 335
formaldehyde   20, 38, 121, 238
formals, cyclic   35
fractionation   527
fragment of initiator   394
- of diffusion jumps   249
fructose   84
fumaronitrile   43, 313, 334
furan   67

gel-effect   250, **396**, 397, 398, 399
generation of primary radicals   482
glass transition point   268, 335
GPC   427
grafting by transfer   **473**
- from   337
- onto   337
group(s), carboxyl   236, 354
-, end   476
-, hydroperoxide   337
-, phenolate   418
-, photoactive   86
-, vinyl   187, 188
growth, isotactic   272
-, stereospecific   **262**, 272
-, syndiotactic   272, 273

half-time of initiation, propagation   242
- of initiator decomposition   79
heat of combustion   31
- of copolymerization   321
- of formation   238
- of hydrogenation   236

–, of polymerization   18, 22, 44, 235, 236, 237, 238, 239, 259, 353
– – –, removal of **259**
– of reaction   44, 200, **235**, 454, 473
– of solvation   200
– of sublimation   200
– of vaporization   44
heptane   456
heterocycles   **33**, 35, 41, 60, 189, 190, 194, 195, 213, 342, 343, 368, 461, **539**
hexafluoropropene   453
1-hexene   53
hindrance, steric   176, 236, 346
homopolymerization   14
humidity of air   431
hydrazine   84
– hydrate   85
hydrogen   207, 520, 523
– bond   238, 419
– sulphide   84
hydrogenation   33
hydrolysis   356
hydroperoxide   83, 97
hydroxide(s) of alkali metals   112, 422
hydroxylamine   84
hyperconjugation   165, 235, 236, 237

imines   196
impurity(ies), inhibiting   61, 404
–, terminating   138, 426
inactivation of centres   **431**, 521
indene   32, 128, 247
indol   67
induction period   107, 122, 154, 402, 404, 426
inhibition   76, 109, 401, 402, 428, 444, 519
inhibition by hydrogen   520
– by organoaluminium compound   519
– period   76, 92, 402, 403, 404, 426, 483
inhibitor(s)   93, 128, 133, 154, 180, 401, 402, 403, 404, 405, 408, 412, 414, 431, 432, 435, 445, 462, 463, 483
–, "ideal"   401
– of radical polymerization   85, 202, 404
inifer, iniferter   **475**
initiation   75, 76, 80, **89**, 91, 105, 107, 109, 114, 116, 117, 121, 128, 133, 145, 149, 153, 174, 185, 189, 280, **281**, 286, 288, 395, 401, 402, 405, 407, 408, 431, 432, 445, 487, 489, 496, 502, 511
– by alkylmetals   **105**

– by fragmentation   **90**
– by radical anions   **116**
–, direct   126, 150
– in plasma   94, **151**
–, mechanochemical   94, 150
– of anionic polymerization   **121**
– of cationic polymerization   128, 132, 152
– of coordination polymerization   153
– of diene polymerization   140, 141
– of non-polymerizing monomers   **151**
– of radical polymerization   **101**
–, rapid   133, 154, 243, 363, 464, 502, 524, 531, 539
–, redox   **83**, 199
–, slow   133, 154, 405, 406, **502**, 531, 534
–, spontaneous   66
–, thermal   **75**, 76, 395, 410, 414, 488, 489
initiator(s)   12, 61, 64, **78**, 92, 95, 98, 99, 100, 105, 107, 109, 111, 114, 119, 122, 124, 126, 129, 147, 153, 154, 194, 217, 221, 386, 394, 450, 454, 458, 482, 484, 500, 531, 541, 550
initiator(s)   281, 284, 288, 289, 396, 405, 409, 419, 422, 424, 443, 448, 452, 461, 475, 476, 487, 499, 502, 503, 504, 505
– dianionic   216
–, dicationic   216
–, difunctional   95, 216, 336,
–, insoluble   **288**
–, multifunctional   95, 97, 98, 130, 335, 400
– of cationic polymerization   **124**, 125, 126, 458
–, optically active   269
–, polymeric   **96**, 97, 98, 335, 400
insertion   272, **339**, 340, 341, 516
–, coordination   339
interaction, chemical   **62**
–, dipole-dipole   306
–, donor-acceptor   94, 140, 143, 145, 152, 275, 333
–, electrostatic   250, 274, 310, 333
–, intermolecular   394
– of electrons   53, 54
– of substitutents   34
–, polymer-polymer   385, 398
–, polymer-solvent   385
–, repulsion   236
–, steric   271
intramolecular cyclization   268
inversion of configuration   342
iodine   150, 218, 250, 352

ionic character 103, 104, 125
ionization 128, 181, 183, 199, 210
ion pair(s) 52, 108, 112, 115, 117, 148, 178, **180**, 181, 183, 184, 186, 190, 191, 195, 197, 198, 200, 247, 266, 339, 514, 515
ion pair(s), contact 123, 128, 182, 183, 184, 185, 198, 219, 220, 221, 363, 367, 368
– –, solvent separated 123, 128, 182, 183, 185, 198, 219, 220, 221, 267, 363, 367, 368, 419
ion(s) 94, 102, 112, 117, 118, 125, 128, 133, 152, **180**, 183, 191, 195, 197, 199, 200, 207, 338, 426, 448
–, acyllactam 133
–, ammonium 461
–, carbenium 51, 191, 265, 460
–, carbonium 191
–, carboxonium 193
–, free 52, 123, 128, 180, 181, 183, 190, 191, 195, 200, 219, 221, 268, 345, 363, 367, 419
–, gaseous 200
–, hydride 194
–, metal 84
–, oxonium 132, 194
–, pair 202, 268
–, phosphonium 461
–, radical 202
–, siloxonium 194, 461
–, sulphonium 198, 461
–, uranyl 89
isobutyl vinyl ether 14, 91, 127, 150, 371
isocyanates 123
isomerization 192, 236, 244, 350, 353, 354, 355, 424, 426, 458
–, cis, trans 277
isoprene 30, 47, 89, 106, 107, 108, 135, 143, 186, 236, 276, 277, 310, 332, 351
ketene 30, 428
ketone(s) 39, 193
–, substituted 114
kinetic analysis 437, 450, 497, 505
kinetics of initiation **153**, 432
– –, copolymerization **328**
– –, polymerization 108, 356, 444
– –, propagation **362**
– –, termination **345**
kneading 150

lactams 25, 35, 60, 61, 121, 133, 190, 196, 197, 245, 343, 352, 354

lactones 35, 123, 124, 189, 194, 342, 343, 361, 421
lecithin 247
length of macromolecules, mean 520
– – –, kinetic, mean 384
– –, sequence, mean 291
Lewis base 208
life time of centres, mean 464
– –, of radicals, mean 180
ligand 34, 218
–, activated 214, **349**
light quantum 90, 144
living polystyrene 68

macroanion(s) 119, 215, 216, 218, 458
macrocation(s) 131, 216, 337, 434, 460, 474
macrocycles 343, **346**, 347
macroester 195, 514, 515
macromers 196, 347, **431**, 473, **474**, 475, 476
–, difunctional 474
macromolecules, branched 231, 451
–, cross linked 30, 451
–, linear 231
macroradical 63, 80, 176, 211, 219, 335, 384, 385, 393, 394, 396, 457, 472, 490, 491, 538
–, acrylonitrile 55
–, electronegative 54
–, electropositive 54
–, methyl methacrylate 176
–, styrene 55
maleic anhydride 43, 66, 67, 347, 301, 302, 334
mass (weight) fraction 530
–, molecular 96, 211, 394, 397, 414, 420, 431, 456, 457, 458, 461, 462, **464**, 473, 481, 514, 524, 550
– –, number average 522, 529
– –, of monomer 519
matrix 246, 252, 266, 338, 507, 508, 509
–, polyelectrolyte 188
–, polymeric 253
mechanism of activated monomer 25
–, living 194, 243, 288, 336, 341, 368, 431, 461
–, monometallic 340
mechanism of deaggregation 106
– –, initiation 140, 141, 432
– –, propagation 45, **338**, 339
– –, termination 388
– –, transfer 448

medium, hydrocarbon   331
-, living   235
-, non polar   112, 123, 128, 180, 280, 345, 418
-, non-solvating   128, 245
-, polar   107, 114, 178, 180, 245, 280
melting point   44
mercaptans   443, 453
metal   423
- alkali   104, 115, 116, 152, 184, 185, 188, 422, 424
- hydrides   424
metals   104, 116
-, electropositive   104, 112, 114
-, non-transition   140
-, transition   22, 83, 133, 134, 136, 137, 138, 140, 141, 142, 153, 203, 206, 207, 208, 212, 218, 219, 270, 271, 272, 273, 274, 275, 278, 279, 339, 340, 341, 348, 349, 359, 428, 429, 431, 548, 549
metathesis (disproportionation)   24, 32, 219, 342, **348**, 349
methacrylamide   57, 371
methacrylonitrile   60, 63, 102, 109, 187, 199, 301, 302, 303, 305, 371, 458, 490, 492
methane   40
method
-, dead stop   418
-, end group   530
-, INIFER   **475**
-, radiochemical tracer   484
methoxide   111
methoxystyrene   53, 54
methyl acrylate   20, 53, 63, 64, 84, 85, 176, 301, 370, 396, 400, 404, 472
p-methyl benzyl vinyl ether   50
3-methyl-1-butene   192
methylene chloride   125, 127, 202, 470
methyl-α-ethyl acrylate   50
methyl isopropenyl ketone   109
methyl methacrylate   20, 29, 47, 53, 63, 64, 78, 84, 86, 87, 89, 90, 91, 93, 96, 100, 109, 111, 112, 172, 173, 174, 176, 177, 189, 214, 237, 238, 254, 260, 266, 301, 302, 305, 310, 313, 316, 318, 320, 332, 342, 370, 389, 390, 392, 396, 417, 433, 448, 450, 455, 456, 457, 472, 474, 476, 550
2-methyl-2-nitrosopropane   432, 433
methyloxirane   59, 111, 198, 213, 215, 260, 349

RS-3-methylpentene   135
4-methyl-1-pentene   354
2-methylpropene   28, 47, 53, 101, 126, 128, 145, 179, 202, 238, 323, 336, 419, 424, 460, 461, 473, 476, 481
α-methylstyrene   47, 66, 115, 119, 127, 128, 143, 184, 201, 202, 235, 237, 238, 268, 301, 302, 327, 360, 371, 458, 459
methylthiiran   124, 188, 189, 371
methyltrichlorosilane   241
methyl vinyl ketone   53, 56
methyl vinyl sulphide   53, 56
micelles   20, 104, 105, 280, 281, 282, 284, 286, 288
micro-Brownian motion   385
micromechanical adhesion   261
micromechanism   369
migration of H atoms   405
milling (grinding)   143, 150, 203
-, vibration   150
modification, physical   143
- of polymers   150
molecularity of the termination reaction   435,
molecular mass distribution   173, 320, 355, 394, 438, 451, 469, 505, 526, **527**, 535, 538, **539**, 542, 543, 544
- - -, the most probable   537
monomer(s)   **27**, 65, 107, 148
-, acrylic   88, 109, 187
-, activated   **190**, 195, 196, 197, 347, **352**
-, alkyne   173
-, conjugated   55, 56, 318, 319
-, cyclic   **31**, 188, 237, 242, 269, 325, 342, 345, 346, 349, 360, 420
-, difunctional   30, 356
-, electrophilic   68, 199, 332
-, heterocyclic   59, 60, 116, 132, 243, 248, 461
-, methacrylate   88, 93, 187, 355
-, multifunctional   338, 350, 356, 527
-, non-polymerizing   68, 153, 444
monomers, nucleophilic   64, 68, 332
- orientation prior to addition   246
-, polar   108, 266, 313, 547
- -, polymer particle   281, 282, 285, 287, 498
-, solubilized   281
-, tetrafunctional   352, 356, 358
-, vinyl   28, 29, 63, 84, 90, 101, 116, 179, 199, 201, 325, 360, 399, 443, 470, 550

naphthalene   116, 117, 177, 313
– metal   201
naphthalenesodium   117
naphthonitrile   117
naphthyl   28
neopentane   152
nitrile(s)   117
nitroaromatics   401
nitrobenzene   127, 419
nitroethylene   236
nitrogen   133
nitromethane   147, 418
nitrosoaromatics   401
nitrostyrene   43, 52, 53, 54, 113, 303
NMR   219, 354, 433, 513
–, ^{31}P   433
nucleophilic attack   54
number of centres, determination   431, 437

occlusion of centre   251, 400
octamethylcyclotetrasiloxane   20, 49, 112, 117, 260, 360, 361, 422
oligomer(s)   111, 173, 285, 405
orbital(s)   143, 144, 170, 209, 549
–, antibonding   114
–, atomic   60, 170
–, frontier   170
–, unoccupied   63, 116, 143, 170
order of polymerization, external   259
– – –, internal   259
– – –, referred to initiator   505
organometal(s)   120, 126, 137, 138, 140, 141, 203, 206, 429, **462**, 463, 464
–, aluminium   104, 105, 134, 135, 136, 137, 139, 208, 210, 431, 516, 517, 519
–, aryllithium   104
–, caesium   104
–, lithium   104, 105, 108, 267
–, magnesium   108, 134, 135, 137, 431
osmometry   530
oxacyclobutane   48
2-oxazoline   195
oxepane   367
oxetane   **34**, 36, 59, 346, 419, 515
oxirane(s)   20, **33**, 41, 59, 112, 117, 121, 189, 196, 213, 260, 261, 334, 342, 346, 347, 352, 371, 421, 515
oxolane   **34**
oxygen   85, 91, **93**, 107, 109, 113, 120, 133, 136, 150, 194, 202, 217, 240, 323, 345, 404, 419, 431
ozone   97

parameter(s), activation   384
–, Arrhenius   173, 384
–, copolymerization   50, 51, 291, **293**, 294, 295, 296, 297, 298, 299, **300**, 301, 302, 304, 305, 306, 308, 311, 312, 313, 318, 329, 471
participation of helix   271
pentadiene   277, 278
pentene   101
permittivity   44, 181, 183, 190, 220, 306, 417
– of medium   181, 304, 363
–, relative   181, 182, 304
peroxides   81, 82, 95, 136
–, decomposition of   79, 93
pH   177
phase, condensed   369
–, crystalline   260
–, isotropic   245
–, nonstationary   154
–, smectic   245
phenanthrene   116
phenylacetylene   424
phenyllithium   108
phosphazene(s)   40, 189, 197, 345
phosphorus   347, 433
– oxychloride   123
photoexcitation   91
photoinitiation   **88**, 89, 93, 174, 488, 550
photopolymerization   90, 91, 476
photoreaction   409, 410
photosensitization   **90**
photosensitizer   88
piston flow   414
pivalolactone   238
placement; isotactic, syndiotactic   360
plasma   94, 151
plasma initiated polymerization   94, 151
plasma polymerization   94, 151
points, branching   255, 453
–, cross-linking   357
–, cyclization   357
polarity of amide bond   60, 61
– of bond   103, 104
– of radical   300
oligoacetaldehyde   35
polyacetals   341
polyaddition   14

polyalkenes   97, 515, 521
polyamides   335, 341, 354, 432
polybutadiene   277, 335
polycombination of zwitterions   147
polycondensation   3, 38, **241**, 431
polydimethylsiloxane   241, 420, 423, 427
polyelectrolytes   188, 246
polyene   405, 424
polyester(s)   343, 361, 431
polyethers   337, 341
polyethylene   203, 431
–, crystalline   355
–, high density   45
–, low density   45, 431
polyimides   354
polyimines   341
polyisoprene   335
polymer(s)
–, amorphous   328, 239
–, asymmetric   269
–, atactic   140, 192, 253, 262
–, branched   338, 452
–, conjugated   39
–, crosslinked   356
–, crystalline   238, 239
–, high-molecular   233
–, inactive   435, 436, 447, 450
–, irregular   21
–, low-molecular   12
–, monodisperse   531
–, regular   203
–, stereoblock   253, 267
–, stereoregular   186, 254, 262, 269
–, stereospecific   17, 186, 268
–, tactic   138, 203, 262
– – isotactic   22, 134, 140, 262, 264, 266
– –, syndiotactic   22, 218, 252, 253, 262, 264, 270
–, telechelic   431
polymeric particle   500
polymerization   416
–, bulk   174, 283, 497
–, chain   14
–, coordination   13, 21, 29, 30, 50, 64, 133, 153, 241, 243, 244, 245, 313, 331, 342, 355, 369, 405, 424, 430, 434, 435, 436, 463, **464**, 473, 515, 523, 530, 547
– –, homogenenous   17
– –, Ziegler-Natta   515
– degree   19, 172, 190, 306, 320, 356, 385, 389, 397, 398, 399, 418, 443, 444, 451, 454, 455, 458, 462, 464, 465, 467, 468, 469, 483, 485, 487, 490, 498, 513, **526**, 528, 534, 535, 536, 538
– –, number average   256, 415, 464, 521, **529**, 530, 531, 533, 537, 538, 541
– – of average sequence   307
– –, of radical   172
– –, mass average   **529**, 530, 537, 538
–, emulsion   19, 20, 62, 251, 259, **280**, 281, 282, 289, 433, 550
–, gas phase   206, 208
–, group transfer   26, **355**, 547
–, heterogeneous   208
–, homogeneous   17
polymerization
– inhibited   402
–, ionic   12, 21, 33, 37, 153, 203, 244, 245, 280, 289, 313, 331, 342, 369, 405, 417, 418, 421, 424, 435, 454, **502**, 511, 535, 547, 549
– –, anionic   41, 50, 51, 57, 64, 105, 109, 111, 112, 121, 124, 146, 189, 199, 200, 215, 353, 419, 421, 424, 427, 457,
– –, cationic   41, 50, 51, 63, 64, 109, 124, 127, 129, 131, 146, 152, 192, 195, 215, 270, 331, 336, 344, 347, 426, 427, 454, 459, 460, 461, 473, 474, 475, 481
–, isomerization   25, 114, 244, **352**, 354
–, liquid phase   212
–, living   12, 22, 116, 192, 243, 244, 256, 259, 335, 362, 363, 395, 401, 417, 419, 427, 459, **464**, **502**, 505
– –, anionic   119, 335, 336
– –, cationic   126, 129, 131, 336, 366, 368, 420, 421, 514
– –, radical   476
– –, with rapid initiation   511, 530
– –, with slow initiation   511
–, migration   14
–, non-stationary   22, 84, 187, 220, 363, 430, 431, 444, 465, 505, **511**, 513
– –, ionic   535
–, photo-initiated   406, 408, 414, 487
–, precipitation   399
–, pseudo-cationic   125, 195
–, radical   41, 50, 51, 63, 99, 102, 133, 146, 164, 242, 243, 244, 252, 280, 288, 289, 342, 369, 383, 398, 400, 401, 405, 406, 408, 435, 436, 443, 453, 458, 464, 473, 475, 482, 484, 492, 511, 535, 538, 547, 549, 550

– –, ideal   394, 450
– –, non ideal   384, 485, 490, **492,** 496
–, ring opening   24, **189, 341,** 342, 343, 346, 347, 349, 352
–, solution   497
–, spatially intermittent   **414**
–, stationary   22, 174, 242, 257, 395, 408, 420, 444, 465
–, stereoelective   268
–, stereoselective   268
polymerization, stereospecific   14, 21, 266
–, suspension   19, 251, 259, 457
– with uncontrolled propagation   21
polymethylmethacrylate   253, 288, 393, 394, 397, 456, 457
poly-α-methylstyrene   459
polyoxirane   97
polyoxymethylene   469
polyperoxides   404
polyphenylacetylene   405
polypropylene   134, 138, 218, 237, 288, 431, 462,
polysiloxanes   337, 341, 360
polystyrene   68, 96, 288, 335, 353, 394, 424, 457, 458, 459, 473, 527
polysulphides   457
polysulphones   180, 360
polyurethanes   335, 431
polyvinylacetate   473
polyvinylchloride   266, 431, 455, 497
post effect   408
potential,
–, half-wave   42
–, ionization   42, 84, 103, 114, 200
–, reduction   42, 150
ppm   21, 241
pre-effect   174, 405, **406,** 408
pressure   95, **239, 390,** 431, 500
process , living   383
–, non-radiative   89
–, stationary   154, 383
processing   231
propadiene   30
propagation   12, 75, 105, 116, 119, 128, 151, 152, 153, 163, 172, 177, 185, 189, 194, 195, 201, 212, 219, **231,** 232, 234, 236, 238, 239, 240, **241,** 251, 254, 256, 259, 269, 270, 273, 280, 288, 289, **338,** 345, 346, 348, 352, 353, 356, 362, 364, 369, 387, 404, 407, 417, 418, 426, 469, 502, 505, 535, 539
–, controlled   186, 261
–, coordination   339, 344
– in emulsion   280, 286, 288
– in living system(s)   **363**
–, isotactic   271, 272
– of heterocycles   346, **514,** 515
– on metal-carbon bond   516
–, syndiotactic   266, 271, 272
propiolactone   148, 199, 238, 325, 334, 344
propene   20, 28, 44, 92, 101, 102, 134, 137, 142, 143, 171, 179, 236, 237, 238, 260, 270, 272, 273, 303, 334, 355, 371, 462, 463, 549
properties,
–, dielectric   44
–, donor-acceptor   **42**
proton   132, 424, 460
proton charge   181
pseudotermination   425
pyridine   39, 94, 140, 334
2-pyrrolidone   60, 61, 123, 238, 303

quantity e, Q 302, 303, 471
– e, q 302, 304
– $\Omega_{1,2}^+, \chi_{1,2}^+$ 305
– r_1, r_2 **301,** 304
quinoline   39, 91, 334
quinone(s)   401

racemic mixture   188
radical(s)   76, 78, 88, 94, 98, 134, 136, 150, 168, 179, 199, 217, 218, 219, 221, 254, 265, 283, 284, 285, 302, 304, 338, 393, 399, 409, 410, 411, 412, 415, 416, 432, 433, 490, 498, 536
–, alkyl   93, 171
–, amine   90
–, anchored (trapped)   399
–, anion   89, 114, **116,** 117, 150, 152, 200, 201, 203
–, benzoyl   80
–, benzoyloxyl   101, 494
–, cation   89, 145, 202
–, free   79
–, frozen   431
–, hot   **174,** 175, 176, 221, 294, 403, 404, 450
– –, excited   99, 403
–, hydroxyl   102

–, initiating   67, 91, 93, 94, 95, 99, 402, 431
–, ion(s)   14, 115, 116, 117, **143**, 144, 145, 149, **199**, 200, 201
–, ion pairs   115, 145
–, isobutyronitrile   85, 86
–, long living   399
–, methyl   56, 101
–, nonpolar   171
–, occluded   400
–, oligomeric   490, 491
–, phenyl   56, 494
–, polar   172
–, polymeric   64, 254, 383, 490, 491
–, polymethylmethacrylate   390
–, polystyrene   390
–, primary   62, 75, 79, 100, 101, 172, 384, 386, 394, 431, 482, 490, 491, 492, 494, 496, 501
–, propagating   399, 431
–, resonance stabilized   455
–, secondary   79, 401, 494, 496
–, stable   180
–, styrene   55, 66
radus –, Stokes   184
range, kinetically non-stationary   451
rate of addition   147, 317, 349
rate of combination   419
– of copolymerization   176, 330, 331, 333
– of copropagation   328
– of depropagation   361
– of diffusion   249, 250
– of inhibitor, initiator consumption   100
– of initiation   78, 86, 92, 101, 106, 123, 128, 129, 153, 154, 243, 328, 395, 401, 403, 404, 410, 412, 415, 432, **482**, 483, 484, 511
– of living polymerization   **530**
– of photochemical initiation   407
– of polymerization   66, 127, 129, 176, 177, 178, 188, 244, 283, 284, 286, 287, 402, 408, 410, 412, 415, 418, 446, 448, 463, 481, 485, 499, 502, 519, 524, 525, 526, 533
– of propagation   115, 123, 154, 242, 244, 284, 361, 362, 387, 401, 465, 490
– of re-initiation   463
–, relative   286, 410
– of termination   154, 255, 262, 284, 329, 387, 395, 396, 401, 409, 415, 416, 465, 476, 485, 490, 511
– of transfer   173, 262, 454, 457, 465, 473, 475
– of transfer agent consumption   446

–, overall reaction   481, 485, 502, 532
reaction, back biting   196, 345, 347
–, bimolecular   101, 247, 384, 419, 430, 465
–, diffusion controlled   384, 494
–, disproportionation   383, 384
–, elementary   484
–, exothermal   403
–, first order   505
–, inhibition   176
–, mechanochemical   94
– of anion with monomer   119
–, polycondensation   **241**, 431
–, stationary   511
–, terminating   405, 429
–, transfer   454
reactivity
 – of monomers   **45**, 50, 58, 291
 – of radicals   **165**, 171, **172**, 176, 192, 404
redistribution of chain lengths   539
redox-initiation   **83**, 84
– system   84, **85**, 92
regeneration of monomers   362
re-initiation   443, 444, 450, 539, 541
relation, Hammet   51, 52
–, isokinetic   166
repulsion, electrostatic   177, 360
– of hydrogen atoms   35
–, steric   273, 360
repulsive efficiency   177
retardation   417, 420, 444, 449, 450, 458, 463
ring opening   38, 40, 260, 342
ROP   **24**, 260, 342, 343
rotating sectors   **408**, 412, 414, 416, 456
rotation, free, – of end group   268
– of double bond   278
rubber
–, cold   472
–, liquid   431
–, natural   30
–, silicone   242
–, vulcanized   452

scattering, light   530
Schiff base   245, 246
segment of siloxane chain   37
sensitization   90
sensitizer   **89**, 90
sequence, – isotactic   246
–, syndioctic   266
series of reactivites   50

silazanes 37
silica 136, 141, 206
silicon 131
siloxanes, cyclic 37, 121, 343, 344
singlet 90
S_N1, S_N2 34
$SnCl_4$ 44, 63, 127
softener 335
solvation 62, 102, 123, 145, 190, 200, 235, 239, 247, 268, 331, 418
– by monomer 195, 461
– of active centres 190, 331
– of ions 133, 181, 192, 200, 245
–, selective 44, 62, 65, 186, 331, 332
solvent(s) 128, 145, 150, 181, 183, 220, 306, 367, 384, 394, 396, 421, 427, 432, 443, 458, 461, 489, 490, 500, 524
– effect 177
– inert 404
–, non-polar 105, 195
–, polar 93, 147, 186, 195
–, solvating 69, 113, 115, 255
space diagram 309
specific gravity (density) 46, 47, 48, 49
spectrophotometry 434, 483, 513
–, NMR, UV 513
spin trapping (conserving) 219, 431, 432, 433
splitting-, resonance 169
stability, thermal, of polymers 338
–, thermooxidative 95
stabilization-, resonance 35, 45, 55, 236, 300, 304, 459, 472
state, energetic 165, 174, 199
–, excited 144, 164, 167, 175, 403
–, isotropic 245, 246
–, nematic 245, 246
state, -non-stationary 405, 406, 502
– of centre, energetic 165
–, oxidation 138, 141, 203, 207, 208, 212, 431, 549
–, singlet 202
–, stationary 125, 133, 242, 243, 284, 307, 313, 315, 316, 329, 357, 405, 411, 412, 415, 446, 449, 467, 484, 501, 502, 511, 516, 535
–, transition 153, 165, 171, 265, 403
–, triplet 202
stationary state conditions 416, 485, 487
statistic, Bernoulli 263, 264, 265, 266, 268
–, Markov 264

– of propagation stereochemistry **262**
stereoblocks 253, 264, 267
stereochemistry of addition 264, 265, 266, 267, 268
strain 31, 32, 35, 37
–, internal 514
–, ring 34, 35, 37, 61, 342, 347, 348, 360
–, steric 235
strength of amide bond 60, 61
structure, cis 134
–, trans 134
styrene 20, 28, 43, 47, 51, 52, 54, 64, 66, 67, 68, 75, 76, 78, 79, 88, 90, 91, 93, 97, 100, 106, 108, 112, 115, 125, 127, 128, 129, 143, 171, 176, 177, 180, 184, 186, 200, 214, 236, 238, 244, 260, 288, 301, 302, 305, 309, 310, 312, 313, 316, 318, 320, 325, 327, 331, 332, 334, 336, 342, 356, 360, 366, 368, 370, 390, 392, 394, 404, 418, 419, 455, 456, 457, 460, 466, 472, 491, 492, 494, 511, 512, 550
styreneoxide (2-phenyloxiran) 59
substituent, electronegative 199, 236
substitution –, nucleophilic 36
–, polar 252, 260
superacid 191, 476
support 207, 208
support, catalyst 134, 151, 206, 207
surface tension 44
suspension 259
system(s), conjugated 119
–, living 211, 457

tacticity 178, 183
– of polymers 183
technique, stopped flow 220
telogen 173, 443, 457, 472, 473
telomer 173, 473
theory, Smith-Ewart **282**, 284, 285
temperature, critical 240
–, glass transition 240, 335
–, isokinetic 166
– of polymerization **244**
– –, –, ceiling 35, 39, 232, 233, 239, 323, 326, 327, 346, 360, 445, 513
– –, –, floor 232, 233, 513
– –, –, limiting 233
termination 12, 151, 152, 153, 192, 199, 211, 212, 231, 243, 244, 254, 286, 347, **383**, 387, 388, **390**, 393, 394, 395, 405, 414, 417, 418, 419, 420, 421, 424, 426, 427, 429, **431**, 433,

434, 436, 445, 449, 452, 457, 459, 462, 464, 469, 500, 501, 502, 506, 512, 531, 532, 535, 536, 539
–, bimolecular   62, 208, 209, 210, 250, 385, 444
– by impurities   426
termination by mutual radical reaction   283, 329, 400, **406, 408**, 444, 499
– by primary radicals   288, **394**, 450, 485, 490, 492
–, exploitation of   **430**
–, hindered   **395**, 396
–, isobasic   422
– of coordination polymerizations   **428**
– of macroradicals   394, 398, 494
–, spontaneous   **430**
terpolymerization   316, 317
tetrachloromethane   456, 457
tetracyanobenzene   202
tetracyanoethylene   424
tetrafluoroethylene   20, 235, 236, 238, 261
tetrahydrofuran   14, 20, 35, 48, 52, 59, 109, 114, 115, 117, 121, 127, 131, 132, 186, 195, 198, 201, 214, 220, 238, 268, 342, 346, 349, 360, 363, 364, 366, 368, 369, 418, 427, 428, 433, 461, 481, 514
tetrahydropyran   59, 368, 369
tetramethylenediamine   124
tetramethylsilane   90
tetraphenylphosphine   275
tetraoxane   35, 346
thermodynamics of copolymerization   **320**
– of propagation   **231**
thermooxidative degradation   97
thermoplastics   452
thietane   36, 461
thiirane   36, 346
thiolactones   123
thiourea   85
thiophene   68
$TiCl_3$ 203, 204, 206, 212, 276, 277, 279, 371, 372, 428, 462, 526
$TiCl_4$ 127, 135, 139, 140, 202, 203, 214, 218, 277, 279, 349, 371, 372, 419, 463, 523
toluene   254, 255, 256, 268, 270
transfer   12, 64, 151, 152, 172, 192, 212, 231, 243, 244, 284, 288, 289, 333, 354, 356, 383, 389, 397, 417, 434, 438, **443**, 444, 452, **453**, 454, 457, 458, 459, 460, 461, 462, 463, 464, 466, 468, 469, 470, 471, 473, 484, 493, 506, 513, 516, 521, 522, 523, 542
–, agents   19, 61, 64, 173, 179, 220, 443, 446, 447, 448, 449, 451, 455, 456, 460, 462, 463, 464, 466, 468, 470, 471, 472, 473, 474, 475, 476, 484, 516, 539, 541, 542, 544
– –, copolymerizing   **469**
– –, difunctional   474
– –, multifunctional   **453**, 457
–, degradative   92, 179, 427, 443, 444, 445, 450, 451, 459, 485, 513
–, electron   51, 83, 89, 90, 108, 114, 116, 117, 143, 144, 147, 152, 163, 173, 199, 200, 405
–, energy   89
–, H^+, H^-, H^* 374, 424, 452, 454, **457**, 458, 460, 461
– in anionic polymerizations   **457**
– in cationic polymerizations   **459**
– in coordination polymerizations   **462**
– in radical polymerizations   **454**
–, inter-, intramolecular   451
– of electron   90, 149, 448
– of hydride anion   424
– of proton   353, 354, 424
– to monomer   138, 399, 455, 456, 458, 461, 466, 475
– to organometal   428, 462
– to polymer   179, 400, 458, 473, 539
– –, –, redistributing   539
transformation of centres   **214**, 336
– –, anion→ cation   **215**
transformation of centres, anion → radical   214, 217
transition, glass   240, 286
triads   262
trialkoxyethylene   43
triethylaluminium   44, 91, 131, 336, 346, 437, 463
triethylamine   140
trinitrobenzene   76
trinitromethane   419
trioxane   17, 20, 25, 35, 38, 49, 59, 193, 238, 251, 311, 331, 346, 469
triphenylmethane   132
triphenylmethyl cation   132
trithian   36
tube reactor   414

unit(s)
–, configurational   21, 262, 269, 270
–, constitutional   21

–, penultimate   186, 237, 279, 290, 294, 326, 360
–, siloxane, difunctional   37
–, stereorepeating   262

vacancy (ies)   139, 140, 205, 516
valerolactone   49, 59
vapour tension   44, 500
vinylacetate   20, 43, 47, 53, 64, 67, 78, 80, 88, 90, 93, 102, 173, 174, 219, 236, 238, 260, 288, 301, 302, 318, 334, 390, 396, 404, 448, 451, 455
vinyl bromide   371
vinyl butyrate   52
N-vinyl carbozole   43, 53, 143, 146, 202, 260, 301, 334, 371, 426
vinyl chloride   20, 29, 47, 50, 53, 78, 92, 93, 97, 108, 171, 180, 214, 219, 236, 238, 260, 288, 301, 302, 316, 318, 331, 360, 371, 399, 400, 404, 487, 488, 489, 497, 499, 501
– compounds   180, 188
2-vinyl cyclopropane   353
vinyl dioxolane   59
vinyl ester   52
vinylether(s)   43, 126, 129, 147, 195, 314, 461
vinyl fluoride   48, 303
– formate   52
vinylidene chloride   20, 46, 50, 89, 108, 235, 237, 238, 301, 302, 320, 331, 455
– cyanide   20, 302
– fluoride   236

vinyl isobutylether   301
vinyl isocyanate   301
vinyl mesitylene   458
vinyl monomers   109, 115, 146
vinyl naphthalene   460
vinyl pivalate   52
vinyl propionate   52
1-vinyl pyrene   260
vinyl pyridine   53, 149, 186, 236, 246, 370
vinyl pyrrolidone   53, 302
vinyl sulphide   43, 67
viscometry   530
viscosity   44, 62, 394, 397, 399, 417
– of medium   398, 490
volume   398
–, activation   391, 455,
–, excluded   385, 490, 491
–, molar   260

wall deposit   251
water   125, 126, 127, 128, 129, 198, 215, 240, 261, 424, 427, 428, 437, 459, 469
wave, reduction   42

yield of catalyst   522
–, quantum   88, 90

z average   530
zip propagation   253
zwitterion   68, 143, 145, 146, 147, **197**, 198, 210, 332, 333